Topics in Applied Physics Volume 54

Topics in Applied Physics Founded by Helmut K. V. Lotsch

1 **Dye Lasers** 2nd Ed. Editor: F. P. Schäfer
2 **Laser Spectroscopy** of Atoms and Molecules. Editor: H. Walther
3 **Numerical and Asymptotic Techniques in Electromagnetics** Editor: R. Mittra
4 **Interactions on Metal Surfaces** Editor: R. Gomer
5 **Mössbauer Spectroscopy** Editor: U. Gonser
6 **Picture Processing and Digital Filtering** 2nd Edition. Editor: T. S. Huang
7 **Integrated Optics** 2nd Ed. Editor: T. Tamir
8 **Light Scattering in Solids** 2nd Edition Editor: M. Cardona
9 **Laser Speckle** and Related Phenomena 2nd Ed. Editor: J. C. Dainty
10 **Transient Electromagnetic Fields** Editor: L. B. Felsen
11 **Digital Picture Analysis** Editor: A. Rosenfeld
12 **Turbulence** 2nd Ed. Editor: P. Bradshaw
13 **High-Resolution Laser Spectroscopy** Editor: K. Shimoda
14 **Laser Monitoring of the Atmosphere** Editor: E. D. Hinkley
15 **Radiationless Processes** in Molecules and Condensed Phases. Editor: F. K. Fong
16 **Nonlinear Infrared Generation** Editor: Y.-R. Shen
17 **Electroluminescence** Editor: J. I. Pankove
18 **Ultrashort Light Pulses** Picosecond Techniques and Applications Editor: S. L. Shapiro
19 **Optical and Infrared Detectors** 2nd Ed. Editor: R. J. Keyes
20 **Holographic Recording Materials** Editor: H. M. Smith
21 **Solid Electrolytes** Editor: S. Geller
22 **X-Ray Optics.** Applications to Solids Editor: H.-J. Queisser
23 **Optical Data Processing.** Applications Editor: D. Casasent
24 **Acoustic Surface Waves** Editor: A. A. Oliner
25 **Laser Beam Propagation in the Atmosphere** Editor: J. W. Strohbehn
26 **Photoemission in Solids I.** General Principles Editors: M. Cardona and L. Ley
27 **Photoemission in Solids II.** Case Studies Editors: L. Ley and M. Cardona
28 **Hydrogen in Metals I.** Basic Properties Editors: G. Alefeld and J. Völkl
29 **Hydrogen in Metals II** Application-Oriented Properties Editors: G. Alefeld and J. Völkl
30 **Excimer Lasers** 2nd Ed. Editor: Ch. K. Rhodes
31 **Solar Energy Conversion.** Solid State Physics Aspects. Editor: B. O. Seraphin
32 **Image Reconstruction from Projections** Implementation and Applications Editor: G. T. Herman
33 **Electrets** Editor: G. M. Sessler
34 **Nonlinear Methods of Spectral Analysis** 2nd Edition. Editor: S. Haykin
35 **Uranium Enrichment** Editor: S. Villani
36 **Amorphous Semiconductors** Editor: M. H. Brodsky
37 **Thermally Stimulated Relaxation in Solids** Editor: P. Bräunlich
38 **Charge-Coupled Devices** Editor: D. F. Barbe
39 **Semiconductor Devices** for Optical Communication. 2nd Ed. Editor: H. Kressel
40 **Display Devices** Editor: J. I. Pankove
41 **The Computer in Optical Research** Methods and Applications. Editor: B. R. Frieden
42 **Two-Dimensional Digital Signal Processing I** Linear Filters. Editor: T. S. Huang
43 **Two-Dimensional Digital Signal Processing II** Transforms and Median Filters. Editor: T. S. Huang
44 **Turbulent Reacting Flows** Editors: P. A. Libby and F. A. Williams
45 **Hydrodynamic Instabilities and the Transition to Turbulence** Editors: H. L. Swinney and J. P. Gollub
46 **Glassy Metals I** Editors: H.-J. Güntherodt and H. Beck
47 **Sputtering by Particle Bombardment I** Editor: R. Behrisch
48 **Optical Information Processing** Fundamentals. Editor: S. H. Lee
49 **Laser Spectroscopy of Solids** Editors: W. M. Yen and P. M. Selzer
50 **Light Scattering in Solids II.** Basic Concepts and Instrumentation Editors: M. Cardona and G. Güntherodt
51 **Light Scattering in Solids III.** Recent Results Editors: M. Cardona and G. Güntherodt
52 **Sputtering by Particle Bombardment II** Sputtering of Alloys and Compounds, Electron and Neutron Sputtering, Surface Topography Editor: R. Behrisch
53 **Glassy Metals II.** Atomic Structure and Dynamics, Electronic Structure, Magnetic Properties Editors: H. Beck and H.-J. Güntherodt
54 **Light Scattering in Solids IV.** Electronic Scattering, Spin Effects, SERS, and Morphic Effects Editors: M. Cardona and G. Güntherodt
55 **The Physics of Hydrogenated Amorphous Silicon I** Structure, Preparation, and Devices Editors: J. D. Joannopoulos and G. Lucovsky
56 **The Physics of Hydrogenated Amorphous Silicon II** Electronic and Vibrational Properties Editors: J. D. Joannopoulos and G. Lucovsky

Light Scattering in Solids IV

Electronic Scattering, Spin Effects, SERS, and Morphic Effects

Edited by M. Cardona and G. Güntherodt

With Contributions by
G. Abstreiter K. Arya M. Cardona S. Geschwind
G. Güntherodt R. Merlin A. Otto A. Pinczuk
R. Romestain B. A. Weinstein R. Zallen R. Zeyher

With 322 Figures

Springer-Verlag
Berlin Heidelberg GmbH 1984

Professor Dr. *Manuel Cardona*

Max-Planck-Institut für Festkörperforschung, Heisenbergstraße 1,
D-7000 Stuttgart 80, Fed. Rep. of Germany

Professor Dr. *Gernot Güntherodt*

Universität zu Köln, II. Physikalisches Institut, Zülpicher Straße 77,
D-5000 Köln 41, Fed. Rep. of Germany

ISBN 978-3-662-31148-6 ISBN 978-3-540-39522-5 (eBook)
DOI 10.1007/978-3-540-39522-5

Library of Congress Cataloging in Publication Data. (Revised for volume 54) Main entry under title: Light scattering in solids. (Topics in applied physics; v. 8, 50) (Series traced differently) Vols. edited by M. Cardona and G. Güntherodt. Includes bibliographies and indexes. 1. Light–Scattering. 2. Raman effect. 3. Semiconductors–Optical properties. 4. Solids–Optical properties. I. Cardona, Manuel, 1934–. II. Güntherodt, G. (Gernot), 1943–. III. Series: Topics in applied physics; v. 8, etc. QC427.4.L53 530.4'1 75-20237

Originally published by Springer-Verlag Berlin Heidelberg New York Tokyo in 1984
Softcover reprint of the hardcover 1st edition 1984

2153/3130-543210

Preface

This volume is the fourth of a series* devoted to light scattering in solids and related phenomena. The first and second volumes (TAP 8 and 50) emphasize general concepts and basic theory, the third one (TAP 51), investigations of specific materials and also Brillouin scattering, while the present one (TAP 54) discusses light scattering by electronic excitations (including magnetic effects), surface-enhanced Raman scattering (SERS), and effects of pressure on phonon spectra. A detailed list of the contents of the whole series can be found in the second edition (paperback) of Vol. 1 (TAP 8).

The reader may be struck by the imbalance in the size of the various contributions in this volume. Two of them (Chaps. 2 and 6) are exceptionally long. They grew to the present size because the fields they cover exploded while they were being written up. The reader may notice this in the structure of these chapters. While Chap. 2 was being written, the *quantum Hall effect* and *modulation doping* were discovered. This gave new impetus to light scattering by two-dimensional electron gases which had been early recognized as an ideal technique for the study of technologically important MOS structures, heterostructures, superlattices, and Schottky barriers. Chapter 6 discusses SERS from an experimental point of view, with particular emphasis on the effects of adatoms and other chemisorption phenomena versus the electromagnetic resonance mechanism, a rather controversial subject which is still the object of considerable current research. Chapter 6 is complemented in Chap. 7 by the theory of chemisorption-induced SERS. Chapters 3–5 discuss various aspects of light scattering by electrons and by phonons in which magnetic interactions are of the essence, including the important family of the rare-earth chalcogenides. Finally, Chap. 8 concerns itself with the dependence of scattering by phonons on hydrostatic pressure. It discusses data obtained mainly with the powerful and elegant diamond-anvil-cell technique.

The editors would like to thank all the authors for their cooperation in bringing this volume together and for the patience of those who complied with the original deadline. One half of the contributions are the fruit of transatlantic collaboration, with all the problems of logistics this involves. In the course of solving them, the editors have come to the realization that in this age, in which hundreds of jetliners cross the Atlantic Ocean daily, there is a lot of room for

* Topics in Applied Physics (TAP) Vols. 8, 50, 51, 54

improvement in the postal service within the Atlantic community. Thank goodness for the telephone and telex!

In view of the impossibility of mentioning explicitly the large number of scientists who directly or indirectly have influenced these volumes, we shall mention a few institutions:

AT & T Bell Laboratories in Murray Hill and Holmdel, NJ, Brown University, the IBM T. J. Watson Research Center in Yorktown Heights, NY, the Ioffe Institute in Leningrad, the Institut für Festkörperforschung der Kernforschungsanlage Jülich, the Max-Planck-Institut für Festkörperforschung in Stuttgart, the University of Pennsylvania in Philadelphia, PA, the Universität zu Köln, the University of Michigan in Ann Arbor, MI, and the Xerox Research Laboratories in Rochester, NY and Palo Alto, CA. Last but not least thanks are due to our secretaries Kerstin Weissenrieder and Suzanne Wood for patient organizational work and skillful typing of large parts of these volumes, and to Mr. B. Hillebrands for text editing and help with the keywords index.

Stuttgart and Köln,
December 1983

Manuel Cardona
Gernot Güntherodt

Contents

Contributors

Abstreiter, Gerhard
Technische Universität München, Physik Department
D-8046 Garching, Fed. Rep. of Germany

Arya, Karamjeet
Physics Department, The City College of the City University of New York
New York, NY 10031, USA

Cardona, Manuel
Max-Planck-Institut für Festkörperforschung, Heisenbergstraße 1
D-7000 Stuttgart 80, Fed. Rep. of Germany

Geschwind, Stanley
AT & T Bell Laboratories, 600 Mountain Avenue
Murray Hill, NJ 07974, USA

Güntherodt, Gernot
Universität zu Köln, II. Physikalisches Institut, Zülpicher Straße 77
D-5000 Köln 41, Fed. Rep. of Germany

Merlin, Roberto
Department of Physics, University of Michigan
Ann Arbor, MI 48109, USA

Otto, Andreas
Universität Düsseldorf, Physikalisches Institut III
Universitätsstraße 1, D-4000 Düsseldorf 1, Fed. Rep. of Germany

Pinczuk, Aron
AT & T Bell Laboratories, 4B-437, Holmdel, NJ 07733, USA

Romestain, Robert
Laboratoire de Spectrometrie Physique, B.P. 68
F-38402 St. Martin d'Hères, France

Weinstein, Bernard A.
Xerox Corporation, Joseph C. Wilson Center for Technology-114
Rochester, NY 14644, USA

Zallen, Richard
Virginia Polytechnic Institute, Blacksburg, VA 24061, USA

Zeyher, Roland
Max-Planck-Institut für Festkörperforschung, Heisenbergstraße 1
D-7000 Stuttgart 80, Fed. Rep. of Germany

1. Introduction

Manuel Cardona and Gernot Güntherodt

...y no puedo enterarme en esta verdad, sino es prouãdola, de manera q̃ la prueua manifieste los quilates de su bondad, como el fuego muestra los del oro...

M. de Cervantes, El Ingenioso Hidalgo don Quixote de la Mancha (J. de la Cuesta, Madrid, 1605) p. 185.

This is the fourth volume of a series devoted to light scattering in solids, with special emphasis on semiconductors. Volume I of the series [1.1] was published in 1975. A second corrected and annotated edition, with cross references to other volumes of the series, appeared in 1982 [1.1]. It includes the list of contents of Vols. I–IV. Volumes II [1.2] and III [1.3] also appeared in 1982.

1.1 Contents of Previous Volumes

Volume I of the series was written at a time when the power of light scattering for studying elementary excitations in solids had been amply demonstrated. However, a large body of results was not yet available. Chapter 1 of that volume is a historical introduction by M. Cardona. Chapter 2, by A. Pinczuk and E. Burstein, briefly discusses the macroscopic and microscopic aspects of the theory of the efficiency for scattering by phonons and by electronic excitations, with emphasis on resonance phenomena. Chapter 3 by R. M. Martin and L. M. Falicov dwells on more detailed aspects of resonant light scattering and the different types of possible resonance profiles. Chapter 4 by M. V. Klein is devoted to electronic Raman scattering, a subject which had received relatively little experimental (but a lot of theoretical) attention at the time the chapter was written. In it the author treats the theory of scattering by single-particle excitations, plasmons, plasmon-LO phonon coupled modes, discrete impurity levels, spin-flip excitations, intervalley fluctuations, acoustic plasmons, etc. Chapter 5 by M. H. Brodsky is devoted to vibrational scattering in amorphous materials, a subject which has picked up considerable technological interest in subsequent years. Chapter 6 by A. S. Pine contains a discussion of Brillouin scattering as applied to semiconductors. It was written prior to the discovery of polariton-mediated resonant Brillouin scattering, a subject which was treated later in [Ref. 1.3, Chap. 7]. Chapter 7 by Y. R. Shen covers stimulated Raman scattering and a few related nonlinear phenomena.

Volume II [1.2] contains a comprehensive article by *M. Cardona* (Chap. 2) on the theory of light scattering by phonons, with an extensive introduction to peripheral concepts such as the theory of phonons, the fluctuation-dissipation theorem, infrared absorption by phonons, etc. The chapter treats both macroscopic and microscopic theoretical concepts and emphasizes, as often in the series, resonance effects.

Chapter 3 of [1.2] by R.K. Chang and M.B. Long is devoted to an important instrumentation topic, namely, that of multichannel detection. Although it may well be the way of the future, this technique has not yet gained general acceptance. Reviews like the one being discussed, may help to propagate the technique. Chapter 4 of [1.2] by H. Vogt discusses coherent Raman and hyper-Raman techniques. It represents an expansion of [Ref. 1.1, Chap. 7] to cover developments in the field between 1975–1982.

Volume III [1.3] contains mainly case studies for families of materials or for classes of phenomena. Chapter 2 by M.S. Dresselhaus and G. Dresselhaus discusses light scattering in graphite intercalation compounds. Chapter 3 of [1.3] by D.J. Lockwood discusses scattering by electronic and magnetic excitations in a wide family of materials: the transition metal halides. Chapter 4 of [1.3] by W. Hayes treats the problem of light scattering by superionic conductors, a family of materials of considerable technological interest. Chapter 5 of [1.3] by M.V. Klein discusses light scattering by phonons and by electronic excitations in metals, using as examples a wide class of transition metal compounds (carbides, nitrides, chalcogenides). The chapter contains a detailed description of the rather elegant, formal theory of light scattering first proposed by Kawabata. This theory enables us to treat, in a compact way, scattering by phonons coupled to other quasiparticles such as electronic excitations, other phonons, etc. The theoretically inclined reader will enjoy the beauty of this presentation. Chapter 6 of [1.3] discusses developments in the field of Brillouin scattering which became possible thanks to the work of the author, J.R. Sandercock, in multipass tandem Fabry-Perot interferometry. It emphasizes scattering by metals and scattering by surface excitations.

Chapter 7 of [1.3] discusses the beautiful phenomenon of resonant (Brillouin) scattering, mediated by polaritons. The phenomenon was predicted in 1972 by Brenig, Zeyher, and Birman and first observed by the authors of this chapter, C. Weisbuch and R.G. Ulbrich, in 1977. It represents one of the rare and most beautiful recent examples of predictive solid-state theory.

The present volume is written in a spirit similar to that of Vol. III: Case studies for families of materials and classes of scattering phenomena. Chapter 2 by A. Pinczuk, G. Abstreiter, and M. Cardona, devoted to electronic excitations in semiconductors, is meant to bring up to date Chap. 4 of [1.1]. It discusses the theory and detailed experimental work concerning scattering by "optical" plasmons and plasmon-LO phonon coupled modes, acoustical plasmons and one-electron excitations (e.g., spin-flip). It also discusses scattering by mixed electronic-vibronic excitations coupled through the deformation potential mechanism and the corresponding renormalization of phonon energies. A large part of Chap. 2 is devoted to scattering by two-dimensional electron gases, including inversion layers, Schottky barriers, heterostructures and superlattices. The development in this field in the past few years has been phenomenal, due in part to the technological interest of these two-dimensional structures.

Chapter 3 by S. Geschwind and R. Romestain is devoted to spin-flip resonant Raman scattering and confined to the great wealth of experimental observations available for CdS. The method is applied to study phenomena as varied as the metal-insulator transition in impurity bands, spin relaxation times and terms linear in k in the energy bands. Chapter 4 by G. Güntherodt and R. Zeyher is concerned with the effects of magnetic order and spin disorder in the scattering by phonons in magnetic seminconductors. Both phenomenological and microscopic theory are presented and illustrated by experimental data for the different magnetic phases of magnetic semiconductors.

Chapter 5 by G. Güntherodt and R. Merlin reviews the very prolific research area of Raman scattering in rare-earth chalcogenides, consisting of magnetic semiconductors, magnetic or superconducting metals and mixed valence compounds. In all these materials the electron-lattice interaction, phonon anomalies and Raman intensities can be described consistently by the concept of phonon-induced local charge deformabilities.

Chapter 6 by A. Otto treats the important topic of surface-enhanced Raman scattering, a subject of great current interest with special conferences devoted solely to it. The emphasis here is to disentangle experimentally the various mechanisms which contribute to this striking phenomenon, including electromagnetic resonances produced by "macroscopic roughness" of the metallic surface as well as chemical effects of roughness on a microscopic scale. In many cases both types of enhancements seem to be of the same order of magnitude and each account for half of the observed effect. Chapter 7 by R. Zeyher and K. Arya presents a theoretical formulation of the problem based, in part, on classical electromagnetic resonances and also on the quantum-mechanical theory of chemisorption. Finally, Chap. 8 by B.A. Weinstein and R. Zallen discusses the most widely studied class of the so-called morphic effects, namely, the effect of hydrostatic pressure on Raman spectra due to phonons. The field has gained great impetus since the development of the ruby manometer for the diamond anvil cell ten years ago. While uniaxial stress is not explicitly treated, a list of relevant references is included.

References

1.1 M.Cardona (ed.): *Light Scattering in Solids*. 1st ed.; *Light Scattering in Solids I*: Introductory Concepts, 2nd ed. Topics Appl. Phys., Vol. 8 (Springer, Berlin, Heidelberg, New York 1975 and 1982)

1.2 M.Cardona, G.Güntherodt (eds.): *Light Scattering in Solids II: Basic Concepts and Instrumentation*, Topics Appl. Phys., Vol. 50 (Springer, Berlin, Heidelberg, New York 1982)

1.3 M.Cardona, G.Güntherodt (eds.): *Light Scattering in Solids III: Recent Results*, Topics Appl. Phys., Vol. 51 (Springer, Berlin, Heidelberg, New York 1982)

2. Light Scattering by Free Carrier Excitations in Semiconductors

Gerhard Abstreiter, Manuel Cardona, and Aron Pinczuk

With 99 Figures

The chapter discusses the applications of light scattering techniques to the investigation of excitations of free carriers in semiconductors. The free carriers are produced either by doping, by thermal, or by optical excitation. Several types of excitations can be studied. In semiconductors with simple, non-degenerate band extrema collective excitations, due to charge density fluctuations, and single particle excitations, related usually to spin density fluctuations, are observed. Semiconductors with multivalley extrema exhibit excitations due to intervalley density fluctuations. In cases where the masses of the particles fulfill the appropriate conditions acoustic plasmons can result. Of particular interest is the resonant behavior of these phenomena for incident laser frequency near electronic interband energy gaps. Such behavior can be used to enhance some of the scattering channels with respect to others and so to discriminate among them.

Changing the laser frequency enables one also to change the scattering wavevector within a limited range. It is then possible to investigate the dependence of scattering response functions on wavevectors. Since these response functions are related to the frequency- and wavevector-dependent dielectric constant $\varepsilon(\boldsymbol{q},\omega)$ the light scattering technique offers the unique opportunity of providing information about the wavevector dependence of $\varepsilon(\boldsymbol{q},\omega)$.

Electronic excitations can be modified by coupling to phonons and vice-versa. Light scattering yields considerable information about these phenomena. Two types of such effects are observed, those due to the coupling of either non-polar phonons or transverse polar phonons to electron systems and those due to the coupling of LO polar phonons to electrons or plasmons. The former are described by means of the deformation potential Hamiltonian, the latter, usually more spectacular, by the Fröhlich Hamiltonian. Investigations of these phenomena yield the real and imaginary parts of the phonon self energies and, in some cases, the Fano asymmetry parameter for the coupling of the phonons to electronic continua. These parameters can often be calculated by using band structure theory and compared with the experimental results.

Light scattering in semiconductors is usually performed with visible laser lines. In most cases the scattering light is absorbed within a penetration depth of the order of 1000 Å. The cross sections are such that phenomena taking place in a layer of as little as 100 Å can be detected. Light scattering is thus not directly sensitive to surface states, localized to a depth of ~5 Å. It can nevertheless reveal information on space charge layers produced by surface pinning of the Fermi energy and electron layers at heterostructure interfaces. Nearly two-dimensional electron gases can be confined to such surfaces and interfaces. The third dimension, perpendicular to the surfaces or interfaces, can exhibit quantum mechanical effects of considerable interest. Because of its sensitivity, such effects can be easily investigated by light scattering. As in the 3-dimensional case, single particle and collective excitations can be observed. These excitations can be coupled to phonons. Their cross section can exhibit resonances for laser frequencies in the neighborhood of electronic interband critical points. The properties of such 2-dim electron gases can be changed by external electric (and magnetic) fields. In the case of semiconductor-atmosphere interfaces the properties of the 2-dim electron systems can be altered by the type of gas (or vacuum) in contact with the surface.

The typical sampling depths of light scattering (~ 1000 Å) enable one also to investigate electron gases in multilayer structures (multiple quantum wells). Also in these cases, single particle and collective excitations are observed.

The present chapter describes phenomena observed in tetrahedrally coordinated semiconductors of the diamond (Ge, Si) and zincblende types (III–V compounds, such as GaAs, InAs, etc.). The wide variety of electronic gaps and band extrema in these materials permits the observation of a large number of resonance phenomena. The fact that some of them (Ge, Si) are non-polar, while others (GaAs, InSb) are polar yields a wide range of electron-phonon interaction phenomena. This family of semiconductors appears to yield model substances for the wide range of phenomena related to scattering of light by free carriers in semiconductors. Their electronic band structures and their lattice dynamics are usually well known, a fact which simplifies theoretical analysis of the results. Because of their technological applications crystal growth and sample preparation pose no major problems.

2.1 Introduction and Historical Survey

There was early interest in the theory of light scattering by plasmas concerning highly diluted gaseous systems [2.1–2.3] and also in the case of solid state electron plasmas [2.4]. The scattering cross section was seen to be related to the spectrum of density fluctuations. After the invention of the laser, it became clear that semiconductor plasmas were excellent candidates for light scattering studies [2.5–7]. While most gaseous plasmas constitute dilute high temperature systems, in semiconductors it is possible to realize cold degenerate plasmas in which electron-electron interactions play a substantial role. The plasma frequencies of free carriers in semiconductors, unlike those of metals, fall in the range accessible to conventional Raman spectroscopy. Moreover, the plasma density can be varied within a wide range by doping and by thermal or optical excitation.

In this theoretical work it was found that at the densities typical of semiconductor plasmas, the one-electron excitations accessible to light scattering are modified by "collective" dynamical screening effects associated with the "longitudinal" polarization of the plasma [2.5]. It was also recognized that band structure effects influence light scattering in a number of ways. One effect is the replacement of the free-electron mass by the effective mass tensor [2.6,7]. Another is the resonant behavior of the light scattering cross section for laser frequencies close to optical interband energy gaps [2.8]. Furthermore, it was also pointed out that nonparabolicity effects should have a substantial influence on selection rules, particularly under magnetic fields [2.8].

It was also recognized that semiconductors can accommodate multicomponent plasmas. These may arise from the multivalley nature of the electron energy bands and also from the simultaneous presence of electrons and holes. In addition to the usual (optic) plasmons, such systems can sustain excitations in which the fluctuation in charge density associated with one type of carriers cancels almost exactly the one associated with other types of carriers. These excitations were discussed by *Platzman* [2.6] and by *McWhorter* [2.7] in the context of light scattering by intervalley density fluctuations and by acoustic plasmons.

Single component, as well as multicomponent systems, have been shown to have a class of excitations which carry no net fluctuation in charge density and therefore have single particle character. These are the spin-density fluctuations, in which the spin of the electrons is changed in the light scattering process through the intervention of spin-orbit coupling [2.9]. Spin-flip transitions in a magnetic field were first considered by *Yafet* [2.10]. Light scattering by single particle excitations without a spin-flip had also been predicted in the case of nonparabolic electron bands [2.11].

The first observation of laser light scattering by a solid-state plasma was reported by *Mooradian* and *Wright* [2.12] in doped n-type GaAs. The spectra displayed scattering by longitudinal charge density excitations, i.e., plasmons coupled to LO phonons. In 1968 *Mooradian* [2.13] reported light scattering by single particle excitations of degenerate free electrons in n-GaAs. This work also opened the possiblity of measuring the distribution of kinetic energies of the free carriers by means of inelastic light scattering [2.14].

The scattering of light by the free electrons of a semiconductor in a magnetic field was first considered by *Wolff* [2.8] and by *Yafet* [2.10]. The first observations were reported by *Slusher* et al. [2.15] in InSb. The spectra were associated with electron spin-flip and Landau-level transitions. Light scattering by collective magneto-plasma modes was reported soon thereafter for n-type InAs [2.16].

This pioneering research on light scattering by semiconductor plasmas has stimulated a steady flow of exciting theoretical and experimental results. Several excellent reviews have covered the developments in the field. *Mooradian* [2.17] has given detailed accounts of his work. *Yafet* [2.18] has reviewed the theory and early experiments in semiconductors, including the work in magnetic fields. *Platzman* and *Wolff* [2.19], in their monograph on solid-state plasmas, have also considered light scattering in semiconductors and have presented a detailed theoretical discussion of magneto-plasmas and multicomponent systems. *Klein* [Ref. 2.20, Chap. 4] as well as *Hayes* and *Loudon* [2.21] have published extensive reviews of experimental and theoretical work in the field. Several aspects of spin-flip light scattering in a magnetic field have been reviewed by *Patel* [2.22a], *Wolff* [2.22b] and *Scott* [2.22c].

Recent years have seen a number of stimulating developments in the light scattering spectroscopy of free carriers in semiconductors. These make up the dominant subject of this chapter. We are concerned with phenomena as rich and diverse as the elementary excitations of multicomponent plasmas, the electron-phonon coupling in heavily doped semiconductors and the frequency and wave-vector dependence of the longitudinal dielectric function. Another active area which we consider here is related to the applications of light scattering to the spectroscopy of free electrons at semiconductor surfaces and interfaces. These include the investigation of surface space-charge layers and of two-dimensional electron systems at semiconductor heterostructures and metal-insulator-semi-conductor interfaces.

Light scattering by density fluctuations in multicomponent carrier systems of semiconductors was first considered theoretically in 1965 [2.6, 7]. These theories predict the presence of *optic* and *acoustic* plasmons as well as of *unscreened* density fluctuations. This and subsequent theoretical work has been reviewed previously ([2.19] and [Ref. 2.20, Chap. 4]). Light scattering by unscreened density fluctuations was first observed in 1977 in heavily doped n-type Si [2.23] and interpreted in terms of intervalley fluctuations. It has been recently suggested that this phenomenon can be used to determine intervalley scattering times [2.24]. The electron-hole plasmas created by photoexcitation with an intense laser source are another kind of multicomponent systems that has been the subject of a number of light scattering studies. Nondegenerate systems have been studied in GaAs [2.25]. Single-particle excitations of photoexcited holes have been observed in Si [2.26] and of photoexcited electrons in GaAs [2.27]. Coupled optical plasmons LO-phonons have been studied in photoexcited GaP [2.28] and GaAs [2.27]. Most recently, the acoustic mode of the electron-hole plasma has been identified in light scattering spectra of photoexcited GaAs [2.29]. The mode was first predicted in 1956 [2.30]. This was the first observation of the acoustic plasmon in a solid-state system.

The q-dependence of coupled plasmons LO-phonons of doped polar semiconductors was investigated extensively by light scattering. The occurrence of plasma dispersion effects in light scattering was initially identified in spectra of nondegenerate [2.31] and degenerate [2.32] plasmas in n-type GaAs. Dispersion effects were also reported in n-type InAs [2.33], n-type InP [2.34] and in n-type GaSb [2.35]. The dispersion of the coupled modes of n-GaAs was studied in considerable detail. It was recognized that the spectra contain features related to Landau-damping of the modes and to wavevector nonconservation due to absorption of the laser light [2.33, 36, 7]. In more recent work [2.38, 39], the spectral line shapes of coupled modes of n-GaAs were interpreted in terms of wavevector and frequency-dependent electron-gas dielectric functions that include the effects of finite temperature and electron relaxation processes.

Besides dispersion effects, the light scattering spectra of coupled plasmon LO-phonon modes are also influenced by surface space-charge layers [2.33]. *Stolz* and *Abstreiter* [2.40] have taken advantage of this effect and demonstrated that Raman spectroscopy is a comparatively easy and sensitive experimental technique to characterize semiconductor surfaces. In more recent work [2.41] they have also been able to observe the plasma mode associated with the surface layer.

Optical phonons also interact with free carriers via the weaker deformation potential mechanism. The effect occurs in polar and nonpolar semiconductors and it involves both interband as well as intraband excitations of the carriers [2.42]. It manifests as small renormalizations of the $q \approx 0$ optical phonons, i.e., changes in phonon frequencies and linewidths. These phenomena were first observed in light scattering spectra of heavily doped p-type Si [2.43,44]. More recently it was also identified in n-type Si [2.45], p-type Ge [2.46], p-type GaAs [2.46] and re-examined in p-type Si [2.47,48]. The most spectacular of these effects are the changes in the line shapes of the light scattering spectra of the renormalized phonons with changes in laser frequency [2.48]. The range of effects observed is rather broad and finds its complement in recent determination of self-energies by means of neutron scattering [2.49]. The Raman measurements, while limited to $q \simeq 0$, are considerably more accurate than neutron data.

The development of tunable dye lasers has made possible the investigation of resonance phenomena in the light scattering cross sections by single-particle and collective excitations. Resonant behavior in the cross sections of the electron-gas excitations was studied in considerable detail in n-type GaAs [2.50,51]. This work led *Burstein* and coworkers [2.52] to propose resonant light scattering as a sensitive method for studies of the elementary excitations of two-dimensional electron systems at semiconductor surfaces and interfaces. This proposal was coincidental with reports of the achievement of high mobility 2-dim electron systems in GaAs-(AlGa)As heterostructures made by molecular beam epitaxy [2.53]. Shortly afterwards came the first observations of light scattering by the electrons confined at these heterostructures [2.54,55]. This success stimulated considerable light scattering work in that system [2.56]. More recent results in GaAs-(AlGa)As heterostructures involve photoexcited 2-dim plasmas [2.57] and the observation of excitations associated with the degrees of freedom for electron motion parallel to the interface [2.58,59]. Successful experiments have also been reported in metal-insulator-semiconductor interfaces [2.60] and in Si-MOS p-type accumulation layers [2.61]. Light scattering has become an extremely useful tool for the investigation of 2-dim electron systems in semiconductors. This is exemplified by the recent work in doping superlattices [2.62] and in Ge-GaAs heterostructures [2.63].

2.2 Light Scattering by Electron Plasmas in Semiconductors

In this section we present some aspects of the theory of light scattering by a free electron gas that are relevant to semiconductors. Most of the material discussed here has already been covered by previous reviews on light scattering by solid-state plasmas [2.17–21], which are also the general references on the field. We have also made use of the conceptual framework outlined in [Ref. 2.64, Chap. 2]. SI units will be used throughout in this section unless otherwise indicated.

2.2.1 Electron-Photon Coupling in Semiconductors

The Hamiltonian that represents the interaction of the incident and scattered light fields with electrons in the semiconductor can be written as [2.18–21,64]

$$H_{\mathrm{ep}} = H''_{\mathrm{ep}} + H'_{\mathrm{ep}}, \tag{2.1}$$

where

$$H''_{\mathrm{ep}} = \frac{e^2}{2m} \sum_j [\boldsymbol{A}(\boldsymbol{r}_j)]^2 \tag{2.2a}$$

and

$$H'_{\mathrm{ep}} = \frac{e}{2m} \sum_j [\boldsymbol{p}_j \cdot \boldsymbol{A}(\boldsymbol{r}_j) + \boldsymbol{A}(\boldsymbol{r}_j) \cdot \boldsymbol{p}_j]. \tag{2.2b}$$

In these equations $\boldsymbol{A}(\boldsymbol{r}_j)$ represents the sum of the vector potentials of the incident and scattered fields with frequencies ω_{L} and ω_{S}, respectively. The quantities e and m are the charge and mass of the free electrons. The summations extend to all the electrons.

Equation (2.2a) is of second order in the fields and therefore leads to light scattering in first-order perturbation theory. It yields the familiar Thomson cross section, proportional to the square of the classical radius of the electron ([2.64] and [Ref. 2.20, Chap. 4]).

H'_{ep} is of first order in the fields. It gives rise to light scattering in second-order perturbation theory. There are two kinds of terms that enter into the second-order contributions of H'_{ep}. They involve the *intraband* and *interband* matrix elements of $\boldsymbol{p}_j$, respectively. The intraband terms represent the solid state analog of the nonrelativistic Compton effect. Such terms being of the order of v_j/c, where the v_j are the electron velocities, can be ignored [2.19,64]. On the other hand, the interband matrix elements do make a contribution that can be important in small effective mass semiconductors [2.8]. In considering their contribution, it is necessary to take into account the fact that the second-order perturbation of H'_{ep} has a *resonant denominator* approximately equal to

$(E_G - \hbar\omega_L)$, where E_G is the optical energy gap. This results in two relatively well-defined situations [Ref. 2.64, Sect. 2.2.2]. One is the *nonresonant* case in which $\hbar\omega_L \ll E_G$. The other is the *resonant* case which $\hbar\omega_L \approx E_G$.

In addition to the enhancement of the scattering cross section, *resonant effects* are anticipated to give rise to light scattering that is *antisymmetric* in the polarizations of the incident and scattered light [2.65]. Light scattering believed to be antisymmetric has been extensively studied in spectra by the electron gases of n-type semiconductor crystals [2.13,17,50,51]. These spectra have been interpreted in terms of spin-density fluctuations. Explicit calculations of the scattering cross sections involve matrix elements of H'_{ep} associated with intermediate electronic states which, due to the spin-orbit interaction, are eigenstates of total angular momentum [2.9,10,18,20,64,65].

The features that are unique to resonant light scattering spectroscopy will be considered in more detail in Sect. 2.3. In this section we shall be largely concerned with a group of phenomena that are common to nonresonant as well as resonant scattering. In most cases, resonant effects can be included simply in the expressions of the light scattering cross section given below by multiplying by the resonant enhancement factor $E_G^4/[E_G^2 - (\hbar\omega_L)^2]^2$ [2.8,9,66,67].

2.2.2 Single-Component Plasmas

We consider here the case in which the semiconductor plasma consists of a single species, which we exemplify as electrons in a single parabolic conduction band with effective mass tensor $\boldsymbol{m}^*$. Effects specifically related to nonparabolicity of the electron energy bands, not taken into account here, have been discussed by *Wolff* [2.11].

In the nonresonant case the contributions to the light scattering cross section associated with H''_{ep} and the interband terms of H'_{ep} can be collected into an effective A^2 light scattering Hamiltonian of the form [2.19,20]

$$\begin{aligned} H''_{\text{eff}} &= \frac{e^2}{2} \sum_j \boldsymbol{A}(\boldsymbol{r}_j) \cdot \frac{1}{\boldsymbol{m}^*} \cdot \boldsymbol{A}(\boldsymbol{r}_j) \\ &= e^2\, \hat{\boldsymbol{\varepsilon}}_L \cdot \frac{1}{\boldsymbol{m}^*} \cdot \hat{\boldsymbol{\varepsilon}}_S \,[\sum_j e^{i\boldsymbol{q}\cdot\boldsymbol{r}_j}]\, A(\omega_L) A(\omega_S), \end{aligned} \tag{2.3}$$

where $\hat{\boldsymbol{\varepsilon}}_L$ and $\hat{\boldsymbol{\varepsilon}}_S$ are the unit polarization vectors of the incident and scattered fields with wavevectors $\boldsymbol{k}_L$ and $\boldsymbol{k}_S$, and

$$\boldsymbol{q} = \boldsymbol{k}_L - \boldsymbol{k}_S \tag{2.4}$$

is the *scattering wavevector*. In addition,

$$A(\omega_L) = \left(\frac{\hbar}{2\varepsilon(\omega_L) V \omega_L}\right)^{1/2} \tag{2.5}$$

and a similar expression for $A(\omega_S)$ are the amplitudes of the vector potentials of the incident and scattered fields. In (2.5), V is the volume of illuminated semiconductor, or *scattering volume*, and $\varepsilon(\omega_L)$ is the dielectric constant at ω_L.

The effective light scattering Hamiltonian can be rewritten as

$$H''_{\text{eff}} = \frac{e^2}{2} V\left(\hat{\varepsilon}_L \cdot \frac{1}{\boldsymbol{m}^*} \cdot \hat{\varepsilon}_S\right) N(-\boldsymbol{q}) A(\omega_L) A(\omega_S), \tag{2.6}$$

where

$$N(\boldsymbol{q}) = \frac{1}{V} \sum_j e^{i\boldsymbol{q}\cdot\boldsymbol{r}_j} \tag{2.7}$$

is the Fourier transform of the electron-density operator. Equation (2.6) shows that light scattering probes the spectrum of density fluctuations of the semiconductor electron plasma, as it occurs in a free electron gas [2.1–3]. Therefore, the differential scattering cross section can be written as [2.4–7, 19–21]

$$\frac{\partial^2 \sigma}{\partial \Omega\, \partial \omega} = r_0^2 \left(\frac{\omega_S}{\omega_L}\right)^2 m^2 V^2 \left(\hat{\varepsilon}_L \cdot \frac{1}{\boldsymbol{m}^*} \cdot \hat{\varepsilon}_S\right)^2 S(\boldsymbol{q},\omega), \tag{2.8}$$

where $r_0 = (e^2/4\pi\varepsilon_0 mc^2)$ is the classical radius of the electron and

$$\omega = \omega_L - \omega_S \tag{2.9}$$

is the *scattering frequency*. $S(\boldsymbol{q},\omega)$ is the *dynamical structure factor*, defined as

$$S(\boldsymbol{q},\omega) = \frac{1}{2\pi} \int_{-\infty}^{\infty} e^{i\omega t} \langle n(\boldsymbol{q},t) n^\dagger(\boldsymbol{q},0)\rangle dt, \tag{2.10}$$

where the angular brackets denote a thermal average over the electron initial states, and $n(\boldsymbol{q},t)$ is the time-dependent (in the Heisenberg sense) version of $N(\boldsymbol{q})$. In the interpretation of absolute scattering intensities, it is often convenient to work with *scattering efficiencies* ($\partial^2 S/\partial\Omega\,\partial\omega$) defined as the ratio of scattered to incident intensities per unit length of traversed material, unit solid angle and unit frequency [2, 64]. $\partial^2 S/\partial\Omega\,\partial\omega$ is obtained by dividing (2.8) by the scattering volume V. Note that (2.8) represents the cross section per volume V, not per electron or per unit cell.

At first sight, (2.8–10) do not appear to be very revealing about the relationship between light scattering spectra and the motion of the free electrons in the semiconductor. Fortunately, this is not the case. The formulation is extremely convenient because the dynamical structure factor $S(\boldsymbol{q},\omega)$ is directly related to the dielectric response of the electron plasma. We note that the effective light scattering Hamiltonian, specified by (2.3–7), can be interpreted as a scalar potential with Fourier components given by

$$\phi(\boldsymbol{q}) = -\frac{V}{2}\left(\hat{\varepsilon}_{\mathrm{L}} \cdot \frac{1}{\boldsymbol{m}^*} \cdot \hat{\varepsilon}_{\mathrm{S}}\right)[eN(-\boldsymbol{q})]A(\omega_{\mathrm{L}})A(\omega_{\mathrm{S}}), \tag{2.11}$$

which drives electron charge-density fluctuations $[eN(-\boldsymbol{q})]$. This way it becomes apparent that our problem is a standard one in the theory of the dielectric response of the electron gas [2.4,68].

The dynamical structure factor will be calculated below in terms of the longitudinal dielectric function by means of the *fluctuation-dissipation theorem*. In carrying out this analysis it is convenient to rewrite (2.10) in terms of the plasma polarization $\boldsymbol{P}(\boldsymbol{q},t)$, related to $n(\boldsymbol{q},t)$ by Poisson's equation

$$n(\boldsymbol{q},t) = \frac{\mathrm{i}\boldsymbol{q} \cdot \boldsymbol{P}(\boldsymbol{q},t)}{-e} = -\frac{\mathrm{i}}{e}\,\boldsymbol{q} \cdot \boldsymbol{P}(\boldsymbol{q},t), \tag{2.12}$$

which enables us to write (2.10) as

$$S(\boldsymbol{q},\omega) = \frac{q^2}{2\pi e^2} \int_{-\infty}^{\infty} \mathrm{e}^{\mathrm{i}\omega t} \langle \boldsymbol{P}(\boldsymbol{q},t)\boldsymbol{P}^\dagger(\boldsymbol{q},0)\rangle dt. \tag{2.13}$$

In the next sections we present the explicit calculation of $S(\boldsymbol{q},\omega)$ for single-component plasmas. Here we shall consider briefly the scattering cross section, which by using (2.13) in (2.8), can be written as

$$\frac{\partial^2\sigma}{\partial\Omega\,\partial\omega} = r_0^2\left(\frac{\omega_{\mathrm{S}}}{\omega_{\mathrm{L}}}\right)^2 m^2 V^2 \left(\hat{\varepsilon}_{\mathrm{L}} \cdot \frac{1}{\boldsymbol{m}^*} \cdot \hat{\varepsilon}_{\mathrm{S}}\right)^2 \cdot \frac{q^2}{2\pi e^2} \int_{-\infty}^{\infty} \mathrm{e}^{\mathrm{i}\omega t} \langle \boldsymbol{P}(\boldsymbol{q},t)\boldsymbol{P}^\dagger(\boldsymbol{q},0)\rangle dt. \tag{2.14}$$

It is interesting to compare (2.14) with the corresponding expression for Raman scattering by lattice vibrations as given, for example, by [Ref. 2.64, Eqs. (2.55,56)]. In the latter, P signifies the *Raman polarizability*, whereas in (2.14) it represents the *polarization* of the free carriers. We focus on the factor ω_{S}^4 which is characteristic of macroscopic formulations of light scattering. It is most striking that the factor occurs in [Ref. 2.64, Eq. (2.55)], but does not appear in (2.14). It has been shown that the ω_{S}^4 factor occurs in microscopic theories of the light scattering cross section, but only under circumstances in which the scattering frequency ω is below that of all the possible intermediate electronic excitations that contribute to the polarizability [2.69]. This condition holds for nonresonant scattering by phonons. It obviously does not hold for scattering by the electron-gas excitations.

2.2.3 Fluctuation-Dissipation Analysis

The calculation of the dynamical structure factor $S(\boldsymbol{q},\omega)$ given by (2.13) is carried out by means of the *fluctuation-dissipation theorem* [2.70]. The applications of the method to inelastic light scattering were described in [2.2, 64]. We consider the specific case of single-component plasmas. Multicomponent plasmas are considered in Sect. 2.2.6.

In order to state the fluctuation-dissipation theorem, we first consider a physical variable $F(\boldsymbol{r},t)$, to be called "generalized force", which produces in the system a change, or response, $X(\boldsymbol{r},t)$ in another physical variable. $X(\boldsymbol{r},t)$ will be called a "generalized displacement". The nomenclature is related to the fact that the variables X and F are "conjugated" in the sense that when multiplied they yield the Hamiltonian associated with the perturbation that occurs when F is acting on the system:

$$H(\boldsymbol{r},t)=F(\boldsymbol{r},t)X(\boldsymbol{r},t). \tag{2.15}$$

The variables F and X can be a true force and a true displacement, as would occur in a mechanical system. In the case of the electron plasma, to evaluate (2.15) we choose as conjugated variables the induced dipole moment $V\boldsymbol{P}(\boldsymbol{q},t)$ and an externally applied ("driving") longitudinal electric field $\boldsymbol{D}(\boldsymbol{q},t)$. This field is introduced here with the only purpose of evaluating the amplitude of the thermal fluctuations of $\boldsymbol{P}(\boldsymbol{q},t)$, which are independent of the magnitude of $\boldsymbol{D}(\boldsymbol{q},t)$. We could also have thought of $\boldsymbol{D}(\boldsymbol{q},t)$ as being associated with the effective light scattering potential Φ given by (2.11) ($\boldsymbol{D}=-\nabla\Phi$). Such an approach leads to a semi-classical derivation of the light scattering cross section, as described in [2.71a].

We introduce next the *linear response* function $T(\boldsymbol{q},\omega)$ that relates the Fourier components of X and F through

$$X(\boldsymbol{q},\omega)=T(\boldsymbol{q},\omega)F(\boldsymbol{q},\omega). \tag{2.16}$$

The fluctuation-dissipation theorem states that

$$\frac{1}{2\pi}\int\limits_{-\infty}^{\infty} e^{i\omega t}\langle X(\boldsymbol{q},t)X^{\dagger}(\boldsymbol{q},0)\rangle dt=\langle X(\boldsymbol{q},t)X^{\dagger}(\boldsymbol{q},0)\rangle_{\omega}$$

$$=\frac{\hbar}{\pi}(1-e^{-\hbar\omega/k_BT})^{-1}\,\mathrm{Im}\,\{T(\boldsymbol{q},\omega)\}. \tag{2.17}$$

Equation (2.17) is the quantum-mechanical version of the theorem. $[1-e^{-\hbar\omega/k_BT}]^{-1}$ is the Bose-Einstein statistical factor. In the context of light scattering, (2.17) describes the Stokes component of the spectra for $\omega>0$ and the anti-Stokes one for $\omega<0$.

In order to calculate the dynamical structure factor defined by (2.13), we set

$$X \equiv V\boldsymbol{P}(\boldsymbol{q},t), \tag{2.18a}$$

$$F \equiv \boldsymbol{D}(\boldsymbol{q},t), \tag{2.18b}$$

and define a linear response function for *purely longitudinal* polarization fluctuations ($\boldsymbol{P} \| \boldsymbol{D} \| \boldsymbol{q}$):

$$P(q,\omega) = \frac{1}{V} T(q,\omega) D(q,\omega). \tag{2.19}$$

Next we introduce the longitudinal dielectric susceptibility of the electron gas $\chi(q,\omega)$ defined as

$$P(q,\omega) = \varepsilon_0 \chi(q,\omega) E(q,\omega), \tag{2.20}$$

where ε_0 is the dielectric constant of vacuum and $E(q,\omega)$ is the *total* longitudinal electric field in the electron plasma given by [2.4,68]

$$E(q,\omega) = D(q,\omega) - \frac{P(q,\omega)}{\varepsilon_0 \varepsilon_\infty}. \tag{2.21}$$

In (2.21), ε_∞ is the background dielectric constant of the semiconductor *relative to vacuum* and the second term in the right side represents the electric field associated with the plasma polarization. From a microscopic point of view, the use of (2.21) is equivalent to treating Coulomb electron-electron interactions, the dominant factor in the response of electron plasmas, within the *random phase approximation* (RPA) [2.4,68].

Equations (2.19–21) lead to the following expression for the response function:

$$T(q,\omega) = \frac{V \varepsilon_0 \varepsilon_\infty \chi(q,\omega)}{\varepsilon(q,\omega)}, \tag{2.22}$$

where

$$\varepsilon(q,\omega) = \varepsilon_\infty + \chi(q,\omega) \tag{2.23}$$

is the *total* dielectric function of the semiconductor. By using (2.17–22) in (2.13), we find that the dynamical structure factor is

$$\begin{aligned} S(q,\omega) &= \frac{\hbar}{\pi} \frac{q^2}{e^2} \frac{\varepsilon_0 \varepsilon_\infty}{V} (1 - \mathrm{e}^{-\hbar\omega/k_B T})^{-1} \operatorname{Im} \left\{ \frac{\chi(q,\omega)}{\varepsilon(q,\omega)} \right\} \\ &= \frac{\hbar}{\pi} \frac{q^2}{e^2} \frac{\varepsilon_0 \varepsilon_\infty^2}{V} (1 - \mathrm{e}^{-\hbar\omega/k_B T})^{-1} \operatorname{Im} \left\{ \frac{-1}{\varepsilon(q,\omega)} \right\}. \end{aligned} \tag{2.24}$$

We next consider two limiting cases that illustrate the properties of semiconductor plasmas which can be studied by inelastic light scattering. One is that of the "highly diluted" plasma; the other is that of the dense, degenerate electron gas. For simplicity we assume that the effective mass is a scalar m^*. Generalization to a tensor mass is straightforward.

2.2.4 Single-Component Plasmas: Highly Diluted Case

In the highly diluted electron gas, the density is so small that the *plasma frequency* ω_P readily satisfies

$$\omega_P = \left(\frac{Ne^2}{\varepsilon_0 \varepsilon_\infty m^*}\right)^{1/2} \ll \omega \tag{2.25}$$

in the frequency range of interest to light scattering. In such highly diluted plasma, the free electrons make a small contribution to the dielectric susceptibility of the semiconductor, i.e.

$$\chi(q,\omega) \ll \varepsilon_\infty . \tag{2.26}$$

Using (2.26) in (2.23,24), we find that the dynamical structure factor is well-approximated by the expression

$$S(q,\omega) = \frac{\varepsilon_0}{V} \frac{\hbar}{\pi} \frac{q^2}{e^2} (1 - e^{-\hbar\omega/k_B T})^{-1} \operatorname{Im}\{\chi(q,\omega)\}. \tag{2.27}$$

Equation (2.27) indicates that the light scattering spectra are directly related to Im $\{\chi(q,\omega)\}$, which is proportional to the density of states for single-particle excitations of the free electron gas [2.4]. Therefore, this result shows that the highly diluted system behaves as a collection of noninteracting electrons. This is not surprising. In a highly diluted system the average separation between the electrons is large and electron-electron interactions that give rise to collective behavior play a minor role.

In the limit of a "degenerate" electron gas, in which the *Fermi energy* is $E_F \gg k_B T$, the energies of possible single particle excitations are within the hatched region of Fig. 2.1. In this figure, k_F represents the *Fermi wavevector* related to the electron density by

$$k_F = (3\pi^2 N)^{1/3} \tag{2.28}$$

and to the Fermi energy by

$$E_F = \hbar^2 k_F^2 / 2m^*. \tag{2.29}$$

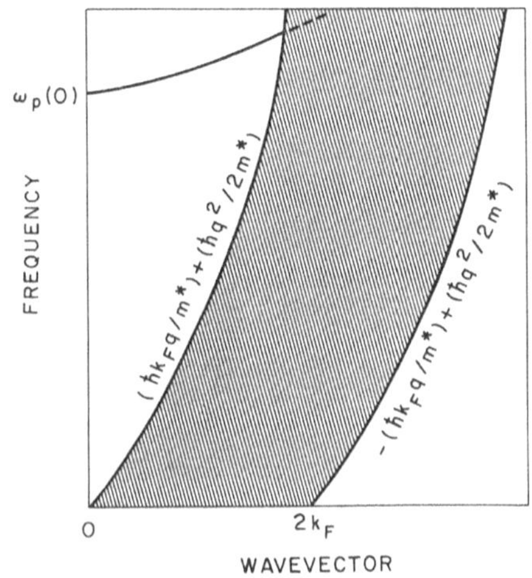

Fig. 2.1. Frequencies of elementary excitations of a degenerate electron gas as a function of wavevector. The hatched region represents the continuum of single-particle excitations (Landau-damping region). The plasma frequency ω_p is shown with a weak q^2-dispersion

In this case, Im $\{\chi_e(q,\omega)\}$ and also $S(q,\omega)$ have the triangular shape that is characteristic of the *Lindhard* dielectric function [2.4, 68, 71b].

In a dilute plasma, the degeneracy condition $E_F \gg k_B T$ can only be satisfied at extremely low temperatures. At these temperatures the free carriers present in the semiconductor, either because of thermal excitation or by light doping, freeze out. In practice, highly diluted semiconductor plasmas are studied at the relatively high temperatures that satisfy $E_F \ll k_B T$. These plasmas have classical (i.e. Maxwellian) behavior. The *imaginary* part of their dielectric susceptibility is given by [2.19, 2.71b]

$$\mathrm{Im}\,\{\chi(q,\omega)\} \simeq \frac{\pi\varepsilon_\infty\omega}{\lambda_D^2 q^3}\left(\frac{m^*}{2\pi k_B T}\right)^{1/2} \exp\left(\frac{-m^*\omega^2}{2q^2 k_B T}\right), \tag{2.30}$$

where λ_D is the *Debye length* given by

$$\lambda_D = \left(\frac{k_B T \varepsilon_0 \varepsilon_\infty}{Ne^2}\right)^{1/2} \tag{2.31}$$

and λ_D^{-1} is the *screening wavevector* of the electron gas in this classical limit. We note that for the limit given by (2.26) to hold, it is required that

$$\lambda_D^{-1} \ll q. \tag{2.32}$$

This condition is fulfilled in GaAs at room temperature for densities $n \lesssim 10^{16}\,\mathrm{cm}^{-3}$. Light scattering by dilute Maxwellian plasmas in GaAs was first observed by *Mooradian* [2.17]. Low-density and high-temperature plasmas were also observed in photoexcited GaAs [2.25, 72].

2.2.5 Single-Component Plasmas: The High Density Case

In the high density plasma, electron-electron interactions play a dominant role. The light scattering cross section, related to the effective Hamiltonian of (2.6) and given by (2.8, 24), is determined by the full longitudinal response function $\mathrm{Im}\{-1/\varepsilon(q,\omega)\}$. The longitudinal dielectric susceptibility of the electron gas is usually calculated within the RPA [2.4, 68]. The conditions under which the RPA is anticipated to be valid can be expressed in terms of the dimensionless parameter r_s:

$$r_s = \left(\frac{3}{4\pi N}\right)^{1/3} \left(\frac{m^* e^2}{4\pi\varepsilon_0\varepsilon_\infty \hbar^2}\right), \tag{2.33}$$

which represents the average distance between the electrons, measured in terms of the effective Bohr radius of free electrons in the semiconductor. The RPA is believed to give a good description of the properties of the electron gas when

$$r_s \ll 1. \tag{2.34}$$

For typical doped or photoexcited semiconductors, $r_s \lesssim 1$. The situation here is better than that in metals, in which $2 \lesssim r_s \lesssim 6$, where the RPA has been extensively used. Light scattering by free carriers in semiconductors is usually interpreted within the theoretical framework of the RPA. *Vasconcellos* and *Luzzi* [2.73] have considered the light scattering cross section of the electron gas beyond the usual RPA with a model calculation of exchange and correlation effects.

Turning now to the specific case of a degenerate semiconductor ($E_F \gg k_B T$), it is clear that we anticipate structure in the light scattering spectra at energies related to those of the maxima in $\mathrm{Im}\{-1/\varepsilon(q,\omega)\}$. The hatched region in Fig. 2.1 corresponds to the range of wavevectors and frequencies where single-particle excitations of the free electrons are possible. This is the region in which the lifetimes of collective modes are limited by decay into single-particle excitations, i.e., *Landau damping* [2.4, 68]. In the case of free electrons occupying states in a single energy band, there is no other loss mechanism within this framework. Losses may occur due to electron-phonon and also electron-impurity interactions. The phenomena associated with these additional losses are usually taken into account by introducing a *collision relaxation time* τ. Their effect can be included in the Lindhard expression for the dielectric susceptibility by means of a procedure formulated by *Mermin* [2.74]. The *Lindhard-Mermin* expression is given by

$$\chi(q,\omega) = \frac{(1+\mathrm{i}\Gamma/\omega)\,[\chi^o(q,\omega+\mathrm{i}\Gamma)]}{1+(\mathrm{i}\Gamma/\omega)\,[\chi^o(q,\omega+\mathrm{i}\Gamma)/\chi^o(q,0)]}, \tag{2.35}$$

where $\chi^o(q,\omega)$ is the Lindhard expression [2.4] and $\Gamma = \tau^{-1}$ is the *collision frequency*.

Outside the Landau-damping region, $\mathrm{Im}\{\chi(q,\omega)\}$ is small and the peaks in $\mathrm{Im}\{-1/\varepsilon(q,\omega)\}$ occur at frequencies and wavevectors that give

$$\mathrm{Re}\{\varepsilon(q,\omega)\}=0. \tag{2.36}$$

Equation (2.36) is the usual condition that determines the frequency $\omega_{\mathrm{P}}(q)$ of the collective plasma oscillation, or *plasmon*, of the electron gas. For small wavevectors, as is often the case in light scattering, the plasma frequency is [2.4, 68]

$$\omega_{\mathrm{P}}^2(q)\simeq\omega_{\mathrm{P}}^2(0)+\frac{3}{5}\,q^2v_{\mathrm{F}}^2\,, \tag{2.37}$$

where $\omega_{\mathrm{P}}(0)$ is given by (2.25) and $v_{\mathrm{F}}=\hbar k_{\mathrm{F}}/m^*$. For larger wavevectors, such that $\omega_{\mathrm{P}}(q)\leqslant qv_{\mathrm{F}}$, the plasmon is inside the Landau-damping regime and careful analysis of the behavior of $\mathrm{Im}\{-1/\varepsilon(q,\omega)\}$ is required. This regime is only marginally reached in light scattering experiments with existing lasers [2.38].

In semiconductors, the best examples of single-component plasmas occur in the n-type doped III-V compounds. These are polar crystals in which we must also consider the coupling of the electron plasma with the longitudinal optical (LO) phonons [2.75–77]. In the RPA, the total longitudinal dielectric function is given by

$$\varepsilon(q,\omega)=\varepsilon_\infty+\chi(q,\omega)+\chi_{\mathrm{L}}(\omega), \tag{2.38}$$

where

$$\chi_{\mathrm{L}}(\omega)=\varepsilon_\infty\,\frac{\omega_{\mathrm{LO}}^2-\omega_{\mathrm{TO}}^2}{\omega_{\mathrm{TO}}^2-\omega^2} \tag{2.39}$$

is the contribution of the polar lattice to the electric susceptibility. In (2.39), ω_{TO} and ω_{LO} are the frequencies of transverse optical (TO) and longitudinal optical (LO) phonons. Provided Landau damping is negligible, the frequencies of the *coupled plasmon* LO-*phonon modes* are determined by (2.40–42) below. In addition, when

$$\omega_{\mathrm{P}}(0)\simeq\omega_{\mathrm{P}}(q)\gg qv_{\mathrm{F}}\,, \tag{2.40}$$

dispersion effects on the plasma frequency are negligible and therefore $\chi(q,\omega)=\chi(0,\omega)$ takes the familiar *Drude* form [2.4]

$$\chi(\omega)=-\varepsilon_\infty\,\frac{\omega_{\mathrm{P}}^2}{\omega^2}\,. \tag{2.41}$$

From (2.36–41) we obtain the coupled mode equation

$$\omega_\pm^2=\tfrac{1}{2}\,\{(\omega_{\mathrm{P}}^2+\omega_{\mathrm{LO}}^2)\pm[(\omega_{\mathrm{P}}^2-\omega_{\mathrm{LO}}^2)^2+4\omega_{\mathrm{P}}^2(\omega_{\mathrm{LO}}^2-\omega_{\mathrm{TO}}^2)]^{1/2}\}. \tag{2.42}$$

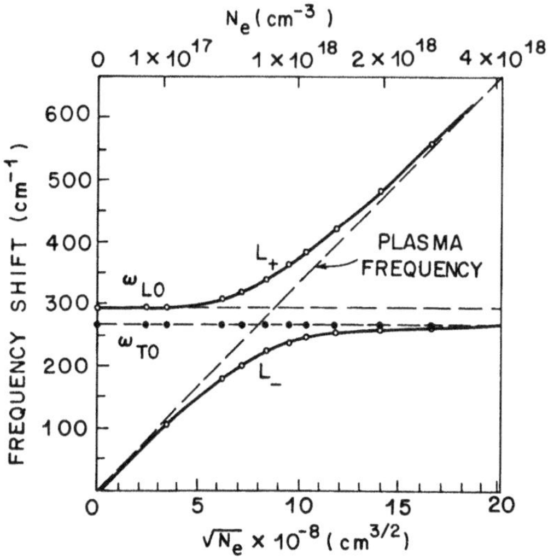

Fig. 2.2

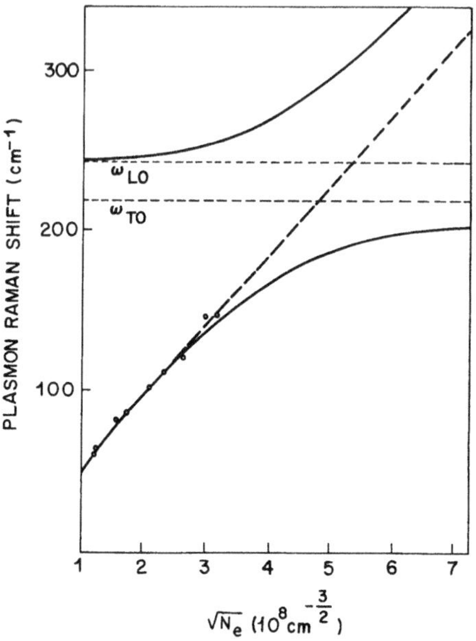

Fig. 2.3

Fig. 2.2. Frequency shifts measured in light scattering spectra of coupled plasmon LO-phonon modes of n-type GaAs. The solid lines are calculated with (2.42). From [2.78]

Fig. 2.3. Coupled plasmon LO-phonon modes in n-InAs. The dots represent the Stokes shifts measured in light scattering spectra. Only the plasmon-like modes were observed. The calculated curves include changes in electron effective mass due to non-parabolicity. From [2.16]

Light scattering by coupled plasmons LO-phonons has been studied in several semiconductors. The case of n-type GaAs [2.12, 17, 31] is particularly interesting because the conduction band is nearly parabolic. Figure 2.2 displays the data reported by *Mooradian* and *McWhorter* [2.78] which show that (2.42) reproduces the density dependence of the coupled modes. In the smaller gap semiconductors, nonparabolicity of the energy bands is an important effect, as first shown by *Slusher* et al. [2.15] in n-type InAs. Figure 2.3 reproduces these results. The curvature measured in the lower coupled mode branch has been explained with (2.42) in terms of changes in m^* with density [2.15].

The light scattering cross section of plasmons also includes additional contributions due to the electro-optic effects associated with the macroscopic electric field of the collective mode. In the case of coupled plasmons LO-phonons, it is also necessary to take into account the partial phonon character of the modes [2.71a, 78, 79]. We shall return to this question in Sect. 2.4 where we consider the light scattering cross section of the coupled modes in the context of the recent applications to studies of the dielectric function of the carriers in semiconductors.

2.2.6 Multicomponent Plasmas

It is quite common for semiconductor plasmas to contain more than one type of mobile carrier. This occurs in many-valley semiconductors, like n-type silicon and germanium. In these cases there are several valleys in which the similar anisotropic effective-mass tensors have their principal axes along nonequivalent directions in wavevector space, the {100} directions in Si and the {111} directions in Ge. For a given orientation of the scattering wavevector, electrons in nonequivalent valleys have different "effective" masses $\hat{\varepsilon}_L \cdot (1/m^*)^{-1} \cdot \hat{\varepsilon}_S$ and, therefore, such a system behaves as a multicomponent plasma. Another important case is that of dense electron-hole plasmas. These types of multicomponent plasmas exist in semimetals, like bismuth, or can be created by photoexcitation. Collective behavior in solid-state electron-hole plasmas was investigated by *Pines* [2.30], by *Nozieres* and *Pines* [2.80] and by *Pines* and *Schrieffer* [2.81]. It was found that, like in the case of a gaseous plasma of electrons and ions, there are *optical* and *acoustical* plasma oscillations. The theory of light scattering by optical and acoustical plasmons of multicomponent plasmas in semiconductors was first considered by *Platzman* [2.6] and by *McWhorter* [2.7]. Detailed calculations applicable to many-valley semiconductors were carried out by *Jha* [2.66, 82] and also by *Tzoar* and *Foo* [2.83, 84]. The theory of light scattering by multicomponent plasmas was extensively reviewed by *Platzman* and *Wolff* [2.19] and by *Klein* [2.20, Chap. 4].

In the discussion of the theory of the light scattering cross section of multicomponent plasmas, it is necessary to generalize the effective electron-photon interaction (2.3) to include several free-carrier species. A convenient way to write the cross section, following *McWhorter* [2.7], is

$$\frac{\partial^2 \sigma}{\partial \Omega \, \partial \omega} = \left(\frac{\omega_S}{\omega_L}\right)^2 V^2 \sum_{ll'} r_{0l} r_{0l'} S_{ll'}(\boldsymbol{q}, \omega), \tag{2.43}$$

where the subscripts l and l' label the electron species. The set of dynamical structure factors is given by

$$S_{ll'}(\boldsymbol{q}, \omega) = \frac{1}{2\pi} \int_{-\infty}^{\infty} e^{i\omega t} \langle n_l(\boldsymbol{q}, t) n_{l'}^{\dagger}(\boldsymbol{q}, 0) \rangle dt \tag{2.44}$$

and the r_{0l} are "effective" Thomson radii of the electrons, given by

$$r_{ol} = m r_0 \, \hat{\varepsilon}_S \cdot \frac{1}{\boldsymbol{m}_l^*} \cdot \hat{\varepsilon}_L . \tag{2.45}$$

The cross section given by (2.43–45) is somewhat similar to that of single-component plasmas. However, it has several novel features that are unique to multicomponent plasmas. These are considered below in the discussion of the dynamical structure factors. We note here that there is an important difference with the isotropic single-component cases, considered in the previous sections, in which the scattering was completely polarized. In the anisotropic effective mass multicomponent plasmas the scattering can become partly depolarized. This occurs whenever $\hat{\varepsilon}_L$ and $\hat{\varepsilon}_S$ are not directed along principal axes of the effective mass tensor.

For purely longitudinal density fluctuations, the set of structure factors $S_{ll'}(\boldsymbol{q},\omega)$ can be expressed in a form that is a generalization of (2.13):

$$S_{ll'}(\boldsymbol{q},\omega)=\frac{q^2}{e_l e_{l'}}\frac{1}{2\pi}\int_{-\infty}^{\infty} e^{i\omega t}\langle P_l(\boldsymbol{q},t)P_{l'}^{\dagger}(\boldsymbol{q},0)\rangle\, dt, \tag{2.46}$$

where the charges associated with each species e_l and $e_{l'}$ are $-e$ for electrons in a conduction band and $+e$ for holes in a valence band. To carry out a fluctuation-dissipation analysis of (2.46), we require a linear response function matrix $T_{ll'}(\boldsymbol{q},\omega)$ that is a generalization to that defined by (2.19):

$$\boldsymbol{P}_l(\boldsymbol{q},\omega)=\frac{1}{V}\sum_{l'} \boldsymbol{T}_{ll'}(\boldsymbol{q},\omega)\boldsymbol{D}_{l'}(\boldsymbol{q},\omega). \tag{2.47}$$

The total electric field is now given by

$$\boldsymbol{E}_l(\boldsymbol{q},\omega)=\boldsymbol{D}_l(\boldsymbol{q},\omega)-\frac{1}{\varepsilon_0\varepsilon_\infty}\sum_{l'}\boldsymbol{P}_{l'}(\boldsymbol{q},\omega). \tag{2.48}$$

In addition, we have

$$\boldsymbol{P}_l(\boldsymbol{q},\omega)=\varepsilon_0\boldsymbol{\chi}_l(\boldsymbol{q},\omega)\boldsymbol{E}_l(\boldsymbol{q},\omega), \tag{2.49}$$

where $\boldsymbol{\chi}_l(\boldsymbol{q},\omega)$ is the dielectric polarizability tensor of the species l.

Equations (2.48,49) constitute a system of linearly coupled equations for the polarizations $\boldsymbol{P}_l(\boldsymbol{q},\omega)$. Since the solution to this system of coupled polarizations has the form of (2.47), they determine the linear response function in terms of the polarizabilities of the species. A simple system of coupled polarizations was considered by *Klein* et al. [2.71a] in their discussion of coupled LO-phonon plasmons. The solution given by these authors can be generalized to any number of electron species and LO phonons. We note that they can be very complex, depending on the effective mass tensor of the different electron species. For example, the case of n-type germanium can be quite involved because the principal axes of the mass tensors of the different axes are not all parallel. For simplicity, we consider here the situation in which all the fluctuating fields and the scattering wave vector $\boldsymbol{q}$ are along principal axes of the mass tensors (this is possible for n-type silicon but not for n-type germanium). In this case, the effective mass and dielectric polarizability tensors associated with each valley are diagonal. For each pair of indices l we have

$$T_{ll'}(\boldsymbol{q},\omega)=V\varepsilon_0[\chi_l(\boldsymbol{q},\omega)\delta_{ll'}-\chi_l(\boldsymbol{q},\omega)\chi_{l'}(\boldsymbol{q},\omega)/\varepsilon(\boldsymbol{q},\omega)], \tag{2.50}$$

where

$$\varepsilon(\boldsymbol{q},\omega)=\varepsilon_\infty+\sum_l \chi_l(\boldsymbol{q},\omega) \tag{2.51}$$

is the total dielectric function for *longitudinal* electric fields along the direction of the corresponding principal axes.

2.2.7 The Simple Two-Component Plasma: Acoustic and Optic Plasmons

In semiconductors, multicomponent plasmas can display a high degree of complexity. Several different species may be present and the response of the system can also reflect electron energy band structure effects like nonparabolicity and warping. In order to reveal some of the new features that occur in light scattering by multicomponent plasmas, we ignore all these complexities and consider a *two-component system* with *isotropic* effective masses m_1 and m_2. In addition, we assume that $m_1 \ll m_2$ and, for the time being, we also ignore the coupling to LO phonons which occurs in the case of polar semiconductors. Such a simple model system was also considered by *Platzman* and *Wolff* [2.19].

Light scattering by this two-component system is dominated by the density fluctuations of the lighter species with effective mass m_1. The expression for the light scattering cross section (2.43) reduces to

$$\frac{\partial^2\sigma}{\partial\Omega\,\partial\omega}\simeq\left(\frac{\omega_{\mathrm{S}}}{\omega_{\mathrm{L}}}\right)^2 V^2 r_{01}^2 S_{11}(q,\omega). \tag{2.52}$$

By means of (2.46, 50, 51) and the fluctuation-dissipation theorem, we obtain

$$S_{11}(q,\omega)=\frac{1}{V^2}\frac{q^2}{e^2}\left(\frac{\hbar}{\pi}\right)(1-\mathrm{e}^{-\hbar\omega/k_{\mathrm{B}}T})^{-1}\,\mathrm{Im}\,\{T_{11}(q,\omega)\}, \tag{2.53}$$

where

$$\begin{aligned}T_{11}(q,\omega)&=V\varepsilon_o[\chi_1(q,\omega)-\chi_1^2(q,\omega)/\varepsilon(q,\omega)]\\&=V\varepsilon_o[\varepsilon_\infty+\chi_2(q,\omega)]\chi_1(q,\omega)/\varepsilon(q,\omega).\end{aligned} \tag{2.54}$$

In (2.54),

$$\varepsilon(q,\omega)=\varepsilon_\infty+\chi_1(q,\omega)+\chi_2(q,\omega). \tag{2.55}$$

Equations (2.52–55) show that the light scattering cross section is expected to have peaks at frequencies which are in the vicinity of the solutions of

$$\mathrm{Re}\,\{\varepsilon(q,\omega)\}=0. \tag{2.56}$$

For the cases in which Landau-damping effects are small, (2.55,56) determine the frequencies of *collective* modes of the simple two-component plasma. Their solutions are somewhat more complex than that of its analog (2.36) for the single-component plasma.

Let us consider first the $q \to 0$ limit. In this case we may write [2.4,68]

$$\mathrm{Re}\{\chi_1(q,\omega)\} = -\frac{\omega_{P1}^2}{\omega^2}\,\varepsilon_\infty, \tag{2.57}$$

$$\mathrm{Re}\{\chi_2(q,\omega)\} = -\frac{\omega_{P2}^2}{\omega^2}\,\varepsilon_\infty, \tag{2.58}$$

where ω_{P1} and ω_{P2} are the plasma frequencies of the two species. Equations (2.55–58) show that in this limit there is only one collective mode, with a frequency given by

$$\omega_{OP} = (\omega_{P1}^2 + \omega_{P2}^2)^{1/2} \simeq \omega_{P1} \tag{2.59}$$

because $\omega_{P1} \gg \omega_{P2}$. We identify this mode as the *optic* plasmon of the system, since it corresponds to the oscillation of the charges of species 1 with respect to the *nearly* immobile charges of species 2. It is clear that optic plasmons of the two-component plasma are similar to plasmons in the isotropic simple component plasmas of doped semiconductors, where the free charge oscillates with respect to the immobile ionized impurities.

We next consider the case in which $q \neq 0$ but is relatively small, so that $\omega_{P1} \gg qv_{F1}$. Here we find two solutions. One is the optical mode which is still given by (2.59). The other occurs at frequencies $\omega \approx \omega_{P2} \ll \omega_{P1}$. Charge fluctuations and macroscopic electric fields at these frequencies are seen by species-one electrons as static. Therefore, instead of (2.57), we now use the static limit of the electron-gas electric susceptibility [2.4,19,68]

$$\mathrm{Re}\{\chi_1(q,\omega)\} = \frac{k_{S1}^2}{q^2}\,\varepsilon_\infty, \tag{2.60}$$

where k_{S1}^{-1} is the screening length of species one. In the $q \ll k_{s1}$ limit, (2.55,56,58,59) yield for this other mode the frequency

$$\begin{aligned}\omega_{AP} &\cong \omega_{P2}[\mathrm{Re}\{\chi_1(q,0)\}]^{1/2}\\ &= \frac{q}{k_{S1}}\,\omega_{P2}\,.\end{aligned} \tag{2.61}$$

Equation (2.61) implies that the charge density fluctuation of the heavier particles, species two, is quasi-statically screened by the lighter species. This mode is identified as the *acoustic* plasma oscillation [2.19,30,80] i.e., the *nearly neutral* excitation in which the two types of particles move in such a fashion that the respective density fluctuations are out of phase and almost cancel each other.

We shall consider here the case of a degenerate system in which k_{S1} is given by the *Fermi-Thomas* expression [2.4, 68]

$$k_{S1} = \frac{\sqrt{3}\,\omega_{P1}}{v_{F1}}, \qquad (2.62)$$

which means that

$$\omega_{AP} = \frac{\omega_{P2} v_{F1}}{\sqrt{3}\,\omega_{P1}}\, q. \qquad (2.63)$$

Equation (2.63) shows that $\omega_{AP} < q v_{F1}$. Therefore, the acoustic plasma mode is *necessarily* subject to Landau damping by the lighter species. Depending on the values of m_1 and m_2, the mode might also be subject to Landau damping by the single-particle excitations of the heavier species. It is easy to see that Landau-damping effects on the acoustic plasma mode are minimized when

$$q v_{F2} < \omega_{AP} < q v_{F1}. \qquad (2.64)$$

In Fig. 2.4 we show the single-particle excitations and collective modes of the two-component plasma when the conditions set by inequalities (2.64) hold. Acoustic plasma waves in Maxwellian solid-state plasmas, not discussed here, have been reviewed in [2.19]. *Ruvalds* and *Kahn* [2.85a] as well as *Fröhlich* [2.85b] have also considered acoustic plasmons and their damping. More recently, *Appel*

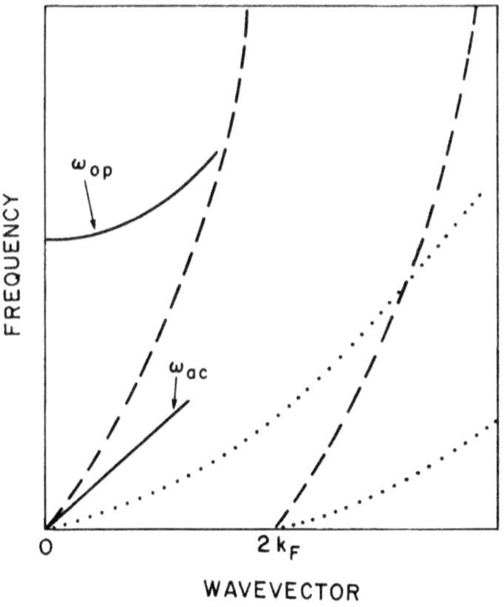

Fig. 2.4. Elementary excitations of a simple two-component plasma. Two single-particle continua exist. The one defined by the dashed lines is for the lighter species; the one between the dotted lines is for the heavier species. Optic (ω_{OP}) and acoustic (ω_{AP}) plasma modes are shown. In this case Landau damping of the acoustic mode is due only to the lighter species

and *Overhauser* [2.86] have studied the acoustic modes of the two-component plasma by solving their coupled Boltzmann equations. Besides the mode described above, two additional overdamped modes are predicted.

We have seen that the total charge-density fluctuation carried by the acoustic plasma oscillation is small. Nevertheless, an examination of (2.47–51) indicates that its light scattering cross section is comparable to that of optic plasmons if Landau damping is weak. This occurs because the light scattering is caused by the screening charge density associated with the lighter species, which is large and nearly cancels that of the heavier species. This aspect of light scattering by acoustic plasmons was emphasized by *Platzman* and *Wolff* [2.19]. They pointed out that light scattering, for the reasons just presented, is an ideal tool to observe the acoustic plasmon. These ideas were confirmed by the recent observation of light scattering by the acoustic plasmon of the electron-hole plasma in photoexcited GaAs [2.29]. A discussion of this work is presented in Sect. 2.4.

2.2.8 The Simple Two-Component Plasma: Neutral Density Excitations

We now consider the simple two-component plasma, but we do not make the approximation $m_1 \ll m_2$. We must in this case take into account the contribution of both types of carriers to the light scattering cross section. By means of (2.43,45) the cross section can be written as

$$\frac{\partial^2 \sigma}{\partial \Omega\, \partial \omega} = \left(\frac{\omega_S}{\omega_L}\right)^2 m^2 V^2 r_o^2 S^{(2)}(q,\omega), \qquad (2.65a)$$

where $S^{(2)}(q,\omega)$ is the dynamical structure factor of the simple two-component plasma, given by

$$S^{(2)}(q,\omega) = [S_{11}(q,\omega)/m_1^2] + [S_{22}(q,\omega)/m_2^2] + [2S_{12}(q,\omega)/m_1 m_2]. \qquad (2.65b)$$

By application of the fluctuation-dissipation theorem in the form of (2.17,50) to the evaluation of the $S_{ll'}(q,\omega)$ defined by (2.46), we obtain

$$S^{(2)}(q,\omega) = \frac{\hbar}{\pi}\frac{q^2}{e^2}\frac{\varepsilon_o}{V}(1-\mathrm{e}^{-\hbar\omega/k_B T})^{-1} \cdot \mathrm{Im}\left\{\frac{1}{\varepsilon}\left[\varepsilon_\infty\left(\frac{\chi_1}{m_1^2}+\frac{\chi_2}{m_2^2}\right)+\chi_1\chi_2\left(\frac{1}{m_1}-\frac{1}{m_2}\right)^2\right]\right\}, \qquad (2.66)$$

where ε is given by (2.55). The factor ε^{-1} represents a "screening" effect that reduces $S^{(2)}(q,\omega)$ and the light scattering cross section except in the case of collective modes when $\mathrm{Re}\{\varepsilon(q,\omega)\} \cong 0$. We note that, as anticipated, $S^{(2)}(q,\omega)$ is smaller when $m_1 = m_2$, and is equivalent to (2.53,54) when $m_1 \ll m_2$.

We focus here on the behavior of $S^{(2)}(q,\omega)$ in the range where $\mathrm{Re}\{\varepsilon(q,\omega)\} \neq 0$ and in the high electron density case when $|\mathrm{Re}\{\varepsilon(q,\omega)\}| \gg \varepsilon_\infty$. Under these conditions we may write

$$S^{(2)}(q,\omega) \simeq \frac{\hbar}{\pi}\frac{q^2}{e^2}\frac{\varepsilon_o}{V}(1-\mathrm{e}^{-\hbar\omega/k_\mathrm{B}T})^{-1}\left(\frac{1}{m_1}-\frac{1}{m_2}\right)^2 \cdot \mathrm{Im}\,\{\chi_1\chi_2/\varepsilon\}. \tag{2.67}$$

We see in (2.67) that when $\chi_1(q,\omega) \approx \chi_2(q,\omega)$, the screening effect is considerably reduced and $S^{(2)}(q,\omega) \sim \mathrm{Im}\,\{\chi_1(q,\omega)\}$. In this frequency and wavevector range which overlaps the single-particle excitations, the cross section is smaller than for the collective modes, but may be sizable provided $m_1 \neq m_2$.

It has been pointed out that this situation is possible in multivalley semiconductors by an appropriate choice of the orientation of the scattering wavevector with respect to that of the valleys. The excitations responsible for this kind of scattering have been identified as *intervalley fluctuations* [2.19]. These are *neutral* excitations in which the charge-density fluctuations of electrons in one valley are cancelled out by the fluctuations in the other valley. For this behavior to occur, the two valleys have to be nonequivalent, hence the condition $m_1 \neq m_2$. We shall return to this class of elementary excitations in the discussion of light scattering by free carriers in heavily doped n-type silicon in Sect. 2.7.

2.3 Resonant Light Scattering by Free Electron Excitations

Experimental studies of light scattering by free electrons in semiconductors are usually carried out with infrared and visible lasers. In semiconductors the energy gap is typically $E_\mathrm{G} \lesssim 2$ eV. Therefore, strictly speaking, the condition $E_\mathrm{G} \gg \hbar\omega_\mathrm{L}$ is invalid. This fact is often of consequence with regard to the character of the measured spectra. In the interpretation of results it is necessary to go beyond the framework of scattering by charge-density fluctuations as presented in the previous section. One of the most important effects, already mentioned, is the appearance of spectra associated with spin-density fluctuations and spin-flip transitions. In the case of free electrons, these two types of excitations are physically equivalent in the absence of an external magnetic field. In the nonmagnetic semiconductors of interest here, these excitations have single-particle character at all carrier densities [2.9,17,18,20]. Additional effects occur when $\hbar\omega_\mathrm{L}$ is very close to optical energy gaps. The most striking ones involve breakdown in the polarization selection rules [2.50] and of the condition of wavevector conservation [2.51,87].

Early theoretical interest in the role of optical interband transitions in light scattering by free electrons in semiconductors was related to the need to understand the scattering mechanisms and the character of the active elementary excitations [2.8,9,88]. The first experimental observations of near-resonance light scattering were reported in n-InSb [2.89,90], n-CdS [2.91] and n-GaAs [2.50]. The *resonance profiles*, i.e., the scattering cross section (or efficiency) vs laser photon energy, were not determined in this seminal work. In fact, in the case of free electron excitations, there have been few studies of resonant profiles. Among these is the work of *Brueck* et al. [2.90] on spin-flip transitions in n-type InSb under high magnetic field, and the research of the present authors in n-type GaAs [2.51]. More recent interest in resonant light scattering by free-electron excitations has been stimulated by the applications of the method to studies of electrons at semiconductor surfaces and interfaces [2.52].

In this section we first give a concise presentation of the theory of the resonant behavior of the light scattering cross section. We then proceed to consider the experimental results in doped III–V compound semiconductors. Two important subjects are not included here. One is the growing field of resonant light scattering by 2-dim electron systems which is considered in Sect. 2.5. The other is the light scattering by spin-flip transitions in a magnetic field. These studies have been the subject of several excellent reviews [2.22].

2.3.1 General Considerations for Light Scattering Cross Sections

a) Electron-Density Fluctuations

We require expressions for the light scattering cross sections by electron-density fluctuations that, unlike those in the previous section, are not based on the approximation $E_G \gg \hbar\omega_L$. These expressions [2.9,66,67] are based on the theoretical work of *Wolff* [2.8] and *Yafet* [2.10]. They take into account the contributions associated with *virtual* interband transitions. The spin-orbit interaction also plays an important role in these light scattering processes, in a manner that is similar to that proposed by *Elliott* and *Loudon* [2.92] for the case of spin waves in magnetic crystals.

Yafet [2.18] and *Klein* [2.20] have given thorough reviews of these theoretical investigations. The presentation we give here is based on the work of *Blum* [2.67]. This formalism, which is similar to that of *Hamilton* and *McWhorter* [2.9], has the advantage that it leads in a natural way to an "effective" light scattering Hamiltonian that is a generalization of the one in (2.6).

We note first that $N(q)$, defined in (2.7), can be expressed in second-quantization notation as [2.4,68]

$$N(\boldsymbol{q}) = \frac{1}{V} \sum_{\boldsymbol{k}} C^{\dagger}_{\boldsymbol{k}+\boldsymbol{q}} C_{\boldsymbol{k}}, \tag{2.68}$$

where $C^{\dagger}_{\boldsymbol{k}}$ and $C_{\boldsymbol{k}}$ are creation and destruction operators associated with the one-electron states occupied by the free carriers in the semiconductor. In writing (2.68), we are labeling the states by their wavevector $\boldsymbol{k}$ and ignoring their spin quantum numbers. This is possible in the framework of the discussion of the previous section, dealing with spin-independent processes.

We describe, following *Blum* [2.67], the scattering of light due to $|\alpha\rangle \rightarrow |\beta\rangle$ electronic excitations, where α and β include spin, by means of the effective Hamiltonian

$$H''_{\text{eff}} = \frac{e^2}{2m} NA(\omega_L)A(\omega_S). \tag{2.69}$$

The operator N can be written as

$$N = \sum_{\alpha,\beta} \gamma_{\alpha\beta} C_\beta^+ C_\alpha, \tag{2.70}$$

where the coefficients $\gamma_{\alpha\beta}$ take into account the H''_{ep} and H'_{ep} type processes (Sect. 2.2). In the absence of an external magnetic field, $\gamma_{\alpha\beta}$ is given by [2.9, 18, 67]

$$\begin{aligned}\gamma_{\alpha\beta} &= (\hat{\boldsymbol{\varepsilon}}_L \cdot \hat{\boldsymbol{\varepsilon}}_S)\,\xi_{\alpha\beta} \\ &+ \frac{1}{m}\sum_{\beta'}\left[\frac{\langle\alpha|\hat{\boldsymbol{\varepsilon}}_S \cdot \boldsymbol{p}\,\mathrm{e}^{-\mathrm{i}\boldsymbol{k}_S\cdot\boldsymbol{r}}|\beta'\rangle\,\langle\beta'|\hat{\boldsymbol{\varepsilon}}_L \cdot \boldsymbol{p}\,\mathrm{e}^{\mathrm{i}\boldsymbol{k}_L\cdot\boldsymbol{r}}|\beta\rangle}{\hbar\omega_L + E_\beta - E_{\beta'}}\right. \\ &\left. + \frac{\langle\alpha|\hat{\boldsymbol{\varepsilon}}_L \cdot \boldsymbol{p}\,\mathrm{e}^{\mathrm{i}\boldsymbol{k}_L\cdot\boldsymbol{r}}|\beta'\rangle\,\langle\beta'|\hat{\boldsymbol{\varepsilon}}_S \cdot \boldsymbol{p}\,\mathrm{e}^{-\mathrm{i}\boldsymbol{k}_S\cdot\boldsymbol{r}}|\beta\rangle}{E_\beta - E_{\beta'} - \hbar\omega_S}\right],\end{aligned} \tag{2.71}$$

where $|\beta'\rangle$ are intermediate states. E_β and $E_{\beta'}$ are one-electron energies and

$$\xi_{\alpha\beta} = \langle\alpha|\mathrm{e}^{\mathrm{i}\boldsymbol{q}\cdot\boldsymbol{r}}|\beta\rangle. \tag{2.72}$$

The first summand in (2.71) corresponds to H''_{ep}, while the second and third ones correspond to H'_{ep}.

The formalism associated with (2.69) enables us to write the light scattering cross section of a single-component plasma as

$$\frac{\partial^2\sigma}{\partial\Omega\,\partial\omega} = r_o^2\left(\frac{\omega_S}{\omega_L}\right) V^2 G(\omega), \tag{2.73}$$

where

$$G(\omega) = \frac{1}{2\pi}\int_{-\infty}^{\infty} \mathrm{e}^{\mathrm{i}\omega t}\langle N(t)N^\dagger(0)\rangle\, dt. \tag{2.74}$$

Equations (2.73, 74) are analogous to (2.8, 10) of the previous section. In fact, the operator N is a generalization of $n(q)$. In the limit $\omega_S \rightarrow \omega_L \rightarrow 0$ we have

$$N \equiv N(q) = m\left(\hat{\boldsymbol{\varepsilon}}_L \cdot \frac{1}{\boldsymbol{m}^*} \cdot \hat{\boldsymbol{\varepsilon}}_S\right) n(q). \tag{2.75}$$

Equations (2.73, 74) are deceptively simple. *Blum* [2.67] has derived general expressions for $G(\omega)$ in terms of linear response functions that are based on the RPA. However, these are extremely difficult to evaluate. The task is simplified considerably when the approximations

$$\omega_L \approx \omega_S, \tag{2.76a}$$

$$E_{\beta'} - E_\beta \simeq E_G, \tag{2.76b}$$

are made in the resonant denominators of (2.70). Equation (2.76a) represents the "quasi-static" approximation, extensively used in the interpretation of resonant Raman scattering by optical phonons [2.64]. In (2.76b), E_G is the value of an *appropriate* optical energy gap. Under resonant conditions, when $\hbar\omega_L \approx E_G$, the use of (2.76b) is clearly invalid. The details of the electron energy level structure play an important role here. Nevertheless, we are making use of (2.76b) because it is the only approach that has led to analytic expressions for the light scattering cross sections. One possible way of going beyond this level of approximation is by means of a series expansion of the resonant denominators in powers of the scattering wave vector q. This procedure is useful in a *qualitative* interpretation of experimental results obtained under extreme resonant conditions [2.20, 50].

In the absence of external fields, the states $|\alpha\rangle$, $|\beta\rangle$, and $|\beta'\rangle$ are represented by the usual electron energy bands of the crystal. When external electric fields are present, the effective-mass approximation often provides a good description of the one-electron states [2.93]. It is also of interest to us that the effective mass approximation gives a good account of the one-electron states in the case of the 2-dim electron systems in space charge layers at semiconductor surfaces and interfaces [2.94].

In this section we are going to assume that the free electrons are in a *parabolic S*-like conduction band. This situation occurs in many n-type III–V compound semiconductors of reasonably large band gap. n-type GaAs and InP are typical examples. In small band gap crystals, like InSb, nonparabolicity effects are important (for a discussion see [Ref. 2.20, Sect. 4.4.3]). We take $|\alpha\rangle$ as the state $|\boldsymbol{k}, s_\alpha\rangle$, with wavevector $\boldsymbol{k}$ and spin index s_α. $|\beta\rangle$ is the state $|\boldsymbol{k}+\boldsymbol{q}, s_\beta\rangle$, as required by wavevector conservation. The set of intermediate levels $|\beta'\rangle$ are all the valence band states that are connected to the conduction band states $|\boldsymbol{k}, s_\alpha\rangle$ and $|\boldsymbol{k}+\boldsymbol{q}, s_\beta\rangle$ by the optical transitions associated with the incident and scattered photons.

In Fig. 2.5 we illustrate *virtual interband* transitions that enter in the evaluation of $\gamma_{\alpha\beta}$. Typical contributions arising from the second term in the right side of (2.71) correspond to the incident photon being absorbed first. In the third term of (2.71), the scattered photon is emitted first. Diagrams like that in Fig. 2.5 help in visualizing the resonant light scattering processes. They are also useful in anticipating the resonant behavior of scattering cross sections. For example, an examination of the diagram indicates that resonant enhancements of light scattering by electron-density fluctuations occur *only* at optical energy gaps that

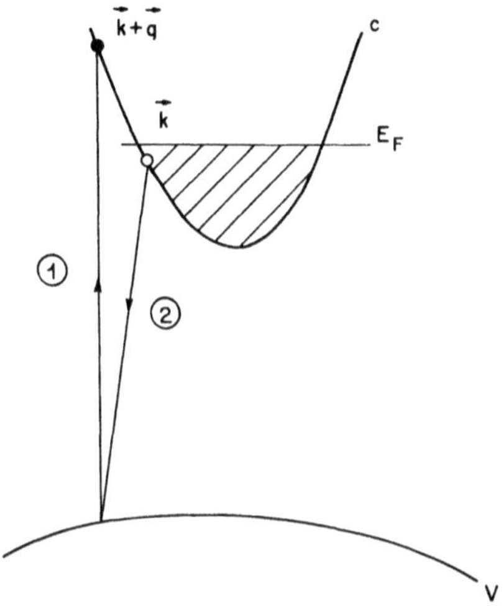

Fig. 2.5. Virtual interband transitions which enter in the evaluation of the contribution of the $\boldsymbol{A}\cdot\boldsymbol{p}$ term of the scattering by free electron excitations [see (2.71)]. The numbers indicate the order of the transitons

involve the states of the free electrons (the appropriate optical gaps mentioned above). These features have been emphasized by *Burstein* et al. [2.52,95] in their consideration of resonant light scattering by 2-dim electron systems.

The explicit expressions of the light scattering cross section depend on details of the electron energy band structure [2.9,66]. We present here the results of *Hamilton* and *McWhorter* [2.9] that apply to n-type III–V semiconductor compounds in which the free electrons occupy conduction band states near the Brillouin zone center (not for Ge, Si and GaP!). The wave functions of the Kane model are used [2.96]. For the conduction band, the states are represented by s-like spin-up and spin-down wave functions:

$$(1/2, 1/2):\ (S)\uparrow, \tag{2.77a}$$

$$(1/2, -1/2):\ (S)\downarrow. \tag{2.77b}$$

The intermediate states are in heavy, light and (spin-orbit) split-off valence bands. Their wave functions are represented by

$$(3/2, 3/2):\ (1/\sqrt{2})\ (X+\mathrm{i}Y)\uparrow, \tag{2.78a}$$

$$(3/2, -3/2):\ (1/\sqrt{2})\ (X-iY)\downarrow, \tag{2.78b}$$

for the heavy valence bands along k_z,

$$(3/2, 1/2):\ (1/\sqrt{6})\ (X+\mathrm{i}Y)\downarrow - \sqrt{2/3}\,Z\uparrow, \tag{2.79a}$$

$$(3/2, -1/2):\ (1/\sqrt{6})\ (X-\mathrm{i}Y)\uparrow + \sqrt{2/3}\,Z\downarrow, \tag{2.79b}$$

for the light valence band along k_z, and

$$(1/2,1/2):\ (1/\sqrt{3})\,(X+\mathrm{i}Y)\downarrow+(1/\sqrt{3})Z\uparrow, \tag{2.80a}$$

$$(1/2,-1/2):\ (1/\sqrt{3})\,(X-\mathrm{i}Y)\uparrow-(1/\sqrt{3})Z\downarrow, \tag{2.80b}$$

for the split-off bands (see (2.148) of [2.64]).

Using these wave functions, *Hamilton* and *McWhorter* [2.9] have obtained [Ref. 2.64, Sect. 2.2.2]

$$\begin{aligned}\gamma_{\alpha\beta}=&\,\hat{\boldsymbol{\varepsilon}}_\mathrm{L}\cdot\boldsymbol{A}\cdot\hat{\boldsymbol{\varepsilon}}_\mathrm{S}\xi_{\alpha\beta}\\&+\mathrm{i}(\hat{\boldsymbol{\varepsilon}}_\mathrm{L}\times\hat{\boldsymbol{\varepsilon}}_\mathrm{S})\cdot\boldsymbol{B}\cdot\boldsymbol{\zeta}_{\alpha\beta},\end{aligned} \tag{2.81}$$

where

$$\boldsymbol{\zeta}_{\alpha\beta}=\langle\alpha|\mathrm{e}^{\mathrm{i}\boldsymbol{q}\cdot\boldsymbol{r}}\boldsymbol{\sigma}|\beta\rangle=\xi_{\alpha\beta}\langle s_\alpha|\boldsymbol{\sigma}|s_\beta\rangle \tag{2.82}$$

and the components of $\boldsymbol{\sigma}$ are the Pauli matrices. The tensor coefficients $\boldsymbol{A}$ and $\boldsymbol{B}$ contain the momentum matrix elements and the resonant denominators [see (2.89,90) below].

Equation (2.81) shows that light scattering by electron-density fluctuations separates into two components. The first term in the right side of (2.81) is symmetric in the polarizations of the incident and scattered light. We shall see below that it is related to *charge-density fluctuations*, like the ones discussed in the previous section. The second term in the right side of (2.81) is *antisymmetric* in $\hat{\varepsilon}_\mathrm{L}$ and $\hat{\varepsilon}_\mathrm{S}$. It is associated with changes in the spin degrees of freedom, giving rise to scattering by *spin-density fluctuations* and *spin-flip transitions* (equivalent in the absence of a magnetic field). The changes in spin are made possible by the spin-orbit interaction [2.92] that leads to intermediate state wave functions like those in (2.79,80) which are not eigenstates of spin or which, like in the case of (2.78), are eigenstates of spin for only one direction of spin quantization. This antisymmetric term is a new feature associated with the resonant light scattering processes that involves the *virtual* optical transitions across the energy gap of the semiconductor. It will be seen below that $\boldsymbol{B}\rightarrow 0$ when $\omega_\mathrm{L}\rightarrow 0$, as required on the basis of general considerations on the properties of antisymmetric light scattering cross sections [2.65].

b) Charge-Density Fluctuations

The effective light scattering Hamiltonian (2.70) associated with charge-density fluctuations is obtained by using the first term in the right side of (2.81) in (2.69). We find

$$H''_\mathrm{eff}=\frac{e^2}{2m}\,(\hat{\boldsymbol{\varepsilon}}_\mathrm{L}\cdot\boldsymbol{A}\cdot\hat{\boldsymbol{\varepsilon}}_\mathrm{S})N(\boldsymbol{q})A(\omega_\mathrm{L})A(\omega_\mathrm{S}), \tag{2.83}$$

where $N(\boldsymbol{q})$ is given by (2.68), for each component of electron spin. Neglecting the anisotropy of the valence bands, the tensor $\boldsymbol{A}$ is written as

$$\boldsymbol{A}=\boldsymbol{I}\left[1+\frac{2P^2}{3m}\left(\frac{E_1}{E_1^2-\hbar^2\omega_L^2}+\frac{E_2}{E_2^2-\hbar^2\omega_L^2}+\frac{E_3}{E_3^2-\hbar^2\omega_L^2}\right)\right], \tag{2.84}$$

where $\boldsymbol{I}$ is the unit dyadic and $P=|\langle S|p_z|Z\rangle|$ is the interband matrix element of momentum. The energies that enter in $\boldsymbol{A}$ are the values of the interband optical energy gaps of the *Kane* model [2.96]. E_1 and E_2 are associated with the heavy and light effective mass valence bands, degenerate at the Γ-point. We can thus write

$$E_1 \simeq E_2 \simeq E_0 . \tag{2.85}$$

In addition, E_3 is the split-off energy gap

$$E_3 = E_0 + \Delta_0 , \tag{2.86}$$

where Δ_0 is the spin-orbit splitting of the top valence band states. We note that in the limit $E_0 \gg \hbar\omega_L$, this expression for $\boldsymbol{A}$ becomes identical to the inverse effective mass tensor of the Kane model, as expected.

The form of the effective light scattering Hamiltonian of (2.83) is similar to that of (2.6). The only difference is the replacement of $(\hat{\varepsilon}_L \cdot \boldsymbol{m}^{*-1} \cdot \hat{\varepsilon}_S)$ by $1/m\,(\hat{\varepsilon}_L \cdot \boldsymbol{A} \cdot \hat{\varepsilon}_S)$. This implies that in a single component plasma under consideration, the scattering cross sections and spectral line shapes are determined by the dynamical structure factor defined in (2.10). This is a consequence of the approximations made in using (2.76). It is clear that when these approximations break down, as occurs under extreme resonance conditions, the spectral line shapes are expected to show a substantial dependence on the photon energy. In semiconductors with a complex energy band structure and in multicomponent plasmas, proper consideration of resonant enhancement factors may be quite complicated, even when not under extreme resonance conditions. Each case requires specific attention.

In the consideration of resonant effects, it is useful to have a more microscopic view of the light scattering processes. This approach provides an interesting physical insight into the results that are otherwise obtained by means of linear response theory. In the case of charge-density fluctuations, this is done by explicitly taking into consideration the Coulomb interaction among the electrons within the RPA. *Jha* [2.88] has carried out an infinite-order perturbation theory calculation. *McWhorter* and *Argyres* [2.97] have calculated the light scattering from charge-density fluctuations by using third-order perturbation theory. In this case the effects of the electron-electron interaction are described, in the RPA, by the macroscopic electric field of (2.21). The interband and intraband electron transitions that enter in this third-order calculation are shown in the diagrams of Fig. 2.6 [2.65,98]. The interband transitions involve the incident and scattered photons as in Fig. 2.5. The intraband transitions are caused by a Fröhlich-like interaction [2.64]

$$H_F = \frac{ie}{q} E(\boldsymbol{q}, \omega) e^{i\boldsymbol{q}\cdot\boldsymbol{r}} \tag{2.87}$$

between the free electrons and the longitudinal potential $(ie/q)E(q, \omega)e^{i\boldsymbol{q}\cdot\boldsymbol{r}}$, set up by the charge-density fluctuations.

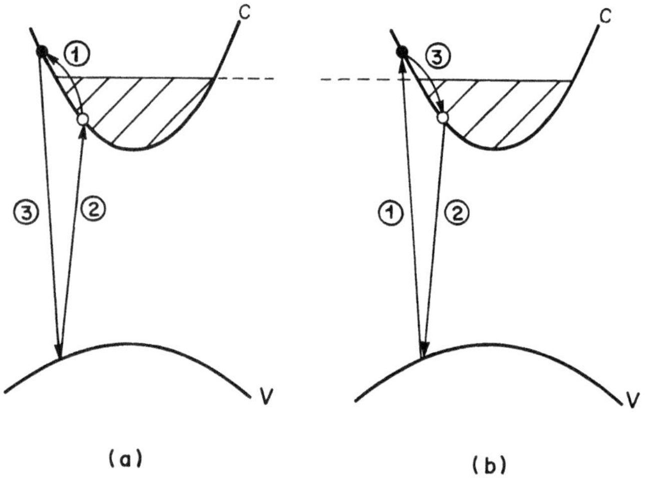

Fig. 2.6a, b. Virtual inter- and intraband transitions which enter in the evaluation of cross sections by collective plasmon-LO phonon excitations [2.65,98]. The interband transitions are caused by photons; the intraband ones by the coupled modes

c) Spin-Density Excitations

We are concerned here with the light scattering processes associated with the second term in the right side of (2.81) that causes changes in the spin of the free electrons. First we consider the terms that originate in the use of σ_z in (2.82). In this case the expression for the effective light scattering Hamiltonian, obtained by means of (2.69,70,81,82), is

$$H''_{\text{eff}} = i \frac{e^2 V}{2m} (\hat{\varepsilon}_L \times \hat{\varepsilon}_S) \cdot \boldsymbol{B} \cdot \hat{\varepsilon}_z \left(\frac{N_\uparrow(q) - N_\downarrow(q)}{2} \right) A(\omega_L) A(\omega_S), \tag{2.88}$$

where $\hat{\varepsilon}_z$ is the unit vector along the z-direction. We see that, even though the σ_z does not change the spin of a given electron, in this case light scattering couples into the spin-density fluctuation [2.20]

$$\frac{1}{2}\,[N_{\uparrow}(q)-N_{\downarrow}(q)], \tag{2.89}$$

where $N_{\uparrow}(q)$ and $N_{\downarrow}(q)$ are the electron-density fluctuations for each value of spin.

The expression for the coefficient $\boldsymbol{B}$ that results from the wave functions in (2.77–80) and the gap energies in (2.85,86) is [2.9] (see also [Ref. 2.20, Eq. (4.80b)])

$$\boldsymbol{B}=\boldsymbol{I}\,\frac{2P^2}{3m}\,\hbar\omega_{\mathrm{L}}\left(\frac{1}{E_0^2-\hbar^2\omega_{\mathrm{L}}^2}-\frac{1}{(E_0+\Delta_0)^2-\hbar^2\omega_{\mathrm{L}}^2}\right). \tag{2.90}$$

We note that this expression for $\boldsymbol{B}$ has the properties that have been anticipated in previous considerations. It tends to zero when $\hbar\omega_{\mathrm{L}}\to 0$, as required for the antisymmetric part of light scattering tensors [2.21, 65], and it vanishes for $\Delta_0\to 0$.

We next consider the terms associated with the use of σ_y in (2.82). These terms correspond to a direct spin-flip of individual electrons, with spin quantized along z. The effective Hamiltonian is, in this case, given by

$$H''_{\mathrm{eff}}=\mathrm{i}\frac{e^2V}{2m}\,(\hat{\varepsilon}_{\mathrm{L}}\times\hat{\varepsilon}_{\mathrm{S}})\cdot\boldsymbol{B}\cdot\hat{\varepsilon}_y\left(\frac{N_{\uparrow}(q)-N_{\downarrow}(q)}{2}\right)A(\omega_{\mathrm{L}})A(\omega_{\mathrm{S}}), \tag{2.91}$$

where $\hat{\varepsilon}_y$ is the unit vector along the y-direction.

This Hamiltonian is very similar to that of (2.88). The only difference is in the relative orientations of the incident and scattered light polarizations with respect to that of the free-electron spin. This difference is immaterial in the absence of an external magnetic field, when the spin quantization direction is arbitrary, and thus the cross sections associated with the Hamiltonians (2.88, 91) are identical.

It is straightforward to write the light scattering cross section that is associated with the effective Hamiltonians of (2.88, 91). The spin-density fluctuations defined by the expression (2.89) have no net fluctuations in charge density [2.20]. This implies that they have no macroscopic electric field of the kind defined by (2.12, 21), valid in the RPA. Therefore, these spin-density fluctuations have single-particle character. The scattering cross section is directly proportional to the density of states of single-particle excitations of the free electrons [2.9, 66, 73], which may be written in terms of the longitudinal dielectric function of the electron gas as

$$\frac{\partial^2\sigma}{\partial\Omega\,\partial\omega}=r_0^2\left(\frac{\omega_{\mathrm{S}}}{\omega_{\mathrm{L}}}\right)^2\frac{V\varepsilon_0}{e^2}\,\frac{\hbar}{\pi}\,q^2\,|(\hat{\varepsilon}_{\mathrm{L}}\times\hat{\varepsilon}_{\mathrm{S}})\cdot\boldsymbol{B}\cdot\hat{\varepsilon}_{\mu}|^2$$
$$\cdot\,(1-\mathrm{e}^{-\hbar\omega/k_{\mathrm{B}}T})^{-1}\,\mathrm{Im}\{\chi(q,\omega)\}, \tag{2.92}$$

where $\hat{\varepsilon}_\mu$ is a unit vector along the x-, y- or z-directions, and z is the axis of quantization of the free-electron spin. Here we encounter unscreened scattering, at single-particle energies, as in the case of nearly neutral density excitations of the two-component plasma discussed in Sect. 2.2.8. We may consider the spin-up and spin-down electrons as two different components of the plasma.

Equation (2.90) applies to scattering by free electrons in zincblende-type semiconductors with a direct gap at the Γ-point, such as GaAs, InP, GaSb, InSb, and InAs. Scattering by spin-density fluctuations should also exist for electrons occupying states at other points in $\boldsymbol{k}$-space, where the bands involved are split by the spin-orbit interaction. Detailed calculations have been carried out in the case of the lead salts [2.66]. The case of electrons in n-type Ge, where they occupy states in four equivalent valleys along the $\{111\}$ directions, is of considerable interest. The relevant optical transitions are here close to the E_1 and $E_1+\Delta_1$ energy gaps [2.52]. The intermediate valence-band states are of the form displayed in the right of (2.78), leading to scattering by spin-density fluctuations described by an effective Hamiltonian similar to that of (2.88), where for each ellipsoid,

$$\boldsymbol{B} \sim \frac{1}{E_1^2-\hbar^2\omega_L^2} - \frac{1}{(E_1+\Delta_1)^2-\hbar^2\omega_L^2} \tag{2.93}$$

with $E_1 \simeq 2.1$ eV and $\Delta_1 \approx 0.2$ eV, and where the components of $\hat{\varepsilon}_L$ and $\hat{\varepsilon}_S$ are perpendicular to the axes of the ellipsoid. One also must sum the contributions due to the four equivalent ellipsoids. Resonant light scattering by single-particle transitions of 2-dim electrons confined in thin Ge layers has been reported [2.63] and interpreted in terms of spin-density fluctuations [2.63, 99]. It is also worth mentioning the case of free electrons in n-type Si, which occupy states in six valleys directed along the $\{100\}$ directions and centered near the X-point (at 85 percent of the zone-edge wavevector). In this case, light scattering by spin-density fluctuations is expected to be weak because the spin-orbit splitting is small in Si and vanishes by symmetry at the X-point [2.100].

d) Coupled Plasmons LO-Phonons

We have seen in Sect. 2.2.5 that the charge-density fluctuations of the free electrons couple to the macroscopic longitudinal electric field of polar (ir-active) optical lattice vibrations in materials which have them (i.e., in GaAs but not in Ge). The resulting coupled modes have mixed plasmon-phonon character [2.77, 79]. Therefore, their light scattering cross section has contributions arising from the electron charge-density fluctuations (cdf), discussed in Sect. 2.3.1b, and also from the LO phonon mechanisms which have been recently reviewed by one of us in [Ref. 2.64, Chap. 2]. In these cases, in the analysis of resonant behavior of the scattering intensities and spectral line shapes, one should anticipate quantum-mechanical interference effects [2.71a]. Such effects have been recently identified in scattering intensities of polarized light spectra from n-GaAs obtained with photon energies near the $E_0+\Delta_0$ optical gap [2.101a].

We consider first the cross section of coupled plasmons LO-phonons in the cdf mechanism. The effective Hamiltonian (2.83) leads to the expression

$$\frac{\partial^2\sigma}{\partial\Omega\,\partial\omega}=r_0^2\left(\frac{\omega_S}{\omega_L}\right)^2 V^2\,(\hat{\varepsilon}_L\cdot \boldsymbol{A}\cdot\hat{\varepsilon}_S)^2\,S(q,\omega). \tag{2.94}$$

In order to evaluate $S(q,\omega)$, the dynamical structure factor of the free electrons, we note that the mixed plasmons LO-phonons of Sect. 2.2.5 constitute a coupled system of longitudinal polarizations. Therefore, we may describe it with the formalism of Sect. 2.2.7 by making the identification $\chi_1\equiv\chi(q,\omega)$ and $\chi_2\equiv\chi_L(\omega)$. Using (2.53–55), we obtain

$$S(q,\omega)=\frac{\hbar}{\pi}\,\frac{q^2}{e^2}\,\frac{\varepsilon_0}{V}\,(1-e^{-\hbar\omega/k_BT})^{-1}\cdot[\varepsilon_\infty+\chi_L(\omega)]\;\mathrm{Im}\left\{\frac{\chi(q,\omega)}{\varepsilon(q,\omega)}\right\}. \tag{2.95}$$

In (2.95) we have neglected the phonon damping and assumed that ε_∞ is real.

The assumption that ε_∞ is real holds for all semiconductors whose lowest electronic absorption edge lies above the frequency of the Raman excitations under consideration. In addition, use of (2.38, 39) enables us to write

$$S(q,\omega)=\frac{\hbar}{\pi}\,\frac{q^2}{e^2}\,\frac{\varepsilon_0\varepsilon_\infty^2}{V}\,(1-e^{-\hbar\omega/k_BT})^{-1}\cdot\left(\frac{(\omega_{LO}^2-\omega^2)}{(\omega_{TO}^2-\omega^2)}\right)^2\;\mathrm{Im}\left\{\frac{-1}{\varepsilon(q,\omega)}\right\}. \tag{2.96}$$

Equations (2.94–96) describe the scattering cross section by the free electron cdf of the coupled plasmon LO-phonon modes. They indicate two characteristic features. One is simple resonant behavior which, under the approximations implicit in (2.76), is given by

$$\frac{\partial^2\sigma}{\partial\Omega\,\partial\omega}\sim\left|\frac{E_G^2}{E_G^2-\hbar^2\omega_L^2-2iE_G\hbar\gamma}\right|^2 S(q,\omega). \tag{2.97}$$

In (2.97), γ is the collision broadening associated with the optical energy gap, generally represented here as E_G. The other feature is a zero in the scattering cross section for $\omega=\omega_{LO}$. This is a characteristic signature of the cdf mechanism because at $\omega=\omega_{LO}$ there are no free-electron density fluctuations in the coupled plasmon LO-phonon system [2.77, 79]. Equations (2.94, 96) can yield explicit expressions for the spectral line shapes associated with the electron charge-density fluctuation. The case in which the free electrons are well described by a Drude susceptibility, including collision damping, is given by (4.23, 24) of [2.20].

The simplest type of light scattering related to the phonon component of the coupled modes is the *deformation potential* mechanism. In this mechanism the lattice vibrations modulate the electron energy band structure and thereby cause a modulation in the electric susceptibility related to interband transitions. An additional contribution arises from the macroscopic electric field that accompanies the coupled modes. This field also produces a modulation in the interband electric susceptibility. Within perturbation theory these effects are described as field-induced intraband and interband transitions. This description leads to two-band and three-band representations of the light scattering cross section ([Ref. 2.20, p. 23] and [2.65]). The three-band electric field contributions can be characterized by a first-order *electro-optic* tensor $\partial\chi(\omega_L)/\partial E$. The combined scattering cross section of coupled modes due to the deformation potential and electro-optical mechanisms can be written as [Ref. 2.20, Eq. (4.25)]

$$\frac{\partial^2\sigma}{\partial\Omega\,\partial\omega}=\frac{\hbar\varepsilon_0}{4\pi V}\left(\frac{\omega_S}{c}\right)^4(1-e^{-\hbar\omega/k_BT})^{-1}\left|\hat{\boldsymbol{\varepsilon}}_L\cdot\frac{\partial\boldsymbol{\chi}(\bar{\omega})}{\partial E}\cdot\hat{\boldsymbol{\varepsilon}}_S\right|$$

$$\cdot\left|\frac{\omega_o^2-\omega^2}{\omega_{TO}^2-\omega^2}\right|\operatorname{Im}\left\{\frac{-1}{\varepsilon(q,\omega)}\right\}, \tag{2.98}$$

where $\bar{\omega}=(\omega_L+\omega_S)/2$ and ω_o, simply a parameter with the dimensions of frequency, is defined as

$$\omega_o^2=\omega_{TO}^2(1+C), \tag{2.99}$$

where

$$C=\frac{e^*\hat{\boldsymbol{\varepsilon}}_L\cdot(\partial\boldsymbol{\chi}/\partial u)\cdot\hat{\boldsymbol{\varepsilon}}_S}{\mu\omega_{TO}^2\hat{\boldsymbol{\varepsilon}}_L\cdot(\partial\boldsymbol{\chi}/\partial E)\cdot\hat{\boldsymbol{\varepsilon}}_S}. \tag{2.100}$$

In (2.100), e^* and μ are the dynamical charge and the reduced mass of the ions, respectively. The dimensionless constant C is the so-called Faust-Henry coefficient. The tensor $\partial\boldsymbol{\chi}/\partial u$ describes light scattering by the deformation potential mechanism. A detailed discussion of these parameters can be found in [2.21] and in [Ref. 2.64, Sect. 2.2].

The resonant behavior of the cross section in (2.98) is contained in the electro-optic tensor $\partial\boldsymbol{\chi}(\bar{\omega})/\partial E$. The parameter C has been found to have only a weak dependence on the laser frequency. On the other hand, both $\partial\boldsymbol{\chi}/\partial E$ and $\partial\boldsymbol{\chi}/\partial u$ display strong resonant behavior at laser frequencies near the optical energy gaps, at interband critical points (see, for example, [Ref. 2.64, Sects. 2.2, 2.3]). Below the lowest absorption edge, $\partial\boldsymbol{\chi}/\partial u$ and $\partial\boldsymbol{\chi}/\partial E$ are real. Immediately above they are almost purely imaginary. The quantum interference between the two mechanisms, deformation potential and electro-optic, is thus contained in the term $(1+C)$ of (2.99). It will be constructive or destructive according to the sign of C.

The two-band electric field contributions are often referred to as the "Fröhlich mechanism" because they are associated with virtual *intraband* transitions of electrons caused by an interaction of the form (2.87). A recent discussion of this mechanism is given in [Ref. 2.64, Sect. 2.2.8]. The scattering amplitude is in this case dependent on the scattering wavevector $\boldsymbol{q}$. These processes are also identified as "forbidden scattering", since its selection rules are different from those of the usual wavevector-independent processes. The leading term in the scattering amplitude is linear in $\boldsymbol{q}$ and the associated cross section goes as q^2. These terms are 90° out of phase with those of the "allowed" term in (2.98) [2.20]. In this case the allowed and forbidden terms do not interfere. Interference is possible, however, if both scattering amplitudes are complex, as it occurs at resonance [2.20, 101a, b].

For the simple case of a single interband optical gap between isotropic parabolic bands, the forbidden Fröhlich contribution can be included in (2.98) by adding to $(\partial\chi/\partial u)$ the following expression [Ref. 2.64, Eq. (2.218)]

$$-\frac{C_F}{8}\frac{m_e-m_h}{M}\left(\frac{\partial^2\chi}{\partial\omega^2}\right)_{\bar{\omega}}|q|, \tag{2.101}$$

where m_e and m_h are the electron and hole masses and $M=m_e+m_h$. The interaction constant C_F, defined by (2.213) of [2.64], is *pure imaginary* and proportional to e*. We note that an ambiguity of sign results in (2.100, 101) depending on whether we take for e^* the charge at the cation (positive) or at the anion (negative). The apparent ambiguity in the relative signs of the several contributions to allowed and forbidden scattering is solved by defining e^*, $(\partial\chi/\partial\mu)$ and C_F consistently. For example, we could take for e^* the charge of the cation and for u the sign of the motion of the cation. The ambiguity in (2.101) is solved in a similar manner. We take $C_F=-\mathrm{i}|C_F|$ if $\boldsymbol{q}$ points along the motion chosen for the cation and negative in the opposite case.

When $\chi(\bar{\omega})$ is real (or pure imaginary), the allowed scattering amplitudes are real (or pure imaginary), while those of the Fröhlich term are imaginary (or real). Quantum interference is only possible when $\bar{\omega}$ is well above the lowest direct interband edge and hence $\chi(\bar{\omega})$ is complex. We also note that light scattering via the cdf mechanism, which is also wavevector dependent [2.71a] and thus includes as a factor in the scattering amplitude the imaginary unit, will display interference phenomena like forbidden Fröhlich scattering [2.101a].

We have seen that scattering due to cdf has a simple resonance given by (2.97). In the phonon mechanisms two resonances appear, an in-going one for $\hbar\omega_L=E_G$ and an out-going one for $\hbar\omega_S=E_G$. If $\omega=\omega_L-\omega_S\lesssim|\omega_L-E_G/\hbar-2\mathrm{i}\gamma|$, both resonances merge into one. We can then invoke the quasi-static approximation (2.76a) on which (2.98–101) are based. For the sake of simplicity we stay within this approximation which predicts a resonance at $E_G\simeq\hbar\bar{\omega}$, and which holds in most cases related to scattering by coupled modes. The resonant behavior is then given by derivatives of the susceptibility $\chi(\bar{\omega})$. We note that at the laser frequencies ($\bar{\omega} \approx \omega_L$), the free carrier and phonon contributions to $\chi(\bar{\omega})$ have usually relaxed away so that only interband contributions to $\chi(\bar{\omega})$ need to be considered.

The resonances of the interband $\chi(\bar{\omega})$ and its derivatives occur at so-called critical points or van-Hove singularities. The singularity in $\chi(\bar{\omega})$ associated with edge excitons has a Lorentzian shape and often leads to a resonance like that of (2.97) (three-band terms) or the square of it (two-band terms) in the electro-optic or deformation-potential mechanism [2.64]. We discuss in what follows the interband type of singularities. Near them the susceptibility $\chi(\bar{\omega})$ has a resonant behavior of the form

$$\begin{aligned} &e^{i\alpha}(E_G - \hbar\bar{\omega} - i\hbar\gamma)^{-1/2} \quad \textit{in one dimension}, \\ &e^{i\alpha} \ln(E_G - \hbar\bar{\omega} - i\hbar\gamma) \quad \textit{in two dimensions}, \\ &e^{i\alpha}(E_G - \hbar\bar{\omega} - i\hbar\gamma)^{1/2} \quad \textit{in three dimensions}, \end{aligned} \tag{2.102}$$

where α depends on the type of critical point and on excitonic-like interactions [2.102]. The one- and two-dimensional situations often occur in three-dimensional crystals when the reduced effective mass is very large in two or one directions, respectively [Ref. 2.64, Sect. 2.3].

There are two types of terms in $(\partial\chi/\partial u)$, the so-called two-band and three-band contributions. The latter have the shape given in (2.102) and thus, in the general case, have much weaker resonant behavior than the former which are proportional to the derivative of (2.102) with respect to ω. For $(\partial\chi/\partial E)$ there are only three-band terms. However in the case of degenerate bands, such as the top valence bands of GaAs (i.e. for the E_0 edge), the three bands should resonate in a way similar to two-band terms. At the $E_0 + \Delta_0$ edge of GaAs, for which extensive light scattering by free carriers has been investigated, only weakly resonant three-band terms are possible. This is a consequence of the nondegenerate character of the Γ_7 and Γ_6 states which form the edge. (The phonon, of Γ_4 symmetry, does not have diagonal matrix elements between nondegenerate states because the product of such states with themselves must have Γ_1 symmetry.)

According to this discussion, the two-band terms should have resonant behavior of the form

$$\begin{aligned} &|E_G - \hbar\bar{\omega} - i\hbar\gamma|^{-3} \quad \textit{in one dimension}, \\ &|E_G - \hbar\bar{\omega} - i\hbar\gamma|^{-2} \quad \textit{in two dimensions}, \\ &|E_G - \hbar\bar{\omega} - i\hbar\gamma|^{-1} \quad \textit{in three dimensions}. \end{aligned} \tag{2.103}$$

The reason why the second derivative of $\chi(\bar{\omega})$ appears in the expression (2.101) that describes the "forbidden", q-dependent, Fröhlich mechanism, as compared with the first derivatives of (2.103), is the forbidden nature of the mechanism. For the scattering cross sections we find the following resonant behavior:

$$\begin{aligned}
&|E_G-\hbar\bar{\omega}-i\hbar\gamma|^{-5} \quad \textit{in one dimension,}\\
&|E_G-\hbar\bar{\omega}-i\hbar\gamma|^{-4} \quad \textit{in two dimensions,}\\
&|E_G-\hbar\bar{\omega}-i\hbar\gamma|^{-3} \quad \textit{in three dimensions.}
\end{aligned} \tag{2.104}$$

The two-dimensional case of (2.103), which occurs at E_1 and $E_1+\Delta_1$ edges of diamond- and zincblende-type semiconductors, should thus resonate in a manner similar to (2.97). We point out, however, that in cdf case (2.97), the energy gap E_G must involve the energy states of the free carriers. This does not happen in GaAs, so that (2.97) does not resonate at the E_1 gap while (2.103) does.

Finally, we should mention that in doped semiconductors the presence of impurities may cause a breakdown of the $\boldsymbol{q}$-conservation selection rule. This effect is particularly important under resonant conditions. *Colwell* and *Klein* [2.103] first considered the case of the resonance with bound-exciton intermediate states. *Gogolin* and *Rashba* [2.104] highlighted the effects of electron-impurity interactions. In the case of coupled modes, evidence of wavevector nonconservation has been found in resonant spectra of n-GaAs [2.51], p-GaAs [2.87, 105] and of n-InP [2.106].

Wavevector nonconservation is particularly important in the cases of the cdf and Fröhlich mechanisms, given the q^2 dependence of their scattering cross sections [see (2.96, 101)]. The q supplied by the impurity can be much larger than that of the photons. The forbidden nature of the phenomena is lifted by the effects of electron-impurity interactions, which can be described within perturbation theory [2.104]. Impurity-induced deformation potential (or electro-optic) scattering, although weaker than that of the q-dependent mechanisms, has also been observed [2.105]. We shall consider further the effects of the breakdown of wavevector conservation in Sect. 2.3.3.

2.3.2 Experimental Results

The most comprehensive experiments on resonant light scattering by free-electron excitations in semiconductors have been carried out in the III-V compounds with the zincblende structure. We thus show in Fig. 2.7 the electron energy band structure of GaAs, which serves as a good example of such a semiconductor with a lowest direct gap. The figure also identifies several direct gaps which produce structure in the optical constants. The lowest are E_0 (the fundamental gap) and its spin-orbit split-off partner $E_0+\Delta_0$. They are followed by the E_1 and $E_1+\Delta_1$ gaps, which are associated with interband critical points along the Λ direction in the Brillouin zone. These are the only gaps for which resonant light scattering has been observed. Limitations in the available lasers and spectrometers have discouraged work at the other resonances.

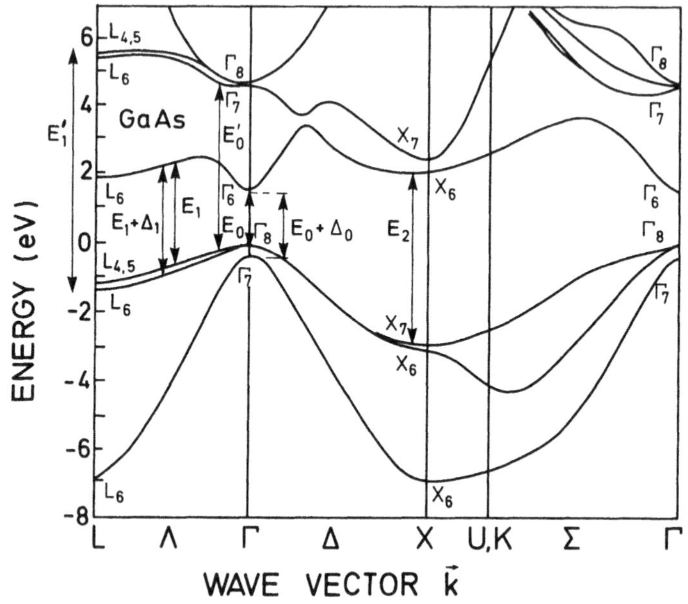

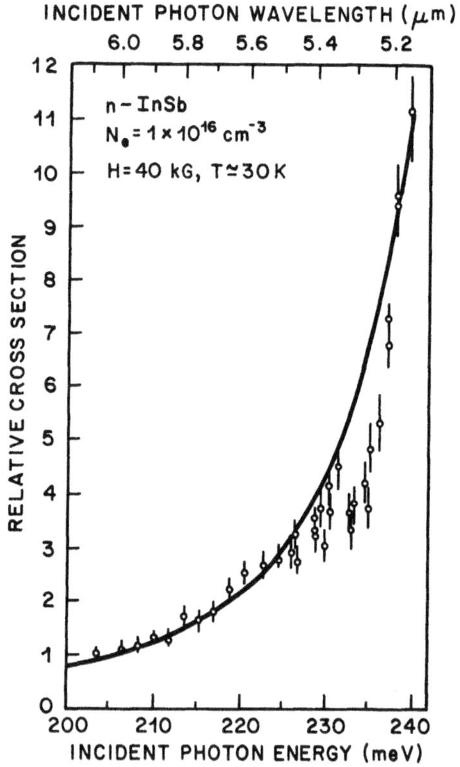

▲ **Fig. 2.7.** Band structure of GaAs (from [2.106b]) showing the main interband critical points which should contribute to Raman resonances (see text)

Fig. 2.8. Resonance of spin-flip electron excitations in a magnetic field near the E_0 gap of InSb. From [2.90]

What might be the nicest example of an E_0 resonance in scattering by free carriers was reported for InSb by *Brueck* et al. [2.90]. These experiments involve spin-flip excitations in a magnetic field. We reproduce the results in Fig. 2.8. A CO gas laser was used for this work. The magnetic field has the effect of enhancing the resonance by reducing from three to one the dimensionality of the E_0 critical point, see (2.102).

In general, the E_0 gap of direct semiconductors is characterized by strong photoluminescence that obscures the observation of inelastic light scattering. This is much less of a problem in semiconductors in which the lowest gap is indirect. Near-resonant light scattering by free electrons has been observed in several semiconductors in this class, including n-GaP [2.107], n-GaAs under large hydrostatic pressure [2.108] and p-channels of Si-MOS devices [2.61]. However, in none of these cases have the resonant profiles been investigated.

The $E_0+\Delta_0$ optical gap, not being the lowest direct one, is only accompanied by weak luminescence. It is therefore very convenient for resonant light scattering studies. It also has the advantage of being only Kramers degenerate, a fact that simplifies the theoretical treatment. In addition, as seen in Sects. 2.3.2a,b, light scattering by *single-particle* (spin-density) and *collective* (charge-density) excitations are allowed at this gap. The first observations of resonant light scattering at a $E_0+\Delta_0$ gap have been reported in n-GaAs, in experiments carried out with He–Ne and Kr^+ lasers [2.32, 50]. More recently, the $E_0+\Delta_0$ resonant profiles have been studied in n-GaAs [2.51] and in n-InP [2.106]. These experiments have been made possible by the development of dye laser systems operating at continuously tunable photon energies that overlap with the corresponding energy gaps: rhodamine 101 [Ref. 2.64, Fig. 2.7] in the case of GaAs and Oxazyne 750 [2.109] in the case of InP.

In the remainder of this section we discuss the studies of the $E_0+\Delta_0$ resonance of light scattering by single-particle and collective excitations of the free carriers. We also consider the resonant behavior of scattering by coupled plasmons LO-phonons near the E_1 and $E_1+\Delta_1$ optical gaps. Due to small penetration depths (~300 Å) and large scattering wavevectors ($q \gtrsim 10^6\ \mathrm{cm}^{-1}$), the light scattering spectra excited near these resonances display a number of challenging features related to surface space-charge layers, Landau damping, and the relaxation of wavevector conservation [2.33].

a) $E_0+\Delta_0$ Gap: Single-Particle Excitations

Figure 2.9 shows spectra of free electrons in doped n-type GaAs obtained with photon energies close to the $E_0+\Delta_0$ gap (1.89–1.91 eV) and also with orthogonal incident and scattered polarizations. According to the discussion in Sect. 2.3.1c, see (2.92), these *depolarized* spectra are associated with spin-density (spin-flip) excitations of the free electrons. The experimental line shape is found to depend slightly on incident photon energy. This effect, perhaps exaggerated by uncertainties in the subtraction of the background luminescence [2.32, 50], emphasizes the approximate character of (2.92). Equations (2.76), which have

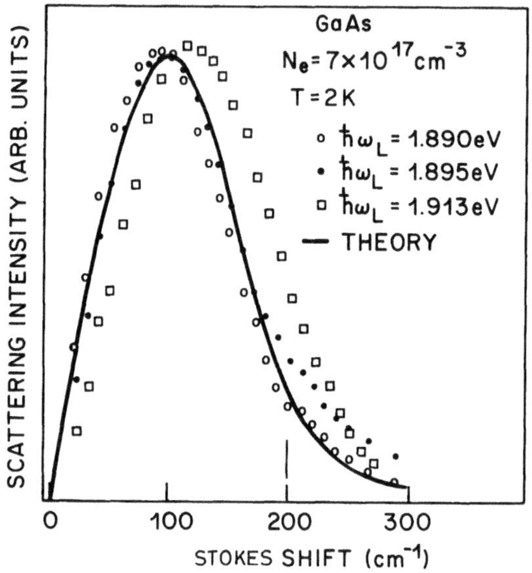

Fig. 2.9. Light scattering spectra of spin-density fluctuations of n-GaAs. Photon energies are close to resonance with the $E_0+\Delta_0$ optical gap from [2.51]. The full line is a fit with (2.35, 92)

been used in the derivation of (2.90,92), can only be valid if $|E_0+\Delta_0-\hbar\omega_L+i\Gamma|>\hbar\omega$, where Γ is the broadening of the optical transitions, not included for simplicity in (2.90). The value of Γ at the $E_0+\Delta_0$ gap is only 8 meV at 4 K [2.110]. Hence, the condition above is not well fulfilled in the resonant spectra of Fig. 2.9, thus explaining the slight dependence of the spectral line shape on laser photon energy. Similar behavior has been found in resonant spectra of n-type InP [2.106a].

We have seen in Sect. 2.3.1c that spin-density excitations have single-particle character. According to (2.92), their light scattering spectral line shapes are expected to be determined by Im $\{\chi(q,\omega)\}$. The solid line in Fig. 2.9 is a fit of the observed line shapes using for the free-electron-gas susceptibility $\chi(q,\omega)$ the Lindhard-Mermin expression (2.35). This susceptibility requires a phenomenological lifetime broadening γ_e, which is the only adjustable parameter required by this analysis. The best fit, shown in Fig. 2.9, corresponds to $\gamma_e = 7$ meV. This value is about 50 percent higher than the one calculated from the Hall mobility of the sample. A similar line shape analysis also gives a satisfactory interpretation of the resonant light-scattering spectra obtained from n-InP [2.106].

There are other calculations of the spectral line shapes. *Voitenko* et al. [2.111] (see also [2.24]) have considered lifetime broadening in a microscopic manner as due to fluctuations in the density of ionized donors. This calculation is expected to be applicable in high-doping situations because the average separation between ionized donors is smaller than the screening length and each electron is subject to the Coulomb potential of more than one ionized impurity. A notable feature of this calculation is that lifetime broadening is not an adjustable

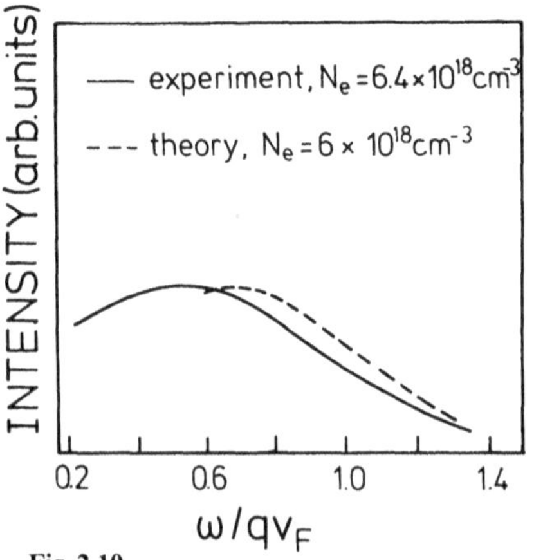

Fig. 2.10

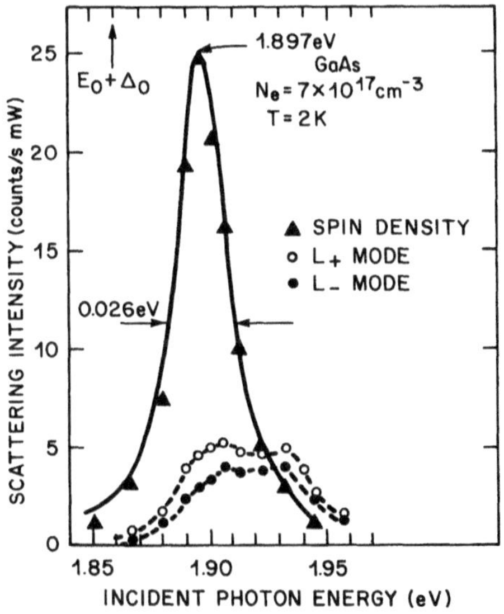

Fig. 2.11

Fig. 2.10. Spin-flip single-particle scattering near $E_0+\Delta_0$ for n-GaAs ($N_e=6.4\times10^{18}$ cm^{-3}, $qv_F=280$ cm^{-1}) [2.115] compared with calculations by *Voitenko* et al. [2.111]

Fig. 2.11. Resonance of the single-particle (spin-density) and the collective mode free-electron excitations near the $E_0+\Delta_0$ gap of GaAs [2.51]

parameter. This aspect of the theory stimulated us to compare its predictions with the spectral line shapes observed in resonant light scattering from highly doped n-GaAs. The comparison is shown in Fig. 2.10. The agreement between theory and experiment is quite satisfactory considering the lack of adjustable parameters. We should also mention here the calculation by *Hertel* and *Appel* [2.112], who considered the possibility of a frequency-dependent broadening parameter.

The resonance profile for the intensities of the spectra of Fig. 2.9 is shown in Fig. 2.11. The solid line is a fit with the Lorentzian expression

$$I(\hbar\omega_L)\sim\left|\frac{1}{(E_0+\Delta_0+\mathrm{i}\Gamma)^2-\hbar\omega_L^2}\right|^2, \tag{2.105}$$

which corresponds to the second term in (2.93). The value of $\Gamma=13$ meV used for the fit is quite reasonable since $\gamma_e=7$ meV and the broadening of the optical gap is 8 meV according to [2.110]. We also note that, as it can be deduced from the diagrams in Fig. 2.5, the resonance has an intrinsic full width of $\sim qv_F$, which for the case in Fig. 2.11 is 23 meV. In [2.51] we determined a *scattering efficiency* (integrated cross section per unit volume) of 8×10^{-3} cm^{-1} for the photon

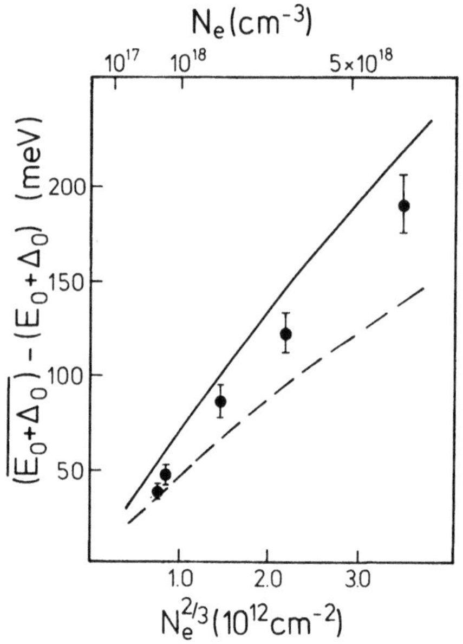

Fig. 2.12

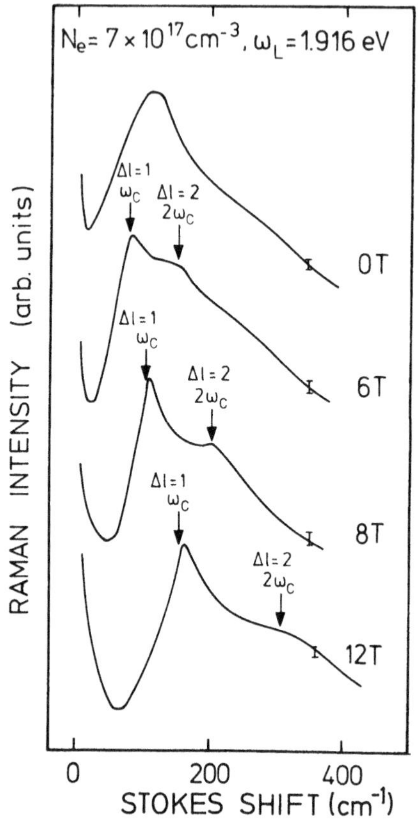

Fig. 2.13

Fig. 2.12. Difference between the resonance energy of single-particle excitations in n-GaAs $\overline{E_0+\Delta_0}$ and the $E_0+\Delta_0$ energy gap versus the 2/3 power of the number of carriers. The dashed curve is the Fermi energy in the conduction band including effects of nonparabolicity. The solid curve includes, in addition, the curvature of the valence band. From [2.115]

Fig. 2.13. Raman spectra of single-particle excitations in a magnetic field in Voigt configuration $(q\perp H)$. The arrows mark the $\Delta l=1$ and $\Delta l=2$ Landau level splittings. From [2.115]

energy at the maximum in the resonance profile ($\hbar\omega_L = 1.897$ eV) by comparison with Raman scattering intensities from transverse optical phonons. The scattering efficiency was also estimated with (2.92,93), the sum rule for $\mathrm{Im}\{\chi(q,\omega)\}$ [2.4] and the accepted value of the interband matrix element of electron momentum [2.104]. We obtained 4×10^{-3} cm^{-1}, in reasonable agreement with the value discussed above.

The maximum in the resonance profile occurs at $\overline{E_0+\Delta_0}$. The bar indicates that this energy is not simply that of the $E_0+\Delta_0$ gap (Fig. 2.11), but is displaced towards higher energies by the analog of the Burstein-Moss shift. This effect simply accounts for the fact that in a semiconductor doped to degeneracy, the excitation gap occurs at a higher energy because the band edge is blocked by the free carriers (Fig. 2.6). Assuming that the wavevector is strictly conserved in optical transitions relevant to these light scattering processes (which might not be the case when impurities are present), the $\overline{E_0 + \Delta_0}$ gap of n-type GaAs is

$$\overline{(E_0 + \Delta_0)} \simeq E_0 + \Delta_0 + E_F \left(1 + \frac{m_e^*}{m_{so}^*}\right), \tag{2.106}$$

where E_F is the Fermi energy of the electrons measured from the bottom of the conduction band and the second term within the brackets arises from $\boldsymbol{k}$-conservation. m_e^* and m_{so}^* are the conduction and split-off valence band effective masses, respectively. If $\boldsymbol{k}$-conservation is relaxed by impurity scattering, nonvertical transitions are possible from the top of the valence band to the Fermi surface of the conduction band and the term m_e^*/m_{so}^* in (2.106) must be omitted.

We plot in Fig. 2.12 the energy shift $\overline{(E_0 + \Delta_0)} - (E_0 + \Delta_0)$ determined in five n-type GaAs samples with different free-carrier densities N_e. If we neglect the conduction band nonparabolicity, E_F is proportional to $N_e^{2/3}$, which is the reason the data are presented as a function of $N_e^{2/3}$. The dashed and solid curves in Fig. 2.12 represent the calculated values of E_F and $E_F(1 + m_e^*/m_{so}^*)$, respectively. In both cases we have included nonparabolicity in E_F [2.113] and taken the $k = 0$ values as $m_e^* = 0.065\,m$ and $m_{so}^* = 0.16\,m$ [2.114]. From the fact that the solid curve fits the experimental points better, we conclude that $\boldsymbol{k}$-conserving nearly vertical transitions are indeed dominant. This result is consistent with the fact that the spectral line shapes have been satisfactorily explained assuming $\boldsymbol{k}$-conservation. In the next sections we are going to encounter situations in which this is not the case. We should also point out that similar behavior has been observed for the intrinsic dielectric constant of a number of semiconductors at E_0 and $E_0 + \Delta_0$. As an example see Fig. 144 of [2.102].

To complete this discussion we show in Fig. 2.13 the effect of an external magnetic field $\boldsymbol{H}$ on the spectra of single-particle spin-density excitations of n-GaAs ($n = 7 \times 10^{17}$ cm^{-3}) [2.115]. These experiments were carried out in the Voigt configuration for which $\boldsymbol{q} \perp \boldsymbol{H}$. We can see that the spectral bands of single-particle excitations split into two components. They correspond to $\Delta l = 1$ and $\Delta l = 2$ transitions between Landau levels at frequencies ω_c and $2\omega_c$, where $\omega_c = eH/m_e^* c$ is the cyclotron frequency of the conduction electrons. The spin-splitting is not resolved because the g-factor is small in GaAs [2.114]. Most puzzling in these results is the observation of the $\Delta l = 1$ transition. Light scattering by this excitation, which has odd parity, is forbidden in the dipole approximation [2.8, 10, 18, 19, 21]. Its observation in small energy gap compounds, e.g., InSb [2.15], is believed to be caused by the nonparabolicity of the energy bands. However, in the larger band gap GaAs such effects are much

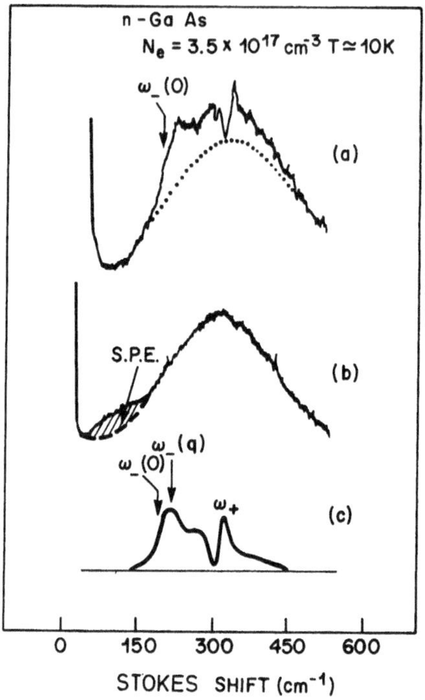

Fig. 2.14

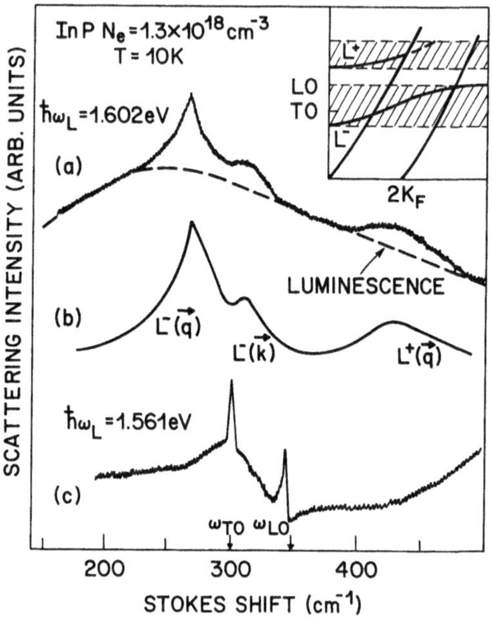

Fig. 2.15

Fig. 2.14a–c. Resonant light scattering spectra of n-GaAs. $\hbar\omega_L = 1.916$ eV (Fig. 2.11). (**a**) Polarized spectrum due to coupled plasmons LO-phonons. The dotted line is the estimated luminescence background as determined from the depolarized spectrum in (**b**) with exclusion of the single-particle excitations (SPE). (**c**) Line shape of the spectrum of coupled modes obtained from (**a**) by subtraction of estimated luminescence showing $\omega_-(q)$ induced by nonconservation of crystal momentum. From [2.32]

Fig. 2.15a–c. Resonant light scattering spectra of coupled plasmons LO-phonons of n-InP. The photon energies are close to the $E_0 + \Delta_0$ gap. (**a**) and (**c**) are measured spectra. (**b**) Spectral line shape determined from (**a**) after subtraction of luminescence. The inset shows the dispersion of the coupled modes. The hatched regions represent the widths of the mode dispersions. From [2.106a]

weaker. In this case resonant light scattering might be due to a breakdown of the dipole approximation. At the present time the theory of resonant light scattering by free electrons in large magnetic fields requires further consideration.

b) $E_0 + \Delta_0$ Gap: Collective Modes

Figure 2.14a shows spectra of n-GaAs obtained with a laser photon energy close to the $E_0 + \Delta_0$ gap and also with parallel incident and scattered polarizations. These *polarized* spectra are assigned to coupled plasmon LO-phonon modes (see Sect. 2.3.1c). The figure also indicates how the luminescence, in this case determined from the depolarized spectrum of Fig. 2.14b, is subtracted to obtain the curve of Fig. 2.14c. This spectrum shows peaks at coupled frequencies $\omega_{\pm}(q)$. They are shifted from the $q=0$ values because at the scattering wavevector $q \simeq 0.7 \times 10^6 \text{ cm}^{-1}$ the plasma frequency, given by (2.37), has substantial dispersion. In the spectrum of Fig. 2.14c there is, in addition, a scattering continuum that exhibits a minimum near the frequency of the $q \approx 0$ LO phonon.

These unusual spectral line shapes are not unique to n-type GaAs. More recently they have also been observed in spectra from doped n-type InP excited near the $E_0 + \Delta_0$ optical gap [2.106a]. Figure 2.15 reproduces these results. The fact that the spectral line shapes depend on the incident photon energies emphasizes their resonant character. Below we shall consider these effects, which are not present in spectra excited off-resonance [2.37, 38, 51]. The dispersions of the coupled modes are discussed in Sect. 2.4.

In [2.51] we proposed that the additional scattering structure observed in spectra excited near the $E_0 + \Delta_0$ gap is caused by resonant light scattering processes in which $\boldsymbol{q}$-conservation breaks down. These processes should allow the observation of coupled modes with wavevectors $k > 2k_F \approx 3\text{–}6 \times 10^6 \text{ cm}^{-1}$. These modes are within the hatched region shown in the inset to Fig. 2.15. Thus, the spectral line shapes associated with these processes are related to a one-dimensional density of states of the coupled modes. This explains qualitatively the major feature of the additional scattering observed in Figs. 2.14c, 15: the minimum that occurs in the vicinity of ω_{LO} is explained by the gap that occurs in the density of states within the frequency range $\omega_{LO} < \omega < \omega_+(0)$.

The dependence of the spectral line shape on incident photon energy (Fig. 2.15) could be interpreted by *different* resonant behavior of the possible light scattering mechanisms involved. For example, it follows from (2.96) that charge-density fluctuations have a minimum (a zero if damping is neglected) at $\omega = \omega_{LO}$. Predominance of this mechanism would explain spectra like those in Figs. 2.14c, 15b. On the other hand, the phonon-like mechanisms discussed in Sect. 2.3.1d are strong for $\omega = \omega_{LO}$. Their strength and a maximum in the density of states should account for the peak at ω_{LO} in the spectrum of Fig. 2.15c.

Breakdown of wavevector conservation has been invoked in resonant Raman scattering by LO phonons. In these processes the intermediate electron states are either excitons bound to impurities [2.103] or free excitons with a relaxation time limited by collisions with impurities [2.104]. In doped semiconductors, breakdown of $\boldsymbol{q}$-conservation can be caused by electron wavevector relaxation processes associated with the ionized dopant impurities [2.51, 87, 105]. The matrix elements for this kind of impurity-assisted resonant light scattering require fourth-order perturbation theory. This is one order higher than the $\boldsymbol{q}$-conserving processes considered in Sects. 2.3.1b, d (see Fig. 2.6 above [Ref. 2.64, Fig. 2.39] and [Ref. 2.65, Figs. 2.4, 5]), so that electron relaxation processes are explicitly included. The interband and intraband electron transitions in two of the many possible fourth-order processes are shown in Fig. 2.16. Excitonic effects in the intermediate states are ruled out since they are screened by the free electrons. As in the cases shown in Fig. 2.6, the nearly vertical interband transitions involve the incident and scattered photons through the H_{ep}

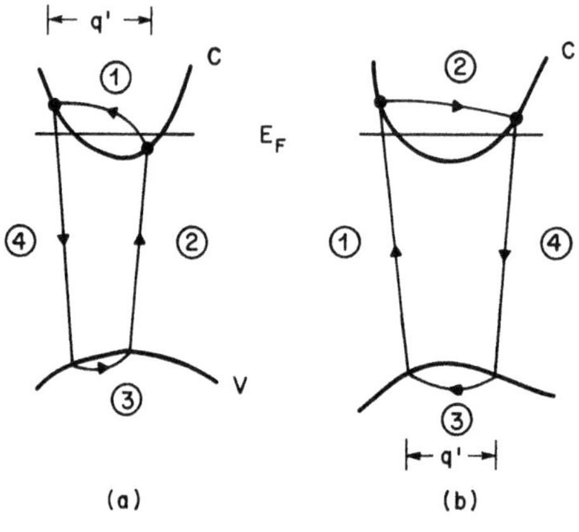

Fig. 2.16. Schematic representation of two fourth-order perturbation terms associated with impurity-induced resonant light scattering by a free electron gas. The wavevector relaxation of q' is due to electron-impurity (Coulomb) interaction

Hamiltonian. The impurity-assisted wavevector relaxation, of magnitude q', involves the Coulomb interaction between the ionized impurities and the free electrons in Fig. 2.16a and that with the intermediate holes in Fig. 2.16b. An evaluation of these matrix elements has not been made. We note, however, that the strength of these processes, like the observation of the additional scattering, depend critically on resonant conditions. This happens because in these matrix elements there is one more interaction and an additional resonant factor.

Breakdown of $\boldsymbol{q}$-conservation should also occur in the resonant light scattering by the spin-density single-particle excitations considered in the previous section. However, because of the intrinsic wavevector dependence of the spectral line shape (Fig. 2.9), any additional scattering is expected to be in the form of a featureless continuum indistinguishable from the background luminescence.

The empty and solid circles shown in Fig. 2.11 represent the resonant profiles of the collective coupled plasmons LO-phonons (labeled L_+ and L_-). Most remarkable is their width (~ 0.06 eV) which is more than twice that of the resonant profile of single-particle spin-density excitations. This result can be interpreted as being caused by two resonant factors in the scattering cross section, one at the incident photon energy with a maximum at $\hbar\omega_L \simeq \overline{(E_0 + \Delta_0)}$, and the other at the scattered photon energy when $\hbar\omega_S \simeq (E_0 + \Delta_0)$. It is useful to recall here that, as seen in Sects. 2.3.1b, 2.3.2a, light scattering by spin-density fluctuations is given by second-order perturbation theory (Fig. 2.5) with a single resonance at $\hbar\omega_L \simeq \overline{(E_0 + \Delta_0)}$. On the other hand, scattering by charge-density fluctuations is given by the third-order processes shown in Fig. 2.6. It is simple to verify that the process of Fig. 2.6a has a resonance at the scattered photon while that of Fig. 2.6b resonates with the incident photon. This double resonance had been hidden in the "quasi-static" approximation (2.76) used in the derivation of (2.97). Unfortunately, these considerations do not explain the observations. It is straightforward to show, as argued in [2.51], that the strength of the scattered photon resonance is about an order of magnitude weaker than the other one. We thus conclude that more work is needed to clarify the double resonance in Fig. 2.11. We should point out that the Fröhlich mechanism (Sect. 2.3.1.d) could make a substantial contribution and thus explain these results [2.64].

c) E_1 and $E_1+\Delta_1$ Gaps

The E_1 and $E_1+\Delta_1$ optical gaps, as can be seen in Fig. 2.7, occur along the $\{111\}$ Λ-directions in the Brillouin zone. In direct gap semiconductors the free carriers occupy states near the Brillouin zone center. In these semiconductors the electron-density light scattering mechanisms, considered in Sects. 2.3.1a–c, are not expected to have resonant behavior at the E_1 and $E_1+\Delta_1$ gaps [2.52]. This can be seen by examination of Figs. 2.5,6 and recalling that the photon wavevectors, typically $k_L \simeq k_S \approx 5\times10^5\ \mathrm{cm}^{-1}$, are much smaller than the Brillouin zone boundaries ($\sim 10^8\ \mathrm{cm}^{-1}$). Thus, in the direct-gap cases only the phonon-like mechanisms discussed in Sect. 2.3.1d, and associated with light scattering by collective coupled modes, show resonant behavior at the E_1 and $E_1+\Delta_1$ optical gaps.

In indirect gap n-type Ge, where the free electrons occupy states near the L-points of the Brillouin zone, the electron density mechanisms resonate near the E_1 and $E_1+\Delta_1$ gaps [2.52]. Until recently the only experimental study of this resonance has been reported in Ge-GaAs interfaces [2.63]. This work will be considered in Sect. 2.5. Below we shall be concerned with the results obtained from direct gap semiconductors. Resonant scattering by plasmons in n-Ge for ω_L near the E_1 gap has just been reported [2.101b].

The resonant profiles of Raman scattering by bare phonons near the E_1 and $E_1+\Delta_1$ optical gaps have been measured in several tetrahedral semiconductors (see, for example, [Ref. 2.64, Figs. 2.46,48]). Similar resonances should exist for Raman scattering by coupled plasmons LO-phonons of the doped polar semiconductors. Although some resonance effects have been reported, the resonance profiles have not been investigated in a systematic fashion. Interest has been diverted by other phenomena that strongly influence Raman scattering near the E_1 and $E_1+\Delta_1$ gaps. The most significant ones are related to the small penetration depths for the laser light (200–500 Å) that are associated with the large absorption. For this reason the Raman scattering spectra are dependent on physical, chemical and mechanical conditions at the sample surfaces. An important effect which occurs at surfaces of the highest perfection, like the ones created by cleavage, is the Fermi-level pinning caused by the presence of surface electron states. This effect leads to the formation of surface space-charge layers in which the free-electron density is very different from that in the bulk.

The importance of space-charge layers was clearly recognized in the first Raman scattering experiments near an E_1 optical gap [2.116a], which were carried out in n-type InSb. In this case it was not possible to observe coupled modes; only the bare optical phonons contributed to the spectra. This behavior was attributed to the presence of a surface depletion layer wider than the penetration

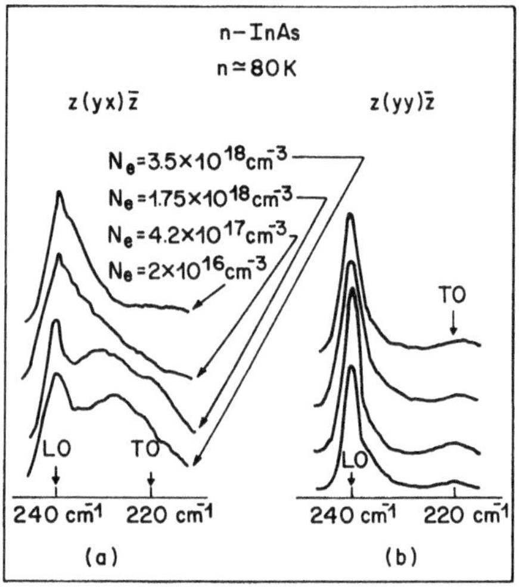

Fig. 2.17

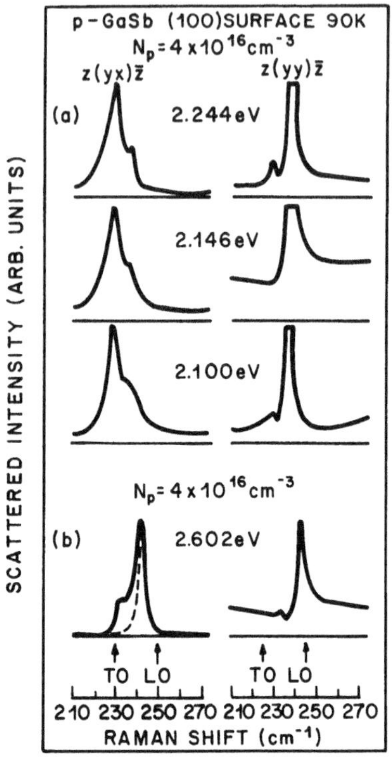

Fig. 2.18

Fig. 2.17a, b. Resonant Raman scattering spectra from n-InAs obtained with photon energies close to the E_1 optical gap. (**a**) Allowed scattering; (**b**) forbidden scattering. From [2.33]

Fig. 2.18. Resonant Raman scattering spectra from p-GaSb surfaces. The photon energies are in the range of the E_1 and $E_1+\Delta_1$ optical gaps. The $z(yx)\bar{z}$ and $z(yy)\bar{z}$ spectra correspond to allowed and forbidden configurations, respectively. From [2.117]

depth of light (≈ 200 Å), even for $N_e \approx 10^{18}$ cm^{-3}. Similar surface conditions prevail in other tetrahedral semiconductors [2.40,41]. In p-type GaAs the depletion-layer effect has been overcome by very heavy doping [2.85, 105]. Some semiconductors support surface accumulation layers that can be investigated by resonant Raman scattering from the coupled modes. This has been the case in n-type InAs [2.33] as well as in p-type GaSb and InSb [2.117]. Thus, the resonant Raman scattering method can be used as a tool for studies of the Fermi level position at surfaces and interfaces. These applications are considered in Sect. 2.6.

The first observations of coupled plasmons LO-phonons in Raman spectra excited with photon energies near an E_1 gap were reported by *Buchner* and *Burstein* [2.33]. The measurements were carried out in *accumulation* space-charge layers at surfaces of n-type InAs. Figure 2.17 reproduces these results. We focus here on the broad band that occurs in the allowed spectrum between the bare LO and TO phonon frequencies. This band has been interpreted as arising from lower branch coupled plasmon LO-phonon modes (the L_- branch in the inset to Fig. 2.15). The spectral line shape is believed to be related to the nonuniform charge densities and electric fields within the surface accumulation layers which, by causing breakdown in q-conservation, allow observation of large wavevector modes.

A particularly interesting $E_1, E_1 + \Delta_1$ resonant behavior of Raman scattering by LO phonons coupled to free-electron excitations has been observed by *Dornhaus* et al. [2.117] in surface space-charge layers of p-GaSb and p-InSb. The basic observations which exemplify the remarkable line shapes that can be obtained in these experiments are shown in Fig. 2.18. It has been suggested that the Fano-type line shape observed in spectra obtained in the *forbidden* configuration is due to coherent interference between Raman scattering by longitudinal excitations and broad emission associated with a recombination of excitons at the E_1 and $E_1 + \Delta_1$ optical gaps. The spectra in the *allowed* configuration have been interpreted, like those from n-InAs, as due to large wavevector coupled modes made Raman active by nonuniform space-charge layers and by the presence of ionized acceptor impurities.

Finally, we should mention that the coupled plasmons LO-phonons of heavily doped p-GaAs were investigated by means of Raman spectra excited with photon energies near the E_1 gap [2.87, 105]. At the doping levels of these samples, surface space-charge layer effects are minimized. The spectral line shapes are in these cases determined by breakdown of $\boldsymbol{q}$-conservation caused by electron-relaxation effects associated with the large densities of ionized impurities. These experiments and their interpretation in terms of calculated line-shape functions will be considered in the following section.

2.4 Scattering by Free Carriers: The Wavevector- and Frequency-Dependent Dielectric Function

In this section we consider the light scattering investigations of the dielectric function of free carriers in semiconductors. Much of this research has been carried out in GaAs and it involves spectra by coupled longitudinal plasmon-phonon modes. The conceptual framework is based on the general and relatively simple applications of the RPA discussed in Sect. 2.2.5. Thus, the impact of these studies is beyond that of their specific cases and the results are also relevant to a large group of doped or photoexcited semiconductors.

2.4.1 Background

The early light scattering studies of coupled plasmon LO-phonon excitations of n-type GaAs [2.12, 2.17, 2.78] were interpreted in terms of simple Drude expressions for the free electron susceptibility, as given by (2.38–42). However, a q-dependence of the plasma frequency was noted in nondegenerate samples [2.31]. Remarkable dispersion effects were later observed at larger q-values in spectra of degenerate n-GaAs obtained with *visible* lasers and a *backscattering* geometry [2.32, 37, 118]. Dispersive behavior has also been reported in other doped semiconductors [2.33–35, 106a, 119, 120]. In these experiments it was found that Landau damping as well as collision damping play nontrivial roles in the spectral line shapes [2.36a, 37]. The interpretations of the spectra are accomplished by numerical fittings of the line shapes by means of longitudinal response functions [2.36b, 38, 39]. Thus, this spectroscopy represents a rather unique tool for the investigation of the wavevector and frequency dependence of the dielectric function of free electrons.

We consider first the values of the wavevector transfer that are accessible by light scattering experiments. In the backscattering geometry the magnitude of the scattering wavevector can be varied by changing the laser wavelength [2.36a, 39]. Because the experiments that concern us are carried out in *opaque* samples, it is necessary to take into consideration the effects associated with absorption of the incident and scattered light. Such effects, like those in metals [2.121], are described in terms of the *complex* refractive index $n(\lambda_L)$ at the laser wavelength λ_L [Ref. 2.65, p. 23]. Thus, there are real and imaginary parts to the scattering wavevector, which for the backscattering geometry are given by

$$\mathrm{Re}\,\{q\} = q = \frac{4\pi}{\lambda_L}\,\mathrm{Re}\,\{n(\lambda_L)\} \tag{2.107a}$$

and

$$\mathrm{Im}\,\{q\} = \alpha = \frac{4\pi}{\lambda_L}\,\mathrm{Im}\,\{n(\lambda_L)\}, \tag{2.107b}$$

where α is the absorption coefficient. Absorption allows coupling to modes with wavevectors in a range $\pm\mathrm{Im}\,\{q\}$ about q, a fact which is of substantial consequence in experiments that probe dispersive phenomena. In Fig. 2.19 we plot the scattering wavevector of GaAs given by (2.107a) as a function of laser wavelength. The hatched region describes the smearing of $\boldsymbol{q}$-conservation caused by absorption. These effects become important at the shorter wavelengths when $\alpha \gtrsim 10^5\ \mathrm{cm}^{-1}$.

We see in Fig. 2.19 that the accessible wavevectors are in the range $0.4 \times 10^6\ \mathrm{cm}^{-1} \leqslant q \leqslant 1.4 \times 10^6\ \mathrm{cm}^{-1}$. Within this range, qv_F is no longer negligible compared to ω_P. This leads to substantial plasma dispersion effects as indicated by (2.37). In the interpretation of light scattering at these wavevectors, we can recognize two aspects which are often complementary. In one, the aim is to determine the electron-gas dielectric function that best describes the spectra [2.36b–40, 72]. In the other, a specific form of the dielectric function is used to obtain electron gas parameters like the plasma frequency and collision

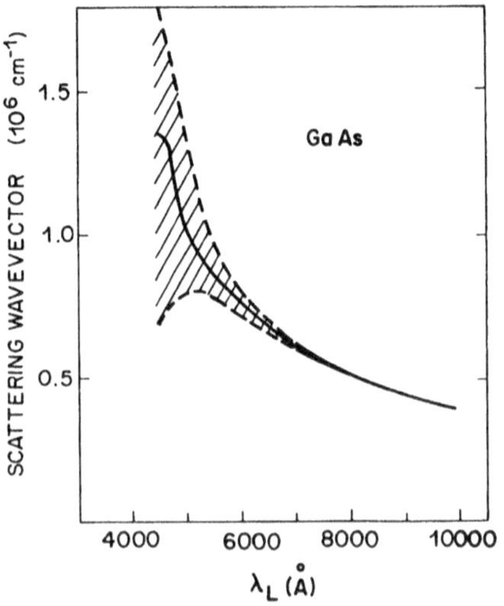

Fig. 2.19. Real part of the backscattering wavevector q obtained in GaAs for various laser wavelengths. The hatched area represents the spread in q (2.107b) due to light absorption (imaginary part of q)

damping [2.35, 71a, 107, 108, 120, 122]. In this section we consider both aspects. We begin with a review of the longitudinal response functions that apply to light scattering. We then proceed with a discussion of the results obtained in doped and photoexcited semiconductors.

2.4.2 Light Scattering Response Functions

The spectral line shape functions that have been employed in light scattering by longitudinal excitations of the coupled free-electron optical-phonon system are derived from scattering cross sections considered in Sect. 2.3. Equations (2.96, 97) apply to charge-density fluctuations. In the case of the phonon-like mechanisms, (2.98–100) correspond to the "allowed" terms and (2.101) are for the "forbidden-Fröhlich" contribution. We are thus led to consider three different longitudinal line shape functions, one for each of the mechanisms. We focus our attention here to the case of a single-component plasma in a simple polar semiconductor, n-type GaAs being one of the most popular examples. Extension to multicomponent plasmas is straightforward by means of response functions similar to those discussed in Sect. 2.2.6, 7.

For charge-density fluctuations, the dependence of the scattering cross section on the frequency shift and the scattering wavevector can be expressed in terms of a line shape function given by [(2.96)]

$$L_C(q,\omega)=q^2(1-e^{-\hbar\omega/k_BT})^{-1}\left(\frac{\omega_{LO}^2-\omega^2}{\omega_{TO}^2-\omega^2}\right)^2 \mathrm{Im}\{-1/\varepsilon(q,\omega)\}, \qquad (2.108)$$

where $\varepsilon(q,\omega)$ is the *total* dielectric function of the coupled plasmon optical-phonon system given by (2.38). For the "allowed" deformation potential and electro-optic contributions, we find from (2.98)

$$L_{\mathrm{A}}(q,\omega)=(1-\mathrm{e}^{-\hbar\omega/k_{\mathrm{B}}T})^{-1}\left(\frac{\omega_{\mathrm{o}}^2-\omega^2}{\omega_{\mathrm{TO}}^2-\omega^2}\right)^2 \mathrm{Im}\left\{\frac{-1}{\varepsilon(q,\omega)}\right\}. \tag{2.109}$$

Finally, in the case of the pure "forbidden" Fröhlich mechanism, we obtain [(2.101)]

$$L_{\mathrm{F}}(q,\omega)=q^2(1-\mathrm{e}^{-\hbar\omega/k_{\mathrm{B}}T})^{-1}\,\mathrm{Im}\left\{\frac{-1}{\varepsilon(q,\omega)}\right\}. \tag{2.110}$$

We consider first the longitudinal response function $\mathrm{Im}\{-1/\varepsilon(q,\omega)\}$. In the case of the coupled plasmon-phonon system, the damping of the optical phonons is negligible compared to that of plasmons (i.e., free electrons). This means that $\mathrm{Im}\{\chi_{\mathrm{L}}(\omega)\}=0$ and $\mathrm{Im}\{\varepsilon(q,\omega)\}=\mathrm{Im}\{\chi(q,\omega)\}$. By means of (2.38, 39) we can then write

$$\mathrm{Im}\left\{\frac{-1}{\varepsilon(q,\omega)}\right\}=\frac{(\omega_{\mathrm{TO}}^2-\omega^2)\,[\mathrm{Im}\,\{\chi(q,\omega)\}]}{\varepsilon_\infty^2 F(q,\omega)+(\omega_{\mathrm{TO}}^2-\omega^2)^2\,[\mathrm{Im}\,\{\chi(q,\omega)\}]^2}, \tag{2.111}$$

where

$$F(q,\omega)=(\omega_{\mathrm{LO}}^2-\omega^2)+\frac{1}{\varepsilon_\infty}\,(\omega_{\mathrm{TO}}^2-\omega^2)\,\mathrm{Re}\,\{\chi(q,\omega)\}. \tag{2.112}$$

In the regime where $\omega \gg q v_{\mathrm{F}}$, the plasma dispersion is given by (2.37) and Landau damping is negligible. Then, $\chi(q,\omega)$ takes a Drude form similar to (2.41) [2.4, 68]:

$$\chi(q,\omega)=-\varepsilon_\infty\,\frac{\omega_{\mathrm{P}}^2(q)}{\omega(\omega+\mathrm{i}\Gamma)}, \tag{2.113}$$

where Γ is the collision frequency of the free carriers. With (2.113) we obtain [2.20, 107]

$$\mathrm{Im}\left\{-\frac{1}{\varepsilon(q,\omega)}\right\}=\frac{1}{\varepsilon_\infty}\,\frac{(\omega_{\mathrm{TO}}^2-\omega^2)^2\omega_{\mathrm{P}}^2(q)\Gamma\omega}{\Delta^2}, \tag{2.114}$$

where

$$\Delta^2=[\omega^2(\omega_{\mathrm{LO}}^2-\omega^2)-\omega_{\mathrm{P}}^2(\omega_{\mathrm{TO}}^2-\omega^2)]^2+\omega^2\Gamma^2(\omega_{\mathrm{LO}}^2-\omega^2). \tag{2.115}$$

In the behavior of $\mathrm{Im}\{-1/\varepsilon(q,\omega)\}$ given by (2.114, 115) we distinguish two different cases. When $\Gamma \ll \omega_{\mathrm{P}}$, the peaks in $\mathrm{Im}\{-1/\varepsilon(q,\omega)\}$ occur at the solutions of

$$\omega^2(\omega_{\mathrm{LO}}^2-\omega^2)-\omega_{\mathrm{P}}^2(q)\,(\omega_{\mathrm{TO}}^2-\omega^2)=0 \tag{2.116}$$

which are the coupled-mode frequencies given in (2.42). The behavior of Im $\{-1/\varepsilon(q,\omega)\}$ is considerably more complex for large plasmon damping. The features are then broad and the peaks no longer occur at the coupled-mode frequencies, but close to the minima of Δ [(2.115)]. For $\Gamma \gtrsim \omega_P$, the expression of Δ given by (2.115) has only one minimum that occurs near the optical phonon frequencies [2.107]. In addition, when the structures in Im $\{-1/\varepsilon(q,\omega)\}$ are broad, we should also identify the zero of (2.111) at $\omega=\omega_{TO}$. This zero occurs because, according to (2.13, 24), Im $\{-1/\varepsilon(q,\omega)\}$ is related to the macroscopic polarization of the longitudinal coupled electron optical-phonon system. This polarization must vanish at $\omega=\omega_{TO}$ [2.76, 79].

In the case of pure plasmons, Landau damping sets in when $qv_F \gtrsim \omega_P$ [2.4, 68]. For scattering wavevectors in the range shown in Fig. 2.19, this regime occurs in n-GaAs in the range $N_e \gtrsim 10^{16}$ cm^{-3} [2.17]. For coupled modes Landau damping depends on the strength of the plasmon component. *Buchner* and *Burstein* [2.33] and also *Murase* et al. [2.36a] observed that Landau damping exists in the spectral range of the lower coupled-mode branch $L_-(q)$. In further work, *Pinczuk* et al. [2.37] studied the q-dependence of light scattering from n-GaAs. The results indicate that Landau damping is important for both branches of the coupled modes (usually $\omega_+ > qv_F$). This is mainly a consequence of collision damping and finite temperatures.

To include Landau damping in the longitudinal response of the coupled modes, *Katayama* et al. [2.36b] considered a Lindhard expression for $\chi(q,\omega)$ [2.123] which assumes $T=0$. *Lemmens* et al. [2.124a] used a similar approach to investigate the zeros of $\varepsilon(q,\omega)$. They also studied the behavior of Im $\{-1/\varepsilon(q,\omega)\}$ in the $T=0$ limit [2.124b]. *Abstreiter* et al. [2.38] represented $\chi(q,\omega)$ with a generalization of the Lindhard function that included collision damping in the relaxation-time approximation of Mermin (see (2.35) and [2.74]). On the other hand, *Richter* and coworkers [2.39] used a hydrodynamical approach that is valid for $\omega > qv_F$ and $q < k_F$.

The *Lindhard-Mermin* approach has proven very successful in the interpretation of light scattering from doped and photoexcited semiconductors [2.29, 38, 51, 106a]. The formulation is based on (2.35), where $\chi^0(q,\omega)$ is the temperature-dependent form of the Lindhard expression given by

$$\chi^0(q,\omega)=\frac{e^2}{2\pi^3 q^2 \varepsilon_0}\int f(k,T)\left(\frac{1}{\hbar^2 q^2/m^* + 2\hbar^2 \boldsymbol{q}\cdot\boldsymbol{k}/m^* - 2\hbar\omega}\right.$$
$$\left.+\frac{1}{\hbar^2 q^2/m^* - 2\hbar^2 \boldsymbol{q}\cdot\boldsymbol{k}/m^* + 2\hbar\omega}\right)d^3k, \qquad (2.117)$$

where $f(k,T)$ is the Fermi distribution function at temperature T. Equations (2.35, 111, 117) have been used in a numerical evaluation of the longitudinal response functions of n-GaAs [2.38]. Figure 2.20 shows the results for Im $\{-1/\varepsilon(q,\omega)\}$ for the case $N_e=7\times10^{17}$ cm^{-3}, $T=80$ K and $\Gamma=55$ cm^{-1}.

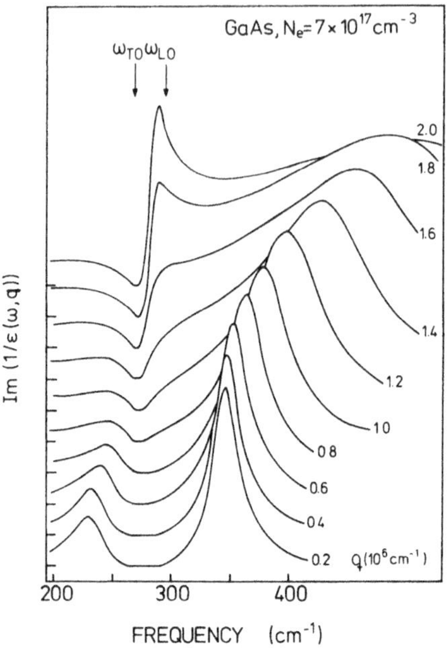

Fig. 2.20. Im $\{-1/\varepsilon\}$ plotted versus frequency ω for different values of a wavevector q obtained with the Lindhard-Mermin approximation for n-GaAs with $N_e = 7 \times 10^{17}$ cm^{-3}. See (2.35, 117)

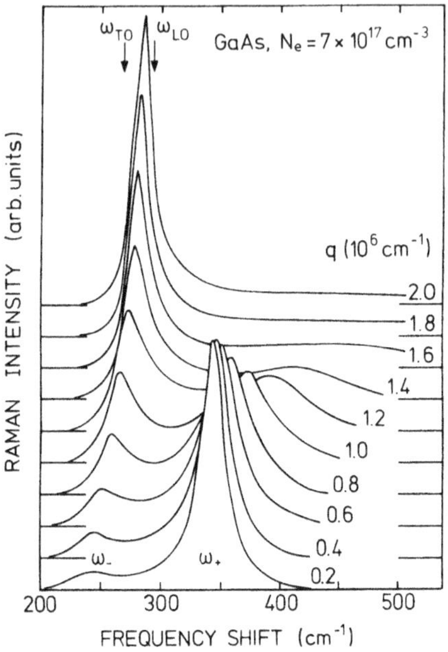

Fig. 2.21. Calculated line shape functions for the allowed contributions to the scattering cross section of n-GaAs with $N_e = 7 \times 10^{17}$ cm^{-3}, according to (2.109)

Other parameters are $\omega_{LO} = 295$ cm^{-1}, $\omega_{TO} = 271$ cm^{-1}, $m^* = 0.072 m_0$ and $\varepsilon_\infty = 11.9$.

In Fig. 2.20 we see that for small values of q, say 0.2×10^6 cm^{-1}, the longitudinal response shows two well-defined peaks at the coupled-mode frequencies of (2.42, 116). With increasing wavevectors there are shifts due to spatial dispersion and broadenings due in part to Landau damping. The effect of Landau damping on the $L_-(q)$ branch is already important at $q = 0.6 \times 10^6$ cm^{-1}. For larger wavevectors, $q \gtrsim 1.6 \times 10^6$ cm^{-1}, the maxima in Im $\{-1/\varepsilon(q,\omega)\}$ associated with the $L_-(q)$ branch occur between ω_{TO} and ω_{LO}, as could be anticipated [2.33]. However, their shapes are distorted by the zero in Im $\{-1/\varepsilon(q,\omega)\}$ at $\omega = \omega_{TO}$.

In Fig. 2.21 we show the "allowed" light scattering response function $L_A(q,\omega)$ defined in (2.109). In the numerical evaluations we have used the results for Im $\{-1/\varepsilon(q,\omega)\}$ shown in Fig. 2.20 and a value of $\omega_0 = 185$ cm^{-1}, which is appropriate to GaAs [2.38]. At the smaller wavevectors we see two well-defined peaks, one for each of the coupled mode branches. In the case of the $L_+(q)$ branch, we notice that broadening due to Landau damping already exists at $q \approx 10^6$ cm^{-1}. This is a wavevector typical of light scattering experiments (Fig. 2.19). Although at these q-values $qv_F < \omega_+(q)$, Landau damping also occurs

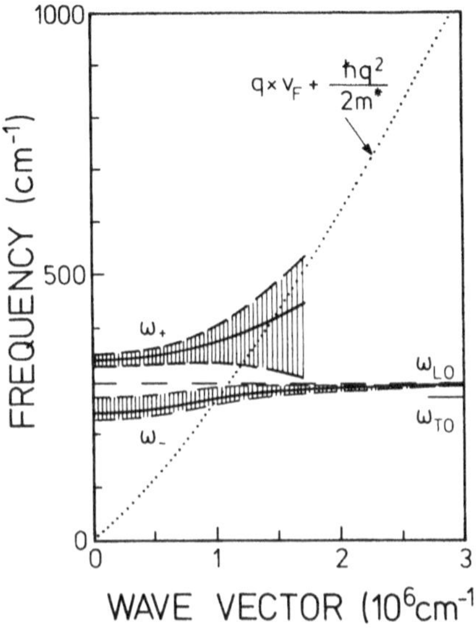

Fig. 2.22

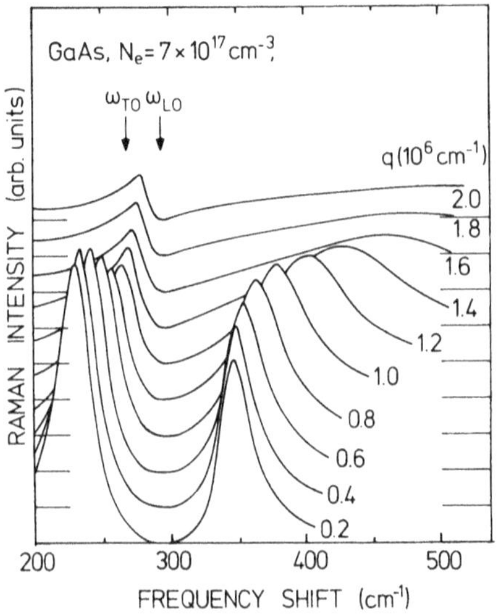

Fig. 2.23

Fig. 2.22. Calculated dispersion of the peak values of the L_+ and L_- modes in n-GaAs (Fig. 2.21). The boundary of single-particle excitations is given by the dotted line. The hatched regions indicate the width of the modes at half maximum

Fig. 2.23. Line shape function for scattering via charge-density fluctuations in n-GaAs with $N_e = 7 \times 10^{17}$ cm^{-3}, calculated with (2.108)

for $L_+(q)$ because the combined effects of finite temperature and collision broadening allow single-particle electron transitions which overlap with $\omega_+(q)$.

In Fig. 2.23 we present the results of a numerical evaluation of the response function of charge-density fluctuations $L_C(q,\omega)$ defined in (2.108). A comparison with the results shown in Figs. 2.20, 21 illustrates the important influence that the detailed light scattering mechanisms have on the spectral line shapes of coupled modes. Figure 2.22 shows the dispersion of the maxima in $L_A(q,\omega)$. We see that the peak associated with the $L_-(q)$ branch occurs between $\omega_-(0)$ and ω_{LO}. Its spectral line shape, shown in Fig. 2.21, is well-behaved since the pre-factor in (2.109) removes the zero of ε^{-1} at ω_{TO}.

We mentioned above the hydrodynamical approach that has also been used in the interpretation of light scattering by coupled modes in n-GaAs [2.39]. This formulation is based on the expression [2.39]

$$\chi(q,\omega) = -\frac{\omega_P^2}{(\omega^2 - \frac{3}{5}q^2 v_F^2) + \mathrm{i}\Gamma\omega} \tag{2.118}$$

which is valid for $\omega > qv_F$ and $q < k_F$. In the regime in which Landau damping is negligible, the results are the same as those obtained with the Lindhard-Mermin theory. Detailed comparisons between the two formulations can be found in [2.39].

2.4.3 Doped Semiconductors

a) n-Type GaAs

We consider here the recent light scattering investigations of the wavevector dependence of coupled plasmon LO-phonon modes in doped n-type GaAs. The presentation is to a great extent based on the result of [2.37, 38, 115].

Figure 2.24 shows backscattering spectra from three n-GaAs samples with different free-electron densities. They were obtained in the "allowed" configuration with $q = 1.02 \times 10^6\ \mathrm{cm}^{-1}$. The spectra show the bands of the L_- and

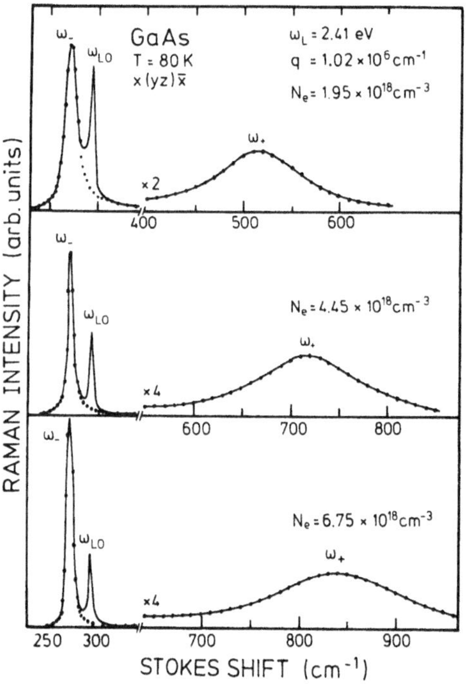

Fig. 2.24

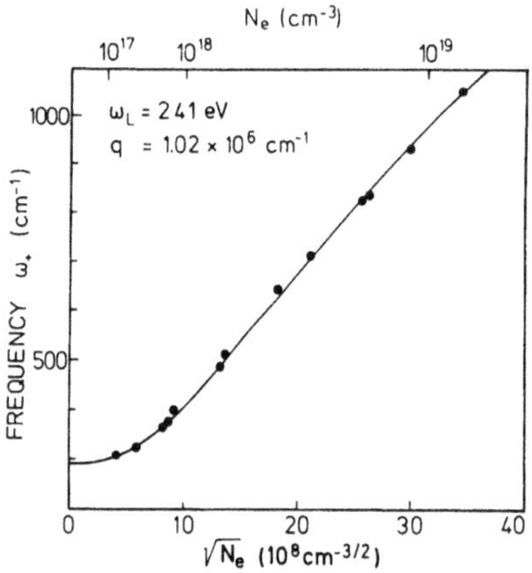

Fig. 2.25

Fig. 2.24. Raman spectra of three different n-GaAs samples obtained in backscattering geometry from (100) surfaces. The dotted lines are spectra of the coupled LO-phonon plasmon modes calculated according to (2.109). From [2.38]

Fig. 2.25. Frequency of the L_+-mode versus $\sqrt{N_e}$ in GaAs. The solid line is extracted from theoretical calculations including the nonparabolicity of the conduction band. From [2.115]

L_+ coupled modes, usually labeled ω_- and ω_+. There is also a peak of LO phonons decoupled from the free carriers. It originates in the surface depletion layer that exists in these samples. Its relative strength is determined by the depletion width (100–400 Å) and the penetration depth of light (~1000 Å). In Sect. 2.6 we shall discuss applications of spectra like those in Fig. 2.24 to the investigation of the Fermi level and electric fields at surfaces of polar semiconductors. We should also mention that this capability to study non-uniform layers can also yield important information on semiconductor structures grown by epitaxy [2.125, 126].

In the spectra of Fig. 2.24 the L_- mode is a heavily screened LO phonon which for $q=1.02\times10^6\ \mathrm{cm}^{-1}$ has a frequency close to ω_{TO}. On the other hand, the L_+ modes can be regarded as the plasmons of the free electrons. This can be appreciated more readily in Fig. 2.25, where the circles show the position of the L_+ band in a group of n-GaAs samples having a wide range of free-electron densities. The line represents the frequency of the L_+ band calculated on the basis of (2.109) and the Lindhard-Mermin expression for $\mathrm{Im}\{-1/\varepsilon(q,\omega)\}$ including the variation of m^* with free-electron concentration. However, for the larger densities ($n\gtrsim2\times10^{18}\ \mathrm{cm}^{-3}$), the Landau damping effects are negligible at $q\approx10^6\ \mathrm{cm}^{-1}$. Thus, the position of the L_+ band is well described by the plasmon-like expression derived within the Drude framework, see (2.113–116) and (2.37–42):

$$\omega_+^2=\frac{\omega_{\mathrm{LO}}^2+\omega_{\mathrm{P}}^2(q)}{2}+\frac{1}{2}\{[\omega_{\mathrm{P}}^2(q)-\omega_{\mathrm{LO}}^2]^2+4\omega_{\mathrm{P}}^2(q)\,[\omega_{\mathrm{LO}}^2-\omega_{\mathrm{TO}}^2]\}^{1/2}. \tag{2.119}$$

The q-dependence of the longitudinal coupled modes was studied by measuring backscattering spectra with different laser wavelengths (Fig. 2.19). Figure 2.26 shows typical spectra obtained in the "allowed" configuration. At these wavelengths the effects of absorption are significant ($\alpha\approx0.3$–$1\times10^5\ \mathrm{cm}^{-1}$) and need to be taken into consideration. This was done by averaging the response functions with a Lorentzian weight factor [2.65]

$$[L(q,\omega)]_\alpha=\int\limits_0^\infty\frac{L(q',\omega)dq'}{(q'-q)^2+\alpha^2}. \tag{2.120}$$

The dotted lines in Fig. 2.26, as well as those in Fig. 2.24, are the spectral line shapes calculated with (2.109, 120). Equations (2.111, 112) were used for $\mathrm{Im}\{-1/\varepsilon(q,\omega)\}$. The *Lindhard-Mermin* expression, given by (2.35, 117), was used for $\chi(q,\omega)$.

The calculated line shapes shown in Fig. 2.26 correspond to $m^*=0.074m_0$ [2.113] and $T=80$ K. The values of the absorption coefficient are those compiled in [2.127]. The fit has two adjustable parameters: ω_0 and Γ. ω_0 was fixed to the value obtained from the fit to the $q=0.87\times10^6\ \mathrm{cm}^{-1}$ spectrum ($\lambda_{\mathrm{L}}=5682$ Å). This is $\omega_0=185\ \mathrm{cm}^{-1}$, which with (2.99) yields $C=-0.53$. The electron damping is $\Gamma=58\ \mathrm{cm}^{-1}$, a value ~50% higher than that calculated from the Hall

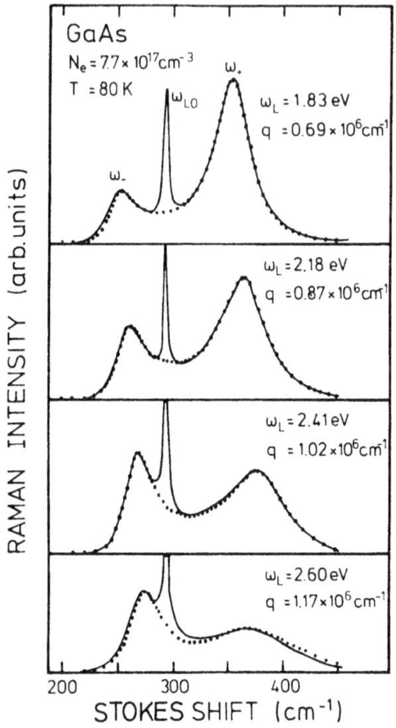

Fig. 2.26. Raman spectra of n-GaAs obtained in backscattering geometry from a (100) surface using different laser excitation energies. The dotted lines are calculated spectra. From [2.38]

mobility. There is good agreement between measured and calculated line shapes. The small discrepancy in the case of $q = 0.69 \times 10^6$ cm^{-1} can be corrected with a minor change in the value of ω_o.

Figure 2.27 shows the dispersions of the coupled modes. The circles are the points measured in a large family of spectra like those in Fig. 2.26. The full lines are the calculated dispersions. They were obtained from the maxima in line shape functions calculated with (2.35, 109, 111, 112, 120). There is good agreement between measured and calculated dispersions. The separation between the L_+ mode and the edge of $T=0$ single-particle excitations is comparable to Γ and kT for $q \gtrsim 0.8 \times 10^6$ cm^{-1}. This implies the existence of significant Landau damping. These effects could not be accounted for with the Drude expression for $\chi(q,\omega)$ [2.37].

Figure 2.28 shows the dispersion of coupled modes in samples having a range of free-electron densities. We have seen above that for $N_e \gtrsim 2 \times 10^{18}$ cm^{-3}, Landau damping of the L_+ mode is not significant. In fact, at these densities the dispersion of this mode is rather well described by (2.37, 119) [2.37]. Quite striking is the downward bending observed in the dispersion of the L_+ mode at $N_e = 7.7 \times 10^{17}$ cm^{-3}. In [2.38] it was concluded that the bending is related to the combined effects of Landau damping and strong absorption. The dashed line in Fig. 2.27 represents the dispersions calculated with an arbitrarily small absorption coefficient ($\alpha = 1.3 \times 10^4$ cm$^{-1} \ll q$). This calculation does not exhibit the

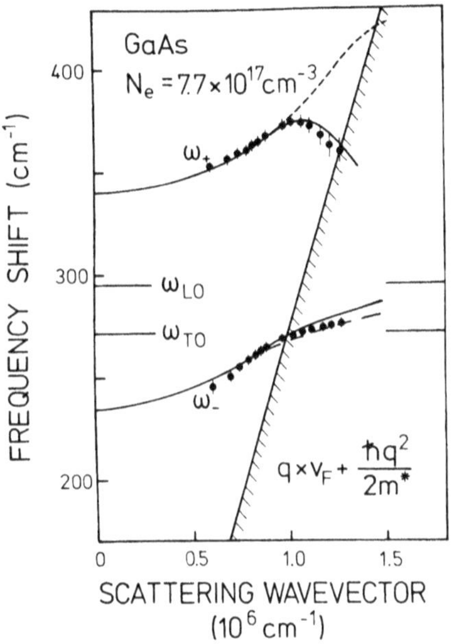

Fig. 2.27

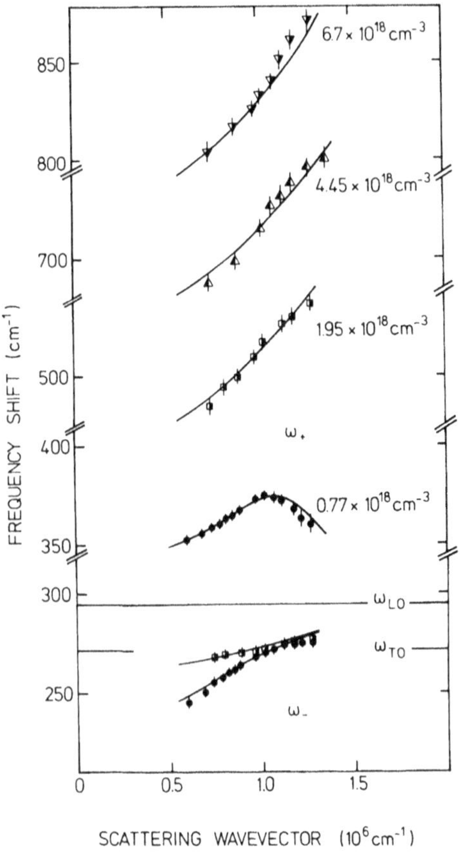

Fig. 2.28

Fig. 2.27. Dispersion of the Raman peaks of coupled LO-phonon plasmon modes of n-GaAs with $N_e = 7.7 \times 10^{17}$ cm^{-3}. The solid and dashed lines are calculated (see text). From [2.38]

Fig. 2.28. Dispersion of the Raman peak of coupled LO-phonon plasmon modes of several n-GaAs samples. From [2.115]

downward bending of the L_+ mode. It indicates that the effect is caused by strong absorption ($\alpha \gtrsim 10^5$ cm^{-1}) which allows the observation of modes in a range of wavevectors, including the small ones. For the smaller wavevectors the L_+ modes are lower in frequency and less affected by Landau damping. For this reason they make a large contribution to the spectra and cause a downward shift of the position of the maximum in scattering intensity.

The downward bending of the dispersion of the L_+ mode was also observed by *Nowak* et al. [2.39] in samples with $N_e \lesssim 6 \times 10^{17}$ cm^{-3}. These authors have shown that the expression of $\chi(q, \omega)$ given by the hydrodynamical theory, Eq. (2.118), cannot explain the effect. This is another indication that the bending is related to Landau damping. In addition, the Lindhard-Mermin fits considered in [2.39] predict downward shifts smaller than the observed ones. The reason for the discrepancies between the results of [2.38, 39] are not fully understood to us. They might be due to the use in [2.39] of the response function $L_C(q,\omega)$ (2.108), which has a zero at ω_{LO} and thus reduces the contributions of smaller wavevectors to spectra weighted with (2.120).

b) n-Type GaAs Under High Hydrostatic Pressure

When n-GaAs is subject to hydrostatic pressures greater than 25 kbar, the free electrons transfer to conduction band minima along the Δ-direction of the Brillouin zone [2.126]. *Pinczuk* et al. [2.108] have used light scattering by the coupled plasmons LO-phonons to investigate the corresponding free-electron behavior up to pressures of 88 kbar. Typical results for the L_+ mode are summarized in Fig. 2.29. For pressures below 27.5 kbar, the spectra are similar to those in Fig. 2.24. At pressures above 32 kbar, at which GaAs has an indirect optical gap [2.129a], the spectra resemble those of n-GaP [2.107]. Only the L_+ mode is resolved, a fact which indicates that damping effects are now very significant ($\Gamma \gtrsim \omega_P$) [2.20, 2.107].

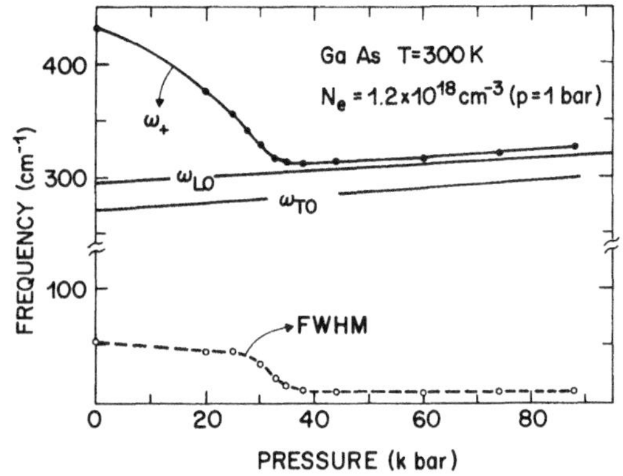

Fig. 2.29. Frequency (solid line and dots) and FWHM (broken line and circles) of the L_+ band measured in Raman spectra of n-type GaAs under large hydrostatic pressures [2.108]

In the pressure range where the free electrons are at the Γ-point ($p \lesssim 27.5$ kbar), the plasma frequency was determined from the position of the L_+ mode by means of (2.119). The optical phonon frequencies measured by *Trommer* et al. [2.129b], also shown in Fig. 2.29, were used in this analysis. For $p \gtrsim 32$ kbar, with the free electrons occupying states in equivalent minima along the Δ-directions, a line-shape analysis of the L_+ band with (2.114, 115) led to a determination of ω_P and Γ. The results are shown in Fig. 2.30.

Figure 2.30 indicates that there are three well-differentiated pressure regimes that are correlated with the transfer of free electrons out of states at the Γ-point minimum. For pressures in the range 1 bar $\lesssim p \lesssim 25$ kbar, the behavior of the

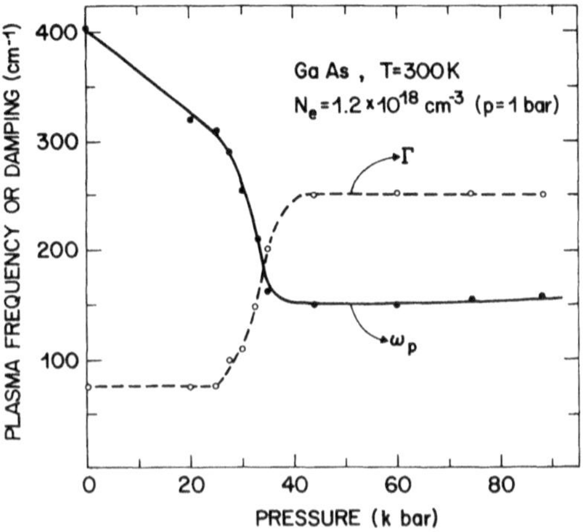

Fig. 2.30. Plasma frequency (solid line and dots) and plasma damping (broken line and circles) of n-GaAs under large hydrostatic pressures obtained from data such as those of Fig. 2.29. From [2.108]

plasma frequency shows a steady increase in the Γ-point electron effective mass caused by the increase of the direct optical gap [2.128]. In the range 25 kbar $\lesssim p \lesssim$ 32 kbar, ω_P undergoes a rapid variation associated with electron transfer. Finally, for $p \gtrsim 32$ kbar, ω_P becomes almost independent of pressure. Its value, about 2.5 times smaller than the one at 1 bar, has been used to obtain the conductivity effective mass at the Δ_{1C} minima of GaAs [2.108].

c) n-Type GaSb

Raman scattering by coupled plasmons LO-phonons of n-GaSb has been reported by *Trommer* and *Ramdas* [2.35]. Interest in these experiments was stimulated by the structure of the conduction band of GaSb in which the L_6 minima (Fig. 2.7) lie only ~90 meV above the absolute minimum at the Γ point. Thus, because of the small conduction band effective mass ($m^* \simeq 0.04 m_0$), the two conduction band minima are populated for densities $N_e \gtrsim 10^{18}$ cm^{-3}. This allows the study of plasma oscillation of a two-carrier system, as well as the process of charge transfer between different conduction band minima.

The measurements were carried out in samples with 10^{18} cm$^{-3} \lesssim N_e \lesssim 3 \times 10^{18}$ cm^{-3}. At these free-electron densities the L_- mode is a heavily screened LO phonon ($\omega_- \approx \omega_{TO}$) and the L_+ mode is plasmon-like. An expression for ω_+ can be obtained within the framework of the simple two-component plasma of Sect. 2.2.7. Equations (2.55, 57–59) lead to an expression like (2.119) for ω_+, where

$$\omega_P^2 = \frac{1}{\varepsilon_0 \varepsilon_\infty} e^2 \left(\frac{N_{e\Gamma}}{m_\Gamma^*} + \frac{N_{eL}}{m_L^*} \right) + \frac{3}{5} q^2 (v_{F,\Gamma}^2 + v_{F,L}^2), \tag{2.121}$$

where the subindices Γ and L label the values of the parameters at the corresponding conduction band minima.

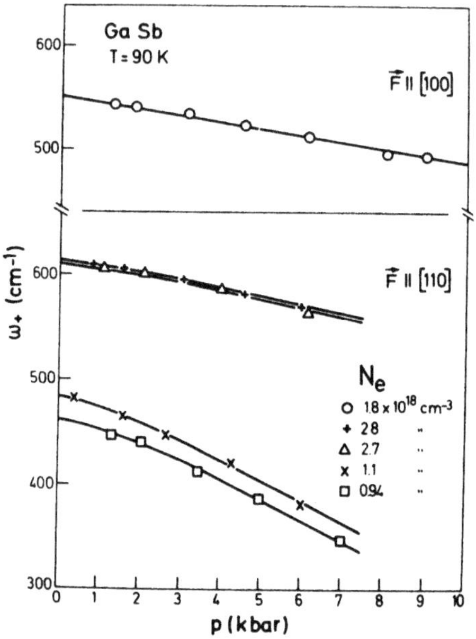

Fig. 2.31. Frequency of the L_+ band measured in n-GaSb under unaxial compressive stress along two different crystallographic axes. From [2.35]

The distribution of free electrons among the conduction band minima was investigated by measuring the position of the L_+ mode as a function of an external compressive force $\boldsymbol{F}$ applied along selected crystallographic directions. The results are shown in Fig. 2.31. The hydrostatic stress component reduces the energy difference between the Γ- and L-point minima. Like in the case considered in the previous section, this reduces the value of $N_{e,\Gamma}$ and increases that of $N_{e,L}$. For $\boldsymbol{F} \| [100]$, the uniaxial stress component does not lift the degeneracy of the eight equivalent L_6 minima. The continuous reduction in ω_+, shown in the upper portion of Fig. 2.31, is related to the transfer of electrons out of the Γ-point minimum and the fact that $m_\Gamma^* < m_L^*$. For $\boldsymbol{F} \| [110]$, the uniaxial stress partially lifts the L-point degeneracy and thus affects the values of $N_{e,\Gamma}$, $N_{e,L}$ and the conductivity effective mass m_L^* that enter in (2.121). This feature of the data shown in the lower part of Fig. 2.31 has been used to estimate shear deformation-potentials at the L point.

d) p-Type GaAs

Olego and *Cardona* [2.105] have recently reported an extensive investigation of Raman scattering from Zn-doped, p-type GaAs. Although the measured spectra show clear evidence of coupling between LO phonons, they could not observe the long-wavelength L_+ and L_- coupled modes given by (2.42). The inability to measure the coupled modes at wavevector $\boldsymbol{q}$ appears to be related to the short relaxation-times and large effective masses associated with the free holes. The hole mobilities measured in the Zn-doped samples indicate that at all values of

N_P, the damping parameter is $\Gamma \gg \omega_P$. Indeed, in overdamped plasmas the response functions related to (2.114,115) show structure only near ω_{LO} [2.107] (see also Sect. 2.4.3b).

The Raman sepctra of heavily doped p-GaAs show bands near the optical-phonon frequencies. For $10^{18}\ \mathrm{cm}^{-3} \lesssim N_p \lesssim 4\times10^{19}\ \mathrm{cm}^{-3}$, the LO phonon bands exhibit considerable broadenings and relatively small downward shifts. At larger densities, the spectra also show the band of the heavily screened LO phonon (with a peak near ω_{TO}). These spectra were assigned to *large wavevector* coupled plasmon LO-phonon excitations [2.87,105]. Breakdown of $\boldsymbol{q}$-conservation was attributed, as in the case of n-GaAs [2.51], to elastic scattering by the ionized dopant impurities.

In order to obtain the total scattering cross section in the case of a non $\boldsymbol{q}$-conserving mechanism, we must fold the q-conserving response functions [given in (2.108–111)] with a "weighting" function $f(q')$ in a way similar to that of (2.120) where $f(q')$ was simply a Lorentzian. In the case of $\boldsymbol{q}$-nonconservation induced by impurities, this function can, to a first approximation, be taken as the square of the Fourier transform of a Yukawa-type impurity potential $V(r)$ [2.87]:

$$V(r) = \frac{1}{r}\exp(-k_{TF}r). \tag{2.122}$$

The function $f(q')$ is then

$$f(q') = \left(\frac{4\pi}{q'^2 + k_{TF}^2}\right)^2. \tag{2.123}$$

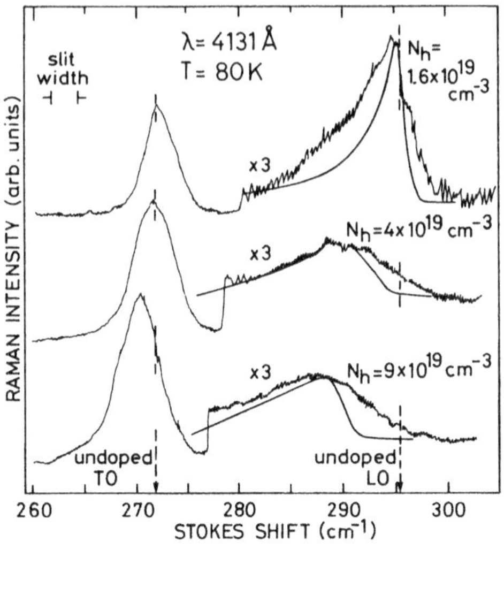

Fig. 2.32. Effect of $\boldsymbol{q}$-nonconservation in the scattering by coupled modes in heavily doped p-type GaAs. The noisy curve is experimental. The smooth ones are theoretical. From [2.87]

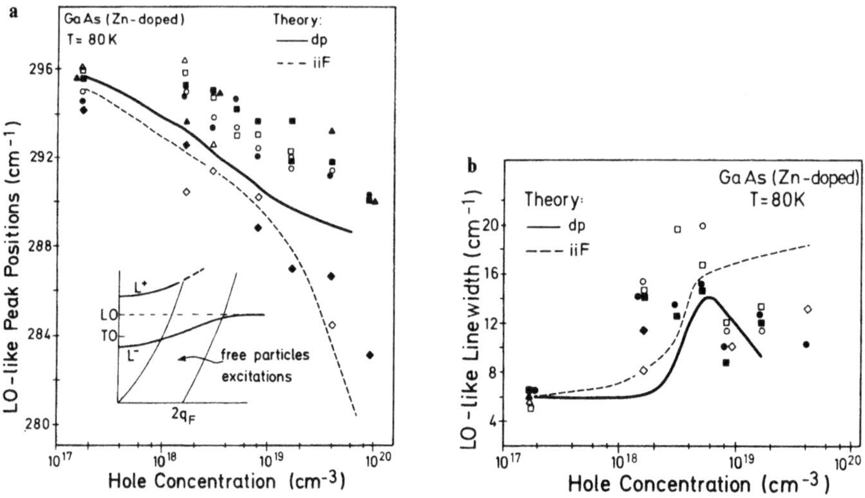

Fig. 2.33a, b. Self-energy effects in the LO phonons of p-GaAs induced by coupling to plasmons with q-nonconservation. (**a**) Frequency shift due to the real part of the self energy; (**b**) linewidth increase due to its imaginary part. From [2.105]

Equations (2.122–124) have been used to interpret Raman scattering from heavily doped p-GaAs [2.87, 105]. Figure 2.32 shows the results for spectra measured with photon energies close to resonance with the E_1 energy gap in the configuration in which the "forbidden" Fröhlich mechanism predominates. The calculated line shapes were obtained with (2.110, 122, 124). Equation (2.111) was used in the evaluation of Im $\{-1/\varepsilon(q,\omega)\}$. The Lindhard expression (2.111) was used for the susceptibility of the heavy holes. The light-hole contribution was neglected. Although Fig. 2.32 shows that the calculation represents the major features of the measured spectra reasonably well, in particular the frequency shift with doping and the asymmetry, we should point out that the corresponding "weighted" response functions

$$\int_0^{q_{\max}} L(q',\omega)\left(\frac{4\pi}{q'^2+k_{\mathrm{TF}}^2}\right)^2 \qquad (2.124)$$

converge rather slowly and that the line shape depends rather critically on the value of $q_{\max}$ chosen as the cutoff[1]. The cutoff can be taken to correspond qualitatively to the decrease in the electron-phonon coupling with increasing q [Ref. 2.64, Sect. 2.2.9]. Detailed calculations of the peak position and linewidths for this type of impurity-induced scattering have been given in [2.105]. In Fig. 2.33 we reproduce these results and compare them with experimental data.

1 (U. Nowak: Private communication)

2.4.4 Photoexcited Plasmas

The absorption of laser light can generate dense ($10^{15}\,\text{cm}^{-3} \lesssim N \lesssim 10^{18}\,\text{cm}^{-3}$) plasmas in semiconductors. Light scattering by these photoexcited plasmas has attracted considerable interest. The first experiments, by *Turtelli* et al. [2.25] in intrinsic GaAs, involved Maxwellian systems. Coupled optical plasmons LO-phonons were reported by *Kardontchik* and *Cohen* [2.28] in photoexcited GaP. In stressed silicon, *Guidotti* et al. [2.26] observed intervalence band transitions of photoexcited holes. These experiments have shown that inelastic light scattering can be applied to the study of photoexcited semiconductors. The method allows the determination of energy level structures as well as plasma properties like temperatures, densities and collective phenomena.

The more recent work has focused on GaAs and related systems. The experiments with photoexcited Maxwellian plasmas were re-examined by *Abramsohn* et al. [2.72] who found that the measured spectral line shapes do not follow the Gaussian behavior predicted by (2.27,30). *Hertel* and *Appel* [2.112] interpreted this behavior in terms of energy-dependent electron-relaxation processes associated with scattering by charged impurities and polar optical phonons. Resonant light scattering by photoexcited 2-dim plasmas was observed in GaAs-(AlGa)As quantum-well heterostructures [2.57,130] and in GaAs doping superlattices [2.62]. Single-particle and collective excitations of degenerate electron-hole plasmas were observed in highly photoexcited GaAs [2.27,29].

Light scattering by photoexcited 2-dim plasmas will be considered in Sect. 2.5. Here we direct our attention to the work on photoexcited electron-hole plasmas in GaAs [2.27,29]. The carriers in this plasma occupy conduction and valence band states near the center of the Brillouin zone. The free electrons are in a nearly parabolic conduction band with $m_e \cong 0.068 m_0$. The free holes are in heavy (V_1) and light (V_2) valence bands, with masses $m_{hh} \simeq 0.62 m_0$ and $m_{lh} \simeq 0.075 m_0$ [2.114,131]. Because there are much fewer light holes and $m_e \ll m_{hh}$, photoexcited plasmas in GaAs have several features in common with the simple two-component plasma considered in Sect. 2.2.7. Light scattering spectra by collective modes of these photoexcited plasmas can be very different from those measured in doped semiconductors. The spectra reveal the multi-component character of the plasma with new structures, like the band of the acoustic oscillation, that were interpreted in terms of electron-gas susceptibilities of electrons and holes [2.29].

In the experiments of *Romanek* et al. [2.27], photoexcitation of GaAs was achieved by means of cavity-dumped ion lasers operating at $\lambda_L = 6471$ Å or $\lambda_L = 5145$ Å. The same light was used to excite the light scattering spectra. With photon energies near the $E_0 + \Delta_0$ optical gap ($\lambda_L = 6471$ Å), single-particle as well as collective excitations were observed. Away from resonance ($\lambda_L = 5145$ Å) only collective modes were observed. Figure 2.34 shows spectra obtained with $\lambda_L = 5145$ Å. The plasma temperatures were determined from the high-energy tail of the luminescence at the fundamental optical gap. At high

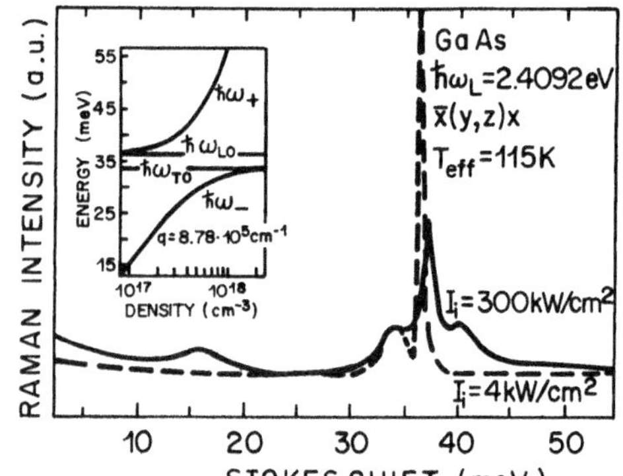

Fig. 2.34

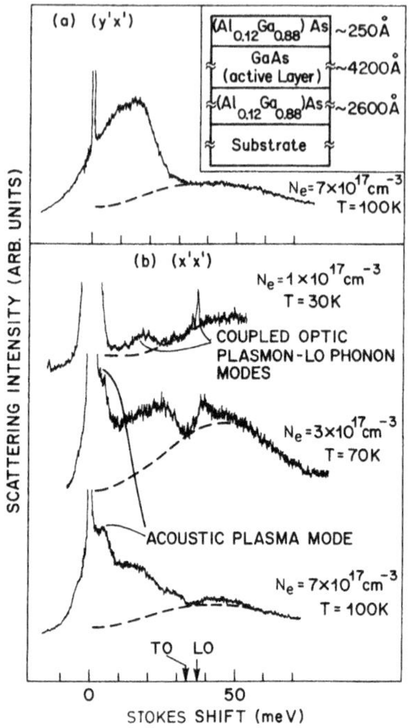

Fig. 2.35

Fig. 2.34. Allowed $[z(xy)\bar{z}]$ Raman spectra from a high purity GaAs sample under high photoexcitation. The inset shows the energies of coupled plasmons LO-phonons calculated at $q=8.78\times10^5\ \mathrm{cm}^{-1}$. From [2.27]

Fig. 2.35a, b. Resonant light scattering spectra from a photoexcited electron-hole plasma in GaAs. The laser photon energy, $\hbar\omega_L = 1.916$ eV, is close to the $E_0+\Delta_0$ optical gap. (**a**) Typical depolarized spectrum; (**b**) polarized spectra for three electron densities. In all cases the dashed lines represent an estimated luminescence background associated with $E_0+\Delta_0$. The inset shows a schematic description of the sample. From [2.29]

photoexcitation levels the spectra show four reasonably well-defined bands. Three of them occur at Stokes shifts (33–40 meV) close to the energies of optical phonons. This led the authors to assign the spectrum to coupled optical plasmons LO-phonons of the photoexcited electron-hole plasma. It was also conjectured that the presence of four bands, as opposed to the two expected ones (Sects. 2.2.5, 7), could indicate two different densities in a spatially nonuniform plasma.

To improve the uniformity of their plasmas, *Pinczuk* et al. [2.29] created them in the microstructures illustrated in the inset to Fig. 2.35. The samples were fabricated by molecular beam epitaxy on (001) GaAs substrates. The two $(Al_{0.12}Ga_{0.88})As$ layers are meant to confine the photoexcited carriers within the GaAs layer with negligible quantization effects [2.132]. The plasma was generated by means of a Kr^+ laser operating cw at $\lambda_L = 6471$ Å and focused to a spot of radius $\sim 10\ \mu m$. The penetration depth of the laser light (~ 3000 Å) is comparable to the thickness of the "active" GaAs layer, and much larger than the thickness of the top $(Al_{0.12}Ga_{0.88})As$ layer.

The laser light that created the photoexcited plasma was also used to obtain the light scattering spectra. The photon energy, 1.916 eV, is close to the $E_0 + \Delta_0$ optical gap at 1.86 eV [2.110]. Under these conditions, the photoexcited electrons are expected to make the dominant contributions to the light scattering spectra. We have seen in Sect. 2.3.2a, in the discussion of Fig. 2.5, that resonant enhancements of light scattering by electron-density fluctuations occur only at optical energy gaps that involve the states of the free carriers. This means that at the $E_0 + \Delta_0$ gap (Fig. 2.7), only the light scattering by density fluctuations of the photoexcited electrons is resonantly enhanced [2.52, 95]. The effect is enhanced further because of the smaller conduction band effective mass [2.96].

Figure 2.35a shows a typical *depolarized* spectrum, reported in [2.29]. Its shape is similar to the ones observed in this configuration in doped n-type GaAs (see Sect. 2.3.2a and [2.50, 51]). Thus, the spectrum is assigned to spin-density fluctuations of the photoexcited electrons. Depolarized spectra like the one in Fig. 2.35a have been used to determine parameters of the photoexcited plasma. Temperatures were obtained from Stokes-anti-Stokes intensities ratios. The plasma densities and electron damping parameters, N_e and Γ_e, were determined from fits of the spectral line shapes. This analysis, similar to that shown in Fig. 2.9, is based on a Lindhard-Mermin expression for Im $\{\chi(q,\omega)\}$ [2.51], in which N_e and Γ_e are the adjustable parameters.

Polarized spectra of the photoexcited electron-hole plasma are shown in Fig. 2.35b for three different densities. These spectra, like the ones from n-type GaAs considered in Sect. 2.3.2b, are assigned to *collective* modes. At the lowest plasma density, $N = 10^{17}$ cm^{-3}, there are two well-defined bands at the energies of the coupled optic plasmon LO-phonon modes (see Sects. 2.2.5, 7). We see in Fig. 2.35b that with increasing plasma density, the polarized spectra show remarkable behavior which is very different from that found in doped n-GaAs. There is a new low-energy band at ~ 5 meV and the spectral line shape of the coupled-optic modes is broadened and distorted.

The low-energy band has been identified as an *acoustic oscillation* of the electron-hole plasma. Its position is in good agreement with that calculated by means of (2.61) with $q = 0.7 \times 10^6$ cm^{-1} and neglecting the light holes. This seems to be the first experimental evidence of an acoustic plasma oscillation in a solid. The unusual behavior of the coupled optic plasmon LO-phonon modes is also of considerable interest. In previous work on photoexcited GaP [2.28] and GaAs [2.27], the observed collective mode energies could not be explained within the coupled-mode framework. By contrast, the spectrum at $N = 10^{17}$ cm^{-3} can be understood by a conventional coupled-mode analysis (Sects. 2.2.6, 2.4.2). The distortion of the spectral line shapes occur at higher densities, when the coupled mode energies overlap with those of intervalence-band transitions ($V_2 \rightarrow V_1$) of the photoexcited holes. Calculations have indicated that coupling to $V_2 \rightarrow V_1$

single-particle excitations has a substantial influence on the energies and damping of collective modes [2.133,134].

This interpretation of light scattering by collective modes of the photoexcited plasma is supported by a calculation of the spectral line shapes. We have seen that in these experiments only the photoexcited electrons contribute to the cross section. Thus, the expression for the light scattering spectrum of charge-density fluctuations can be written as (see (2.52,53) above or [Ref. 2.19, Eqs. (37.3–5)])

$$\begin{aligned} L_e(q,\omega) &= (1-e^{-\hbar\omega/k_BT})^{-1}\,\mathrm{Im}\,\{T_e(q,\omega)\} \\ &\propto (1-e^{-\hbar\omega/k_BT})^{-1} I(q,\omega), \end{aligned} \tag{2.125}$$

where $T_e(q,\omega)$ is the response function of the electrons and $I(q,\omega)$ represents the line shape function of the reduced spectrum. $T_e(q,\omega)$ can be written as (see (2.54) or [Ref. 2.19, Eq. (37.11)])

$$T_e(q,\omega) = V\varepsilon_0[\chi_e(q,\omega) - \chi_e^2(q,\omega)/\varepsilon(q,\omega)], \tag{2.126}$$

where $\chi_e(q,\omega)$ is the susceptibility of the free electrons and $\varepsilon(q,\omega)$ is the total longitudinal dielectric function. $\varepsilon(q,\omega)$ can be written as a generalization of (2.38,55):

$$\varepsilon(q,\omega) = \varepsilon_\infty + \chi_e(q,\omega) + \chi_h(q,\omega) + \chi_L(\omega). \tag{2.127}$$

In (2.127) $\chi_h(q,\omega)$ is the susceptibility of the photoexcited holes.

In [2.29], $\chi_e(q,\omega)$ was represented by a *finite temperature Lindhard-Mermin* dielectric function (Sect. 2.4.2). Intraband and intervalence-band contributions to $\chi_h(q,\omega)$ were considered. The intraband terms of heavy and light holes were also evaluated with Lindhard-Mermin functions. The intervalence-band susceptibility was approximated by [2.133,134]

$$\chi_h^{\mathrm{inter}}(q,\omega) = -\left(\frac{\omega_H}{\omega}\right)^{1/2}\left[f(\omega) + i\,\frac{\pi}{2}\right], \tag{2.128}$$

where

$$\omega_H = \frac{4m_{lh}e^4}{\hbar^3\pi^3 c\varepsilon_0}$$

and

$$f(\omega) = \frac{1}{2}\ln\left|\frac{\omega^{1/2}+\omega_M^{1/2}}{\omega^{1/2}-\omega_M}\,\frac{\omega^{1/2}-\omega_m^{1/2}}{\omega^{1/2}+\omega_m^{1/2}}\right| - \arctan\left[\frac{\omega^{1/2}(\omega_M^{1/2}-\omega_m^{1/2})}{\omega+(\omega_m\omega_M)^{1/2}}\right]. \tag{2.129}$$

In (2.129) ω_m and ω_M are lower and upper cutoff frequencies for the spectrum of vertical $V_2 \to V_1$ transitions. These expressions of $\chi_h^{\mathrm{inter}}(q,\omega)$ are strictly valid

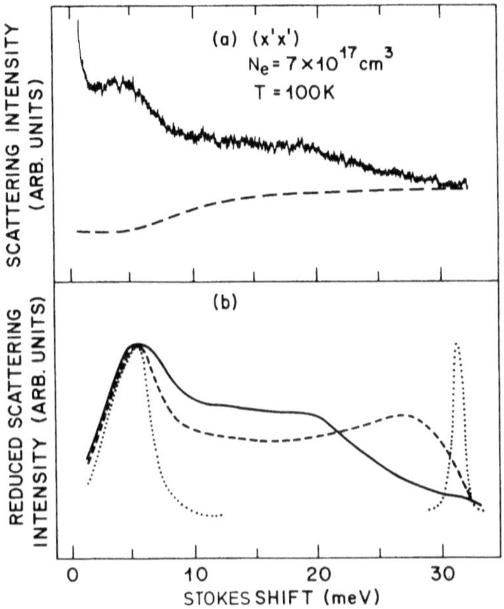

Fig. 2.36a, b. Comparison between measured and calculated light scattering spectra of a photoexcited electron-hole plasma in GaAs. (**a**) Low energy portion of the measured spectrum. The dashed line is the estimated luminescence; (**b**) full line is the measured spectrum reduced by the Bose factor. The dashed line was calculated including the interband susceptibility of holes, the dotted line without including it. From [2.29]

only for $T=0$ and they also neglect the warping of the valence bands (see also Sect. 2.7.2b).

Figure 2.36 compares measured and calculated light scattering spectra by charge-density fluctuations of the photoexcited plasma. Only the energy range of the acoustic plasma oscillation and the lower coupled mode are shown in the figure. At the plasma density and temperature shown, the photoexcited electrons are degenerate but the photoexcited holes are nearly Maxwellian. $I(q,\omega)$ was evaluated numerically as described above. In the case of the inter-valence-band susceptibility, because the holes are nondegenerate, ω_m was set to zero and $\omega_M = k_B(m_{hh}/m_{lh})$.

Figure 2.36b indicates that the evaluation of $I(q,\omega)$ reproduces the major features of the reduced light scattering spectra. The theory is particularly accurate in the region of the acoustic plasma oscillation. Its scattering is strong, despite its near neutrality, because the cross section of electrons is much larger than that of holes. An analysis of the several contributions to $I(q,\omega)$ indicates that the spectral line shape is modified by Landau damping by single-particle excitations of electrons as well as holes, though the peak position is well approximated by (2.61). It is also clear from Fig. 2.36b that the spectral line shape of collective modes is expected to be strongly affected by $V_2 \rightarrow V_1$ transitions. The relatively small differences between measured and calculated line shapes are probably due to the approximations involved in the evaluation of $\chi_h^{\text{inter}}(q,\omega)$.

2.5 Light Scattering by Two-Dimensional Electron Systems

The two-dimensional systems of interest here occur at interfaces of semiconductor microstructures, where potential discontinuities and space-charge electric fields *quantize* the electron motion normal to the interfaces into *discrete* energy levels. Since the electrons are free to move in a plane parallel to the interface, each of the discrete energy levels gives rise to a two-dimensional *subband*. Electrons in subband states have many properties in common with those of an idealized two-dimensional electron gas [2.94]. The early interest in the field focused on inversion and accumulation layers at metal-insulator-semiconductor (MIS) and metal-oxide-semiconductor (MOS) structures. At the present time, considerable attention is being devoted to semiconductor heterostructures.

Experimental and theoretical research of 2-dim electron systems at semiconductor interfaces has been extensive during the last ten years [2.94,135,136]. These systems are ideal for studies of free-electron behavior under conditions of reduced dimensionality. The recent observations of the quantized Hall effect [2.137,138] and the anomalous magneto-transport behavior in the extreme quantum limit [2.139] are among the most exciting new developments in the realm of solid state physics. Further interest is stimulated by the relevance of the subject to the technology of modern solid-state electronics.

In 1978 *Burstein* et al. [2.52] proposed *resonant* inelastic light scattering as a sensitive method for the investigation of the elementary excitations of 2-dim electron systems at semiconductor interfaces. It was noted that resonant light scattering from n-GaAs (Sects. 2.3.1, 2) indicates that the excitations of $\simeq 5\times 10^{11}$ cm^{-2} carriers, typical of 2-dim semiconductor plasmas, should be readily observable. It was also emphasized that, as is the case for 3-dim systems, the method should yield separate spectra of single-particle and collective excitations. This feature was expected to lead to the determination of energy level structures and collective electron-electron interactions.

The proposal was soon followed by the observations of *resonant* light scattering by *intersubband excitations*, between discrete energy levels, of electrons in GaAs-(Al_xGa_{1-x})As heterostructures [2.54,55,140–143]. These results confirmed predictions of [2.52,95], and also yielded substantial information on novel [2.53] heterostructures made by molecular beam epitaxy. In the last three years the light scattering method has been applied to studies of 2-dim electron systems at several semiconductor interfaces and heterostructures. This section is devoted to what we believe are the highlights of this research.

2.5.1 Resonant Light Scattering

We consider first the mechanisms and selection rules for resonant light scattering by 2-dim electron systems in semiconductors. The discussion presented here avoids detailed formulas for the cross sections and emphasizes the general ideas

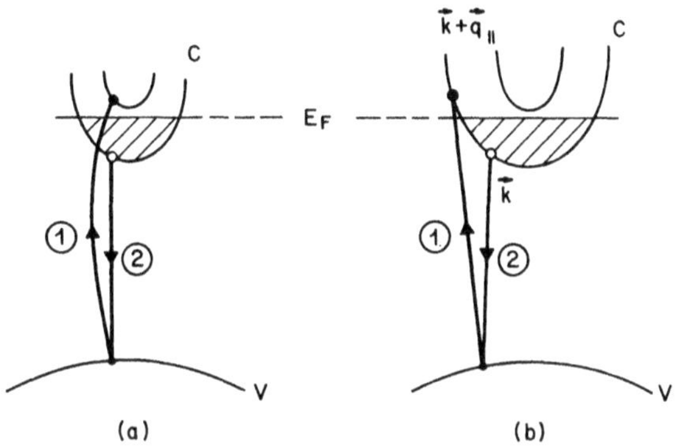

Fig. 2.37. Schematic diagrams showing the virtual valence-to-conduction band transitions that contribute to resonant light scattering by a 2-dim plasma. (**a**) Vertical intersubband excitations; (**b**) intrasubband excitations ($q_{\|}$ is the in-plane component of the scattering wave vector)

that are relevant to the interpretation of experimental results. The reader will find further theoretical discussions in [2.95, 144, 145].

Burstein et al. [2.52, 95] pointed out that within the effective mass approximation, the mechanisms and selection rules for resonant light scattering by a 2-dim semiconductor plasma are similar to those of 3-dim systems considered in Sect. 2.3. Thus, resonant light scattering due to electron-density fluctuations can be described in terms of diagrams analogous to those in Fig. 2.5. In Fig. 2.37 we show the diagrams that correspond to light scattering by *intersubband* and *intrasubband* excitations. These diagrams lead to the conclusion that resonant enhancements in light scattering by electron-density fluctuations are expected only at photon energies near optical gaps that involve the states occupied by the free electrons [2.52, 95] (Sect. 2.3.1a). Table 2.1 lists optical gaps accessible in several semiconductors. In the cases of GaAs, InP and p-Si, the gaps are at the Γ-point of the Brillouin zone. In n-Ge it is at the L-point and in n-Si close to the X-point.

In the case of *vertical* intersubband excitations, as in Fig. 2.37a, only changes in electron motion normal to the plane are involved. The energies of the excitations are expected to be close to the spacings between the discrete energy

Table 2.1. Optical energy gaps for resonant light scattering by electron-density fluctuations

System	Optical gap	E_G [eV]
n-Si	E_2	4.35
p-Si	E_0'	3.35
n-Ge	E_1	2.20
	$E_1+\Delta_1$	2.40
n-GaAs	E_0	1.52
	$E_0+\Delta_0$	1.86
p-GaAs	E_0	1.52
n-InP	E_0	1.43
	$E_0+\Delta_0$	1.56

levels. On the other hand, the intrasubband transitions are in-plane excitations. Their energy spectrum is a continuum, which for degenerate plasmas is similar to that shown in Fig. 2.1. In the case of in-plane components of the scattering wavevector that are $q_{\|} \leqslant 2k_F$, the energy spectrum extends from zero to $q_{\|}v_F$. We note that *nonvertical* intersubband transitions, not considered explicitly in Fig. 2.37, are also possible. They should have an intrinsic width of the order of $q_{\|}v_F$.

The polarization selection rules are also expected to be similar to those for 3-dim plasmas considered in Sect. 2.3.1. When the spin-orbit splitting is reasonably large (0.1 eV or larger), as in all cases considered in Table 2.1 with the exception of those involving n-type Si, we anticipate that light scattering by *spin-density* fluctuations, besides that due to *charge-density* fluctuations, should occur in *depolarized* spectra ($\hat{\varepsilon}_L \perp \hat{\varepsilon}_S$). The elementary excitations associated with these spectra have *single-particle* character. In the case of charge-density fluctuations the spectra are *polarized*, with $\hat{\varepsilon}_L \| \hat{\varepsilon}_S$, and the elementary excitations are expected to have *collective* character. The collective behavior is due to the macroscopic electric fields associated with the fluctuating electron-charge densities.

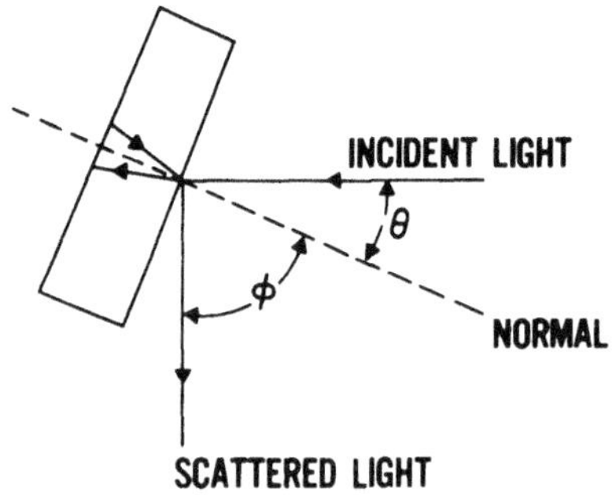

$\theta + \phi = 90°$

Fig. 2.38. Typical (*near*) backscattering geometry used in the investigations of the electronic transitions described here

In polar semiconductors, where collective excitations couple strongly to LO phonons [2.95], light scattering by coupled modes can also occur through the phonon-like mechanisms. These light scattering processes are similar to those of 3-dim semiconductor plasmas considered in Sect. 2.3.1d. Of particular interest is the fact that they are expected to have resonant behavior at *all* optical energy gaps, not only those related to states where the carriers are located. This feature is of importance in direct semiconductors with small energy gaps, like InSb and InAs, where the E_0 and $E_0 + \Delta_0$ optical gaps are not accessible by means of instrumentation currently available for light scattering spectroscopy.

The resonant light scattering spectra are usually obtained in the back-scattering configuration, with light propagating inside the sample along a direction close to the normal to the plane of the 2-dim system. Figure 2.38 describes a convenient form of backscattering geometry. The in-plane and normal components of the scattering wavevector are given by

$$q_{\|} = \frac{2\pi}{\lambda_L} (\sin\theta - \cos\theta), \tag{2.130a}$$

$$q_{\perp} \simeq \frac{4\pi}{\lambda_L} n \left(1 - \frac{1}{4n^2}\right). \tag{2.130b}$$

The values of the refractive index n are between 3 and 5 in the semiconductors of interest here. Thus, in typical cases, the range of $q_{\perp}$ is similar to that shown in Fig. 2.19. For the angle of incidence, $\theta = 45°$, $q_{\|} = 0$. The requirement of $\boldsymbol{q}$-conservation implies that for this case only intersubband excitations can be expected in the spectra. In practice, the smallest values of $q_{\|}$ attainable in a backscattering geometry are $\simeq 10^4\ \mathrm{cm}^{-1}$. In these cases, $q_{\|} v_F$ is too small to be resolved in typical light scattering spectra. The largest values of $q_{\|}$, which correspond to $\theta = 0°$ or $\theta = 90°$, are still about an order of magnitude smaller than those of $q_{\perp}$. This is an intrinsic feature of the backscattering geometry.

2.5.2 GaAs-(Al$_x$Ga$_{1-x}$)As Heterostructures

The first observations of resonant light scattering by 2-dim electron systems were carried out in *doped* GaAs-(Al$_x$Ga$_{1-x}$)As heterostructures [2.54, 55, 140, 141]. In this seminal work the spectral bands are broad, with line shapes that reflect the sample quality at the time. Refinements in the technique of molecular beam epitaxy as well as in the *modulation-doping* procedure [2.53] led to the achievement of extremely high mobilities in multiple quantum wells [2.146] and in single heterojunctions [2.147, 148]. Figures 2.39, 40 show the structure of the

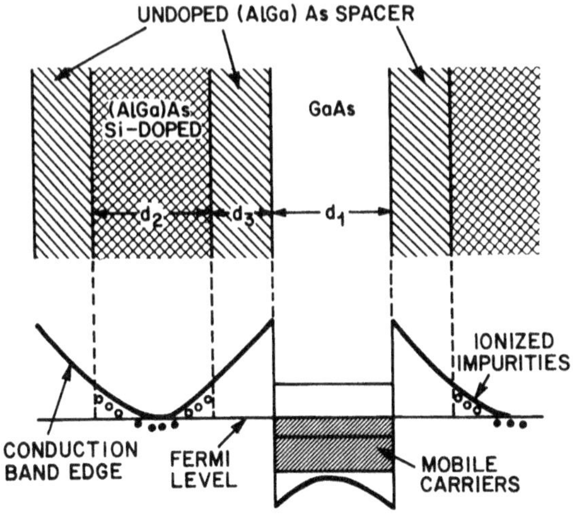

Fig. 2.39. Sequence of layers and structure of the conduction band edge in the modulation-doped multiple quantum well GaAs-(AlGa)As heterostructures. An energy diagram is also shown

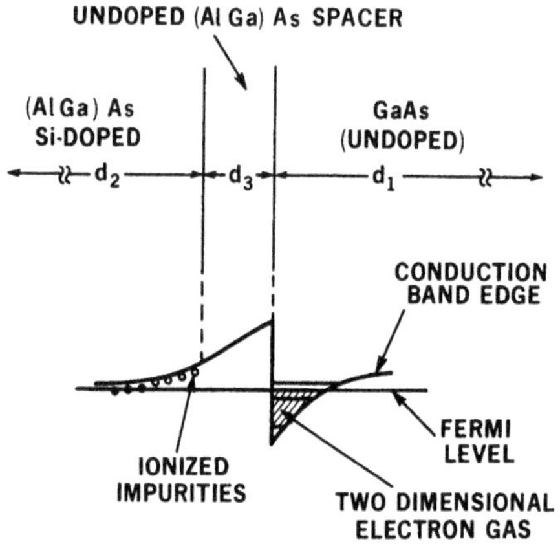

Fig. 2.40. Sequence of layers and structure of the conduction band edge in selectively doped GaAs-(AlGa)As single heterojunctions. An energy diagram is also shown

samples. In these modulation-doped heterostructures, the spatial separation between free electrons and the parent donors enhances the mobility. This is due to the strong reduction of scattering of the mobile electrons by the ionized impurities. The undoped (AlGa)As spacer, with thickness d_3, separates further the free electrons from the ionized donors.

Resonant light scattering by 2-dim electron systems in GaAs-(AlGa)As heterostructures has been extensively investigated. The first intersubband transitions have been observed in GaAs-(AlGa)As single heterostructures [2.54]. The results are shown in Fig. 2.41. Besides the spin-flip single-particle excitations of GaAs bulk carriers ($\simeq$ below $\boldsymbol{q} \cdot \boldsymbol{v}_{\mathrm{F}}$), one identifies single-particle intersubband excitations of the electrons confined at the interface (S.E.). It is also demonstrated that with a Schottky barrier arrangement one is able to control the 2-dim carrier density. The energies of the single-particle excitations have been compared with self-consistent calculations in [2.140]. They are found to be in good agreement with the subband splitting ω_{01}.

The first experimental evidence for the separation of single-particle and collective excitations were found in the inelastic light scattering spectra of [2.55]. Figure 2.42 shows the results obtained from two multiple heterostructures. All the relatively wide features in these spectra were assigned to intersubband excitations of free electrons in the GaAs quantum wells. These experiments were also the first to provide spectroscopic evidence of confined electron states in these superstructures.

We discuss in some detail various results obtained later with better quality samples. In the case of *intersubband spectroscopy*, this work has revealed energy-level structures and resonant behavior of scattering intensities, collective Coulomb interactions, correlations with electron mobilities and the effects of photoexcitation at high intensities. We also consider here two reports of scattering by excitations with in-plane motion. One is the observation and

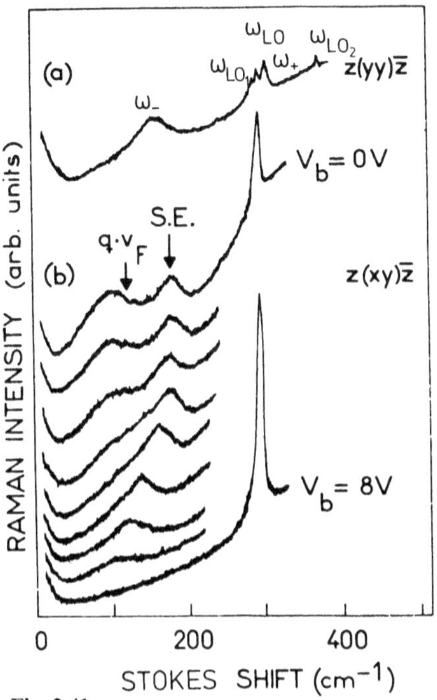

Fig. 2.41

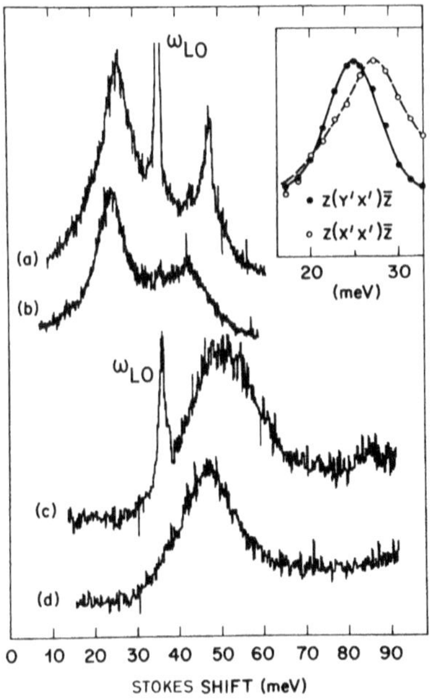

Fig. 2.42

Fig. 2.41. Resonant light scattering spectra from a selectively doped GaAs-($Al_xGa_{1-x}As$) single heterostructure as a function of applied voltage V_b across a Ni-Schottky barrier on the GaAs layer. Photon energies are close to the $E_0 + \Delta_0$ optical gap of GaAs. The peak labeled S.E. corresponds to single-particle intersubband excitations of the 2-dim electron system confined at the interface. From [2.54]

Fig. 2.42a–d. Resonant light scattering spectra from two modulation-doped multiple GaAs-(Al_xGa_{1-x})As heterostructures. The free electrons occupy states in the conduction subbands of the GaAs well (of thickness d_1). Photon energies are close to the $E_0 + \Delta_0$ optical gap of GaAs. (**a**) and (**b**) are from a sample with $d_1 = 400$ Å, $x = 0.28$ and $N_s = 1.6 \times 10^{12}$ cm^{-2}. (**c**) and (**d**) are from the other sample with $d_1 = 221$ Å. $x = 0.24$ and $N_s = 3.1 \times 10^{12}$ cm^{-2}. The small differences between the positions of bands in spectra (**a**) and (**b**) have been assigned to collective effects (Sect. 2.5.3b). From [2.55]

measurement of the dispersion of the plasmons in multiple heterostructures. The other involves transitions between Landau levels of electrons in a large magnetic field normal to the interface.

a) Intersubband Spectroscopy

Figure 2.43 shows results obtained from modulation-doped multiple quantum wells (MQW). These backscattering spectra, reported by *Pinczuk* et al. [2.142], are the first in which there is clear separation between single-particle and

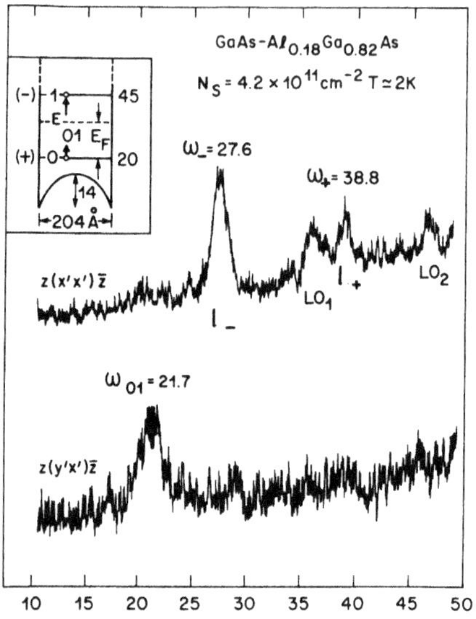

Fig. 2.43

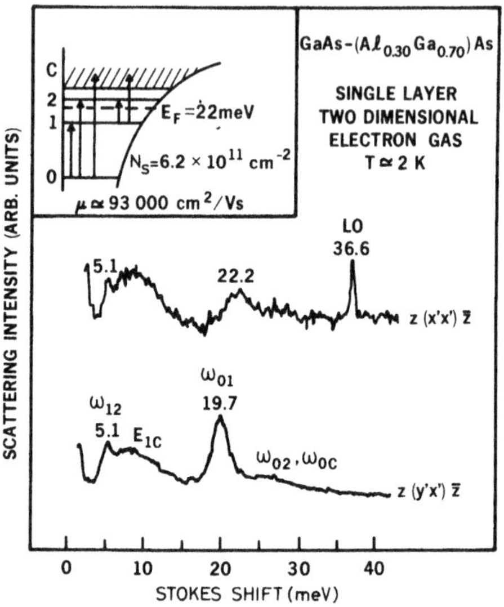

Fig. 2.44

Fig. 2.43. Light scattering spectra of a modulation-doped multiple quantum well GaAs-($Al_{0.18}Ga_{0.82}$)As heterostructure. The inset shows calculated energies of the lowest quantum well states, the band bending and the Fermi energy. The GaAs well width is 204 Å. Only the energy range of the lowest intersubband excitation is shown. The LO_1 and LO_2 peaks are from the ($Al_{0.18}Ga_{0.82}$)As layers. From [2.142]

Fig. 2.44. Light scattering spectra of a modulation-doped GaAs-($Al_{0.3}Ga_{0.7}$)As single heterojunction ($d_3 = 100$ Å). The inset shows the Fermi level and the assignment of the intersubband transitions. C indicates a quasi-continuum of energy levels. From [2.56]

collective excitations. *Abstreiter* et al. [2.143] later reported similar results. Figure 2.44 shows spectra obtained from modulation-doped single heterojunctions. In this case separation between the two types of spectra is also achieved [2.56].

In Figs. 2.43, 44, z and $\bar{z}$ are the [001] and [00$\bar{1}$] propagation directions of the incident and scattered light both normal to the layers. x' and y' are the light polarizations. They correspond to the [110] and [1$\bar{1}$0] directions in the plane of the layers. The depolarized $z(y'x')\bar{z}$ spectrum of Fig. 2.43 shows a single peak at $\omega_{01} = 21.7$ meV. On the basis of the selection rules considered in Sect. 2.5.2, it is assigned to the lowest spin-density intersubband excitation of single-particle character. In the polarized $z(x'x')\bar{z}$ spectrum the peaks at $\omega_+ = 38.8$ meV and $\omega_- = 27.6$ meV, labelled I_+ and I_-, are assigned to the coupled collective intersubband-LO phonon excitations proposed by *Burstein* et al. [2.95].

Single-particle intersubband energies were measured in $z(y'x')\bar{z}$ spectra from a number of samples [2.56, 140, 141, 143]. In all cases there is good agreement with calculated subband spacings [2.140, 149, 150]. The differences in energy between single-particle and collective excitations reveal in a direct way the macroscopic electric fields (or depolarization fields) due to *resonant screening* phenomena associated with intersubband excitations. Quantitative interpretation of collective effects in these polar semiconductors requires an analysis of electron-LO phonon coupling.

b) Intersubband Spectroscopy: Collective Electron LO-Phonon Modes

Macroscopic electric fields associated with intersubband excitations were first predicted by *Chen* et al. [2.151]. These phenomena, often called depolarization field effects, had been invoked in the interpretation of subband optical absorption in Si-MOS structures [2.152, 153] and, more recently, in InSb and InAs MIS structures [2.154]. Light scattering spectra like those shown in Figs. 2.43, 44 have made possible a well-defined experimental identification of these effects, based on the differences between the spectra due to single-particle and collective excitations. We present below a quantitative analysis for the case of results obtained in the MQW GaAs-(AlGa)As heterostructures. The analysis yields the Coulomb matrix elements for the intersubband transitions.

In the MQW heterostructures, the envelope functions of the lowest-lying levels are largely confined to the GaAs wells [2.149, 150]. Since the macroscopic electric fields of vertical intersubband excitations are also confined, coupling between excitations in different wells is negligible. Under these circumstances, an extension of the treatment of *Dahl* and *Sham* [2.155] to the case of isolated quantum wells gives the collective intersubband excitation energies as solutions of [2.56]

$$\det\left|\delta_{ij,kl}-\frac{\varepsilon_\infty}{\varepsilon_L(\omega)}\frac{\omega_P^2(ij,kl)}{\omega^2-\omega_{kl}^2}\right|=0, \tag{2.131}$$

where ω_{kl} is the energy associated with the vertical transition $k\to l$ between subbands k and l. In (2.131), $\delta_{ij,kl}=1$ when the transitions $k\to l$ and $i\to j$ are identical ($i=k$ and $j=l$), and is zero otherwise. $\omega_P(ij,kl)$ is a set of plasma energies that describe the macroscopic electric field effects, including the couplings among the different intersubband excitations of one well. $\varepsilon_L(\omega)$ is a longitudinal dielectric function that accounts for the screening of the macroscopic electric field by the polar lattice, see (2.39) and (2.134) below.

The general expression for $\omega_P^2(ij,kl)$ is

$$\omega_P^2(ij,kl)=2[N_s(k)-N_s(l)]e^2\omega_{kl}L(ij,kl)/\hbar\varepsilon_\infty\varepsilon_0, \tag{2.132}$$

where $N_s(k)$ and $N_s(l)$ are the areal densities of electrons in subbands k and l and $L(ij,kl)$ are the matrix elements of the Coulomb interaction expressed as a length. In the case of an isolated quantum well, $L(ij,kl)$ can be written as [2.56]

$$L(ij,kl)=\int_{-\infty}^{\infty}dz\left[\int_{-\infty}^{z}dz'\,\xi_i(z')\xi_j(z')\right]\left[\int_{-\infty}^{z}dz''\,\xi_k(z'')\xi_l(z'')\right], \tag{2.133}$$

where the $\xi(z)$ are the envelope functions that describe the subband states within the effective mass approximation [2.94, 149, 150]. The off-diagonal terms of the $L(ij, kl)$ matrix describe the coupling between the $i \rightarrow j$ and $k \rightarrow l$ transitions. Since the envelope functions are even or odd with respect to the mirror planes of the quantum wells, nonzero elements of $L(ij, kl)$ exist only for the cases in which the transitions $i \rightarrow j$ and $k \rightarrow l$ have the same parity. That is, there is coupling only among collective intersubband excitations that have identical parity. It is easy to see from (2.133) why collective effects vanish for the case of spin-density excitations. In this case the orthogonality of the spin components of ξ_l and ξ_k forces $L(ij, kl)$ to vanish.

The coupling between collective intersubband excitations and the polar lattice is determined by $\varepsilon_L(\omega)$. For excitations fully confined within the quantum wells, $\varepsilon_L(\omega)$ can be taken as the lattice dielectric function of bulk GaAs, (2.39):

$$\varepsilon_L(\omega) = \varepsilon_\infty(\omega_{LO}^2 - \omega^2)/(\omega_{TO}^2 - \omega^2), \tag{2.134}$$

where $\omega_{TO} = 33.7$ meV and $\omega_{LO} = 36.6$ meV. Equations (2.131–134) describe a system of coupled collective modes between intersubband excitations and longitudinal-optical lattice vibrations. They have been used with considerable success in the interpretation of collective intersubband excitations of odd [2.56, 142, 143] and even [2.56, 143] parities.

As an example of this analysis of coupled electron LO-phonon excitations in a 2-dim electron system, we consider the *odd parity* I_+ and I_- excitations shown in the $z(x'x')\bar{z}$ spectra of Fig. 2.43. In this case coupling to other intersubband excitations of odd parity, at much higher energy, can be neglected [2.142]. Thus, we find that ω_+ and ω_- are the solutions of

$$1 - \left(\frac{\omega_{TO}^2 - \omega_\pm^2}{\omega_{LO}^2 - \omega_\pm^2}\right) \cdot \left(\frac{\omega_P^2}{\omega_\pm^2 - \omega_{01}^2}\right) = 0, \tag{2.135}$$

where

$$\omega_P^2 = 2 N_s e^2 \omega_{01} L_{01} / \hbar \varepsilon_\infty \varepsilon_0 . \tag{2.136}$$

The only adjustable parameter in (2.135, 136) is the Coulomb matrix element L_{01}. Equations (2.135, 136) when used in conjunction with the values of ω_{01}, ω_+ and ω_- obtained from the spectra in Fig. 2.43 yield an adjusted value of $L_{01} = 15$ Å [2.142]. A similar analysis of light scattering spectra has been carried out in a number of samples for the cases of *odd parity* ω_{01} and *even parity* ω_{02} transitions [2.56, 142, 143, 156]. Evaluations of the Coulomb matrix elements by means of subband envelope functions obtained from simple model calculations are in good agreement with those obtained from the light scattering spectra [2.56, 142, 143, 150, 156].

c) Intersubband Spectroscopy: Correlation with Transport Properties

One of the important questions in the light scattering spectroscopy of intersubband excitations is the relation to transport properties of the 2-dim electron system. *Pinczuk* et al. [2.157] have shown that there is a correlation between resonant light scattering spectra and the in-plane electron mobilities. There is, in addition, an unexpected effect which takes the form of a dependence of the spectral linewidths on incident photon energies in the samples of "lower" mobility ($\mu \lesssim 40{,}000\ \mathrm{cm^2/V\,s}$).

The systematic part of these studies has been carried out in the four similar modulation-doped MQW GaAs-($Al_{0.12}Ga_{0.88}$)As heterostructures in which the electron mobilities were found to be directly related to the thickness d_3 of the undoped (AlGa)As) spacer (Fig. 2.39) [2.146]. In Table 2.2 we reproduce some of the relevant sample parameters. We see that the electron mobilities do increase monotonically with d_3. Figure 2.45 shows depolarized $z(y'x')\bar{z}$ spectra obtained from three of these samples. The observed bands are assigned to single-particle transitions as indicated in the figure. The widths of all these bands decrease with increasing Hall mobility and with d_3. Even more striking is the way the spectral widths depend on laser photon energies in the range of the $E_0 + \Delta_0$ optical gap. Figure 2.46 shows the FWHM of the ω_{01} transition in spectra of the three samples with the lowest Hall mobilities. In the case of the sample with the lowest mobility of these three, the FWHM peaks for photon energies close to the maximum in resonant enhancement.

To interpret these results it was proposed [2.157] that resonant inelastic light scattering by the intersubband excitations is affected by wavevector relaxation processes related to those that limit the electron mobilities. These relaxation processes, due to scattering of the electrons by the Coulomb potential of the ionized donors, should depend on d_3. Under these circumstances, as in the cases discussed in Sect. 2.3.3b, the condition of wavevector conservation breaks down for resonant light scattering. The intersubband transitions active in backscattering spectra are no longer necessarily vertical as in Fig. 2.47a, but may occur with an in-plane wavevector transfer $\boldsymbol{q}'$ as shown in Fig. 2.47b. The energies of such nonvertical transitions are shifted from the vertical subband spacing as shown in the figure. Since the shifts form a continuum, the spectral bands are expected to show considerable width. The resonant inelastic light

Table 2.2. Sample parameters of MQW devices used in the experiments described here

Sample	d_1 [Å]	d_2 [Å]	d_3 [Å]	x	N_s [$\mathrm{cm^{-2}}$]	Hall mobility [$\mathrm{cm^2/V\,s}$]
9–23–80(3)	255	302	0	0.12	10×10^{11}	12.500
9–25–80(1)	245	290	51	0.12	8.8×10^{11}	28.000
9–25–80(2)	244	287	99	0.12	6.8×10^{11}	62.000
9–25–80(3)	250	293	151	0.12	5.9×10^{11}	93.000

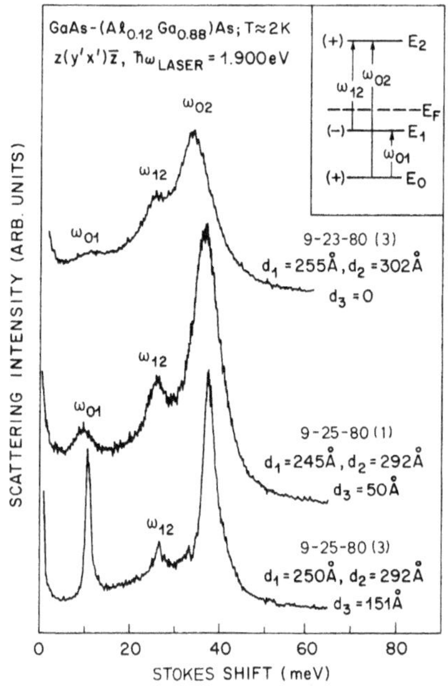

Fig. 2.45. Depolarized light scattering spectra from three modulation-doped multiple GaAs-$(Al_{0.12}Ga_{0.88})As$ heterostructures. Sample parameters are listed in Table 2.2. The inset shows the assignments of the single-particle intersubband excitations. From [2.56, 157]

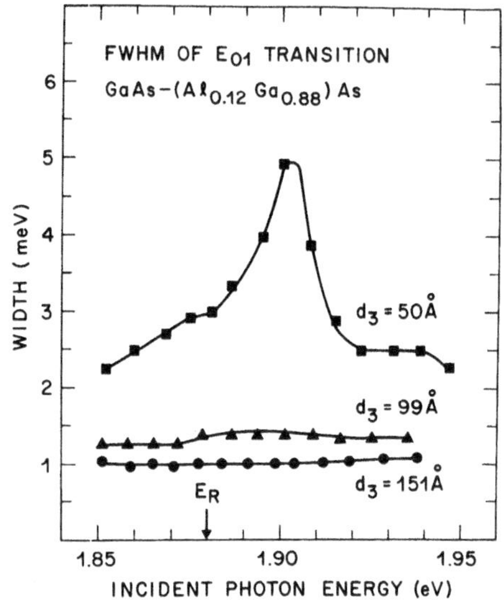

Fig. 2.46. FWHM of the ω_{01} excitation band as a function of incident photon energy for three of the samples listed in Table 2.2. $E_R = 1.88$ eV is the photon energy of the center of the resonant profile (Fig. 2.48). From [2.56, 157]

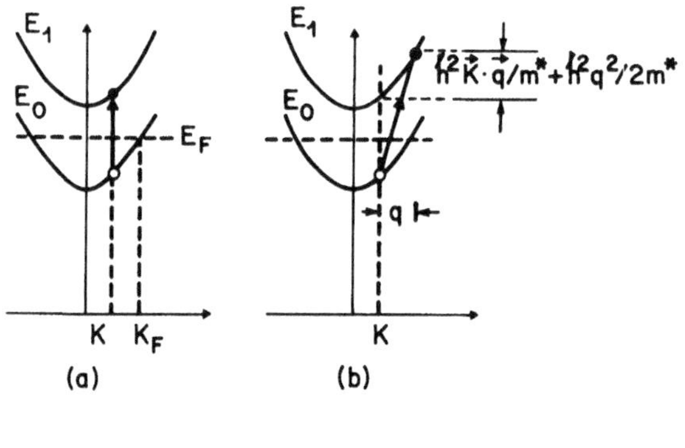

Fig. 2.47a, b. Schematic representation of intersubband transitions ω_{01} in the two-dimensional wavevector space of the conduction subbands. (**a**) Vertical transitions ($q_{\|} = 0$); (**b**) nonvertical transitions ($q_{\|} \neq 0$). From [2.57, 157]

scattering with the breakdown of $\boldsymbol{q}$-conservation is expected to show more pronounced resonant behavior than that with $\boldsymbol{q}$-conservation, as is the case for electron systems (Sect. 2.3.1b and Fig. 2.16).

d) Intersubband Spectroscopy: Resonant Enhancements

Measurements of the resonant enhancements of light scattering by intersubband excitations have been reported in single [2.54, 140, 158] and multiple [2.55, 141, 143, 159] GaAs-(AlGa)As heterostructures. These results, obtained with photon energies across the $E_0 + \Delta_0$ optical gaps of the GaAs layers, display the general behavior anticipated by *Burstein* et al. [2.52, 95]. In the discussion presented below we focus on the modulation-doped multiple quantum well heterostructures. The underlying physics of these phenomena is similar in single and multiple heterostructures. However, the analysis is simplified in the case of quantum wells because the relevant intermediate valence band states are also confined to the GaAs layers [2.95].

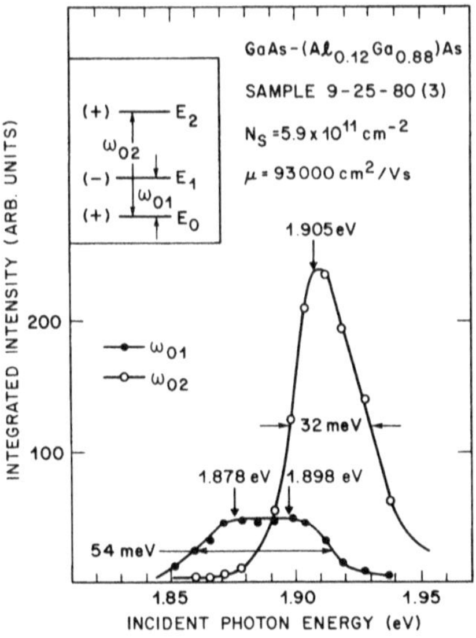

Fig. 2.48. Resonant enhancement of light scattering by single-particle spin-density intersubband excitations of a modulation-doped multiple GaAs-($Al_{0.12}Ga_{0.88}$)As heterostructure (sample parameters are listed in Table 2.2). From [2.159]

Figure 2.48 shows resonant profiles measured in $z(y'x')\bar{z}$ spectra from sample 9-25-80(3) (Table 2.2) [2.159]. We choose to focus here on the results from this sample because, as seen in Sect. 2.5.2b, the wavevector appears to be strictly conserved in its light scattering spectra. The laser photon energies are in the range of the $E_0 + \Delta_0$ optical gap of GaAs (Table 2.1). A typical spectrum, due to single-particle spin-density intersubband excitations, can be seen in the lower part of Fig. 2.45. The resonant profiles in Fig. 2.48 have several aspects of considerable interest. Particularly striking, in our opinion, is the relative magnitude of the *integrated* intensities of ω_{01} and ω_{02} excitations at their maxima in resonant enhancement. We can see that the intensities of the *even parity* ω_{02} excitations are much larger than those of the *odd parity* ω_{01} excitations. Parity, as already mentioned in Sect. 2.5.2b, refers here to the mirror symmetry of the quantum wells.

Interpretation of resonant profiles requires consideration of the light scattering matrix element [2.95, 141, 159]. An expression can be obtained from (2.71, 81, 90) by using the effective mass approximation [2.94] to write the wave functions of the relevant conduction and valence band states. In the case of modulation-doped MQW GaAs-(AlGa)As heterostructures, the envelope functions are largely confined within the GaAs layers and coupling between different wells can be neglected.

The resonant light scattering matrix element for spin-density intersubband excitations in $z(x'y')\bar{z}$ spectra can be written as

$$\gamma_{\mathrm{if}}(\boldsymbol{k}) \simeq (\hat{\boldsymbol{e}}_{\mathrm{L}} \times \hat{\boldsymbol{e}}_{\mathrm{S}}) \frac{p^2}{3m} \sum_{\nu} \frac{\langle \phi_{\mathrm{f}}(z)|\mathrm{e}^{\mathrm{i}k_{\mathrm{L}}z}|\phi_{\nu}(z)\rangle \langle \phi_{\nu}(z)|\mathrm{e}^{-\mathrm{i}k_{\mathrm{S}}z}|\phi_{\mathrm{i}}(z)\rangle}{E(k,\nu) - \hbar\omega_{\mathrm{L}}}, \tag{2.137}$$

where $\phi_{\mathrm{i}}(z)$ and $\phi_{\mathrm{f}}(z)$ are the envelope functions of the initial and final conduction subbands, $\phi_{\nu}(z)$ represent the envelope functions of the intermediate valence subbands and $\boldsymbol{k}$ is the two-dimensional wavevector common to these three states. Implicit in (2.137) is the assumption that spin-density intersubband excitations have single-particle character and thus, each vertical transition at wavevector $\boldsymbol{k}$ is an independent excitation of the 2-dim electron system. The expression for the resonant energy $E(k,\nu)$ is

$$E(k,\nu) = E_{\mathrm{G}} + E_{\mathrm{f}} + E_{\nu} + \hbar^2 k^2/2\mu, \tag{2.138}$$

where E_{G} is the $E_0 + \Delta_0$ gap of bulk GaAs. E_{f} and E_{ν} are the $k=0$ energies of the final and intermediate subband states measured, respectively, from the bottom of the conduction band and the top of the valence band of GaAs. μ is the reduced mass of the conduction and spin-orbit split-off valence bands ($\mu^{-1} = m_{\mathrm{e}}^{-1} + m_{\mathrm{so}}^{-1}$).

The selection rules that apply to the parities of the initial and final states are determined by the product of matrix elements associated with subband states in (2.137), which can be written as

$$\begin{aligned} &\langle \phi_{\mathrm{f}}(z)|\mathrm{e}^{\mathrm{i}k_{\mathrm{L}}z}|\phi_{\nu}(z)\rangle \langle \phi_{\nu}(z)|\mathrm{e}^{-\mathrm{i}k_{\mathrm{S}}z}|\phi_{\mathrm{i}}(z)\rangle \\ &\quad = \langle \phi_{\mathrm{f}}(z)|\phi_{\nu}(z)\rangle \langle \phi_{\nu}(z)|\phi_{\mathrm{i}}(z)\rangle \\ &\qquad + \mathrm{i}k_{\mathrm{L}} \langle \phi_{\mathrm{f}}(z)|z|\phi_{\nu}(z)\rangle \langle \phi_{\nu}(z)|\phi_{\mathrm{i}}(z)\rangle \\ &\qquad - \mathrm{i}k_{\mathrm{S}} \langle \phi_{\mathrm{f}}(z)|\phi_{\nu}(z)\rangle \langle \phi_{\nu}(z)|z|\phi_{\mathrm{i}}(z)\rangle + \dots \end{aligned} \tag{2.139}$$

The first term in the right side of (2.139) represents the dipole approximation. It is clear that this term is different from zero only for pairs of initial and final states that have the same parity. This implies that, as expected, there is no light scattering by odd parity excitations within the dipole approximation. Therefore, the existence of substantial light scattering by the odd parity ω_{01} excitations in the results of Fig. 2.48 is rather surprising. Light scattering by odd parity excitations could possibly be explained by the higher-order terms (quadrupole) in (2.139). We note that $k_L \simeq k_S \simeq 3.5 \times 10^5\ \mathrm{cm}^{-1}$ is fairly large and comparable to values of d_1^{-1} of typical samples. However, at the present time there is no quantitative evaluation of this suggestion.

Finally, we note that (2.138) leads to a simple interpretation of the different positions of the maxima in the resonant profiles of the ω_{01} and ω_{02} transitions. The maximum of the enhancement of light scattering by ω_{02} occurs at a higher energy than for ω_{01} because its final state is also at a higher energy. We also note that the larger width of the resonant profile of ω_{01} suggests, as indicated in Fig. 2.48, that two confined valence subband states make a similar contribution to its resonant light scattering matrix element.

e) Intersubband Spectroscopy: Photoexcited Plasmas

Resonant light scattering by photoexcited electrons was first reported by *Pinczuk* et al. [2.57] in *undoped* GaAs-(AlGa)As MQW heterostructures and by *Zeller* et al. [2.130] in modulation-doped heterostructures. These experiments made possible light scattering studies of 2-dim systems in which the free-electron density can be varied by changing the photoexcitation intensity.

Figure 2.49 shows results obtained from the undoped heterostructures [2.57]. In order to achieve carrier photoexcitation, the laser beam, originating

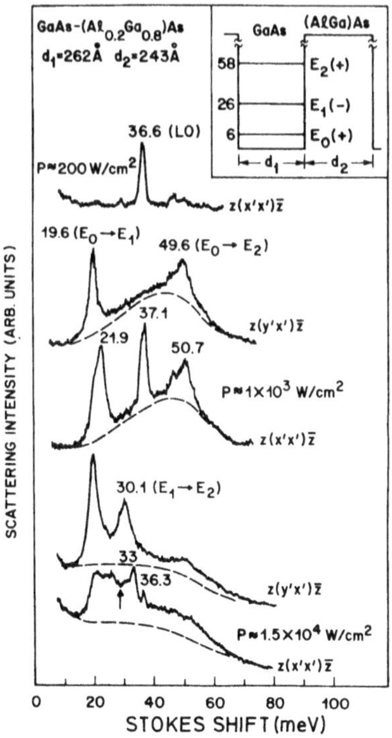

Fig. 2.49. Light scattering spectra from an *undoped* multiple quantum well GaAs-($Al_{0.2}Ga_{0.8}$)As heterostructure for three incident laser power densities. The laser photon energy (1.916eV) is close to the $E_0 + \Delta_0$ optical gap. The dotted line is the estimated $E_0 + \Delta_0$ luminescence. The inset shows calculated finite quantum-well energy levels. From [2.58]

from a Kr^+ laser operating at 6471 Å, was focused to a spot of radius ~10 μm. This beam also excited the light scattering spectra. The incident photon energy $\hbar\omega_L = 1.916$ eV is close to resonance with the $E_0 + \Delta_0$ optical gaps of the GaAs layers.

At a relatively low power density of ~200 W/cm², the spectrum shows only the peak of the LO phonon of GaAs. This is an indication that at these power densities, carrier photoexcitation is low. When the power density is increased to ~10^3 W/cm², new bands appear. The assignments of the new bands are similar to those in Sect. 2.5.2a. The bands in the $z(y'x')\bar{z}$ spectrum are assigned to the single-particle spin-density excitations of photoexcited *electrons* as indicated in Fig. 2.49. The bands in the $z(x'x')\bar{z}$ spectrum are assigned to the corresponding collective excitations coupled to LO phonons. Excitations related to the photoexcited holes were not observed. Their scattering cross section is believed to be much smaller because it is not enhanced at the $E_0 + \Delta_0$ optical gap (Table 2.1) [2.52, 95].

At a higher power density of ~1.5×10^4 W/cm² there is a new band, at 30.1 meV, in the $\bar{z}(y'x')\bar{z}$ spectrum. This has been interpreted as arising from transitions of photoexcited electrons that populate the next higher conduction subband at large power densities. At these densities the $z(x'x')\bar{z}$ spectra, assigned to collective excitations, are rich in structure related to the occupation of more than one conduction subband. The dip in the scattering intensity just below the ω_{01} transition, indicated by an arrow, has been attributed to the coupling of collective excitations associated with *odd parity* ω_{01} and ω_{12} transitions.

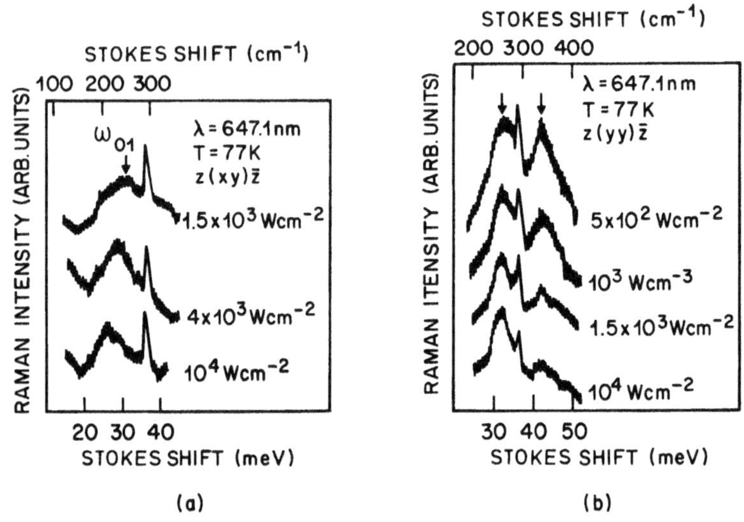

Fig. 2.50a, b. Light scattering spectra from a modulation-doped multiple GaAs-($Al_{0.15}Ga_{0.85}$)As heterostructure for several photoexcitation levels. $d_1 = 170$ Å, $d_3 = 50$ Å and $N_s = 5 \times 10^{11}$ cm^{-2}. **(a)** Single-particle spin-density intersubband excitation; **(b)** collective coupled charge density-LO phonon excitations. From [2.130]

One of the striking aspects of the results of Fig. 2.49 is the observation that in this range of photoexcited carrier densities (10^{11}–10^{12} cm^{-2}), there is no identifiable change in the single-particle energies measured in $z(y'x')\bar{z}$ spectra. This has been considered an indication that photoexcitation *does not* cause substantial changes in the potential distribution within the quantum wells. This means that photoexcited electrons and holes become confined in the GaAs layers with identical densities, since spatial separation between positive and negative charges would create a large electrostatic potential that would modify the spacing between the subbands.

Zeller et al. [2.130] have studied the effects of photoexcitation in modulation-doped heterostructures. Their results, shown in Fig. 2.50, indicate different behavior from that found in undoped heterostructures. Figure 2.50a shows that photoexcitation decreases the spacing between the two lowest subbands. This suggests substantial changes in the electrostatic potential caused by different spatial distributions of the densities of photoexcited electrons and holes. The smearing of the ω_+, seen in Fig. 2.50b, reflects the nonuniform photoexcitation density due to strong absorption of laser light at $\lambda_L = 647.1$ nm.

f) Spectroscopy of In-Plane Motion: Landau Level Excitations

Worlock et al. [2.58] and *Tien* et al. [2.160] investigated resonant inelastic light scattering from modulation-doped GaAs-(AlGa)As MQW heterostructures under strong magnetic fields normal to the layers. For this orientation of magnetic field, the electron motion in the plane of the layers is also quantized into Landau levels and the energy spectrum of the 2-dim electron system is fully discrete [2.94].

Typical depolarized $z(y'x')\bar{z}$ spectra are shown in Fig. 2.51. The photon energies overlap with the $E_0 + \Delta_0$ optical gaps of the GaAs layers. The band labeled $\hbar\omega_c$ has been assigned to $\Delta l = 1$ Landau-level excitations of single-particle character. The assignment is based on the linear dependence of its energy on magnetic field. Similar spectra have been obtained from several samples having a range of free-electron densities and layer thicknesses. Figure 2.52 shows the measured values of $\hbar\omega_c$ as a function of magnetic field. The slope of the straight line gives an electron effective mass of $m^* = 0.068 \pm 0.003$. The polarized $z(x'x')\bar{z}$ spectra show bands at positions somewhat higher than the cyclotron energies. These effects have been tentatively interpreted as arising from collective (i.e., magnetoplasma) effects.

We have seen in Sect. 2.3.2a that light scattering by the $\Delta l = 1$ excitations, having odd parity, is forbidden in the dipole approximation. In the experiments described above [2.58, 160] carried out with the scattering geometry shown in the inset to Fig. 2.51, the in-plane component of the wave-vector is $q_{\parallel} \simeq 0.9 \times 10^5$ cm^{-1}. Breakdown of the dipole approximation might be the reason for the observation of light scattering by $\Delta l = 1$ excitations [2.160a].

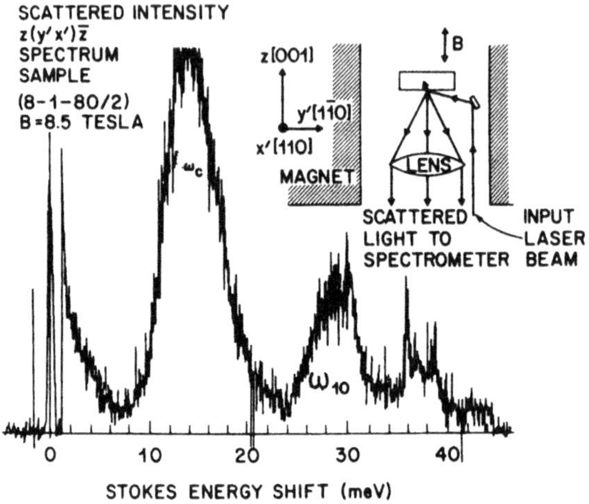

Fig. 2.51. Depolarized light scattering spectrum of a modulation-doped multiple GaAs-($Al_{0.25}Ga_{0.75}$)As heterostructure under a magnetic field of 8.5 Tesla normal to the 2-dim electron system. The band at 14 meV (ω_c) is assigned to $\Delta l = 1$ Landau-level transitions. The one at 29 meV is a ω_{01} intersubband excitation. Photon energies are close to the $E_0 + \Delta_0$ optical gap of GaAs. The inset shows the sample arrangement in the magnet. From [2.58, 160]

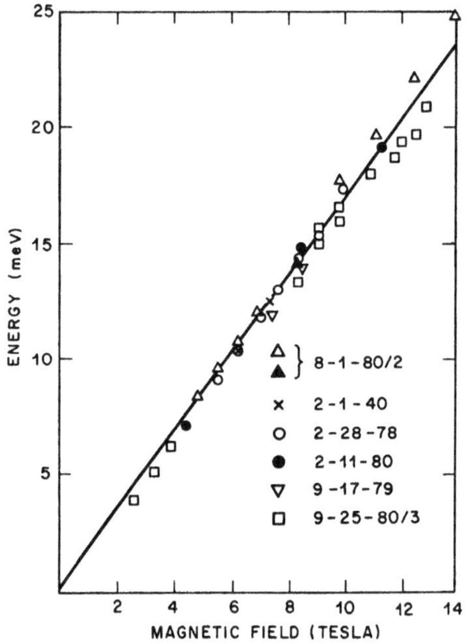

Fig. 2.52. Plot of the energies of $\Delta l = 1$ single-particle Landau-level transitions as a function of a magnetic field constructed from data like those in Fig. 2.51. Results from six different samples are shown. The straight line gives an effective mass ratio of 0.068 ± 0.002. From [2.58, 160]

g) Spectroscopy of In-Plane Motion: Plasma Oscillations

Olego et al. [2.59] have recently reported inelastic light scattering by *plasma* oscillations in modulation-doped GaAs-(AlGa)As MQW heterostructures. The oscillations are the collective modes associated with the in-plane 2-dim motion of the electrons confined in the GaAs layers. Like the plasmons of the idealized layered electron gas [2.161], the plasma oscillations of layered semiconductor microstructures are expected to show large dispersive effects [2.162]. In

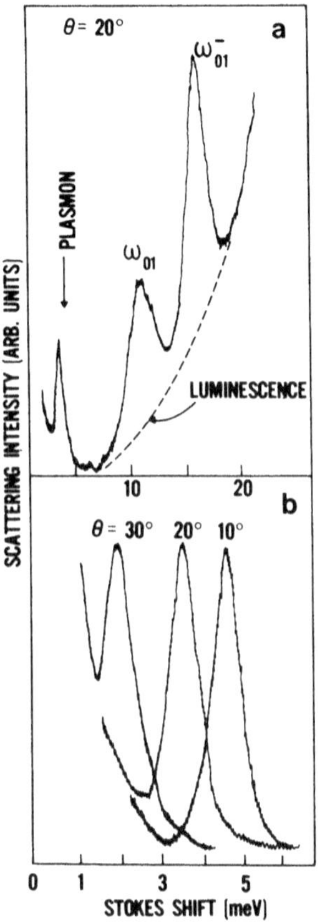

Fig. 2.53. (**a**) *Nearly* backscattering spectrum showing scattering by plasmons in a modulation-doped multiple GaAs-($Al_{0.20}Ga_{0.80}$)As heterostructure; (**b**) plasma line for different angles of incidence. From [2.59]

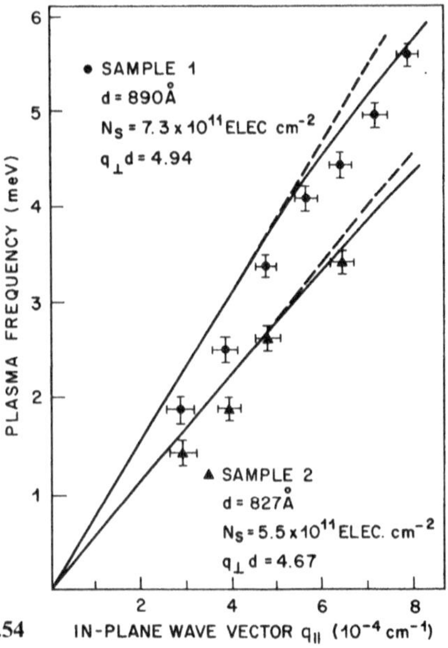

Fig. 2.54

Fig. 2.54. Dispersion of the plasma frequency measured in two samples. Sample 1 is the same as in Fig. 2.53. Sample 2 is listed in Table 2.2 as 9-25-80 (3). From [2.59]

multilayer systems or superlattices, these effects manifest themselves as strong dependences of the plasma frequencies on $q_\parallel$ and $q_\perp$ [2.163, 164]. The light scattering experiments have confirmed the predictions of the theoretical models.

These experiments were carried out with photon energies ($\sim$1.58 eV) close to the fundamental optical gap E_0 of GaAs (Table 2.1). This allowed larger penetration depths ($\sim$2 μm) than those near the $E_0+\Delta_0$ gap while retaining the benefits of resonant enhancements and relatively low luminescence backgrounds. Figure 2.53a shows a spectrum from a sample with $d_1=262$ Å, $d_2=317$ Å, $d_3=163$ Å and $N_s=7.3\times10^{11}$ cm^{-2}. The ω_{01} and ω_{01}^- bands are associated with single-particle and collective intersubband excitations. There is, in addition, a *new* low-energy peak at 3.5 meV. Its most remarkable feature, displayed in Fig. 2.53b, is that its energy depends on the angle of incidence. Through (2.130) this implies a dependence on $q_\parallel$. Such dispersive behavior led *Olego* et al. [2.59] to assign the peaks to plasma oscillations of the MQW heterostructures.

Figure 2.54 shows the dispersion of these plasmons measured in two samples. These results were interpreted by means of a plasma frequency of the form [2.161, 163, 164]

$$\omega_P = \left(\frac{2\pi N_s e^2}{\varepsilon_M m^*} q_\parallel \frac{\sinh q_\parallel d}{\cosh q_\parallel d - \cos q_\perp d} \right)^{1/2}, \tag{2.140}$$

where m^* is the effective mass, ε_M is the background dielectric constant of the multilayers and $d = d_1 + d_2 + 2d_3$. The full lines in Fig. 2.54 correspond to (2.140). The agreement between measured and calculated values is good; small differences can be explained by the uncertainties in the values of the sample parameters.

The dashed lines in Fig. 2.54 are the extrapolations of the *linear* behavior predicted for relatively low $q_\parallel$ [2.59], a dispersion similar to that of acoustic plasmons in two-component 3-dim systems (Sect. 2.2.7). However, while acoustic plasmons of 3-dim electron gases are Landau-damped [2.81], in the experiments of [2.59], $\omega_P > q_\parallel v_F$ and Landau damping should be absent. This absence suggests that investigations of lifetime effects in multilayer plasmas are possible. Plasma oscillations of single-layer 2-dim system with little damping have been observed by means of infrared absorption [2.165, 166] and emission [2.167] from grating structures. The observation of the plasmon and the determination of its dispersion demonstrate novel applications of the light scattering method to studies of 2-dim electron systems of semiconductors.

2.5.3 Ge-GaAs Heterostructures

Merlin et al. [2.63] have reported inelastic light scattering from Ge-GaAs heterojunctions fabricated by molecular beam epitaxy. The differences in resonant behavior of the light scattering intensities (Table 2.1) was used in the assignment of spectral features as arising either from Ge or GaAs layers. The procedure allowed the identification of light scattering associated with the excitations of a 2-dim electron system at the Ge side of the interface.

The samples used in this work consist of a thin layer of Ge (~300 Å) grown on top of a 2 μm layer of GaAs. These heterojunctions are fabricated on (001) GaAs substrates. The Ge is *unintentionally* doped with As at levels of ~1.5 × 10^{18} cm^{-3}. Two samples were investigated in detail. In *sample (1)*, the GaAs layer is n-type (Ge doped) with a free-electron density of $N_e = 1.4 \times 10^{18}$ cm^{-3}. For *sample (2)*, the GaAs layer is nominally undoped with $N_e \sim 6.6 \times 10^{15}$ cm^{-3}.

Figure 2.55 shows spectra obtained from the selectively doped sample (1). The relatively broad structures seen in these spectra *are not* observed in sample (2), in which the GaAs layer has much lower free-electron density. The resonant profile shown in the inset of Fig. 2.55, for the better resolved depolarized $z(x'y')\bar{z}$ spectrum, indicates a well-defined resonance for photon energies close to the E_1 optical gap of Ge. This resonant behavior unambiguously identifies the

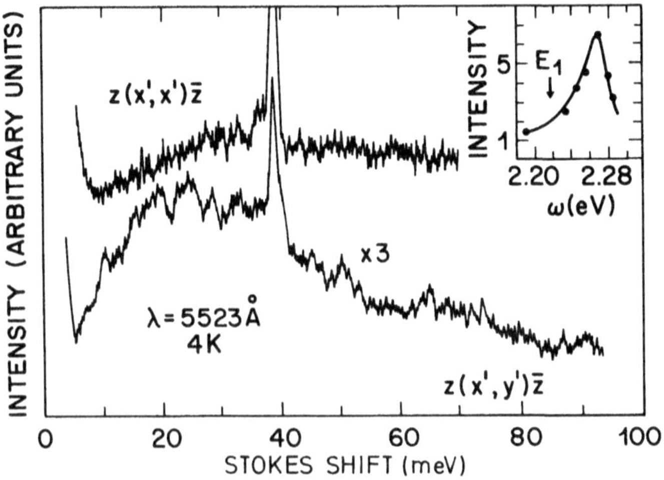

Fig. 2.55. Resonant light scattering from a selectively doped Ge-GaAs heterostructure (sample 1). Typical polarized and depolarized spectra are shown. The sharp band at 38.5 meV is the optical phonon of the Ge layer. The broad features are assigned to excitations of the 2-dim electron system. The inset shows the resonant profile of the integrated intensity measured in $z(x'y')\bar{z}$ spectra. From [2.63]

scattering as associated with the Ge layer. In addition, the fact that they are observed only in the sample in which the GaAs layer is heavily n-type indicates that the spectra involve excitations of electrons that have transferred from the adjacent n-GaAs. Such a transfer of charge and the subsequent formation of an *accumulation layer* at the Ge side of the interface is expected because the bottom of the conduction band of Ge lies ~0.2 eV below that of GaAs [2.168].

This interpretation was supported by a calculation of the line shape of the $z(x'y')\bar{z}$ spectra assigned to spin-density fluctuations. The inset to Fig. 2.56 shows the energy level scheme calculated in real space for the accumulation layer. The most important result of the calculation is the prediction of a *single* occupied bound state (E_0) at 18 meV below the bottom of the conduction band of Ge. This is taken as an indication that the spectra originate from transitions from the 2-dim electron system to a 3-dim free-electron-like continuum. To provide further support for this interpretation, the $z(x'y')\bar{z}$ spectrum was fitted to a broadened 1-dim density of states:

$$N(\omega)=\mathrm{Re}\,\{\omega-\omega_0+\mathrm{i}\Gamma\}^{-1/2}. \quad (2.141)$$

This joint density of states occurs when the 2-dim and 3-dim subbands have the same effective masses for in-plane motion. As can be seen in Fig. 2.56, there is good agreement between calculated and measured spectra. The parameters of the fit are $\omega_0=15$ meV (close to the energy calculated for the bound state) and Γ

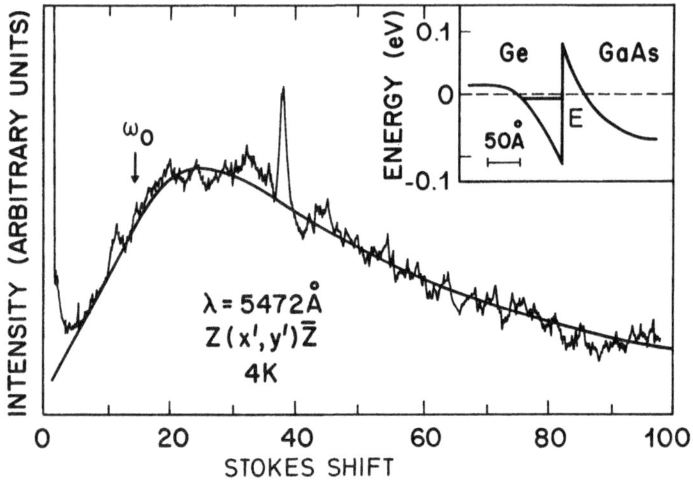

Fig. 2.56. Depolarized spectrum from the selectively doped Ge-GaAs heterostructure (sample 1). $\omega_L = 2.266$ eV, near the maximum in resonant enhancement. The solid line is a fit with (2.141). The inset shows the calculation of the conduction band edge at the heterojunction. *E* represents the only bound state. Stokes shift in meV. From [2.63]

$= 16.5$ meV. The large value of Γ is consistent with the rather low carrier mobilities in these heterostructures.

2.5.4 Periodic GaAs Doping Multilayer Structures

a) Description of the System

An important application of resonant inelastic light scattering by 2-dim electrons is the investigation of periodic doping multilayer structures, so-called "nipi" crystals. This type of semiconductor superlattice was first proposed and analyzed theoretically by *Döhler* [2.169]. It is composed of a periodic sequence of ultrathin n- and p-doped layers of GaAs, possibly with intrinsic layers in between, and exhibits various novel and exciting properties which are caused by purely space-charge-induced potential wells. In Fig. 2.57 the spatial variation of the conduction and valence band edges of a np-superlattice is shown schematically for the ground state and for an excited state. A doping superlattice is simply an alternation of p–n and n–p junctions of an otherwise homogeneous semiconductor. In the ground state the structure contains no free carriers when the concentration of the donors N_D times the thickness of the n-type layer d_n is equal to the number of acceptors N_A times the thickness of the p-type layers d_p (compensated nipi-crystal). The electrons from the donors are attracted by the acceptors in the p-type layers resulting in a periodic rise and fall of the conduction and valence band. The magnitude of this variation depends on the distribution of donors and acceptors and the thicknesses of the n- and p-type

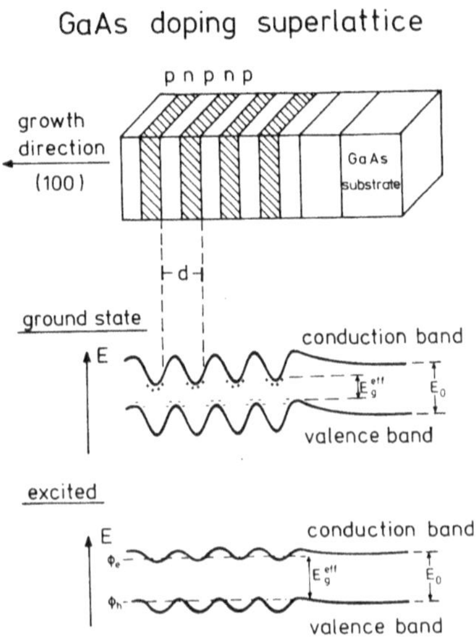

Fig. 2.57. Schematic illustration of a periodic doping multilayer structure. Also shown is the periodic modulation of the conduction and valence band eges in the ground state and under excitation. ϕ_e and ϕ_n are the quasi-Fermi energies in the conduction and in the valence band, respectively

layers. The periodic potential is obtained from Poisson's equation:

$$\frac{d^2V}{dz^2} = -\frac{e}{\varepsilon}\,[N_D(z) - N_A(z)]. \tag{2.142}$$

In the case of a constant doping concentration, integration of (2.142) yields a potential $V(z)$ which is periodic in z [2.169]. The potential fluctuations due to the statistical distribution of the doping impurities in the individual layers have been neglected in the above description.

Such periodic doping multilayer structures exhibit the special feature of a semiconductor with an "indirect gap in real space". The lowest conduction band states are shifted in the direction of the superlattice by half a period from the highest states in the valence band. Consequently, excited electrons and holes are separated in space and may have recombination lifetimes orders of magnitude larger than homogeneous bulk crystals. The reduced effective band gap results in absorption tails far below the band gap of the unmodulated material [2.170, 171]. The long recombination lifetime of excited electron-hole pairs leads to large deviations of electron and hole concentrations from equilibrium, which results in a strong tunability of the effective energy gap and of the absorption coefficient.

In order to calculate the periodic potential for the excited case, the free carriers have to be taken into account. *Ruden* and *Döhler* [2.172] have solved the Schrödinger and Poisson equations self-consistently for a realistic doping structure, where they take into account the Coulomb interaction of the free carriers in the Hartree approximation and the exchange and correlation effects in the density functional formalism. In the case when the superlattice potential is

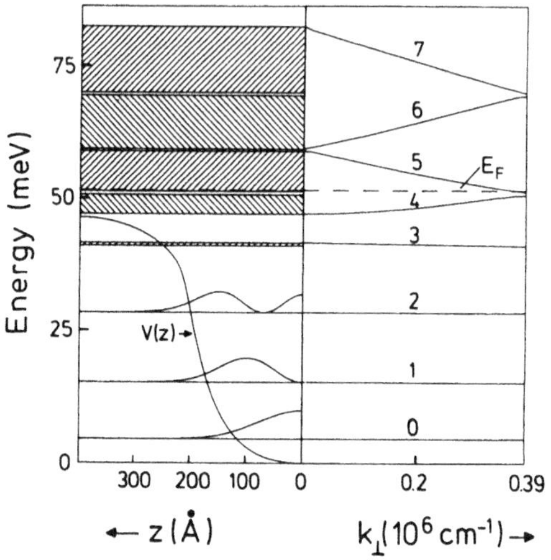

Fig. 2.58. Energy level diagram of an np multilayer structure with $d_n = d_p = 40$ nm and $N_D = N_A = 10^{18}$ cm^{-3}. To the right we show the dependence of the eigenstates on $k_\perp$, to the left the potential and the charge densities of bound states as a function of z together with the allowed energies. From [2.173]

deep, the electrons are strongly bound to one layer and the superlattice potential can be treated as a series of independent potential wells in which the carriers behave quasi two-dimensionally. The binding of the electrons leads to quantization into two-dimensional subbands with dispersion:

$$E_n(\boldsymbol{k}_\parallel) = E_n + \frac{\hbar^2 k_\parallel^2}{2m}, \tag{2.143}$$

where E_n is the bottom of the nth subband and $k_\parallel$ is the wavevector parallel to the layers. The highly excited case, in which so many electrons are created that states with energy near or even above the top of the superlattice potentials are occupied, has been treated by *Zeller* and coworkers [2.173]. In this case the two-dimensional nature of the system is less pronounced, since electrons can easily tunnel between the wells. Results of theoretical calculations for a highly excited sample with $d_n = d_p = 40$ nm and $N_D = N_A = 10^{18}$ cm^{-3} are shown in Fig. 2.58 [2.173]. The left side shows the self-consistent potential $V(z)$, the energy levels for $k_\parallel = 0$ and the corresponding charge distributions. The right side shows the dispersion $E_n(k)$ in the direction perpendicular to the layers.

In order to grow such doping superlattices, a crystal growth technique with the possibility of producing exactly tailored doping profiles is necessary. In recent years rapid progress has been made in obtaining abrupt doping profiles in GaAs using molecular beam eptitaxy (MBE) [2.174]. With this technique *Ploog* and coworkers [2.170] were recently able to grow GaAs doping superlattices of the required layer smoothness and the desired doping profile. The first experimental evidence for the validity of the basic theoretical ideas was the observation of absorption tails extending far into the gap of the pure material [2.170, 171]. Shortly thereafter followed the direct observation of the tunable effective band gap and the quantization of photoexcited carriers in space-charge-induced potential wells using photoluminescence and resonant inelastic light scattering experiments [2.62]. Very recently both single-particle and collective intersubband excitations concomitant with the transition from quasi two-dimensional to quasi three-dimensional behavior have been reported [2.173, 175]. In the following these results are discussed in more detail.

b) Tunable Effective Energy Gap Photoluminescence

The tunability of the effective band gap of various periodic GaAs doping multilayers has been demonstrated unambiguously by *Döhler* et al. [2.62] and *Jung* et al. [2.176] as well as by *Zeller* et al. [2.173] using photoluminescence experiments. The measured luminescence energy shows a strong dependence on the excitation intensity. Photoluminescence spectra obtained with a nipi-crystal having $N_D = N_A = 10^{18}$ cm^{-3} and $d_n = d_p = 40$ nm are shown in Fig. 2.59 [2.62]. The position of the asymmetric luminescence line is found to shift strongly as a function of excitation intensity. Such behavior has been observed in various samples with different doping concentrations and/or different layer thicknesses.

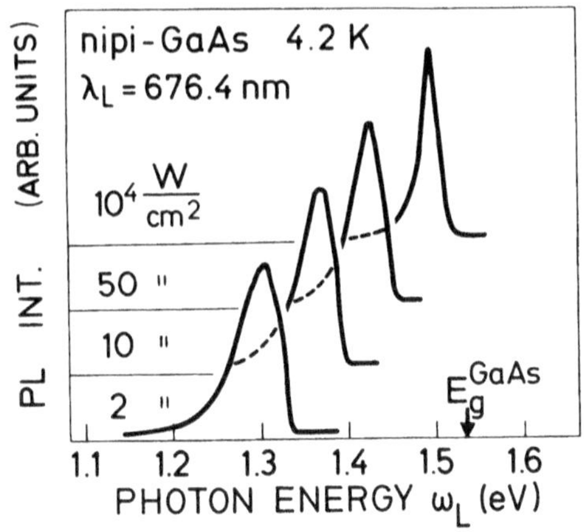

Fig. 2.59. Photoluminescence spectra from a GaAs doping superlattice for several exciting power densities. From [2.62]

This dominant luminescence process has been interpreted as the recombination of electrons in the conduction subbands with holes in the acceptor impurity band [2.62, 173, 176], i.e., transitions across an indirect gap in real space. The recombination energies reflect the effective energy gap of the studied nipi crystals which depends strongly on the excitation intensity or on the concentration of photoexcited carriers. The position of the photoluminescence line consequently also reflects the carrier concentration in the sample. *Döhler* et al. [2.62] and *Zeller* et al. [2.173, 175] made use of this carrier concentration for the interpretation of resonant inelastic light scattering experiments discussed below.

c) Single-Particle and Collective Excitations

Resonant inelastic light scattering experiments similar to those described in Sects. 2.5.1–3 have been used to get direct information on the quantization of photoexcited carriers in doping superlattices [2.62, 173, 175]. Spin-flip single-particle intersubband excitations have been studied using different power densities of the $\omega_L = 1.916$ eV laser line of a Kr^+ laser. In order to obtain information on the carrier density, the photoluminescence spectrum has been measured simultaneously with each Raman spectrum. At low excitation intensities (below 10^3 W cm^{-2}), several distinct peaks have been observed on top of a hot luminescence background (Fig. 2.60). These peaks have been identified as $\Delta = 1$, $\Delta = 2$, and $\Delta = 3$ intersubband transitions of photoexcited electrons in the conduction band of the nipi superlattice. $\Delta = 1$ stands for all possible transitions between nearest subbands, $\Delta = 2$ between second nearest, and so on. Consequently, each peak consists of several transitions between different subbands depending on the quasi-Fermi energy. As shown in Fig. 2.60, similar experiments have been performed using different samples with different superlattice periods and/or doping concentrations [2.173]. The observation of spin-flip single-particle intersubband transitions is taken as direct evidence for the quantization of photoexcited carriers in purely space-charge-induced potential wells in periodic doping multilayer structures.

The simultaneously measured photoluminescence allows a correlation of the experimentally determined subband splittings with the carrier concentrations.

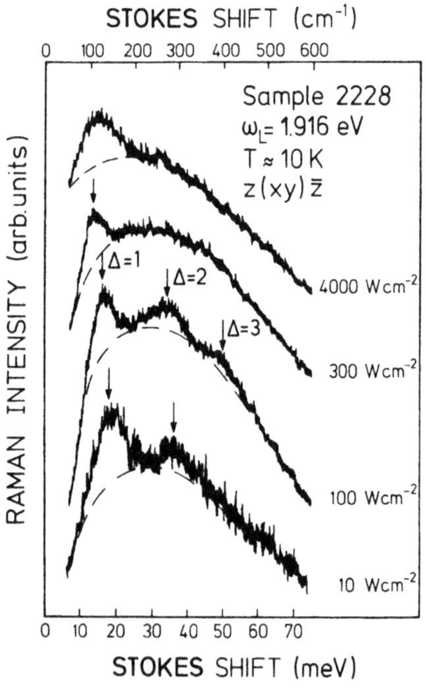

Fig. 2.60. Single-particle scattering by intersubband electron excitations in GaAs nipi-structures for several incident laser power densities. $\Delta = 1$ represents transitions between nearest levels, $\Delta = 2$ between second-nearest and $\Delta = 3$ between third-nearest levels. From [2.173]

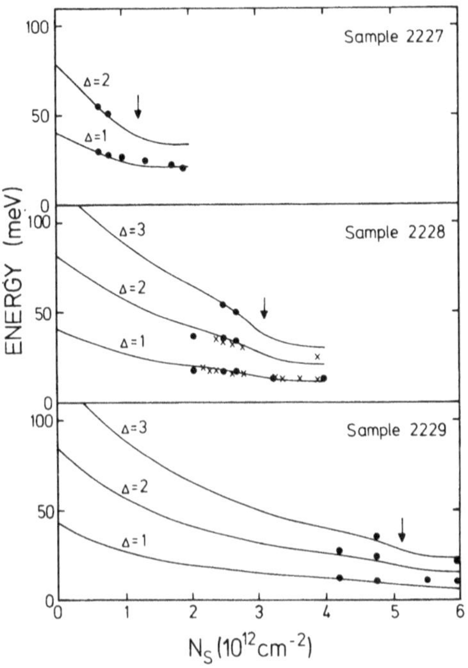

Fig. 2.61. Comparison of the energies of the $\Delta = 1,2,3$ peaks (points and crosses) of three different GaAs nipi-structures with self-consistent calculations of the energy separations between first, second and third-neighbor intersubband transitions (solid lines). For carrier densities higher than those marked by arrows, the Fermi energy lies above the top of the potential wells. From [2.173]

The splitting energies have been compared with the results obtained by means of self-consistent subband calculations [2.62, 173]. This is shown in Fig. 2.61 for three different samples. The solid lines labeled $\Delta = 1$, $\Delta = 2$ and $\Delta = 3$ are the theoretical results of averaging over all possible subband transitions. They have been obtained by weighting the excitation energies by the number of electrons which can make the transitions. The energies obtained in this way are in excellent agreement with experiment.

Zeller and coworkers [2.173, 175] also investigated the change from quasi two-dimensional to quasi three-dimensional behavior of the photoexcited electrons in the highly excited regime. The arrows in Fig. 2.61 mark the carrier concentration where the Fermi energy reaches the top of the potential wells. At these high densities the subband splittings get smaller and the higher occupied states show considerable dispersion in the direction perpendicular to the layers (see also Fig. 2.58). This has been observed by studying the single-particle excitations in the highly excited case. While at low laser-power densities distinct subband transitions are observed, the individual peaks merge at high power densities into one broad single-particle excitation band which has a line shape similar to that obtained for a homogeneously doped GaAs single crystal of comparable carrier concentration (Fig. 2.14).

In order to take the broad quasi-continuum states into account, the imaginary part of the electronic susceptibility Im $\{\chi(q,\omega)\}$, which is directly proportional to the single-particle excitation spectrum, has been calculated

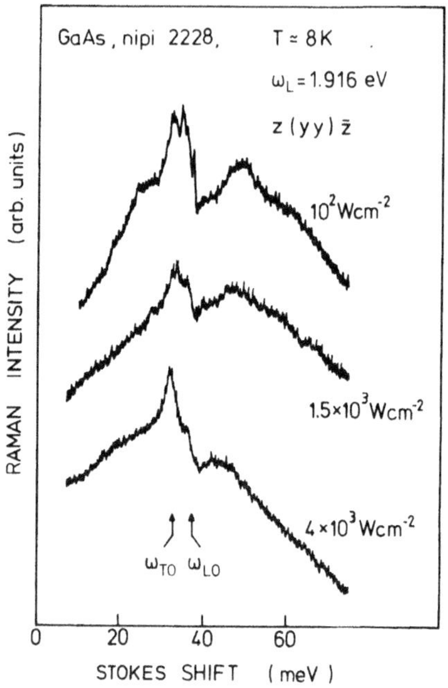

Fig. 2.62. Raman spectra of collective electron excitations coupled to LO phonons in a nipi-structure. From [2.175]

[2.175]. Using realistic parameters for the scattering wavevector, a strong transfer of oscillator strength from the sharp intersubband transitions to the broad continuum was found when the carrier concentration was sufficiently increased. This fact is borne out by the experiments.

Raman spectra of collective excitations, which in principle contain information on depolarization shifts and Coulomb matrix elements, have also been observed recently [2.175] (Fig. 2.62). Because of the occupation of several subbands already at relatively low power densities, and the coupling to LO phonons, the observed collective excitations exhibit a complicated structure. At low excitation intensities, three peaks which represent several L_- modes have been observed below the LO phonon mode. A broad L_+ mode has been identified above the LO phonon. The positions of these modes are found to be in good agreement with calculations performed by *Ruden* and *Döhler* [2.172]. At high excitation intensities, the coupled-mode spectra change and look very similar to coupled phonon-plasmon modes of homogeneously doped n-GaAs Again the behavior with increasing power density is concomitant with the transition from a quasi two-dimensional to a quasi three-dimensional electron system.

2.5.5 Metal-Insulator-Semiconductor Structures

Apart from the extensive work on GaAs-based heterojunctions and multilayer structures, resonant electronic light scattering has been used very recently as a tool to investigate surface accumulation layers in metal-insulator-semiconductor (MIS) structures. In these samples the carrier concentration is easily tunable by simply changing the gate voltage. Thus far, results have been reported on MIS structures based on p- and n-type InAs [2.60, 177], InP [2.177, 178] and Si [2.61]. In light scattering experiments, the space-charge field is usually altered by screening due to the photoexcited electron-hole pairs in the space-charge region. This is especially important for the case of inversion layers, where an additional contribution to the total surface charge comes from the depletion of the bulk donor states. The extra field is partly screened by photoexcited carriers. The decrease in depletion field results in an apparent shift of the threshold voltage. This effect has been studied for Si inversion layers by means of low temperature capacitance measurements [2.61, 179]. For the intense laser radiation used in light scattering experiments it was found that the surface-carrier concentration increased by as much as the total depletion charge [2.61]. Consequently, one studies quasi-accumulation layers with quasi-Fermi energies for holes and for electrons. This has to be taken into account when comparing results obtained from light scattering experiments with far-infrared absorption measurements or with calculated subband splittings.

a) Electrons at InAs Surfaces

The first Raman experiments with InAs MIS structures were performed by *Ching* et al. [2.60]. They obtained spectra at $T=77$ K for various positive and negative gate voltages for both n- and p-InAs substrates. At positive gate voltages they observed a narrow peak at 236 cm^{-1} (29.5 meV) just below the peak due to the unscreened LO phonon. This peak disappeared for negative gate voltages, i.e., when there were no electrons at the InAs surface. Their results are shown in Fig. 2.63. The peak at 236 cm^{-1} (29.5 meV) is attributed to scattering by modes involving collective electron-LO-phonon excitations. The experiments were

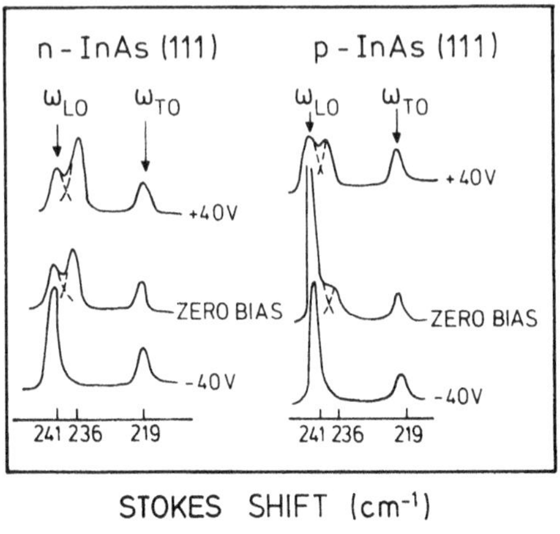

Fig. 2.63. Scattering by LO and TO phonons and by mixed surface intersubband excitations in n and p-type InAs MIS structures with various surface gate voltages and for laser frequencies near the E_1 gap. From [2.60]

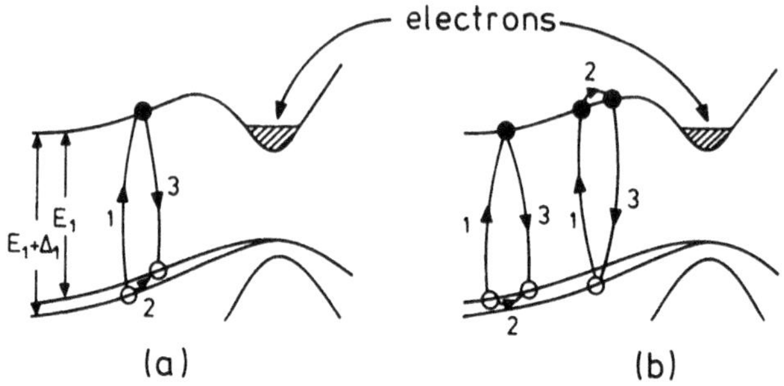

Fig. 2.64. Scattering mechanisms involved in the collective modes of Figs. 2.62, 65. The interaction of the electrons with the collective modes (deformation potential and Fröhlich) is represented by "2". (**a**) Electro-optic and 3-band deformation potential mechanism; (**b**) two-band deformation potential and Fröhlich mechanism

performed with laser energies close to the E_1 gap of InAs. At this resonance, direct optical interband transitions involve no carrier-occupied states in the conduction band (see Sect. 2.3.1). Consequently, carrier-density scattering mechanisms are negligible (i.e., nonresonant).

In polar semiconductors, however, the collective excitations are coupled to LO phonons. Hence, as discussed in Sects. 2.3.1, 2.5.2b, they can take part in resonant scattering (deformation potential, electro-optic, Fröhlich intraband electric-field-induced mechanisms). Resonance enhancement by the LO phonon scattering mechanisms occurs at all optical gaps independent of where the carrier-occupied states are. The transitions involved in these scattering processes are shown in Fig. 2.64. According to [2.101b] such resonant transitions are also possible for pure plasmons.

The mode observed below the LO phonon (Fig. 2.63) is interpreted as the low-frequency coupled LO-phonon intersubband mode. The frequency of this mode is nearly independent of the surface carrier concentration N_s when the subband splitting ω_{01} is much larger than ω_{LO}. As shown by infrared subband spectroscopy and Shubnikov-deHaas measurements [2.180], the subband splitting is already much larger than the energy of the LO phonon at surface-carrier densities $N_s \gtrsim 10^{12}\,\mathrm{cm}^{-2}$. The low-frequency coupled mode was only observed when it was close to the LO phonon. The high-frequency mode was not detected in [2.60]. These experiments have been repeated and verified by *Tränkle* and *Abstreiter* [2.177], who also carefully searched for the high-frequency LO phonon plus intersubband coupled mode. In these measurements a sensitive differential detection technique was applied. The gate voltage was chopped at low frequencies (1.5 to 4 Hz) between flat band and a value corresponding to the desired surface-carrier concentration N_s. The difference Raman signal, as detected with a lock-in amplifier, gives the surface-electron-induced changes of the Raman spectrum directly. This chopping technique also avoids problems arising from long-time threshold drifts which may occur in these samples when a high gate voltage is steadily applied. The results obtained for $N_s \simeq 3.5$

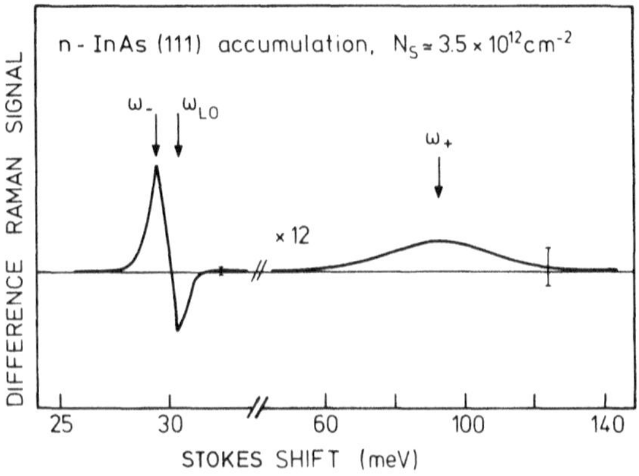

Fig. 2.65. Differential Raman spectrum obtained with a chopped gate voltage (between flat band $V_g = -15$ V and $V_g = +25$ V which corresponds to $N_s = 3.5 \times 10^{12}$ cm^{-2}) at a n-InAs surface. The spectrum shows a positive signal at the ω_- and ω_+ coupled modes (subband excitations plus LO phonons) and a negative signal at ω_{LO}. From [2.177]

$\times 10^{12}$ cm^{-2} are shown in Fig. 2.65. A strong positive signal is observed at $\omega_- = 29.5$ meV which is due to the low-frequency coupled mode discussed above. At the frequency of the LO phonon (30.4 meV) a signal of opposite sign is observed, indicating the decrease in intensity of the pure LO mode when the surface electron layer is turned on. The spectrum in the range 50 meV is shown strongly enlarged. It exhibits a very weak positive signal which peaks around 90 meV. This mode represents the high-frequency coupled LO-phonon intersubband excitation ω_+, equivalent to the depolarisation-shifted resonance $\tilde{E}_{01}$ observed by *Reisinger* and *Koch* [2.180].

The experiments with InAs MIS structures are the only ones performed so far where collective excitations of the surface electron system are observed using laser excitation lines for which the interband transitions do not involve the carrier-occupied states directly. The results of [2.60, 177] demonstrate the wide applicability of resonant inelastic light scattering as a tool for studying electronic properties of surface space-charge layers in semiconductors.

b) Hole Accumulation Layers in Si

The most widely studied surface space-charge layers in semiconductors are accumulation and inversion layers on (100) Si surfaces. An excellent review of these studies has been published recently by *Ando* et al. [2.94]. Despite this extensive work, until recently no successful inelastic light scattering experiments had been published for Si surface layers. This is probably due to the experimental difficulty in fulfilling the resonance condition for carriers in Si. For electrons in the conduction band the direct optical interband transitions are far in the uv and, consequently, are not accessible to conventional Raman scattering setups (Table 2.1). For holes, however, the resonant energy is connected with the direct E_0' gap at $k \simeq 0$, which is about 3.4 eV. Direct interband transitions at this energy involve hole states in the valence band, which makes resonant light scattering via carrier-density mechanisms possible. In the schematic diagram of Fig. 2.66, the

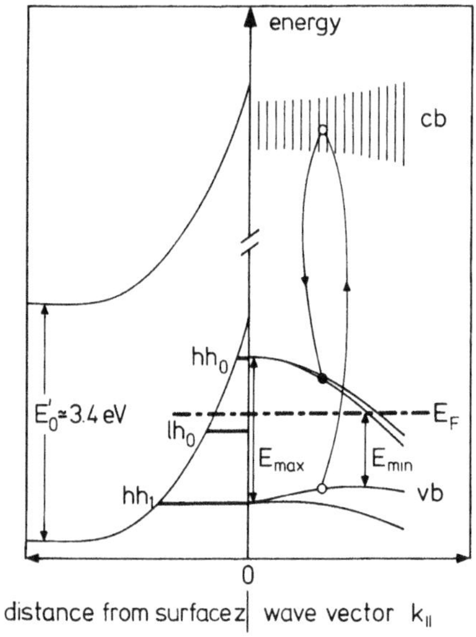

Fig. 2.66. Schematic diagram of a two-step intersubband electronic Raman process for p-type Si. From [2.61]

transitions involved in a resonant two-step scattering process are shown. The $k_{\|}$ dispersion of the hole subbands indicates nonparabolic behavior. The subband structure of hole subbands is considerably more complicated than that of electrons. In the effective mass approximation the subband energies are derived from the Luttinger-Kohn Hamiltonian taking spin-orbit interaction into account. The six p-like valence bands are mixed by the surface electric field. The hole subband structure was calculated by *Bangert* et al. [2.181] and independently by *Ohkawa* and *Uemura* [2.182].

There have been some successful attempts recently to study hole space-charge layers in Si using resonant inelastic light scattering techniques. *Abstreiter* et al. [2.61] observed hole intersubband transitions on (100) Si surfaces using the 3.48 eV emission line of a Kr^+ laser. This laser line is apparently close enough to the E_0' gap of Si to fulfill the resonance condition. These experiments are the first spectroscopic study of hole space-charge layers as a function of energy. In earlier infrared absorption experiments by *Kamgar*, *Kneschaurek* and coworkers [2.171, 183], the excitation energy was fixed. In order to match the ir laser energy with the subband splitting, the surface-carrier concentration was changed by the gate voltage. The resulting absorption spectrum is a complicated function of the carrier concentration and therefore it is difficult to obtain information on line shapes and linewidths using this method.

In [2.61], resonant inelastic light scattering experiments have been performed at various fixed values of N_s set by means of the applied gate voltage. A relatively broad excitation line, whose peak shifts with increasing negative gate voltage to higher energies, was found. A typical spectrum for $N_s = 0$ and $N_s \simeq 3.5 \times 10^{12}\ \mathrm{cm}^{-2}$ ($V_g = -25\ V$) is shown in Fig. 2.67. The peak of the hole excitation line at $N_s \simeq 2.5 \times 10^{12}\ \mathrm{cm}^{-2}$ is at about 40 meV. The observed bands have a

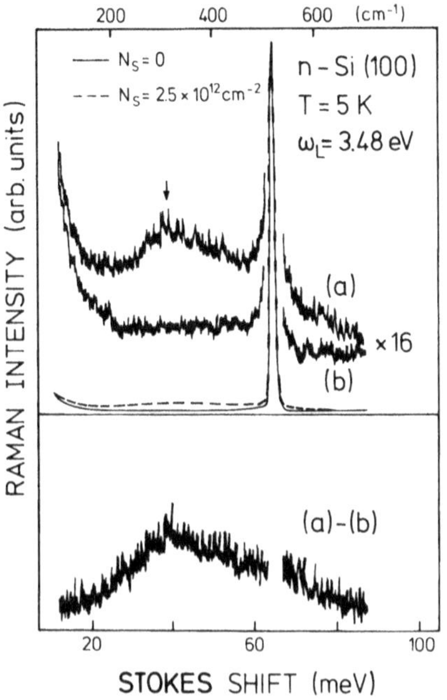

Fig. 2.67. Raman spectra of an MOS structure on Si with ω_L near the E_0' gap with surface hole densities $N_s = 2.5 \times 10^{12}$ cm^{-2} and $N_s = 0$. In the lower part the difference of the two spectra is shown. The arrow indicates the peak of the intersubband excitations. From [2.61]

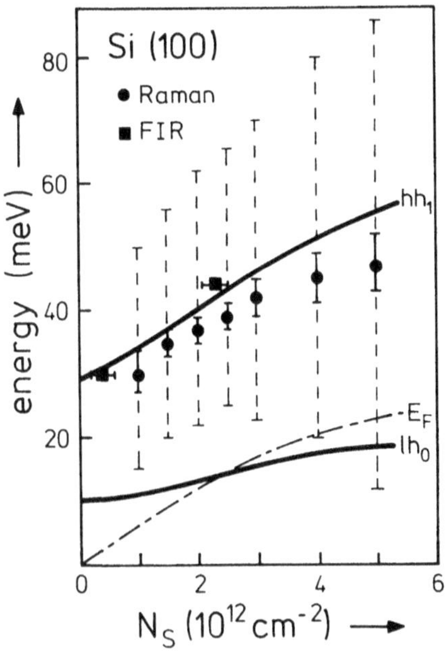

Fig. 2.68. Peak energies and widths of the intersubband excitations of Fig. 2.67 (and similar ones for different N_s) compared with the theoretical calculations [2.181] of the heavy hole ground state and the first excited heavy subband hole hh_1. Also given is the excitation energy from the heavy hole to the light hole ground state lh_0. From [2.61]

relatively sharp onset at the low-energy side and a long tail to the high-energy side. They broaden considerably with increasing N_s. The energy of the peak values has been compared with the subband calculations of *Bangert* et al. [2.181]. It is possible to relate the lower subbands to the heavy hole or light hole valence band by analyzing the Bloch part of the total wave function. *Bangert* has made this analysis for the lowest three subbands. The comparison between theory and experiment (Fig. 2.68) suggests that the observed hole excitation lines are mainly transitions between the heavy hole ground state and the first excited heavy hole subband. These calculations, however, do not include many-body corrections as well as depolarization shifts which should be present in the experiments. The calculations have been performed for inversion layers and thus include the effects of the depletion field on the subband energies. In the experiments this contribution is screened by the photo-excited electron-hole pairs as discussed in the beginning of this paragraph. Considering all these uncertainties, one must conclude that the excellent agreement shown in Fig. 2.68 may be partly fortuitous. The reported experiments, however, are the first ones of this type and it is hoped that both new experimental and theoretical results will lead to a better understanding of the hole subband structure at Si-surfaces.

c) Electron Accumulation Layers in InP

InP has a band structure very similar to that of GaAs. The spin-orbit split $E_0+\Delta_0$ band gap at low temperature is $\simeq 1.6$ eV. At these energies the light scattering mechanisms and selection rules are the same as those extensively studied for GaAs. Until recently, however, no good tunable MIS structures or heterojunctions were available on the basis of InP. Contrary to the case of GaAs, several groups have recently been able to fabricate MIS transistor devices on InP. [2.184–186]. The most extensive studies of the physical properties of surface electron subbands have been performed by *Cheng* and *Koch* [2.187] using mobility, magneto-conductivity and cyclotron resonance measurements. But no spectroscopic information on the subband energies was obtained in these experiments.

Abstreiter et al. [2.178] have performed resonance enhanced Raman scattering experiments using InP MIS structures fabricated in a way similar to that described by *Cheng* and *Koch*. The laser lines chosen are the emission lines of a Kr^+-laser pumped cw oxazine-dye laser, whose energies are in the region of the $E_0+\Delta_0$ gap of InP. Both polarized spectra, obtained with parallel incident and scattered light polarizations, as well as depolarized spectra in which the polarizations are perpendicular to each other have been observed. The depolarized spectra are assigned to spin-flip single-particle excitations while the polarized spectra show features related to the collective coupled LO-phonon intersubband excitations. These assignments are similar to those made for GaAs heterojunctions. In Fig. 2.69 typical spectra obtained for InP n-accumulation layers are shown. The single-particle excitations (Fig. 2.69a) shift to higher energies with increasing positive gate voltage (increasing surface-carrier concentration). In the polarized spectra (Fig. 2.69b) only the low frequency mode (ω_-) is observed clearly. The intensity of the high frequency mode (ω_+) is low and its position overlaps with the tail of the photoluminescence spectrum of the sample.

The measured energies of the coupled modes and single-particle excitations can be satisfactorily compared with self-consistent calculations for this subband system as done in Fig. 2.70. However, the increase of the subband splittings with N_s is found to be smaller than theoretically predicted. Part of this discrepancy is due to the nonparabolicity of the conduction band which has not been taken into account in the self-consistent theory. From the positions of the measured single-particle peaks and the collective modes, the size of the depolarization shift and the Coulomb-matrix elements can be extracted. The experimental results are in good agreement with the values calculated from the theoretical subband wave functions in this system. The experiments on InP surface accumulation layers are

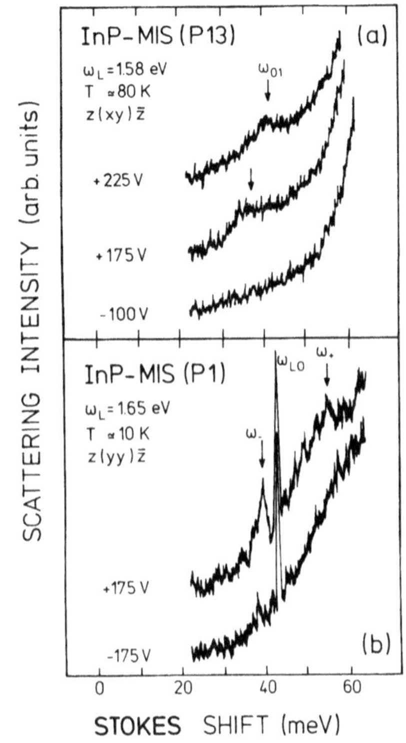

Fig. 2.69a, b. Typical single-particle (**a**) and collective coupled-mode (**b**) spectra obtained for intersubband excitations in MIS accumulation layers of InP. The arrows mark the excitation energies. From [2.178]

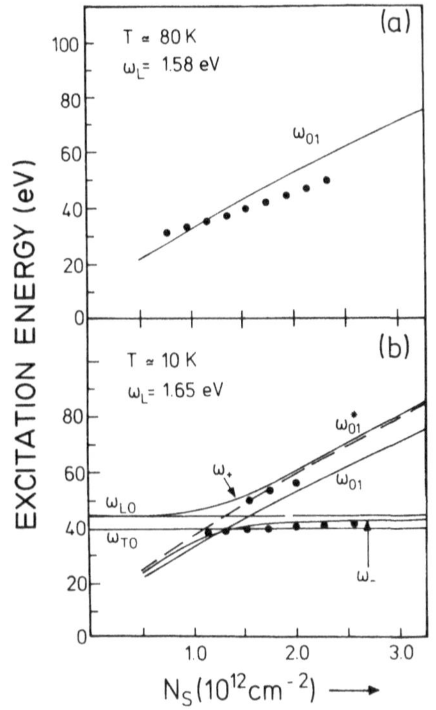

Fig. 2.70. Excitation energies obtained from spectra of the type of Fig. 2.69 as a function of surface density N_s (dots) compared with calculated results. From [2.178]

the first ones which measure both single-particle excitations and collective excitations in a MIS structure with a wide tunability of carrier concentration and subband splitting.

2.6 Barriers on Semiconductor Surfaces

The problems addressed in this section deal with properties of semiconductor surfaces which are strongly influenced by the nature and energy of surface states. In doped semiconductors the carrier concentration near the surface is usually changed due to the presence of space-charge layers. The nature of these regions is determined by the Fermi energy at the surface which depends on the density distribution and character of surface or interface states. The natural space-charge layer consequently reflects properties of the surface. For example, in GaAs exposed to air, the Fermi level is pinned around midgap. Both n- and p-type

GaAs are therefore depleted towards the surface. It has been shown, however, that no Fermi level pinning exists at clean, ultrahigh-vacuum cleaved (110) surface of GaAs [2.188]. In other semiconductors, such as InAs or ZnO, one finds natural surface accumulation, depletion or even inversion layers, depending on the chemical treatment [2.189–191].

For accumulation layers the Fermi level moves deeper into the conduction band in n-type semiconductors. Inversion layers, on the other hand, are characterized by the bulk minority carriers which dominate at the surface. They are separated from the bulk by a depletion layer. Carriers at those surfaces form a quasi two-dimensional system. The electronic structure of such layers, especially the quantization perpendicular to the surface, has been discussed in Sect. 2.5, together with the electronic properties of multilayers and superlattices. Here we restrict the discussion to the nature of space-charge layers of clean and dirty semiconductor surfaces. The understanding of Fermi level pinning by surface states is important in technological application of Schottky barrier and metal-insulator-semiconductor devices. In the following sections we discuss a few inelastic light scattering experiments which have been used to gather information on semiconductor surface space-charge layers.

2.6.1 Electric-Field-Induced Raman Scattering

Electric-field-induced Raman scattering by phonons was studied in paraelectrics such as $KTaO_3$ and $SrTiO_3$ using externally applied fields [2.192–194]. The effect of an electric field on the scattering of electromagnetic radiation by optical phonons in centrosymmetric crystals has been discussed by *Burstein* et al. [2.195].

In polar crystals the applied electric field produces a relative displacement of the atoms in the unit cell and deforms the charge distribution of the atoms. The center of inversion is removed, therefore the conventional selection rules are violated. Raman scattering by normally forbidden optical phonons can become allowed. The so-called atomic displacement mechanism is usually responsible for electric-field-induced scattering by optical phonons in paraelectric crystals.

At the surface of semiconductors, an internal electric field often exists which is associated with a surface space-charge layer. The depletion layer, for example, can be described in the framework of a Schottky barrier model. In this approximation the dependence of the surface electric field on the barrier height ϕ_B and on the carrier concentration N is given by

$$F_s = \left(\frac{2\phi_B N}{\varepsilon_S \varepsilon_0}\right)^{1/2}, \qquad (2.144)$$

where ε_S is the static dielectric constant of the semiconductor. Unlike in the case of an externally applied field in insulators, here the field is not constant over the

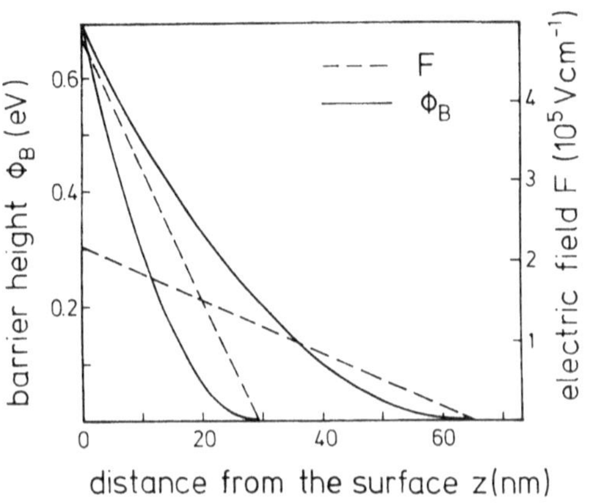

Fig. 2.71. Potential and field distribution at a GaAs surface barrier for a fixed barrier height $\phi_B = 0.7$ eV and for two different bulk electron concentrations $N_e = 2 \times 10^{17}$ cm^{-3}(larger penetration) and $N_e = 1 \times 10^{18}$ cm^{-3} (smaller penetration)

depletion layer width z_d, given by

$$z_d = \left(\frac{2\varepsilon_S \varepsilon_0 \phi_B}{e^2 N} \right)^{1/2} . \tag{2.145}$$

The potentials and the electric fields of depletion layers in GaAs as calculated for a fixed barrier height ϕ_B, and two different carrier concentrations N_e are shown in Fig. 2.71. The electric field is strongest at the surface and decreases linearly to zero over the width z_d. Electric-field-induced Raman scattering, therefore, becomes important when the semiconductors studied are opaque, i.e., when the light penetration depth is smaller or comparable to the depletion width.

The first evidence for light scattering induced by the internal electric field of a space-charge layer has been reported by *Pinczuk* and *Burstein* [2.116, 196]. They studied the resonance-enhanced Raman scattering by $q \simeq 0$ LO phonons in backscattering geometry from (110) surfaces of several n-type InSb single crystals. The photon energies used in those experiments were close to the E_1 gap of InSb, where the optical penetration depth is of the order of 500 Å. Scattering by LO phonons is forbidden in the applied configuration for InSb. *Pinczuk* and *Burstein* did observe an LO phonon mode which increased in intensity with increasing carrier concentration. In a surface depletion layer the electric field F_s is proportional to $N^{1/2}$ (2.144). This results in a stronger electric-field-induced LO scattering intensity in highly doped samples for a fixed barrier height, in qualitative agreement with the observations.

The atomic displacement mechanism is thought to have played little role in those experiments as the enhancement only applies to LO phonons. *Gay* et al. [2.197] discussed a different effect of an electric field specific to scattering by LO phonons. This mechanism is related to the Franz-Keldysh effect [Ref. 2.64, Sect. 2.2.8]. The excited electron-hole pairs (excitons) are spatially separated in the electric field. This mechanism is believed to be responsible for the surface-field-induced resonant Raman scattering by LO phonons in III–V compound semiconductors. The Franz-Keldysh-type theory leads to the result that the scattering intensity near the E_1 gap with scattering wavevector q parallel to the electric field is proportional to the square of the field in the low field region [Ref. 2.64, Eq. (2.227)]. At very strong electric fields the scattering intensity is expected to saturate or even decrease.

Electric-field-induced resonant Raman scattering by LO phonons has also been observed in cubic IV–VI semiconductors such as PbTe [2.198]. In these crystals all the atoms are at a center of symmetry; consequently, the $q \simeq 0$ optical-phonon modes are Raman inactive. Using metal-semiconductor interfaces, surface electric fields can be produced. The observed LO phonon modes in backscattering geometry of so-prepared surfaces show a resonance enhancement at the E_2 optical gap. It is believed that both the atomic displacement and the Franz-Keldish-type mechanisms are responsible for the field-induced Raman scattering in the (paraelectric) IV–VI semiconductors [2.198].

It is often difficult to separate electric-field-induced Raman scattering from other types of forbidden LO-phonon scattering, especially when the exciting photon energies are in resonance with direct gaps. One such mechanism is scattering by LO phonons with finite q-vectors. Intraband scattering of electrons or holes becomes allowed for finite q-vectors of the LO phonons via the Fröhlich interaction [Ref. 2.64, Sect. 2.2.8]. It has been shown that in this case the Raman tensor is proportional to the second derivative of the electric susceptibility with respect to the photon energy [Ref. 2.64, Eq. (2.218)]. In [2.104] it has been proposed that also the presence of impurities in the crystal induces forbidden scattering (Fig. 2.16). The electron is scattered elastically by the impurities so as to provide the necessary momentum change to make the electron LO-phonon coupling allowed. *Shand* et al. [2.199] first tried to separate the electric-field-induced part of forbidden LO phonon scattering in CdS from the other mechanisms, especially the wavevector-dependent scattering. They measured the difference in forbidden LO scattering with and without an applied electric

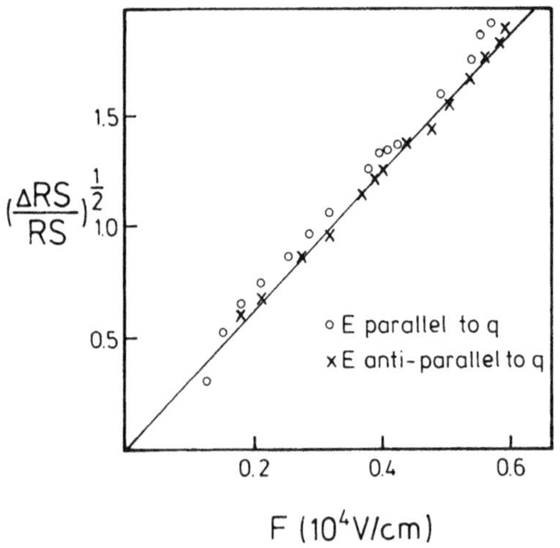

Fig. 2.72. Dependence of the square root of the field-induced Raman signal on the magnitude of the applied field in CdS. From [2.199]

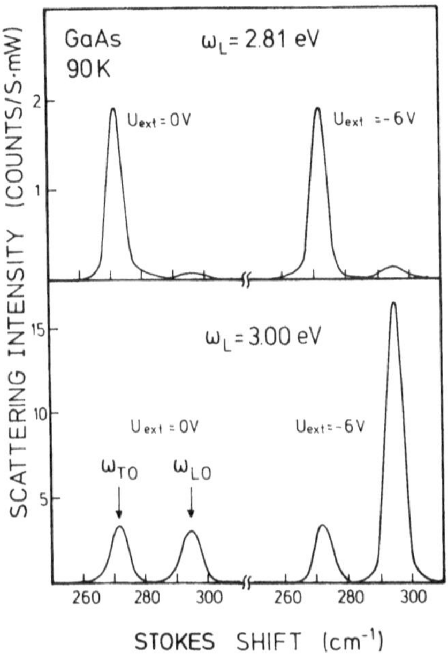

Fig. 2.73. Dependence of the Raman scattering by phonons in GaAs on an applied Schottky barrier potential for two laser frequencies. From [2.200]

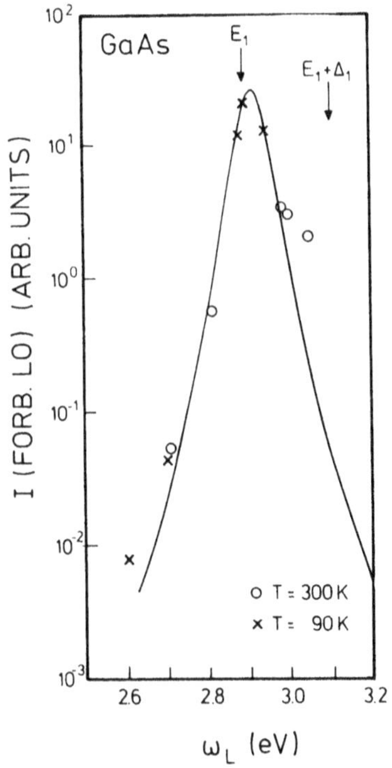

Fig. 2.74. Resonance of the forbidden, field-induced scattering by LO phonons in GaAs in the neighborhood of the E_1, $E_1+\Delta_1$ gaps. From [2.200]

field. The scattering induced by the field was indeed found to be proportional to the square of the applied field. Their results are shown in Fig. 2.72. *Trommer* et al. [2.200] used a Schottky barrier arrangement with various n-type GaAs (110) surfaces of different doping concentrations in order to quantitatively separate the three processes: electric-field-induced scattering, impurity induced scattering and scattering at finite q via the Fröhlich interaction. In those experiments the surfaces were covered with semitransparent Ni-films in order to apply an external voltage. The electric field was determined by means of barrier capacitance measurements. At zero bias the Schottky barrier height was found to be $\simeq 0.8$ eV. The excitation energies $\hbar\omega_L$ were chosen close to the E_1 gap of GaAs where the absorption coefficient $\alpha = 7\times10^5$ cm^{-1}. Thus, except for the very highly doped samples, the optical skin depth was always much smaller than the Schottky barrier width z_d. The experimental results show a field independent scattering intensity due to the other forbidden scattering mechanisms, in reasonable agreement with theory. The electric field induced part follows the predicted F^2 behavior at small electric field. At high electric field a flattening of the I(LO) vs F^2 response has been observed. This is in qualitative agreement with

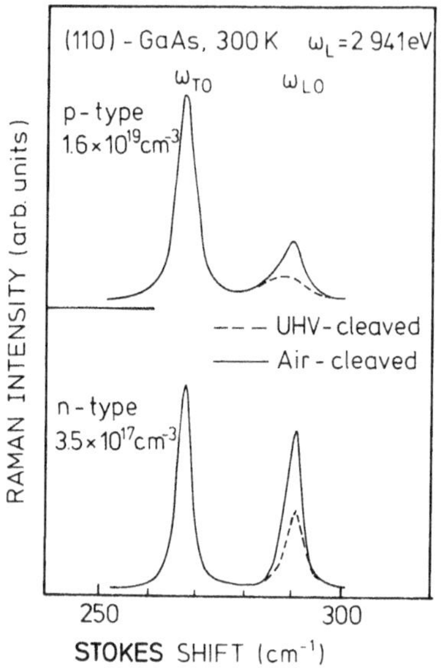

Fig. 2.75. Electric-field-induced forbidden LO-phonon scattering in air exposed GaAs as compared with cleaved GaAs in which the field is absent. From [2.40]

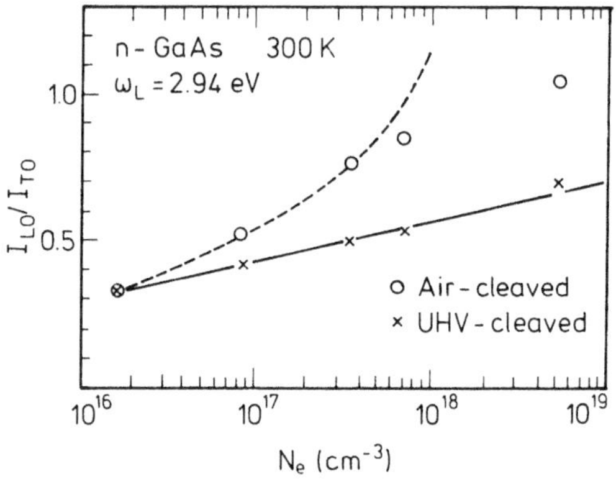

Fig. 2.76. Intensity of the forbidden LO-phonon scattering in air-cleaved and UHV-cleaved GaAs versus bulk electron concentration N_e. From [2.41]

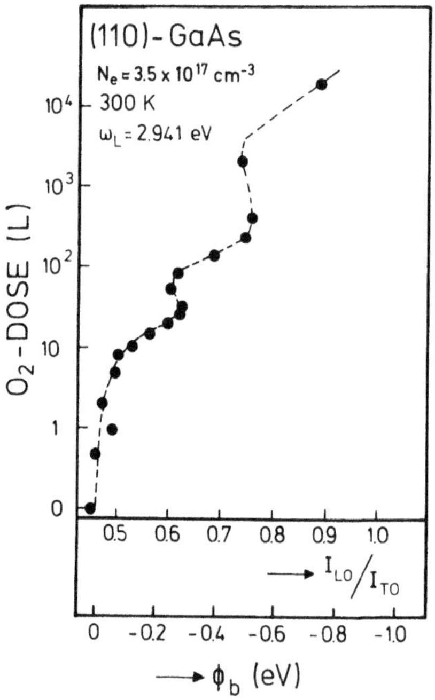

Fig. 2.77. Dependence of the forbidden LO scattering intensity on oxygen exposure and equivalent dependence of the surface barrier ϕ_B in GaAs. From [2.201]

the calculations of *Gay* et al. [2.197]. The screening of the surface depletion layer due to the photoexcited carriers, however, has not been ruled out. In Fig. 2.73 Raman spectra taken from a sample with $N_e = 7.45 \times 10^{17}$ cm^{-3} are shown with 0 V and -6 V bias at the Schottky barrier. To eliminate field independent terms, the differences between LO scattering intensities with bias voltage on and off are plotted versus laser-excitation energy in Fig. 2.74. The resonance enhancement seems to be much stronger than that expected for the q-dependent scattering, in agreement with the discussion in [Ref. 2.64, Sect. 2.2.8d].

Recently, electric-field-induced Raman scattering has been used to study GaAs surfaces cleaved in ultra-high vacuum and the formation of a surface depletion layer under oxygen exposure [2.40, 41, 201]. Other techniques like contact-potential-difference measurements [2.188], uv-excited photoemission spectroscopy [2.202] and photoemission yield spectroscopy [2.203] have shown that on the clean "well-cleaved" surface of GaAs, there are no surface states in the band gap and consequently no surface band bending. *Stolz* and *Abstreiter* [2.40] have used the measurement of symmetry forbidden LO-phonon Raman scattering, resonantly excited at the E_1 gap of GaAs, to detect by means of its surface electric field dependence changes in the surface band bending. In Fig. 2.75 results are shown for two different samples (n- and p-type). In the backscattering configuration used, TO phonon scattering is allowed and is independent of the electric field. Therefore, the spectra have been normalized to the TO intensity. The observed difference of the LO intensity between air-cleaved and clean UHV-cleaved samples is thought to be due to the surface electric field. In Fig. 2.76 the ratio I_{LO}/I_{TO} is shown for different n-type GaAs samples.

This technique has been applied to study the influence of small oxygen exposures on surface band bending [2.201]. Starting with a clean surface, cleaved in UHV, a stepwise increase of the LO intensity is observed with increasing oxygen exposure (Fig. 2.77). This has been analyzed as a stepwise shifting of the surface Fermi level which demonstrates that several discrete physical changes and/or surface chemical reactions take place during oxygen exposure at coverages less than 1 % of a monolayer. This work has been extended recently by *Schäffler* and *Abstreiter* [2.204]. They studied the very early stages of Schottky barrier formation for several adsorbates (O_2, CO, Ag) at different temperatures as well as the influence of laser radiation on the chemisorption. Their results show that part of the step-like behavior found in earlier work might be due to laser-induced chemical surface reactions.

Recently, the question has been solved of why an LO-like mode still exists, even under flat-band conditions where, in principle, the LO phonon should be screened by the bulk carriers present up to the surface. A careful study of the line shape leads to the conclusion that the observed structure is due to the wavevector dependence of coupled phonon-plasmon modes, as discussed in Sect. 2.4 ([2.41]). It has been shown [2.41] that the L_--mode reaches the unscreened LO phonon frequency for q-vectors larger than the Thomas-Fermi screening wavevector. In the presently discussed experiments a large scattering wave-

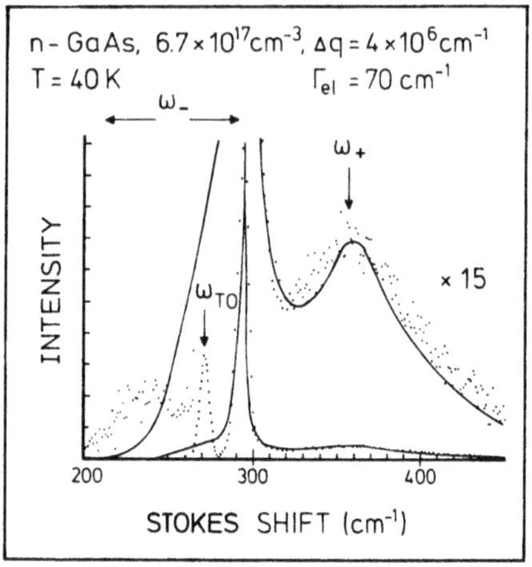

Fig. 2.78. Scattering by unscreened LO phonons in n-type GaAs under flat band conditions due to spread in the wavevector produced by strong absorption of the light. From [2.41]

vector is provided by the strong absorption of the incident and scattered light ($\Delta q = 2\pi\alpha \simeq 4 \times 10^6\ \mathrm{cm}^{-1}$). In Fig. 2.78 an experimental spectrum has been fitted with the relaxation time approximation of the Lindhard-Mermin dielectric function. The charge-carrier concentration N_e determined from this fit is $7 \times 10^{17}\ \mathrm{cm}^{-3}$, in excellent agreement with the measured bulk value. The constant carrier concentration close to the surface directly demonstrates the flat-band condition in well-cleaved GaAs surfaces.

2.6.2 Unscreened LO Phonons and Coupled Modes

The coupling of LO phonons with collective plasma excitations of the carrier system can be used to study the surface depletion width, even at surfaces where scattering by LO phonons is allowed [for example, GaAs (100) surfaces]. The intensity ratio of the unscreened LO phonon, which originates from the surface depletion layer, and the coupled LO-phonon plasmon modes from the bulk directly reflects the depletion layer width. To make use of this information one has to perform Raman scattering experiments with laser excitation lines which have a penetration depth larger than the surface depletion-layer width to be studied. As shown in Sect. 2.4 and in [2.125], the intensity ratio of the L_--coupled phonon-plasmon mode and the unscreened LO phonon from the surface depletion layer in n-GaAs depends critically on the carrier concentration and on the exciting laser energy (Figs. 2.24, 26). In these experiments, wavevector nonconservation is not important because of the relatively small absorption coefficient and the off-resonance condition. Changes in the intensity ratio have also been reported using metal-semiconductor Schottky barriers and externally applied electric fields [2.118].

In Fig. 2.79 spectra obtained from highly doped n-GaAs with a Ni-Schottky barrier on top of the (100) surface [2.205] are shown. The data were obtained

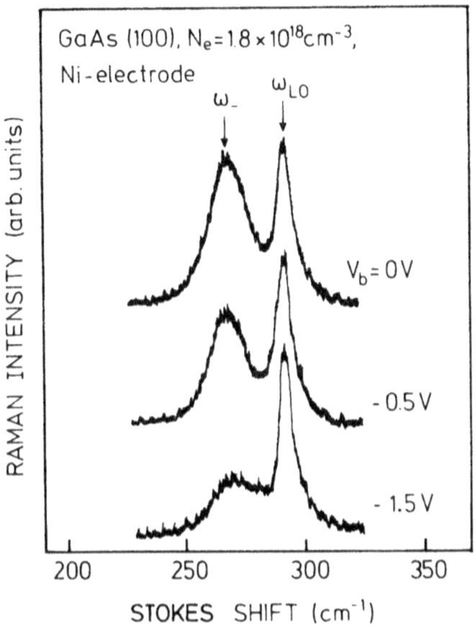

Fig. 2.79. Dependence of the ω_- mode and of the ω_{LO} peak produced in a depletion layer on GaAs with a Ni-electrode, as a function of applied backward bias at 80 K. From [2.205]

with the $\omega_L = 2.41$ eV laser emission line of an Ar^+ laser using backscattering geometry. One observes a strong decrease of the coupled LO-phonon plasmon mode L_- when the backward bias voltage is increased. At the same time, the LO phonon mode gains in intensity due to the increased depletion layer width.

Recently *Pinczuk* et al. [2.206] used this method to obtain quantitative information on surface space-charge layers and Schottky barrier formation in

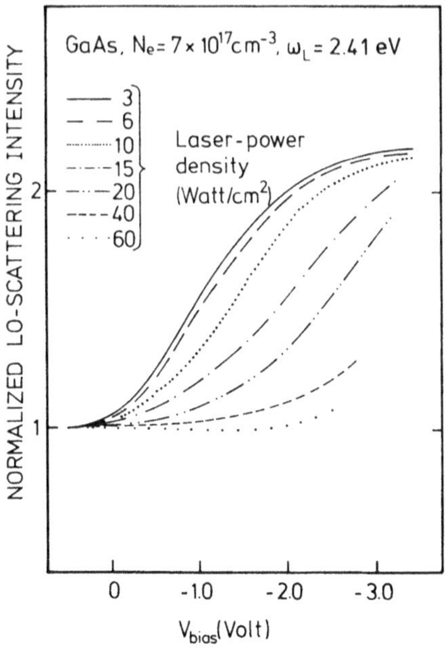

Fig. 2.80. Scattering intensity by LO phonons vs bias voltage at a GaAs Schottky barrier for various incident laser powers. From [2.205]

InP. Their experiments yielded information similar to that found in photoemission and photovoltaic measurements.

In all the experiments just discussed, one has to consider carefully the effect of photoinjected carriers on the intensity ratio. This is usually done by studying the dependence of the spectra on laser power. As an example, in Fig. 2.80, the LO intensity in n-GaAs is plotted versus backward bias voltage for various laser power densities [2.205]. At low power densities (less than 10 W/cm^2) the LO mode increases in intensity with increasing bias voltage and tends to saturate at high V_{bias}. With higher power density the increase is less and less pronounced and vanishes almost completely for more than 60 W/cm^2. This behavior is due to the screening of the depletion field by the photoexcited electron-hole pairs and depends on various parameters like doping-concentration surface-recombination, surface field, and absorption coefficient. Consequently, for a quantitative interpretation of surface barrier heights, a careful study of the effect of photoexcited carriers in each system under investigation is necessary.

2.6.3 Resonance Effects in InAs

Buchner et al. [2.207] used another feature of the resonance behavior of electric-field-induced Raman scattering. They studied A (111) and B $(\bar{1}\bar{1}\bar{1})$ surfaces of InAs at the E_1 gap. In III–V compound semiconductors, the A(111) surface terminates ideally in group III atoms, while the B$(\bar{1}\bar{1}\bar{1})$ surface terminates in group V atoms. It is expected that surface states associated with these atoms lie close either to the conduction band or to the valence band, respectively. Therefore, the space-charge fields on the two surfaces are different. *Buchner* et al. used resonance-enhanced Raman scattering by LO phonons at the E_1 gap of InAs to probe the difference in the character of the two surfaces of various n- and p-type samples in air. In backscattering from (111) surfaces, both the allowed and the forbidden, field-induced or wavevector dependent scattering tensors are diagonal. Therefore they can interfere either constructively or destructively, depending on their relative signs. Another interesting aspect is the fact that the resonance peaks at the E_1 energy gap of InAs are shifted with respect to each other by about 70 meV as shown by *Renucci* et al. [2.208].

The overlap of the two resonance curves, however, is strong enough that one can obtain information on both the relative strengths of the allowed and forbidden scattering as well as about the interference between them. *Buchner* et al. used several laser lines of an Ar^+ and a Kr^+laser to study the resonance behavior of both LO and TO phonon scattering in various n- and p-doped InAs samples. Results for two different p-type InAs samples are shown in Fig. 2.81. R_f and R_a mark the resonance position of forbidden and allowed scattering. Scattering by TO phonons is only due to allowed processes; the corresponding resonance curve is also shown in the figure. The scattering by LO phonons exhibits different features depending on the carrier concentration and on the surface direction. For the sample with the lower concentration, the resonance

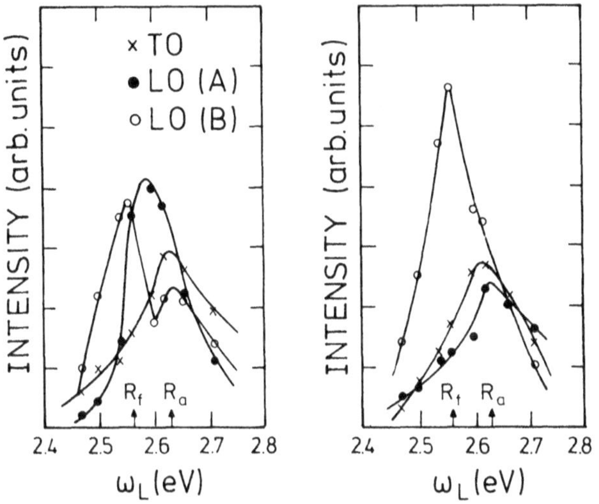

Fig. 2.81. Resonance of the TO and LO phonon scattering at A and B [(111) and $(\bar{1}\bar{1}\bar{1})$] surfaces of two p-type InAs samples showing interferences for the LO scattering between allowed and forbidden scattering amplitudes. R_f and R_a designate the peaks of the forbidden and allowed cross sections, respectively. Left side $N_p = 3.6 \times 10^{17}\,\mathrm{cm}^{-3}$; right side $N_p = 5.8 \times 10^{18}\,\mathrm{cm}^{-3}$

curve has a peak between the forbidden and allowed resonance positions for scattering from the A surface and a dip for scattering from the B surface. The peak is attributed to the constructive interference, while the dip indicates destructive interference. The same behavior is found for n-type samples with relatively low carrier concentrations. From these results it has been concluded that the surface electric field points in opposite directions at the A and B surfaces of both n and p-type InAs. For higher carrier concentration samples, scattering by LO phonons shows strong forbidden resonance at B surfaces. Scattering from the A surfaces, however, shows different behavior for n- and p-type samples. While for n-InAs the forbidden LO scattering is very strong, it shows allowed behavior for p-InAs. This indicates that no surface electric field exists at the A surface of highly doped p-type samples. In n-type samples, on the other hand, both A and B surfaces are depleted towards the surface. This appears to be quite different from the situation at (100) surfaces where one always finds an electron accumulation layer at the surface [2.191].

2.7 Light Scattering in Heavily Doped Silicon and Germanium

Germanium and silicon are nonpolar materials, i.e., their optical phonons at $k \simeq 0$ ($\Gamma_{25'}$ or T_{2g} symmetry) are not infrared active. Consequently, they do not split at $k \simeq 0$ into LO and TO components. There is, in this case, no Fröhlich scattering mechanism. Also, no LO phonon-plasmon coupled modes, of the type discussed in Sects. 2.3, 4, exist. The electron-phonon interaction is of the

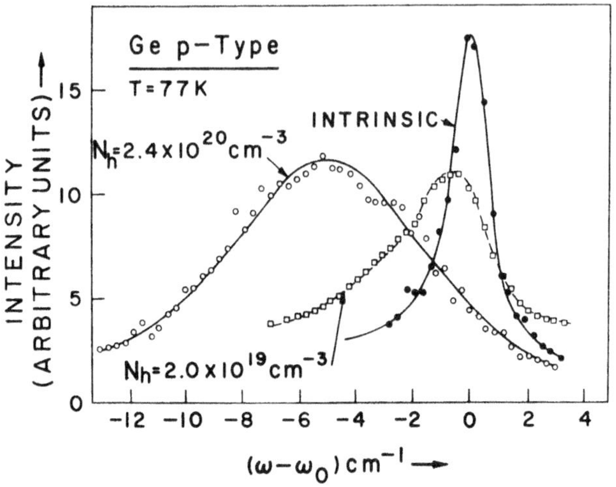

Fig. 2.82. One-phonon Raman spectra of pure and heavily doped p-type Ge showing the real (frequency shift) and imaginary (frequency broadening) parts of the self-energy due to free carriers. From [2.44]

deformation potential type and, to first order, does not depend on the wavelength of the phonons. Plasmon-phonon coupling is negligible but the phonons can couple with electronic excitations in doped semiconductors via the deformation potential mechanism. These effects are usually small but can be *directly* observed in Raman scattering. An example is shown in Fig. 2.82 for p-Ge: the Raman phonon shifts to lower frequencies and broadens [2.209, 210]. These phenomena are interpreted as the formation of a quasiparticle consisting of a phonon dressed with single-particle electronic excitations. The quasiparticle has a self-energy. Its real part is the frequency shift, its imaginary part the additional broadening introduced by the doping.

Similar effects are observed (in ir absorption) for nondegenerate materials when the transitions between the ground and an excited state of an impurity level is close to the energy of the optical phonons [2.211]. Self-energy effects have been also observed due to free holes for the localized vibrational modes of boron in Si [2.212].

Germanium and silicon also offer other interesting features when compared with the III–V semiconductors. Germanium has a multivalley conduction band structure with 4 equivalent minima at the *L*-point (L_1 symmetry, edge of the *BZ* along $\{111\}$, see Fig. 2.7). Silicon has six Δ_1 minima (near edge of the BZ along $\{100\}$), a situation similar to that for GaP and AlSb. Hence, in these materials it should be possible to observe unscreened low-frequency excitations of the type described in Sect. 2.2.8 (intervalley density fluctuations). They were predicted by *Platzman* [2.6, 19] and observed by *Chandrasekhar* et al. in n- and p-type Si [2.23, 47]. Also, n-Ge offers the possibility of observing spin-density-fluctuation-type single-particle excitations (Sect. 2.3.1c) resonant at the E_1 gap.

2.7.1 Scattering by Intervalley Density Fluctuations

The basic physics of these scattering processes is contained in ([Ref. 2.20, p. 23] and [2.67]). These equations can be generalized to n-electron valleys by replacing the scattering form factors $S^{(2)}(q,\omega)$ by

$$S^{(n)}(q,\omega) \simeq \frac{\hbar}{\pi}\frac{q^2}{e^2}\frac{\varepsilon_0}{V}(1-\mathrm{e}^{-\hbar\omega/k_\mathrm{B}T})\cdot\sum_{i<j}^{\nu}\left[\hat{\boldsymbol{e}}_\mathrm{L}\cdot\left(\frac{1}{\boldsymbol{m}_i}-\frac{1}{\boldsymbol{m}_j}\right)\cdot\hat{\boldsymbol{e}}_\mathrm{S}\right]^2\cdot\mathrm{Im}\left\{\frac{\chi_i\chi_j}{\varepsilon(q,\omega)}\right\}$$

$$\varepsilon(q,\omega) = \varepsilon_\infty + \sum_i^n \chi_i\,, \tag{2.146}$$

where $\hat{\boldsymbol{e}}_\mathrm{L}$ and $\hat{\boldsymbol{e}}_\mathrm{S}$ are the polarization vectors of the incident and scattered light which were implicitly included in the masses in (2.67).

In order to evaluate (2.146), we must use the appropriate expression for $\chi_i(\boldsymbol{q},\omega)$. The simplest possible ansatz is the Drude susceptibility:

$$\chi_i(\boldsymbol{q},\omega) = -\frac{N_i e^2}{q^2\omega(\omega+\mathrm{i}\tau^{-1})\varepsilon_0}\,\boldsymbol{q}\cdot\frac{1}{\boldsymbol{m}_i}\cdot\boldsymbol{q}, \tag{2.147}$$

where τ is some average scattering time dominated by the intravalley momentum relaxation time. The scattering efficiency obtained with (2.146, 147) has a rather complicated tensorial dependence as a function of the orientation of $\hat{\boldsymbol{e}}_\mathrm{L}$, $\hat{\boldsymbol{e}}_\mathrm{S}$ and $\boldsymbol{q}$:

$$\frac{\partial^2 S}{\partial\Omega\,\partial\omega} \propto \sum_{i<j}^{\nu}\left[\hat{\boldsymbol{e}}_\mathrm{L}\cdot\left(\frac{1}{\boldsymbol{m}_i}-\frac{1}{\boldsymbol{m}_j}\right)\cdot\hat{\boldsymbol{e}}_\mathrm{S}\right]^2\left(\boldsymbol{q}\cdot\frac{1}{\boldsymbol{m}_i}\cdot\boldsymbol{q}\right)\left(\boldsymbol{q}\cdot\frac{1}{\boldsymbol{m}_j}\cdot\boldsymbol{q}\right). \tag{2.148}$$

Equation (2.148) implies that the dependence of the scattering efficiency on scattering configuration and crystal orientation is given by an *eighth-rank tensor*. The few experimental observations available support a fourth-rank tensor type of dependence, namely, that of (2.148) *after* removing the terms involving $\boldsymbol{q}$ [2.23]:

$$\frac{d^2 S}{d\Omega\,d\omega} \propto \sum_{i<j}^{\nu}\left[\hat{\boldsymbol{e}}_\mathrm{L}\cdot\left(\frac{1}{\boldsymbol{m}_i}-\frac{1}{\boldsymbol{m}_j}\right)\cdot\hat{\boldsymbol{e}}_\mathrm{S}\right]^2. \tag{2.148a}$$

A heuristic explanation of this anomaly has been suggested in [2.23]: in (2.148), the terms involving $\boldsymbol{q}$ should by symmetrized in order to take into account the small mean free path l (note that $ql \ll 1$). In other words, the electrons do not see the scattering $\boldsymbol{q}$ vector l but this $\boldsymbol{q}$ vector smeared out by an isotropic l^{-1}.

The collision time τ in the susceptibility of (2.147) was introduced in (2.23) by replacing ω with $\omega+\mathrm{i}\tau^{-1}$. We have mentioned in Sect. 2.2.5 that this is not the correct way of including a collision time; it leads to a local nonconservation of the charge [2.74]. The correct way of introducing a phenomenological relaxation time is given in (2.35). In the Drude approximation we must replace in this equation for each valley (at $T \ll E_F$)

$$\chi_i^0(q,\omega) = -\frac{N_i e^2}{q^2(\omega + i\tau^{-1})\varepsilon_0}\left(\boldsymbol{q}\cdot\frac{1}{\boldsymbol{m}_i}\cdot\boldsymbol{q}\right)$$

$$\chi^0(q,\omega) = \frac{q_{TF}^2}{q^2} = \frac{1}{q^2}\left(\frac{3N_i e^2}{E_F\varepsilon_0}\right). \tag{2.149}$$

In this manner we obtain [Ref. 2.23, Eq. (25)] which can be rewritten as

$$\chi_i = -\frac{N_i e^2}{q^2(\omega + i\tau^{-1})\,[\omega - i\tau^{-1}\frac{1}{3}(ql_i)^2]\varepsilon_0}\left(\boldsymbol{q}\cdot\frac{1}{\boldsymbol{m}_i}\cdot\boldsymbol{q}\right), \tag{2.150}$$

where l_i is the electron mean free path along the $\boldsymbol{q}$ direction. Typical values of l_i in heavily doped semiconductors are 10 Å, hence $ql_i \simeq 10^{-3}$, and for $\tau \simeq 10^{-14}\ \mathrm{s}^{-1}$ the $(ql_i)^2$ terms in (2.150) are negligible at all frequencies for which the experiments are performed ($\omega > 1\ \mathrm{cm}^{-1}$).

It has been shown in [2.23] that (2.150) leads to cross sections for intervalley density fluctuations which are at least one order of magnitude too low when compared with the experimental ones. There are also some serious difficulties with the ω-dependence of the scattering cross section predicted by (2.146, 150) which should diverge like ω^{-1} for $\omega \to 0$, a fact not observed experimentally. In order to remedy these difficulties, *Ipatova* et al. [2.24] introduced *intervalley* scattering into the problem. The scattering is then due to charge fluctuations between the various valleys. Such fluctuations are negligible for $\omega \neq 0$ if the intervalley scattering time τ_{inter} is much larger than ω^{-1} (the case considered so far). In order to obtain the scattering cross section for this mechanism, we must go back to (2.43) and evaluate the structure factors $S_{ll'}$ with the help of (2.44) and the fluctuation-dissipation theorem. Let us apply to the various valleys, labeled by l, the potentials $V_l(t) = U_l \exp(i\omega t)$.

The response function $\Pi_{ll'}(\omega)$ between the electron density in valley l and the potential in valley l' can then be used to calculate $S_{ll'}$ according to

$$S_{ll'}(\omega) = -\frac{\hbar}{\pi}\,(1 - e^{-\hbar\omega/k_B T})^{-1}\ \mathrm{Im}\,\{\Pi_{ll'}(\omega)\}. \tag{2.151}$$

In order to calculate the response function $\Pi_{ll'}(\omega)$, we write the linearized kinetic equation for the charges δn_l induced at an energy ε by the potentials $V_{l'}$:

$$-i\omega\delta n_l(\varepsilon,\omega) - \sum_{l'} I_{ll'}\delta n_{l'}(\varepsilon,\omega)$$

$$= \frac{\partial f^{(0)}}{\partial E_F}\sum_{l'} I_{ll'}(\varepsilon)\,U_\beta(\omega), \tag{2.152}$$

where $f^{(0)}$ is the Fermi function, E_F the Fermi energy. The collision integral $I_{ll'}$ is the inverse of the corresponding intervalley scattering time for $l \neq l'$. For $l = l'$ it has been defined as

$$I_{ll} = -\sum_{l \neq l'} I_{ll'} \tag{2.153}$$

so as to obtain no response when all U_β's are equal. Equation (2.152) leads to the response function

$$\Pi_{ll'}(\omega) = \int \frac{2d^3\boldsymbol{k}}{(2\pi)^3} \{\delta_{ll'} + \mathrm{i}\omega[-\mathrm{i}\omega\delta_{ll'} - I_{ll'}(\varepsilon)]\}^{-1} \frac{\partial f^0}{\partial E_F}. \tag{2.154}$$

In (2.154) the term in square brackets is a matrix of dimensionality $v = 3$ for n-Si, and the exponent -1 means that the matrix must be inverted. For the case of n-Si we must consider three nonequivalent electron valleys. The collision integral $I_{12} = I_{13} = 2\tau_{\text{inter}}^{-1}$ [the factor of two represents the fact that there are two equivalent $(k_0, -k_0)$ valleys to scatter into] and $I_{11} = -4\tau_{\text{inter}}^{-1}$. We note that only f-type scattering between valleys at right angles can contribute to light scattering. We thus neglect g-type scattering between the k_0 and $-k_0$ valleys. The diagonal and off-diagonal terms of $[-\mathrm{i}\omega\delta_{ll'} - I_{ll'}(\varepsilon)]^{-1}$ are

$$\begin{aligned} [-\mathrm{i}\omega\delta_{ll'} - I_{ll'}(\varepsilon)]^{-1}_{\text{diagonal}} &= \frac{\mathrm{i}}{\omega} + \frac{4\tau_{\text{inter}}^{-1}}{\mathrm{i}\omega(-\mathrm{i}\omega + 6\tau_{\text{inter}}^{-1})} \\ [-\mathrm{i}\omega\delta_{ll'} - I_{ll'}(\varepsilon)]^{-1}_{\text{off-diagonal}} &= \frac{2\mathrm{i}\tau_{\text{inter}}^{-1}}{\omega(-\mathrm{i}\omega + 6\tau_{\text{inter}}^{-1})}. \end{aligned} \tag{2.155}$$

We thus obtain for the response function $\Pi_{ll'}$

$$\begin{aligned} \Pi_{ll} &= \frac{4\tau_{\text{inter}}^{-1}}{(-\mathrm{i}\omega + 6\tau_{\text{inter}}^{-1})} \int \frac{d^3\boldsymbol{k}}{4\pi^3} \frac{\partial f^0}{\partial \zeta} \\ \Pi_{ll'} &= \frac{-2\tau_{\text{inter}}^{-1}}{(-\mathrm{i}\omega + 6\tau_{\text{inter}}^{-1})} \int \frac{d^3\boldsymbol{k}}{4\pi^3} \frac{\partial f^0}{\partial \zeta}, \qquad l \neq l'. \end{aligned} \tag{2.156}$$

The integrals in (2.156) are calculated for a pair of equivalent $(k_0, -k_0)$ valleys. The scattering efficiency thus becomes

$$\begin{aligned} \frac{\partial^2 S}{\partial\omega\,\partial\Omega} &= \frac{\hbar r_0^2}{\pi}\left(1 - \mathrm{e}^{-\frac{\hbar\omega}{k_B T}}\right)^{-1} \frac{2\omega\tau_{\text{inter}}^{-1}}{\omega^2 + \left(\dfrac{6}{\tau_{\text{inter}}}\right)^2} \\ &\cdot \sum_{i<j}^{v}\left[\hat{\boldsymbol{e}}_L \cdot \left(\frac{1}{\boldsymbol{m}_i} - \frac{1}{\boldsymbol{m}_j}\right) \cdot \hat{\boldsymbol{e}}_S\right]^2 \cdot \int \frac{d^3\boldsymbol{k}}{4\pi^3} \frac{\partial f^0}{\partial E_F}. \end{aligned} \tag{2.157}$$

This cross section has the tensorial dependence on orientation found experimentally, i.e., no explicit dependence on $\boldsymbol{q}$. The dependence of the scattering efficiency on ω given in (2.157) is also in agreement with experiment [2.23]. From (2.146, 150, 157) it is easy to see that the ratio of the efficiency of the interband mechanism to that for intraband scattering is only approximately given by (at $T \simeq 0$, $\omega < \tau_{\rm intra}^{-1}$, $6\tau_{\rm inter}^{-1}$)

$$\frac{\sigma_{\rm inter}}{\sigma_{\rm intra}} \simeq \frac{(\hbar\omega)^2}{36 E_{\rm F}(\hbar^2 q^2/2m^*)} \frac{\tau_{\rm inter}}{\tau_{\rm intra}}, \tag{2.158}$$

where m^* is an average effective mass. For typical values of $\omega = 100\ {\rm cm}^{-1} \simeq 0.012$ eV, $E_{\rm F} = 0.1$ eV, $m^* = 0.2$ and $q = 5 \times 10^5\ {\rm cm}^{-1}$, (2.158) becomes

$$\frac{\sigma_{\rm inter}}{\sigma_{\rm intra}} \simeq \frac{1}{5} \frac{\tau_{\rm inter}}{\tau_{\rm intra}}. \tag{2.159}$$

The intervalley scattering time is always longer than its intravalley counterpart (for heavily doped Si a factor of 10 to 100); hence (2.159) shows that the intervalley scattering mechanism is dominant in scattering by intervalley fluctuations.

An attempt to generalize this treatment by including intervalley and intravalley scattering in the form of a diffusion term has been presented in [2.213].

a) n-Type Si

Low frequency scattering due to multivalley electronic excitations has been observed for n-Si with $N_{\rm e} = 1.5 \times 10^{19}\ {\rm cm}^{-3}$ and with 1.5×10^{20} electrons/cm^3. Typical spectra are shown in Fig. 2.83. In this figure, data taken with a $\rm Kr^+$ laser (6471 Å) are presented with and without a uniaxial stress (15 kbar) applied along [001]. The stress has the function of splitting the $[00\zeta]$ valley from the $[0\zeta0]$ and $[\zeta00]$ mates. The $[00\zeta]$ valleys become the lowest and the multicomponent system is, in this manner, transformed into a one-component system (note that $[00\zeta]$ actually means the pair of $[00\zeta]$ and $[00\bar{\zeta}]$ valleys). The low-frequency tail should decrease strongly upon application of the pressure if due to multicomponent carrier scattering. Figure 2.83 shows that this is indeed the case.

The selection rules observed for this scattering are those which correspond to

$$\sum_{i<j}^{\nu} \left[\hat{e}_{\rm L} \cdot \left(\frac{1}{\boldsymbol{m}_i} - \frac{1}{\boldsymbol{m}_j}\right) \cdot \hat{e}_{\rm S}\right]^2 \tag{2.160}$$

for effective mass ellipsoids along the $\{\zeta 00\}$ direction as found in n-Si ($m_{\parallel} = 0.91\ m_0$, $m_{\perp} = 0.19\ m_0$). It is easy to see that this effective mass tensor leads in (2.160) to a Raman tensor of Γ_{12} (i.e., $E_{\rm g}$) symmetry (see Table 2.1 in [2.64]).

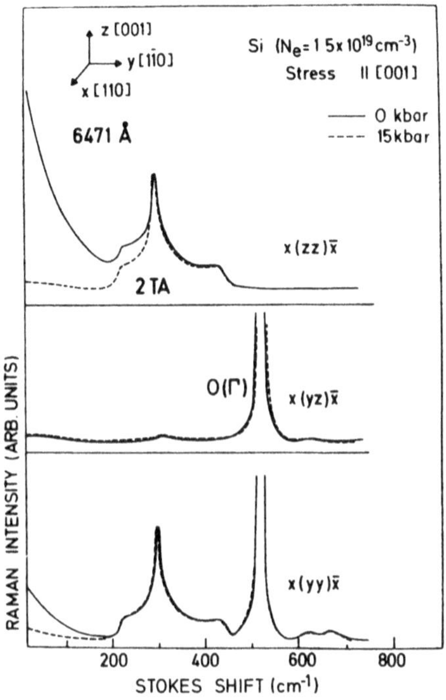

Fig. 2.83. Raman spectra of heavily doped n-type Si showing a low-frequency intervalley fluctuations tail in the configurations which contain the Γ_{12} component $(x(zz)\bar{x}$ and $x(yy)\bar{x})$. Also shown is the effect of uniaxial stress along (001). From [2.23]

Hence, the effect can be observed in scattering configurations in which the Raman phonon ($\Gamma_{25'}$) is absent. It is, however, possible to choose configurations in which both the Raman phonon and the multivalley tail appear [Ref. 2.64, Table 2.2]. This fact is important in order to determine the absolute scattering efficiency of the multivalley tail [2.23].

Figure 2.84 shows fits to the experimental data at large ω based on the intravalley theory of (2.146, 150), assuming $\omega \ll \tau^{-1}$ (dashed line). Under this assumption the ω-dependence of the efficiency on frequency reduces to

$$\frac{\partial^2 S}{\partial\omega\,\partial\Omega} \propto \frac{1}{\omega}\,(1-\mathrm{e}^{-\hbar\omega/k_\mathrm{B}T}). \tag{2.161a}$$

At low frequencies this fit gives higher efficiencies than found experimentally. A good fit (solid line) is obtained with the frequency dependence of (2.157), i.e., with the function

$$\hbar\omega\,[\omega^2+(6\tau_{\mathrm{inter}}^{-1})^2]^{-1}\,(1-\mathrm{e}^{-\hbar\omega/k_\mathrm{B}T})^{-1}. \tag{2.161b}$$

The fit yields $\tau_{\mathrm{inter}} = 5\times10^{-13}$ s. This scattering time is much shorter than the values reported for pure Si ($\sim 10^{-10}$ s), probably due to additional scattering by the ionized impurities.

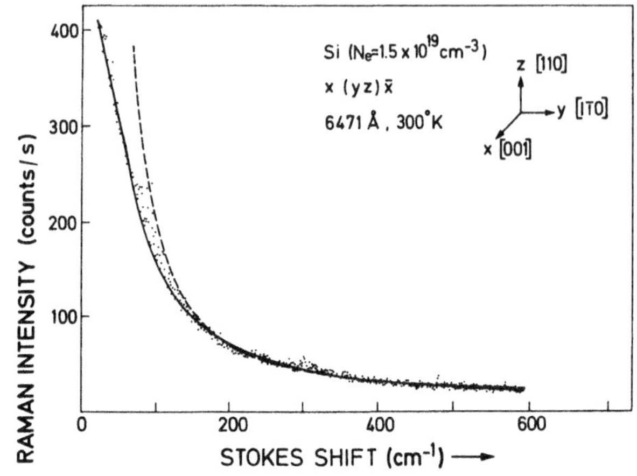

Fig. 2.84. Scattering spectrum due to intervalley fluctuations in n-type Si (dots) fitted with (2.161b) (solid line) and (2.161a) (dashed line). From [2.23]

We should mention that a line shape similar to that of (2.161b) had already been suggested in [2.23] along phenomenological grounds based on (2.150). The scattering efficiency was written in [2.23] as

$$\frac{\partial^2 S}{\partial\omega\,\partial\Omega}=r_0^2\,\frac{\hbar}{\pi}\,(1-\mathrm{e}^{-\hbar\omega/k_{\mathrm{B}}T})^{-1}\left(\frac{N_{\mathrm{e}}}{\nu^2}\right)\sum_{i<j}^{\nu}\left[\hat{e}_{\mathrm{L}}\cdot\left(\frac{1}{\boldsymbol{m}_i}-\frac{1}{\boldsymbol{m}_j}\right)\cdot\hat{e}_{\mathrm{S}}\right]^2 \cdot\frac{\hbar\omega}{A_0[1+(\hbar\omega/E_{\mathrm{a}})^2]\,(2E_{\mathrm{F}}/3)\hbar\tau_{\mathrm{intra}}^{-1}}, \tag{2.161c}$$

where $\nu=3$ is the number of inequivalent valleys and A_0 is a dimensionless phenomenological parameter which, for $N_{\mathrm{e}}=1.5\times10^{19}\,\mathrm{cm}^{-3}$, a laser line at 6471 Å and room temperature, is found experimentally to be equal to 0.01. This value was obtained by comparing the scattering efficiency with that of the Raman phonon calculated with [Ref. 2.64, Eq. (2.134a)] for a Raman polarizability $a=50\,\text{Å}^2$ [Ref. 2.64, Fig. 2.47]. Note, however, that in [2.23], a value of $a=22\,\text{Å}^2$ was used. This value, which is known to be incorrect, leads to a larger $A_0(A_0\simeq0.05)$. By comparing (2.157) with (2.161c), we obtain for $kT\ll E_{\mathrm{F}}$:

$$A_0=6\frac{\tau_{\mathrm{intra}}}{\tau_{\mathrm{inter}}}=6\frac{1.5\times10^{-14}}{5\times10^{-13}}=0.18 \tag{2.162}$$

More careful absolute scattering efficiency measurements are called for in order to clarify the discrepancy between the value of A_0 calculated with (2.162) and the experimental one $A_0=0.01$.

The dependence of the efficiency for scattering by intervalley fluctuations on laser frequency has also been investigated in [2.23]. A dependence $\sim\omega_{\mathrm{L}}^2$ would result from (2.146, 150). Instead, a frequency-independent efficiency is found. This is in agreement with (2.157, 161a). We should point out, however, that this efficiency should increase when approaching the lowest gap which enters into the $\boldsymbol{k}\cdot\boldsymbol{p}$ expression for the effective masses. For n-Si this gap is 4.3 eV, well above the largest photon energy (2.5 eV) used in [2.23].

b) p-Type Si

A low-frequency scattering tail which can be attributed to free carriers has also been observed for p-Si (Fig. 2.85). Contrary to the case of n-Si where the scattering had Γ_{12} symmetry, the symmetry of the corresponding Raman tensor in p-Si is $\Gamma_{25'}$, i.e., the same as for the Raman phonon. In fact, as shown in Fig. 2.85, the intervalley tail may contribute to the Fano interference of the $\Gamma_{25'}$ phonon (Sect. 2.6.2). In p-Si most of the holes are located in the heavy-hole band which is strongly anisotropic (Fig. 2.86). Hence, the angular dependence of the scattering efficiency is given by the generalization of (2.148a) to carriers with masses continuously variable as a function of $\boldsymbol{k}$:

$$\frac{\partial^2 S}{\partial\omega\,\partial\Omega}\propto\iint\left[\hat{e}_{\mathrm{L}}\cdot\left(\frac{1}{m_i(\boldsymbol{k})}-\frac{1}{m_i(\boldsymbol{q})}\right)\cdot\hat{e}_{\mathrm{S}}\right]^2 d\boldsymbol{k}\,d\boldsymbol{q}. \tag{2.163}$$

The masses which enter (2.163) can be obtained by differentiation of the dispersion relation of the heavy-hole band [2.47]:

$$E(\boldsymbol{k})=Ak^2-\sqrt{B^2k^4+C^2(k_x^2k_y^2+k_y^2k_z^2+k_z^2k_x^2)}\,. \tag{2.164}$$

Rather than embarking on the formidable task of evaluating (2.164), the authors of [2.47] preferred to approximate the constant energy profile of Fig. 2.86 by 6 ellipsoids along the $\{110\}$ directions, which contain 98 % of the carriers, and an isotropic sphere with 2 % of the carriers. With this model the integral (2.163) goes over into a sum over 7 valleys which yields Raman tensors of all possible symmetries ($\Gamma_{25'}$, Γ_{12}, Γ_1). That of $\Gamma_{25'}$ symmetry, however, is dominant (cross section $\geqslant 20$ times larger than the others), a fact which is borne out by experiment (Fig. 2.85).

The line shape of the intervalley tail of Fig. 2.85 has not been analysed in detail. The deviations from the shape of (2.161b) may be due to the inevitable interference with the Raman phonon (same symmetry!). Also, the "intervalley" scattering time of (2.161b) is in our case an *intravalley* time (within heavy-hole band), connecting parts of the heavy-hole bands with different masses. Thus there may also be a distribution of scattering times modifying the frequency dependence of (2.161b).

The dependence of the scattering efficiency on ω_{L} should be, in our case, different from n-Si as the heavy-hole mass is determined by the $\Gamma_{25'}\rightarrow\Gamma_{15}$ gap, called E_0' (Fig. 2.7), which equals 3.4 eV. Hence, at 2.6 eV (the maximum ω_{L} was

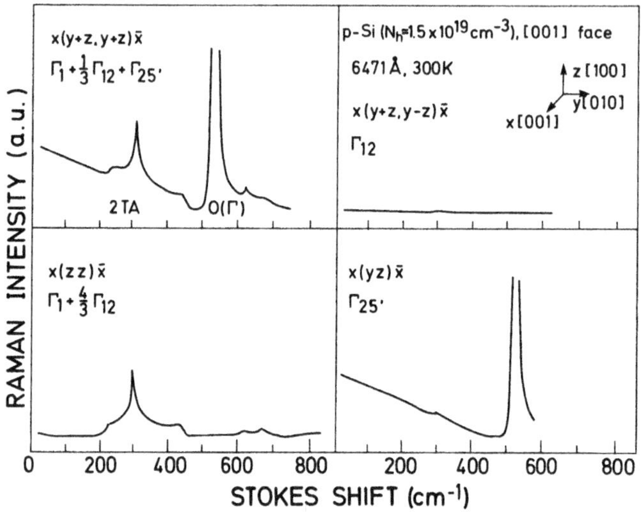

Fig. 2.85. Raman spectra of heavily doped p-type Si showing the low-frequency tail due to the warping of the heavy-hole band in the spectra of $\Gamma_{25'}$ symmetry. From [2.47]

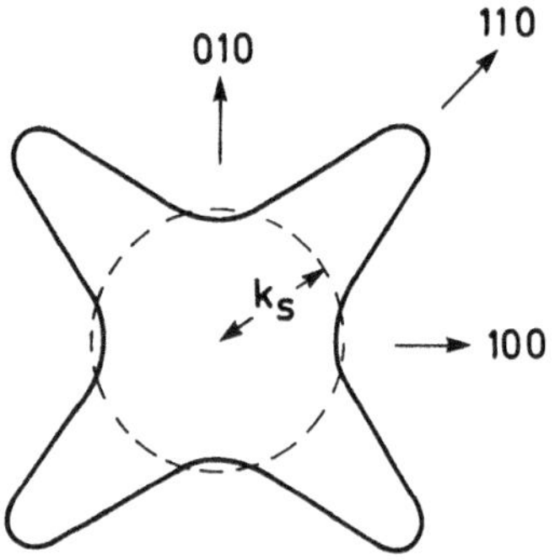

Fig. 2.86. Constant energy surface for heavy holes in p-type Si (solid line). For the calculations of [2.47], this surface is approximated by six ellipsoids along the $\{110\}$ directions and a sphere (dashed line in the figure). From [2.47]

used in [2.47]), resonance enhancement can be observed. This resonance enhancement follows the theoretical prediction based on (2.71)

$$\frac{\partial^2 S}{\partial\omega\,\partial\Omega}\propto\left|1-\frac{E_0E_\mathrm{P}}{E_0^2-(\hbar\omega_\mathrm{L})^2}-\frac{E_0'E_\mathrm{P}'}{E_0'^2-(\hbar\omega_\mathrm{L})^2}\right|^2, \tag{2.165}$$

with $E_0=4.1$ eV, $E_0'=3.4$ eV, $E_\mathrm{P}=21.6$ eV, $E_\mathrm{P}'=14.4$ eV [Ref. 2.47, Fig. 9].

We have considered, so far, n- and p-Si doped beyond the degeneracy limit ($\sim 5\times10^{18}$ cm^{-3}) so as to have no carrier freeze-out at low temperatures. Below this limit electronic transitions between impurity levels are observed. In n-Si the effect discussed here continuously goes over into transitions between the ground state and the valley-orbit-split impurity states. The selection rules (Γ_{12}

symmetry) are the same [2.214]. For n-Ge, valley-orbit transitions of $\Gamma_{25'}$ symmetry have been observed [2.215]. The corresponding intervalley fluctuations should have the same symmetry but have not been observed.

2.7.2 Interaction Between Raman Phonons and Electronic Continua

The intraband continua just described, and also *interband* continua, overlap sometimes with the Raman phonon. In these cases electron-phonon interaction couples the phonon to the electronic continuum and quantum-mechanical interference may result. Examples of these interferences observed for p-type Si are shown in Fig. 2.87. They manifest themselves in an asymmetry of the otherwise symmetric (in the absence of free carriers) phonon line with an enhancement of one flank and a decrease, even a minimum (antiresonance), on the other. This is expected from a quantum-mechanical mixing of the continuum with the discrete state. Coupling in first-order perturbation theory suffices to explain the interference; it breaks down, however, very close to the discrete line. The theory of these phenomena (to all orders) is presented in [Ref. 2.20, Sect. 4.3.1] from a rather elementary point of view. An elegant and formal discussion of this theory based on the Green's functions treatment of *Kawabata* [2.216] is given in [Ref. 2.232, Sect. 5.4.2]. Similar discussions can be found in [2.217–219]. We shall briefly review here the most salient aspects of this theory.

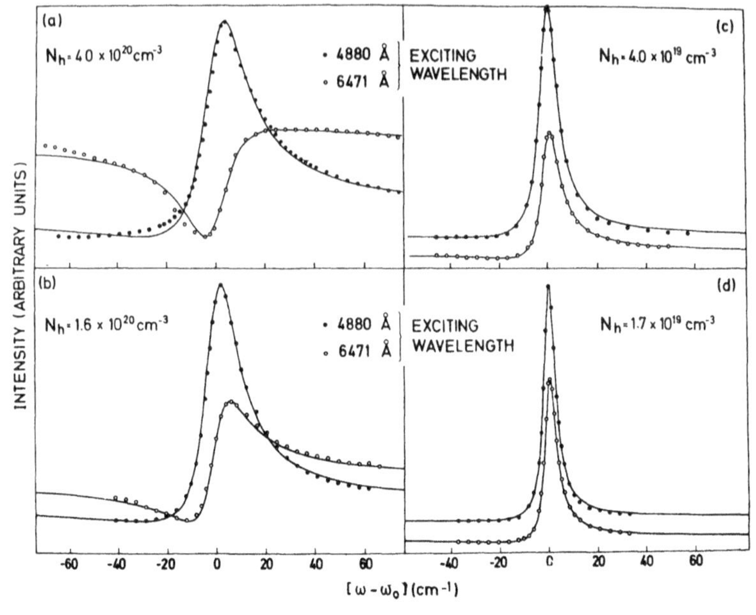

Fig. 2.87. One-phonon Raman spectra of four heavily doped p-type Si samples at two different laser wavelenghts at room temperature. The line shapes have been fitted with the Fano-Breit-Wigner expression (2.168). From [2.48]

Scattering or absorption of radiation by a discrete excitation interacting with an overlapping continuum is described by the so-called Fano-Breit-Wigner profile which can be written in any of the following equivalent forms:

$$\frac{\partial^2 S}{\partial\omega\,\partial\Omega} \propto \frac{|Q+\varepsilon|^2}{1+\varepsilon^2} \tag{2.166}$$

$$= \frac{Q_i^2}{1+\varepsilon^2} + \frac{(Q_r+\varepsilon)^2}{1+\varepsilon^2} \tag{2.167}$$

$$= 1 + \frac{Q_i^2+Q_r^2}{1+\varepsilon^2} + \frac{2Q_r\varepsilon-1}{1+\varepsilon^2} = \sigma_0+\sigma_1+\sigma_2\,. \tag{2.168}$$

In these equations ε represents a reduced frequency

$$\varepsilon = \frac{\omega-\bar{\omega}_0}{\Gamma}\,,$$

where Γ is a linewidth parameter and $Q=Q_r+iQ_i$ a complex line-shape parameter. Notice that of the three components in (2.168), σ_0 represents the continuum without interaction, σ_1 is the imaginary part of a Lorentzian function (i.e., the non-interfering part of the discrete excitation) and σ_2 the interfering part of the discrete excitation. Hence, only the real part of Q contributes to the interference. Note that for $Q_r \to \infty$, a noninteracting profile (σ_1) is found while for $Q=0$, one finds a scattering "window" around $\bar{\omega}_0$ (Fig. 2.87a). It is not possible to obtain separate values of Q_i and Q_r with (2.168) if there is an *additional* noninteracting scattering continuum ($\tilde{\sigma}_0 \neq 1$), as is usually the case. It is then customary to fit the Fano profiles with (2.168) plus a constant under the assumption that Q is *real* ($=\tilde{Q}$). The relationship between $\tilde{Q}$ and the complex Q is then

$$\frac{2\tilde{Q}}{\tilde{Q}^2-1} = \frac{2Q_r}{Q_r^2+Q_i^2-1}\,, \qquad \tilde{\sigma}_0 = \frac{\tilde{Q}}{Q_r}\,. \tag{2.169}$$

A complex Q is thus equivalent to keeping Q_{real} ($=\tilde{Q}$) and adding a constant background to σ_0.

The Fano-Breit-Wigner profiles just mentioned are not privy to light scattering. They appear often in atomic and nuclear physics [2.220a]. In the case of light scattering, the parameters Q, Γ and $\bar{\omega}_0$ are given by [2.220b]

$$\begin{aligned} &Q = [V(T_p/T_e) + V^2R(E)]/[\pi V^2\varrho(E)] \\ &\hbar\Gamma = \pi V^2\varrho(E) + \hbar\gamma \\ &\hbar\bar{\omega}_0 = \hbar\omega_0 - V^2R(\varepsilon)\,, \end{aligned} \tag{2.170}$$

where V is the matrix element for the interaction between the discrete excitation and the continuum (actually a quasi-continuum: its eigenstates are normalized to one). The matrix element V is assumed to be constant in this expression.

In (2.170), $\varrho(E)$ is the density of continuum states, $\pi^{-1}R(E)$ its Hilbert transform, ω_0 the discrete frequency in the absence of electron-phonon interaction, T_p and T_e the scattering amplitudes for the decoupled phonon and the electronic continuum, respectively, and γ the phonon linewidth in the absence of interaction with the continuum. This linewidth is usually determined by anharmonic decay into two acoustic phonons [2.221]. Note that the expression for $\bar{\omega}_0$ in (2.170) is only approximate, for $V^2\varrho(E) \ll \hbar\omega_0$. The exact expression can be found in [Ref. 2.232, Eq. (5.24)].

The expressions for Γ and $\bar{\omega}_0$ in (2.170) contain a line shift and a broadening which are independent of the asymmetry parameter Q. Hence we can view the scattering as produced by a quasiparticle consisting of a phonon mixed with an electronic excitation. The self-energy of the phonon due to the mixing is

$$\Sigma(E) = V^2[-R(E) + \pi i \varrho(E)]. \tag{2.171}$$

The real part of Σ represents the frequency shift of ω_0 and its imaginary part the broadening due to the interaction with the continuum. They may exist even for symmetric profiles (Q_r or $\langle Q_r \rangle = 0$).

The scattering amplitude T_p in (2.170) is basically related to the Raman polarizability a (for definition see [Ref. 2.64, Eq. (2.67)] through the square root of [Ref. 2.64, Eq. (2.134a)]. It will be real or complex depending on a. Below the absorption edge a is real; above it, it is complex. In order to investigate whether Q is real or complex, we must also investigate the reality of V and T_e.

We note that the intraband electronic backgrounds seldom lead to asymmetric profiles because for them, often $Q \simeq \infty$. Let us consider, for instance, the intraband background of electrons in a nondegenerate minimum. In order to produce effects of the type discussed, the phonon should couple the minimum with itself; hence only those phonons with $q=0$ which are completely symmetric with respect to the group of the q-vector of the minimum can couple them. For a minimum at $q=0$ (e.g., GaAs), the Γ_{15} Raman phonon does not couple and, in this case, neither asymmetry nor self-energy arises. For minima off $q=0$, intraband coupling is possible (e.g., Ge, but not for Si). Self-energies do then result. The Q-parameter, however, is, in this case, usually either pure imaginary (below the gap) or complex, since T_e is pure imaginary (Sect. 2.3.1d). Hence, asymmetric Fano profiles have been only positively identified for phonons interacting with *interband* continua. Figure 2.88a, b show band sketches for the cases which have been extensively studied: n- and p-type Si and p-type Ge.

The case of interband electronic scattering has not been discussed so far in this chapter. It can occur in intrinsic semiconductors (across the gap) or in doped semiconductors with subsidiary extrema close to those where the carriers are placed. Because of the orthogonality of the periodic parts of the Bloch functions in *different* bands, the interband excitations are not accompanied by charge-

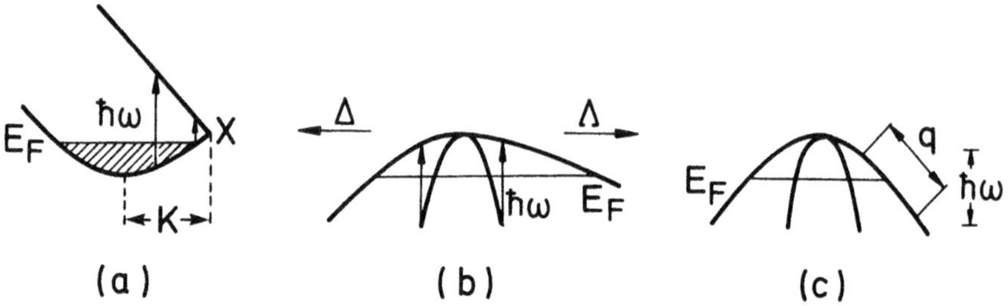

Fig. 2.88a–c. Diagrams of band extrema leading to phonon self-energies and Fano line shapes. (**a**) n-type Si; (**b**) intervalence-band transitions, p-type Ge and Si; (**c**) intravalence-band transitions, p-type Ge and Si

density fluctuations the way intraband ones are. Thus, screening does not play a role and the spectrum of single-particle excitations is seen in light scattering. The expressions for the scattering cross section are given in [Ref. 2.64, Sect. 2.2.2] and [Ref. 2.232, Sect. 5.4.1]. The scattering amplitudes which enter in the expression for Q (2.170) are obtained from the square root of the scattering efficiencies, whereby care must be taken to recover possible phase factors which disappeared in the efficiencies.

Raman scattering can thus be used to investigate the self-energies of optical phonons at $q \simeq 0$ in interaction with electronic excitations. These self-energies have often (but not necessarily) a negative real part, i.e., they lead to a *softening* of the phonons. The maximum softening attainable is usually small; in the extrinsic case it is limited by the solubility of impurities ($\sim 1\%$). In some intrinsic cases, however, it may be large enough to make $\bar{\omega}_0 = 0$ and hence induce phase transitions [2.222]. Similar effects have been postulated to explain laser annealing phenomena in cases in which the melting point is not reached [2.223]. Large carrier concentrations are, in principle, possible in nonequilibrium conditions (pulsed laser excitations) and in samples implanted with doping ions and laser annealed[2].

Acoustic and optic phonons of large q are not accessible to first-order Raman scattering. Only some of them are accessible to rather weak second-order scattering. Acoustic phonons of $q \simeq 0$ can be detected by means of either ultrasonic propagation (small q) or Brillouin scattering (larger q). Ultrasonic techniques have been extensively used for measuring self-energies due to doping in Si and Ge [2.224]. Very recently, *Pintschovius* et al. [2.49] have been able to measure with inelastic neutron scattering the real parts of the self-energies in heavily doped n- and p-Si for acoustic phonons along various directions through the whole Brillouin zone. We show an example of these results in Fig. 2.89. In this case the self-energy is due to intraband coupling. An interesting fact appears: TA phonons along high symmetry directions (Fig. 2.89) do not have any component of A_1 symmetry. Hence they produce intraband excitations which *are not*

---

2 A. Compaan, G. Contreras, M. Cardona, A. Axmann: J. Physique, **C5**, 197 (1983); G. Contreras, A. K. Sood, M. Cardona, A. Compaan: Solid State Commun. **49**, 303 (1984)

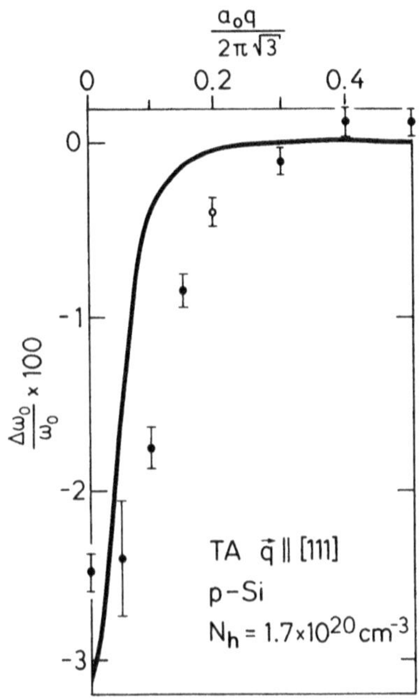

Fig. 2.89

Fig. 2.89. Real part of the self-energy of transverse acoustic phonons versus q-vector as determined by neutron scattering in p-type Si at 300 K. The solid line has been obtained theoretically. From [2.49]

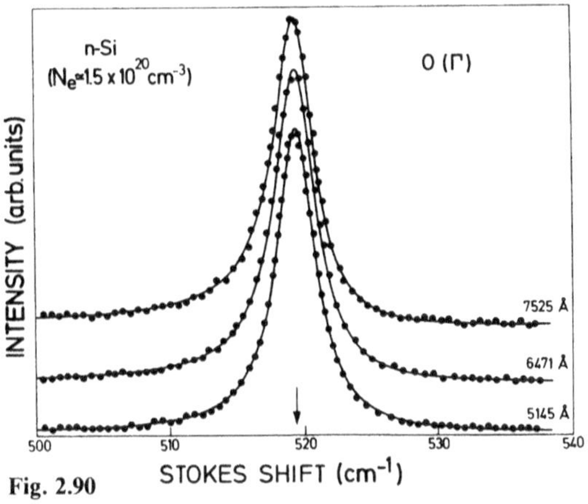

Fig. 2.90

Fig. 2.90. One-phonon Raman spectra of heavily doped n-type Si at different laser wavelengths (330K). The solid line is a fit to the Fano-Breit-Wigner expression (2.168). From [2.45]

screened by the dielectric constant of the free carriers (the free carriers can only efficiently screen excitations of A_1 symmetry). For LA phonons, however, the excitations have partly A_1 character and are thus partly screened. The screened part does not contribute to the self-energy at $q \simeq 0$ but yields a contribution at $q \neq 0$, with a maximum in the neighborhood of k_{TF} (the inverse of the Thomas-Fermi screening radius).

a) n-Type Si

We show in Fig. 2.90 the Raman phonons observed with different laser lines for n-Si doped with phosphorus ($n = 1.5 \times 10^{20}$ cm^{-3}) [2.45, 225]. The profile obtained for $\lambda_L = 7525$ Å is slightly asymmetric. The others are less asymmetric. The former has been fitted with (2.168) yielding $Q = -13 \pm 1$ (Table 2.3). The relatively high value of $|Q|$ is responsible for the weak asymmetry as compared with the p-Si case (Fig. 2.87). Note that the asymmetry of the curves in Fig. 2.90 (antiresonance at high frequencies) is opposite to that in Fig. 2.87 (antiresonance at low frequencies); the values of Q are negative in the former case and positive in the latter.

The conduction band extrema of n-Si (Fig. 2.88a) are rather simple, a fact which enables us to calculate *microscopically* all the parameters of the Fano

Table 2.3. Experimental and calculated values of the asymmetry parameter $\tilde{Q}$ for several exciting wavelengths [2.45] in n-type Si with $N_e = 1.5 \times 10^{20}$ cm^{-3}

Exciting wavelength [Å]	$\tilde{Q}$ (exp.)	$\tilde{Q}$ (theor.)
5145	-50 ± 10	-46^{+18}_{-10}
6471	-24 ± 2	-21^{+8}_{-4}
6764	-20 ± 2	-16^{+6}_{-4}
7525	-13 ± 1	-12^{+5}_{-3}

profile $(\bar{\omega}_0 - \omega_0, \Gamma, Q)$. We follow the line of [2.45]. The matrix element of the electron-phonon interaction coupling the Δ_1 minimum with the $\Delta_{2'}$ band is written in terms of the deformation potential D_0 [eV]:

$$V = (2/a_0) D_0 (\hbar/4MN\omega_0)^{1/2}, \tag{2.172}$$

where a_0 is the lattice constant, M the atomic mass of Si and N the number of primitive cells per unit volume. In [2.45] it was shown that D_0 can be estimated by the pseudopotential technique. A simple calculation which can be performed by hand yields $D_0 \simeq -8.1$ eV. This deformation potential has also been calculated by *Vogl* and *Pötz* [2.226] with the tight-binding method. They find for Si $D_0 = -6.6$ eV, in reasonable agreement with the pseudopotential result (note that the definition of the deformation potential in [2.226] differs from that used here by a factor of 2, $D_0 = d_{x0}/2$).

The bands involved in the electron-phonon interaction (Fig. 2.88a) can be considered to be parabolic with a longitudinal mass $m_l = 0.9 m_0$ and a transverse mass $m_t = 0.2 m_0$ [2.114]. The density of states for k-conserving excitations is

$$\varrho(E) = \frac{1}{4\pi^2} \frac{m_l m_t}{K\hbar^4} \left[E_F - \frac{\hbar^2}{2m_l} \left(K - \frac{E m_l}{2K\hbar^2} \right)^2 \right], \tag{2.173}$$

where K gives the position of the lowest minimum measured from the edge of the Brillouin zone (Fig. 2.88a). "Negative" values of $\varrho(E)$ obtained for large and small E with (2.173) correspond to transitions forbidden by the Pauli exclusion principle. By using (2.172, 173) it is easy to find with (2.170) the imaginary self-energy for $N_e = 1.5 \times 10^{20}$ cm^{-3}: $\Gamma = \pi V^2 \varrho(E) = 0.14 \pm 0.04$ cm^{-1}, which compares well with the average experimental value $\hbar(\Gamma - \gamma) = 0.21 \pm 0.14$ cm^{-1}. The real self-energy is also given in (2.170). For our bands it can be explicitly written as (for $T \simeq 0$)

$$\bar{\omega}_0 - \omega_0 = \Delta_0 = \frac{1}{2\pi} \frac{D_0^2}{a_0^2} \frac{m_l m_t}{MN\omega_0 K\hbar^2} P \int_{\omega_{\min}}^{\omega_{\max}} \left(\frac{\omega}{\omega_0 - \omega} - \frac{\omega}{\omega_0 + \omega} \right) \varrho(\omega)\, d\omega$$

$$\text{with } \varrho(\omega_{\min}) = \varrho(\omega_{\max}) = 0. \tag{2.174}$$

This expression can be analytically integrated, The result can be found in [Ref. 2.45, Eq. (10a)].

From (2.174) we obtain for $N_i = 1.5 \times 10^{20}\,\mathrm{cm}^{-1}$, $\Delta\omega = -0.65\,\mathrm{cm}^{-1}$, in excellent agreement with the experimental result $\Delta\omega_0 = -0.5 \pm 0.2\,\mathrm{cm}^{-1}$ [2.45].

The self-energy just calculated has been obtained by treating the first-order electron-phonon coupling Hamiltonian in second-order perturbation theory. There is another contribution which arises from the second-order electron-phonon Hamiltonian given in [Ref. 2.64, Eq. (2.234)]. This contribution is

$$\Delta\omega_0^{(2)} = \frac{D_1^{(2)} N_e}{3 a_0^2 N M \omega_0}. \tag{2.175}$$

The second-order electron-phonon coupling constant $D_1^{(2)}$ can be easily estimated with the pseudopotential method used in [2.45]. We find for the Δ_1 minimum of Si

$$\begin{aligned} D_1^{(2)} &= \frac{a^2}{4} \langle \Delta_1 | \nabla^2 V | \Delta_1 \rangle \\ &= -2\pi^2 [4\beta^2 v_8 + 3\sqrt{2}\alpha\beta v_3] \simeq -130\,\mathrm{eV}. \end{aligned} \tag{2.176}$$

In (2.176) v_3 and v_8 are pseudopotential form factors and $\alpha = -0.46$, $\beta = 0.89$ are coefficients of the expansion of the wave functions in $\{111\}$ and $\{200\}$ plane waves, respectively. Using the value of $D_1^{(2)} = -130\,\mathrm{eV}$, (2.175) yields $\Delta\omega_i^{(2)} = -0.17\,\mathrm{cm}^{-1}$, smaller than the contribution of (2.174) ($-0.65\,\mathrm{cm}^{-1}$) but not negligible. This contribution is, however, usually not taken into account in the literature. It is equivalent to the correction referred to as "adiabatic" in [2.227].

The same method can be used for estimating the asymmetry parameter Q. An analytical calculation [2.45] yields a Q which depends on the laser frequency, in agreement with experiment (Table 2.3). Such dependence arises from the fact that the scattering efficiency for phonons resonates at the E_0' gap (3.4 eV) while that for electronic excitations is not resonant at this frequency; its first resonance should occur for $E_2 = 4.4$ eV. The negative sign of Q predicted by theory is also in agreement with the data of Fig. 2.90.

b) p-Type Si

Typical data for Raman spectra of heavily doped p-type Si have been shown in Fig. 2.87. The self-energy and Fano asymmetry effects are much stronger than in the case of n-Si (Fig. 2.90). The electronic continuum is clearly seen, especially in Figs. 2.87a,b. This electronic continuum corresponds to transitions from the light to the heavy hole bands (Fig. 2.88). The calculation of the Raman spectrum for this continuum of electronic excitations is not as easy as in the case of n-Si because of the band degeneracy at $\boldsymbol{k} = 0$ and the ensuing warping of the valence bands, see (2.164). Also, the matrix element of the electron-phonon interaction cannot be assumed to be a constant, independent of the $\boldsymbol{k}$ of the electrons, in the way done in (2.174). For a given phonon direction, these matrix elements depend strongly on the direction of $\boldsymbol{k}$. This dependence is obtained by finding the Bloch functions for a finite $\boldsymbol{k}$ as a linear combination of Bloch functions for $\boldsymbol{k}=0$. This involves the diagonalization of a 6×6 matrix which, if the doping is not too high so that the spin-orbit-split valence band can be neglected, reduces to a 4×4 matrix (parabolic case valid if $E_F \ll \Delta_0$). Complicated analytic expressions for the electron-phonon matrix element as a function of the direction of $\boldsymbol{k}$ have been given by *Lawaetz* [2.228]. Following this author, the squared matrix element for the electron-phonon interaction between two phonon subbands λ and λ' is written as

$$|H_{\lambda\lambda'}(\boldsymbol{k})|^2 = \frac{\hbar d_0^2 a_0}{16 M\omega_0 V}\, g_{\lambda\lambda'}(\boldsymbol{k}), \tag{2.177}$$

where the $\boldsymbol{k}$-dependence is contained in the function $g_{\lambda\lambda'}(\boldsymbol{k})$ and d_0 is the deformation potential defined in [Ref. 2.64, Eq. 2.188)]. This deformation potential, which lies around $+30$ eV for most zincblende and germanium-like semiconductors, also plays a very important role in the theory of resonant Raman scattering by phonons [2.64]. The expressions for $g_{\lambda\lambda'}$ as a function of $\boldsymbol{k}$ are rather complicated in the general case which includes explicitly the spin-orbit split-off valence band. In the parabolic case, $g_{\lambda\lambda'}(\boldsymbol{k})$ is given by simple analytic expressions [Ref. 2.228, Eq. (B27)]:

$$\begin{aligned} &g_{11} = g_{22} = \frac{\gamma_2}{3(\gamma_1+\gamma_2)}, \qquad g_{33}=0 \\ &g_{12} = 1 - g_{11}, \qquad g_{13} = g_{23} = 1, \end{aligned} \tag{2.178}$$

where

$$\begin{aligned} &\gamma_1(k) = B^2(1-3\mu) \\ &\gamma_2(k) = D^2\mu \\ &\qquad \mu = (k_x^2k_y^2 + k_y^2k_z^2 + k_z^2k_x^2)k^{-4}. \end{aligned}$$

In (2.178) the subindices 1 and 2 represent the heavy- and the light-hole band, respectively, while 3 represents the split-off band. The band parameter D is related to the A, B, C of (2.164) through $D^2 = C^2 + 3B^2$.

It is of interest to evaluate the angular averages of g_{11}, g_{22} and g_{12}. Taking into account that for Si $B=4.2$, $D=5.0$,

$$g_{22}=g_{11}=\frac{\mu}{3}\left(\frac{D^2}{B^2}\right)\left[1+\left(\frac{D^2}{B^2}-3\right)\mu\right]^{-1}$$

$$\langle g_{22}\rangle=\langle g_{11}\rangle\simeq\frac{1}{3}\left(\frac{D^2}{B^2}\right)\langle\mu\rangle\left[1+\left(\frac{D^2}{B^2}-3\right)\langle\mu\rangle\right]^{-1}=0.3$$

$$\langle g_{12}\rangle\simeq 0.7. \tag{2.179}$$

We thus realize that the average interband electron-phonon-interaction matrix element, which is the same for the light as for the heavy-hole band, is much stronger than the intraband counterparts. Intraband effects on the self-energy will be negligible except if the frequency and wavevector of the phonon involved lie near the boundary of the Landau-damping region (Fig. 2.1). In this case vanishing energy denominators produce a logarithmic singularity in the real part of the self-energy. The condition for this to happen in backscattering is (Fig. 2.1)

$$\omega_0=\frac{\hbar k_F q}{m^*}+\frac{\hbar q^2}{2m^*}\simeq\frac{\hbar k_F 4\pi n}{m^*\lambda_L}. \tag{2.180}$$

For the heavy-hole band of Si ($m^*_{hh}=0.52$), the only one which may yield observable intraband effects, (2.180) is satisfied at $\lambda=1200$ Å for a typical $N_h=10^{20}\,\mathrm{cm}^{-3}$, a wavelength in the vacuum-uv and beyond the possibilities of present lasers. For Ge, however, ω_0 is smaller and so is $m^*_{hh}=0.35$. The index n in the visible is larger than for Si. Hence, (2.180) can be fulfilled at standard laser wavelengths. In this case intraband self-energy effects seem to have been observed (Sect. 2.7.2d and [2.46]).

Because of the complicated warped and degenerate nature of the valence bands at $\boldsymbol{k}=0$, the interband spectrum for excitations from the light hole to the heavy-hole band must be obtained by numerical integration [(2.178, 179) do not apply in this case]. This has recently been done by *Kanehisa* et al. [2.229]. We show the results of this calculation for $N_h=1.13\times10^{20}$ and $1.60\times10^{20}\,\mathrm{cm}^{-3}$ in Fig. 2.91 for two polarization configurations. This figure indicates that the corresponding Raman tensor must have mainly $\Gamma_{25'}$ symmetry as hardly any scattering is obtained for $\hat{e}_L\|\hat{e}_S\|$ [100]. The maximum of this electronic scattering occurs rather close to the Raman phonon frequency (520 cm^{-1}). Hence, strong self-energy and Fano interference effects are expected.

In order to calculate the corresponding imaginary self-energy we must, in principle, also perform a numerical three-dimensional integration using (2.164) for all possible interband transitions between occupied light-hole and empty heavy-hole states. This has been done in [2.229]. We show in Fig. 2.92 the results of this calculation performed for $d_0=27$ eV, compared with experimental data (all at $T=77$ K). The agreement is good for high dopings. At lower dopings the real part of the calculated self-energy reverses sign, a fact which is not observed experimentally. The sign reversal corresponds to the fact that at low dopings the frequencies of most electronic excitations lie below ω_0 (hence, $\Delta\omega_0>0$), while at

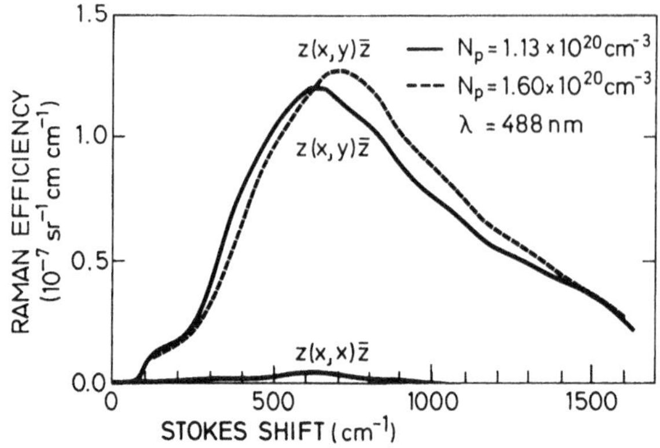

Fig. 2.91. Calculated Raman scattering spectra by inter-valence-band electronic excitations for two heavily doped p-type silicon samples. From [2.230]

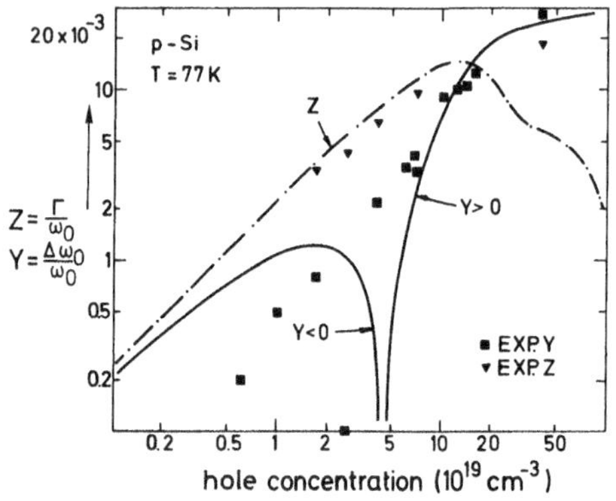

Fig. 2.92. Real and imaginary parts of the phonon self-energy, normalized to the phonon frequency ω_0, for p-type Si at 77 K. From [2.229]. The lines are calculated while the points are experimental

high dopings the opposite is true ($\Delta\omega_0 < 0$). The failure to observe this sign reversal experimentally may be due to $\boldsymbol{k}$-nonconservation induced by the impurities. This $\boldsymbol{k}$-nonconservation would shift the electronic excitation spectrum to higher frequencies and thus possibly make $\Delta\omega_0 < 0$ for all frequencies.

An approximate evaluation of the self-energy is also possible without numerical integration by using an average electron-phonon matrix element of the type given in (2.177) and the density of states corresponding to parabolic bands. For details see [2.48].

It remains now to discuss the values of Q, their sign and their increase with increasing photon energy. The sign found for Q in p-type Si is in all cases positive [2.48, 230]. This is contrary to the case of n-Si. It has been shown in [2.230] that a positive sign should indeed result from the positive sign of d_0 and the positive sign of the Raman polarizability a [Ref. 2.64, Table 2.8]. In fact, according to [2.230], the positive sign of Q requires the sign of a to be positive; no other experimental determinations of the sign of a are available for nonpolar semiconductors [Ref. 2.64, Sect. 2.1.18g].

Table 2.4. Fano parameters of the optic phonon and B^{11} local mode in Si for different exciting frequencies and hole concentrations, [2.212]

Carrier concentration $[cm^{-3}]$	Exciting wavelength [Å]	Optic phonon			B^{11} local mode		
		$\tilde{Q}_P$	$\Gamma_P [cm^{-1}]$	$\omega_P + \Delta\omega_P$ $[cm^{-1}]$	$\tilde{Q}_L$	$\Gamma_L [cm^{-1}]$	$\omega_L + \Delta\omega_L$ $[cm^{-1}]$
4×10^{20}	4579	4.54	12.4 ± 0.5	512 ± 2	1.82	8.0 ± 1.5	615 ± 2
	5145	2.57	11.8 ± 0.5	511 ± 2	1.16	7.4 ± 1.5	614 ± 2
	6471	0.55	12.6 ± 0.5	509 ± 2	0.15	10.0 ± 1.5	613 ± 2
			avg. 12.3 ± 0.5	avg. 510.7 ± 1.5		avg. 8.5 ± 1.8	avg. 614 ± 2
1.5×10^{20}	4880	6.0	9.2	516.7	2.7	7.3	614.8
	6471	1.8	8.5	514.8	0.93	6.4	614.2
			avg. 8.85 ± 0.5	avg. 515.7		avg. 6.9 ± 0.6	avg. 614.6
6×10^{19}	4880	10	6.2	519.7	3.2	6.4	615.8
1.5×10^{19}	4880	44	3.1	520.2	8	6.0	617.1
	6471	20.0	2.65	520.2	5.8	5.9	617.3
			avg. 2.88	avg. 520.2		avg. 5.95	avg. 617.2
5×10^{18}	4880	200	1.8	520	33	6.8 ± 1.0	618
Pure Si (10^{13})			1.5	520.0		5.5	620

The next interesting feature of the Fano Q's of p-Si is their increase with increasing ω_L, (Table 2.4), i.e., the asymmetry of the Raman profile becomes smaller with increasing ω_L. The ratio $\beta = Q(2.54\,\text{eV})/Q(\omega_L = 1.92\,\text{eV})$ equals 2 ± 0.6 nearly independently of hole concentration [2.48]. It is easy to attribute this change with ω_L to a different dependence on ω_L of $T_p \sim a$ and T_e in (2.170). The resonant denominators of T_p and T_e, however, are expected to be nearly the same; for T_p we expect a resonance like $|\hbar\omega_L - E_1|^{-1}$ for $E_1 \simeq 3.4\,\text{eV}$ ([Ref. 2.64, Sect. 2.2.4], 2-dimensional critical point) and similarly for the electronic scattering amplitude (Fig. 2.93). We must therefore examine the prefactors in the scattering efficiency, such as the ω_L^4 dependence of the cross section for phonon scattering (i.e., $T_p \propto \omega_L^2$). This prefactor does not exist for the interband electronic scattering. It leads to $Q \propto \omega_L^2$, i.e., to $\beta = 1.8$, in rather good agreement with experiment.

No calculations of the magnitude of Q have appeared. The difficulty lies in the fact that the parameters V and T_e of (2.170) actually vary with the initial $\boldsymbol{k}$ of the participating electron and hence (2.170) becomes meaningless. The correct expression for Q in this case has been derived in [2.219]. It is given in terms of a tensor $R_{yz}^{(x)}$ by

$$Q = -\frac{\mathrm{Re}\,\{R_{yz}^{(x)}\}}{\mathrm{Im}\,\{R_{yz}^{(x)}\}}, \tag{2.181}$$

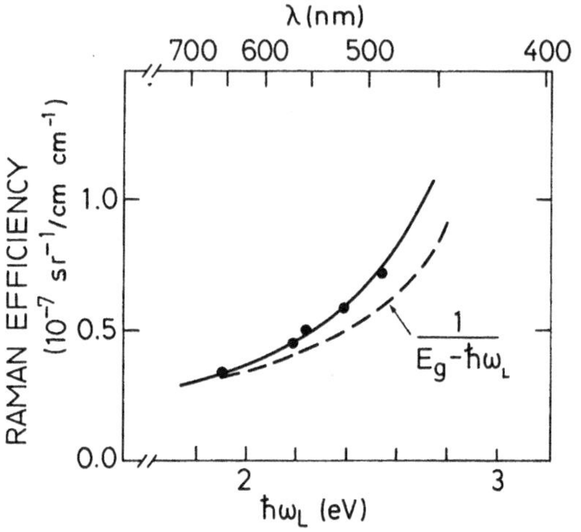

Fig. 2.93. Resonance of the intervalence-band light scattering in p-Si (see Fig. 2.91). From [2.230]

where y and z are incident and scattered photon polarization indices and (x) is the phonon polarization. The tensor $R_{yz}^{(x)}$ is given by

$$R_{yz}^{(x)} = \sum_{\mathrm{if}\boldsymbol{n}} \frac{p_{\mathrm{im}}^{y} p_{\mathrm{mf}}^{z}}{\hbar\omega_{\mathrm{L}} - E_{\mathrm{m}} + E_{\mathrm{i}}} H_{\mathrm{fi}}(\boldsymbol{k}) \frac{f(E_{\mathrm{i}}) - f(E_{\mathrm{f}})}{\hbar\omega - E_{\mathrm{f}} + E_{\mathrm{i}} + \mathrm{i}\delta}, \tag{2.182}$$

where p_{im}^{y} represent matrix elements of linear momentum, $H_{\mathrm{fi}}(\boldsymbol{k})$ those of the electron-phonon Hamiltonian (2.177) and n labels the bands to which the i and f states belong. If the matrix elements $H_{\mathrm{fi}}(\boldsymbol{k})$ in (2.182) are constant, (2.182) can be easily transformed into (2.170). Note that since (2.182) is summed over all i, f and m, it contains both T_{p} and the product VT_{e}.

To our knowledge a numerical evaluation of (2.182) has not been performed. Using for $H_{\mathrm{fi}}(\boldsymbol{k})$ the independently averaged values given in (2.178), it is possible to write Q as

$$Q \simeq \left(\frac{dS_{\mathrm{p}}/d\Omega}{\pi\Gamma(\partial^2 S_{\mathrm{e}}/\partial\omega\,\partial\Omega)} \right)^{1/2} \frac{\Delta\omega_0}{\Gamma}, \tag{2.183}$$

where S_{p} and S_{e} are the phonon and electronic scattering efficiencies, respectively. We evaluate (2.183) at 2.54 eV for $N_{\mathrm{h}} = 1.6 \times 10^{20}\,\mathrm{cm}^{-3}$. From Fig. 2.91 we find $\partial^2 S_{\mathrm{e}}/\partial\omega\,\partial\Omega = 1 \times 10^{-7}\,\mathrm{sr}^{-1}\mathrm{cm\,cm}^{-1}$, from Fig. 2.47 and [Ref. 2.64, Eq. (2.134a)] $\partial S_{\mathrm{p}}/d\Omega = 2 \times 10^{-4}\,\mathrm{sr}^{-1}\,\mathrm{cm}^{-1}$. From Table 2.4 we find for $N_{\mathrm{h}} = 1.5 \times 10^{20}\,\mathrm{cm}^{-3}$, $\Gamma = 8.85\,\mathrm{cm}^{-1}$ and $\Delta\omega_0 = -4.3\,\mathrm{cm}^{-1}$. By replacing these values into (2.183) we obtain $Q = 8.2$ in reasonable agreement with the experimental value ($Q = 6.0$) given in Table 2.4.

It is interesting to note that the frequency shifts $\Delta\omega_0$ obtained for p-Si are much larger than the corresponding self-energy shifts for n-Si. The reason is partly due to the fact that d_0 (p-Si) is larger than the corresponding deformation potential D_0 of n-Si. Also, the average energy denominators for the interband coupling are $\sim E_F \lesssim 0.1$ eV for p-Si while for n-Si they are ~ 0.5 eV. Hence, the self-energies for the p-type material should be one order of magnitude larger than in the n-type case. This is confirmed by experiment. We note that the contribution of the second-order electron-phonon coupling (2.175) is also negligible for the p-type material. In this case $D_1^{(2)} \simeq 470$ eV [Ref. 2.64, Eq. (2.237)], which is about 3 times larger than for n-Si. The contribution of the first-order electron-phonon Hamiltonian is, however, more than 10 times larger than for n-Si.

We conclude with some qualitative considerations concerning the magnitude of the real part of the self-energy. In p-Si one may naively say that the phonon softening due to the doping results from the fact that each hole breaks a bond. The total number of bonds is $2N_a$ (N_a: number of *atoms* per unit volume: 5×10^{22} cm^{-3} per Si). We would then expect a softening

$$\frac{\Delta\omega_Q}{\omega_0} \simeq -\frac{1}{2}\frac{N_h}{2N_a}. \tag{2.184}$$

For $N_h = 4 \times 10^{20}$ cm^{-3} (2.184) yields $\Delta\omega_0 \simeq -1$ cm^{-1}. Instead, $\Delta\omega_0 = -10$ cm^{-1} is observed [Ref. 2.49, Table 1]. Hence there must be an enhancement factor which makes the holes at the Γ point more effective than the "average" valence electron which is removed when breaking a bond. Some light is shed on this question by considering that the shift arises from an expression of the type

$$\Delta(\omega_0) \simeq -\frac{M^2}{\Delta E} N_h, \tag{2.185}$$

where M is a matrix element of the electron-phonon interaction and ΔE an average gap for the electronic excitations. The frequency ω_0 can also be viewed as related to electronic transitions from the valence to conduction bands through an expression of the type (2.185) with N_h replaced by $4N_a$ and ΔE by E_G, the average excitations gap of Si (Penn gap ~ 4.5 eV). We thus find

$$\frac{\Delta\omega_0}{\omega_0} \simeq -\frac{E_G}{\Delta E}\left(\frac{N_h}{4N_a}\right). \tag{2.186}$$

For a typical $\Delta E = 0.2$ eV and $N_h = 4 \times 10^{20}$ cm^{-3}, (2.186) yields $\Delta\omega_0 = 24$ cm^{-1}, somewhat larger than found experimentally.

The parameters $\Delta\omega_0$, Γ and Q have been studied in [2.48] as a function of uniaxial stress. The triply-degenerate phonon splits into a singlet and doublet which have different self-energies and Q's. Similar effects have also been observed and discussed for n-Si in [2.45]. In [2.26] equivalent studies were performed with carriers produced by laser excitation, their concentration being

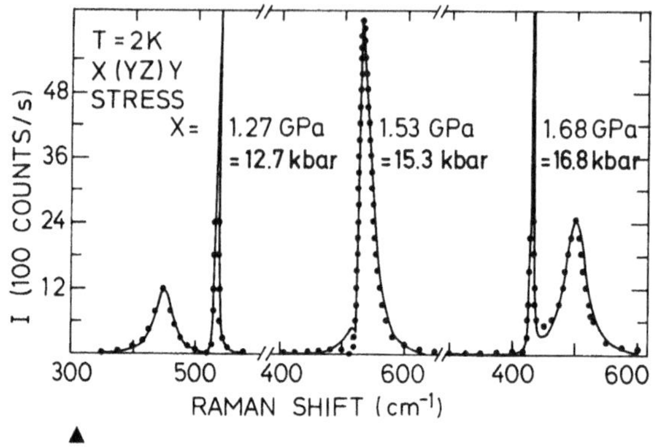

Fig. 2.94. Raman scattering by phonons (sharp peaks) and by intervalence-band transitions (broad peaks) in photo-excited silicon under uniaxial stress. From [2.232]

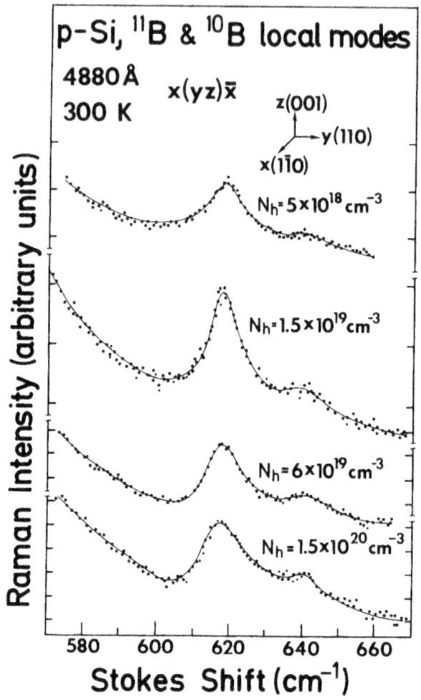

Fig. 2.95. Scattering by local modes of boron (isotopes B^{10} and B^{11}) in p-type silicon. From [2.212]

enhanced through strain confinement. The Fano interference can be enhanced by shifting the light-hole to heavy-hole interband continuum through the application of a uniaxial stress. Typical examples of the results are shown in Fig. 2.94. The sign of the parameter Q is positive, as expected for the interband hole continuum.

c) Local Vibrational Modes of B in Si

Most of the p-type Si available is doped substitutionally with boron. This element produces localized vibrational modes in Si due to its light mass. Two natural isotopes of B exist, ^{11}B and ^{10}B. Their natural abundance ratio is 5:1. Hence, if the doping is performed with natural B, two local modes appear at ~ 620 and 640 cm^{-1}, respectively, the former five times stronger than the latter (Fig. 2.95). Doping with isotopically pure B is possible through ion implantation.

Like the Raman phonon, the local modes due to boron find themselves in the middle of the interband background of Fig. 2.95. Real and imaginary self-energies and a Fano interference results. We present in Table 2.4 the self-energy and Q-parameters of these modes, and also of the Raman phonons, obtained from a fit to the spectra of several B-doped samples in [2.212]. We note that the sign of $\tilde{Q}_L$, the Q-parameter of the local modes, is the same as for the Raman phonons (positive). However, the self-energies and the magnitudes of Q for the local mode and for the phonons are not the same.

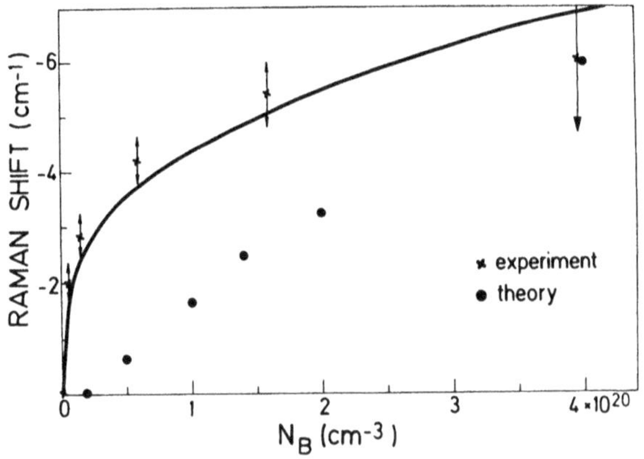

Fig. 2.96. Dependence of the real part of the self-energy of local modes of boron in silicon on boron concentration N_B. Experiment: crosses and error flags. Dots: calculations based on deformation potential interaction. Solid line: fit with $\Delta\omega_0 \propto N_B^{1/3}$. From [2.212]

The dependence of the real part of the self-energy on the hole concentration ($\equiv$B concentration) is shown in Fig. 2.96. In this figure we also display the results of a calculation based on an appropriate deformation potential type of treatment (dots). Although the magnitude of the observed shifts is explained by the calculation, the detailed shape of the dependence of $\Delta\omega_0$ on N_B is not; $\Delta\omega_0$ rises steeply at low N_h and saturates at $N_B \simeq 4 \times 10^{20}$ cm^{-3}. The experimental trend is fitted rather well by $\Delta\omega_0 \propto N_B^{1/3}$. This suggests screening by the free holes as a possible source of the observed self-energies; the Thomas-Fermi radius is proportional to $N_h^{-1/6}$, hence the Fourier component of the potential for small k is proportional to $N_h^{-1/3}$.

The potential squared multiplied by the hole concentration gives the dependence of $\Delta\omega_0 \propto N_h^{1/3}$. A recent suggestion by *Szabo* and *Genzel* [2.231] attributes the self-energy of Fig. 2.96 to some form of electrostatically coupled mode between the ir-active local mode and the hole plasma. Model calculations of this effect reproduce the shape of $\Delta\omega_0(N_h)$ found experimentally. The magnitude of the calculated effect, however, seems to be too small to explain that of Fig. 2.96. This model is actually equivalent to an *intra*band self-energy due to electrostatic coupling. The interband term seems to be at least partly responsible for the observed effect.

Another interesting feature of the Raman spectra of the local modes of B in Si is their strength. Their cross section per atom is much larger than for the Raman phonon ($\sim$150 times for $N_h = 8 \times 10^{18}$ cm^{-3}). This ratio of cross sections decreases with increasing N_B approximately like $N_B^{-2/3}$! Since the Raman cross section is also related to electron-phonon coupling, it is natural to conjecture that the cause of its peculiar dependence on N_B is the same as that of the anomalous dependence of $\Delta\omega_0$ on doping. It is thus possible to introduce a coupling constant which depends on $N_B \simeq N_h$ and to fit both the dependence on doping of $\Delta\omega_0$ and that of the cross section. This coupling constant may be related to the dynamical charge of the B ions. The details of this relationship, however, remain obscure. We may, nevertheless, say that the enhancement in the cross section is much stronger than that in the self-energy.

d) p-Type Ge, p-Type GaAs

Phonon self-energies and Fano interferences have been observed for heavily doped (with Ga) p-type Ge with N_h between 2×10^{19} and 2×10^{20} cm^{-3}. The results obtained for the real and imaginary parts of the self-energy and the corresponding Q parameters are given in Table 2.5 for several ω_L. We note that the Q parameters are large (small asymmetry) and increase with increasing ω_L, a fact which was also observed for p-Si (Fig. 2.87) and which, for Ge, must be due to the resonance of the phonon Raman tensor near E_1 and $E_1 + \Delta_1$ (Fig. 2.9 of [2.232]) and also to the fact that the imaginary part of this Raman tensor increases with increasing frequency. This makes Q complex. The Q's of Table 2.5 were obtained under the assumption that they are real (we should call them $\tilde{Q}$, according to [2.169]). This would certainly be a good approximation for Si with ω_L below the direct gap (3.4 eV) but not for Ge near 2 eV (the direct gap of Ge lies around 0.8 eV). We note that $\tilde{Q}$ is related to the "true" complex Q through (2.169). An increase in Q_i leads to an increase in the effective $|\tilde{Q}|$.

Table 2.5. Shifts, broadenings, and asymmetry parameters of the Raman lines for heavily doped p-Ge for different incident laser energies $\hbar\omega_L$. The experimental error in the determination of the shifts and broadening is ± 0.2 cm^{-1} [2.46]

Hole concentration [cm^{-3}]	$\hbar\omega_L$ [eV]	Shifts $\hbar\Delta\omega$ [cm^{-1}]		Broadenings $\Delta\Gamma$ [cm^{-1}]		Fano asymmetry parameter $\tilde{Q}$	
		300 K	77 K	300 K	77 K	300 K	77 K
2.4×10^{19}	1.83	−0.6	−0.8	1.4	1.1		
	1.92	−0.9	−0.7	1.2	1.3		-25 ± 5
	2.18	−0.1	−0.5	2.2	2		-50 ± 10
	2.41	0.9	0.9	3	2.8		-150 ± 20
	2.54	1	1.3	3.5	4.7		
	2.71	1	0.9	4.2	4.9		
5.5×10^{19}	1.83	−2.2	−2.3	2.4	2.1		
	1.92	−2.1	−2.5	2.4	2.5	-15 ± 2	-9 ± 2
	2.18	−2	−2	3.4	3.1	-50 ± 10	-14 ± 2
	2.41	−1.2	−1.1	3.8	3.5	-80 ± 15	-30 ± 5
	2.54	−1.1	−0.5	5.2	4.9		
	2.71	−0.5	−1.2	6.7	6.8		
2×10^{20}	1.83	−6.8	−7.0	5.6	5.6		
	1.92	−6.5	−6.0	5.4	4.7		-30 ± 10
	2.18	−6.8	−7.4	5.6	6.0		
	2.54	−8.9	−9.1	7.0	8.9		-50 ± 15
	2.71	−8.4	−9.0	8.8	10.2		

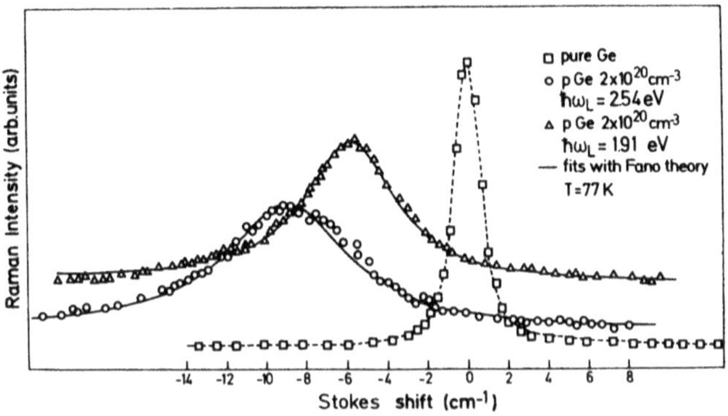

Fig. 2.97. Raman spectra due to optical phonons ($q=0$) of pure and heavily doped p-type germanium showing the frequency-dependent self-energy. From [2.46]

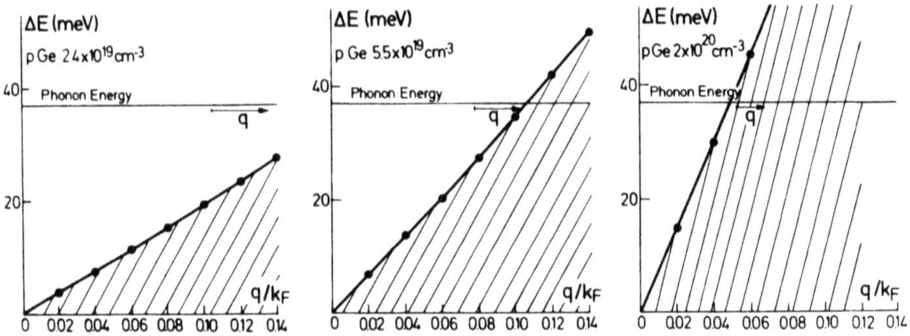

Fig. 2.98. Schematic diagram of single-particle excitations in p-type Ge with three different doping levels, together with the range of q-vectors available for light scattering. From [2.46]

An interesting fact is the negative sign of Q for p-Ge, contrary to the sign found for p-Si. The sign of V is supposed to be the same for p-Ge as for p-Si [2.230], while T_e may reverse sign since the laser frequency lies above the lowest resonant gap (E_0); T_p is complex, but below E_1 ($=2.1$ eV) the imaginary part is small and the real part should be positive like that of Si [Ref. 2.232, Sect. 2.2.6]. Hence, the sign reversal in Q_r may be related to a sign reversal in T_e. We notice that the same sort of anomaly in $\tilde{Q}$ exists for p-GaAs [2.46].

Maybe the most interesting peculiarity of phonons interacting with the continuum of hole excitations in p-Ge is the fact that the phonon self-energy $\Delta\omega_0$ depends on laser frequency (Fig. 2.97, Table 2.5). This dependence has been explained in [2.46,233] as arising from an interaction with the *intra*band continuum (Landau-damping region). We show in Fig. 2.98 the continuum of intraband excitations for the heavy holes in p-Ge. The low-q boundary of this continuum is given by

$$\omega \simeq \frac{\hbar k_F q}{m_{hh}^*}. \tag{2.187}$$

For p-Ge, $m^*_{\rm hh} = 0.34m_0$ and the q of "visible" lasers in backscattering comes close to the free-hole excitation boundary. Hence, large dispersion effects in the self-energy are expected. The real part of the self-energy is given by [cf. (2.174)],

$$\hbar\Delta\omega_0(q) = V^2 P \int \varrho(E', q)\left(\frac{1}{\hbar\omega_0 - E'} - \frac{1}{\hbar\omega_0 + E'}\right) dE'$$
$$= -\frac{q^2\varepsilon_0}{e^2} V^2 \chi(\omega_0, q), \tag{2.188}$$

where P represents the principal part of the integral, $\varrho(E, q)$ is the density of states for electronic excitations and V is the average matrix element of the electron-phonon interaction obtained by averaging (2.177). The electron-phonon coupling does not need to be screened because the phonon does not have the symmetry of the crystal. The function $\chi(\omega_0, q)$ is the Lindhard dielectric constant in the absence of scattering. In [2.46] the integral of (2.188) was replaced by the approximate expression

$$P \int_0^{\Delta_{\rm max}(q)} \varrho(E, \boldsymbol{q})\left(\frac{1}{\hbar\omega_0 - E} - \frac{1}{\hbar\omega_0 + E}\right) dE$$
$$\simeq \frac{m^{*2}_{\rm hh}}{\pi^2\hbar^4 q}\,[\Delta_{\rm max}(\boldsymbol{q}) - \hbar\omega_0 \ln|\hbar\omega_0 - \Delta_{\rm max}(\boldsymbol{q})|$$
$$+ \hbar\omega_0 \ln(\hbar\omega_0)], \tag{2.189}$$

where $\Delta_{\rm max}(\boldsymbol{q}) = E_{\rm F}(2q/k_{\rm F} + q^2/k^2_{\rm F})$.

Figure 2.99 shows the real parts of the self-energy obtained experimentally for two p-Ge samples with several laser lines, corresponding to different values of q. The solid line is a fit with (2.189) for $d_0 = 32$ eV. We note that for Ge [$B = 8.5$,

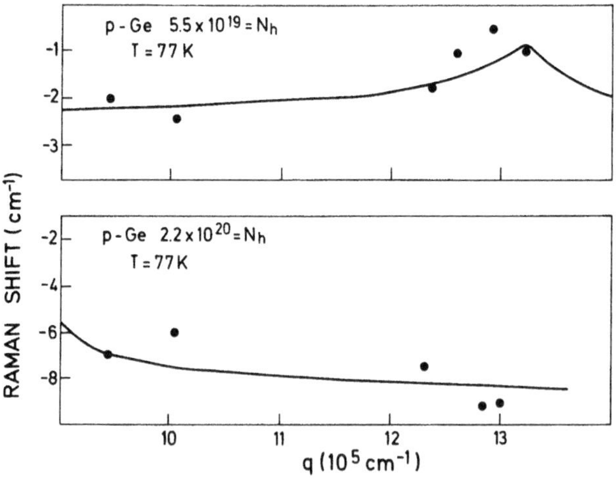

Fig. 2.99. Dispersive real part of the phonon self-energy observed for two heavily doped p-type Ge samples. From [2.46]

$D = 20$ according to (2.228)] $\langle g_{11} \rangle \simeq 0.24$, while in [2.46] a less accurate average ($\langle g_{11} \rangle \simeq 0.15$) was used. Hence, the d_0 required for the fit should be corrected to be $d_0 = 32\ (0.15/0.24)^{1/2} = 25$ eV.

The theory just discussed, while on the surface reasonable, has one serious difficulty. The hole mean free path in heavily doped p-Ge is rather small ($l \sim 60$ Å). Hence, one would not expect the strict wavevector conservation required by the Lindhard $\chi(\omega_0, q)$ to hold. In fact, $2\pi/l$ is much larger than the magnitude of the scattering vector and one should use in (2.188) an l-dependent χ which is not available. In any case, since $q \ll 2\pi/l$ it seems that this would mostly eliminate the dependence of $\Delta\omega_0$ on ω_L and lead to results in disagreement with experiment. More elaborate calculations, including explicitly the mean free path l, are obviously needed in order to clarify the nature of the observed dispersion in $\Delta\omega_0$. A recent effort in this direction can be found in [2.234].

Acknowledgement. We would like to thank Dr. I. P. Ipatova for a critical reading of the manuscript.

References

2.1 E. E. Salpeter: Phys. Rev. **120**, 1528 (1960)
2.2 M. N. Rostoker, N. Rosenbluth: Phys. Fluids **5**, 776 (1962)
2.3 D. F. Du Bois, V. Gilinsky: Phys. Rev. **133**, A1308 (1964)
2.4 D. Pines: *Elementary Excitations in Solids* (Benjamin, New York 1963)
P. Grosse: *Freie Elektronen in Festkörpern* (Springer, Berlin, Heidelberg, New York 1979)
2.5 P. M. Platzmann, N. Tzoar: Phys. Rev. **136**, A11 (1964)
2.6 P. M. Platzman: Phys. Rev. **139**, A379 (1965)
2.7 A. L. McWhorter: *Physics of Quantum Electronics* (McGraw-Hill, New York 1966) p. 111
2.8 P. A. Wolff: Phys. Rev. Lett. **16**, 225 (1966)
2.9 D. C. Hamilton, A. L. McWhorter: In *Light Scattering Spectra of Solids*, ed. by G. B. Wright (Springer, Berlin, Heidelberg, New York 1969) p. 309
2.10 Y. Yafet: Phys. Rev. **152**, 858 (1966)
2.11 P. A. Wolff: In [2.9] p. 273
2.12 A. Mooradian, G. B. Wright: Phys. Rev. Lett. **16**, 999 (1966)
2.13 A. Mooradian: Phys. Rev. Lett. **20**, 1102 (1968)
2.14 A. Mooradian: In [2.9] p. 285
2.15 R. E. Slusher, C. K. N. Patel, P. A. Fleury: Phys. Rev. Lett. **18**, 77 (1967)
2.16 C. K. N. Patel, R. E. Slusher: Phys. Rev. Lett. **21**, 1563 (1968); Phys. Rev. **167**, 413 (1968)
2.17 A. Mooradian: In *Festkörperprobleme IX* (Pergamon-Vieweg, Braunschweig 1969) p. 74; also *Laser Handbook*, Vol. II, ed. by F. T. Arecchi, E. O. Schulz-Du Bois (North-Holland, Amsterdam 1972) p. 1309
2.18 Y. Yafet: In *New Developments in Semiconductors*, ed. by P. R. Wallace, R. Harris and M. J. Zuckerman (Nordhoff, Leyden 1973) p. 469
2.19 P. M. Platzman, P. A. Wolff: *Waves and Interactions in Solid State Plasmas*, (Academic Press, New York 1973)
2.20 M. Cardona (ed.): *Light Scattering in Solids*, 2nd ed., Topics Appl. Phys., Vol. 8 (Springer, Berlin, Heidelberg, New York 1982)
2.21 W. Hayes, R. Loudon: *Scattering of Light by Crystals*, (Wiley, New York 1978)
2.22a C. K. Patel: In *Fundamental and Applied Laser Physics*, ed. by M. S. Feld, A. Javan, N. A. Kurmit (Wiley, New York 1973)

2.22b P.A.Wolff: In *Physics of Quantum Electronics*, Vol. 4, ed. by S.F.Jacobs, M.Sargent III, M.O.Scully, C.T.Walker (Addison-Wesley, Reading, MA 1976) pp. 305 – 324
2.22c J.F.Scott: In [Ref. 2.22b, pp. 325 – 341], also Rep. Prog. Phys. **43**, 61 (1980)
2.23 M.Chandrasekhar, M.Cardona, E.O.Kane: Phys. Rev. ***B*16**, 3579 (1977)
2.24 I.P.Ipatova, A.V.Subashiev, V.A.Voitenko: Solid State Commun. **37**, 893 (1981)
2.25 R.S.Turtelly, A.R.B.Castro, R.C.C.Leite: Solid State Commun. **16**, 969 (1975) A.R.Vasconcellos, R.S.Turtelli, A.R.B. De Castro: Solid State Commun. **22**, 97 (1977)
2.26 D.Guidotti, Shui Lai, M.V.Klein, J.P.Wolfe: Phys. Rev. Lett. **43**, 1950 (1979)
2.27 K.M.Romanek, H.Nather, E.O.Göbel: Solid State Commun. **39**, 23 (1981)
2.28 J.E.Kardontchik, E.Cohen: Phys. Rev. Lett. **42**, 669 (1979)
2.29 A.Pinczuk, Jagdeep Shah, P.A.Wolff: Phys. Rev. Lett. **47**, 1487 (1981)
2.30 D.Pines: Can. J. Phys. **34**, 1379 (1956)
2.31 B.Tell, R.J.Martin: Phys. Rev. **167**, 381 (1968)
2.32 A.Pinczuk, L.Brillson, E.Burstein, E.Anastassakis: *Light Scattering in Solids*, ed. by M.Balkanski (Flammarion, Paris 1971) p. 115
2.33 S.Buchner, E.Burstein: Phys. Rev. Lett. **33**, 908 (1974)
2.34 V.I.Zemski, E.L.Irchenko, D.N.Mirlin, I.I.Reshina: Solid State Commun. **16**, 221 (1975)
2.35 R.Trommer, A.K.Ramdas: In *Physics of Semiconductors 1978*, ed. by B.L.H.Wilson (The Institute of Physics, London 1979) p. 585
2.36a K.Murase, S.Katayama, S.Ando, H.Kawamura: Phys. Rev. Lett. **33**, 1481 (1974)
2.36b S.Katayama, K.Murase, H.Kawamura: Solid State Commun. **16**, 945 (1975)
2.37 A.Pinczuk, G.Abstreiter, R.Trommer, M.Cardona: Solid State Commun. **21**, 959 (1977)
2.38 G.Abstreiter, R.Trommer, M.Cardona, A.Pinczuk: Solid State Commun. **30**, 703 (1979)
2.39 W.Richter, U.Nowak, A.Stahl: J. Phys. Soc. Japan **49** (Suppl. A), 703 (1980); U.Nowak, W.Richter, G.Sachs: Phys. Stat. Sol. **108**(b), 131 (1981)
2.40 H.J.Stolz, G.Abstreiter: Solid State Commun. **36**, 857 (1980)
2.41 H.J.Stolz, G.Abstreiter: J. Vac. Sci. Technol. **19**, 380 (1981)
2.42 M.Cardona: J. Phys. Soc. Japan **49**, (Suppl. A), 23 (1980)
2.43 R.Beserman, M.Jouanne, M.Balkanski: *Proc. of the Eleventh Intern. Conf. on the Physics of Semiconductors* (Polish Scientific Publications, Warsaw 1972) p. 1181
2.44 F.Cerdeira, M.Cardona: Phys. Rev. B**5**, 1140 (1972)
2.45 M.Chandrasekhar, M.Renucci, M.Cardona: Phys. Rev. B**17**, 1623 (1978)
2.46 D.Olego, M.Cardona: Phys. Rev. B**23**, 6592 (1981)
2.47 M.Chandrasekhar, U.Rössler, M.Cardona: Phys. Rev. B**22**, 761 (1980)
2.48 F.Cerdeira, T.A.Fjeldly, M.Cardona: Phys. Rev. B**8**, 4734 (1973)
2.49 L.Pintschovius, J.A.Verges, M.Cardona: J. Physique C**6**, 942 (1981); also Phys. Rev. B**26**, 5658 (1982)
2.50 A.Pinczuk, L.Brillson, E.Burstein, E.Anastassakis: Phys. Rev. Lett. **27**, 317 (1971)
2.51 A.Pinczuk, G.Abstreiter, R.Trommer, M.Cardona: Solid State Commun. **30**, 429 (1979)
2.52 E.Burstein, A.Pinczuk, S.Buchner: In [Ref. 2.35, p. 1231]
2.53 R.Dingle, H.L.Störmer, A.C.Gossard, W.Wiegmann: Appl. Phys. Lett. **33**, 665 (1978) H.L.Störmer, R.Dingle, A.C.Gossard, W.Wiegmann, R.A.Logan: In [Ref. 2.35, p. 557]
2.54 G.Abstreiter, K.Ploog: Phys. Rev. Lett. **42**, 1308 (1979)
2.55 A.Pinczuk, H.L.Störmer, R.Dingle, J.M.Worlock, W.Wiegmann, A.C.Gossard: Solid State Commun. **32**, 1001 (1979)
2.56 A.Pinczuk, J.M.Worlock: Surf. Sci. **113**, 69 (1982)
2.57 A.Pinczuk, J.Shah, A.C.Gossard, W.Wiegmann: Phys. Rev. Lett. **46**, 1307 (1981)
2.58 J.M.Worlock, A.Pinczuk, Z.J.Tien, C.H.Perry, H.L.Störmer, R.Dingle, A.C.Gossard, W.Wiegmann, R.Aggarwal: Solid State Commun. **40**, 867 (1981)
2.59 D.Olego, A.Pinczuk, A.C.Gossard, W.Wiegmann: Phys. Rev. B**25**, 7867 (1982)
2.60 L.Y.Ching, E.Burstein, S.Buchner, H.H.Wieder: In [Ref. 2.39, p. 951]
2.61 G.Abstreiter, U.Claessen, G.Tränkle: Solid State Commun. **44**, 673 (1982)
2.62 G.H.Döhler, H.Künzel, D.Olego, K.Ploog, P.Ruden, H.J.Stolz, G.Abstreiter: Phys. Rev. Lett. **47**, 864 (1981)
2.63 R.Merlin, A.Pinczuk, W.T.Beard, C.E.E.Wood: J. Vac. Sci. Technol. **21**, 516 (1982)

2.64 M.Cardona, G.Güntherodt (eds.): *Light Scattering in Solids* II, Topics Appl. Phys., Vol. 50 (Springer, Berlin, Heidelberg, New York 1982)
2.65 A.Pinczuk, E.Burstein: In [Ref. 2.20, p. 23]
2.66 S.S.Jha: Phys. Rev. **182**, 815 (1969)
2.67 F.A.Blum: Phys. Rev. B**1**, 1125 (1970)
2.68 D.Pines, P.Nozieres: *The Theory of Qunatum Liquids* (Benjamin, New York 1966)
2.69 R.Zeyher, H.Bilz, M.Cardona: Solid State Commun. **19**, 57 (1976)
2.70 L.D.Landau, E.M.Lifshitz: *Statistical Physics* (Pergamon Press, Oxford 1968)
2.71a M.V.Klein, B.N.Ganguly, P.J.Colwell: Phys. Rev. B**6**, 2380 (1972)
2.71b E.M.Lifshitz, L.P.Pitaevskii: *Physical Kinetics* (Pergamon Press, Oxford 1981), p. 129
2.72 D.A.Abramsohn, K.T.Tsen, R.Bray: Phys. Rev. B**26**, 6571 (1982)
2.73 A.R.Vasconcellos, R.Luzzi: Solid State Commun. **17**, 501 (1975)
2.74 N.D.Mermin: Phys. Rev. B**1**, 2362 (1970)
2.75 I.Yokota: J. Phys. Soc. Japan **16**, 2075 (1961)
2.76 F.Stern: *Solid State Physics* **15**, 362 (Academic Press, New York 1963)
2.77 B.B.Varga: Phys. Rev. **137**, A1896 (1965)
2.78 A.Mooradian, A.L.McWhorter: Phys. Rev. Lett. **19**, 849 (1967)
2.79 E.Burstein, A.Pinczuk, S.Iwasa: Phys. Rev. **157**, 611 (1967)
2.80 P.Nozieres, D.Pines: Phys. Rev. **109**, 1062 (1958)
2.81 D.Pines, J.R.Schrieffer: Phys. Rev. **124**, 1387 (1961)
2.82 S.S.Jha: Phys. Rev. **179**,764 (1969)
2.83 N.Tzoar, E.N.Foo: In [Ref. 2.32, p. 119]
2.84 E.N.Foo, N.Tzoar: Phys. Rev. B**6**, 4553 (1972)
2.85a J.Ruvalds, L.M.Khan: Phys. Lett. **70**A, 477 (1979); S.K.Sinha, C.M.Varma: Phys Rev. B**28**, 1663 (1983)
2.85b H.Fröhlich: J. Phys. C**1**, 544 (1968)
2.86 J.Appel, A.W.Overhauser: Phys. Rev. B**26**, 507 (1982)
2.87 D.Olego, M.Cardona: Solid State Commun. **32**, 375 (1979)
2.88 S.S.Jha: Nuovo Cimento **58**B, 331 (1969)
2.89 F.A.Blum, A.Mooradian: In *Proc. 12th Intern. Conf. Physics of Semiconductors*, ed. by S.Keller, J.C.Hensel, F.Stern (USAEC, Oak Ridge, Tenn. 1970) p. 755
2.90 S.R.J.Brueck, A.Mooradian, F.A.Blum: Phys. Rev. B**7**, 5253 (1973)
2.91 J.F.Scott, T.C.Damen, J.Ruvalds, A.Zawadovski: Phys. Rev. **83**, 1295 (1971)
2.92 R.J.Elliott, R.Loudon: Phys. Lett. **3**, 189 (1963)
2.93 J.Callaway: In *Quantum Theory of the Solid State*, Part B (Academic Press, New York 1974)
2.94 T.Ando, A.Fowler, F.Stern: Revs. Mod. Phys. **54**, 437 (1982)
2.95 E.Burstein, A.Pinczuk, D.L.Mills: Surf. Sci. **98**, 451 (1980)
2.96 E.O.Kane: J. Phys. Chem. Solids **1**, 249 (1957)
2.97 A.L.McWhorter, P.N.Argyres: In [Ref. 2.9, p. 325]
2.98 E.Burstein, A.Pinczuk: In *The Physics of Opto-Electronic Materials*, ed. by W.A.Albers (Plenum Press, New York 1971) pp. 33–79
2.99 R.Merlin: Private communication
2.100 M.Cardona, K.L.Shaklee, F.H.Pollak: Phys. Rev. **154**, 696 (1967)
2.101a C.Y.Chen: Phys. Rev. B**27**, 1436 (1983); J.Henendez, M.Cardona: Phys. Rev. Lett. **51**, 1297 (1983)
2.101b Note that the cdf mechanism can also be construed as an effect of the Fröhlich Hamiltonian (2.87) of the plasmons acting as either step 1 or 3 of Fig. 2.6. In the mechanism usually referred to as forbidden Fröhlich, however, this field acts as step 2 (see Figs. 2.4 of [2.65] and 2.39 of [2.64], also F.Cerdeira, N.Mestres, M.Cardona: Solid State Commun. (in press)
2.102 M.Cardona: In *Modulation Spectroscopy*, (Academic Press, New York 1969)
2.103 P.J.Colwell, M.V.Klein: Solid State Commun. **8**, 2095 (1970)
2.104 A.A.Gogolin, E.I. Rashba: Solid State Commun. **19**, 1177 (1976)
2.105 D.Olego, M.Cardona: Phys. Rev. B**24**, 7217 (1981)
2.106a D.Olego, A.Pinczuk, A.A.Ballman: Solid State Commun. **45**, 941 (1983)
2.106b J.R.Chelikowsky, M.L.Cohen: Phys. Rev. B**14**, 556 (1976)

2.107 D.T.Hon, W.L.Faust: Appl. Phys. **1**, 241 (1973)
2.108 A.Pinczuk, S.G.Louie, B.Welber, J.C.Tsang, J.A.Bradley: In [Ref. 2.35, p. 1191]
A.Pinczuk, B.Welber, J.C.Tsang, J.A.Bradley: Unpublished
2.109 G.D.Aumiller: Optics Commun. **41**, 115 (1982)
2.110 D.E.Aspnes, A.A.Studna: Phys. Rev. B**7**, 4605 (1971)
2.111 V.A.Voitenko, I.P.Ipatova, A.V.Subashiev: In *Light Scattering in Solids*, ed. by J.L.Birman, H.Z.Cummins, K.Rebane, (Plenum Press, New York 1979) p. 83
2.112 P.Hertel, J.Appel: Phys. Rev. B**26**, 5730 (1982)
2.113 D.Julienne, F. Le Saos, A.Fortini, Ph.Bauduin: Phys. Rev. B**13**, 2576 (1976)
2.114 Landolt-Börnstein Tables, Vol. 17a, ed. by O.Madelung (Springer, Berlin, Heidelberg, New York 1982)
2.115 G.Abstreiter: Unpublished
2.116 A.Pinczuk, E.Burstein: Phys. Rev. Lett. **21**, 1073 (1968); also in [Ref. 2.89, p. 727]
2.117 R.Dornhaus, R.L.Farrow, R.K.Chang: Solid State Commun. **35**, 123 (1980)
2.118 G.Abstreiter, A.Pinczuk, R.Trommer, R.Tsu: In *Proc. 13th Intern. Conf. on Physics of Semiconductors*, ed. by G.Fumi (Tipografia Marves, Rome 1976) p. 779
2.119 E.L.Ivchenko, D.N.Mirlin, I.I.Reshina: Sov. Phys. Solid State **17**, 1510 (1976)
2.120 W.Richter, A.Krost, U.Nowak, E.Anastassakis: Z. Phys. B**49**, 191 (1982)
2.121 D.L.Mills, A.A.Maradudin, E.Burstein: Phys. Rev. Lett. **21**, 1178 (1968)
2.122 D.Olego, T.Y.Chang, E.Silberg, E.A.Caridi, A.Pinczuk: Appl. Phys. Lett. **41**, 476 (1982)
2.123 T.Lindhard: Dan. Mat. Fys. Medd. **28**, 881 (1954)
2.124a L.F.Lemmens, J.T.Devreese: Solid State Commun. **14**, 1339 (1974)
2.124b L.F.Lemmens, F.Brosens, J.T.Devreese: Solid State Commun. **17**, 337 (1975)
2.125 G.Abstreiter, E.Bauser, A.Fischer, K.Ploog: Appl. Phys. **16**, 345 (1979)
2.126 P.Parayanthal, F.H.Pollak, J.M.Woodall: Appl. Phys. Lett. **41**, 961 (1982)
2.127 B.O.Seraphin, H.E.Bennett: In *Semiconductors and Semimetals*, **3**, 499 (Academic Press, New York 1967);
D.E.Aspnes, A.A.Studna: Phys. Rev. B**27**, 985 (1983)
2.128 G.D.Pitt, J.Lees: Phys. Rev. B**2**, 4144 (1970)
2.129a B.Welber, M.Cardona, C.K.Kim, S.Rodriguez: Phys. Rev. B**12**, 5729 (1975)
2.129b R.Trommer, E.Anastassakis, M.Cardona: In *Light Scattering in Solids*, ed. by M.Balkanski, R.C.C.Leite, S.P.S.Porto (Flammarion, Paris 1976), p. 396
2.130 Ch.Zeller, G.Abstreiter, K.Ploog: Surf. Sci. **113**, 85 (1982)
2.131 W.F.Brinkman, T.M.Rice: Phys. Rev. B**7**, 1508 (1973)
2.132 R.Dingle, A.C.Gossard, W.Wiegmann: Phys. Rev. Lett. **33**, 827 (1974)
2.133 D.L.Mills, R.F.Wallis, E.Burstein: In [Ref. 2.32, p. 107]
2.134 M.Combescot, P.Nozieres: Solid State Commun. **10**, 301 (1972)
2.135 H.L.Störmer: J. Phys. Soc. Japan **49** (Suppl. A), 1013 (1980)
2.136 L.L.Chang, L.Esaki: Surf. Sci. **98**, 70 (1980)
2.137 K. von Klitzing, G.Dorda, M.Pepper: Phys. Rev. Lett. **45**, 494 (1980)
2.138 D.C.Tsui, H.L.Störmer, A.C.Gossard: Phys. Rev. B**25**, 1405 (1982)
2.139 D.C.Tsui, H.L.Störmer, A.C.Gossard: Phys. Rev. Lett. **48**, 1559 (1982)
2.140 G.Abstreiter: Surf. Sci. **98**, 117 (1980)
2.141 A.Pinczuk, J.M.Worlock, H.L.Störmer, R.Dingle, W.Wiegmann, A.C.Gossard: Surf. Sci. **98**, 126 (1980)
2.142 A.Pinczuk, J.M.Worlock, H.L.Störmer, R.Dingle, W.Wiegmann, A.C.Gossard: Solid State Commun. **36**, 43 (1980)
2.143 G.Abstreiter, Ch.Zeller, K.Ploog: In *Proc. 8th Intern. Symp. on GaAs and Related Compounds*, ed. by H.W.Thim (The Institute of Physics, London 1981), p. 741
2.144 S.Das Sarma: Appl. Surf. Sci. **11/12**, 535 (1982)
2.145 T.Ando: J. Phys. Soc. Japan **51**, 3893 (1982)
2.146 H.L.Störmer, A.Pinczuk, A.C.Gossard, W.Wiegmann: Appl. Phys. Lett. **38**, 691 (1981)
2.147 H.L.Störmer, A.C.Gossard, W.Wiegmann, K.Baldwin: Appl. Phys. Lett. **39**, 912 (1981)
2.148 J.C.M.Hwang, A.Kastalsky, H.L.Störmer, V.G.Keramidas: 2nd Intern. Symp. on Molecular Beam Epitaxy, Tokyo, Japan (1982), unpublished
2.149 T.Ando, S.Mori: J. Phys. Soc. Japan **47**, 1518 (1979)

2.150 G.Fishman: Phys. Rev. B**27**, 7611 (1983)
2.151 W.P.Chen, Y.J.Chen, E.Burstein: Surf. Sci. **58**, 263 (1976)
2.152 S.J.Allen Jr., D.C.Tsui, B.Vinter: Solid State Commun. **20**, 425 (1976)
2.153 T.Ando: Solid State Commun. **21**, 133 (1977)
2.154 H.R.Chang, H.Reisinger, F.Schäffler, J.Scholz, K.Wiesinger, F.Koch: J. Phys. Soc. Japan **49**A, 955 (1980)
2.155 D.Dahl, L.J.Sham: Phys. Rev. B**16**, 651 (1977)
2.156 G.Fishman, A.Pinczuk, J.M.Worlock, H.L.Störmer, A.C.Gossard, W.Wiegmann: J. Physique **12**, C**6**, 305 (1981)
2.157 A.Pinczuk, J.M.Worlock, H.L.Störmer, A.C.Gossard, W.Wiegmann: J. Vac. Sci. Technol. **19**, 561 (1981)
2.158 A.Pinczuk, J.M.Worlock, H.L.Störmer, A.C.Gossard, W.Wiegmann, R.Dingle: Bull. Am. Phys. Soc. **26**, 254 (1981)
2.159 A.Pinczuk, J.M.Worlock, H.L.Störmer, A.C.Gossard, W.Wiegmann: To be published
2.160 Z.J.Tien, J.M.Worlock, C.H.Perry, A.Pinczuk, R.L.Aggarwal, H.L.Störmer, A.C.Gossard, W.Wiegmann: Surf. Sci. **113**, 89 (1982)
2.161 A.L.Fetter: Ann. Phys. **88**, 1 (1974)
2.162 S.Das Sarma, A.Madhukar: Phys. Rev. B**23**, 805 (1981)
2.163 S.Das Sarma, J.J.Quinn: Phys. Rev. B**25**, 1603 (1982)
2.164 W.L.Bloss, E.M.Brody: Solid State Commun. **43**, 523 (1982)
2.165 S.J.Allen,Jr., D.C.Tsui, P.A.Logan: Phys. Rev. Lett. **38**, 980 (1977)
2.166 T.N.Theis, J.P.Kotthaus, P.J.Stiles: Solid State Commun. **26**, 603 (1978)
2.167 R.Höpfel, G.Lindemann, E.Gornik, G.Stangl, A.C.Gossard, W.Wiegmann: Surf. Sci. **113**, 118 (1982)
2.168 E.A.Kraut, R.W.Grant, J.R.Waldrop, S.P.Kowalczyk: Phys. Rev. Lett. **44**, 1620 (1980)
2.169 G.H.Döhler: Phys. Stat. Sol. (b), **52**, 79 and 533 (1972), and J. Vac. Sci. Technol. **16**, 851 (1979)
2.170 K.Ploog, A.Fischer, H.Künzel: J. Electrochem. Soc. **128**, 400 (1981)
2.171 G.H.Döhler, H.Künzel, K.Ploog: Phys. Rev. B**25**, 2616 (1982)
2.172 P.Ruden, G.H.Döhler: Phys. Rev. B**27**, 3538 and 3547 (1983)
2.173 Ch.Zeller, B.Vinter, G.Abstreiter, K.Ploog: Phys. Rev. B**26**, 2124 (1982)
2.174 See for example, K.Ploog: In *Crystals: Growth, Properties, and Applications*, **3**, 73 (Springer, Berlin, Heidelberg, New York 1980)
2.175 Ch.Zeller, B.Vinter, G.Abstreiter, K.Ploog: Physica B**117**/**118**, 729 (1983)
2.176 H.Jung, G.H.Döhler, H.Künzel, K.Ploog, P.Ruden, H.J.Stolz: Solid State Commun. **43**, 291 (1982)
2.177 G.Tränkle, G.Abstreiter: Unpublished
G.Tränkle: Diploma thesis, Techn. Univ. München (1981)
2.178 G.Abstreiter, R.Huber, G.Tränkle, B.Vinter: Solid State Commun. **47**, 651 (1983)
2.179 P.Kneschaurek, A.Kamgar, J.F.Koch: Phys. Rev. B**14**, 1610 (1976)
2.180 H.Reisinger, F.Koch: Solid State Commun. **37**, 429 (1981)
2.181 E.Bangert, K.von Klitzing, G.Landwehr: In *Proc. 12th Intern. Conf. Phys. Semicond.*, ed. by M.H.Pilkuhn (Teubner, Stuttgart 1974) p. 714, and E.Bangert: Unpublished (1975)
2.182 F.J.Ohkawa, Y.Uemura: Suppl. Progr. Theoret. Phys. **57**, 164 (1975)
2.183 A.Kamgar, P.Kneschaurek, W.Beinvogl, J.F.Koch: In *Proc. 12th Intern. Conf. Phys. Semicond.*, ed. by M.H.Pilkuhn (Teubner, Stuttgart 1974) p. 709
2.184 D.L.Lile, D.A.Collins, L.G.Meiners, L.Messick: Electron Lett. **14**, 657 (1978)
2.185 D.Fritzsche: In *Insulating Films on Semiconductors*, ed. by G.G.Roberts, M.J.Morant (Institute of Physics, Bristol 1979) p. 258
2.186 K. von Klitzing, Th.Englert, E.Bangert, D.Fritzsche: J. Appl. Phys. **51**, 5893 (1980)
2.187 H.C.Cheng, F.Koch: Solid State Commun. **37**, 911 (1981) and Phys. Rev. B**26**, 1989 (1982)
2.188 A.Huijser, J. van Laar: Surf. Sci. **52**, 202 (1975), and J. van Laar, A.Huijser: J. Vac. Sci. Technol. **13**, 76 (1976)
2.189 G.Heiland, E.Mollwo, F.Stöckmann: In *Solid State Phys.* **8**, 191 (Academic Press, New York 1959)
2.190 Y.Goldstein, A.Many, D.Eger, Y.Grinshpan, G.Yaron, M.Nitzan: Phys. Lett. **62**A, 57 (1977)

2.191 D.C.Tsui: Phys. Rev. B**4**, 4438 (1971); B**8**, 2657 (1973); B**12**, 5739 (1974)
2.192 V.Dvorak: Phys. Rev. **159**, 652 (1967)
2.193 J.M.Worlock: In [Ref. 2.9, p. 411]
2.194 P.A.Fleury, J.M.Worlock: Phys. Rev. **174**, 613 (1968)
2.195 E.Burstein, A.A.Maradudin, E.Anastassakis, A.Pinczuk: Helv. Phys. Acta **41**, 730 (1968)
2.196 A.Pinczuk, E.Burstein: In [Ref. 2.9, p. 42]
2.197 J.G.Gay, J.D.Dow, E.Burstein, A.Pinczuk: In [Ref. 2.32, p. 33]
2.198 L.Brillson, E.Burstein: Phys. Rev. Lett. **27**, 808 (1971)
2.199 M.L.Shand, W.Richter, E.Burstein, J.G.Gay: J. Nonmetals **1**, 53 (1972)
2.200 R.Trommer, G.Abstreiter, M.Cardona: In *Proc. Intern. Conf. on Lattice Dynamics*, ed. by M.Balkanski, (Flamarion, Paris 1977) p. 189
2.201 H.J.Stolz, G.Abstreiter: In *Proc.* 15th *Intern. Conf. Phys. Semicond.*, Kyoto (1980); J. Phys. Soc. Japan **49**, Suppl. A, 1101 (1980)
2.202 P.Pianetta, J.Lindau, P.E.Gregory, C.M.Garmer, W.E.Spicer: Surf. Sci. **72**, 298 (1978)
2.203 C.D.Thurault, G.M.Guichar, C.A.Sebenne: Surf. Sci. **80**, 273 (1979)
2.204 F.Schäffler, G.Abstreiter: *Surface Studies with Lasers*, Springer Ser. Chem. Phys., Vol. 33 (Springer, Berlin, Heidelberg, New York, Tokyo 1983), p. 131
2.205 G.Abstreiter: Verhandl. DPG VI **12**, 81 (1977)
2.206 A.Pinczuk, A.A.Ballman, R.E.Nahory, M.A.Pollak, J.M.Worlock: J. Vac. Sci. Technol. **16**, 1168 (1979)
2.207 S.Buchner, L.Y.Ching, E.Burstein: Phys. Rev. B**14**, 4459 (1976)
2.208 M.A.Renucci, J.B.Renucci, M.Cardona: Phys. Stat. Sol. (b) **49**, 625 (1972)
2.209 F.Cerdeira, T.A.Fjeldly, M.Cardona: Phys. Rev. B**8**, 4734 (1972)
2.210 D.Olego, M.Cardona: Phys. Rev. B**23**, 6592 (1981)
2.211 H.R.Chandrasekhar, A.K.Ramdas, S.Rodriguez: Phys. Rev. B**14**, 2417 (1976)
2.212 M.Chandrasekhar, H.R.Chandrasekhar, M.Grimsditch, M.Cardona: Phys. Rev. B**22**, 4825 (1980)
2.213 I.P.Ipatova, A.V.Subashiev, V.A.Voitenko: J. Raman Spectr. **10**, 221 (1981)
2.214 K.Jain, S.Lai, M.V.Klein: Phys. Rev. **13**, 5448 (1976)
2.215 J.Doehler: Phys. Rev. B**12**, 2917 (1975)
2.216 A.Kawabata: J. Phys. Soc. Japan **30**, 68 (1971)
2.217 F.Bechstedt, K.Peuker: Phys. Stat. Sol. (b) **72**, 743 (1975)
2.218 M.Balkansi, K.P.Jain, R.Besserman, M.Jouanne: Phys. Rev. B**12**, 4328 (1975)
2.219 I.P.Ipatova, A.V.Subashiev: In *The Theory of Light Scattering in Solids*, ed. by V.M.Agranovitch, J.L.Birman (Nauka, Noscow 1976) p. 248; ibid. Sov. Phys. Solid State **18**, 1251 (1976)
2.220a A.Nitzan: Mol. Phys. **27**, 65 (1973)
2.220b U.Fano: Phys. Rev. **124**, 1866 (1961)
2.221 T.R.Hart, R.L.Aggarwal, B.Lax: Phys. Rev. **1**, 638 (1970); J.Menéndez, M.Cardona: Phys. Rev. (in press)
2.222 N.Kristoffel, P.Konsin: Phys. Stat. Solidi **28**, 731 (1969); H.Kawamura, K.Murase, S.Nishikawa, S.Nishi, S.Katayama: Solid State Commun. **17**, 341 (1976)
2.223 R.Biswas, V.Ambegaokar: Phys. Rev. B**26**, 1980 (1982)
2.224 R.B.Keyes: In *Solid State Physics*, **20**, 37 (Academic Press, New York 1967); T.A.Fjeldly, F.Cerdeira, M.Cardona: Phys. Rev. B**8**, 4723 (1973)
2.225 M.Jouanne, R.Besserman, I.Ipatova, A.Subashiev: Solid State Commun. **16**, 1047 (1975)
2.226 P.Vogl, W.Pötz: Phys. Rev. **24**, 2025 (1981)
2.227 E.G.Brovman, Yu.M.Kagan: In *Dynamical Properties of Solids*, Vol. 1, ed. by G.K.Horton, A.A.Maradudin (North-Holland, Amsterdam 1974) p. 191
2.228 P.Lawaetz: Unpublished report, Technical University of Denmark (1978)
2.229 M.A.Kanehisa, R.F.Wallis, M.Balkanski: Phys. Rev. B**25**, 7619 (1982)
2.230 M.Cardona, F.Cerdeira, T.Fjeldly: Phys. Rev. **10**, 3433 (1974)
2.231 R.Szabo, L.Genzel: Unpublished
2.232 M.Cardona, G.Güntherodt (eds.): *Light Scattering in Solids* III, Topics Appl. Phys., Vol. 51 (Springer, Berlin, Heidelberg, New York 1982)
2.233 I.P.Ipatova, A.V.Subashiev: Sov. Phys. JETP **39**, 349 (1974)
2.234 I.P.Ipatova, A.V.Subashiev, V.A.Shchukin: Sov. Phys. Solid State **24**, 1932 (1982)

3. High Resolution Spin-Flip Raman Scattering in CdS

Stanley Geschwind and Robert Romestain

With 28 Figures

This review of spin-flip Raman scattering (SFRS) in CdS covers a variety of experimental findings, many of which were a direct result of the use of very *high resolution* techniques.

3.1 Introductory Comments

3.1.1 Historical Background

SFRS from conduction electrons was first suggested by *Yafet* [3.1] following a suggestion by *Wolff* [3.2] on light scattering from Landau levels in a semiconductor. SFRS is most conveniently illustrated in terms of noninteracting bound donor electrons. The electric field of incident laser light of frequency ω_L interacts via spin-orbit coupling with a spin in a magnetic field H_0 and induces a spin flip $|\downarrow\rangle \rightarrow |\uparrow\rangle$ with the frequency of the scattered light shifted to $\omega_L - \omega_0$ (Stokes), where $\omega_0 = g\beta H_0/\hbar$, the Zeeman frequency. Similarly, a transition $|\uparrow\rangle \rightarrow |\downarrow\rangle$ shifts the light to higher frequency (anti-Stokes) $\omega_L + \omega_0$. The frequency of the scattered light is therefore tunable by a magnetic field and was first observed by *Slusher* et al. [3.3] for conduction electrons in InSb which have a g-value of -50. This led to development of a class of tunable far-infrared lasers using stimulated SFRS (see reviews by *Patel* [3.4] and *Colles* and *Pidgeon* [3.5]). It was first studied for bound donors in CdS by *Thomas* and *Hopfield* [3.6]. The wide application of conventional SFRS to the study of the properties of donors and acceptors in semiconductors has been summarized in a number of fine earlier reviews [3.7, 8].

3.1.2 Role of High Resolution Fabry-Perot Spectroscopy in SFRS

In almost all of the SFRS work referred to above, the scattered Raman radiation was spectrally analyzed using grating spectrometers whose resolution were typically 0.1 to 0.5 cm^{-1}. Even with multipass grating instruments, the unrejected residual laser light makes it difficult to observe Raman shift of less than 1 cm^{-1}. Thus, fairly high magnetic fields were needed as well. By contrast, high-resolution Fabry-Perot techniques afford a resolution of $< 10^{-3}$ cm^{-1}, and SFRS has been studied in CdS in fields as low as 160 Gauss [3.9] corresponding to Raman shifts of $\sim 0.01\ cm^{-1}$. This high resolution has led to a wide variety of experiments and findings in CdS, many of which would have been more difficult if not impossible with grating instruments. Examples of these are the measurement of the k-linear term in the conduction band, the observation of pure spin diffusion, field-induced dispersion in the amorphous antiferromagnet and SFRS from microwave-induced coherent states. These will be among the topics in high resolution SFRS reviewed in more detail below.

3.1.3 Experimental Procedure

The experimental arrangement used in high resolution SFRS is almost identical to that used in Brillouin scattering (except for the addition of the external magnetic field). A simple schematic diagram of such a system is illustrated in Fig. 3.1. In our experiments, a single longitudinal mode cw Ar^+ laser was used as the source, allowing operation at a number of discrete lines. The 4880 Å line is most convenient in terms of the large SFRS cross section in CdS at this wavelength and was most often used. The scattered light is spectrally analyzed by a Fabry-Perot which is piezoelectrically scanned at rates anywhere from a few to ten Hz. The light transmitted through the Fabry-Perot is processed by photon counting equipment and a multichannel analyzer whose sweep rate is synchronized with that of the Fabry-Perot. The sample of CdS is contained in a low temperature dewar with optical access and is immersed in either pumped liquid He for the lowest temperature or in a temperature regulated He gas flow for temperatures from ~ 5 to 70 K. Various scattering angles are obtained by use of prisms adjacent to the sample which is mounted on a rotating stage.

As in all inelastic light scattering, it is vital to reduce the amount of stray laser light reaching the detection system. This can be done by use of an iodine filter as first demonstrated by *Devlin* et al. [3.10]. This technique is, however, limited to working at the single laser wavelength of 5145 Å and requires computer correction for other iodine absorption lines within a $1\ cm^{-1}$ region [3.11]. A

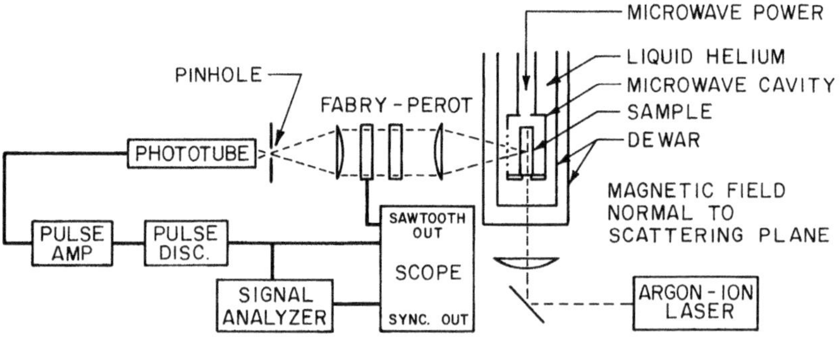

Fig. 3.1. Schematic diagram of apparatus for high resolution SFRS. The microwave apparatus is added for the experiments on Raman scattering from coherent spin states described in Sect. 3.8.1

more general technique to reduce stray laser light is to use a multipass Fabry-Perot pioneered by *Sandercock* [3.12]. There are many excellent reviews on modern Fabry-Perot spectroscopy including multipass and tandem operation and the most recent ones by *Sandercock* [3.13a, b] and *Dil* et al. [3.14] contain references to other work as well. Further information may be found in a brochure published by Burleigh Instruments, USA who manufacture much of the needed equipment.

While CdS was an extremely convenient choice for these studies because its band gap is in the visible region and close to a number of lines of low power cw argon ion lasers, the continued refinement of cw single frequency dye lasers should result in the extension of high resolution SFRS to other semiconductors as well.

3.2 Review of Spin-Flip Raman Scattering

3.2.1 Classical Picture and Role of Spin-Orbit Coupling

A classical picture of SFRS is illustrated in Fig. 3.2. The spin precesses in an external magnetic field H_0 at the Zeeman frequency ω_0 and drags around a charge cloud tied to it by spin orbit (s-o) coupling as shown. Thus, the incident light sees a polarizability α modulated at ω_0 and a polarization is induced with frequencies $\omega_L \pm \omega_0$ which then radiates at these frequencies. This is analogous to vibrational scattering at $(\omega \pm \omega_v)$ due to the modulation of α by a vibration

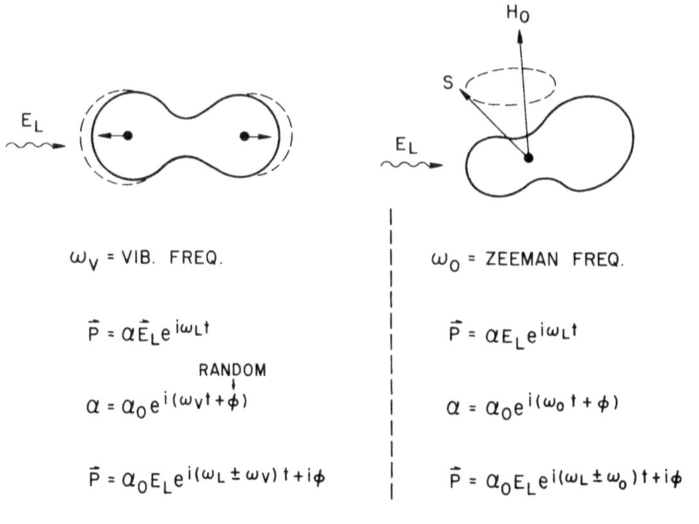

Fig. 3.2. Classical picture of SFRS in analogy with Raman scattering from molecular vibrations. Spin-orbit coupling drags a charge cloud around at spin precessional frequency ω_0, thus modulating the polarizability and giving rise to sidebands at $\omega_L \pm \omega_0$

frequency at ω_v. The s-o coupling is the handle via which the electric field of the light couplies to the spin. The essential role of s-o coupling in SFRS will be further emphasized below in Sect. 3.3.2.

3.2.2 Cross Section by Semiclassical Treatment: Raman Dipole $D^{(2)}$

In order to obtain a quantitative picture of SFRS, it is very convenient to work in the semiclassical description of interaction between light and matter [3.15]. Treating the field classically is also very convenient for the coherent phenomena to be discussed later. The Raman scattering is assumed to occur between two states $|a\rangle$ and $|b\rangle$ of the ground state manifold where $|a\rangle$ and $|b\rangle$ may correspond, for example, to the two spin states $|\downarrow\rangle$ and $|\uparrow\rangle$ of shallow donors in CdS with $S=1/2$. The electric field $\boldsymbol{E}_\mathrm{L}\cos\omega_\mathrm{L}t$ of the incident laser beam admixes electronic excited states $|n\rangle$ into the states $|a\rangle$, $|b\rangle$ so that the modified states are given in first-order time-dependent perturbation theory by

$$|\psi_a\rangle=\exp(-\mathrm{i}\omega_a t-\mathrm{i}\phi)\times\left(|a\rangle-\sum_n\frac{|n\rangle\langle n|e\boldsymbol{r}\cdot\boldsymbol{E}_\mathrm{L}|a\rangle}{2(E_n-E_a-\hbar\omega_\mathrm{L})}\exp(-\mathrm{i}\omega_\mathrm{L}t)\right),\tag{3.1}$$

where the nonresonant term in $\exp(+\mathrm{i}\omega_\mathrm{L}t)$ has been omitted for simplicity. $|\psi_b\rangle$ is given by a similar expression without the phase factor $\exp(-\mathrm{i}\phi)$ which describes the relative phase between the two states $|a\rangle$ and $|b\rangle$. The matrix element $\langle\psi_b|e\boldsymbol{r}|\psi_a\rangle$ describes the dipole emitting Raman radiation at frequencies $\omega_\mathrm{L}\mp\omega_{ab}$.

It is convenient to introduce an effective Raman dipole operator $\boldsymbol{D}^{(2)}$ which operates in the 2 dimensional manifold of the unperturbed states $|a(t)\rangle=|a\rangle\exp(-\mathrm{i}\omega_a t-\mathrm{i}\phi)$ and $|b(t)\rangle=|b\rangle\exp(-\mathrm{i}\omega_b t)$ such that the matrix element

$$\langle\psi_b(t)|e\boldsymbol{r}|\psi_a(t)\rangle=\langle b(t)|\boldsymbol{D}^{(2)}|a(t)\rangle=\boldsymbol{D}^{(2)}_{ab}.\tag{3.2}$$

Since $\boldsymbol{D}^{(2)}$ operates in the 2-dim manifold of $|a\rangle$, $|b\rangle$, it can be expressed by a pseudospin $S=1/2$ with Pauli matrices $\bar{\sigma}_j$ and the identity matrix $\tilde{I}$ as

$$D^{(2)}_k=\sum_{ij}E_{\mathrm{L}i}\bar{\sigma}_j(\alpha_{ijk}\mathrm{e}^{-\mathrm{i}\omega_\mathrm{L}t}+\mathrm{c.c.})+E_{\mathrm{L}i}\tilde{I}(\alpha^0_{ik}\mathrm{e}^{-\mathrm{i}\omega_\mathrm{L}t}+\mathrm{c.c.}),\tag{3.3}$$

where c.c. denotes the complex conjugate. While the donor site symmetry in CdS is C_{3v}, for convenience sake we will, for the time being, use the great simplification in (3.3) which results in cubic symmetry and in Sect. 3.3.1 return to a consideration of the lower symmetry. For cubic symmetry, in the case where $|a\rangle$ and $|b\rangle$ are time reversed states of each other,

$$\boldsymbol{D}^{(2)}=\boldsymbol{\sigma}\times\boldsymbol{E}_\mathrm{L}(\alpha\mathrm{e}^{-\mathrm{i}\omega_\mathrm{L}t}+\mathrm{c.c.})+(\alpha_0\boldsymbol{E}_\mathrm{L}\boldsymbol{I}\mathrm{e}^{-\mathrm{i}\omega_\mathrm{L}t}+\mathrm{c.c.}),\tag{3.4}$$

where

$$\alpha = e^2 \sum_n \frac{\mathrm{i}}{2} \frac{\langle a|x|n\rangle \langle n|z|b\rangle}{E_n - E_a - \hbar\omega_L}. \tag{3.5}$$

α^0 involves similar matrix elements and corresponds to the usual polarizability which gives rise to the forward coherent Rayleight scattering and index of refraction of light and will be omitted in further discussions of $\boldsymbol{D}^{(2)}$ [Ref. 3.16, Sect. 2.2.2]. If $|a\rangle$ and $|b\rangle$ correspond to spin states $|\downarrow\rangle$ and $|\uparrow\rangle$, then clearly the excited states $|n\rangle$ must contain admixtures of $|\downarrow\rangle$ and $|\uparrow\rangle$ for the matrix elements to be nonvanishing. This admixture comes about from spin-orbit coupling.

Off-diagonal elements of $\boldsymbol{D}^{(2)}$ involve $\langle b(t)|\sigma_{x,y}|a(t)\rangle = \exp(\mathrm{i}\omega_{ba}t - \mathrm{i}\phi) \cdot \langle b|\sigma_{x,y}|a\rangle$ so that $D_{ab}^{(2)}$ contains the Raman frequencies $(\omega_L \pm \omega_{ba})$. Following the semiclassical treatment of fluorescence [3.15], one obtains the power emitted at the Raman frequency during each spin-flip transition by inserting the dipole $D_{ab}^{(2)} + D_{ab}^{(2)*}$ in the classical radiation formula [Ref. 3.16, Eq. (2.1)]. Due to the randomness of the phase factor ϕ, each dipole radiates incoherently so that the total power is proportional to their number. The power radiated per center into the solid angle $d\Omega$ in a plane perpendicular to the dipole is $\omega^4|\alpha|^2 E_L^2 n_S/2\pi c^3$ (cgs units are used), where n_S is the index of the medium at $\omega = (\omega_L \pm \omega_{ba})$. Since the scattering is inelastic, this radiated power must be multiplied by $\omega_L/(\omega_L \pm \omega_{ba})$, and dividing by the incident flux $cn_L E_L^2/8\pi$ gives the spontaneous photon Raman scattering section per center in the crystal:

$$\left(\frac{d\sigma}{d\Omega}\right)_{\mathrm{sp}} = 4|\alpha|^2 \frac{\omega_L(\omega_L \pm \omega_{ba})^3}{c^4} \frac{n_S}{n_L}. \tag{3.6}$$

Because the frequency shifts we deal with are so small $(\omega_{ba} \ll \omega_L)$, $n_L/n_S \simeq 1$ and the cross section is basically $4|\alpha|^2\omega_L^4/c^4$.

Note that in (3.6) the c.c. part in the expression for $D^{(2)}$ in (3.3) does not enter since it would represent another process whereby light at the Raman frequency is absorbed while the laser frequency is emitted. However, both processes are taken into account if one writes the interaction of the Raman dipole $\boldsymbol{D}^{(2)}$ with the Raman light field $\boldsymbol{E}_S \cos\omega_S t$ as an effective Hamiltonian (cubic symmetry is assumed):

$$H_{\mathrm{SF}}^{(2)} = \left(\frac{\alpha}{2} \mathrm{e}^{-\mathrm{i}(\omega_L - \omega_S)t} + \mathrm{c.c.}\right) \boldsymbol{\sigma} \cdot (\boldsymbol{E}_L \times \boldsymbol{E}_S). \tag{3.7}$$

This Hamiltonian appears in the early description of SFRS in semiconductors [3.1, 7] and it will be used in the description of the Raman echo experiments in Sect. 3.8. It clearly shows the selection rules for cubic symmetry since $\boldsymbol{E}_L \times \boldsymbol{E}_S$ must have a component perpendicular to the magnetic field so that σ has off-diagonal elements corresponding to a spin flip.

3.2.3 Momentum Representation for Delocalized Electrons

When electrons are no longer localized at donor sites but are described by plane waves $|\pm 1/2, \exp(\mathrm{i}\boldsymbol{k}\cdot\boldsymbol{r})\rangle$, it is more convenient to use matrix elements of momentum $\boldsymbol{p}$ instead of the dipole $\boldsymbol{D}=e\boldsymbol{r}$ since the former are well known for interband transitions. It is straighforward to go from one to the other since

$$m(E_n-E_a)\,\langle n|\boldsymbol{r}|a\rangle = -\mathrm{i}\hbar\langle n|\boldsymbol{p}|a\rangle. \tag{3.8}$$

With the matrix elements in α thus modified, one recovers the usual expressions for $(d\sigma/d\Omega)$ and $H_{\mathrm{SF}}^{(2)}$ for conduction electrons [3.1, 7].

3.2.4 SFRS in Terms of Quantization of Radiation Field

We use the same form of the interaction Hamiltonian as in (3.1), i.e.,

$$H=-e\boldsymbol{r}\cdot\boldsymbol{E}, \tag{3.9}$$

where $\boldsymbol{E}$ is now the field operator in a normalization volume V expressed in terms of the annihilation and creation operators a and a^+ as

$$\boldsymbol{E}=\mathrm{i}\sum_{k,\gamma}\sqrt{\frac{2\pi\hbar\omega_k}{\varepsilon_k V}}\,\hat{e}_{k,\gamma}\,(a_{k_\gamma}\mathrm{e}^{\mathrm{i}\boldsymbol{k}\cdot\boldsymbol{r}}-a_{k_\gamma}^{+}\mathrm{e}^{-\mathrm{i}\boldsymbol{k}\cdot\boldsymbol{r}}). \tag{3.10}$$

ε_k is the dielectric constant of the medium at ω_k and $\hat{e}_{k,\gamma}$ is a unit polarization vector. Consider the scattering from a single polarization of the incident field to a single polarization of the Raman light. The initial state $|I\rangle=|a,\bar{n}_\mathrm{L},\bar{n}_\mathrm{S}\rangle$ is one in which the electron center is in state $|a\rangle$, with $\bar{n}_\mathrm{L}$ quanta in the incident laser field and $\bar{n}_\mathrm{S}$ quanta in the Raman field. In the final state $|F\rangle=|b,\bar{n}_\mathrm{L}-1,\bar{n}_\mathrm{S}+1\rangle$, one incident photon has been scattered into a Raman photon with the center making a transition to state $|b\rangle$. The rate for this process is given by the golden rule as

$$dw=\frac{2\pi}{\hbar^2}\sum_N\left|\frac{\langle F|e\boldsymbol{r}\cdot\boldsymbol{E}|N\rangle\,\langle N|e\boldsymbol{r}\cdot\boldsymbol{E}|I\rangle}{E_N-E_I-\hbar\omega_\mathrm{L}}\right|^2\varrho(\omega_\mathrm{S})d\Omega, \tag{3.11}$$

where the density of states per unit solid angle $\varrho(\omega_\mathrm{S})=\varepsilon_\mathrm{S}^{3/2}\omega_\mathrm{S}^2 V/(2\pi c)^3$; $\varrho(\omega_\mathrm{S})$ for a single polarization and $d\Omega$ is the element of scattering solid angle. $\boldsymbol{E}$ includes both the incident and scattered fields. There are two types of intermediate states $|N\rangle$, i.e., $|n,\bar{n}_\mathrm{L}-1,\bar{n}_\mathrm{S}\rangle$ and $|n,\bar{n}_\mathrm{L},\bar{n}_\mathrm{S}+1\rangle$ corresponding, respectively, to absorption of the incident photon or emission of the Raman photon, where $|n\rangle$ are the excited states of the electronic center. Just as we did in the semiclassical description earlier, we will neglect the nonresonant term corresponding to the second process. Using $a|\bar{n}\rangle=\bar{n}^{1/2}|\bar{n}-1\rangle$ and $a^+|\bar{n}\rangle=(\bar{n}+1)^{1/2}|\bar{n}+1\rangle$, the scattering rate per unit solid angle in the dipole approximation [exp $(i\boldsymbol{k}\cdot\boldsymbol{r})\sim 1$] is found to be

$$\frac{dw}{d\Omega}=4|\alpha|^2\frac{\varepsilon_S^{1/2}}{V\varepsilon_L c^3}\omega_L\omega_S^3\bar{n}_L(\bar{n}_S+1), \tag{3.12}$$

where α is the 2nd order matrix element given earlier by (3.5). This is the rate of scattering per center into modes in normalization volume V and into unit solid angle. It depends upon the intensity of the incident beam $I_L\sim\bar{n}_L/V$ and the number of Raman photons $\bar{n}_S$.

The differential scattering cross section

$$\frac{d\sigma}{d\Omega}=\frac{dw/d\Omega}{\text{incident flux}}=\frac{dw/d\Omega}{\bar{n}_L c/V\varepsilon_L^{1/2}} \tag{3.13}$$

or

$$\left(\frac{d\sigma}{d\Omega}\right)=4|\alpha|^2\frac{\omega_L\omega_S^3}{c^4}\left(\frac{\varepsilon_S}{\varepsilon_L}\right)^{1/2}(\bar{n}_S+1). \tag{3.14}$$

The $\bar{n}_S$ corresponds to stimulated scattering and its effect will be treated in Sect. 3.8. When $\bar{n}_S\to 0$, (3.14) gives the same result for $(d\sigma/d\Omega)_{sp}$ as derived earlier in (3.6). Corrections to (3.14) for an anisotropic medium including the effects of transmission through the surfaces are given by *Lax* and *Nelson* [3.17].

3.2.5 SFRS as Measuring the Transverse Spin Susceptibility $\chi_+(\boldsymbol{q},\omega)$

The differential cross section for light scattering from $\omega_L,\boldsymbol{k}_L$ to $\omega_S,\boldsymbol{k}_S$ from an elementary excitation of frequency $\omega=\omega_L-\omega_S$ and wave vector $\boldsymbol{q}=\boldsymbol{k}_L-\boldsymbol{k}_S$ is generally expressed as [3.18] (see also [Ref. 3.16, Eqs. (2.55, 56)] where the assumption $\omega_S\simeq\omega_R$ and $n_S\simeq n_L$ has been made)

$$\frac{d^2\sigma}{d\Omega\, d\omega_S}=\frac{\omega_L\omega_B^3}{c^4}\frac{n_S}{n_L}\frac{\langle\hat{\varepsilon}_S\cdot\boldsymbol{P}_S^*(\boldsymbol{k}_S)\hat{\varepsilon}_S\cdot\boldsymbol{P}_S(\boldsymbol{k}_S)\rangle_{\omega S}}{|\boldsymbol{E}_L|^2}, \tag{3.15}$$

where $\hat{\varepsilon}_S$ is a unit vector in the direction of polarization of the scattered light, n_S and n_L the indices of refraction, $\boldsymbol{P}_S(k_S)$ is the induced dipole moment per unit volume at wave vector $\boldsymbol{k}_S$ and $\langle\ \rangle_{\omega_S}$ refers to the ω_S component. For noninteracting fixed point scatterers and spontaneous emission, (3.15) has no q-dependent effects and it reduces to (3.6) multiplied by the number of scatterers. However, in the presence of interactions between the spins or if the spins are in motion, then q-dependent effects are generally present and the $\boldsymbol{q}$-dependence of $\boldsymbol{\sigma}$ in (2.4) should be recognized. Substituting for $\boldsymbol{P}$ in (3.15), $\boldsymbol{P}=\boldsymbol{D}^{(2)}=\alpha\boldsymbol{\sigma}\times\boldsymbol{E}$ from (3.4) we find [Ref. 3.16, Eqs. (2.55, 56)]

$$\frac{d^2\sigma}{d\Omega\, d\omega_S}=\frac{|\alpha|^2}{(4\pi)^2}\frac{\omega_L\omega_S^3}{c^4}\frac{n_S}{n_L}S_+(q,\omega), \tag{3.16}$$

where

$$S_+(q,\omega)=\int_{-\infty}^{\infty} e^{-i\omega t}\langle\sigma_q^-(t)\sigma_{-q}^+(0)\rangle \frac{dt}{2\pi}. \tag{3.17}$$

$\langle\ \rangle$ here is the thermal average over the exact many-body states of the interacting system. The spin-spin correlation function $S_+(q,\omega)$ is related to the imaginary part of the transverse spin susceptibility $\mathrm{Im}\{\chi_+(q,\omega)\}$ by the fluctuation-dissipation theorem [3.19]:

$$S_+(q,\omega)_{\mathrm{eq}}|_{\mathrm{Stokes}}=[\bar{n}(\omega)+1]\ \mathrm{Im}\{\chi_+(\boldsymbol{q},\omega)\} \tag{3.18}$$

$$S_+(q,\omega)^*_{\mathrm{eq}}|_{\substack{\mathrm{Anti-}\\ \mathrm{Stokes}}}=\bar{n}(\omega)\ \mathrm{Im}\{\chi_+(\boldsymbol{q},\omega)\} \tag{3.19}$$

$$\bar{n}=(e^{\hbar\omega/k_BT}-1)^{-1}. \tag{3.20}$$

The SFRS cross section thus effectively measures the (q,ω) component of the transverse spin susceptibility. The $\sigma_z(t)\sigma_z(0)$ correlation function, which does not involve spin flip, measures $\chi_{\|}(q,\omega)$ (Chap. 2). Both $\chi_{\|}(\mathrm{q},\omega)$ and $\chi_{\perp}(q,\omega)$ are accessible in light scattering from spin excitations although $\chi_{\|}(q,\omega)$ will correspond to quasielastic scattering and will appear under the laser line, making it more difficult to observe. The q-dependence of the scattering will be returned to in later sections.

3.2.6 Multiple Spin-Flip Raman Scattering

In addition to SFRS at the Zeeman frequency ω_0, spontaneous scattering has also been seen at integral multiples of the Zeeman frequency [3.20,21]. Initial observations seemed to indicate that the integral multiple relationship was not exact, i.e., $\omega(\Delta S=2)-2\omega(\Delta S=1)$ was not equal to zero but of the order of $0.1\ \mathrm{cm}^{-1}$ [3.20]. It was suggested that this shift corresponded to a binding energy of two $\Delta S=1$ excitations. However, subsequent high-resolution examination [3.21] of the frequencies of the $\Delta S=1$, 2 and 3 lines indicated that they were integral multiples of each other to better than 1 part in 2×10^3 and no pair excitation binding energy was observed.

The ratios, R, of the intensity of adjacent sidebands in a given sample is a constant, i.e., $R=I_{\omega_0}/I_{2\omega_0}\simeq I_{2\omega_0}/I_{3\omega_0}$, although R increases with the concentration, N_D, of bound donors. It has been suggested by *Wolff* [3.22] that these higher sidebands are due to the *near* fields of the Raman dipole $D_S^{(2)}\exp[i(\omega_L-\omega_0)t]$ which falls off as $1/r^3$, inducing a dipole at a donor at a distance r which then oscillates at $\exp[i(\omega_L-2\omega_0)t]$, etc. This suggestion is borne out by the polarizations of the incident and scattered light. For example, calling $\boldsymbol{E}_L$ the laser field $\boldsymbol{E}(\Delta S=1)\perp\boldsymbol{E}_L$ while $\boldsymbol{E}(\Delta S=2)\perp\boldsymbol{E}(\Delta S=1)$ and $E(\Delta S=2)\|\boldsymbol{E}_{\mathrm{inc}}$, etc. According to this model, one would also expect $R\sim N_D^2$. For a sample with N_D

$=2\times10^{17}$ cm^{-3}, we found $R=0.03$, and for a sample with $N_D=8\times10^{16}$ cm^{-3}, $R\simeq0.012$. However, there are serious corrections in measuring the intensities due to the severe dichroic absorption at 4880 Å so that it cannot be claimed that this feature was accurately verified.

Recently, using a tuneable dye laser in resonance with the I_2 exciton, *Oka* and *Cardona* [3.23] observed multiple spin flips as high as $\Delta S=5$ in the SFRS spectrum of CdS. These multiple spin flips were seen as sidebands on an LA phonon. This phonon line was nondispersive and interpreted as scattering via the bound exciton of the donor. P. Hu (private communication) has also observed sidebands as high as $\Delta S=8$, unconnected with phonons, using a pulsed dye laser. However, this was strictly *stimulated* SFRS, which is easily seen when the laser frequency is close to the I_2 exciton as the resonant cross section becomes extremely large, and is unrelated to the *spontaneous* multiple SFRS described at the beginning of this section.

3.3 Excited States Contributing to SFRS in CdS

3.3.1 SFRS Selection Rules for C_{3v} Symmetry

We now return to the general expression (3.3) for the Raman dipole $\boldsymbol{D}^{(2)}$ and consider the form appropriate to the donor site symmetry, C_{3v}. $\boldsymbol{\sigma}$ transforms as an axial vector and $\boldsymbol{E}$ as a polar vector. In the notation of *Koster* et al. [3.24], E_z and σ_z transform, respectively, according to the Γ_1 and Γ_2 representations of C_{3v}, where z is along the c-axis while each pair E_x, E_y and σ_y, $-\sigma_x$ transform according to the two-dimensional representation Γ_3. All possible products $E_j\sigma_K$ span the reducible representation

$$(\Gamma_1+\Gamma_3)\otimes(\Gamma_2+\Gamma_3)=\Gamma_1+2\Gamma_2+3\Gamma_3\,. \tag{3.21}$$

Since $\boldsymbol{D}^{(2)}$ is a polar vector whose components transform as Γ_1 and Γ_3, only these terms appear and the components of the Raman dipole will have four coefficients, one for the Γ_1 component and 3 for the Γ_3 component. A straightforward group theoretical reduction of (3.3) yields:

$$\Gamma_1\rightarrow D_z^{(2)}=\alpha(\sigma_xE_y-\sigma_yE_x), \tag{3.22a}$$

$$\Gamma_3\begin{cases}D_x^{(2)}=-\beta(\sigma_2E_y)+\gamma(\sigma_yE_z)+\delta(-\sigma_xE_x+\sigma_yE_y), & (3.22\text{b})\\ D_y^{(2)}=+\beta(\sigma_zE_x)-\gamma(\sigma_xE_z)+\delta(\sigma_xE_y+\sigma_yE_x). & (3.22\text{c})\end{cases}$$

The symmetry of the polarizability coefficients [3.25] requires in (3.3)

$$\alpha_{ijk}=\alpha^*_{kji} \tag{3.23}$$

so that

$$\alpha^* = -\gamma, \quad \beta = -\beta^* \quad \text{and} \quad \delta = \delta^*. \tag{3.24}$$

In addition, since for spin flip we are dealing with time reversed states $|a\rangle$ and $|b\rangle$, α as well as β is pure imaginary and δ is real so that there are only three independent coefficients given by expressions similar to (3.5). For cubic symmetry, $\alpha = \beta$ so that $D^{(2)}$ reduces to the form in (3.4). Note that the δ term appears as a contribution to a *symmetric* Raman or susceptibility tensor χ_{ij} $(D_i = \chi_{ij}E_j)$[1], i.e., $D_x = +\delta\sigma_x E_y$ and $D_y = +\delta\sigma_y E_x$, in contrast to β and α which are components of an *anti-symmetric* tensor. However, such symmetric terms in the χ_{ij} tensor must be *even* in σ or H or their product [3.25, 26a]. The δ term, unlike α and β, must therefore be linear in H_0 and is of order $\mu H_0/(E_n - E_g)$ smaller than α or β, and is therefore quite negligible as will be illustrated in Sect. 3.2. It is associated with a higher-order Cotton-Mouton effect [3.26b]. One does occasionally see spurious XX or YY scattering corresponding to the δ-term which may be as large as 20 % of the α or β terms. However, its origin is not intrinsic and is due to strong depolarization of the light at 4880 Å because of crystal strains. Equations (3.22–24) describe the selection rules for SFRS for the donor C_{3v} site symmetry in CdS. The magnitudes of the coefficients depend upon details of the excited states and how near the laser frequency is to the intermediate states as described in the next section.

3.3.2 Role of Bound Excitons in SFRS from Bound Donors

The band structure of CdS is shown in Fig. 3.3a. The conduction band is s-like and the p-like valence band is first split by spin-orbit (s-o) coupling ($\delta' = 57$ meV) into $J = 1/2$ and $J = 3/2$ components with the $J = 3/2$ further split by the trigonal field ($\Delta' = 16$ meV) into $m_J = \pm 3/2$ and $m_J = \pm 1/2$. The bound donor ground state is made up of conduction band states near $k = 0$. The binding energy of the isolated donor is approximately 28 meV. We are primarily interested in excited states of the donor whose separation from the ground state falls in the vicinity of the laser light. These excited states will be excitons made up of holes in the A, B, C valence bands (Fig. 3.3a) and an electron in the conduction band, bound to a neutral donor as shown in Fig. 3.3b. This bound exciton complex is analogous to a hydrogen molecule in which the two electrons pair up with opposite spins into a singlet and one of the heavy nuclei is replaced by a p-like hole with $L = 1$ and $S = 1/2$. In this respect, the neutral donor exciton is analogous to an alkali atom with an inverted s-o coupling. The wave functions of the A, B, and C

1 D_i is a component of the dipole or induced polarization and not to be confused with the displacement or induction.

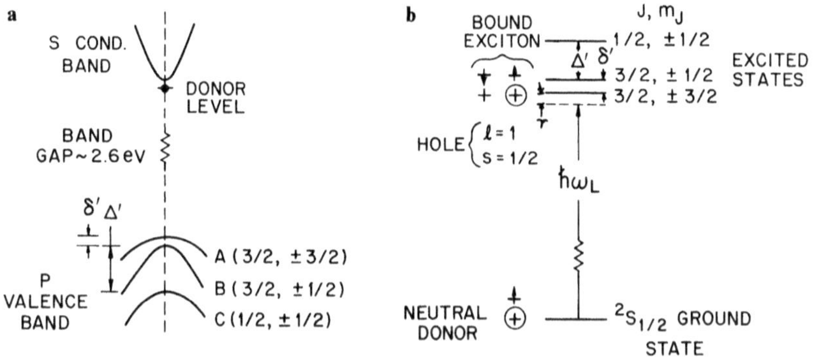

Fig. 3.3. (**a**) Band structure of CdS. (**b**) Bound excitons as excited states of donor in the region of the laser light (see text)

excitons given by *Hopfield* [3.27] are

$$\begin{aligned}
C:&\quad \begin{matrix} -\sin\theta Y_{1,1}|\downarrow\rangle + \cos\theta Y_{1,0}|\uparrow\rangle \\ -\sin\theta Y_{1,-1}|\uparrow\rangle + \cos\theta Y_{1,0}|\downarrow\rangle \end{matrix} \\
B:&\quad \begin{matrix} \cos\theta Y_{1,1}|\downarrow\rangle + \sin\theta Y_{1,0}|\uparrow\rangle \\ \cos\theta Y_{1,-1}|\uparrow\rangle + \sin\theta Y_{1,0}|\downarrow\rangle \end{matrix} \\
A:&\quad \begin{matrix} Y_{1,1}|\uparrow\rangle \\ Y_{1,-1}|\downarrow\rangle \end{matrix}\,.
\end{aligned} \tag{3.25}$$

The $Y_{n,m}$'s are the spherical harmonics corresponding to the *p*-wave function of the hole and the quantization is along the *c*-axis:

$$\cos^2\theta = \frac{1}{2}\left(1 - \frac{1 - \dfrac{3\Delta}{\zeta}}{\left[\left(1 - \dfrac{3\Delta}{\zeta}\right)^2 + 8\right]^{1/2}}\right) \tag{3.26}$$

and for $\Delta \to 0$ reduces to the familiar coupling coefficients for $L=1$ and $S=1/2$ for cubic or spherical symmetry. The energy separation of the excitons are

$$\begin{aligned}
E_B - E_A = \Delta' &= \frac{\zeta+\Delta}{2} - \frac{1}{2}\left[\left(\frac{\zeta}{3} - \Delta\right)^2 + \frac{8}{9}\zeta^2\right]^{1/2} \\
E_c - E_A = \delta' &= \frac{\zeta+\Delta}{2} + 1/2\left[\left(\frac{\zeta}{3} - \Delta\right)^2 + \frac{8}{9}\zeta^2\right]^{1/2}.
\end{aligned} \tag{3.27}$$

The donor ground state wave functions are

$$\begin{aligned}&\cos\frac{\phi}{2}\,|\uparrow\rangle+\sin\frac{\phi}{2}\,|\downarrow\rangle\\&-\sin\frac{\phi}{2}\,|\uparrow\rangle+\cos\frac{\phi}{2}\,|\downarrow\rangle,\end{aligned}\tag{3.28}$$

where ϕ is the angle between the c-axis and the external magnetic field H_0. Note that H_0 is, in practice, sufficiently small compared to the axial crystal field that it does not affect the c-axis quantization of the exciton states. The significance of s-o coupling in SFRS may be seen as follows. Using the wave functions (3.25, 26), it may be easily shown that the intermediate state sum of the contributions to the matrix elements of $\boldsymbol{rr}$ from the A, B, C excitons cancel each other. For example, for XY scattering the contributions from A, B and C are proportional, respectively, to $\sin\phi$, $-\sin\phi\cos^2\theta$ and $-\sin\phi\sin^2\theta$. Thus there would be no SFRS except for the inequality of the energy denominators in (3.5) arising from s-o splitting. As a consequence, when $\hbar\omega_L$ is significantly less than the bandgap E_g, the effective cross section $\sigma_{eff}\sim\sigma_0\zeta/E_g$, where σ_0 is the Thomson cross section. However, for 4880 Å $\hbar\omega_L$ is <5 meV from the A exciton which is much less than the s-o coupling of 57 meV, and the magnitude of the s-o coupling is effectively infinite. The differential cross section in this one-level approximation [3.6] becomes

$$\frac{d\sigma}{d\Omega}=\frac{3}{32\pi}f^2\sigma_0\left(\frac{\omega_L}{\omega_L-\omega_A}\right)^2,\tag{3.29}$$

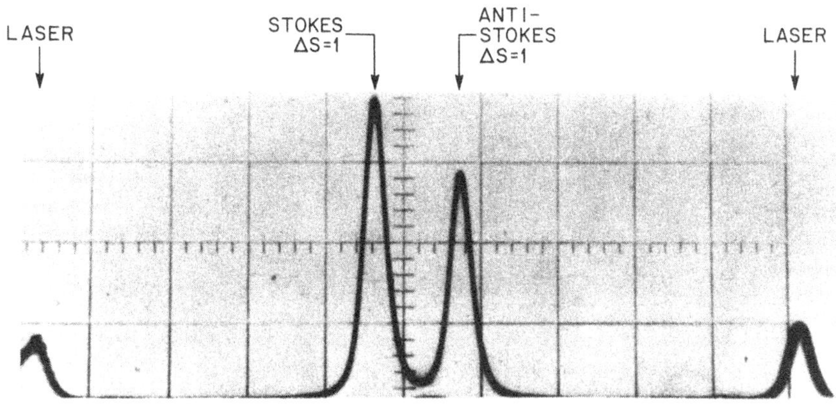

Fig. 3.4. Strong SFRS at 4880 Å in CdS from bound donors at 2 K and in a field of 10 kg. The two laser lines are different interference orders of the Fabry-Perot spectrometer with the Stokes belonging to the one on the extreme right and the anti-Stokes to the one on the left. The free spectral range is approximately 2 cm^{-1}

where $f \sim 10$ is the oscillator strength of the A exciton. In this near-resonant situation the cross-sections are large ($\geqslant 10^{-18}$ cm^{-2}) and the SFRS intensity is very often comparable to the stray elastically scattered light and may be easily observed on an oscilloscope without signal averaging as seen in Fig. 3.4. Further from resonance, at 5145 Å, the cancellation effect from the B and C excitons becomes pronounced and the cross section is much reduced. In addition, since at 5145 Å $E_a - \hbar\omega_L \gg \Delta$, the trigonal field splitting is obscured and the selection rules appear the same as those for cubic symmetry.

The direct experimental determination of the absolute cross section is difficult, especially at 4880 Å where very strong but little understood surface absorption occurs in polished samples of CdS. In addition, there is strong linear dichroic absorption at 4880 Å which drastically alters the relative intensities of light polarized parallel and perpendicular to the c-axis with changing path length. However, the coefficients α_{ij} in (3.22) which reflect the cross section may be measured by Faraday rotation which is unaffected by absorption, as will be described in Sect. 3.5.1.

3.3.3 Excited States for Scattering from Delocalized Electrons

When the concentration N of donors exceeds a critical value N_c such that the mean separation d between donors is $d \sim 4a_H$ (a_H is the donor Bohr radius), an insulator-metal (IM) transition results (Sect. 3.4). In CdS, $N_c \simeq 1.0 \times 10^{18}$ cm^{-3}. The screening of the electrostatic interaction with increasing n prevents the binding of excitons as well as the binding of electrons to donors. The donor impurity band broadens, eventually merges with the conduction band and the Fermi level appears above the bottom of the conduction band as indicated in Fig. 3.5. In this case of extreme degeneracy, the excited states are unbound electron-hole pairs of much higher energy, spread over a continuum. With the oscillator strength no longer concentrated in a sharply bound exciton level, the

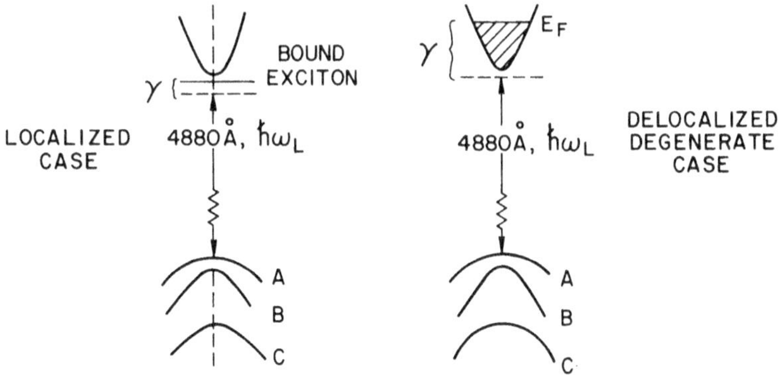

Fig. 3.5. Comparison of the resonance denominator γ in SFRS for the localized and the delocalized degenerate case (see text)

near resonance of 4880 Å light is absent and there is a considerable decrease in cross section. The selection rules now look more like those for systems with cubic symmetry. A further reduction in cross section per spin occurs since the Fermi statistics prevent a spin reversal except for electrons within kT of the Fermi level. Since for $n \simeq 10^{18}$ the Fermi temperature is approximately 300 K, this reduction is quite significant at low temperatures. This case has been treated in detail in Chap. 2.

It should be noted, however, that just beyond the IM transition a vestige of electron-hole binding still remains due to a many-body effect as described by *Girvin* [3.28]. Thus, at $N \sim 10^{18}$ one still observes a resonant feature in the optical emission spectrum at the position of the A exciton [3.29]. At still higher concentrations this feature broadens and becomes asymmetric, shifts to higher energy and begins to reflect the density of states in the impurity band. At $n > 4 \times 10^{18}$, the exciton feature has been sufficiently suppressed so that the cross section even at 4880 Å is drastically reduced and is found to be several orders of magnitude smaller than for bound donors.

3.3.4 Polariton Effects in CdS

When the frequency of the incident laser is near the intrinsic free exciton of the crystal, the proper description of the light in the crystal is that of an excitonic polariton, i.e., a mixed mode of photon and exciton [Ref. 3.30, Chap. 7]. In this case, SFRS from bound donors involves the scattering of an incident polariton $(\boldsymbol{k}_\mathrm{L}, \omega_\mathrm{L})$ from a neutral donor into a polariton $(\boldsymbol{k}_\mathrm{S}, \omega_\mathrm{S})$ with a simultaneous spin flip [3.31]. Close to the intrinsic exciton, the polariton is largely exciton-like whereas far below resonance it is primarily photon-like. However, as indicated by *Hopfield* [3.31], it is often not appreciated how exciton-like the polariton is, even at what at first glance appears to be a significant separation from the exciton. Hopfield has shown that for frequency ω such that

$$|\hbar(\omega_\mathrm{ex} - \omega)| \lesssim \sqrt{\hbar\omega_\mathrm{ex} E_\mathrm{LT}}, \tag{3.30}$$

the polariton has a considerable exciton-like component. E_LT is the exciton-photon coupling which is 15 cm^{-1} in CdS and $\hbar\omega_\mathrm{ex} = 20{,}589$ cm^{-1} so that at 4880 Å the polariton is largely exciton-like. This considerably modifies the picture of SFRS for bound donors in at least two ways. First the appropriate SFRS matrix element corresponds to the scattering of a free exciton off the donor. Second, since the polariton is highly dispersive in this region, the group velocity v_g rather than the phase velocity c/n (n: real part of refractive index) enters once in the flux and once in the density of states. *Hopfield* [3.31] has given the total cross section in the single-resonance level approximation and it is recast here in the following from:

$$\sigma \simeq r_0^2 \left(\frac{\omega}{\omega - \omega_\mathrm{res}}\right)^2 \left(\frac{c}{v_\mathrm{g}}\right)^2 \frac{n_\mathrm{L}^2 E_\mathrm{B}}{2E_\mathrm{H}} \left(\frac{m_\mathrm{e}}{m}\right)^3, \tag{3.31}$$

where m_e is the free electron mass, m the intrinsic exciton mass, E_B the binding energy of the donor, E_H the binding energy of hydrogen (13.6 eV), v_g the group velocity $\partial\omega/\partial k$ of the polariton, c the free space velocity of light, n_L the index of refraction and r_0 the Thomson radius of the electron. However, if we treat the Raman polarizability matrix elements as a phenomenological tensor relationship between $\boldsymbol{P}$ (our dipole $\boldsymbol{D}$) and $\boldsymbol{E}$ in the medium, i.e., $\boldsymbol{P}=\tilde{\alpha}\boldsymbol{E}$, then we may calculate $(d\sigma/d\Omega)$ classically ([3.18] and Eq. (3.15)): no group velocity factors appear and (3.6) still holds even in the extreme polariton region. This is so because the energy density in a dispersive region is given by $nE^2c/4\pi v_g$ [3.25] and the energy velocity by v_g so that the Poynting vector is still given by $ncE^2/4\pi$. While indeed the calculation of α will involve the polariton picture as detailed by *Hopfield* [3.31], one can still extract α experimentally from the measured oscillator strength or lifetime of the bound exciton. This will be the approach used below in Sect. 3.5 in discussing the magnitude of $(d\sigma/d\Omega)$.

3.4 The Insulator-Metal (IM) Transition in CdS Studied by SFRS

3.4.1 The Insulator-Metal Transition

As the spacing between a regular lattice of hydrogen-like atoms is decreased, the screening of the nuclei by the electrons increases and at a critical separation the electrons will no longer have a bound state and will delocalize. This electronic transition, first proposed by *Mott* [3.32], is also viewed in the *Hubbard* [3.33] picture as the decrease in kinetic energy through delocalization winning out over the correlation energy needed to put two electrons on the same site in the delocalized or metallic state. This *Mott-Hubbard* transition [3.34] occurs when the separation d between centers is such that $d \sim 4a_H$, where a_H is the Bohr radius.

The radii a_H^* of shallow donors in semiconductors is given by $a_H^* \simeq a_H\varepsilon/m^*$, where a_H is the Bohr radius of hydrogen, ε the dielectric constant and m^* the effective mass of the donor electron. The usual large ε and small m^* result in a large a_H^*. In CdS, for example, $a_H^* \sim 25$ Å so that $d \sim 4a_H^* = 100$ Å. Thus, at a donor concentration of parts per million level, one would expect to observe the IM transition (at zero temperature without any thermal activation), putting aside for a moment the question of the random placement of the donors. This is illustrated for CdS in Fig. 3.6 where at a critical concentration N_c, between 5×10^{17} and 10^{18} cm^{-3} a drop of eight decades in resistivity ϱ is observed. Similar ϱ vs N curves, indicating this so-called impurity banding, are observed for many doped semiconductors where N_c scales as $\sim(a_H^*)^{-1/3}$ [3.35]. The precise pinpointing of the transition is a delicate question because it sharpens up even more at lower T and ideally, this type of measurement should be carried out at as low a temperature as possible, as has been done in Si:P by *Rosenbaum* et al. [3.36] and *Thomas* et al. [3.37]. Experimentally, however, a problem still remains since the carrier concentration, as observed in a Hall measurement, seems to become

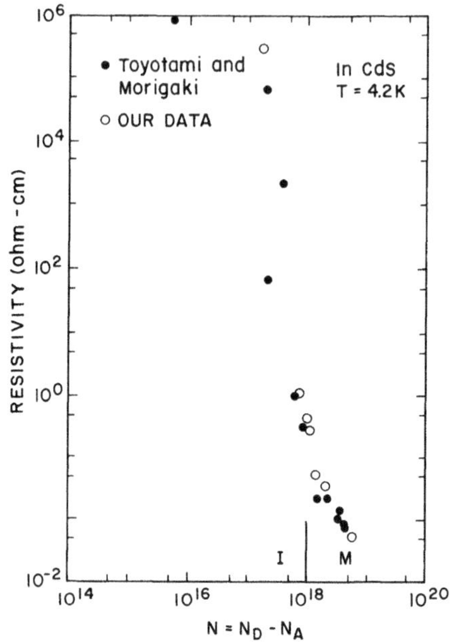

Fig. 3.6. Resistivity vs uncompensated donor concentration in In-doped CdS showing the insulator-metal (IM) transition

independent of temperature at a value of donor concentration that is slightly below the concentration at which ϱ becomes independent of T [3.38]. A most reasonable value for N_c in CdS seems to be $N_c \simeq 1.0 \pm 0.1 \times 10^{18}\,\mathrm{cm}^{-3}$.

The random spacing of the donors does complicate the nature of the transition, however, due to variable overlap between the centers as well as random potentials which could bring about *Anderson* localization [3.39]. The Hall effect problem cited may have its origin in this randomness. A central question in the entire subject is the relative importance of electron correlation compared to Anderson localization. Recent reviews of this problem have been presented by *Mott* [3.34, 40] and *Thouless* [3.41] as well as in several conference proceedings [3.42]. Our main concern in this section is to consider how the electron delocalization manifests itself in the SFRS.

3.4.2 Charge Diffusion in Terms of a Collisionally-Narrowed Doppler Width

Information on electron dynamics appears in the SFRS linewidths. When the electrons are bound to the donors and there is no spin diffusion, i.e., in the limit of very low donor concentration, the optical linewidth will be the same as the EPR linewidth which may be due to a T_1 or T_2 spin lifetime process or to inhomogeneous broadening T_2^*. On the other hand, when the electron delocalize in passing through the IM transition, then superimposed on the Zeeman shift will be a Doppler shift $\Delta\omega_{\mathrm{Dop}} = \boldsymbol{q} \cdot \boldsymbol{v}$, where $\boldsymbol{v}$ is the electron velocity and $\boldsymbol{q} = \boldsymbol{k}_{\mathrm{L}} - \boldsymbol{k}_{\mathrm{S}}$ is the light scattering vector. The line shape will then reflect the electron velocity distribution as was shown for GaAs by *Mooradian* [3.43] and *Hamilton* and *McWhorter* [3.44]. In CdS, however, carrier mobilities are so low that the mean collision time $\tau_c \ll 1/\Delta\omega_{\mathrm{Dop}}$. One therefore observes a collisionally-narrowed Lorentzian line, centered at the Zeeman frequency, whose full width at half maximum intensity is given by the familiar motional narrowing expression [3.45]

$$\Delta\omega_{\mathrm{F.W.}} = 2(\Delta\omega_{\mathrm{Dop}})^2\tau_c\,. \tag{3.32}$$

Of course, this diffusional linewidth will be superimposed on any intrinsic EPR width cited above. Equation (3.32) may be rewritten as

$$\begin{aligned}\Delta\omega_{\mathrm{F.W.}} &= 2(\boldsymbol{q}\cdot\boldsymbol{v})^2_{\mathrm{av}}\tau_c = \frac{2}{3}\,v^2_{\mathrm{av}}\tau_c q^2\\ \Delta\omega_{\mathrm{F.W.}} &= 2D_c q^2,\end{aligned} \tag{3.33}$$

where D_c is the diffusion constant, $v^2_{\mathrm{av}}\tau_c/3$. For a degenerate and nondegenerate noninteracting electron gas, respectively, D_c is given by

$$D_c = 2\mu E_F/3e, \quad \text{degenerate,} \tag{3.34}$$

$$D_c = kT\mu/e, \quad \text{nondegenerate,} \tag{3.35}$$

where E_F is the Fermi energy, μ is electron mobility, $e\tau_c/m^*$, T the temperature, k is Boltzmann's constant and e the electronic charge.

3.4.3 Distinction Between Spin and Charge Diffusion

One should consider diffusion, however, in more general terms than that of noninteracting mobile electric charges as just outlined. Recall from Sect. 3.2.4 that in SFRS, one observes the transverse spin susceptibility $\chi_\perp(q,\omega)$. *Wolff* et al. [3.46] have grafted on to the Bloch equation of motion for the transverse magnetization $M_+(q,\omega)$ in an external field $H_0 + h_+(q,\omega)$ a diffusion term $-D_s\nabla^2 M_+$, i.e.,

$$\frac{dM_+}{dt} + \mathrm{i}\omega_s M_+ + \frac{(M_+ - \chi_0 h_+)}{T_2} - D_s\nabla^2 M_+ = \mathrm{i}\mu g M_0 h_+\,, \tag{3.36}$$

where T_2 is the spin lifetime. Solving for $\chi_+(q,\omega) = M_+(q,\omega)/h_+(q,\omega)$, they demonstrated that the damping term or half linewidth for SFRS is given by

$$\Delta\omega = \frac{1}{T_2} + D_s q^2\,. \tag{3.37}$$

The subscript "s" in D_s emphasizes that it is always spin diffusion that one measures because $\chi(q,\omega)$ in SFRS measures spin-density fluctuations.

Charge diffusion would be seen directly in so-called single-particle scattering [3.47] which, unfortunately, is too weak here to be seen because of screening. For mobile charges in a semiconductor, $D_s = D_c$ if electron correlation and many-body effects may be neglected. One anticipates that this would be the case on the very metallic side of the IM transition. In general, however, correlation effects might result in $D_s \neq D_c$ [3.46]. Moreover, pure spin diffusion may exist without any attendant charge transport, i.e., for bound donor electrons, as will be seen in Sect. 3.7.2a.

3.4.4 Experimental Results on Diffusive Linewidths

We now turn to an examination of experimental results on D_s through the IM transition. For $N \simeq 10^{18}\,\mathrm{cm}^{-3}$ and $m^* = 0.2\,m_e$, the Fermi temperature T_F is $\simeq 300$ K so that at He temperatures, the electron gas is degenerate and (3.34) applies. Since $E_F = (3N/8\pi)^2/3\hbar^2/m^*$ and the resistivity $\varrho = (Ne\mu)^{-1}$, D_c may be expressed in terms of the easily measured transport parameters N and ϱ. $q = (2\pi n/\lambda)\sin\theta/2$, where θ is the scattering angle so that (3.33) becomes

$$\Delta\nu_{\mathrm{F.W.}} = \frac{\Delta\omega_{\mathrm{F.W.}}}{2\pi} = \Delta\nu_0 \sin^2\frac{\theta}{2} = \frac{4.2\times10^5}{\varrho N^{1/3}} \sin^2\left(\frac{\theta}{2}\right)\,[\mathrm{cm}^{-1}], \tag{3.38}$$

where the following parameters have been used: $\lambda = 4880$ Å, the index of refraction $n = 3$ and $m^* = 0.2$. A Lorentzian line shape and a $\sin^2(\theta/2)$ dependence upon the light scattering angle are hallmarks of diffusion and this q^2-dependence is illustrated in Fig. 3.7 for small angle scattering for a metallic sample above the IM transition. For geometries where neither $\boldsymbol{q}$ nor $\boldsymbol{H}_0$ are parallel to the c-axis, the effective q is modified due to the linear k-term in CdS (Sect. 3.6) and this has been accounted for in Fig. 3.7 where the effective q is plotted. Just above the IM transition for $\theta \gtrsim 90°$, $\Delta\omega$ vs q varies more slowly than q^2 and for larger H is asymmetric. This behavior could be connected with a "coherence length" in the spin-flip excitation (Sect. 3.7) arising from correlations in the electron gas. Therefore, in examining the experimental data for a fit to (3.38), the coefficient $\Delta\nu_0$ is determined from scattering angles $\theta < 90°$ where the $\sin^2(\theta/2)$ behavior is observed. The results are listed in Table 3.1 for a range of donor concentrations N. Samples A and B are insulating, i.e., the donors are frozen out at low temperatures so that (3.38) is not applicable. In addition, within experimental accuracy, no variation of $\Delta\nu$ with θ is observed in sample A. While an angular dependence is observed in sample B, it is connected with pure spin diffusion without charge transport as will be described in Sect. 3.7.2a. In samples C to F, the observed $\Delta\nu_0$ agrees with (3.38) within a factor of two or so, indicating that $D_s \simeq D_c$ and is dominated by particle motion. The factor of two or so agreement is quite reasonable as the SFRS probes a region of the sample which is 100 μ in size and variations as much as 50 % in $\Delta\nu$ are occasionally seen

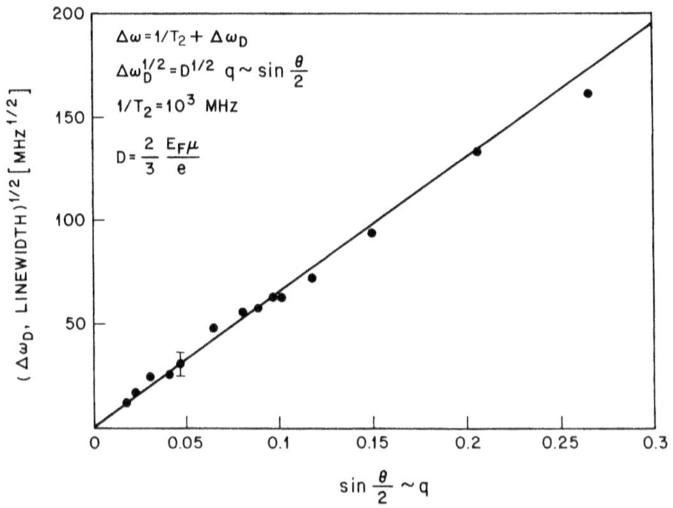

Fig. 3.7. Small angle SFRS in a degenerate metallic sample of CdS showing the diffusive motion of carriers via $q^2(\sim \sin^2 \theta/2)$ dependence on linewidth

Table 3.1. Comparison of SFRS linewidths with predictions of (3.38). $\Delta\nu_0$ is the coefficient of $\sin^2(\theta/2)$ in (3.38)

N A,B-room temperature C to F: 1.7 K	ϱ at 1.7 K [Ω cm]	$\Delta\nu_0$ [cm^{-1}] calculated*	$\Delta\nu_0$ measured*
A 8×10^{16}	$>10^7$	Not	0.003
B 2.3×10^{17}	10^7	applicable	0.016
C 7×10^{17}	0.48	0.98	0.87
D 1.1×10^{18}	0.05	8.1	4.1
E 1.9×10^{18}	0.033	10.2	6.8
F 4×10^{19}	0.0014	88	28

in the same sample. In addition, slight inhomogeneities in N corresponds to large inhomogeneities in ϱ which are averaged in a resistivity measurement. Thus, in the metallic region, defined by an absence of activation in the Hall carrier concentration, the observed spin diffusion corresponds to charge diffusion as given by (3.38) even when μ varies with T as it does in sample C. However, when there is significant activation of N, at finite T one has a mixture of localized electrons interacting with conduction electrons as well as hopping conductivity. The relationship between $D(T)$ and $\varrho(T)$, $N(T)$ in this mixed region, is still a difficult unresolved problem. Earlier measurements of $D(T)$ in CdS with $N \simeq 5 \times 10^{17}$ cm^{-3} by *Scott* et al. [3.20], later interpreted by *Wolff* et al. [3.46], were probably in this mixed region since the IM transition occurs at $N \simeq 10^{18}$ cm^{-3}.

3.5 Relationship Between Spin Faraday Rotation and SFRS

3.5.1 Spin Faraday Rotation and Raman Dipole

Measurements of absolute Raman scattering cross section ($d\sigma/d\Omega$) are always cumbersome experiments. Apart from the necessity of measuring absolute values of radiated power, care must be taken in correcting for absorption and reflection of both laser and scattered beams in the crystal. These effects are especially severe near the band edge of semiconductors where it is often desirable to work in order to obtain resonant enhancement of the cross section, as described in Sect. 3.3.2. However, a very simple relation links the cross section to the Faraday rotation (FR). Close to resonance this FR becomes very large and allows an easy determination of the SFRS cross section.

In the semiclassical treatment of SFRS, a light field $E_{\mathrm{L}} \cos \omega_{\mathrm{L}} t$ acting upon the donor induces a dipole

$$\boldsymbol{D}^{(2)} = \boldsymbol{\sigma} \times \boldsymbol{E}_{\mathrm{L}}(\alpha \mathrm{e}^{\mathrm{i}\omega_{\mathrm{L}} t} + \alpha^* \mathrm{e}^{-\mathrm{i}\omega_{\mathrm{L}} t}) + (\alpha_0 \boldsymbol{E}_{\mathrm{L}} \tilde{I} \mathrm{e}^{-\mathrm{i}\omega_{\mathrm{L}} t} + \mathrm{c.c.}), \tag{3.4}$$

where for simplicity we restrict ourselves to the cubic form. We saw in Sect. 3.2 that associated with σ_x and σ_y were off-diagonal elements $\boldsymbol{D}_{ab}^{(2)}$ of this dipole between the spin states $a(\downarrow)$ and $b(\uparrow)$ representing a dipole oscillating at $\exp[\mathrm{i}(\omega_{\mathrm{L}} \mp \omega_{ba})t]$ emitting Stokes and anti-Stokes radiation. On the other hand, the component of σ along the magnetic field σ_z has diagonal elements only and gives rise to dipoles radiating at the same frequency ω_{L}, but whose polarization is rotated by 90° relative to $\boldsymbol{E}_{\mathrm{L}}$, i.e., for $E_{\mathrm{L}} \| y, x$ (recall that α is pure imaginary),

$$D_{aa}^{x} = -\langle a|\sigma_z|a\rangle E_y(\alpha \mathrm{e}^{\mathrm{i}\omega_{\mathrm{L}} t} + \mathrm{c.c.}), \tag{3.39}$$

$$D_{aa}^{y} = +\langle a|\sigma_z|a\rangle E_x(\alpha \mathrm{e}^{\mathrm{i}\omega_{\mathrm{L}} t} + \mathrm{c.c.}), \tag{3.40}$$

with similar expressions for D_{bb}. These dipoles radiate with the phase of the incident electric field so that emission is constructive only along the direction of propagation of the incident beam and the effect will be a simple rotation of the plane of polarization, i.e., a Faraday rotation. A gyrotropic relationship between $\boldsymbol{D}^{(2)}$ and $\boldsymbol{E}$, expressed by the $\boldsymbol{\sigma} \times \boldsymbol{E}$ term in (3.4), and its leading to a Faraday rotation, was discussed by *Landau* and *Lifschitz* [3.25]. We see, moreover, that the same constant α is associated with both the off-diagonal and diagonal elements of $\boldsymbol{D}^{(2)}$ so that the SFRS cross section can be determined from the Faraday rotation (FR). This connection was emphasized by *Pershan* et al. [3.48] and *Shen* and *Bloembergen* [3.49a] in estimating cross sections for light scattering from spin waves from Faraday rotation measurements. It was further elaborated by *Le Gall* et al. [3.26, 27], who in addition considered the connection between two-magnon scattering and the Cotton-Mouton effect. It was first applied to spins in a semiconductor by *Romestain* et al. [3.49b]. The spin FR may be calculated by considering the volume polarizability for left and right circularly polarized light associated with (3.39, 40):

$$\chi_{\pm} = \pm 2\mathrm{i}\alpha\langle\sigma_z\rangle N + 2\alpha_0 N + \chi_{\mathrm{B}}, \qquad (3.41)$$

where χ_{B} is the background of the pure crystal;

$$\varepsilon_{\pm} = 1 \pm \mathrm{i}8\pi\alpha\langle\sigma_z\rangle N + 8\pi\alpha_0 N + 4\pi\chi_{\mathrm{B}}, \qquad (3.42)$$

where $\langle\sigma_z\rangle N$ is the difference in density between the up and down spins. The FR per unit length ϕ/l is

$$\frac{\phi}{l} = \frac{\pi(n_+ - n_-)}{\lambda} = \frac{\pi(\varepsilon_+ - \varepsilon_-)}{(n_+ + n_-)\lambda} \simeq \mathrm{i}\alpha\,\frac{8\pi^2 N\langle\sigma_z\rangle}{n\lambda}, \qquad (3.43)$$

where $n \simeq (n_+ + n_-)/2$ and λ is the free space wavelength of the light. Comparing (3.43) with (3.6), the SFRS cross section may be expressed in terms of the spin FR as

$$\left(\frac{d\sigma}{d\Omega}\right) = \left(\frac{\phi}{l}\right)^2 \frac{n^2}{N^2\langle\sigma_z\rangle^2\lambda^2}. \qquad (3.44)$$

3.5.2 Wavelength Dependence of SFRS Cross Section in CdS Determined from Spin Faraday Rotation

Figure 3.8 shows the FR for a crystal with $N_{\mathrm{D}} = 7 \times 10^{16}$ bound donors [3.49a]. Figure 3.8a displays the transmitted intensity vs magnetic field H_0 at 4880 Å when the analyser is kept fixed, so that the separation between transmission maxima corresponds to a rotation of 180°. In Fig. 3.8b the rotation vs H_0 corresponding to Fig. 3.8a is plotted. In addition to the spin Faraday rotation $\sim\langle\sigma_z\rangle$, there is an intrinsic background rotation η which is due to band to band transitions. Since η is essentially temperature independent, it may be determined by extrapolating the observed rotation to high temperature where $\langle\sigma_z\rangle = 0$. η also depends on wavelength and is a fractionally small correction at 4880 Å. While η is smaller in absolute value at 4965 Å, it is a fractionally larger correction since the spin Faraday rotation is so much smaller. Table 3.2 summarizes the results on the cross section determinations at different wavelengths and clearly displays the resonant enhancement of α as the wavelength approaches the A exciton. R is the rotation in deg/kG mm and $(d\sigma/d\Omega)_{\mathrm{FR}}$ is determined from (3.44) by estimating $\langle\sigma_z\rangle$ at a given H_0. The FR is measured with $H_0 \| C$ but SFRS with $H_0 \| C$ does not involve the A exciton while the FR does. The A-exciton does, however, enter in SFRS for $H_0 \perp C$ so that $(d\sigma/d\Omega)_{\mathrm{FR}}$ refers to the SFRS cross section with $H_0 \perp C \perp E_{\mathrm{L}}$, and $E_{\mathrm{S}} \| H_0$. Of course E_{S} and E_{L} may be interchanged here. $(d\sigma/d\Omega)_{\mathrm{RS}}$ is the direct measurement of the Raman cross section. $(d\sigma/d\Omega)_{\mathrm{theor}}$ is the calculation of the cross section using the schematic wave functions (3.25, 28), the known positions of the A, B and C excitons and assuming isolated

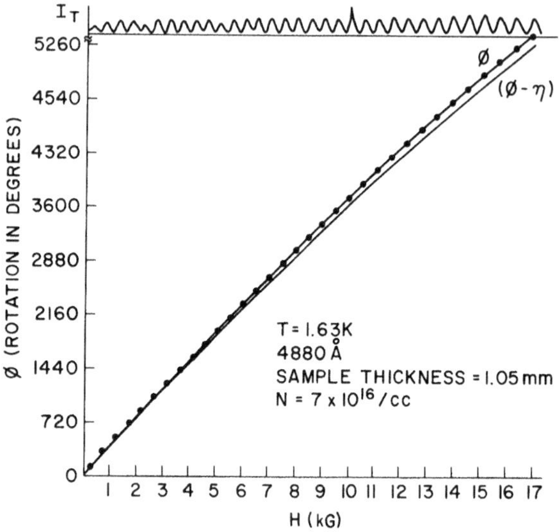

Fig. 3.8. Faraday rotation (FR) in CdS. (a) Transmitted intensity I_T vs H with fixed position of Polaroid analyser. Separation of peaks correspond to a rotation of 180°. (b) Points on curve labeled ϕ are plot of peaks in (a) and $(\phi-\eta)$ is spin FR where η is the background rotation in the pure crystal (see text and [3.49b])

Table 3.2. Values of measured Faraday Rotation (FR) induced by bound donor electron spins, and spin-flip cross sections as determined from FR, $(d\sigma/d\Omega)_{FR}$; Raman scattering, $(d\sigma/d\Omega)_{RS}$ and theoretical calculations. FR data are for sample containing 7×10^{16} In/cm^3 and for $T=1.63$ K

λ [Å]	Spin FR [deg/kG mm]	$\left(\frac{d\sigma}{d\Omega}\right)_{FR}$ [cm^2]	$\left(\frac{d\sigma}{d\Omega}\right)_{RS}$ [cm^2]	$\left(\frac{d\sigma}{d\Omega}\right)_{theor}$ [cm^2]
4880	363	1.9×10^{-18}	4×10^{-18} [3.6]	4.4×10^{-19}
4965	8.8	9.9×10^{-22}	3×10^{-20} [3.6]	6.4×10^{-22}
5145	2 ± 1	4×10^{-23}	10^{-25} [3.20]	1.8×10^{-23}

noninteracting donors. In this calculation, the essential matrix element of the type $\langle n|er|g\rangle$ is obtained from the measured lifetime of the A exciton, $\tau\simeq0.5\times10^{-9}$ s [3.50]. In the earlier calculation of $(d\sigma/d\Omega)_{FR}$ [3.21], the value of $\langle\sigma_z\rangle$ was probably overestimated since the antiferromagnetic exchange interaction between donors was neglected. *Hu* et al. [3.51] have recently measured the spin FR in several samples at many more wavelengths using a tuneable dye laser. The observed variation of the spin FR with wavelength agrees with the theoretically expected variation using the A, B and C excitons but seriously disagrees in magnitude, being more than a factor of 10 larger than the calculated value. This discrepancy is still an unresolved problem and may have its origin in the estimates of $\langle\sigma_z\rangle$, determination of N, lifetime τ, or possible omission of other excited states contributing to the spin FR as well as neglect of the interaction between donors which give rise to a coherent spin excitation (Sect. 3.7.2). Nevertheless, (3.44) is quite general and the spin FR should provide the most reliable experimental determination of SFRS cross section.

3.5.3 Measurement of Donor Susceptibility by Faraday Rotation

Since $(\phi-\eta)$ is proportional to $\langle\sigma_z\rangle$, the specific rotation R is proportional to the magnetic susceptibility χ, i.e.,

$$R=\frac{d}{dH}[(\phi-\eta)/l]\sim\chi. \tag{3.45}$$

Since ϕ/l can be made extremely large by selecting a wavelength to resonantly enhance α, one has a method of selectively measuring the χ of the donors at ppm levels unobscured by other impurities. For example, even though the concentration of Mn^{2+} impurities in the sample of Fig. 5.1 is greater than the donor concentration, their contribution to the FR is estimated to be at least four orders of magnitude smaller. The application of this technique to the measurement of the temperature dependence of χ for interacting bound donors (a model amorphous antiferromagnet) is described in Sect. 3.7. In the extremely metallic region, $N \geqslant 4\times10^{18}$, the rotation ϕ shows a weaker wavelength dependence, does not saturate with H and is independent of temperature as one would expect for a Pauli susceptibility. In the metallic region just beyond the IM transition, however, R (and hence χ) displays both a temperature-dependent and a temperature-independent component [3.52], as seen in Fig. 3.9. Similar behavior is also observed in a sample with $N_D\sim2\times10^{18}$ for which both N and the resistivity ϱ are temperature independent. The existence of these two compo-

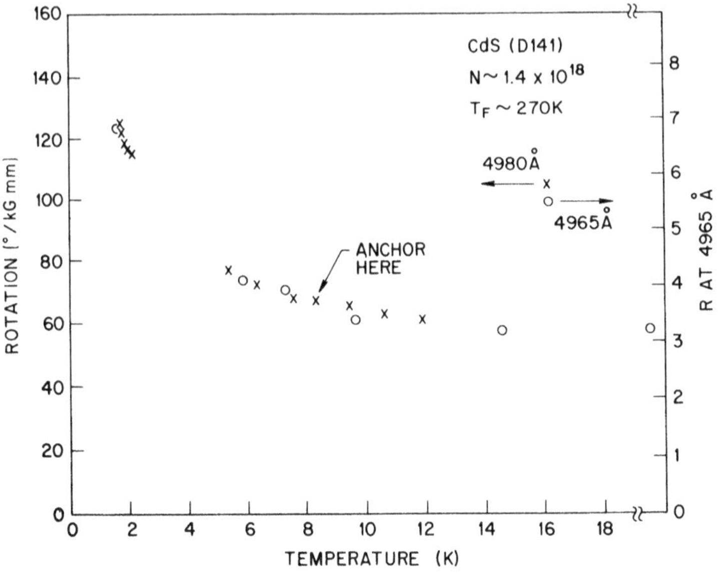

Fig. 3.9. Spin Faraday rotation ($\sim\chi_s$ the spin susceptibility) vs temperature for sample just above IM transition. The scale for 4965 Å is adjusted so that the data points for the two wavelengths coincide at 8.5 K (see text)

nents of χ, which have also been seen in P:Si [3.53], has been interpreted as evidence for the simultaneous coexistence of spatially separate insulating and metallic phases. However, the dynamic behavior of the spins seems homogeneous since the SFRS linewidth displays only a single Lorentzian component $\Delta\omega$ with strong diffusive character, i.e., $\Delta\omega = Dq^2$. It has been suggested [3.52, 38] that we are indeed dealing with a single phase and that the temperature dependence of χ in these samples is due to electron correlation in the metallic region [3.55]. The observation of saturation in the temperature-dependent component at still lower temperatures would support the single phase hypothesis.

Large values of α and hence of FR are expected in other direct band-gap semiconductors but α is expected to be orders of magnitude smaller in indirect gap semiconductors. We can therefore anticipate similar measurements of FR in direct band-gap semiconductors using tuneable dye lasers.

3.5.4 Measurement of Donor Relaxation T_1 by Faraday Rotation

Experiments have been performed using both FR and SFRS techniques to monitor the population of the spin levels during or after irradiation with a pulse of resonant microwaves. FR has proven to be an extremely sensitive tool for detecting the EPR of the donor even at very low concentration ($N_D \sim 10^{16}$ cm^{-3}) on very small samples [3.56] and allows precise determination of the spin-lattice relaxation time. The main result is that T_1 shortens dramatically well before the Mott transition, its value ranging from 40 ms at $N_D \sim 10^{16}$ cm^{-3} to a few µs when $N_D \sim 10^{17}$ cm^{-3}. This speedup is probably due to the strong antiferromagnetic coupling between the spins (Sect. 3.7) and its modulation by the phonons.

3.6 Determination of the k-Linear Term in the Conduction Band of CdS by SFRS

3.6.1 Origin of the k-Linear Term

In a polar (or more precisely, pyroelectric) crystal like CdS, the crystallographic structure allows for the existence of a permanent electric field $\boldsymbol{E}$ along the c-axis. Mobile charges with velocity $v = \hbar k/m$ experience an effective magnetic field H which couples to their spin $\boldsymbol{S}$ in a manner very similar to the spin-orbit coupling. The electron energy measured from the conduction band minimum becomes [3.57]

$$E = \frac{\hbar^2 k^2}{2m^*} + \lambda(\boldsymbol{k} \times \boldsymbol{c}) \cdot \boldsymbol{S}, \tag{3.46}$$

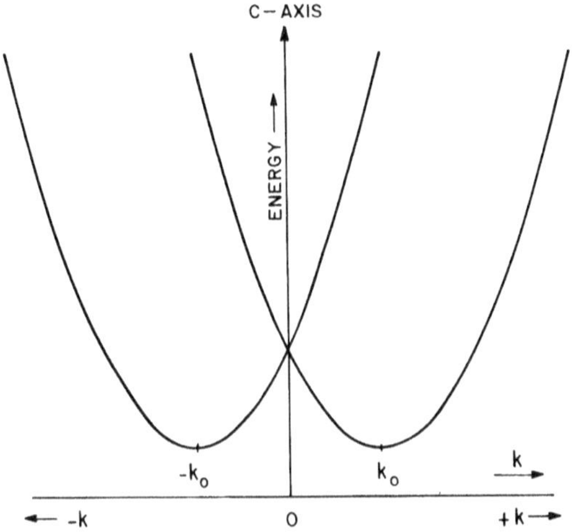

Fig. 3.10. Effect of k-linear term in CdS in lifting spin degeneracy away from $k=0$ in conduction band

where $\boldsymbol{c}$ is a unit vector along the c-axis and λ is proportional to the internal electric field (Fig. 3.10). The two-fold conduction band degeneracy is then lifted for each value of $\boldsymbol{k}$, while time reversal invariance still requires $(\boldsymbol{k}, m_s)$ and $(-\boldsymbol{k}, -m_s)$ to be degenerate.

This k-linear term arises from admixture of the other bands through third-order perturbation involving the $\boldsymbol{k}\cdot\boldsymbol{p}$ interaction, spin orbit coupling $\zeta\boldsymbol{L}\cdot\boldsymbol{S}$ and the average crystal electric field $\boldsymbol{E}$ and is given by [3.58]

$$\lambda(\boldsymbol{k}\times\boldsymbol{c})\cdot\boldsymbol{S}=\sum_{\{p\}}\left\langle s\left|\frac{\hbar}{m}\,\boldsymbol{k}\cdot\boldsymbol{p}\right|p\right\rangle\frac{\langle p|\zeta\boldsymbol{L}\cdot\boldsymbol{S}|p\rangle}{(E_p-E_s)^2}\,\langle p|eE\boldsymbol{c}\cdot\boldsymbol{r}|s\rangle, \tag{3.47}$$

where $|s\rangle$ refers to the conduction band and $|p\rangle$ represents opposite parity band states of energy E_p. This effect is much larger than the simple relativistic effect of the magnetic field $H=\boldsymbol{v}\times E\boldsymbol{c}$. While a clear contribution to λ comes from the valence band intermediate state $|p\rangle$, it will be suggested below that other bands must also participate.

Terms of this type, odd in k, have been proposed as a relaxation mechanism for conduction electron spins in noncentrosymmetric crystals [3.59, 60]. The k-linear term has been discussed as well by *Hopfield* [3.61] and measured indirectly via anomalies in the exciton reflectivity spectra of CdS by *Hopfield* and *Thomas* [3.58] and *Mahan* and *Hopfield* [3.62]. *Koteles* and *Winterling* [3.63] have also recently measured λ in the valence band of CdS using resonant Brillouin scattering [Ref. 3.31, Chapt. 7]. However, the first measurement of λ in any conduction band was made using SFRS in CdS [3.64a], as described in the next section.

3.6.2 Appearance of the k-Linear Term in Diffusional SFRS Linewidth

The term in λ in (3.46) implies that the spin will be quantized along an axis $\boldsymbol{k} \times \boldsymbol{c}$ defined by the $\boldsymbol{k}$ vector. For free electrons, however, the momentum collision rate in CdS is so rapid that this effective internal magnetic field ($\sim \lambda \boldsymbol{k} \times \boldsymbol{c}$) is spatially randomized at a rate greater than any spin precessional frequency so that the spins quantize along the external magnetic field H_0. The energy of the two spin states $m_s = \pm 1/2$ are given by $[g\mu_B H_0 + \lambda(\boldsymbol{k} \times \boldsymbol{c}) \cdot \boldsymbol{h}_0] m_s$, where $\boldsymbol{h}_0$ is a unit vector along H_0. Assuming that the momentum transfer $\boldsymbol{q}$ for the transitions $k, \mp 1/2 \rightarrow \boldsymbol{k} \pm \boldsymbol{q}, \pm 1/2$ is small compared to $\boldsymbol{k}$, it is clear that the frequency shift $\Delta\omega$ will be given by

$$\hbar\Delta\omega = \frac{\hbar^2 \boldsymbol{k}}{m^*} \cdot \left[\boldsymbol{q} \pm \frac{\lambda m^*}{\hbar^2} (\boldsymbol{c} \times \boldsymbol{h}_0)\right] \pm g\mu_\beta H_0, \tag{3.48}$$

where $(\hbar^2 \boldsymbol{k}/m^*) \cdot \boldsymbol{q}$ is the Doppler shift or the transfer of kinetic energy described in Sect. 3.4.2. The $\mp$ signs refer to the Stokes or anti-Stokes processes, respectively. In other words, the Doppler shift appears modified by replacement of $\boldsymbol{q}$ by $\boldsymbol{q} \pm \boldsymbol{q}_0$, where $\boldsymbol{q}_0 = \lambda m^*(\boldsymbol{c} \times \boldsymbol{h}_0)/\hbar^2$. We have seen in Sect. 3.4.2 that this term is motionally narrowed and results in a diffusional linewidth $\Delta\omega = Dq^2$ at least for small values of q. The net effect of the k-linear term on SFRS in CdS is thus seen in the q-dependence of the linewidth where $\Delta\omega = D(\boldsymbol{q} \pm \boldsymbol{q}_0)^2$. For a given geometry, the Stokes and anti-Stokes lines will appear with different linewidths: this is confirmed by experiment for a sample with mobile donor electrons, as seen in Fig. 3.11 for a sample with $N_D \simeq 7 \times 10^{17}\,\mathrm{cm}^{-3}$. Reversing

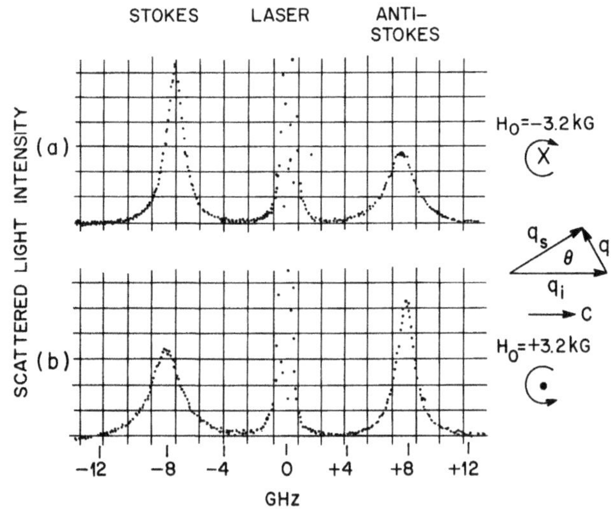

Fig. 3.11. (a) Asymmetry between Stokes and anti-Stokes linewidths at scattering angle $\theta = 30°$ due to k-linear term; **(b)** reversal of asymmetry with field reversal. $N_D = 7 \times 10^{17}\,\mathrm{cm}^{-3}$ and $T = 1.6\,\mathrm{K}$ [3.64]

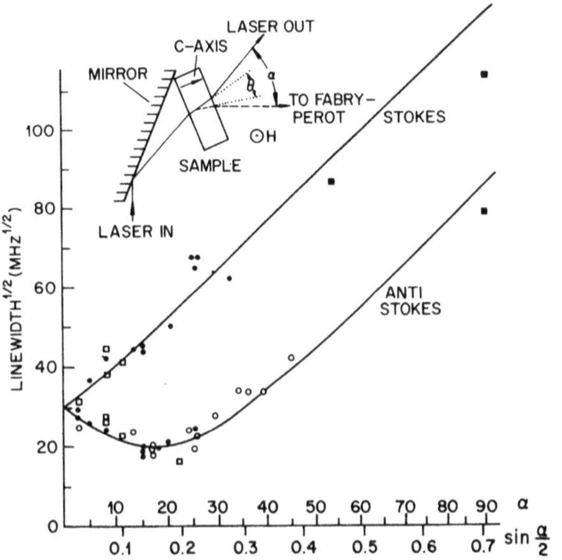

Fig. 3.12. Asymmetry between Stokes and anti-Stokes linewidths vs scattering angle for CdS with $N = 1.8 \times 10^{18}\,\mathrm{cm}^{-3}$. Minimum at $\alpha = 18°$ determines k-linear term (see text) [3.64a]

either the magnetic field or the direction of the crystal c-axis reverses the asymmetry. It has been verified that indeed the linewidth asymmetry varies as $(\boldsymbol{q} \times \boldsymbol{c}) \cdot \boldsymbol{h}_0$ for different geometries, vanishing when $\boldsymbol{c} \| \boldsymbol{H}$ or $\boldsymbol{c} \| \boldsymbol{q}$.

Since the observed SFRS lines have an additional linewidth contribution due to the lifetime of the spin states (T_1 and T_2 relaxation times), it is necessary to study $\Delta\omega$ for different values of q in order to determine λ. To optimize the asymmetry, a geometry in which $\boldsymbol{q}$ remains perpendicular to $\boldsymbol{c}$ and $\boldsymbol{h}_0$ was chosen and is shown in the inset of Fig. 3.12. The sample and mirror rotate together about an axis perpendicular to the figure. This arrangement has the additional feature that $2k_0 \sin(\alpha/2) = 2nk_0 \sin(\theta/2) = |q|$, where n is the index of refraction and k_0 is the free space wave vector of light (not to be confused with the k vector of the electrons). Thus q is determined independently of n. In Fig. 3.12 $(\Delta\omega)^{1/2}$ is plotted vs $\sin \alpha/2$ for an even more degenerate sample of CdS than in Fig. 3.11, with $N_D = 1.8 \times 10^{18}\,\mathrm{cm}^{-3}$. It shows very clearly the existence of a minimum in $\Delta\omega$ obtained for a nonzero value of q. In effect, the linewidth at $q = 0$, obtained for instance in a microwave experiment, has a component due to the k-linear term which can be exactly cancelled by the Doppler effect by appropriate choice of scattering geometry corresponding to $q = q_0$. Thus one has a remarkable situation in SFRS that may be termed Doppler narrowing of a line.

The solid lines in Fig. 3.12 represent a best fit to $\Delta\omega$ of the form $\Delta\omega = \Delta\omega_0 + [(4D/2\pi)k_0^2 \cdot (\sin \alpha/2 \pm \sin 18°/2)^2]$ with $\Delta\omega_0 = 400$ MHz and $(4D/2\pi)k_0^2 = 22.5$ GHz. The discrepancy at higher values of q is linked to a breakdown of diffusional behavior at small distances as described below. The minimum at $\alpha = 18°$ corresponds to a value of $\lambda_c = 1.6 \times 10^{-10}$ eV cm (the subscript c refers to the conduction band). It is noteworthy that the same value of λ_c has been obtained for samples with widely different concentrations and D values by using the effective mass m^* of the conduction band, even when the impurity band is well separated from it; this point will be clarified in Sect. 3.6.4.

3.6.3 Comparison of λ in Conduction and Valence Bands

Equation (3.47) could, in principle, be used for comparison with the experimental value of λ_c. However, there exists, to our knowledge, no precise determination of the crystal structure of CdS which would allow knowledge of the internal electric field E, or even of its sign. A crude point-charge model [3.64a] requires a 3% distortion of the tetrahedron, which is very reasonable. A more crucial test seems to be a comparison with the experimental value of λ_v observed in the valence band by *Mahan* and *Hopfield* [3.62] or more recently by *Koteles* and *Winterling* [3.63] who found $\lambda_v = 6.7 \times 10^{-10}$ eV cm. Since the valence band is already split by spin-orbit coupling ζ, one does not require a 3rd order perturbation and one expects a ratio $(\lambda_c/\lambda_v) \simeq (\zeta/E_g)$ if the conduction band value comes mainly through the admixture of the valence band [3.61]. The experimental ratio λ_c/λ_v, is 0.25, much larger than $\zeta/E_g = 0.02$. Yafet has suggested in a private communication [3.64b] that the higher conduction band, derived from the $5p$ states of Cd^+ ion, may be most effective here. This band is 5.6 eV above the s-band and has a spin-orbit splitting of 0.3 eV which is mostly centered on the same ion as the conduction band, i.e., the Cd, so that one may expect the matrix elements of $\boldsymbol{k} \cdot \boldsymbol{p}$ and $\boldsymbol{E} \cdot \boldsymbol{r}$ to be larger than for the valence band. This, in addition to the fact that $\zeta(5p, \mathrm{Cd})/(E_{5p} - E_{5s})^2 \gtrsim \zeta(\text{valence band})/E_g^2$, may explain the unexpectedly large value of λ_c.

In this connection it is interesting to note a recent determination of λ_c in CdSe. *Rashba* [3.65] has emphasized the influence of the k-linear term in mixing odd-parity states into the time-reversed spin states thus giving rise to a direct electric dipole matrix element between the time reversed spin states, calling such a transition a "combined resonance". *Dobrowolska* et al. [3.66] have recently measured the direct electric dipole matrix element between donor spin levels in CdSe which also has the Wurzite structure. The crystal was doped with several percent of Mn to provide a large splitting of the donor $m_s = \pm 1/2$ states through the exchange interaction between Mn and donor electrons. In moderate magnetic fields this splitting is sufficiently large (in the far infrared), bringing the $1s, m_s = +1/2$ donor state close to its $2p$ state so that the mixing effect of the k-linear term is large, resulting in a large dipole moment matrix element between the $m_s = \pm 1/2$ states. By measuring the strength of this dipole moment by the infrared absorption $(1s, -1/2 \rightarrow 1s, +1/2)$, they arrived at a value $\lambda_c(\mathrm{CdSe}) = 10^{-10}$ eV cm. They suggest that this value is consistent with $\lambda_c(\mathrm{CdSe})/\lambda_c(\mathrm{CdS}) \simeq \zeta(\mathrm{S})/\zeta(\mathrm{Se})$. This result would imply that it is the mixing-in of the valence band and the ζ on the sulfur or selenium that are important rather than any excited states of the Cd, in which case the difficulty of how to explain the near equality of λ_c and λ_v still remains. Perhaps closer examination of λ_c in

CdSe by SFRS and λ_v by resonant Brillouin scattering may help resolve this question.

3.6.4 Generalization to Bound Donors with Spin Diffusion

The derivation of the shift as presented in Sect. 3.6.1 applies to mobile charges in a conduction band. The experimental results presented so far prove that the derivation is valid in impurity bands as well. Experiments have also been performed on a more dilute sample with $N_D = 2.3 \times 10^{17}\,\mathrm{cm}^{-3}$. Here the electrical transport data [3.54] show a decrease in conductivity of more than seven orders of magnitude between 40 and 1.6 K. Clearly the donor electrons in this sample are frozen out at 1.6 K and one has bound donors, so that charge diffusion should be vanishingly small. Yet SFRS experiments (Fig. 3.13) show a

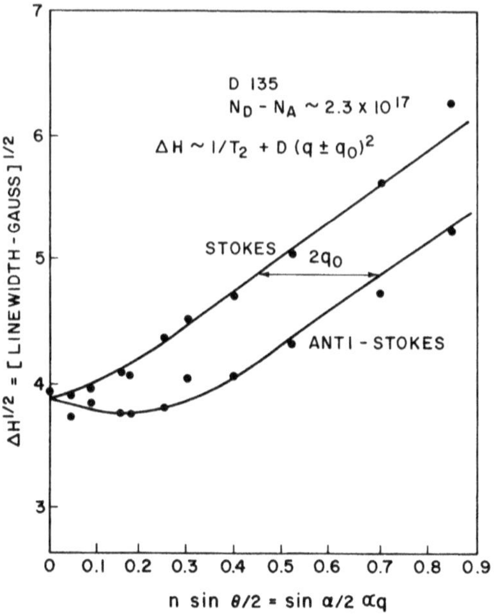

Fig. 3.13. Evidence of k-linear term (asymmetry in widths of Stokes and anti-Stokes) and spin diffusion (variation of widths with q^2) for bound donors; $N = 2.3 \times 10^{17}$. 1 Gauss $= 2.5$ MHz, corresponding to a donor $g \simeq 1.8$ [3.54]

definite variation of $\Delta\omega$ vs q, corresponding to diffusion. This pure spin diffusion is due to the strong exchange interaction between bound donors (Sect. 3.7.2a). That the effect of k-linear term is still present for localized electrons, as clearly seen in Fig. 3.13, is at first sight puzzling. We therefore present a more general demonstration that within certain approximations its effect amounts to a shift in the q-space of momentum transfer for any system of electrons interacting either with a static potential (localized donors) or with each other (electrostatic interaction leading either to antiferromagnetic interaction or to charge delocalization).

In the effective mass approximation, one can write a generalized many-body Hamiltonian:

$$\mathscr{H}_0=\sum_{ij}-\frac{\hbar^2}{2m^*}\nabla_i^2+V_{ij}+V_i, \tag{3.49}$$

where V_{ij} represents electrostatic interactions between electrons and V_i, the pontential of ionized impurities, depends only on space coordinates and not on spin. This Hamiltonian describes either a set of interacting bound donor impurities, an impurity band, or a strongly doped metallic-like sample. The addition of a k-linear term can be represented in this approximation by the Hamiltonian

$$\mathscr{H}_1=-\mathrm{i}\sum_i \lambda(\boldsymbol{c}\times\boldsymbol{S}_i)\cdot\nabla_i. \tag{3.50}$$

Let

$$\boldsymbol{q}_0^i=\frac{\lambda m^*}{\hbar^2}(\boldsymbol{c}\times\boldsymbol{S}_i)=q_0(\boldsymbol{c}\times\boldsymbol{S}_i). \tag{3.51}$$

Now if

$$\mathscr{H}_0\psi_0=E_0\psi_0, \tag{3.52}$$

then it is easily shown that

$$(\mathscr{H}_0+\mathscr{H}_1)\psi_0\exp\left(-\mathrm{i}\sum_i \boldsymbol{q}_0^i\cdot\boldsymbol{r}_i\right)$$

$$=\left(E_0-\sum_i\frac{\hbar^2 q_0^{i2}}{2m^*}\right)\psi_0\exp\left(-\mathrm{i}\sum_i \boldsymbol{q}_0^i\cdot\boldsymbol{r}_i\right). \tag{3.53}$$

Thus, aside from the shift in energy, the effect of the k-linear term is to multiply the original wave function by the phase factor $\exp(-\mathrm{i}\sum_i \boldsymbol{q}_0^i\cdot\boldsymbol{r}_i)$.

The SFRS is described by a matrix element

$$\langle\psi_\mathrm{f}|S^{\pm}\mathrm{e}^{+\mathrm{i}\boldsymbol{q}\cdot\boldsymbol{r}}|\psi_\mathrm{i}\rangle \tag{3.54}$$

$$=\left\langle\psi_{0_f}\exp\left(\mp\mathrm{i}\sum_i \boldsymbol{q}_0^i\cdot\boldsymbol{r}_i\right)\middle|S^{\pm}\mathrm{e}^{\mathrm{i}\boldsymbol{q}\cdot\boldsymbol{r}}\middle|\psi_{0_i}\exp\left(\pm\mathrm{i}\sum_i \boldsymbol{q}_0^i\cdot\boldsymbol{r}_i\right)\right\rangle. \tag{3.55}$$

We now make the assumption that the coherence length of any spin excitation is very small compared to $1/q_0$ as well as to the length of our light probe $1/q$, i.e., that we are in a diffusive regime. We may then expand the exponentials to first order. In the practical case where $\boldsymbol{c}$, $\boldsymbol{q}$ and $\boldsymbol{H}$ are at right angles to each other along x, y and z, respectively (which maximizes the effect of the k-linear term), it is seen by using the commutation rules for the components of S that this amounts to replacing $S^{\pm}(1 \pm iqy)$ by $S^{\pm}(1 \pm \mathrm{i}(q \mp q_0)y] + q_0 z S_Z$. The alternate signs refer to Stokes and anti-Stokes scattering, respectively. The zS_Z involves a transition at zero frequency and is unimportant. Thus, the main effect of the k-linear term is to replace the momentum transfer q by $q \mp q_0$, and any q-dependent results will contain this feature. This is true for any dispersive shifts that might be present [3.67]; see Sect. 3.7.2b), as well as in a diffusive linewidth whose widths will be given by $D(q \pm q_0)^2$, as seen in Fig. 3.13. [Note, however, that at large q, where the condition for the expansion of the exponentials in (3.55) fails, diffusive behavior is no longer seen ($\Delta\omega$ is becoming independent of q). At the same time, the difference between Stokes and anti-Stokes linewidths become smaller and they no longer follow $D(q \pm q_0)^2$. See Sect. 3.7.2a and Fig. 3.18.] In summary, as long as the coherence length of the spin excitation is small compared to $1/q$ and $1/q_0$ and V_{ij} is independent of spin coordinates, this result regarding the shift in q-space is valid and will be seen both for charge and pure spin diffusion.

3.7 Bound Donors as Model Amorphous Antiferromagnets

3.7.1 Static Properties Studied by Faraday Rotation

A very active topic in condensed matter physics is that of random magnetic systems. Bound shallow donors are an example of one class of such systems called amorphous antiferromagnets. The donors in their $1S$ ground state interact with each other via the familiar antiferromagnetic hydrogenic-like exchange $J_{ij}\boldsymbol{S}_i \cdot \boldsymbol{S}_j$, where $J_{ij} > 0$ and its asymptotic behavior has been given by *Herring* and *Flicker* [3.68] as

$$J_{ij}(r_{ij}) = 1.636 r_{ij}^{5/2} \exp(-2r_{ij}/a_{\mathrm{H}}^*) \,\mathrm{Ry}, \tag{3.56}$$

where r_{ij} is the donor separation, a_{H}^* is the donor Bohr radius and Ry is the appropriately scaled Rydberg. In CdS, for a donor concentration $N \sim 8 \times 10^{16}\,\mathrm{cm}^{-3}$, $a_{\mathrm{H}}^* = 28$ Å and Ry $= 24$ meV. The large donor orbits compared to the crystalline-lattice spacing, coupled with their dilution (a few ppm) and random placement on the crystalline lattice sites, make this system a model "essentially amorphous" antiferromagnet and one which is well characterized. The term antiferromagnet is used here only to describe the positive sign of J_{ij} rather than to imply any kind of antiferromagnetic ordering involving sublattices. The tendency of any two donors to align antiparallel is frustrated by the random placement of neighboring donors and if the spins order, they could only do so in the sense of spin glasses [3.69] with spins frozen in random orientations and a net moment of zero, as illustrated in Fig. 3.14. There would

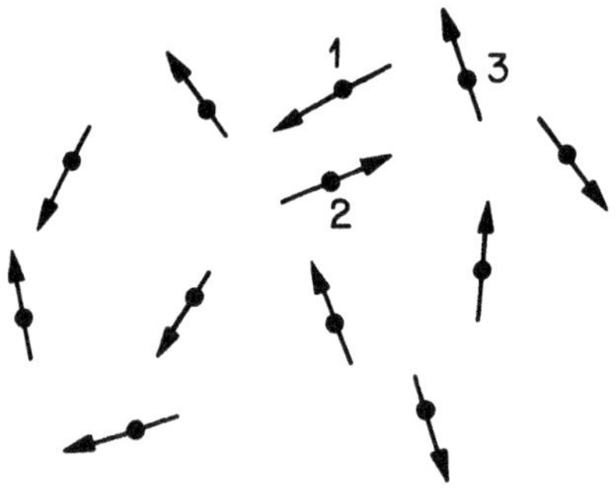

Fig. 3.14

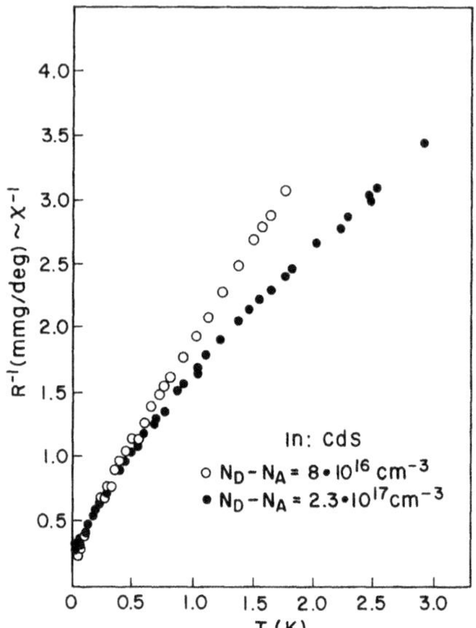

Fig. 3.15

Fig. 3.14. Schematic diagram of random frozen orientations of spins in a spin glass

Fig. 3.15. Inverse Faraday rotation susceptibilities for two concentrations of neutral bound In donors in CdS as a function of temperature. No sign of ordering appears [3.70]

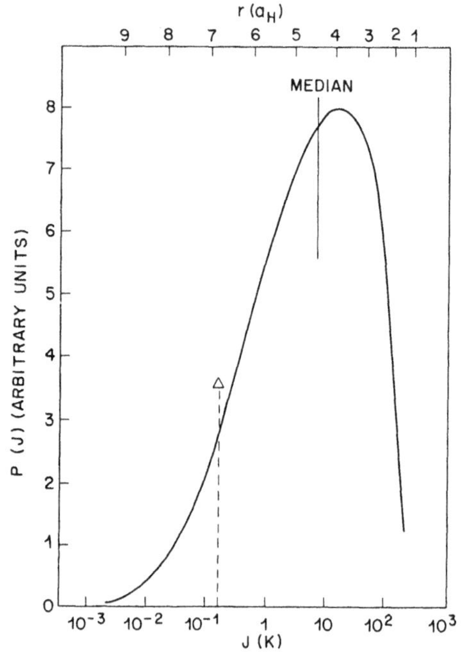

Fig. 3.16. Distribution of closest-neighbor donor-donor exchange in CdS for $N=8\times10^{16}\,cm^{-3}$. $\Delta=0.150$ K is the cluster percolation threshold (see text) [3.71]

be no such frustration in a random ferromagnet ($J_{ij} < 0$). The hallmark of ordering in a prototype spin glass such as Mn:Cu is an abrupt change in the susceptibility χ at a finite temperature $T_g \gg$ Presumably, below T_g the spins begin to freeze in, such that an order parameter q exists [3.69] given by $q = \lim_{t \to \infty} \langle \boldsymbol{S}_i(t) \cdot \boldsymbol{S}_i(0) \rangle$, where t is time and the average is over all spins. To examine whether such ordering exists in the amorphous antiferromagnet, *Kummer* et al. [3.70] studied the susceptibility of n-CdS using the Faraday rotation method described earlier. Special techniques using optical fibers to couple the light into and out of the sample in a dilution refrigerator were used to study χ down into the millikelvin range. The results for χ^{-1} are plotted in Fig. 3.15 for two samples in the insulating region where donor freezeout is complete. The antiferromagnetic coupling is manifest in the negative Curie-Weiss constant extrapolated from the slope of χ^{-1} vs T at higher T, i.e., the tangent intercepts the negative T axis. χ^{-1} continues to curve downward with decreasing T with no sign of any minimum corresponding to the cusp-like peak in χ seen in spin glasses or in $\chi_{\parallel}$ in normal antiferromagnets. This downward slope in χ^{-1} persists to temperatures far below the median closest-neighbor exchange of ~ 7 K shown in Fig. 3.16, where the distribution of closest-neighbor exchanges, $P(J_{ij})$, is plotted for $N \sim 8 \times 10^{16}$ cm^{-3}. The distribution of closest-neighbor exchange interactions is calculated from (3.56) assuming random placement of the donors. It is noteworthy that the distribution of J's spans many decades, i.e., the relatively narrow Poisson distribution of closest distances is converted into a very broad distribution of J's due to the exponential dependence of J_{ij} upon r_{ij}. *Walstedt* et al. [3.71] have done a computer simulation of randomly placed donors and have calculated χ for $N = 8 \times 10^{16}$ cm^{-3} using a cluster approximation where clusters of up to 8 spins are diagonalized exactly and the intercluster couplings are treated by a molecular field approximation. They obtained excellent agreement with the data shown in Fig. 3.15. While the cluster calculation says nothing about ordering, it does yield a value for the temperature and for magnetic percolation T_p, i.e., that value of T such that an infinite connected network of bonds exists with $J_{ij} > kT_p$. For the 8 $\times 10^{16}$ sample, $T_p \sim 0.1$ K while for the 2.3×10^{16}, $T_p \sim 1$ K. Thus, it can also be stated that no magnetic transition is found at a temperature more than an order of magnitude below the percolation threshold for the higher concentration sample. This is in contrast to RKKY spin glasses, where the transition temperature T_g is only slightly below T_p. Essentially similar behavior is observed in P doped Si using SQUID techniques to measure χ [3.72], although a leveling off in χ at temperatures below 20 mK is seen experimentally but still no peak in χ. This failure of the amorphous antiferromagnet to show any signs of ordering at finite temperature, in contrast to other spin glasses, may be due to several factors. For one, the short-range interaction, i.e., the exponential fall off with distance, gives a broad distribution of J_{ij}'s with many weak couplings (this effectively renormalizes the exchanges downward), in contrast to the long range $1/r^3$ fall off in the RKKY interaction and narrower distribution of couplings in metallic spin glasses. Also, the fact that the donor spin is $S=1/2$ (contrasted with $S=5/2$ of Mn in Cu:Mn) may result in very large zero-point fluctuations which prevent ordering [3.73]. These and other possible reasons for lack of ordering in the amorphous antiferromagnet are discussed in greater detail by *Walstedt* et al. [3.71], *Andres* et al. [3.72] and by *Bhatt* and *Lee* [3.74] using a renormalization group technique.

Thus, the Faraday rotation, by allowing one to measure the static susceptibility χ of donors at the ppm level in CdS, has provided us for the first time with important information on this well-characterized $S=1/2$ amorphous antiferromagnet. One would anticipate that it will prove a powerful technique in examining the χ of donors in other direct band-gap semiconductors. We will examine the dynamic properties of the amorphous antiferromagnet as given by $\chi(q, \omega)$ which is studied by SFRS in Sect. 3.7.2.

3.7.2 Dynamics of the Amorphous Antiferromagnet

Even though as seen above in the discussion of the static susceptibility, the amorphous antiferromagnet does not show any long-range ordering analogous to a spin glass like freezing, it is nonetheless of interest to examine how the exchange interactions are manifest in the excitations of this system at temperatures low compared to typical J_{ij}'s. We therefore now examine the dynamical behavior of the amorphous antiferromagnet as revealed by $S(q, \omega)$ in SFRS. In discussing $S(q, \omega)$, it is appropriate to consider two regimes of external magnetic field H_0. The first is the low-field case when H_0 is significantly less than kT as well as most exchange couplings J_{ij}. The second region is where H_0 becomes comparable to typical J_{ij}'s.

a) Low-Field Regime: Pure Spin Diffusion

In low fields (less than 0.1 Tesla), the Stokes and anti-Stokes lines in the SFRS spectrum peak at $\omega_L \mp \omega_0$ where $\omega_0 = g\beta H_0/\hbar$, i.e., the Zeeman frequency. The peak positions do not shift as a function of scattering angle within the experimental accuracy of ~ 6 MHz. However, the linewidth does vary with q and is given by $\Delta\omega = 1/T_2 + Dq^2$, where T_2 is some residual EPR linewidth. This is illustrated in Fig. 3.17 for a sample with donor concentration $N = 2.3 \times 10^{17}$, a factor of four below the IM transition[2]. As discussed in Sect. 3.6.4, transport measurements on this sample at low temperatures [3.54] verify that the donor electrons are frozen out, so that this diffusive behavior is unambiguously associated with bound donors and corresponds to pure spin diffusion without

2 As discussed in Sect. 3.6.4, the maximal effect of the linear k-term appears in certain geometries, i.e., when $\boldsymbol{H} \perp \boldsymbol{c} \perp \boldsymbol{q}$ and effectively shifts q to $q \pm q_0$ for the Stokes and anti-Stokes lines (where q_0 is given) by (3.51) so that their diffusive widths are given by $D(q \pm q_0)^2$. The q_0 term when it appears has been corrected for in all the data presented in this section.

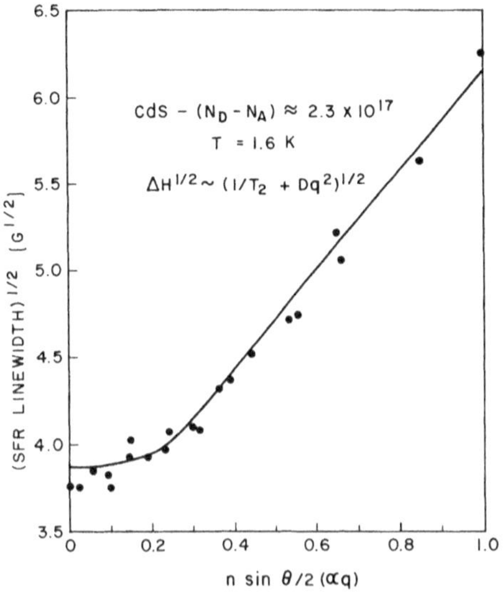

Fig. 3.17. Pure spin diffusion for bound donors in CdS with $N = 2.3 \times 10^{17}$. D is the diffusion constant. 1 Gauss = 2.5 MHz, corresponding to $g \simeq 1.8$ of donors in CdS

any attendant charge transport. This diffusion arises from the transverse terms $J_{ij}S_i^+S_j^-$ in the exchange interaction. A rough estimate of the diffusion constant might be given by $D \simeq \langle r \rangle^2/\tau = \langle r_{ij}^2 J_{ij} \rangle / h$. Taking $r_{ij} \simeq 100$ Å and $J_{ij} \sim 10$ K, one obtains an estimated value $D \sim 0.25\,\mathrm{cm}^2/\mathrm{s}$ which is almost two orders of magnitude larger than the experimental value $D = 2 \times 10^{-3}\,\mathrm{cm}^2/\mathrm{s}$ for this sample. There are several reasons for this large overestimate. For one, the use of averages in this random system is a questionable procedure because diffusion will be controlled by the weak lines in the chain of J_{ij}'s. Secondly, there is a general reduction of effective exchanges between spins, one of which is already tightly coupled to other spins. Finally, account must be taken of a coherence length in the spin system. While the calculation of spin diffusion in this random system is clearly a formidable problem, SFRS has until now been the only source of experimental values of D for these systems and should provide a stimulus for further theoretical work.

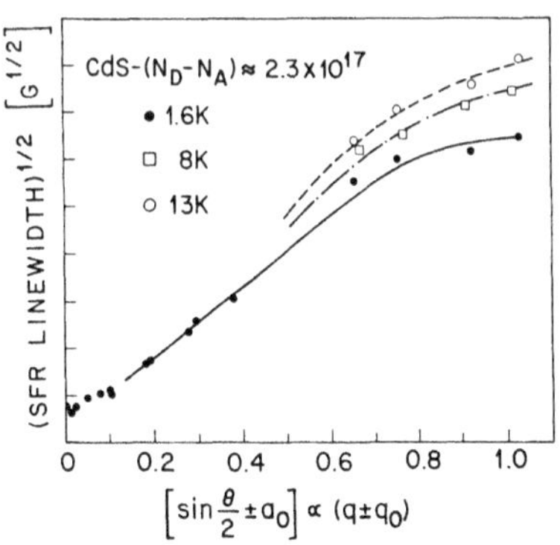

Fig. 3.18. Breakdown of spin diffusion at large $q(\theta \gtrsim 90°)$ and partial recovery with increasing temperature

At larger angles the linewidth varies more slowly with q and above $\theta=90°$ at $T=1.6$ K is almost independent of $\boldsymbol{q}$, as seen in Fig. 3.18. It should be noted that at $\theta=90°$ with a CdS index of 2.9 at 4880 Å, $q \sim 5.3\times 10^5\ \mathrm{cm}^{-1}$ so that $\boldsymbol{q}\cdot\boldsymbol{r}_{ij}\simeq 0.50$, i.e., the wavelength of our probe extends only over a small number of donor distances such that diffusive behavior will no longer pertain. An alternate possibility is that this value of q, where diffusive behavior breaks down, measures a coherence length in the spin excitations [3.54].

b) High-Field Case: Field-Induced Exchange Stiffness and Dispersion

In contrast to low fields, at high field (52 kG), *Carlson* et al. [3.67] observed a shift from the Zeeman frequency ν_0 in the peak position of spin-flip excitation

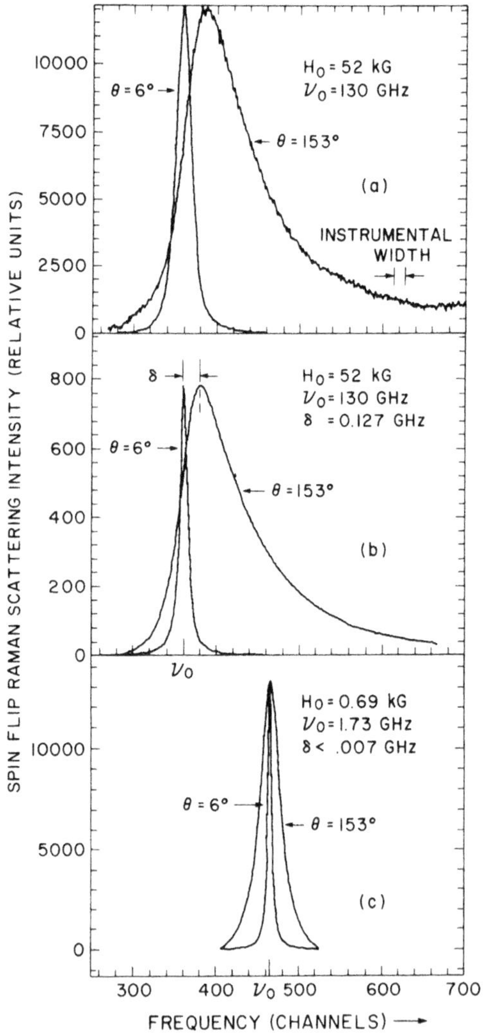

Fig. 3.19a–c. Dispersion (i.e., variation of peak position vs $q \sin\theta/2$ displayed for two values of θ) in SFRS excitation at high field. (**a**) Raw data; (**b**) same as in (**a**) but smoothed and corrected for instrumental width; (**c**) low-field SFRS showing no shift in peak with q [3.67]

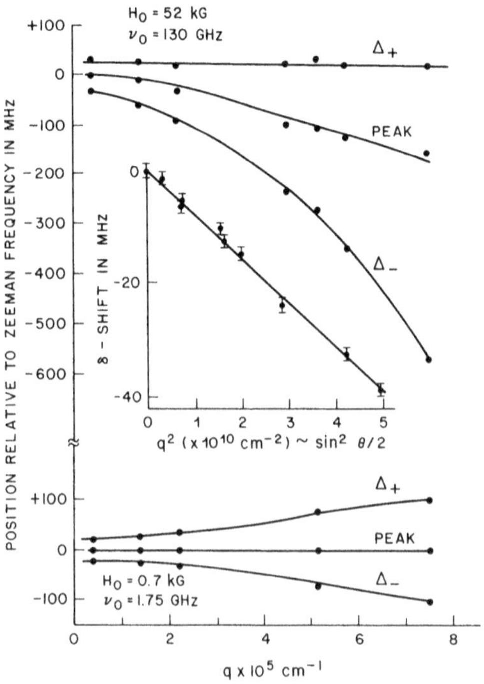

Fig. 3.20. Field-induced negative dispersion in the amorphous antiferromagnet. $\Delta_\pm$ are the half intensity points (see text) [3.67]

with increasing q. This is illustrated for the Stokes line in Fig. 3.19 for two scattering angles $\theta=6^\circ$ ($q\sim 0$ Å^{-1}) and $\theta=153^\circ$ ($q=0.0075$ Å^{-1}) for the same sample as in Figs. 3.17, 18. Figure 3.19a is the raw data and Fig. 3.19b shows the $S(q,\omega)$ corresponding to Fig. 3.19a, smoothed and corrected for the instrumental width. In addition to the shift of the peak, a severely asymmetric line shape is also observed. The anti-Stokes line (corrected for the linear k-term when present) is a mirror image about the laser line of the Stokes, but much reduced in intensity because of the Boltzman factor. Note that the observed shifts, for example, of the Stokes scattered light to higher frequency, i.e., towards the laser line, implies a decrease from ν_0 in the frequency of the spin-flip excitations, i.e., a negative dispersion. For comparison, the low-field data for these angles are presented in Fig. 3.19c. The contrasting behavior of $S(q,\omega)$ in high and low fields is summarized in Fig. 3.20 for different values of q. $\Delta_\pm$ are the half intensity points and the data as presented includes the intrinsic width $1/T_2$. This field-induced negative dispersion may be understood as follows. In high fields, the individual spins acquire induced moments along H_0 which interact antiferromagnetically. Turning a spin over in H_0 costs the Zeeman energy $h\nu_0$, but since the reversed spin is now oppositely oriented to the average direction of the others, the exchange energy is lowered due to the antiferromagnetic coupling and hence the excitation is less than $h\nu_0$. If the coupling between spins were ferromagnetic, the dispersion would be in the opposite direction. The system of exchange coupled spins in n-CdS resembles a regular antiferromagnet in a field in its paramagnetic phase which has a dispersion relation with ω decreasing as q increases [3.75]. However, because of the random placement of the donors, a given $\boldsymbol{q}$ will correspond to a spectrum of excitations at different frequencies with some component at q. The spread of frequencies which project on q increases with increasing q with most of them lying below their most probable value, giving rise to the increased width and asymmetry observed.

A very rough estimate of the magnitude of the dispersion would be given by $\Delta\nu \sim \langle S_{iz} J_{ij} r_{ij}^2 \rangle q^2$, where $\langle\ \rangle$ refer to some ill-defined averages in this random system. The observed shift clearly goes as q^2 as seen in the inset of Fig. 3.20, and a discussion of its magnitude may be found in [3.67]. The dispersion is also found experimentally to increase with increasing donor concentration reflecting the increase in J_{ij}. In summary, the amorphous antiferromagnet shows no exchange stiffness or dispersion in the absence of a field, but as the spins are polarized in increasing field, a field-induced exchange stiffness appears giving rise at long wavelength to dispersive hydrodynamic-like spin wave modes. In closing this section, it should be noted that the observation, both of the pure spin diffusion and the dispersion at the q-values of light as reported here would have been clearly difficult, if not impossible, with a grating spectrometer, but is easily seen with the instrumental resolution of 7.5 MHz (or 3×10^{-4} meV) at the particular Fabry-Perot free spectral range used here.

3.8 Coherence Effects in SFRS and Stimulated SFRS

3.8.1 Scattering from Coherent States in CdS

Spontaneous Raman scattering is an incoherent process, characterized by the random phase factor ϕ appearing between the initial and final states in (3.1). The total power emitted is proportional to the density N of spins and is described by a Raman cross section per spin. We have seen that Faraday rotation, on the contrary, is coherent in that each dipole radiates with a well-defined phase linked to the phase of the incident light wave. For a small rotation ϕ, proportional to $N\langle\sigma_z\rangle$, the power radiated in the direction of the incident beam with the polarization at right angles to the incident polarization is proportional to $\sin^2\phi \sim N^2$. Faraday rotation is linked to the existence of a longitudinal magnetization $\mu_B\langle\sigma_z\rangle$, and one can ask what is the related phenomenon for a transverse magnetization $\mu_B\sigma_T$ generated, for instance, by resonant microwaves at frequency ω_0, and a beam propagating perpendicular to the field. Since σ_T oscillates at ω_0, the radiated power will also be modulated and one expects side bands at frequency $\omega_L \pm \omega_0$.

Microwave radiation of frequency ω_0, near resonant with the Zeeman levels $|a(t)\rangle$ and $|b(t)\rangle$ of the donor ground state, induces a coherent superposition $|c(t)\rangle$ of these states, i.e.,

$$|c(t)\rangle = \zeta|a(t)\rangle + \mu \exp(\mathrm{i}\omega_0 t)|b(t)\rangle. \tag{3.57}$$

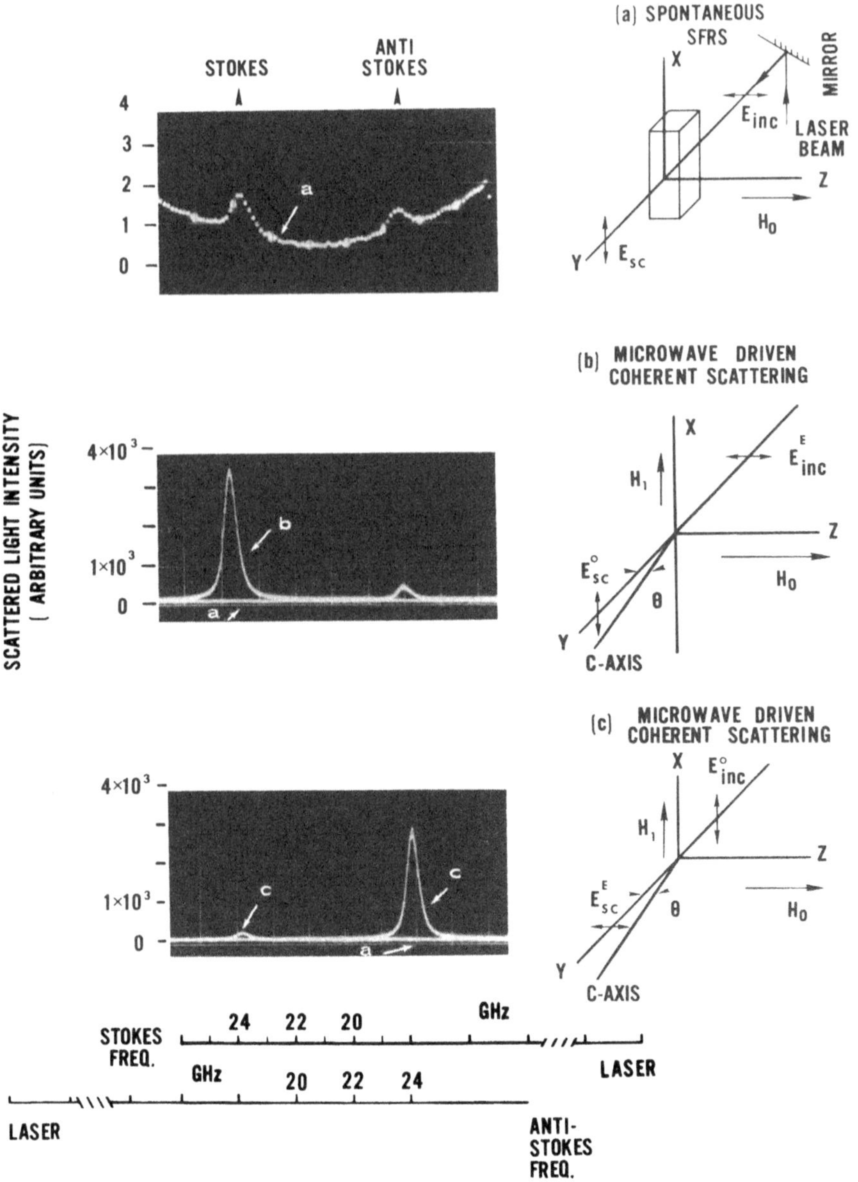

Fig. 3.21a–c. Forward Raman scattering from microwave-induced coherent spin states. (**a**) Spontaneous forward SFRS; (**b**) microwave-driven coherent scattering; (**c**) change of Stokes to anti-Stokes ratio with change of polarizations of E_{inc} and E_{S} due to phase matching (see text) [3.80]

ζ and μ are related to the components of $\langle\boldsymbol{\sigma}\rangle$ which are in turn given by the solution of the Bloch Eqs. [3.76]. The magnitude of the transverse component of precessing spin is $\sigma_{\mathrm{T}}=|\zeta\mu^*|/2$. Using (3.4) one sees that the state $|c\rangle$ displays an oscillating Raman dipole

$$\begin{aligned}\boldsymbol{D}_{cc}^{(2)}&=\langle c(t)|\boldsymbol{D}^{(2)}|c(t)\rangle\\&=(\zeta\mu^*\langle a|\boldsymbol{\sigma}|b\rangle \mathrm{e}^{-\mathrm{i}\omega_{\mathrm{L}}t}\times \boldsymbol{E}_{\mathrm{L}}+\mathrm{c.c.})\,(\alpha\mathrm{e}^{-\mathrm{i}\omega_{\mathrm{L}}t}+\mathrm{c.c.}).\end{aligned}\tag{3.58}$$

σ_{T} will have a spatial variation $\exp[\mathrm{i}(\boldsymbol{k}_{\mathrm{L}}\pm\boldsymbol{k}_0)\cdot\boldsymbol{r}]$, where $\boldsymbol{k}_0$ is the wave vector of the microwaves. The dipoles will emit cooperatively at $\omega_{\mathrm{S}}=\omega_{\mathrm{L}}\pm\omega_0$ provided that the phase matching condition $\Delta\boldsymbol{k}=\boldsymbol{k}_{\mathrm{L}}-\boldsymbol{k}_{\mathrm{S}}-\boldsymbol{k}_0=0$ is satisfied where $\boldsymbol{k}_{\mathrm{S}}$ is the wave vector of the scattered light. If N is the number of centers per unit volume, $ND_{cc}^{(2)}$ is inserted as a source term in Maxwell's equation and one has for the geometry shown in Fig. 3.21

$$-\frac{\partial^2 E_{\mathrm{S}}}{\partial y^2}-\left(\frac{\varepsilon_{\mathrm{s}}'\omega^2}{c^2}+\mathrm{i}k_{\mathrm{S}}k''\right)E_{\mathrm{S}}=-\frac{4\pi\omega^2}{c^2}ND_{cc}^{(2)},\tag{3.59}$$

where $k''=k_{\mathrm{S}}\varepsilon''/\varepsilon'$ is the background absorption of the crystal. We seek a solution $E_{\mathrm{S}}=E_{\mathrm{S}}(y)\exp[\mathrm{i}(k_{\mathrm{S}}y-\omega_{\mathrm{S}}t)]$. Neglecting $\partial^2E(y)/\partial y^2$ compared to $k\,\partial E(y)/\partial y$, one finds

$$2\mathrm{i}k_{\mathrm{S}}\left(\frac{\partial E_{\mathrm{S}}(y)}{\partial y}+\frac{k''}{2}E_{\mathrm{S}}\right)=-\frac{4\pi\omega^2}{c^2}D_{cc}^{(2)}N\mathrm{e}^{\mathrm{i}\Delta ky}.\tag{3.60}$$

Note that $D_{cc}^{(2)}$ has a factor $\exp(-k''y/2)$ via E_{L}. Integration of (3.60) over the interaction length L and expressing α^2 in terms of $d\sigma/d\Omega$ via (3.6) leads to the ratio of the sideband power P_{S} at ω_{S} to the transmitted laser power P_{L}:

$$\frac{P_{\mathrm{S}}}{P_{\mathrm{L}}}=\frac{1}{2}\left(\frac{d\sigma}{d\Omega}\right)\frac{\lambda^2N^2\sigma_{\mathrm{T}}^2}{\varepsilon_{\mathrm{s}}'}\left(\frac{1-\cos(\Delta k\cdot L)}{\Delta k^2}\right),\tag{3.61}$$

where λ is the free space wavelength. For $\Delta k\cdot L\ll 1$, the factor in brackets approaches $L^2/2$. A more precise solution of (3.59), taking proper account of boundary conditions [3.77] also reveals a wave at ω_{S} reflected back into free space with amplitude E_{B} so that E_{S} into the medium actually starts off at $y=0$ with this amplitude E_{B}. However, the ratio of the maximum intensity of the transmitted wave E_{S} to E_{B} is of the order of $(k/\Delta k)^2$ so that E_{B} is quite negligible.

3.8.2 Experimental Observation of SFRS from Coherent States and Phase Matching

The experimental arrangement used by *Romestain* et al. [3.80] to observe SFRS from coherent states is shown schematically in Fig. 3.21. The CdS crystal is now mounted inside a microwave cavity with optical access and tuned to

$\omega_0 \simeq 24$ GHz (Fig. 3.1). The Raman light was detected along the laser beam and the conversion efficiency was large enough (up to 6% in best cases) that this posed no problem. Enough cw microwave power was available to saturate σ_T, i.e., $\sigma_T^{sat} \simeq \langle\sigma_z\rangle(T_2/T_1)^{1/2}/2$. The estimation of $\langle\sigma_z\rangle$ is prone to error due to the exchange couplings discussed in Sect. 3.7, but the best estimate for this sample with $N \simeq 2\times10^{17}$ and a Zeeman frequency of 24 GHz is $\langle\sigma_z\rangle \simeq 0.1$. $T_2 = 5\times10^{-9}$ (see below) and $T_1 = 5\times10^{-6}$ (Sect. 3.5.2) so that $(\sigma_T^{sat})^2 \simeq 2.5\times10^{-6}$. Using $(d\sigma/d\Omega)_{sp} \simeq 2\times10^{-18}\,\mathrm{cm}^2$ at 4880 Å, $\varepsilon_s' \simeq 9$, $L = 1$ mm, and assuming $\Delta k \cdot L \ll 1$, one finds $P_S/P_L \simeq 0.13$ which is not too far from the observed value of 0.06, considering the uncertainties in several of the quantities in (3.61). Depletion of the laser beam and the interaction between Stokes and anti-Stokes waves require a more sophisticated treatment [3.77, 78]. In addition, P_S/P_L is probably further reduced due to the difficulty of obtaining proper phase matching on a converging beam.

Indeed, the phase matching condition is quite severe since one cannot neglect the dispersion $(dn/d\omega)\omega_0$ when one is very close to the band gap, even though $\hbar\omega_0$ is as small as 0.8 cm^{-1}. The phase matching condition $\Delta\boldsymbol{k} = \boldsymbol{k}_L - \boldsymbol{k}_S - \boldsymbol{k}_0$ may be written as

$$|\Delta\boldsymbol{k}| \simeq \omega_0\left(\frac{\partial k_L}{\partial\omega} - \frac{n_\mu}{c}\right) = \omega_0\left(\frac{1}{v_g} - \frac{n_\mu}{c}\right), \tag{3.62}$$

where v_g is the group velocity at ω_L and n_μ is the index of refraction at the microwave frequency ω_0. Near the band gap, the dispersion in pure CdS is quite large with $v_g \simeq c/50$ at 4880 Å compared to a phase velocity of $c/3$. This value of v_g may be determined from the parameters of the polariton dispersion in CdS [3.79] and the expression for v_g in terms of these parameters [3.31]. Since $n_\mu \simeq 2.5$, one may neglect n_μ/c compared to $1/v_g$ so that $\Delta k \simeq (k_L - k_S) = \omega_0/v_g$. However, as the main selection rule is $\boldsymbol{E}_S \perp \boldsymbol{E}_L$, this dispersion may be compensated for by the birefringence $b(\theta)$ of CdS, which for propagation at angles θ close to the c-axis may be expressed as $b(\theta) = (n_o - n_e)\theta^2$, where n_o and n_e are, respectively, the ordinary and extraordinary indices. Thus, phase matching will occur when $\theta \simeq (c\omega_0)^{1/2}\,[v_g\omega_L(n_o - n_e)]^{-1/2}$, provided the higher frequency wave of the two waves ω_L and ω_S propagates as an extraordinary wave. Using our measured value at 4880 Å of $(n_o - n_e) = 0.22$, we find $\theta \simeq 5.4°$. This estimated value is larger than previously reported [3.80] due to a smaller value of v_g but is still close to the observed θ within experimental error, given the uncertainty in v_g. At this angle, the Stokes wave is phase matched if $\boldsymbol{E}_L$ is in the extraordinary polarization (Fig. 3.21b). If $\boldsymbol{E}_L$ is changed to ordinary polarization, the anti-Stokes is phase matched as seen by the reversal of intensities in Fig. 3.21c.

When the magnetic field H_0 is swept through resonance, the power P_S is emitted at the constant frequency $\omega_L \pm \omega_0$ (whereas the spontaneous SFRS frequency varies as H_0) and varies with changing σ_T as indicated by (3.61). In contrast to conventional EPR measurements, which detect a signal proportional to a transverse component of the spin in the rotating frame $\sigma_{x'}'$, in phase with the

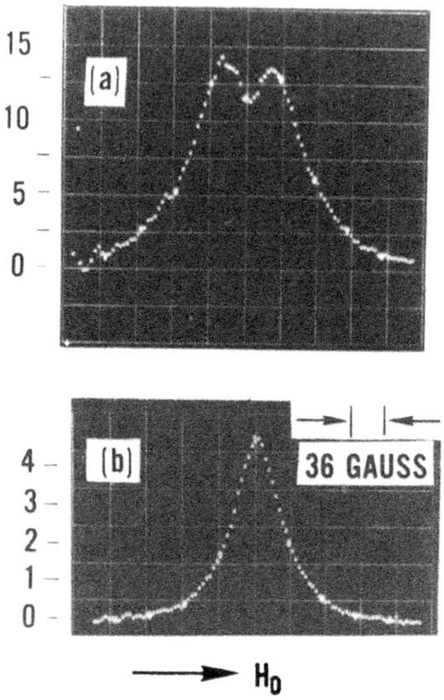

Fig. 3.22a,b. Magnetic field dependence of scattering from coherent states. (**a**) High microwave power; (**b**) Low power (see text) [3.80]

rotating rf field, H_1 (dispersion) or σ'_y 90° out of phase (absorption), the signal here is proportional to $\sigma_{\mathrm{T}}^2 = \sigma_x'^2 + \sigma_y'^2$. Assuming a homogeneous line, σ_{T}^2 can be calculated using the Bloch equations [3.76]. This dependence upon σ_{T}^2 becomes quite evident in the saturation of the EPR line, i.e., when $(\gamma_0 H_1) \cdot (T_1 T_2)^{1/2} > 1$, at which point a double peak appears in σ_{T}^2 centered about $H_0 = \omega_0/\gamma_0$ as H_0 is swept through resonance. Figure 3.22 shows the magnetic field dependence of P_{S}. At low microwave power, the line has a Lorentzian line shape with $T_2 = 5 \times 10^{-9}$ s. At saturation, a dip appears in the center confirming the model.

The coherent scattering discussed above may also be described in terms of the nonlinear mixing of the two frequencies ω_{L} and ω_0. The dipole $D_{cc}^{(2)}$ appearing in (3.58) is proportional to $E_{\mathrm{L}} \cos \omega_{\mathrm{L}} t$ and to $H_1 \cos \omega_0 t$ through the $\zeta\mu^*$ term. Hence one can define a second-order susceptibility $\chi(\omega_{\mathrm{L}}, \omega_0)$ [3.77] such that

$$D(\omega_{\mathrm{L}} \pm \omega_0) = \chi(\omega_{\mathrm{L}}, \omega_0) E(\omega_{\mathrm{L}}) H_1(\omega_0). \tag{3.63}$$

As usual, χ has resonant factors linked with ω_{L} through α in (3.4) and with ω_0 through the magnetic-field dependence. It should be noted, however, that the emission at $\omega_{\mathrm{L}} \pm \omega_0$ depends upon σ_{T} and will therefore persist for a phase memory time T_2 even after the microwave field H_1 is shut off. This will be very evident in the discussion of the Raman echo in Sect. 3.8.4.

3.8.3 SFRS from Coherent States Viewed as Modulation of Faraday Rotation

As suggested at the beginning of Sect. 3.8.1, one may view the scattering from coherent spin states as a modulated Faraday rotation, as illustrated in Fig. 3.23. The coherence between states $|a\rangle$ and $|b\rangle$ is represented by the transverse spin

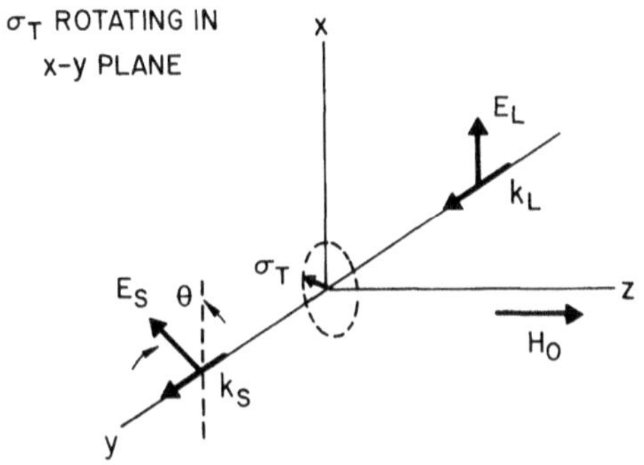

Fig. 3.23. SFRS from coherent state viewed as modulation of Faraday rotation θ by coherently precessing spins θ is proportional to projection of σ_r along y (see text)

component $\sigma_{\rm T}$ precessing at frequency ω_0 in the x-y plane about H_0. The direction of polarization of light, which is incident along the y-direction, experiences a Faraday rotation through an angle θ about y by an amount proportional to the y-component of $\sigma_{\rm T}$ given by

$$\theta=\phi\,\frac{\sigma_{\rm T}\sin\omega_0 t}{\langle\sigma_z\rangle_{\rm eq}}=\theta_0\sin\omega_0 t, \qquad (3.64)$$

where ϕ is the normal spin Faraday rotation corresponding to $\langle\sigma_z\rangle_{\rm eq}$ for a beam propagating along z and $\theta_0=\phi\sigma_{\rm T}/\langle\sigma_z\rangle_{\rm eq}$. We have assumed phase matching as before. Thus, the Faraday rotation for a beam propagating along y is modulated at ω_0 with

$$E_x=E_{\rm L}\cos(\omega_{\rm L}t)\cos(\theta_0\sin\omega_0 t), \qquad (3.65)$$

$$E_z=E_{\rm L}\cos(\omega_{\rm L}t)\sin(\theta_0\sin\omega_0 t). \qquad (3.66)$$

For *small* θ_0, which corresponds to the assumption of small depletion of the pump beam,

$$E_z=-\frac{E_{\rm L}\theta_0}{2}\,[\sin(\omega_{\rm L}-\omega_0)t-\sin(\omega_{\rm L}+\omega_0)t] \qquad (3.67)$$

and the ratio of the single sideband power $P_{\rm S}$ to the power P_0 transmitted at $\omega_{\rm L}$ is given by

$$\frac{P_{\rm S}}{P_0}=\frac{1}{4}\left(\phi\,\frac{\langle\sigma_{\rm T}\rangle}{\langle\sigma_z\rangle_{\rm eq}}\right)^2=\frac{1}{4}\left(\frac{d\sigma}{d\Omega}\right)_{\rm sp}\left(\frac{NL\sigma_{\rm T}\lambda}{n}\right)^2, \qquad (3.68)$$

in agreement with (3.61) with $\Delta k\to 0$, when ϕ is expressed in terms of $(d\sigma/d\Omega)_{\rm sp}$ via (3.44). For general θ_0, (3.65, 66) may be expanded as

$$E_x=E_{\rm L}\cos\omega_{\rm L}t\,\{{\rm J}_0(\theta_0)+2\,[{\rm J}_2(\theta_0)\cos 2\omega_0 t+{\rm J}_4(\theta_0)\cos 4\omega_0 t+\ldots]\}, \qquad (3.69)$$

$$E_z = 2E_L \cos\omega_L t\,[J_1(\theta_0)\sin\omega_0 t + J_3(\theta_0)\sin 3\omega_0 t + \dots], \tag{3.70}$$

where J_n are the Bessel functions of order n. However, (3.69, 70) are only valid for relatively low-frequency modulation of σ_T such that phase matching occurs at all the sidebands with significant amplitude $J_n(\theta_0)$. As many as eight harmonics have been seen in CdS on both the anti-Stokes and Stokes side for large tipping angles achieved in stimulated SFRS (Sect. 3.8.4) using higher powered pulsed excitation of the laser light [3.81]. However, they are not describable by (3.69, 70) but are generated by successive scattering from each sideband to the next and nonlinear interaction of all these waves and the medium [3.78].

3.8.4 Stimulated SFRS

Equation (3.12) describes the rate of scattering per center into all modes in unit solid angle in normalization volume V. The scattering rate into a *single* mode at the center of the Raman line, which we will call w^0, is found by dividing by the total number of modes, $\varrho(\omega)\cdot\Delta\omega_{\text{eff}}$, where $\Delta\omega_{\text{eff}}$ is the effective Raman linewidth. For a Lorentzian line $\Delta\omega_{\text{eff}} = \pi\Delta\omega/2$, where $\Delta\omega$ is the angular width of the line between half intensity points. Therefore,

$$w^0 = \frac{(dw/d\Omega)}{\varrho(\omega)\cdot\pi\Delta\omega/2}. \tag{3.71}$$

Substituting from (3.12) and (3.6) and considering N_a scattering centers in the state $|a\rangle$ in volume V gives

$$w_{ab}^0 = A_{ab}\bar{n}_L(\bar{n}_S + 1)N_a, \tag{3.72}$$

where

$$A_{ab} = \left(\frac{d\sigma}{d\Omega}\right)_{\text{sp}} \frac{2\cdot(2\pi)^3 c^4}{\varepsilon_S^{3/2}\varepsilon_L^{1/2}\pi V^2\omega_S^2\Delta\omega}. \tag{3.73}$$

The Raman photons are also being depleted by the reverse process

$$w_{ba}^0 = A_{ba}\bar{n}_S(\bar{n}_L + 1)N_b. \tag{3.74}$$

Since $A_{ab} \simeq A_{ba} \simeq A$, we have

$$\dot{\bar{n}}_S = w_{ab}^0 - w_{ba}^0 = A\bar{n}_L\bar{n}_S(N_a - N_b) + A(\bar{n}_L N_a - \bar{n}_S N_b). \tag{3.75}$$

For $\bar{n}_S$ and $\bar{n}_L \gg 1$, the second term in (3.75) may be neglected and, assuming negligible depletion of the laser beam, i.e., $\bar{n}_L \simeq$ const, the solution of (3.75) describes an exponential growth of the Stokes signal with distance along the direction of the laser beam, and at a distance z into the crystal,

$$\bar{n}_S(z) = \exp[A\bar{n}_L(N_a - N_b)t] = \exp[A\bar{n}_L(N_a - N_b)z\varepsilon_S^{1/2}/c]. \tag{3.76}$$

If $\bar{n}_L$ is related to the intensity of the laser beam by $I_L = \bar{n}_L \hbar\omega_L c/V\varepsilon_L^{1/2}$, then

$$\bar{n}_S(z) = \bar{n}_S(0)e^{g_S z}, \tag{3.77}$$

where the gain constant g_S is given by (compare with [Ref. 3.16, Eq. (2.133)])

$$g_S = \frac{2}{\pi} \frac{c^2\left(\frac{N_a - N_b}{V}\right) I_L \left(\frac{d\sigma}{d\Omega}\right)_{sp}}{\varepsilon_S h \nu_S^2 \nu_L \Delta\nu}. \tag{3.78}$$

If this buildup of $\bar{n}_S$ due to the SFRS exceeds the background attenuation $\exp(-\gamma Z)$ of the beam by the crystal, there will be a net exponential growth of $\bar{n}_S$. As a typical example, for bound donors in CdS with the following parameters: $(N_a - N_b)/V \sim 10^{17}$ cm^{-3}, $(d\sigma/d\Omega)_{sp} \sim 4 \times 10^{-18}$ cm^2 for $\lambda_L = 4880$ Å, $\varepsilon_S = 9$. $\Delta\nu_L \sim 10^8$ Hz, and with a laser power as little as 5 mW focused to a 50 μm beam diameter, i.e., $I_L \sim 250$ W/cm^2, one finds $g \sim 46$ cm^{-1}! compared to a $\Gamma \sim 20$–30 cm^{-1} at this wavelength. Stimulated SFRS is therefore easily seen under these conditions, as illustrated in Fig. 3.24. The 1st Stokes Raman signal is sufficiently strong to generate a 2nd Stokes, which in turn generates a 3rd Stokes. The threshold pump power for SFRS in CdS is, of course, extremely dependent upon pump wavelength via $d\sigma/d\Omega$ and, for example, is orders of magnitude smaller at 4925 Å [3.20].

Note also the appearance of a 1st order stimulated anti-Stokes in Fig. 3.24, whose origin is not contained in the theory outlined above. As a matter of fact, the equation for anti-Stokes generation corresponding to (3.75) is

$$\dot{\bar{n}}_{AS} = -A\bar{n}_L\bar{n}_{AS}(N_a - N_b) + A(\bar{n}_L N_b - \bar{n}_{AS} N_a). \tag{3.79}$$

Since $(N_a - N_b)$ is positive, the term corresponding to stimulation is now negative and acts in a direction to impede further growth of $\bar{n}_{AS}$, i.e., one has a negative gain $(-g_S)$ or stimulated anti-Stokes *absorption*. The energy of the anti-Stokes wave is used to amplify the laser pump. This is the situation as long as the Stokes and anti-Stokes are uncoupled as expressed in the separate equations (3.75) and (3.76). Note that there are no phase matching considerations in this situation.

This behavior of the anti-Stokes light in n-CdS is illustrated in Fig. 3.25 where the laser intensity $\bar{n}_L$ is kept fixed and instead, g_S is varied through $(N_a - N_b)$ by increasing the external magnetic field H_0. The initial decrease of $\bar{n}_{AS}$ is due to the decrease in N_b, but an even faster decrease ensues due to the stimulated absorption $-A\bar{n}_L\bar{n}_{AS}(N_a - N_b)$ as $(N_a - N_b)$ increases with H_0. The

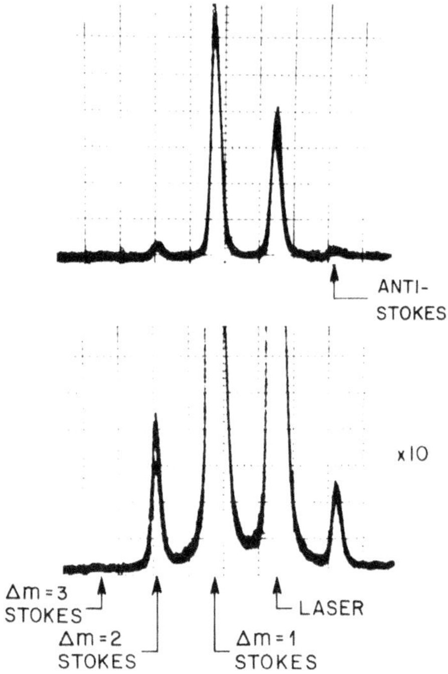

Fig. 3.24.

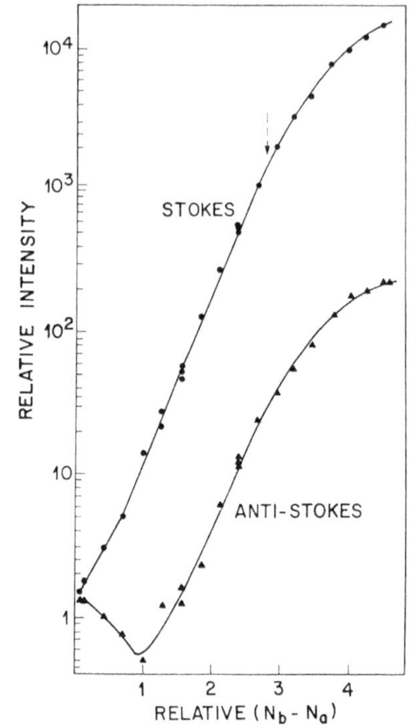

Fig. 3.25.

Fig. 3.24. Stimulated SFRS from bound donors in CdS seen with 5 mW of pump power

Fig. 3.25. Variation of stimulated SFRS by varying gain with magnetic field by varying difference of spin populations. Note initial *negative* gain of anti-Stokes line

eventual turnaround and increase in anti-Stokes is due to the interaction of the increasing stimulated Stokes field with the laser light to coherently drive the spin system (Sect. 3.8.5) and subsequent scattering of the laser field from this coherent state (Sect. 3.8.1). One, in effect, has a coupling between the Stokes and anti-Stokes via the Raman susceptibility and the problem is most conveniently treated by the classical Maxwell's equations with the coherent Raman polarization as the source terms [3.77, 78]. The arrow in the figure corresponds to the point at which the laser beam has been depleted by approximately 5 % and the 1st Stokes gain shows signs of decreasing. At this point the coupled equations for the higher-order Raman waves and coherent medium (spin) excitation become nonlinear and the solutions very complicated [3.77]. The reader is referred to the general review article by *Shen* [3.78] in this series on stimulated Raman scattering for further details.

3.8.5 Raman Echo

The irreversible decay time of the induced coherence in a two-level system, called T_2, is an important parameter which measures stochastic interactions with the environment. T_2 is usually studied by spin echo or photon echo experiments in which the coherence is directly induced by resonant radiation at ω_{ba}.

It would be advantageous to use a nonresonant method for inducing the coherence in those cases where sufficiently intense resonant radiation is not available (as in the far infrared), or its use is otherwise inconvenient. This may be done by two-wave light mixing or stimulated Raman scattering [3.82], as is readily apparent from the form of the SFR Hamiltonian (3.7) i.e., $\mathscr{H}=\alpha\boldsymbol{\sigma}\cdot(\boldsymbol{E}_{\mathrm{L}}\times\boldsymbol{E}_{\mathrm{S}})\,\exp\left[\mathrm{i}(\omega_{\mathrm{L}}-\omega_{\mathrm{S}})t\right]$. It is seen that $\alpha\boldsymbol{E}_{\mathrm{L}}\times\boldsymbol{E}_{\mathrm{S}}/g\beta$ acts as an effective oscillating magnetic field of frequency $(\omega_{\mathrm{L}}-\omega_{\mathrm{S}})$ which will generate a coherent state measured by σ_{T}.

If the pump fields E_{L} and E_{S} are applied for a time Δt, which is short compared to relaxation effects, then analogous to magnetic resonance, the coherence is given by

$$\sigma_{\mathrm{T}}=\langle\sigma_z\rangle_{\mathrm{eq}}\sin\theta\cos\left[(\boldsymbol{k}_{\mathrm{L}}-\boldsymbol{k}_{\mathrm{S}})\cdot\boldsymbol{r}\right],$$

where θ is the "tipping angle" given by

$$\theta=\frac{1}{\hbar}\,|\alpha_{\mathrm{L}}|E_{\mathrm{L}}E_{\mathrm{S}}\Delta t.$$

With moderate laser power applied for several nanoseconds, tipping angles of the order of $\pi/2$ can be easily obtained. With the coherence generated this way, one may shut off the pump beams E_{L} and E_{S} and probe the time decay of the coherence, with a cw laser at another frequency ω_{p}, by observing the decay of the sidebands at $\omega_{\mathrm{p}}\pm\omega_{ba}$, as described in Sect. 3.8.1a. However, one thereby measures the free induction decay of σ_{T}, which is the Fourier transform of the linewidth which would be seen in spontaneous Raman scattering. In an inhomogeneously broadened line, this decay T_2^* is considerably different from the T_2 of the constituent homogeneous packets. To unambiguously measure T_2, an echo experiment is needed, i.e., a Raman echo experiment [3.83].

In the first Raman echo experiments which were done in CdS [3.84], coherence was established between the donor spin levels by a 5 ns dye laser pulse at 4905 Å operating simultaneously in two longitudinal modes ω_{L} and ω_{S}, separated by 32 GHz. The external field was set to 12.75 kG to adjust the Zeeman splitting to 32 GHz. However, it is clearly more convenient to obtain ω_{S} automatically from ω_{L} by stimulated SFRS (Sect. 3.8.2), allowing operation at any field value and thus the study of T_2 as a function of field or ω_{ba}. This more versatile method [3.85] is illustrated in Fig. 3.26. A 30 ns laser pulse (I) from a mode-locked argon ion laser at frequency ω_{L} ($\lambda=4880$ Å) is incident on the

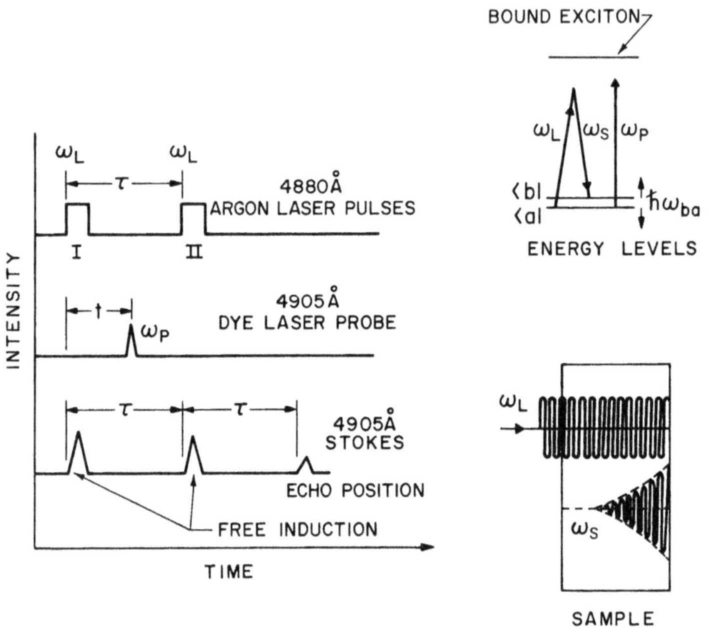

Fig. 3.26. Schematic diagram of generation and probing of Raman echo from donor spins in CdS; see text [3.84, 85]

sample at $t=0$. Via stimulated SFRS, a coherent light signal at the Raman frequency ω_S builds up farther down the crystal which, in concert with ω_L, induces coherence between $|a\rangle$ and $|b\rangle$.

A 5 ns pulse from a dye laser at ω_p ($\lambda=4905$ Å) is used to probe σ_T at a variable time t by observing the magnitude of the Stokes radiation $\omega_p-\omega_{ab}$ due to σ_T. ω_p is sufficiently attenuated so that it does not stimulate any SFRS and thus acts solely as a probe. If the probe pulse is applied in a time domain close to the shutoff of pulse *I*, the free induction decay will be measured. After dephasing of σ_T, pulse *II* is applied at time τ resulting in the rephasing of σ_T, or echo, at time 2τ, which is again probed by ω_p. By varying τ and synchronizing the timing of the probe so that it occurs at 2τ, the echo amplitude is measured as a function of τ, yielding T_2. The samples of CdS used in these experiments had an optical path length of 3 mm, a donor concentration of $2\text{–}4\times10^{16}\ \mathrm{cm}^{-3}$ and were cooled to 1.6 K. With a pump power intensity of $10^4\ \mathrm{W/cm^2}$, the Raman gain was sufficient to produce an echo which could be seen directly on the oscilloscope as shown in Fig. 3.27. The echo signal as a function of τ was studied for a continuous range of frequencies ω_{ab} from 10 to 46 GHz. It is plotted in Fig. 3.28 for two widely different values of dc magnetic field H_0. T_2 is found to be, within experimental error, independent of H_0 at this concentration. We emphasize that a similar study by conventional electron-spin echo techniques would have required different microwave setups for the different frequency ranges.

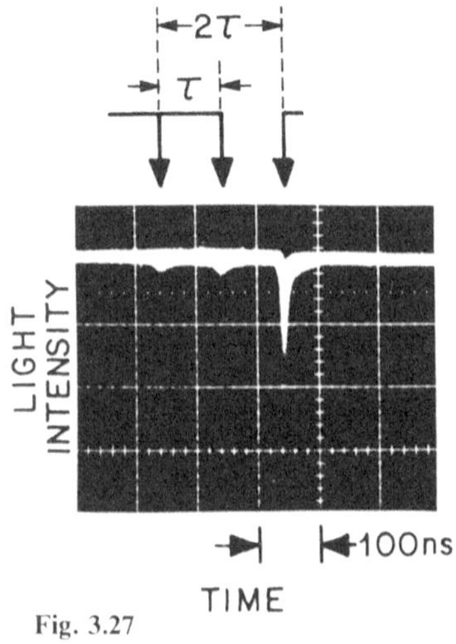

Fig. 3.27

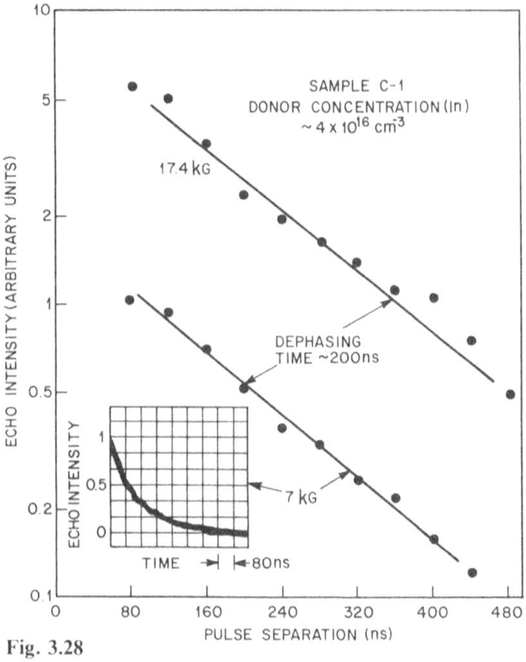

Fig. 3.28

Fig. 3.27. Observation of Raman echo in CdS with $N = 4 \times 10^{16}$ cm^{-3}. The first two pulses separated by τ are leakage of 4880 Å light corresponding to pulses I and II in Fig. 3.26. The echo at 2τ is probed at 4905 Å (see text) [3.84, 85]

Fig. 3.28. Measurement of T_2 of donors in CdS by echo decay as a function of pulse separation τ (see text) [3.85]

In addition to the measurement of T_2 by the 2 pulse echo method, the spin-lattice relaxation or T_1 of the donors was measured by *Hu* et al. [3.85] using a 3 pulse sequence which generates a "stimulated" echo.

With picosecond techniques, one can visualize the extension of the Raman echo to shorter time scales and to a wide variety of two-level system not directly accessible by a one-photon process.

Acknowledgements. We to wish thank our colleagues Y. Yafet, L. R. Walker, and P. Wolff for many helpful discussions during the course of the work described in this chapter. We especially wish to thank G. E. Devlin for his extensive experimental assistance related to the work reviewed here.

References

3.1 Y.Yafet: Phys. Rev. **152**, 855 (1966)

3.2 P.A.Wolff: Phys. Rev. Lett. **16**, 225 (1966)

3.3 R.E.Slusher, C.K.N.Patel, P.Fleury: Phys. Rev. Lett. **18**, 77 (1967)

3.4 C.K.N.Patel: In *Laser Spectroscopy*, ed. by R.G.Brewer, A.Mooradian (Plenum, New York 1974)
3.5 M.J.Colles, C.R.Pidgeon: Rep. Prog. Phys. **38**, 329 (1975)
3.6 D.G.Thomas, J.J.Hopfield; Phys. Rev. **175**, 1021 (1968)
3.7 Y.Yafet: In *New Developments in Semiconductors*, ed. by P.R.Wallace, R.Harris, M.J.Zuckermann (Nordhoff, Leyden 1973)
3.8 J.F.Scott: In *Laser Applications to Optics Spectroscopy*, ed. by S.F.Jacobs et al. (Addison-Wesley, Reading, MA 1975)
3.9 S.Geschwind, R.Romestain, G.E.Devlin: In *Physics of Semiconductors 1978*, ed. by B.L.H. Wilson (The Institute of Physics, London 1979) p. 1013
3.10 G.E.Devlin, J.L.Davis, L.Chase, S.Geschwind: Appl. Phys. Lett. **19**, 138 (1971)
3.11 K.B.Lyons, P.Fleury: J. Appl. Phys. **47**, 4898 (1976)
3.12 J.R.Sandercock: In *Light Scattering in Solids*, 2nd ed., ed. by M.Balkanski (Flammarion, Paris 1971) p. 9
3.13a J.R.Sandercock: In *Proc. 7th Intern. Conf. on Raman Spectroscopy*, ed. by W.Murphy (North-Holland, Amsterdam 1980) p. 364
3.13b J.R.Sandercock: In *Light Scattering in Solids III*, ed. by M.Cardona, G.Güntherodt, Topics Appl. Phys., Vol. 51 (Springer, Berlin, Heidelberg, New York 1982) p. 173
3.14 J.G.Dil, N.C.J.A. van Hijningen, F. van Dorst, R.M.Aarts: Appl. Opt. **20**, 137A (1981)
3.15 A.S.Davydov: *Quantum Mechanics* (Pergamon Press, New York 1965) p. 316
3.16 M.Cardona, G.Güntherodt (eds.): *Light Scattering in Solids II*, Topics Appl. Phys., Vol. 50 (Springer, Berlin, Heidelberg, New York 1982)
3.17 M.Lax, D.F.Nelson: In *The Theory of Light Scattering in Condensed Matter*, ed. by V.M.Agranovich, J.L.Birman (Plenum Press, New York 1976) p. 371
3.18 W.Hayes, R.Loudon: *Light Scattering in Solids* (Wiley, New York 1978)
3.19 H.B.Callen, T.A.Welton: Phys. Rev. **83**, 34 (1951)
3.20 J.F.Scott, T.C.Damen, P.A.Fleury: Phys. Rev. **6**, 3856 (1972)
3.21 R.Romestain, S.Geschwind, G.E.Devlin: Private communication (1974)
3.22 P.A.Wolff: Private communication (1974)
3.23 Y.Oka, M.Cardona: J. Physique **42**, C6, 459 (1981)
3.24 G.F.Koster, J.O.Dimmock, R.G.Wheeler, H.Statz: *Properties of the Thirty Two Point Groups* (MIT Press, Cambridge, MA 1963)
3.25 L.D.Landau, L.M.Lifschitz: *Electrodynamics of Continuous Media* (Pergamon Press, Oxford 1960)
3.26a H.LeGall, J.P.Jamet: Phys. Stat. Sol. (b) **46**, 467 (1971)
3.26b H.LeGall, Vien Tran Khanh, B.Desormiere: Phys. Stat. Sol. (b) **47**, 591 (1971)
3.27 J.J.Hopfield: Phys. Chem. Solids **15**, 97 (1960)
3.28 S.M.Girvin: Phys. Rev. B**17**, 1877 (1978)
3.29 R.Romestain, S.Geschwind, G.E.Devlin: (Private communication, Fig. 1 in Ref. [3.28])
3.30 M.Cardona, G.Güntherodt (eds.): *Light Scattering in Solids* III, Topics Appl. Phys., Vol. 51 (Springer, Berlin, Heidelberg, New York 1982)
3.31 J.J.Hopfield: Phys. Rev. **182**, 945 (1969)
3.32 N.F.Mott: Pur. Phys. Soc. A**62**, 416 (1949); Philos. Mag. **6**, 287 (1961)
3.33 J.Hubbard: Proc. Roy. Soc. A**277**, 237 (1964); A**281**, 401 (1964)
3.34 N.F.Mott: *Metal Insulator Transition* (Taylor and Francis, London 1974)
3.35 P.PEdwards, M.J.Sienko: Phys. Rev. B**17**, 2575 (1978)
3.36 T.F.Rosenbaum, K.Andres, G.A.Thomas, R.N.Bhatt: Phys. Rev. Lett. **45**, 1723 (1980)
3.37 G.A.Thomas, Y.Ootuka, S.Kobayashi, W.Sasaki: Phys. Rev. **24**, No. 8, 4886 (1981)
3.38 S.Geschwind, R.E.Walstedt, R.Romestain, V.Narayanamurti, R.B.Kummer, R.Feigenblatt, G.E.Devlin: Philos. Mag. B**42**, 961 (1980)
3.39 P.W.Anderson: Phys. Rev. **109**, 1492 (1958)
3.40 N.F.Mott: In *The Metal Non-Metal Transition*, ed. by L.R.Freedman, D.P.Tunstall (SUSSP Publications, Edinburgh 1978)
3.41 D.J.Thouless: In [3.4a]
3.42a L.R.Friedman, D.P.Tunstall (eds.): *The Metal Non-Metal Transition* (SUSSP Publications, Edinburgh 1978)

3.42b R.C.Dynes, P.A.Lee: Science (1984) To be published
3.43 A.Mooradian: Phys. Rev. Lett. **20**, 1102 (1968)
3.44 D.C.Hamilton, A.L.McWhorter: In *Light Scattering in Solids*, ed. by G.B.Wright (Springer, Berlin, Heidelberg, New York 1969) p. 309
3.45 R.H.Dicke: Phys. Rev. **89**, 472 (1953)
3.46 P.A.Wolff, J.G.Ramos, S.Yuen: *Theory of Light Scattering in Condensed Matter*, ed. by B.Bendow, J.L.Birman, V.M.Agranovich (Plenum Press, New York 1976) p. 574
3.47 P.M.Platzmann, N.Tsoar: Phys. Rev. **182**, 510 (1069)
3.48 P.S.Pershan, J.P. van der Ziel, L.D. Malstron: Phys. Rev. **143**, 574 (1966)
3.49a Y.R.Shen, N.Bloembergen: Phys. Rev. **143**, 372 (1966)
3.49b R.Romestain, S.Geschwind, G.E.Devlin: Phys. Rev. Lett. **35**, 803 (1975)
3.50 C.H.Henry, K.Nassau: Phys. Rev. B**1**, 1628 (1970)
3.51 P.Hu: Private Communication
3.52 S.Geschwind, R.Romestain, G.E.Devlin: J. Physique **37**, C4, 313 (1976)
3.53 J.D.Quirt, J.R.Marko: Phys. Rev. B**7**, 3892 (1973)
3.54 S.Geschwind, R.E.Romestain, G.E.Devlin, R.Feigenblatt: In *Light Scattering in Solids*, ed. by J.L.Birman, H.Z.Cummins, K.Rebane (Plenum Press, New York 1979) p. 189
3.55 W.F.Brinkman, T.M.Rice: Phys. Rev. B**2**, 4302 (1970)
3.56 R.Romestain: J. Phys. **13**, 1097 (1980)
3.57 R.C.Casella: Phys. Rev. Lett. **5**, 371 (1960)
3.58 J.J.Hopfield, D.G.Thomas: Phys. Rev. **132**, 563 (1963)
3.59 T.Shimizu, K.Morigaki: J. Phys. Soc. Japan **28**, 1468 (1970)
3.60 M.I.Dyakanov, V.I.Perel: Sov. Phys. Solid State **13**, 3581 (1972)
3.61 J.J.Hopfield: J. Appl. Phys. Suppl. **32**, 2277 (1961)
3.62 G.D.Mahan, J.J.Hopfield: Phys. Rev. **135**, A428 (1964)
3.63 E.S.Koteles, G.Winterling: Phys. Rev. Lett. **44**, 948 (1980)
3.64a R.Romestain, S.Geschwind, G.E.Devlin: Phys. Rev. Lett. **39**, 1583 (1977)
3.64b Y.Yafet: Private communication
3.65 E.I.Rashba: J. Magnestism Magn. Mat. **11**, 63 (1979)
3.66 M.Dobrowska, H.D.Drew, J.K.Furdyna, T.Ichigui, P.A.Wolff: Phys. Rev. Lett. **49**, 845 (1982)
3.67 N.W.Carlson, S.Geschwind, L.R.Walker, G.E.Devlin: Phys. Rev. Lett. **49**, 165 (1982)
3.68 C.Herring, M.Flicker: Phys. Rev. A**134**, 362 (1964)
3.69 S.F.Edwards, P.W.Anderson: J. Phys. F**5**, 1965 (1975)
3.70 R.B.Kummer, R.E.Walstedt, S.Geschwind, V.Narayanamurti, G.E.Devlin: Phys. Rev. Lett. **40**, 1098 (1978)
3.71 R.E.Walstedt, R.B.Kummer, S.Geschwind, V.Narayanamurti, G.E.Devlin: J. Appl. Phys. **50**, 1700 (1979)
3.72 K.Andres, R.N.Bhatt, P.Goalwin, T.M.Rice, R.E.Walstedt: Phys. Rev. **24**, 244 (1981)
3.73 R.A.Klemm: J. Phys. C**13**, L755 (1977)
3.74 R.N.Bhatt, P.Lee: Phys. Rev. Lett. **48**, 344 (1982)
3.75 F.B.Anderson, H.B.Callen: Phys. Rev. **136**, A1068 (1964)
3.76 A.Abragam: *Principles of Nuclear Magnetism* (Oxford University Press, Oxford 1961) C.P.Slichter: *Principles of Magnetic Resonance*, Springer Ser. Solid-State Sci., Vol. 1 (Springer, Berlin, Heidelberg, New York 1980)
3.77 N.Bloembergen: *Non-Linear Optics* (Benjamin, New York 1965)
3.78 Y.R.Shen: In *Light Scattering in Solids I*, ed. by M.Cardona, Topics Appl. Phys., Vol. 8 (Springer, Berlin, Heidelberg, New York 1982)
3.79 G.Winterling, E.S.Koteles: Solid State Commun. **23**, 95 (1977)
3.80 R.Romestain, S.Geschwind, G.E.Devlin, P.A.Wolff: Phys. Rev. Lett. **33**, 10 (1974)
3.81 P.Hu: Private communication (1978)
3.82 J.A.Giordmaine, W.Kaiser: Phys. Rev. **144**, 676 (1966)
3.83 S.R.Hartman: IEEE J. *Quantum Electron.* QE-**4**, 802 (1968)
3.84 P.Hu, S.Geschwind, T.M.Jedju: Phys. Rev. Lett. **37**, 1357 (1976)
3.85 P.Hu, S.Geschwind, T.M.Jedju: In *Coherence and Quantum Optics IV*, ed. by L.Mandel, E.Wolf (Plenum Press, New York 1978)

4. Spin-Dependent Raman Scattering in Magnetic Semiconductors

Gernot Güntherodt and Roland Zeyher

With 19 Figures

During the past two decades elementary excitations of the lattice and spin systems of solids, i.e., phonons and magnons, respectively, have been the subject of extensive investigations by means of light scattering [4.1–4a, b]. A new "degree of freedom" is added to the usual phonon Raman scattering in solids by investigating magnetic ordering materials, in particular, magnetic semiconductors which exhibit interactions between the phonon and spin systems [4.5–21]. These simultaneous spin-phonon excitations [4.9, 11–21] offer the intriguingly new possibility of probing the order and dynamics of the spin system by measuring the temperature and magnetic field dependence of the phonon Raman scattering. Investigations in this direction have addressed the magnetic and semiconducting Cd-Cr spinels [4.22–26], the europium chalcogenides [4.5–21] and magnetic insulators such as VI_2 [4.27, 28]. The major aim of light scattering investigations of these model substances has been the study of the basic, microscopic scattering mechanisms, of the spin correlations in the various magnetic phases and of the critical behavior near magnetic phase transitions. More recent interests are directed towards applying these scattering mechanisms and the concepts developed to studies of critical exponents of magnetic phase transitions, of spin correlations in spin glasses like $Eu_xSr_{1-x}S$, of valence fluctuations in mixed valence materials and of spin superstructures in antiferromagnets.

4.1 Outline

Experimental investigations of spin-dependent Raman scattering from phonons had been initiated by the early work on Cd-Cr spinels [4.22]. The strong increase of the Raman intensity of some of the phonon modes upon cooling through the Curie temperature T_C had been found to follow a spin correlation function. Subsequent phenomenological theory expressed the scattering intensity in terms of spin correlation functions by taking into account a modulation of the (Cr)d electron transfer matrix element by the displacements of the magnetic (Cr) as well as intervening nonmagnetic (S, Se) ions [4.23]. However, changes in phonon Raman scattering intensity with laser frequency observed in nonmagnetic $CdIn_2S_4$ [4.29] started a systematic search for resonance effects in Raman scattering of Cd-Cr spinels which will be described in more detail in Sect. 4.7. Besides the "standard" resonance Raman scattering associated with valence to conduction band transitions in these spinels, an additional temperature-dependent resonance enhancement occurs in the magnetic Cd-Cr spinels with the onset of ferromagnetic order due to the band splitting into spin-polarized subbands [4.24–26]. Consequently, in this class of magnetic semiconductors, the magnetic order influences only indirectly the phonon Raman scattering intensity via resonance effects due to spin splittings of the electronic band structure.

On the other hand, the actual spin-dependent phonon Raman scattering due to a simultaneous excitation of the spin and phonon systems has been found in the other prominent class of magnetic semiconductors, the europium chalcogenides. In these NaCl structure-type materials, the symmetry forbidden first-order Raman scattering in the paramagnetic phase and its overtones have been a puzzle for almost a decade. While this scattering intensity was shown to be subject to quenching with the onset of ferromagnetic order or in an external magnetic field [4.5–7], the most striking evidence for magnetic-phase-dependent Raman scattering was found for meta-magnetic EuSe [4.8]. The different magnetic spin superstructures have been found to show up in Brillouin-zone folding effects of the phonon branches. Shortly thereafter, the earlier conjecture [4.5] of spin-disorder induced Raman scattering from phonons in the paramagnetic phase of the europium chalcogenides could quantitatively be proved by resonance Raman scattering and its symmetry analysis [4.13, 17, 18]. The appearance of the antisymmetric Raman tensor component gave direct proof of simultaneous one phonon-one spin excitations due to spin-orbit coupling in the excited intermediate state [4.13]. A detailed discussion of Raman scattering in the paramagnetic and magnetically-ordered phases of the europium chalcogenides will be given in Sects. 4.4, 5, respectively. At the same time these investigations could rule out the interpretation in terms of a hot luminescence model [4.10, 30, 31] which was based on a comparison of multiphonon Raman scattering in europium chalcogenides and in the diamagnetic ytterbium chalcogenides. For the latter case it could be shown by the difference in multiphonon frequencies observed in optical absorption and Raman scattering [4.32] that a hot luminescence model does not apply (see also Chap. 5). The important role of magnetic-phase-dependent Raman scattering for studying spin fluctuations near phase transitions, order parameters and critical exponents has been emphasized [4.12, 15, 19] and will be discussed in Sect. 4.6.

Brillouin-zone folding effects of the phonon branches in the antiferromagnetic phase of layered structure VI_2 revealed another new spin-phonon excitation mechanism [4.27, 28]. A detailed analysis of the phonon eigenvectors of the new Raman-active modes in the antiferromagnetic phase gave evidence for a phonon modulation of the antiferromagnetic exchange interaction between the V^{2+} spins. This analysis made clear that such a net nonzero exchange modulation by phonons, conjectured previously for the Cd-Cr spinels [4.22, 23, 33], can occur only for very complex spin structures and therefore has escaped experimental observation. Section 4.8 deals with Raman scattering in the vanadium dihalides.

We would like to mention here another interaction of phonons with localized electrons in unfilled $4f$ shells in terms of "magnetic phonon splittings" observed in paramagnetic rare-earth trifluorides and trichlorides [4.34, 35]. The magnetic-field dependent splittings of doubly-degenerate (E_g) optical phonon states result from magnetoelastic interactions between the lattice deformation due to optic phonons and the multipole moments of the rare-earth charge distribution [4.36].

In $CsCoBr_3$ the temperature dependence of the scattering intensity of the E_{1g} phonon near the Néel temperature has been attributed to a spin-dependent polarizability mechanism [4.37]. The scattering intensity is found to follow a spin correlation function involving interchain Co^{2+} pairs.

Concepts of phonon-pseudospin couplings have been used successfully to describe orientational disorder of NH_4^+ molecules in the (pseudospin) disorder-induced phonon Raman scattering of ammonium halides [4.38].

In Sect. 4.2 we present a phenomenological treatment of spin-dependent phonon Raman scattering with application to the various magnetic phases of the europium chalcogenides. It accounts for the symmetry selection rules and the wave vector dependence of the scattered intensity. The symmetry independent coupling constants in the scattering cross section and their frequency dependence are derived in Sect. 4.3 in the framework of the microscopic theory considering the detailed electronic structure and the underlying fundamental interactions. These two sections serve as a basis for the discussion of the experimental results of the europium chalcogenides in Sects. 4.4–6.

4.2 Phenomenological Theory

The differential Raman cross section can be written as [4.39]

$$\frac{\partial^2\sigma}{\partial\Omega\,\partial\omega'}=\frac{\omega'^3\omega}{c^4}\frac{1}{2\pi}\int_{-\infty}^{\infty}dt\exp[\mathrm{i}(\omega'-\omega)t]$$

$$\cdot\sum_{\gamma_1\ldots\gamma_4}\langle\alpha^*_{\gamma_1\gamma_2}(0)\,\alpha_{\gamma_3\gamma_4}(t)\rangle\,e^*_{\gamma_1}e'_{\gamma_2}e_{\gamma_3}e'^*_{\gamma_4}\,. \tag{4.1}$$

The frequencies and polarizations of the incident and scattered light are denoted by $\omega, \boldsymbol{e}$ and $\omega', \boldsymbol{e}'$, respectively. c is the velocity of light. $\alpha_{\gamma_1\gamma_2}(t)$ is the polarizability tensor at time t with the indices of Cartesian coordinates γ_1 and γ_2. $\langle\;\rangle$ denotes an average over the canonical ensemble.

In the phenomenological treatment of spin-dependent Raman scattering one assumes that α is a function of the displacements $\boldsymbol{u}_\gamma(l\kappa)$ of the nuclei and of the spins $\boldsymbol{S}_\gamma(l\kappa)$. A possible dependence of α on laser frequency is neglected. l is a cell index, κ counts the atoms in the primitive cell and γ is an index of Cartesian coordinates. In the following, $(l\kappa\gamma)$ will be abbreviated by λ or μ. Electronic states do not have to be considered explicitly in this approach because the initial and final electronic states are assumed to be the same (usually the ground state). This phenomenological approach gives complete information on the various scattering possibilities and their selection rules. If the functional dependence of α on $\boldsymbol{u}$ and $\boldsymbol{S}$ is known or can be parameterized in terms of a few unknown constants, the different symmetry components of the scattering cross section can be calculated as function of the scattered frequency. On the other hand, the dependence of the scattered spectrum on the incident frequency (which probes the electronic states) as well as the magnitude and the relative weight of the various scattering possibilities can be explained only within a microscopic theory.

The expansion of α in terms of the displacements of the nuclei and of the spins can be written as:

$$\alpha_{\gamma_1\gamma_2} = \bar{\alpha}_{\gamma_1\gamma_2} + \alpha^{(1,0)}_{\gamma_1\gamma_2} + \alpha^{(1,1)}_{\gamma_1\gamma_2} + \alpha^{(1,2)}_{\gamma_1\gamma_2} + \dots \quad (4.2)$$

$\bar{\alpha}_{\gamma_1\gamma_2}$ contains all contributions in the Taylor expansion which are not of first order in the displacements. This term describes optical refraction, spin-flip scattering (Chap. 3), usual second-order Raman scattering, etc., and is of no interest in the following. The remaining terms in (4.2) are linear in the displacements $\boldsymbol{u}$. $\alpha^{(1,0)}$ describes first-order Raman scattering, $\alpha^{(1,1)}$ one phonon-one spin-flip scattering, $\alpha^{(1,2)}$ one phonon-two spin-flip scattering, etc. These are the terms we are interested in in the following.

There are a few general restrictions on the expansion coefficients in (4.2) due to the following facts:

a) The expansion coefficients in front of the displacements and spins in (4.2) must be a direct sum of irreducible tensor components with respect to the space group of the crystal. As a result, many coefficients are interrelated or zero by symmetry. For instance, the second term in (4.2) is zero for the EuX lattice because of spatial inversion symmetry.

b) Under time inversion, $\alpha_{\gamma_1\gamma_2}(t)$ transforms into $\alpha^*_{\gamma_1\gamma_2}(-t)$; within the adiabatic approximation, the time dependence occurs only in the displacement and spin variables. This means that the above transformation yields directly conditions for the expansion coefficients in front of the displacement and spin variables in (4.2) and that taking the conjugate complex is equivalent to interchanging γ_1 and γ_2. Furthermore, spins change their signs under time inversion whereas the displacements are unchanged. This implies that in the general term,

$$\alpha^{(i,j)}_{\gamma_1\gamma_2} = \sum_{\substack{\lambda_1\dots\lambda_i \\ \mu_1\dots\mu_j}} \alpha^{(i,j)}_{\gamma_1\gamma_2}(\lambda_1\dots\lambda_i, \mu_1\dots\mu_j)\boldsymbol{u}(\lambda_1)\dots\boldsymbol{u}(\lambda_i)\cdot \boldsymbol{S}(\mu_1)\dots\boldsymbol{S}(\mu_j) \quad (4.3)$$

the coefficient $\alpha^{(i,j)}_{\gamma_1\gamma_2}(\lambda_1\dots\lambda_i, \mu_1\dots\mu_j)$ is symmetric (antisymmetric) in γ_1, γ_2 if j is even (odd). In a microscopic treatment (Sect. 4.3) $\alpha_{\gamma_1\gamma_2}$ depends also on ω and ω'. We will show that the above selection rules remain valid if the difference between ω and ω' is neglected (quasistatic approximation).

c) $\alpha_{\gamma_1\gamma_2}$ must be invariant against infinitesimal translations in space.

In addition to the above rigorous constraints one, in general, also has to use less rigorous arguments to cut down the number of independent parameters appearing in the expansion (4.3). In our case the underlying microscopic interactions (electron-ion interaction, spin-orbit interaction, etc.) are effectively short-ranged. As a result, one expects that the coefficients $\alpha^{(i,j)}_{\gamma_1\gamma_2}(\lambda_1 \ldots \lambda_i, \mu_1 \ldots \mu_j)$ are only nonzero if the cluster of involved lattice sites is small. It is also plausible that the coefficients decrease if one considers first self-terms, then nearest-neighbor terms, then second-neighbor terms and so on. Using this classification scheme we discuss in the following the main contribution to $\alpha^{(1,1)}$ and $\alpha^{(1,2)}$ [4.9, 12–14, 18–20, 23, 40]. We denote the Eu ions by $\kappa=1$, the chalcogenide ions by $\kappa=2$ and take an Eu ion as origin.

$\alpha^{(1,1)}_{\gamma_1\gamma_2}$: this term is linear both in the displacements and in the spins and is given by (4.3) for $i=j=1$. From (b) we have the property

$$\alpha^{(1,1)}_{\gamma_1\gamma_2}(\lambda, \mu) = -\alpha^{(1,1)}_{\gamma_2\gamma_1}(\lambda, \mu). \tag{4.4}$$

Equation (4.4) implies that in the tensor product of Cartesian coordinates $\Gamma^-_{15} \times \Gamma^-_{15} = \Gamma^+_1 + \Gamma^+_{12} + \Gamma^+_{15} + \Gamma^+_{25}$, only the antisymmetric component Γ^+_{15} can be observed. Thus, $\alpha^{(1,1)}_{\gamma_1\gamma_2}$ is only nonzero if the tensor product of Γ^+_{15} (the irreducible representation of the spins) and Γ_Q (the irreducible representation of the displacement pattern around the site of a spin) contains Γ^+_{15} (the antisymmetric representation of Cartesian coordinates). This means that only phonons with the local symmetries Γ^+_1, Γ^+_{12}, Γ^+_{15}, and Γ^+_{25} around the site of a spin can occur in the Raman spectrum.

Writing out the various components explicitly we obtain

$$\begin{aligned}\alpha^{(1,1)}_{\gamma_1\gamma_2} = \sum_{l,n} \begin{bmatrix} 0 & -1 & 0 \\ 1 & 0 & 0 \\ 0 & 0 & 0 \end{bmatrix} \Big\{ & A(n, \Gamma^+_1) S^z_l Q_l(n, \Gamma^+_1) \\ & + A(n, \Gamma^+_{12}) S^z_l Q_l(n, \Gamma^+_{12}, 3z^2 - r^2) + A(n, \Gamma^+_{25}) \frac{1}{\sqrt{2}} [S^x_l Q_l(n, \Gamma^+_{25}, zx) \\ & + S^y_l Q_l(n, \Gamma^+_{25}, yz)] + A(n, \Gamma^+_{15}) \frac{1}{\sqrt{2}} [S^x_l Q_l(n, \Gamma^+_{15}, l_y) \\ & - S^y_l Q_l(n, \Gamma^+_{15}, l_x)] + \text{cyclic permutations of (xyz)} \Big\}. \end{aligned} \tag{4.5}$$

The 3×3 matrix in (4.5) is due to the Cartesian coordinates. $Q(n, \Gamma^+_{12}, 3z^2 - r^2)$ denotes the nth irreducible displacement pattern (phonon coordinate) of local symmetry Γ^+_{12} with respect to site l which transforms like $3z^2 - r^2$. The other symmetry coordinates are defined in an analogous way. The corresponding symmetry independent coupling constants are denoted, equivalently, by $A(n, \Gamma)$.

Equation (4.5) is completely general. According to our classification scheme we expect that the largest coupling constants in (4.5) are caused by the longitudinal displacements of nearest neighbors. There are three coupling coefficients which correspond to the representations Γ_1^+, Γ_{12}^+, and Γ_{25}^+. Furthermore, one can conjecture that the Γ_1^+ coupling should be the dominant one (see p. 218). The comparison with experiment shows, indeed, that for all EuX crystals, only the Γ_1^+ coupling is needed to explain the experimental spectra. Omitting all the other couplings, one obtaines after a Fourier transformation

$$\alpha_{\gamma_1\gamma_2}^{(1,1)} = \begin{bmatrix} 0 & -1 & -1 \\ 1 & 0 & -1 \\ 1 & 1 & 0 \end{bmatrix} \sum_{\boldsymbol{k}j} \frac{2A(\Gamma_{15}^+)}{\sqrt{6}} \sqrt{\frac{2\hbar}{N\omega(\boldsymbol{k}j)M_2}}\, \sigma_{\boldsymbol{k}}(\Gamma_{15}^+)$$

$$\cdot \sum_{\gamma=1}^{3} \sin\,(k_\gamma a)\, e_\gamma(2,\boldsymbol{k}j)\,(a_{\boldsymbol{k}j} + a_{-\boldsymbol{k}j}^+). \tag{4.6}$$

$\sigma_{\boldsymbol{k}}(\Gamma_{15}^+)$ is defined by

$$\sigma_{\boldsymbol{k}}(\Gamma_{15}^+) = \frac{1}{\sqrt{N}} \sum_l S_l^z \exp[i\boldsymbol{k}\cdot\boldsymbol{x}(l1)], \tag{4.7}$$

where N is the number of primitive cells. $e_\gamma(2,\boldsymbol{k}j)$ is the eigenvector of the chalcogenide ion in γ-direction due to the phonon with wave vector $\boldsymbol{k}$ and branch j. The frequencies of the phonons are denoted by $\omega(\boldsymbol{k}j)$, the creation and destruction operators by $a_{\boldsymbol{k}j}^+$ and $a_{\boldsymbol{k}j}$, respectively. M_2 is the mass of the chalcogen ion and a is half the cubic lattice constant.

$\alpha_{\gamma_1\gamma_2}^{(1,2)}$: The simplest and most important case is obtained by taking the two spins at the same Eu ion. Time inversion symmetry allows only the symmetric representations Γ_1^+, Γ_{12}^+, and Γ_{25}^+ for the polarizability tensor. The tensor product of two spins decomposes into the representations Γ_{12}^+ and Γ_{25}^+ (the basis vector for the Γ_1^+ representation is $\sim \boldsymbol{S}^2$; it does not fluctuate and therefore should be omitted). As a result, all phonons with an even representation with respect to the site of a spin can occur in the Raman spectra. Using, however, similar arguments as in the case of $\alpha^{(1,1)}$, one expects that the coupling to the full-symmetric longitudinal displacement pattern of nearest-neighbor ions should be the dominant one (an expectation which again is supported by experiment). Then only the Γ_{12}^+ and Γ_{25}^+ scattering components are nonzero. Writing out the symmetry components explicitly one obtains for this case:

$$\alpha_{\gamma_1\gamma_2}^{(1,2)} = \sum_{\boldsymbol{k}j} \left\{ \begin{bmatrix} 1 & 0 & 0 \\ 0 & 0 & 0 \\ 0 & 0 & 0 \end{bmatrix} A(\Gamma_{12}^+)\sigma_{\boldsymbol{k}}(\Gamma_{12}^+) \right.$$

$$\left. + \begin{bmatrix} 0 & 1 & 0 \\ 1 & 0 & 0 \\ 0 & 0 & 0 \end{bmatrix} A(\Gamma_{25}^+)\sigma_{\boldsymbol{k}}(\Gamma_{25}^+) \right\} \frac{2}{\sqrt{6}} \sqrt{\frac{2\hbar}{N\omega(\boldsymbol{k}j)M_2}}$$

$$\cdot \sum_{\gamma=1}^{3} \sin\,(k_\gamma a)\, e_\gamma(z,\boldsymbol{k}j)\,(a_{\boldsymbol{k}j} + a_{-\boldsymbol{k}j}^+), \tag{4.8}$$

with

$$\sigma_{\boldsymbol{k}}(\Gamma_{12}^{+})=\frac{2}{3\sqrt{N}}\sum_{l}(2S_l^{x^2}-S_l^{y^2}-S_l^{z^2})\ \exp[\mathrm{i}\boldsymbol{k}\cdot\boldsymbol{x}(l1)] \tag{4.9}$$

and

$$\sigma_{\boldsymbol{k}}(\Gamma_{25}^{+})=\frac{1}{\sqrt{N}}\sum_{l}(S_l^xS_l^y+S_l^yS_l^x)\exp[\mathrm{i}\boldsymbol{k}\cdot\boldsymbol{x}(l1)]. \tag{4.10}$$

$A(\Gamma_{12}^{+})$ and $A(\Gamma_{25}^{+})$ are the two independent coupling constants. The case where the two spins belong to nearest and second-nearest neighbor Eu ions has been worked out in [4.14]. It seems however, that the dominant term given in (4.8) is sufficient to explain the available experimental spectra in the EuX.

Inserting (4.6, 8) into (4.1), one obtains for the independent scattering components of the cross section the following expressions assuming Stokes scattering:

$$\frac{\partial^2\sigma^{\Gamma}}{\partial\Omega\,\partial\omega'}=\frac{\omega'^3\omega}{c^4}\sum_{\boldsymbol{k}j}|A(\Gamma)X(\boldsymbol{k}j)|^2[1+n(\boldsymbol{k}j)]\mathrm{S}(\Gamma,\boldsymbol{k},\omega'-\omega+\omega(\boldsymbol{k}j)), \tag{4.11}$$

for $\Gamma=\Gamma_{15}^{+}$, Γ_{12}^{+} and Γ_{25}^{+}, and

$$\frac{\partial^2\sigma^{\Gamma_1^+}}{\partial\Omega\,\partial\omega'}=0. \tag{4.12}$$

S is a spin correlation function defined by

$$S(\Gamma,\boldsymbol{k},\omega)=\frac{1}{2\pi}\int_{-\infty}^{\infty}dt\exp(\mathrm{i}\omega t)\,\langle\sigma_{\boldsymbol{k}}(\Gamma)\sigma_{-\boldsymbol{k}t}(\Gamma)\rangle \tag{4.13}$$

and $X(\boldsymbol{k}j)$ is defined by

$$X(\boldsymbol{k}j)=\sqrt{\frac{8\hbar}{\sqrt{6}\,N\omega(\boldsymbol{k}j)M_2}}\sum_{\gamma=1}^{3}\sin\ (k_\gamma a)\ e_\gamma(2,\boldsymbol{k}j). \tag{4.14}$$

$n(\boldsymbol{k}j)$ is the thermal occupation number of the phonon $(\boldsymbol{k}j)$. According to (4.7, 9, 10) the spin-correlation function in (4.11) is for $\Gamma=\Gamma_{15}^{+}$, a two-spin and for $\Gamma=\Gamma_{12}^{+}$ and Γ_{25}^{+}, a four-spin correlation function.

Two important results follow immediately from (4.11, 12). Independent of the exact form for the spin-correlation function (in particular, independent of the magnetic phase), the scattered spectra should contain broad bands because of the $\boldsymbol{k}j$ summation. Furthermore, the largest contributions are expected in Γ_{15}^{+}

symmetry (first order in the spin), followed by those in Γ_{12}^+ and Γ_{25}^+ symmetries (second order in the spin). The contribution of Γ_1^+ symmetry should be rather small and nonzero only because of higher-order processes. This is just the opposite of what is usually observed for the broad bands of second-order Raman scattering (e.g., alkali halides, III–V and II–VI compounds).

In order to be able to discuss the spectrum as a function of the scattered frequency, explicit expressions for the spin correlation functions are needed. The leading scattering component involves the two-spin correlation function which has the following form for a ferromagnet in the classical limit [4.12, 19, 20, 42].

Paramagnetic phase $T > T_C$:

$$S(\gamma,\gamma',\mathbf{k},\omega)=\delta_{\gamma\gamma'}\delta(\omega)\,\frac{S(S+1)}{3}\,\frac{f(T/T_C)}{\frac{T-T_C}{T_C}+g(\mathbf{k})}\,, \tag{4.15}$$

$$g(\mathbf{k})=(J(0)-J(\mathbf{k}))/J(0)\,; \tag{4.16}$$

$J(\mathbf{k})$ is the Fourier transform of the exchange coupling constants for the lattice under consideration.

Ferromagnetic phase $T \lesssim T_C$:

$$S(\gamma,\gamma',\mathbf{k},\omega)=\delta_{\gamma\gamma'}\delta(\omega)\left[\langle\sigma_{\mathbf{k}}^{\Gamma_{15}^+}\rangle\langle\sigma_{-\mathbf{k}}^{\Gamma_{15}^+}\rangle\delta_{\mathbf{k},0}\delta_{\gamma z} + \frac{S(S+1)}{3}\,\frac{f(T/T_C)}{(T-T_C)/T_C\,\delta_{\gamma z}+g(\mathbf{k})}\right]. \tag{4.17}$$

Spin wave regime $T \ll T_C$:

$$S(\gamma,\gamma',\mathbf{k},\omega)=\delta_{\gamma\gamma'}\left[\delta(\omega)\langle\sigma_{\mathbf{k}}^{\Gamma_{15}^+}\rangle\,\langle\sigma_{-\mathbf{k}}^{\Gamma_{15}^+}\rangle\delta_{\mathbf{k},0}\delta_{\gamma z} + \frac{S}{3}\sum_{+-}\delta(\omega\mp\Omega(\mathbf{k}\lambda))\,[n(\mathbf{k}\lambda)+\tfrac{1}{2}\pm\tfrac{1}{2}]\right], \tag{4.18}$$

where $\Omega(\mathbf{k}\lambda)$ is the frequency of the magnon with wave vector $\mathbf{k}$ and branch λ; $n(\mathbf{k}\lambda)$ is the corresponding thermal occupation number.

The slightly generalized two-spin correlation function $S(\gamma,\gamma',\mathbf{k},\omega)$ is defined by (4.13) with $\sigma_{\mathbf{k}}(\Gamma)$, $\sigma_{-\mathbf{k}t}(\Gamma)$ replaced by $\sigma_{\mathbf{k}}^{\gamma}$, $\sigma_{-\mathbf{k}t}^{\gamma'}$, respectively. $\sigma_{\mathbf{k}}^{\gamma}$ is given by the right-hand side of (4.7) with z replaced by γ. It is assumed in (4.17, 18) that the spontaneous magnetization is parallel to the z direction. Equations (4.15, 17) are mean-field expressions which have been slightly improved so that the sum rule [4.41].

$$\sum_{\mathbf{k},\gamma}\left[\int d\omega\, S(\gamma,\gamma',\mathbf{k},\omega)-\delta_{\gamma z}\delta_{\mathbf{k}0}|\langle\sigma_{\mathbf{k}}^{\Gamma_{15}^+}\rangle|^2\right]=S(S+1) \tag{4.19}$$

is fulfilled. The function $f(T/T_C)$ is determined by the condition (4.19).

Equations (4.15–18) also hold for the two-sublattice antiferromagnetic phases of EuSe and EuTe if $\boldsymbol{k}$ is replaced by $\boldsymbol{q}_L - \boldsymbol{k}$, where $\boldsymbol{q}_L$ is the vector to the L-point in reciprocal space.

In spite of their crudeness, expressions (4.15, 17), together with (4.18), account for the gross features of the scattered spectrum as a function of frequency and temperature. At high temperatures $T \gg T_C$ (for the EuX room temperature fulfills already this condition), S is independent of $\boldsymbol{k}$ and the scattered spectrum consists of a symmetry-projected one-phonon density of states weighted by the form factor $X^2(\boldsymbol{k}j)$ in (4.11). Because of the sine function in $X(\boldsymbol{k}j)$ [see (4.14)], a broad peak of mainly LO phonons near the L-point is obtained. For $T \lesssim T_C$ the $\boldsymbol{k}$ dependence of S, which enhances the contributions of those $\boldsymbol{k}$ values which are near the wave vector describing the magnetically-ordered state below T_C, must be taken into account. The same is essentially true for $T < T_C$ if an average over the domains is performed and if the projected one-phonon density of states is replaced by a projected one phonon-one magnon density. In addition to the broad features, sharp lines occur for $T < T_N$ due to the first term in (4.17) describing Bragg scattering.

Another quantity of interest is the temperature dependence of the integrated scattered spectrum, in particular its Γ_{15}^+ component $\sigma^{\Gamma_{15}^+}(T)$. Omitting the trivial phonon occupation number $n(\boldsymbol{k}j)$, this quantity varies only slowly with T for $T \gg T_C$. Near the phase transition one expects an increase of $\sigma^{\Gamma_{15}^+}(T)$ because of the diverging static susceptibilities S of (4.17). In the case of a ferromagnet this increase is, however, very small: the values of S for small $\boldsymbol{k}$'s are enhanced. These values, however, cannot contribute substantially to the total cross section because of the form factor $X(\boldsymbol{k}j)$ and, in addition, because of phase space arguments. Near the transition point to the two-sublattice antiferromagnetic phase the increase may, however, be well pronounced: S is enhanced for wave vectors near the L-point which have a large form factor $X(\boldsymbol{k}j)$ and a large phase space contribution. For $T \ll T_C$ the spin wave expression (4.18) yields (after a domain average) $\sigma^{\Gamma_{15}^+}(T=0)/\sigma^{\Gamma_{15}^+}(T=\infty) = (S+1)^{-1}$ [4.12]. This means that $\sigma^{\Gamma_{15}^+}(T)$ quenches rapidly for $T < T_C$ and assumes a small but finite value (due to phonon-magnon scattering) at $T=0$. Numerical calculations for $\sigma^{\Gamma_{15}^+}(T)$ [4.12, 14, 19] confirm the above qualitative arguments. Reference [4.14] also presents results for scattering from single domains as well as the dependence on an external applied magnetic field.

The assumed frequency dependence of S in (4.16, 18) is rather crude; even in the limit $T \to \infty$, the scattering from the disordered spins is not exactly elastic but quasi-elastic [4.42a]. The dynamic response of the spin system (dynamic spin disorder) can be taken into account by including in (4.16) a relaxation or spectral-shape function [4.42a]. This can be approximated for ferromagnets at large wave vectors by either a Gaussian [4.12, 42a] or a damped-harmonic oscillator function [4.42b]. The complicated frequency dependence of a realistic S-function for intermediate temperatures does not greatly influence the Raman spectra. At low temperatures the frequency shift due to the magnons should, however, be observable. In the critical region $T \sim T_C$, the mean field value 1 for the critical coefficient γ (near T_C the temperature dependence in (4.15) should be written as $[(T-T_C)/T_C]^{-\gamma}$ [4.43]) must be replaced by a more realistic value. Critical scattering will be discussed in Sect. 4.6.

Reference [4.14] also considers the temperature and magnetic field dependence of the Γ_{12}^+ and Γ_{25}^+ scattering components which involve the four-spin correlation function. The results obtained are qualitatively the same as for the Γ_{15}^+ component discussed above.

4.3 Microscopic Theory

The phenomenological theory presented in the previous section was based on a formal expansion of the polarizability tensor in terms of ion displacements and spins. The aim of a microscopic treatment is to explain these formal expansion coefficients in terms of the underlying fundamental interactions (electron-ion interaction, spin-orbit interaction, etc.) and to study their frequency dependencies. In particular, one wants to obtain answers to the following questions: Why is the antisymmetric component of the scattering stronger than the symmetric ones? Why have these spin-dependent contributions only been seen under resonance conditions (i.e., when the incident frequency is in the neighborhood of strong electronic transitions) so that originally hot luminescence models have been proposed to account for them [4.30].

In order to calculate the nonlinear polarizability α in the frequency region near the fundamental gap, the ground as well as the low-lying excited states must be specified. The energetically lowest electronic states result from the $4f^7(^8S)$ ground state of Eu^{2+} in the EuX crystals. A basis for these states can be written as

$$\prod_l |O_l\rangle|7/2, M_l\rangle. \tag{4.20}$$

$|O_l\rangle$ is the orbital part of the ground state of an Eu^{2+} ion at the site l consisting of filled f orbitals. $|7/2, M_l\rangle$ is a 7/2 spin function with projection M_l. The spin degeneracy of the states (4.20) is partially lifted by the Heisenberg interaction between spins at different Eu sites. For the calculation of α, these splittings are negligible, however (the corrections are of the order of the magnon energy over the energy of the fundamental gap). In the following we will abbreviate the states in (4.20) by $|a\rangle$ and their (approximate) eigenenergies by ε_0.

A basis for the excited states near the fundamental gap is

$$|l'n, lm; \{M_l\}\rangle = c_{dn}^+(l')c_{fm}(l)|0\rangle|1/2, ml\rangle|3, M_l\rangle \prod_{l' \neq l} |7/2, M_{l'}\rangle. \tag{4.21}$$

c_{dn}^+ and c_{fm} are creation and annihilation operators for one-particle states: $c_{dn}^+(l')$ creates a d electron at the site l' with band index n; $c_{fm}(l)$ annihilates a f electron at site l with the magnetic quantum number m (we assume that the

f-bands are dispersionless). $|O\rangle$ denotes the orbital ground state of the Eu ions and is given by $\prod_l |O_l\rangle$. $|3, M_l\rangle$ is a $S=3$ spin function with the projection $M_l = -3\ldots3$ formed from the spins of the six remaining f electrons at the Eu ion l. $|1/2, m_l\rangle$ is a spin 1/2 function with projection $m_l = \pm 1/2$ due to the spin of the d electron.

According to the model of *Kasuya* [4.44], the first residual interaction to be diagonalized in the orbital space of the states (4.21) is

$$H_d^{\text{hopp}} + H_{d-f}^{\text{Coul}}. \tag{4.22}$$

H_d^{hopp} consists of hopping terms for the d electrons which cause a finite width and splittings of the d bands. In lowest order H_d^{hopp} splits the five d states into a triplet of Γ_{25}^+ and a doublet of Γ_{12}^+ symmetry. The doublet is about 1.5 eV higher in energy than the triplet and thus may be neglected. n therefore assumes in the following only the values $\pm 2, 0$. The Hamiltonian H_{d-f}^{Coul} represents the screened Coulomb interaction between the d electrons and the f holes. Interaction (4.22) is approximately taken into account by introducing correlated electron-hole operators $B_{\lambda lm}^+$:

$$B_{\lambda lm}^+ = \sum_{l',n} \beta_{\lambda lm}(l'n) c_{dn}^+(l') c_{fm}(l). \tag{4.23}$$

The amplitudes β are chosen such that (4.23) is diagonal in the subspace with a fixed value for m. An explicit dependence of β on m is neglected. The corresponding eigenenergies of (4.23) are independent of l and (approximately) m and are denoted by E_λ.

A second residual interaction is due to the d–f exchange. Applying a mean field-type approximation to the orbital part and using the basis (4.21), this interaction has the form

$$H_{d-f}^{\text{Exch}} = -I_0 |\beta_{\lambda lm}(ln)|^2 \boldsymbol{s} \cdot \boldsymbol{S}_F - I_1 \sum_{l' \neq l} |\beta_{\lambda lm}(l'n)|^2 \boldsymbol{s} \cdot \boldsymbol{S}_{l'}. \tag{4.24}$$

The first term in (4.24) describes the interaction of the spin $\boldsymbol{s}$ of the d electron ($|\boldsymbol{s}| = 1/2$) with the spin $\boldsymbol{S}_F$ ($|\boldsymbol{S}_F| = 3$) of the f^6 core corresponding to the case where electron and hole are sitting at the same site. The second term in (4.24) describes the interaction of $\boldsymbol{s}$ with the ground state spin $\boldsymbol{S}_l$, with $l' \neq l$ corresponding to the case where the electron and the hole are sitting at different sites. The second contribution is rather small in the paramagnetic phase because of the l' summation in (4.24) and thus may be neglected to a first approximation (the influence of this term has been discussed in [4.45]. The first term in (4.24) can be diagonalized by forming eigenfunctions for $\boldsymbol{S}_0 = \boldsymbol{s} + \boldsymbol{S}_F$, namely, $S_0 = 7/2$ and $S_0 = 5/2$ functions. The $S_0 = 5/2$ functions are optically inactive and more than 1 eV in energy above the $S_0 = 7/2$ functions [4.44] and thus may be dropped.

Finally, the spin-orbit interaction H^{so} of the f electrons must be diagonalized in the subspace of $S_0 = 7/2$ functions. Using the fact that the $S_0 = 7/2$ spin function of seven spin-1/2 particles is totally symmetric with respect to interchanging two particles, the single spin operators can be replaced by the total spin operator S_0 with spherical components S_{0z}, $S_{0\pm}$. H^{so} then becomes

$$H^{so} = \frac{\lambda_f}{7} \sum_{lm} [c^+_{fm}(l) c_{fm}(l) S_{0z}(l) + \tfrac{1}{2} \sqrt{12 - m(m+1)}\, c^+_{fm+1}(l) c_{fm}(l) S_{0+}(l)$$
$$+ \tfrac{1}{2} \sqrt{12 - m(m-1)}\, c^+_{fm-1}(l) c_{fm}(l) S_{0-}(l)] . \tag{4.25}$$

Diagonalizing H^{so} in the space of $S_0 = 7/2$ functions yields the new eigenstates

$$|l\lambda;\, JN_l;\, \{M_{l'}\}\rangle = \sum_{m_l, M_l} \left\langle {J \atop N_l} \middle| {3 \atop m_l} {7/2 \atop M_l} \right\rangle B^+_{\lambda l, m_l} |0\rangle\, |\tfrac{7}{2} M_l\rangle \prod_{l' \neq l} |\tfrac{7}{2} M_{l'}\rangle \tag{4.26}$$

and eigenenergies

$$E(\lambda, J) = E_\lambda - \frac{\lambda_f}{56} [4J(J+1) - 111] . \tag{4.27}$$

J assumes the values 1/2...13/2 and $\left\langle {J \atop N_l} \middle| {3 \atop m_l} {7/2 \atop M_l} \right\rangle$ denotes a Clebsch-Gordan coefficient. We abbreviate the states (4.26) and the energies (4.27) in the following by

$$|\beta\rangle, \quad \varepsilon_\beta . \tag{4.28}$$

The most resonant term for the transition amplitude of the Raman process γ, $\{\alpha\} \to \gamma'$, $\{\alpha'\}$, Qj is given by

$$T^{\gamma\gamma'}_{\alpha\alpha'}(\omega, Qj) = \sum_{\beta, \beta'} \frac{G_\gamma(\alpha, \beta) F_{Q_j}(\beta, \beta') G^*_{\gamma'}(\alpha', \beta')}{(\hbar\omega + i\eta - \varepsilon_\beta)\ [\hbar\omega + i\eta - \hbar\omega(Qj) - \varepsilon_{\beta'}]} . \tag{4.29}$$

$G_\gamma(\alpha, \beta)$ denotes the electronic transition element $\alpha \to \beta$ due to an electric field of polarization γ and unit strength. F is the electron-phonon coupling written in the basis (4.28). The microscopic cross section associated with (4.29) can be written in the form of (4.1) if the polarizability $\sum_{\nu=0}^{\infty} \alpha^{(1\nu)}$, which is linear in the displacements, and T are identified as follows:

$$\sum_{\nu=0}^{\infty} \alpha^{(1\nu)}_{\gamma_1\gamma_2}(Qj) = \sqrt{\frac{\hbar}{2\omega(Qj)}} \sum_{\alpha, \alpha'} T^{\gamma_1\gamma_2}_{\alpha\alpha'}(\omega, Qj)\, |\alpha\rangle \langle\alpha'| . \tag{4.30}$$

The first factor on the right-hand side of (4.30) is due to the fact that $\alpha^{(1\nu)}$ is defined as a first derivative with respect to the normal coordinates and not with respect to the coordinate $(a^+_{Qj} + a_{-Qj})$. In the phenomenological treatment $\sum_{\nu=0}^{\infty} \alpha^{(1\nu)}$ is expanded in terms of powers of spins. The connection between the resulting expansion coefficients and the microscopic terms on the right-hand side of (4.30) can be obtained in the general case by expanding the spin operators in terms of the projection operators $|\alpha\rangle\ \langle\alpha'|$ and identifying the prefactors of identical projection operators on the left and right-hand side in (4.30).

The identification of formal expansion coefficients and microscopic matrix elements becomes rather straighforward in our case if we use a few simplifications in the microscopic amplitude T. We assume that the f electrons couple only to the full-symmetric longitudinal displacement pattern of nearest neighbors and neglect the coupling of d electrons to displacements. $F_{Qj}(\beta\beta')$ then becomes diagonal in β and β'. Furthermore, it is convenient to expand T in terms of powers of λ_f and to calculate the contributions to $\sum_{\nu=0}^{\infty} \alpha^{(1\nu)}$ only from the first two terms. Such a procedure is valid outside the very resonant region. It turns out that the term $\sim\lambda_f$ contributes to $\alpha^{(1,1)}$ and the term $\sim\lambda_f^2$ to $\alpha^{(1,2)}$. Neglecting, furthermore, the small phonon energy in the second denominator in (4.29) (adiabatic or quasistatic approximation), expressing the G-matrix elements in terms of a reduced matrix element and Clebsch-Gordan coefficients and inserting numerical values for the Clebsch-Gordan coefficients, one finds for the irreducible coupling constants in (4.11) the following expressions [4.18]:

$$|A(\Gamma_{15}^{+})|^2 = 3.36\ \lambda_f^2 |R_1(\omega)|^2 , \tag{4.31}$$

$$|A(\Gamma_{12}^{+})|^2 = 0.202\ \lambda_f^4 |R_2(\omega)|^2 , \tag{4.32}$$

$$|A(\Gamma_{25}^{+})|^2 = 0.710\ \lambda_f^4 |R_2(\omega)|^2 \tag{4.33}$$

with

$$R_\nu(\omega) = (-1)^{\nu+1} \frac{1}{\nu!} \frac{\partial^{\nu+1}}{\partial(\hbar\omega)^{\nu+1}} \chi(\omega) . \tag{4.34}$$

$\chi(\omega)$ is the resonant term of the complex linear susceptibility. It can be either calculated within the above model for the electronic states and the residual interactions or can be deduced from the experimental absorption spectrum. Strictly speaking, (4.31–33) are not valid in the very resonant region $|\hbar\omega - E_\lambda| \lesssim \lambda_f$ because the expansion in terms of powers of λ_f breaks down. Numerical calculations show, however, that the difference between a smooth interpolation through the resonant region using (4.31–33) [4.13] and the exact curve [4.20] is small.

The microscopic theory of the cross section presented above reproduces essentially all the features of phenomenological theory. The only difference which is of a more conceptual nature is due to the fact that the phenomenological approach, in contrast to the microscopic one, is based on the adiabatic approximation. As a result, the phonon energy appears in the second denominator in (4.29) which can, however, be safely neglected in the EuX because of the large damping constant $\eta \approx 1000\ \mathrm{cm}^{-1}$ [4.20]. Otherwise all the selection rules as well as the cross sections of (4.11–14) are also valid in the microscopic approach. In addition, the microscopic description allows us to calculate the leading contributions to the symmetry independent coupling constants $A(\Gamma_{15}^{+})$, $A(\Gamma_{12}^{+})$ and $A(\Gamma_{25}^{+})$ in terms of Clebsch-Gordan coefficients and fundamental interactions such as the spin-orbit coupling constant λ_f. Moreover, it allows us to determine the frequency dependence of the coupling constants (i.e., resonance effects).

The influence of the spin-orbit interaction of the d electrons on $\alpha^{(1,1)}$ has been considered in [4.14]. Inserting appropriate values for EuS and EuSe, the corresponding contributions are about ten times smaller than those considered above. In [4.14] the contributions of various other mechanisms to $\alpha^{(1,1)}$ are also shown to be very small.

4.4 Scattering in the Paramagnetic Phase of EuX (X = O, S, Se, Te)

4.4.1 Selection Rules and Scattering Intensity

The europium chalcogenides are the model class of magnetic semiconductors showing unusual interrelated electronic, optical and magnetic properties [4.46]. They crystallize in the NaCl structure and are excellent examples of the Heisenberg exchange, being described by two exchange constants $J_1(>0)$ for the nearest-neighbor (nn) ferromagnetic indirect superexchange interaction and $J_2(<0)$ for the next-nearest neighbor (nn) antiferromagnetic superexchange interaction. While EuO and EuS order ferromagnetically, the comparable magnitude of J_1 and J_2 in metamagnetic EuSe gives rise to a rather complex magnetic phase diagram. EuTe is a type-II (MnO) antiferromagnet.

First-order Raman scattering in the fcc (NaCl structure) europium chalcogenides is symmetry forbidden. However, the Raman spectra exhibit in the paramagnetic phase a rather broad first-order peak which extends over the frequency range between the zone-center TO and LO phonons and a series of overtones of this peak [4.5, 6, 47]. The ("fundamental") peak was identified as a first-order scattering process [4.5, 7, 48a] and as consisting predominantly of a one-phonon density of states from the LO phonon branch [4.48a].

As an example we show in Fig. 4.1 the Raman spectrum of EuS at 300 K. The first-order Raman scattering of $Eu^{32}S$ (upper curve) near $\omega_0 = 240\ \mathrm{cm}^{-1}$ is shifted to $235\ \mathrm{cm}^{-1}$ upon substituting ^{32}S by ^{34}S (see inset). The shifted frequency scales directly with the square root of the anion mass ratio. This anion mass dependence of the peak position ω_0 of the first-order Raman scattering has been confirmed for the entire EuX (X = O, S, Se, Te) series at 300 K [4.17, 48a]. Hence, optic phonons from the zone boundary are shown to be the dominant contribution to the first-order Raman scattering intensity. Obviously, this fact seems to contradict momentum conservation in the scattering process. The clue

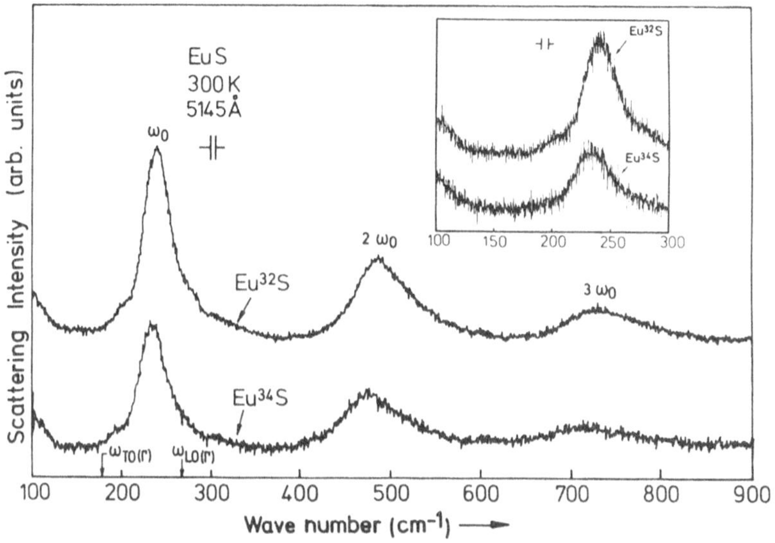

Fig. 4.1. Raman spectrum (unpolarized) of EuS at 300 K for two different S isotopes (^{32}S and ^{34}S)

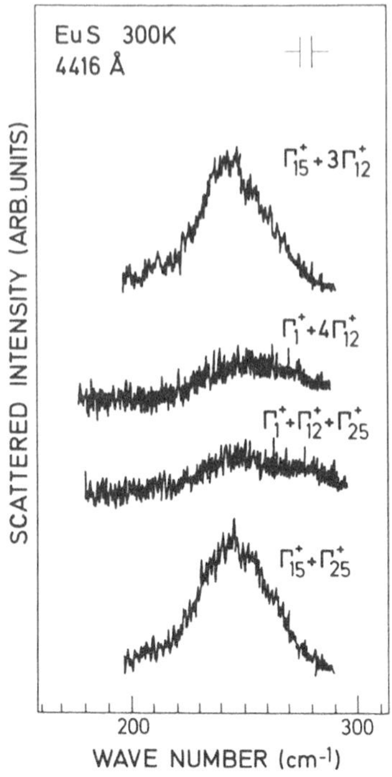

Fig. 4.2. Symmetry analysis of the first-order Raman scattering from a (001) face of EuS at 300 K for 4416 Å laser excitation, using linear and circular polarization

in resolving this long-standing puzzle came from polarized Raman scattering experiments [4.13,17].

The symmetry analysis of the Raman scattering intensity of a cleaved (001) face of EuS at 300 K under 4416 Å laser excitation is shown in Fig. 4.2. (The polarized Raman scattering under resonant 5145 Å laser excitation is shown in Chap. 5 of this volume). Consistency in the analysis was achieved by using linear as well as circular polarized light and by considering, besides the symmetric Raman tensor components $\Gamma_1^+(A_{1g})$, $\Gamma_{12}^+(E_g)$ and $\Gamma_{25}^+(T_{2g})$, also the antisymmetric component $\Gamma_{15}^+(T_{1g})$. The dominant contribution from the antisymmetric $\Gamma_{15}^+(T_{1g})$ Raman tensor component is evident from Fig. 4.2, the three other components being practically negligible. This result gives direct proof of the participation of the spin component in the scattering process as follows. As pointed out in Sect. 4.2, the polarizability tensor $\alpha^{(1,1)}$ of the one phonon-one spin scattering process is only nonzero if the product of the irreducible representations of the spins (Γ_{15}^+) and the phonons (Γ_Q) around the site of a spin contains the antisymmetric (Γ_{15}^+) representation of the Cartesian coordinates. From the classification scheme of coupling constants in (4.5) we expect the participation of phonon coordinates of local symmetry Γ_1^+, Γ_{12}^+ and Γ_{25}^+. The dominance of the Γ_1^+ component is plausible by considering [4.17] that the laser excitation gives rise to an optical transition $4f^7(\mathrm{Eu}^{2+}) \rightarrow 4f^6 5d^1(\mathrm{Eu}^{3+})$ and hence to a 14% reduction in ionic radius of Eu^{2+}. The resulting isotropic compression of the Eu ion (see also Chap. 5) couples to the locally fully symmetric (Γ_1^+) displacements of the surrounding octahedron of chalcogen ions.

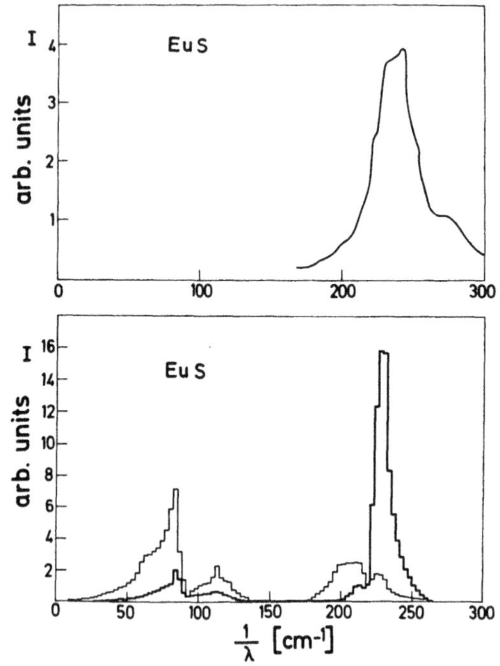

Fig. 4.3. First-order Raman scattering of EuS in the paramagnetic phase. Top: measured $\Gamma_{25}^+ + \Gamma_{15}^+$ spectrum at 300 K using 5145 Å laser excitation. Bottom: calculated temperature-weighted one-phonon density of states (*thin lines*) and calculated spin-disorder induced one-phonon Raman scattering intensity (*thick lines*)

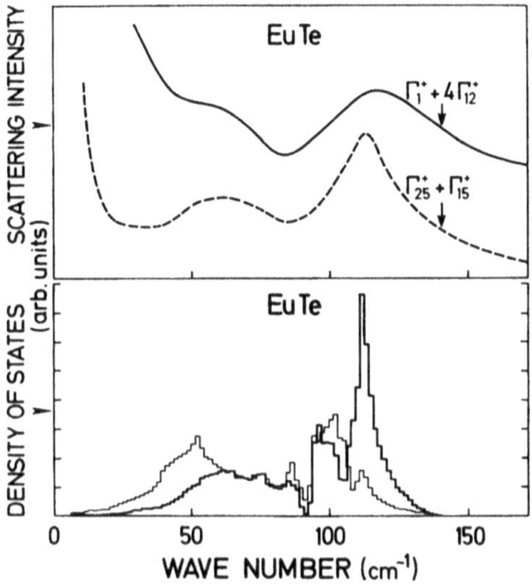

Fig. 4.4. First-order Raman scattering of EuTe in the paramagnetic phase. Top: experiment at 300 K using 5145 Å laser excitation. Bottom: calculated temperature-weighted one-phonon density of states (*thin lines*) and calculated spin-disorder induced one-phonon Raman scattering intensity (*thick lines*)

Therefore, the dominant antisymmetric (Γ_{15}^+) Raman tensor component observed in Fig. 4.2 reflects directly the symmetry of the spin component and proves the simultaneous one phonon-one spin excitation according to (4.3). This excitation also restores momentum conservation in the scattering process: the disorder in the spin system in the paramagnetic phase gives rise to the observation of a one-phonon density of states, which is weighted by the electron-phonon matrix element of Γ_1^+ symmetry.

In Fig. 4.3 (upper part) we show the measured one-phonon Raman spectrum of a cleaved (001) face of EuS at 300 K under resonant 5145 Å laser excitation in the $\Gamma_{25}^+ + \Gamma_{15}^+$ scattering configuration. The temperature-weighted one-phonon density of states of EuS derived from the calculated phonon dispersion curves [4.18] is shown in the lower part of Fig. 4.3 (thin line). Since at high temperatures $T \gg T_C$, the spin-correlation function $S(\Gamma, \boldsymbol{k}, \omega)$ is $\boldsymbol{k}$-independent, the scattered intensity is obtained from (4.11, 14) by multiplying the one-phonon density of states by the square of the form factor $|X(\boldsymbol{k}j)|^2$ representing the electron-phonon matrix element, which according to (4.6) is proportional to $\left|\sum_{\gamma=1}^{3} \sin(k_\gamma a)\right|^2$. Hence, the dominant contributions to the symmetry-projected one-phonon density of states arise from the zone boundary, particularly from the LO phonons near the L point ("breathing mode"). The calculated spin-disorder induced one-phonon Raman scattering intensity of EuS shown in Fig. 4.3 (lower part, thick line) reproduces about one third of the experimental FWHM $= 30\ \mathrm{cm}^{-1}$. Another third of the experimental FWHM can be explained by including broadening due to dynamic spin disorder as discussed in Sect. 4.4.2. This additional dynamic-spin disorder broadening of about $10\ \mathrm{cm}^{-1}$ has been estimated by assuming a Gaussian [4.12, 42a] for the spectral shape function and considering *nn* and *nnn* exchange interactions [4.48b].

In the case of EuO this estimate is in good agreement with the spin fluctuation rate measured by neutron scattering for wave vectors near the zone boundary [4.42b]. The FWHM of the spin-disorder induced Raman scattering from optic phonons of EuO at $T_C = 69$ K can be explained by the weighted one-phonon density of states [4.18] broadened by dynamic spin disorder [4.48b]. The 30% larger experimental FWHM at 300 K points to anharmonic phonon effects. The latter may also explain the 30% discrepancy between experimental and calculated FWHM of the spin-disorder induced optic-phonon scattering of EuS at 300 K. For this type of Raman scattering the nonnegligible broadening effect due to dynamic spin disorder has been emphasized by *Safran* et al. [4.12].

As another representative example we show in Fig. 4.4 (upper part) the measured one-phonon Raman spectra of EuTe at 300 K. The spectrum in the $\Gamma_{25}^{+} + \Gamma_{15}^{+}$ scattering configuration shows good agreement with the calculated spin-disorder induced one-phonon Raman spectrum in the lower part (thick line). In particular the contributions form the LO as well as the LA phonon branches are clearly exhibited in both the experimental and theoretical spectra. The results for the cases of EuO and EuSe are described in [4.18].

4.4.2 Coupling Constants and Resonance Enhancement

The different contributions from the one phonon-one spin or two-spin excitation processes in (4.2) to the scattering cross section, as given by the symmetry independent coupling constants $|A(\Gamma)|^2$ in (4.11). can be determined quantitatively from the resonance enhancement described in (4.31–34). A systematic analysis of the resonance enhancement of first-order Raman scattering has been performed for EuS and EuSe [4.13, 17] between 1.5 and 2.8 eV using all available lines of Ar^+ and Kr^+ lasers. The resonance enhancement of the antisymmetric (Γ_{15}^{+}) and symmetric ($\Gamma_{12}^{+}, \Gamma_{25}^{+}$) Raman tensor components of EuS at 300 K is shown in Fig. 4.5 as a function of exciting laser frequency. The Γ_1^{+} component has been found to be zero within experimental errors, in agreement with the conclusions in Sect. 4.2 [(4.12)]. The solid and dashed lines represent the theoretical resonance curves obtained by using for the coupling constants A^2 in (4.11) those from the microscopic theory in (4.31–33). Hence the resonance enhancement is determined by the spin-orbit coupling constant, the Clebsch-Gordan coefficients, and the frequency dependence in terms of susceptibility derivatives. The latter have been calculated using experimental data of the complex linear electric susceptibility $\chi(\omega)$ [4.13, 18]. The theoretical resonance curves have all been scaled to the experimental data by the same arbitrary factor.

The Γ_{15}^{+} component in Fig. 4.5 shows a resonance enhancement over more than two orders of magnitude and gives the dominant contribution to first-order Raman scattering compared to the Γ_{12}^{+} and Γ_{25}^{+} components. The latter two are weaker in intensity because they appear in a higher-order [one phonon-two spin (4.8)] process compared to Γ_{15}^{+} [one phonon-one spin (4.6)]. The resonance curve

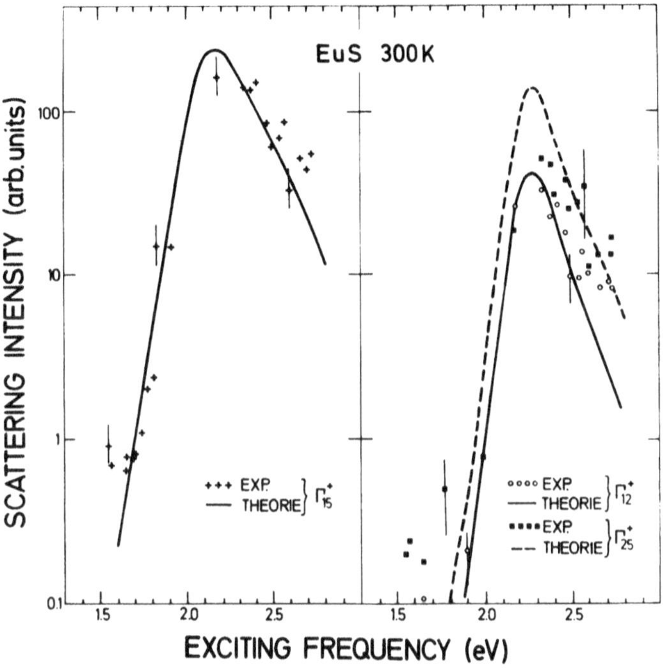

Fig. 4.5. Resonance enhancement of first-order Raman scattering of EuS at 300 K for the antisymmetric (Γ_{15}^+) and symmetric (Γ_{12}^+, Γ_{25}^+) Raman tensor components. The full symmetric (Γ_1^+) component is zero within experimental errors. The *solid* and *dashed lines* are the result of the microscopic theory (see text)

of the Γ_{12}^+ and Γ_{25}^+ components in Fig. 4.5 is narrower, i.e., more resonantly enhanced compared to that of the Γ_{15}^+ component, in agreement with the higher-order susceptibility derivative in (4.32–33) compared to (4.31). The ratio of the Γ_{12}^+ and Γ_{25}^+ components is determined only by Clebsch-Gordan coefficients without any adjustable parameter. On the other hand, the ratio of the Γ_{15}^+ and Γ_{12}^+ or Γ_{25}^+ components contains the spin-orbit coupling constant λ_f according to (4.31–33). Hence we have obtained $\lambda_f \sim 0.3$–0.4 eV [4.18]. For comparison, the free ion value of Eu^{2+} is $\lambda_f = 0.17$ eV [4.44]. The good agreement supports our theoretical model description of the scattering mechanism.

4.4.3 Second-Order Raman Scattering

Contrary to the symmetry forbidden first-order Raman scattering from phonons in the EuX series, the second-order scattering is allowed and appears for nonresonant laser excitation [4.17]. As an example we show the second-order Raman spectrum of EuS for 7993 Å laser excitation in the $\Gamma_1^+ + 4\Gamma_{12}^+$ configuration (upper part of Fig. 4.6, solid line). This spectrum with a cutoff near the $2\mathrm{LO}(\Gamma)$ frequency (534 cm^{-1} [4.49]) and a broad maximum near the $2\mathrm{LO}(L)$

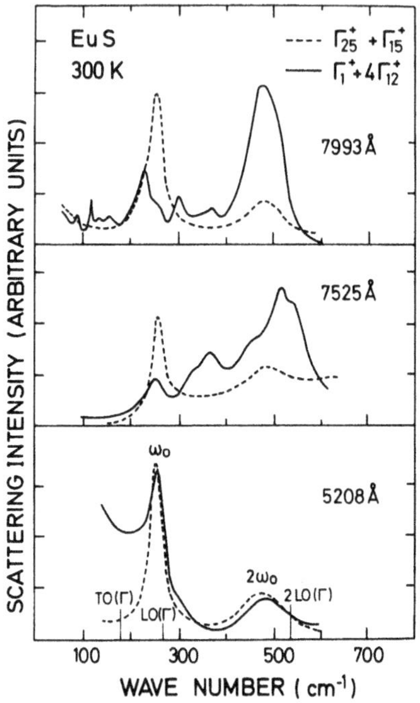

Fig. 4.6. Polarized Raman spectra of EuS at 300 K for different laser excitations below (7993 Å), near (7525 Å) and above (5208 Å) the fundamental absorption edge (7522 Å)

frequency (480 cm^{-1} [4.18, 48a]), is very similar to that of SrO [4.50]. The $\Gamma_1^+ + 4\Gamma_{12}^+$ spectrum shows qualitative agreement with the calculated two-phonon density of states [4.18] concerning peak positions, but not intensities. Further comparisons throughout the EuX series are given in [4.18]. The rich, detailed structure of the $\Gamma_1^+ + 4\Gamma_{12}^+$ spectrum disappears as the exciting laser line approaches the maximum of the $4f^7 \to 4f^6 5d(t_{2g})$ absorption band near 5400 Å [4.51], as seen in Fig. 4.6 for the 7525 Å and 5208 Å laser lines. For the latter the $\Gamma_1^+ + 4\Gamma_{12}^+$ and the $\Gamma_{25}^+ + \Gamma_{15}^+$ spectra become very similar, both exhibiting the typical $n\omega_0 = n\omega_{LO(L)}$ overtone sequence. This illustrates again that the Γ_{12}^+ component (solid line in Fig. 4.6) becomes resonant in a much narrower excitation energy range compared to the Γ_{15}^+ component (dashed line), which shows resonance enhancement even for 7993 Å laser excitation below the fundamental gap (7522 Å) of EuS.

The Raman scattering experiments in EuS, EuSe and EuTe using laser excitations below the fundamental gap have also helped to clarify the controversy of hot luminescence [4.10, 30, 31] versus Raman scattering as the origin of multiphonon overtone scattering. The appearance of the $n\omega_0 = n\omega_{LO(L)}$ overtone sequence of EuS in the $\Gamma_{25}^+ + \Gamma_{15}^+$ configuration for 7993 Å excitation, which lies $3\omega_{LO(L)}$ below the fundamental gap, cannot be explained by a hot luminescence model. Multiphonon scattering in rare-earth chalcogenides will be discussed in Chap. 5.

4.5 Raman Scattering in the Magnetically-Ordered Phases of EuX (X = O, S, Se, Te)

The effect of magnetic order on the phonon Raman scattering in EuX has been investigated quite extensively in numerous works [4.5–21, 30, 40, 48a, 52–55]. In EuO, EuS and EuSe, the spin-disorder induced scattering in the paramagnetic phase is subject to quenching upon cooling below the magnetic ordering temperature. The quenching is enhanced in an external magnetic field. In EuS the temperature-dependent intensity-quenching in the vicinity of the Curie temperature follows a spin correlation function [4.7, 52]. On the other hand, in the two-sublattice antiferromagnet EuTe, the first-order scattering intensity in the paramagnetic phase increases upon cooling below the Néel temperature [4.16, 55].

4.5.1 Ferromagnetic Phase

In Fig. 4.7 we show the temperature dependence of the scattering intensity of EuO in the $\Gamma_{25}^{+}+\Gamma_{15}^{+}$ scattering configuration. Cooling below the Curie temperature $T_C = 69$ K results in a strong quenching of the scattering intensity. At the same time a shift of the peak position near 411 cm^{-1} at 300 K to about 445 cm^{-1} at 5 K is observed. Near T_C one actually expects the scattering intensity to increase because of the divergence in the spin-correlation fucntion S in (4.17) due to small-k fluctuations becoming critical. However, the enhancement of wave vectors near $\boldsymbol{k}=0$, which describes the magnetically-ordered state below T_C (spin ordering wave vector), is suppressed by the large wave vector weighting function $X^2(\boldsymbol{k}j)$. The remaining, weak scattering intensity for $T < T_C$ consists of a projected one-phonon and one-magnon density of states, respectively, involving weighting factors for large (L point) and small (Γ point) wave vectors. Consequently, the region in k-space contributing substantially to the spectrum is much less restricted to the L point than for $T > T_C$. This explains why the main maximum at 5 K in Fig. 4.7 occurs at a smaller frequency than $\omega_{LO}(\boldsymbol{k}_L)+\omega_m(-\boldsymbol{k}_L)$, with $\omega_m(\boldsymbol{k}_L) = 44$ cm^{-1} the magnon frequency [4.56]. The total shift of the peak position of the first-order Raman scattering of EuO and EuS, respectively, from 300 K to 4.2 K by 17 % and 25 % is much larger than expected from magnetostriction [4.17]. The nonlinear shift near T_C and its magnitude for $T < T_C$ are direct proof of simultaneous phonon-magnon excitations.

The integrated scattering intensity of EuS as function of temperature is displayed in Fig. 4.8 by the open squares [4.7] and open circles [4.52]. As discussed above and as described by the (mean field) Ornstein-Zernike form of the two-spin correlation function [4.15, 19, 57, 58] (dashed line in Fig. 4.8), the scattering intensity should show a maximum near T_C. Good agreement with the experimental data, however, is obtained by using the normalized Ornstein-Zernike form of the correlation function [4.15, 19, 58] (solid line), which obviously describes the $k \neq 0$ ("large") wave vector fluctuations much better than the unnormalized version.

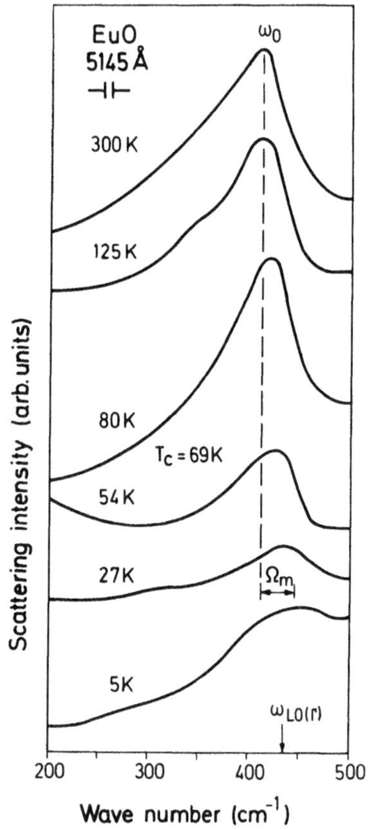

Fig. 4.7.

Fig. 4.7. Temperature dependence of first-order Raman scattering of EuO above and below the Curie temperature $T_C = 69$ K in the $\Gamma_{25}^+ + \Gamma_{15}^+$ configuration. Ω_m is the magnon frequency

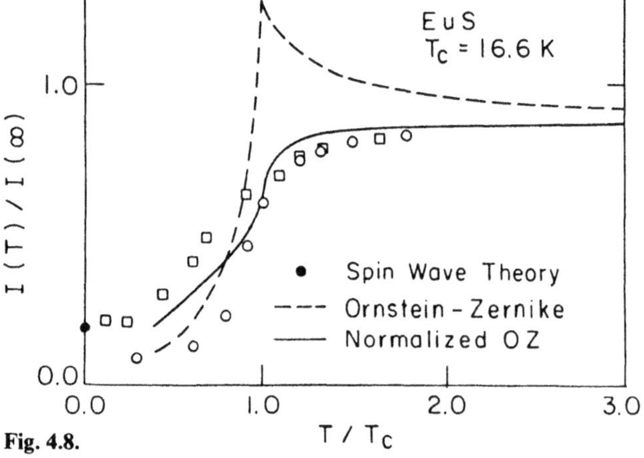

Fig. 4.8.

Fig. 4.8. Calculated integrated, normalized intensity of spin-disorder induced one-phonon Raman scattering of EuS as a function of reduced temperature [4.15, 19]. Data points: open squares from [4.7], open circles from [4.52]. Spin wave result from [4.12]

4.5.2 Magnetic "Bragg" Scattering from Spin Superstructures

Besides the inelastic phonon-magnon scattering, one expects also elastic, i.e., so-called "Bragg" scattering from spin superstructures as described by the first term in (4.17). This scattering is observed through zone folding effects for the phonon branches. Because of simultaneous spin-phonon excitations, the symmetry of the spin system (magnetic unit cell) determines that of the phonon system (chemical unit cell). First observations of spin superstructure-dependent new phonon modes have been reported by *Silberstein* et al. [4.8] for EuSe. In their work, first-order Raman spectra of EuSe were measured at 4 K in the type-I antiferromagnetic phase (AF-I) and as a function of a magnetic field in the ferrimagnetic and ferromagnetic phases. More detailed measurements for EuSe in a zero applied magnetic field as a function of temperature [4.11], also including the AF-II phase below 1.8 K [4.15, 19, 21, 59], have been performed by *Silberstein* et al. (see also Chap. 5). The assignment of the lines is in agreement with the predictions of the one phonon-one spin scattering mechanism [4.12, 19, 40]. Selection rules for the one phonon-two spin mechanism have been given in [4.21].

In Fig. 4.9 the first-order Raman spectra of EuSe are shown as a function of temperature for the four-sublattice AF-I, the three-sublattice ferrimagnetic and the two-sublattice AF-II phases (see inset) [4.19, 59]. With the spin superstructures occurring in the [111] direction, the 176 cm^{-1} peak in the AF-I phase arises from scattering of LO phonons near $1/2\boldsymbol{k}_L$, whereas the 169 cm^{-1} peak in the ferrimagnetic phase originates from LO phonons near $2/3\boldsymbol{k}_L$ [4.8, 11, 18, 19, 21, 59]. The 153 cm^{-1} peak in the AF-II phase is due to LO phonons near $\boldsymbol{k}_L$. On the basis of the zone folding effects observed in the various magnetic phases of EuSe, the phonon dispersion curves have been constructed for the [111] direction [4.11, 21]. The experimentally determined phonon frequencies for selected $\boldsymbol{k}$ values in the [111] direction of EuSe are in good agreement with the calculated phonon dispersion curves using a shell model with macroscopic input parameters [4.18]. This calculation shows good agreement with that based on a breathing shell model [4.40].

As another example of magnetic "Bragg" scattering we show in Fig. 4.10 the temperature and magnetic-field dependent Raman spectra of EuTe covering the AF-II phase. For nonresonant 6471 Å laser excitation, the $\Gamma_1^+ + 4\Gamma_{12}^+$ spectrum at 300 K exhibits second-order scattering. Cooling below the Néel temperature $T_N = 9.8$ K gives rise to the appearance of a new mode at 113 cm^{-1}. Because of the two-sublattice spin superstructure in the [111] direction and the dominant full symmetric (Γ_1^+) electron-phonon coupling (see Sect. 4.1), this mode has been attributed to the LO(L) phonon frequency [4.16, 60a]. This is consistent with the mode found at 112 cm^{-1} and its interpretation as a LO(L) phonon by *Schmutz* et al. [4.61]. The experimental values agree well with the calculated LO(L) phonon frequency [4.18, 40]. The scattering intensity of the LO(L) phonon mode at 2 K in Fig. 4.10 decreases as a function of external magnetic field. It is quenched at the critical field $H_c = 6.5$ T, which corresponds to the second-order phase

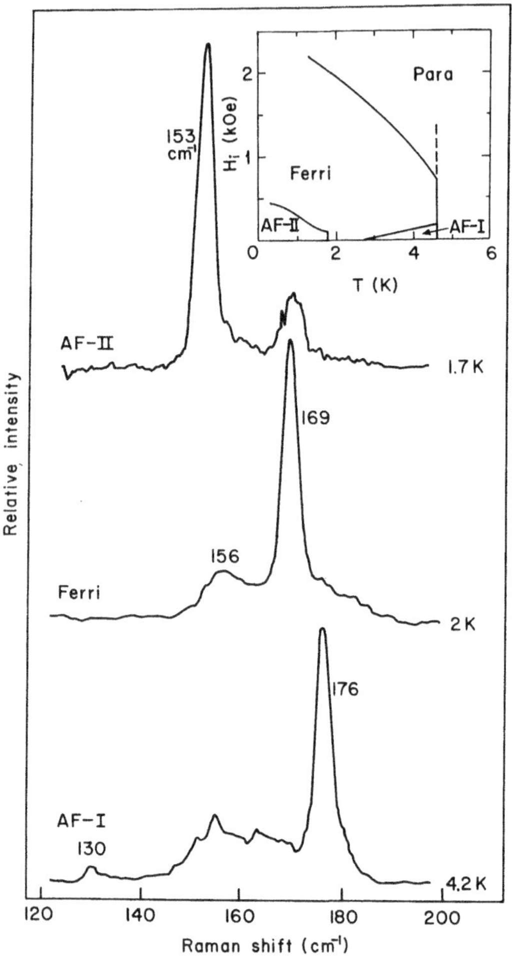

Fig. 4.9

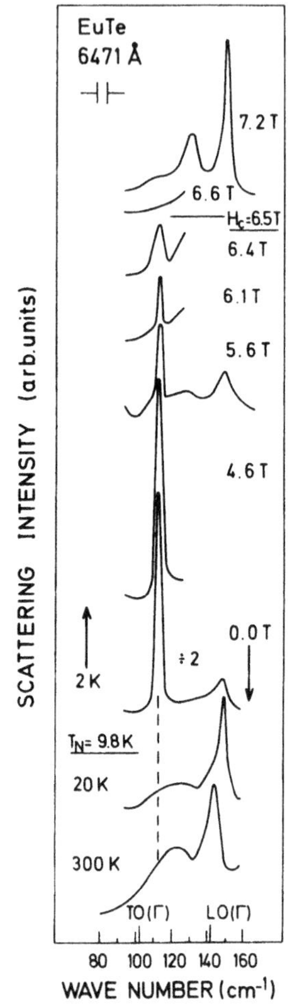

Fig. 4.10

Fig. 4.9. Zero-field Raman spectra of EuSe as a function of several temperatures in the four-sublattice antiferromagnetic (AF-I), three-sublattice ferrimagnetic and two-sublattice antiferromagnetic (AF-II) phases [4.11,21]. Inset: magnetic phase diagram of EuSe [4.73]

Fig. 4.10. Raman spectra of EuTe under nonresonant 6471 Å laser excitation in the $\Gamma_1^+ + 4\Gamma_{12}^+$ configuration as a function of temperature and applied magnetic field. The zone-center TO and LO phonon frequencies at 300 K are indicated

transition from the spin-flop to the spin-parallel aligned paramagnetic or "ferromagnetic" phase [4.62] at an actual sample temperature of 3.5 K [4.16]. The critical behavior of light scattering near this phase transition will be discussed in Sect. 4.6.

4.5.3 Resonant Raman Scattering

Under resonant laser excitation both terms of the spin correlation function in (4.17), i.e., elastic as well as inelastic scattering, can contribute to the Raman intensity. The most pronounced effect is expected for the two-sublattice antiferromagnetic phase of EuSe or EuTe, since both the magnetic "Bragg" scattering and the phonon-magnon scattering give maximum contributions for wave vectors near the L point of the Brillouin zone. The spin-disorder induced Raman spectrum of EuTe at 300 K under resonant 5145 Å laser excitation is shown in Fig. 4.11. Contrary to the cases of ferromagnetic EuO and EuS discussed in Sect. 4.5.1, no quenching of the scattering intensity, but rather an enhancement is found upon cooling below the Néel temperature $T_N = 9.8$ K. However, the narrow 113 cm^{-1} magnetic "Bragg" peak found for nonresonant excitation in Fig. 4.10 is not observed atop the broad continuum in Fig. 4.11. This is due to the strong resonance enhancement of the second term in (4.17). Besides the form factor $X^2(\boldsymbol{k}j)$, also the function $g(\boldsymbol{k})$ peaks for wave vectors near $\boldsymbol{k}_L$, since $J(0)$ in (4.16) has to be replaced by $J(\boldsymbol{k}_L)$. Moreover, the phonon-magnon shift $\omega(\boldsymbol{k}_L) + \omega_m(-\boldsymbol{k}_L)$ is negligible since $\omega_m(\boldsymbol{k}) < 4\ cm^{-1}$ [4.63a]. For 4765 Å laser excitation of EuTe at 1.8 K, the Raman spectrum has been reported to exhibit both LO(L) and LO(Γ) phonons, respectively, near 112 cm^{-1} and 145 cm^{-1} [4.61].

The nonresonant and resonant Raman scattering of EuTe in Figs. 4.10 and 4.11, respectively, has recently been interpreted by *Ousaka* et al. [4.63b] in terms of a modulation of the $5d$ spin-orbit coupling by lattice displacements. This different kind of electron-phonon coupling is believed to dominate in EuTe over the usually considered $4f^6$–$5d^1$ electron-phonon coupling in the excited

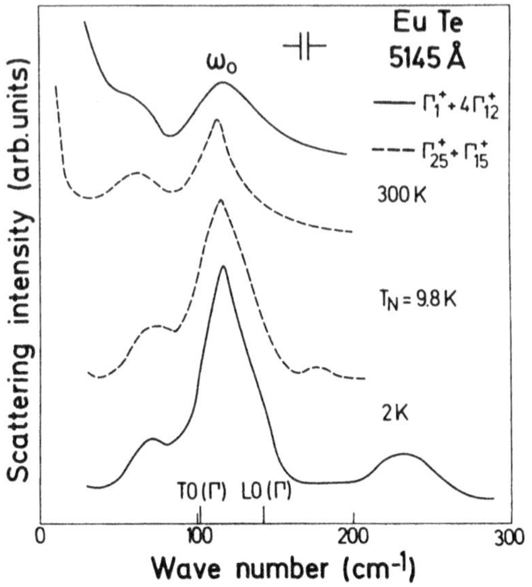

Fig. 4.11. Raman spectra of EuTe under resonant 5145 Å laser excitation above and below the Néel temperature. The zone-center TO and LO phonon frequencies at 300 K are indicated

intermediate state (see above). However, our experimental results do not confirm the conclusions of this model calculation. In nonresonant Raman scattering of EuTe the LO(L) phonon near 113 cm^{-1} does not only appear for $\boldsymbol{e} \perp \boldsymbol{e}'$, but is also observed for $\boldsymbol{e} \| \boldsymbol{e}'$, as shown in Fig. 4.10 [4.60a, b]. Moreover, no increase of the scattering intensity is observed in a magnetic field (Fig. 4.10). In [4.18] the very general form of the electron-phonon interaction has been considered by the difference between the f- and d electron-phonon coupling. A consistent description of the Raman data of the EuX series could be given by omitting the d electron-phonon coupling. Furthermore, the $4f^6$ spin-orbit coupling term is by a factor six larger than that of the $5d$ electron [4.63b]. The dominant coupling between the localized $4f$ hole and lattice displacements is also demonstrated for a variety of other, i.e., nonmagnetic or intermediate-valent rare earth compounds in Chap. 5.

An interesting feature of the resonant Raman scattering of EuTe in the antiferromagnetic (AF-II) phase develops as a function of magnetic field. The unpolarized Raman spectrum of EuTe at 2 K in Fig. 4.12 exhibits scattering intensity near $\omega_0 = \omega_{\rm LO}(\boldsymbol{k}_{\rm L})$ and its overtone $2\omega_0$. As a function of magnetic field, a peak gradually evolves near $\omega_{\rm LO}(\boldsymbol{k}=0)$ and $2\omega_{\rm LO}(\boldsymbol{k}=0)$. For fields above the critical value $H_{\rm c} = 6.5$ T, the LO(Γ) phonon scattering and its overtone become resonantly enhanced and dominate the spectra. The integrated peak intensity of the LO(Γ) phonon scattering as a function of applied magnetic field

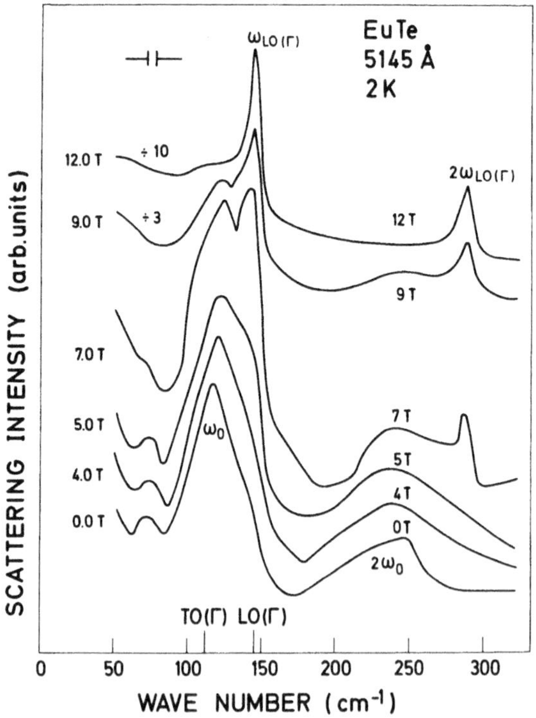

Fig. 4.12. Raman spectra (unpolarized) of EuTe for resonant 5145 Å laser excitation as a function of applied magnetic field $H \| [001]$. The zone-center TO and LO phonon frequencies at 300 K are indicated

is shown in Fig. 4.13 [4.8]. A saturation is indicated near 120 kOe. Such "forbidden" LO(Γ) phonon scattering can be allowed under resonance conditions due to the Fröhlich electron-phonon interaction associated with finite wave-vector effects of the phonons [4.64]. The essential condition for its occurrence here is that the magnetic field establishes full translational symmetry, determined by $\boldsymbol{k}=0$ spin-ordering wave vectors describing the ferromagnetic phase. Hence, there is no difference between the "forbidden" resonance Raman scattering from LO(Γ) phonons and their overtones in the ferromagnetic phases of europium chalcogenides and in diamagnetic ytterbium chalcogenides [4.32]. A more detailed discussion of this point will be given in Chap. 5.

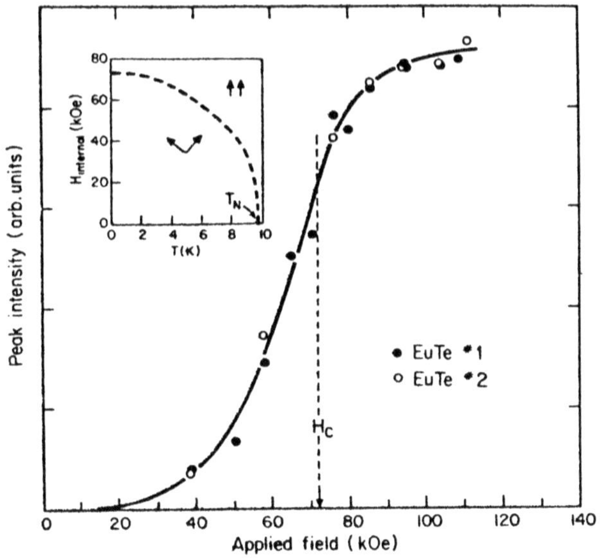

Fig. 4.13. Integrated peak intensity of "forbidden" LO(Γ) phonon scattering vs applied magnetic field ($H \| [100]$) in EuTe at 2 K using 5145 Å laser excitation [4.8]. Inset: magnetic phase diagram of EuTe [4.62,68]

The resonance Raman scattering of LO(Γ) phonons as a function of exciting laser frequency has been investigated in EuS, EuSe and EuTe [4.8, 21]. Unfortunately, the resonance curve could not be measured for a single material over a wide frequency range. Instead, for a fixed, limited excitation frequency range, the systematic shifts in energy of the electronic band structure from EuS to EuTe had to be employed in order to assemble the resonance curve. Strong resonance enhancement has been found near the E_1 and E_1'' magnetoreflectance peaks [4.65]. From the fit of the resonance curve by an excitonic enhancement mechanism [4.66], strong support has been concluded in favor of the localized magnetic exciton model as compared to the one-electron band model [4.21].

4.6 Spin Fluctuations near Magnetic Phase Transitions

The wave vector dependence of Raman scattering in the different magnetic phases of EuX has given evidence for spin-fluctuation effects near magnetic phase transitions. A prominent example concerns the transition from the paramagnetic into the four-sublattice AF-I phase of EuSe [4.19, 21, 59, 67]. The inset [4.11, 21] of Fig. 4.14 shows at $T=32$ K $> T_N=4.6$ K, the spin-disorder induced Raman scattering near 151 cm^{-1} due to LO(L) phonons. Upon cooling towards T_N, a broad shoulder develops above this maximum which appears at $T = T_N$ near 176 cm^{-1}. This shoulder has been correlated with large fluctuations near the (AF-I) spin-ordering wave vector $\boldsymbol{k}_L/2$ [4.19, 21, 59]. The Raman line shapes have been calculated using both the quasielastic and mean-field approximations for the two-spin correlation function [4.19, 59] and are shown in Fig. 4.14 for $T > T_N$ and $T = T_N$. They are in good qualitative agreement with the experimental observations. Hence, the Raman spectrum near T_N is not only determined by the form factor $X^2(\boldsymbol{k}j)$ with its maximum near $\boldsymbol{k}_L$, but also by the large spin fluctuations near $\boldsymbol{k}_L/2$.

The above example shows that the spin-disorder induced Raman scattering is sensitive to spin fluctuations. On the other hand, it has been pointed out by *Safran* et al. [4.15, 19, 59] that the sharp lines due to magnetic "Bragg" scattering measure the magnetic-order parameter. From the first term on the right side of (4.17 or 18), it is evident that the intensities of the "Bragg" peaks associated with

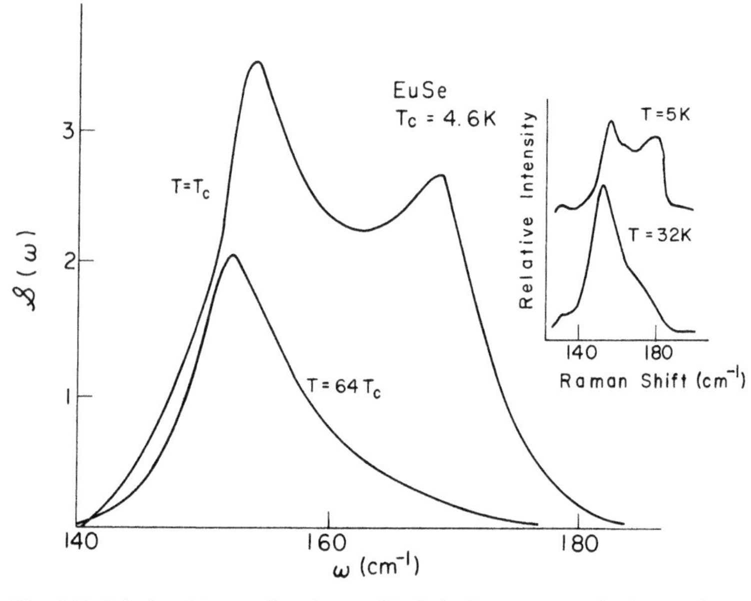

Fig. 4.14. Calculated Raman line shapes of EuSe in the paramagnetic phase and near the transition into four-sublattice AF-I phase [4.19, 59]. Inset: measured spin-disorder induced one-phonon Raman scattering in the paramagnetic phase ($T=32$ K) and near the paramagnetic to AF-I phase transition ($T=5$ K) [4.11, 21]

the one phonon-one spin mechanism are proportional to the square of the staggered magnetization $N = \langle \sigma_k(\Gamma_{15}^+) \rangle$. Thus, the "Bragg" peak intensity is a direct measure of the temperature dependence of the order parameter, which varies as $N \sim (T_N - T)^\beta$ near second-order phase transitions and is discontinuous at T_N for first-order transitions. Of particular interest is the second-order spin-flop to spin-parallel aligned paramagnetic ("ferromagnetic") transition of EuTe occurring at the critical field given by $H = H_c(T)$ [4.62, 68]. Near $H_c(T)$ the scattering intensity is given by $I \sim N^2 \sim [H_c^2(T) - H^2]^{2\beta}$, where H is the applied magnetic field. Hence, a determination of the critical exponent β should be feasible. This is of particular relevance in view of predictions by renormalization group calculations of a quantum crossover from $\beta = 1/2$ mean-field behavior near $T = 0$ [4.69] to $\beta = 0.36$ critical behavior for $0 < T < T_N$ near $H_c(T)$ [4.70, 71]. Experimental attempts in this direction [4.60a] have pointed out the experimental difficulties in accurately determining the decreasing scattering intensity for H approaching H_c.

An intriguingly interesting spin-fluctuation effect has been observed in EuSe for $T \approx T_N$ in terms of an unexpected, sharp "Bragg" peak near 118 cm^{-1} [4.21, 67]. This mode is quenched upon cooling below $T_N = 4.6$ K, i.e., in the four-sublattice AF-I phase. It also disappears at $T > T_N$ in an applied magnetic

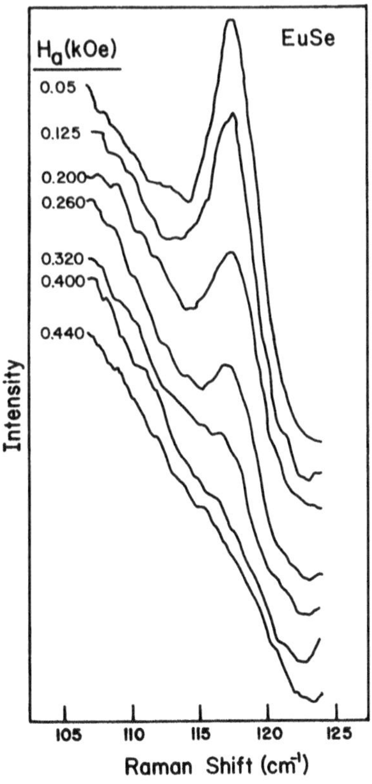

Fig. 4.15. Magnetic field dependence of the 118 cm^{-1} "spin-fluctuation mode" in EuSe at 5.0 K using 6764 Å laser excitation and right-angle geometry $y(xx+xy)z$ [4.21, 67]. The curves with identical intensity scale have been shifted vertically for clarity

field up to 0.4 kOe (Fig. 4.15). The scattering intensity is strongly polarized for parallel electric field vectors of incident and scattered photons. This mode near 118 cm^{-1} has been identified as the LA(L) phonon, in good agreement with the calculated phonon dispersion curves [4.18, 40].In connection with AF-I-phase spin fluctuations, this mode can only be allowed for the one phonon-two spin mechanism [4.21]. This mechanism, which is usually related to a modulation of the exchange interaction by vibrations of the magnetic ions (see also Sect. 4.8), has been considered in the framework of a spin-lattice coupling model [4.72]. The latter has been used to explain the magnetostriction and the stabilization of the spin structures of EuSe. In this model the attraction between spin-parallel (111) planes and the repulsion between spin-antiparallel (111) planes is consistent with the lattice vibrations associated with the LA(L) phonon mode and provides its strong coupling to the AF-I spin ordering scheme. Very small applied magnetic fields have been considered to align the spins, leading to a quenching of the LA(L) mode scattering [4.21, 67].

4.7 Cadmium-Chromium (Cd-Cr) Spinels ($CdCr_2X_4$, X = S, Se)

The observation of a strong intensity increase for some of the Raman lines of $CdCr_2S_4$ and $CdCr_2Se_4$ upon cooling below the Curie temperature [4.22, 74] had originally stimulated the search for spin-dependent Raman scattering from phonons in magnetic semiconductors. The theoretical description [4.23] had been based on a phenomenological concept by *Moriya* [4.75] in expanding the spin-dependent polarizability tensor of a magnetic material in terms of spin components. To account for the spin-dependent Raman scattering from phonons, the polarizability tensor had been further expanded in powers of the ionic displacements with respect to the equilibrium positions [4.23]. This concept laid the basis for all phenomenological treatment of spin-phonon excitations in magnetic semiconductors. For the case of one-phonon Raman scattering in Cd-Cr spinels, only the spin-independent term and the phonon-modulated isotropic nearest-neighbor exchange interaction have been considered [4.23]. The last term has been attributed to the variation of the d electron transfer energy due to the relative displacement of the ions. This intermediate-state interaction had been considered under the assumption that the absorption edge corresponds to a charge transfer transition from the (Cr) d level to the conduction band.

However, strong changes in the scattering intensity of phonon modes in nonmagnetic $CdIn_2S_4$ [4.29] as a function of temperature (fixed laser frequency) or laser frequency (fixed temperature) indicated that magnetic-order induced changes in the electronic band structure of Cd-Cr spinels may simply give rise to resonance effects. This was then demonstrated for $CdCr_2S_4$, for which the resonance Raman scattering has been related to the magnetic-order induced "red shift" in the absorption spectrum [4.24]. Figure 4.16 shows the Raman spectra of $CdCr_2S_4$ at 15 K for various excitation wavelengths. In addition to the

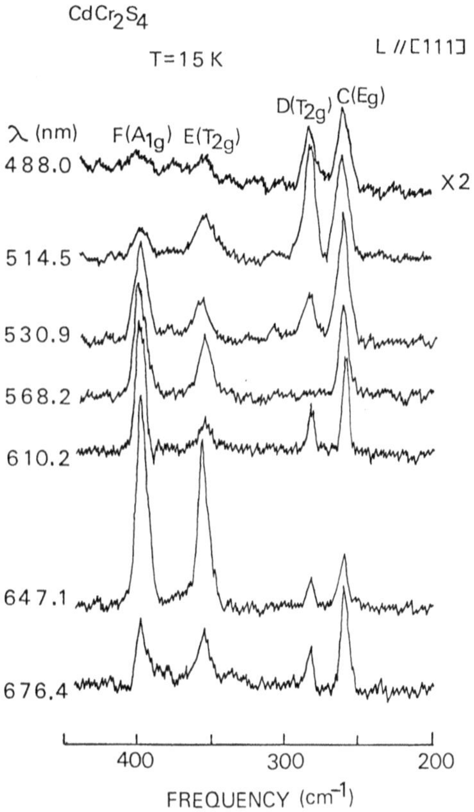

Fig. 4.16. Raman spectra of $CdCr_2S_4$ at 15 K for various excitation wavelengths and parallel incident and scattered electric-field vectors. From [4.24]

four phonon lines shown, another one (A) appears near 105 cm^{-1}, constituting altogether the 12 Raman-active modes: A_{1g}, E_g, $3T_{2g}$. Obviously, the scattering intensity of the individual phonon modes varies strongly as a function of excitation wavelength. Unfortunately, in the analysis of the integrated intensities, only their ratio with respect to that of line C has been considered. This normalization to a constant intensity of the C line, however, has to be handled with caution: the integrated C-line intensity itself exhibits a well-pronounced resonance near 2.0 eV excitation energy [4.25]. *Koshizuka* et al. [4.24] demonstrated for $CdCr_2S_4$ that for 6471 Å laser excitation, the F/C intensity ratio has a temperature dependence which follows qualitatively a spin-correlation function $\langle S_1 \cdot S_2 \rangle / S^2$, but that a temperature-independent intensity ratio is obtained for 5145 Å excitation. Similar effects have also been observed in $CdCr_2Se_4$ for the D/C- and F/C-line intensity ratios [4.76] and for the integrated C- and D-line intensities in $HgCr_2Se_4$ [4.77]. In the latter case it has been pointed out that the temperature-dependent variations of the Raman intensities for different excitation energies cannot be fitted by either the spin correlation function or by its sum with a spin-independent term predicted by theory [4.23].

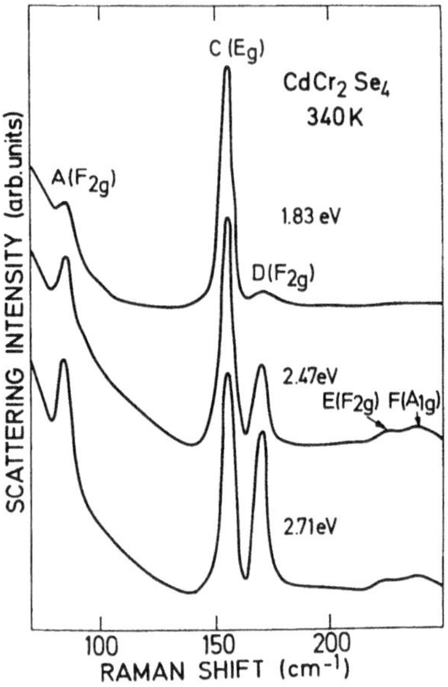

Fig. 4.17. Unpolarized Raman sepctra of $CdCr_2Se_4$ at 340 K for various laser excitation energies. The intensities have been normalized to that of the C line [4.25]

In Fig. 4.17 we show the Raman spectra of $CdCr_2Se_4$ at 340 K for three different excitation energies. While the integrated intensity of line C does not change much, those of line D and line A increase strongly with increasing excitation energy. The behavior of line D in Fig. 4.17 is qualitatively similar to the increase of the D/C-line intensity ratio upon cooling below $T_C = 130$ K for 6328 Å (1.96 eV) laser excitation [4.22]. Hence, a (qualitative) equivalence exists between resonance effects occurring at a fixed temperature as function of excitation energy and at fixed excitation energy but with varying temperatures.

Systematic studies of the resonance Raman scattering using intensity calibration with respect to CaF_2 have been performed for $HgCr_2Se_4$ [4.77] and $CdCr_2Se_4$ [4.25]. The resonance curve of $HgCr_2Se_4$ at 300 K for lines C and D exhibits two maxima near 1.96 and 2.35 eV excitation energy, with an increase toward lower photon energies. Although this resonance curve has not been corrected for absorption and reflection losses, it actually shows a strong similarity to the calculated $|d\varepsilon/d\hbar\omega|^2$ curve of $CdCr_2Se_4$ [4.25] (see below).

The resonance in the Raman efficiency vs photon energy for the C- and D-lines of $CdCr_2Se_4$ is shown in Fig. 4.18 for temperatures above and below $T_C = 130$ K [4.25]. The resonance curve at 340 K shows a maximum near 2.0 eV with a halfwidth of 0.2–0.3 eV, which increases to about 0.5 eV at $T = 110$ K. The arrows in Fig. 4.18 indicate the position of a prominent structure in the thermoreflectance spectrum near 2.03 eV at 340 K which undergoes a splitting for $T \lesssim T_C = 130$ K [4.78]. The broadening of the resonance curve has been ascribed to the magnetic-order induced exchange splitting of the electronic band structure. The temperature-dependent changes of the Raman intensities are thus

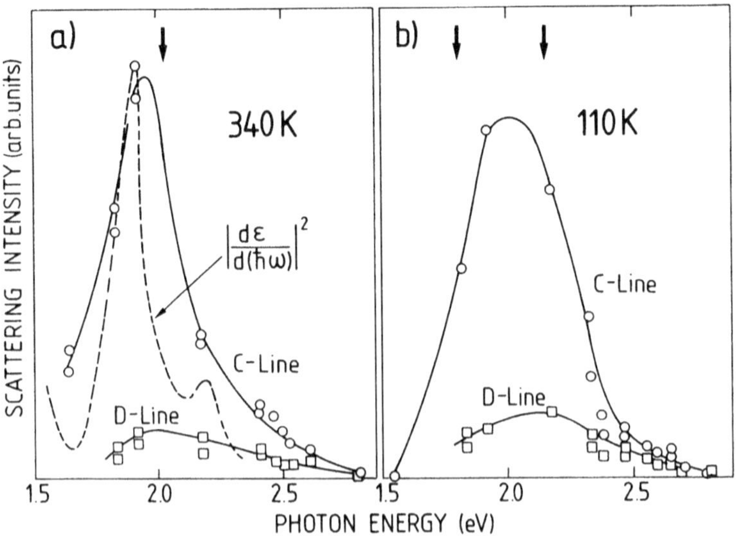

Fig. 4.18a, b. Resonant Raman scattering of the C (E_g) and D (F_{2g}) phonons of $CdCr_2Se_4$ at 340 K (**a**) and 110 K (**b**) [4.25]. The arrows indicate the prominent structure in the thermoreflectance spectra near 2.03 eV at 340 K (**a**) which undergoes a splitting for $T \lesssim T_C = 130$ K (**b**) [4.78]

attributed to temperature-induced changes of the resonance conditions, depending on the exciting photon energy.

The resonance curve at 340 K in Fig. 4.18 shows good agreement with the $|d\varepsilon/d\hbar\omega|^2$ curve (dashed line) calculated using optical data for $\varepsilon(\omega)$ from *Itoh* et al. [4.79]. This indicates that a two-band scattering process takes plane [4.1, 4a].

The maximum of the resonance curve at 340 K coincides with the direct interband gap near 2.0 eV [4.79–86]. A gap of 1.8 eV between a p-valence band and an s-conduction band has been recently identified by photoluminescence [4.87]. Since the exchange splitting of the s-conduction band has been reported to be 3.5 meV [4.88], the broadening of the resonance curve in the ferromagnetic phase (Fig. 4.18) must be due to the exchange splitting of the p-valence band. Such a spin-dependent electronic band structure has been proposed by *Goodenough* [4.89]. The relatively large exchange splitting of the p-valence band may be due to its hybridization with the $(Cr^{3+})3d^3$ states, which have been identified near 1.45 eV below the top of the p-valence band [4.90].

Resonance Raman scattering of $CdCr_2S_4$ has been studied near the "red-shifting" transition ($\sim$1.9 eV) using circular polarized laser excitation in the Faraday configuration [4.91, 93]. The magnetic-circular polarization effects of the phonon Raman scattering intensities have been related to the magnetic-circular dichroism of the electronic states involved in the resonance.

4.8 Vanadium Dihalides

Raman scattering in the 3*d*-metal dihalides with layered structure has been investigated quite extensively as can be seen from a recent review by *Lockwood* [4.94]. Besides the characterization of the Raman-active phonons, these studies were aimed at detecting magnons and at exploring the ground state and crystal field levels of the 3*d*-metal ion by electronic Raman scattering. Investigations of the layer-structure vanadium dihalides VX_2(X=Cl, Br, I), in particular, have focussed on the influence of magnetic order on the phonon Raman scattering [4.27, 28]. The vanadium dihalides crystallize in a layered structure of the CdI_2 type, consisting of hexagonal sheets of V^{2+} ions sandwiched between hexagonal sheets of anions. While VI_2 orders antiferromagnetically at $T_N = 15$ K, VCl_2 and VBr_2 do not show any long-range magnetic order down to 4.2 K.

The antiferromagnetic insulator VI_2 has been of particular interest with respect to spin-dependent phonon Raman scattering. In Fig. 4.19 we show the Raman spectrum of VI_2 under 4765 Å laser excitation as a function of temperature. At 300 K the two Raman-active (E_g, A_{1g}) phonons are observed. Cooling below the Néel temperature $T_N = 15$ K results in the abrupt appearance

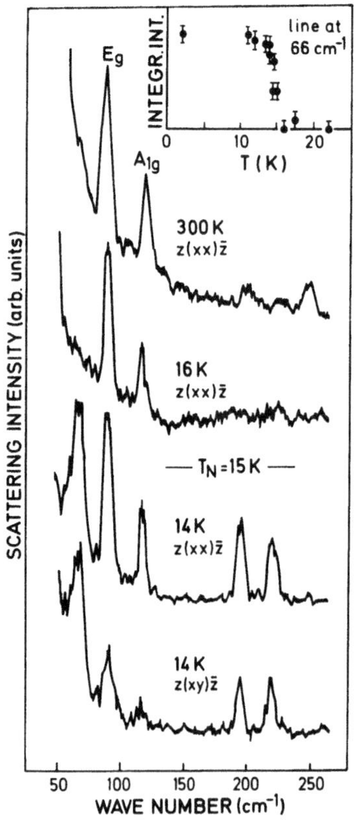

Fig. 4.19. Raman spectra of VI_2 for 4765 Å laser excitation as a function of temperature. The inset shows the temperature dependence of the integrated intensity of the 66 cm^{-1} line in $z(xx)\bar{z}$ geometry [4.27]

of three new lines at 66, 195 and 220 cm^{-1}. These have been attributed to inelastic scattering from zone-boundary phonons which become Raman active due to zone-folding effects induced by the elastic ("Bragg") magnetic scattering from the antiferromagnetic spin superstructure. The microscopic spin-dependent electron-phonon interaction has been identified as a modulation of the magnetic exchange interaction by selected phonon modes. The interaction Hamiltonian is given by [4.27, 28]

$$H_{\mathrm{ep}} \sim \boldsymbol{w}(+|\boldsymbol{q}\lambda) \sum_n \frac{1}{r_n}\left(\frac{\partial J_n}{\partial r_n}\right)\left(\sum_i^{(n)} \boldsymbol{r}_{i,l}\boldsymbol{S}_i\boldsymbol{S}_l\right), \tag{4.35}$$

where $\boldsymbol{w}$ is the phonon polarization vector for wave vector $\boldsymbol{q}$ and branch index λ associated with the vanadium site $(+)$, J_n is the exchange constant between spins assumed to depend only on the V-V distance $|\boldsymbol{r}_n| = |\boldsymbol{r}_i - \boldsymbol{r}_l| = |\boldsymbol{r}_{i,l}|$, with l the cell index $(l = 1, 2, \ldots N)$ and i running over the nth shell of spins at a distance r_n. $\boldsymbol{S}_i$ is the V^{2+} ion spin. The net nonzero phonon modulation of the exchange interaction has been determined by inspection of the spin arrangement.

The periodicities of the antiferromagnetic spin superstructure have been determined by neutron scattering [4.95, 96] as $|\boldsymbol{a}_m| = 2|\boldsymbol{a}|$, $|\boldsymbol{b}_m| = \sqrt{3}|\boldsymbol{a}|$ and $|\boldsymbol{c}_m| = 2|\boldsymbol{c}|$, with a and c the hexagonal lattice constants. The points in the crystallographic BZ subject to folding into the zone center due to the interaction in (4.35) are $\boldsymbol{M} = (0, 2\pi/\sqrt{3}\mathrm{a}, 0)$, $\boldsymbol{K}' = (\pi/a, 0, 0)$ and $\boldsymbol{A} = (0, 0, \pi/c)$. The scalar product in (4.35) selects only A_u modes at the M point whose eigenvectors $\boldsymbol{w}(+|M, A_u)$ are polarized along the x-axis. The displacement pattern of the A_u acoustic and optic modes reveal indeed a net nonzero modulation of the exchange interaction [4.28]. The calculated frequencies of the two A_u modes at the M point at 65 cm^{-1} and 223 cm^{-1} [4.27, 28, 97] are in good agreement with two of the three new Raman frequencies (66 and 220 cm^{-1}) in Fig. 4.19. Because of the reduction of the D_{3d} point group symmetry of the crystallographic unit cell to the C_{2h} subgroup of the magnetic unit cell, the A_u modes have then even symmetry, actually being transformed into a mixture of A_g and B_g modes. The origin of the third line at 195 cm^{-1} could not yet be clarified unambiguously. There exists an intriguing coincidence of this line with the $E_{u,x}$ optical mode at the A point (199 cm^{-1}). The activation of this mode due to the interaction in (4.35), however, depends on the stacking of spins along the c-axis and would call within the same magnetic unit cell [4.95] for a slightly different spin orientation [4.27]. The 195 cm^{-1} line may also be due to the combination $A_{1g}(\Gamma) + A_u(M)$.

The observation of only three new modes out of 36 Raman-active modes for the antiferromagnetic spin superstructure of VI_2 emphasizes that the phonon-modulated exchange interaction is an apparently rare spin-dependent phonon Raman scattering process, despite its simplicity. This process, previously proposed for the magnetic-order dependent phonon Raman scattering in ferromagnetic Cd-Cr spinels [4.23], is obviously related to rather complex spin superstructures, although its coupling strength is comparable to the orbital electron-phonon interaction [4.27].

Raman scattering in VBr_2 and VCl_2 has been attributed to a one-phonon density of states induced by the long-range spin disorder which is weighted by a q-dependent spin-phonon coupling associated with local magnetic order [4.28]. The shift of the observed spectra to lower frequencies with respect to the calculated, unrenormalized one-phonon density of states has been explained by a self-energy renormalization due to the spin-phonon coupling mechanism.

4.9 Conclusions

In this chapter we have emphasized the important role of simultaneous spin-phonon excitations for the wealth of diverse phenomena observed in light scattering in magnetic semiconductors. A systematic description of this diversity has been given by a phenomenological theory with respect to symmetry selection rules and wave vector dependence of the scattering intensity, and by a microscopic theory concerning coupling constants and their frequency dependence. Various spin-dependent phonon Raman scattering mechanisms have been discussed and shown to exist in different materials and/or magnetic phases. So far scattering mechanisms associated with spin-orbit coupling and phonon-modulated ion-ion exchange interactions have been identified experimentally. In rare-earth compounds, the strength of the spin-orbit coupling mechanism exceeds even that of the exciton-LO phonon Fröhlich interaction. This is evidenced by the spin-disorder induced scattering in the paramagnetic phase, which even under resonance conditions, dominates the "forbidden" LO(Γ) phonon scattering.

Other possible scattering mechanisms have been discussed theoretically [4.14, 23]. Among these are (i) the phonon modulation of the spin-orbit coupling constant λ, depending on the degree of localization of the electron wave function, (ii) the phonon-modulated ion-position dependent exchange interaction between an excited conduction electron and an ion spin, and (iii) the phonon modulation of the transfer energy of electrons from the partially filled, magnetic, inner shells. Generally speaking, any term of the spin-dependent Hamiltonian which is a function of the ion positions involves phonons and consequently spin-phonon excitations. The "internal" modulation of the electric susceptibility of magnetic semiconductors by phonons and spin excitations makes resonant Raman scattering a useful tool for studying their electronic structure. In particular, the frequency dependence of the scattering cross section has been found to be proportional to the square of the magnitude of the derivative of the spin-dependent electric susceptibility.

The $4f$ electron-phonon interaction investigated by Raman scattering in the europium chalcogenides has served as essential input for the understanding of the lattice dynamics and Raman intensities of valence-fluctuating rare-earth compounds described in Chap. 5. Hence it was also feasible to identify in the magnetic semiconductor Eu_3S_4 the breathing mode of the S ions with respect to the valence fluctuating Eu ion [4.98] which was expected to be strongly involved in the $Eu^{2+} \leftrightarrow Eu^{3+}$ "valence hopping" [4.99].

While Raman studies of magnetic semiconductors have been quite extensive, there have been only a few measurements of Brillouin scattering. Bulk magnons have been investigated in $CrBr_3$ [4.100]. The first observation of surface magnons has been reported for EuO [4.101]. These have also been observed in EuS [4.102]. A most desirable application of Brillouin scattering would be to study quasielastic scattering from spin fluctuations in magnetic semiconductors.

The spin-disorder induced Raman scattering observed in magnetic semiconductors is another example of disorder-induced scattering phenomena due to either orientational disorder [4.103], structural disorder [4.104] or defects [4.105–107] in solids. Magnetic "Bragg" scattering from spin superstructures has its analogy in zone-folding effects observed in superlattices [4.108].

Acknowledgements. We are very grateful to R. Merlin, W. Kress, P. Grünberg, and H. Bilz for many stimulating discussions. It is a particular pleasure to thank R. Merlin for the fruitful cooperation in large parts of the experimental work, and P. Grünberg for his initiative and participation in the early stages of the experiments. We would like to express our gratitude to W. Kress for his expertise and help with the lattice dynamical calculations. The cooperation of A. Frey, M. Iliev, E. Anastassakis, E. Zirngiebl, F. Canal, and G. Abstreiter is gratefully acknowledged. We are obliged to K. Fischer, W. Zinn, and F. Holtzberg for providing us with the samples. We thank H. Hirt, G. Wolff, and P. Wurster for expert technical assistance. We are indebted to S. Wood for help with the manuscript.

Note Added in Proof

As already mentioned in the preface to this chapter, Raman scattering in the spin glass system $Eu_xSr_{1-x}S$ is of particular interest for observing $q \neq 0$ spin correlations as a result of simultaneous spin (q) – phonon $(-q)$ excitations. This goal, the first Raman measurements in a spin glass and their quantitative interpretation, has been achieved during the completion of this article [4.48b, 109]. $Eu_xSr_{1-x}S$ exhibits a rather unusual magnetic phase diagram [4.110]: ferromagnetism breaks down for $x \leq 0.51$; spin glass behavior is observed at low temperatures for $0.13 \leq x \leq 0.65$, and a reentrant phase boundary between the ("frustrated") ferromagnet and the spin glass is found for $0.51 \leq x \leq 0.65$. Raman measurements in all magnetic phases of $Eu_xSr_{1-x}S$ ($x = 0.65$, 0.54, 0.40, 0.05) show a *new* type of $q \neq 0$ magnetic scattering at low tempeartures. The intensity is quantitatively reproduced using an Ornstein-Zernike ansatz [4.12, 15] for the spin correlation function in (4.11), modified by an adjustable exponent y [4.111], which describes the spin system from uncorrelated ($y = 0$) via statistically coupled ($y = 2$) to ferromagnetically aligned ($y > 2$) spins. The systematic variations of y as a function of composition x and temperature in high magnetic fields up to 7 T yield a consistent, microscopic understanding of spin correlations in correspondence with the magnetic phase diagram and in good agreement with neutron scattering results [4.111, 112].

References

4.1 M.Cardona (ed.): *Light Scattering in Solids I*, 2nd ed., Topics Appl. Phys., Vol. 8 (Springer, Berlin, Heidelberg, New York 1982)
4.2 W.Hayes, R.Loudon: *Scattering of Light by Crystals* (Wiley, New York 1978)
4.3 S.Nakajima, Y.Toyozawa, R.Abe: *The Physics of Elementary Excitations*, Springer Ser. Solid-State Sci., Vol. 12 (Springer, Berlin, Heidelberg, New York 1980)
4.4a M.Cardona, G.Güntherodt (eds.): *Light Scattering in Solids II*, Topics Appl. Phys., Vol. 50 (Springer, Berlin, Heidelberg, New York 1982)
4.4b M.Cardona, G.Güntherodt (eds.): *Light Scattering in Solids III*, Topics Appl. Phys., Vol. 51 (Springer, Berlin, Heidelberg, New York 1982)
4.5 J.C.Tsang, M.S.Dresselhaus, R.L.Aggarwal, T.B.Reed: Phys. Rev. **B9**, 984 (1974)
4.6 J.C.Tsang, M.S.Dresselhaus, R.L.Aggarwal, T.B.Reed: Phys. Rev. **B9**, 997 (1974)
4.7 A.Schlegel, P.Wachter: Solid State Commun. **13**, 1865 (1973)
4.8 R.P.Silberstein, L.E.Schmutz, V.J.Tekippe, M.S.Dresselhaus, R.L.Aggarwal: Solid State Commun. **18**, 1173 (1976)
4.9 N.Suzuki: J. Phys. Soc. Jpn. **40**, 1223 (1976)
4.10 J.Vitins, J. Magnetism Magn. Mat. **5**, 212 (1977)
4.11 R.P.Silberstein, V.J.Tekippe, M.S.Dresselhaus: Phys. Rev. **B16**, 2728 (1977)
4.12 S.A.Safran, G.Dresselhaus, B.Lax: Phys. Rev. **B16**, 2749 (1977)
4.13 R.Merlin, R.Zeyher, G.Güntherodt: Phys. Rev. Lett. **39**, 1215 (1977)
4.14 O.Sakai, M.Tachiki: J. Phys. Chem. Solids **39**, 269 (1978)
4.15 S.A.Safran, R.P.Silberstein, G.Dresselhaus, B.Lax: Solid State Commun. **29**, 339 (1979)
4.16 G.Güntherodt: J. Magnetism Magn. Mat. **11**, 394 (1979)
4.17 G.Güntherodt, R.Merlin, P.Grünberg: Phys. Rev. **B20**, 2834 (1979)
4.18 R.Zeyher, W.Kress: Phys. Rev. **B20**, 2850 (1979)
4.19 S.A.Safran: J. Physique Coll. C-5, **41**, 223 (1980)
4.20 Y.Ousaka, O.Sakai, M.Tachiki: J. Phys. Soc. Jpn. **48**, 1269 (1980)
4.21 R.P.Silberstein: Phys. Rev. **B22**, 4791 (1980)
4.22 E.F.Steigmeier, G.Harbeke: Phys. Cond. Matter **12**, 1 (1970)
4.23 N.Suzuki, H.Kamimura: J. Phys. Soc. Jpn. **35**, 985 (1973)
4.24 N.Koshizuka, Y.Yokoyama, T.Tsushima: Solid State Commun. **18**, 1333 (1976)
4.25 M.Iliev, G.Güntherodt, H.Pink: Solid State Commun. **27**, 863 (1978)
4.26 N.Koshizuka, S.Ushioda, T.Tsushima: Phys. Rev. **B21**, 1316 (1980)
4.27 G.Güntherodt, W.Bauhofer, G.Benedek: Phys. Rev. Lett. **43**, 1427 (1979)
4.28 W.Bauhofer, G.Güntherodt, E.Anastassakis, A.Frey, G.Benedek: Phys. Rev. **B22**, 5873 (1980)
4.29 N.Koshizuka, Y.Yokoyama, H.Hiruma, T.Tsushima: Solid State Commun. **16**, 1011 (1975)
4.30 J.Vitins, P.Wachter: Solid State Commun. **17**, 911 (1975)
4.31 J.Vitins, P.Wachter: Physica **86–88B**, 213 (1977)
4.32 R.Merlin, G.Güntherodt, R.Humphreys, M.Cardona, R.Suryanarayanan, F.Holtzberg: Phys. Rev. **B17**, 4951 (1978)
4.33 W.Baltensperger: J. Appl. Phys. **41**, 1052 (1970)
4.34 G.Schaack: Solid State Commun. **17**, 505 (1975)
4.35 G.Schaack: Z. Physik **B26**, 49 (1977)
4.36 P.Thalmeier, P.Fulde: Z. Physik **B26**, 323 (1977)
4.37 I.W.Johnstone, L.Dubicki: J. Phys. **C11**, 121 (1980)
4.38 S.R.Andrews, R.T.Harley, I.R.Jahn, W.F.Sherman: J. Phys. **C15**, 4679 (1982)
4.39 R.A.Cowley: In *The Raman Effect*, Vol. 1, ed. by A.Anderson (Dekker, New York 1973)
4.40 Y.Ousaka, O.Sakai, M.Tachiki: Solid State Commun. **23**, 589 (1977)
4.41 R.Loudon: Adv. Phys. **13**, 423 (1964)
4.42a W.Marshall, R.D.Lowde: Rep. Progr. Phys. **31**, 705 (1971)
4.42b H.A.Mook: Phys. Rev. Lett. **46**, 508 (1981)

4.43 H.Stanley: *Introduction to Phase Transitions and Critical Phenomena* (Clarendon Press, Oxford 1971)
4.44 T.Kasuya: Crit. Rev. Solid State Sci. **3**, 131 (1972)
4.45 S.Abiko: Z. Physik B**39**, 53 (1980)
4.46 P.Wachter: In *Handbook on the Physics and Chemistry of Rare Earth*, Vol. 2, ed. by K.A.Gschneider, Jr., L.Eyring (North-Holland, Amsterdam 1978) p. 507
4.47 R.K.Ray, J.C.Tsang, M.S.Dresselhaus, R.L.Aggarwal, T.B.Reed: Phys. Lett. **37**A, 129 (1971)
4.48a P.Grünberg, G.Güntherodt, A.Frey, W.Kress: Physica B**89**, 225 (1977)
4.48b E.Zirngiebl: Diplomarbeit, Universität Köln (1983) unpublished
4.49 J.D.Axe: J. Phys. Chem. Solids **30**, 1403 (1969)
4.50 K.H.Rieder, B.A.Weinstein, M.Cardona, H.Bilz: Phys. Rev. B**8**, 4780 (1973)
K.H.Rieder, R.Migoni, B.Renker: Phys. Rev. B**12**, 3374 (1975)
4.51 G.Güntherodt: Phys. Cond. Matter **18**, 37 (1974)
4.52 V.J.Tekippe, R.P.Silberstein, M.S.Dresselhaus, R.L.Aggarwal: Phys. Lett A**49**, 295 (1974)
4.53 S.A.Safran, B.Lax, G.Dresselhaus: Solid State Commun. **19**, 1217 (1976)
4.54 G.Güntherodt: In Proc. 13th Intern. Conf. Phys. Semicond., ed. by F.G.Fumi (Tipografia Marves, Rome 1976) p. 291
4.55 R.Merlin, R.Zeyher, G.Güntherodt: In *Physics of Semiconductors*, ed. by B.L.H.Wilson, IOP Conf. Ser. No. 43, 145 (1978)
4.56 L.Passel, O.W.Dietrich, J.Als-Nielsen: AIP Conf. Proc. **5**, 1251 (1971)
4.57 D.S.Ritchie, M.E.Fisher: Phys. Rev. B**5**, 2668 (1972)
4.58 S.Alexander, J.S.Helman, I.Balberg: Phys. Rev. B**13**, 304 (1976)
4.59 S.A.Safran, G.Dresselhaus, R.P.Silberstein, B.Lax: J. Magnetism Magn. Mat. **11**, 403 (1979)
4.60a G.Güntherodt, G.Abstreiter, W.Bauhofer, G.Benedek, E.Anastassakis: J. Magnetism Magn. Mat **15–18**, 777 (1980)
4.60b G.Güntherodt, R.Merlin, G.Abstreiter: J. Magnetism Magn. Mat. **15–18**, 821 (1980)
4.61 L.Schmutz, G.Dresselhaus, M.S.Dresselhaus: J. Magnetism Magn. Mat. **11**, 412 (1979)
4.62 N.F.Oliveira, S.Foner, Y.Shapiro, T.B.Reed: Phys. Rev. B**5**, 2634 (1972)
4.63a S.Maekawa: J. Phys. Soc. Jpn **34**, 1477 (1973)
4.63b Y.Ousaka, O.Sakai, M.Tachiki: J. Phys. Soc. Jpn **52**, 1034 (1983); and J. Magnetism Magn. Mat. **31–34**, 583 (1983)
4.64 A.Pinczuk, E.Burstein: In [Ref. 4.1, Chap. 2]; see also [Ref. 4.4a, Sect. 2.2.8]
4.65 C.R.Pidgeon, J.Feinleib, W.J.Scouler, J.Hanus, J.O.Dimmock, T.B.Reed: Solid State Commun. **7**, 1323 (1969)
4.66 S.A.Safran, G.Dresselhaus, M.S.Dresselhaus, B.Lax: Physica **89**B, 229 (1977)
4.67 R.P.Silberstein, S.A.Safran, M.S.Dresselhaus: J. Magnetism Magn. Mat. **11**, 408 (1979)
4.68 T.R.McGuire, M.W.Shafer: J. Appl. Phys. **35**, 984 (1964)
4.69 J.A.Hertz: Phys. Rev. B**14**, 1165 (1976)
4.70 S.K.Ma: *Modern Theory of Critical Phenomena* (Benjamin, Reading, MA 1976)
4.71 M.E.Fisher: AIP Conf. Proc. **24**, 273 (1975)
4.72 H.Callen, M.A.deMoura: Phys. Rev. B**16**, 4121 (1977)
4.73 R.Griessen, M.Landolt, H.R.Ott: Solid State Commun. **9**, 2219 (1971)
4.74 G.Harbeke, E.F.Steigmeier: Solid State Commun. **6**, 747 (1968)
4.75 T.Moriya: J. Phys. Soc. Jpn. **23**, 490 (1967)
4.76 N.Koshizuka, Y.Yokoyama, T.Tsushima: Solid State Commun. **23**, 967 (1977)
4.77 M.N.Iliev, E.Anastassakis, T.Arai: phys. stat. sol. (b) **86**, 717 (1978)
4.78 S.G.Stoyanov, M.N.Iliev, S.P.Stoyanova: phys. stat. sol. (a) **30**, 133 (1975)
4.79 T.Itoh, N.Miyata, S.Narita: Japan. J. Appl. Phys. **12**, 1265 (1973)
4.80 R.K.Ahrenkiel, F.Moser, S.Lyu, C.R.Pidgeon: J. Appl. Phys. **42**, 1452 (1971)
4.81 H.Fujita, Y.Okada, F.Okamoto: J. Phys. Soc. Jpn. **31**, 610 (1971)
4.82 K.Sato: J. Phys. Soc. Jpn. **43**, 719 (1977)
4.83 S.G.Stoyanov, M.N.Iliev, S.P.Stoyanova: phys. stat. sol. (a) **30**, 133 (1975)
4.84 K.Sato, T.Teranishi: J. Phys. Soc. Jpn **29**, 523 (1970)
4.85 A.Amith, S.B.Berger: J. Appl. Phys. **42**, 1472 (1971)

4.86 S.G.Stoyanov, M.N.Iliev, S.P.Stoyanova: Solid State Commun. **18**, 1389 (1976)
4.87 S.S.Yao, F.Pellegrino, R.R.Alfano, W.J.Miniscalco, A.Lempicki: Phys. Rev. Lett. **46**, 558 (1981)
4.88 S.S.Yao, R.R.Alfano: Phys. Rev. Lett. **49**, 69 (1982)
4.89 J.B.Goodenough: J. Phys. Chem. Solids **30**, 261 (1969)
4.90 W.J.Miniscalco, B.C.McCollum, N.G.Stoffel, G.Magaritondo: Phys. Rev. B**25**, 2947 (1982)
4.91 N.Koshizuka, Y.Yokoyama, T.Okuda, T.Tsushima: J. Appl. Phys. **49**, 2183 (1978)
4.92 N.Koshizuka, Y.Yokoyama, T.Okuda, T.Tsushima: J. Phys. Soc. Jpn. **45**, 1439 (1978)
4.93 N.Koshizuka, S.Ushioda, T.Tsushima: Phys. Rev. B**21**, 1316 (1980)
4.94 D.J.Lockwood: In [Ref. 4.4b, Chap. 3]
4.95 S.R.Kuindersma, C.Haas, J.P.Sanchez, R.Al: Solid State Commun. **30**, 403 (1979)
4.96 P.J.Brown, K.R.A.Ziebeck, C.Esribe: J. Magnetism Magn. Mat. **15–18**, 515 (1980)
4.97 A.Frey, G.Benedek: Solid State Commun. **32**, 305 (1979)
4.98 G.Güntherodt, W.Wichelhaus: Solid State Commun. **30**, 1147 (1981)
4.99 J.Mulak, K.W.H.Stevens: Z. Physik B**20**, 21 (1975)
4.100 J.R.Sandercock: Solid State Commun. **15**, 1715 (1974)
4.101 P.Grünberg, F.Metawe: Phys. Rev. Lett. **39**, 1561 (1977)
4.102 P.Grünberg: J. Magnetism Magn. Mat. **15–18**, 766 (1980)
4.103 C.F.Wang, R.B.Wright: J. Chem. Phys. **56**, 2124 (1972); **57**, 4401 (1972); **58**, 1411 (1973)
4.104 M.H.Brodsky: In [Ref. 4.1, Chap. 5]
4.105 W.Spengler, R.Kaiser, H.Bilz: Solid State Commun. **17**, 19 (1975)
W.Spengler, R.Kaiser: Solid State Commun. **18**, 881 (1976)
4.106 E.Anastassakis, H.Bilz, M.Cardona, P.Grünberg, W.Zinn: In *Light Scattering in Solids*, ed. by M.Balkanski et al. (Flammarion, Paris 1976) p. 367
4.107 G.Güntherodt, P.Grünberg, E.Anastassakis, M.Cardona, H.Hackfort, W.Zinn: Phys. Rev. B**16**, 3504 (1977)
4.108 C.Colvard, R.Merlin, M.V.Klein, A.C.Gossard: Phys. Rev. Lett. **45**, 298 (1980)
4.109 E.Zirngiebl, G.Güntherodt, H.Maletta: to be published
4.110 H.Maletta, P.Convert: Phys. Rev. Lett. **42**, 108 (1979)
4.111 H.Maletta, G.Aeppli, S.M.Shapiro: J. Magnetism Magn. Mat. **31–34**, 1367 (1983)
4.112 H.Maletta, G.Aeppli, S.M.Shapiro: Phys. Rev. Lett. **48**, 1490 (1982)

5. Raman Scattering in Rare-Earth Chalcogenides

Gernot Güntherodt and Roberto Merlin

With 25 Figures

Among the various classes of rare-earth[1] (RE) compounds, the fcc (NaCl-type) RE monochalcogenides (REX, with X denoting the chalcogen) are the simplest from a structural and magnetic-exchange point of view. Some of these compounds, e.g., EuS and SmS, have received much attention during the past decade because of their many unique properties owed to the presence of partially filled inner $4f$ shells of the RE within the forbidden gap and close to the conduction band. Moreover, their structure affords conceptual simplicity. Although closely related in their chemical aspects, the REX show strongly differing electronic, magnetic and optical properties along with the filling of the $4f$ shell: REX can be ferromagnetic semiconductors (e.g., EuS), metallic antiferromagnets (e.g., GdS), superconductors (e.g., LaS) or exhibit valence fluctuations after semiconductor-metal transformation (e.g., SmS). The occurrence of two different *atomic* configurations for the RE ($4f^n\,5d^1\,6s^2$ and $4f^{n+1}6s^2$), together with the nonbonding character of the $4f$ shell, are responsible for this seemingly irregular pattern most of the properties of the REX appear to follow. These dissimilar features of the $4f$ states are a product of the *lanthanide contraction* (i.e., the small, but regular decrease of the $4f$ radii from Ce towards Yb), of an increasing $4f$ binding energy from Ce towards the heavier rare earths and of the strong Hund's rule correlation effects near the half-filled ($4f^7$) and filled ($4f^{14}$) $4f$ shell, favoring the $4f^{n+1}6s^2$ configuration [5.1–2]. The sudden changes in the properties of the $4f$ wave functions result in the near equality of the $4f$, $5d$ and $6s$ binding energies, despite the localization of the $4f$ states inside the filled $5s$ and $5p$ orbitals of the Xe core [5.1]; the latter being the reason why the $4f$ states largely retain their highly correlated, atomic nature after compound formation. These particular properties of the $4f$ shell, apart from being the cause of the most interesting aspects of RE research, are also the source of the problems posed by the determination of the electronic structure of these materials. These problems arise because of the inadequacy of the conventional methods of band calculation to handle correlated many-electron states.

The combination of Raman scattering with the great diversity of properties displayed by the REX has resulted in a very prolific area of research as judged, for instance, by the large number of publications in the field. These studies have

1 As it is customary, the term *rare earth* is used here to refer to the lanthanides (i.e., the elements La through Lu), and the elements *scandium* and *yttrium*.

both contributed to the understanding of these materials and have revealed new scattering phenomena. In this chapter we review the experiments and the theory of Raman scattering in REX with some emphasis placed on the material side of the subject. Our treatment of some topics, particularly Sect. 5.2.1, complements certain aspects covered in Chap. 4 of this volume, where the discussion of the scattering mechanisms in EuX is emphasized. Here we want to present the full range of scattering phenomena which are exhibited by the family of REX compounds.

The scope of the chapter is to give first a brief introduction to the electronic, optical and magnetic properties of the REX, considering those few aspects which are relevant to the discussion of Raman scattering (Sect. 5.1). A more comprehensive presentation of the many experimental and theoretical aspects of the REX and a compilation of original references can be found in a number of recent conference proceedings [5.3a, b–5a, b] and review articles [5.2, 6–12]. In Sect. 5.2.1 we review the Raman work on EuX [5.13–53], while the results on YbX [5.22–24, 26, 33–34, 54–56] are discussed in Sect. 5.2.2. Both families of compounds allow for case studies with the scattering cross section being determined by electronic transitions involving the $4f$ states. Most of the unique features of the phonon scattering in EuX and YbX [e.g., spin-disorder induced or multiple $LO(\Gamma)$ phonons] are related to *resonant* behavior. Because of a number of reasons (e.g., the most relevant electronic transitions occur outside the frequency range of gas lasers), *resonant* Raman studies have not yet been reported for the other REX. Section 5.3 deals with the case of metallic (trivalent) REX [5.57–69] with their (nonresonant) spectra characterized by first-order (defect-induced) and second-order phonon features. In Sect. 5.4 we discuss semiconductor-metal transitions, i.e., precursor effects of valence instabilities as evidenced by phonon anomalies and electronic Raman scattering from $4f$ multiplet levels near configuration crossover [5.60–67]. Raman scattering in intermediate valence REX materials [5.61–62, 66–74] is described in Sect. 5.5. Much of the emphasis here is given to the interrelationship between phonon anomalies and Raman intensities as described by an electron-lattice interaction model [5.70]. For completeness, Sect. 5.6 discusses the Raman results in the higher RE chalcogenides, such as RE_3X_4 and RE_2X_3 [5.33, 75–79a, b]. Of particular interest are compounds exhibiting thermally activated valence fluctuations (i.e., Eu_3S_4 and Sm_3S_4). Section 5.7 closes this article with a summary and some comments about possible areas of interest for future work.

5.1 An Overview of the Properties of Rare-Earth Monochalcogenides

The REX divide into semiconductors (divalent RE ions) and either trivalent or mixed-valent metals. TmSe is the only case known of REX exhibiting intermediate valence at atmospheric pressure [5.80–81]. TmTe [5.82], as well as the known monochalcogenides of Sm, Eu and Yb [5.9], are all semiconductors at atmospheric pressure. Under pressure, most of these compounds undergo (except for structural changes) phase transitions into a mixed-valent phase characterized by the appearance of metallic conductivity, nonmagnetic behavior and a large reduction in lattice parameter [5.2, 11b, 83–87]. The rest of the REX (including the monochalcogenides of Sc and Y) are all metals with the RE ions in the trivalent state.

Why is, e.g., EuS a semiconductor while, e.g., PrS is a metal and why do the semiconducting REX become metallic under pressure? Two quite different paths lead to answers to these questions. A calculation of the electronic structure of the REX is a rather difficult task for it requires a combination of atomic and band calculations under one single algorithm. Although this difficulty can be overcome as shown by the work of *Herbst* et al. [5.88] for RE metals, a systematic investigation of the electronic properties of the REX has not yet been reported. An almost trivial and perhaps unexpected way to attack this problem is based on simple considerations from thermodynamics, as largely inspired by the work of *Johansson* and *Rosengren* [5.89, 90] on RE metals. The important point of the argument is the established fact that as the $4f$ electrons do not contribute to the chemical bonding, the energy of the $4f^{n+1} \rightarrow 4f^{n}5d^{1}$ atomic transition $\varepsilon_n(f\rightarrow d)$ governs the systematics of the cohesive energy changes in RE systems [5.89–95]. In Fig. 5.1 we show the double periodic behavior exhibited by $\varepsilon_n(f\rightarrow d)$ when plotted as a function of the atomic number of the RE [5.1]. As indicated by the energy scale, most of the lanthanides adopt the $4f^{n+1}6s^2$ or "*divalent*" configuration in the atomic state; only for La, Ce, Gd and Lu ($n=14$, not shown in Fig. 5.1) is the $4f^{n}5d^{1}6s^{2}$ "*trivalent*" configuration favored. The cohesive energy of the REX displays a somehow different double periodicity [5.93–94][2]. For those elements in which the configuration is the same for the RE in the solid phase and in the atomic state (e.g., La, Eu, or Yb), the cohesive energy E_c is approximately constant across the series, adopting different values for the semiconductors ($E_c \approx E_s^0$) and the metals ($E_c \approx E_m^0$). For the other RE, which are found in the "divalent" configuration as free atoms but form metals and therefore become trivalent, E_c varies in such a way that the sum $E_c + \varepsilon_n(f\rightarrow d)$ remains nearly constant and equal to E_m^0. These trends are simply a result of the nonbonding properties of the $4f$ shell [5.89–93]. What is significant to our discussion is the remarkable consistency in the difference $\Delta = E_m^0 - E_s^0$ of ~ 2.1 eV per RE atom for all the chalcogenide series. This result (based only on experimental data), together with the fact that $\delta = \varepsilon_n(f\rightarrow d) - \Delta$ determines whether the divalent ($\delta > 0$) or the trivalent phase ($\delta < 0$) is more stable, permits us to establish a very intuitive picture of the systematics of the REX. As indicated in Fig. 5.1, the horizontal line at $\Delta = 2.1$ eV is the predicted boundary between divalent, semiconducting and trivalent, metallic REX. This prediction is

2 Cohesive energy data are available for the sulfides [5.93], selenides [5.94] and only a few oxides and tellurides. In the latter cases, the cohesive energy can be inferred from the data of the dissociation energy of the gaseous REX [5.95] using the empirical fact that the difference of both quantities does not change much within a chalcogenide series [5.93].

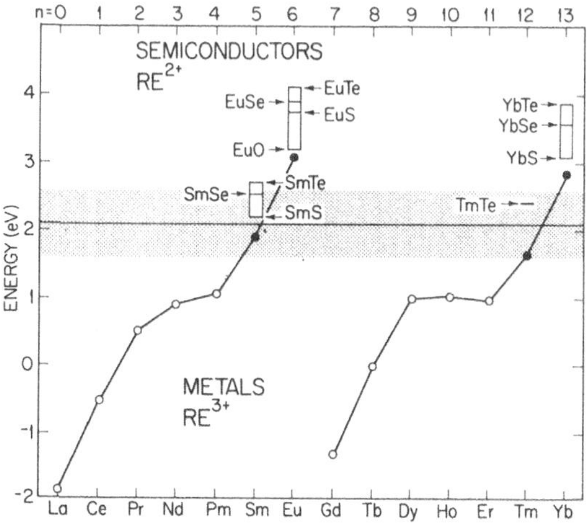

Fig. 5.1. Energy difference $\varepsilon_n(f\to d)$ between the lowest levels of the $4f^{n+1}6s^2$ and $4f^n5d^16s^2$ atomic configurations as a function of n [5.1]. The horizontal line at $\Delta=2.1$ eV indicates the nearly constant difference in the free energy of a trivalent metal and a divalent semiconductor, while the dotted region accounts for the uncertainties of this magnitude. The position of the energy gaps of the semiconducting REX [5.9] is given relative to Δ

observed in the trends exhibited by the electronic properties of these materials, especially if one takes into account the small variations of Δ across the series and the uncertainties in the cohesive energy results (which add up to ± 0.5 eV). The majority of the REX are trivalent metals because the gain in binding energy overcomes the energy cost of the $4f$ promotion. Only in the case of Sm, Eu, Tm, and Yb, and as a result of the increasing importance of the Hund's rule coupling as we approach the half and totally filled $4f$ shell [5.1], is $\varepsilon_n(f\to d)$ large enough to stabilize the semiconducting phase. (For the monochalcogenides of Tm, the actual value of δ is probably very small, as suggested by the fact that TmTe is a semiconductor [5.82] while TmS is a metal).

The bulk moduli of the semiconducting and trivalent metallic REX vary only slightly across a chalcogen series ($\sim 20\%$) with an average value for the metals being larger by a factor of approximately two[3]. This indicates that the free energy of the semiconducting phase increases more rapidly with pressure and therefore that a semiconductor-*trivalent* metal transition should eventually take place under pressure. Actually, this integer-valence change is never observed. The reason is that it is energetically more advantageous for the material to undergo first a transition into a different metallic phase where the RE valence, instead of being 3+, adopts noninteger values. The semiconductor-*mixed-valent* metal transition can be abrupt, as in the case of SmS; or gradual, as it occurs in SmSe, where the RE valence appears to change continuously from 2+ towards 3+ [5.11b, 83–87]. In addition, this transition can be induced by the substitution of either the RE^{2+} ions by smaller size cations (e.g., Gd^{3+}, Y^{3+}) or the chalcogen by larger size anions (e.g., As^{3-}), which introduce "lattice pressure" [5.9, 10, 11b, 99–103]. In the state of intermediate valence, the atomic-like $4f$

3 References to the bulk modulus data are given in [5.11b, 85–87, 96–98].

levels and the $s-d$ conduction band coexist at the Fermi level giving rise to a hybridization of the $4f^{n+1}(2+)$ and $4f^n5d^1(3+)$ configurations and therefore, to interconfigurational fluctuations characterized by a frequency roughly proportional to the $4f$ bandwidth [5.2]. This rather unique situation leads to pronounced anomalies of the transport, magnetic, thermal, elastic and phonon properties of the intermediate-valence compounds [5.2, 4, 5a, b, 11b]. In particular, the softening of certain branches of the phonon dispersion curves [5.104–106] has been ascribed to the $f-d$ hybridization and phonon-induced $4f^{n+1} \leftrightarrow 4f^n5d^1$ excitations [5.107, 108]. These phonon anomalies are mostly a consequence of the large difference ($\sim 15\%$) in the ionic radii associated with the divalent and trivalent ionic configurations, resulting in a substantial volume contraction accompanying the promotion of $4f$ electrons into $5d$ states. This effect is also responsible for some of the unique features of the (resonant) Raman spectra of EuX and YbX, where the $4f^{n+1} \rightarrow 4f^n5d^1$ optical transitions essentially determine the scattering cross section [5.42, 43, 54] (see Chap. 4).

The magnetic and some optical properties of the REX are governed by the $4f$ electron states. A partially filled $4f$ shell carries a magnetic moment when the ground state total angular momentum J is different from zero. This is the case of, e.g., $Eu^{2+}(4f^7, {}^8S_{7/2})$ and $Tm^{2+}(4f^{13}, {}^2F_{7/2})$ and, consequently, collective ordering is observed in the semiconducting EuX and TmX. Accounting for a totally filled $4f^{14}$ shell, the monochalcogenides of Yb^{2+} are diamagnetic. On the other hand, the semiconducting SmX show van Vleck paramagnetism, a result of the small separation (≈ 285 cm^{-1}), between the two lowest lying $J=0$ and $J=1$ levels of the $4f^6({}^7F_J)$ configuration [5.109, 110]. The metallic $RE^{3+}X$ should all be magnetically ordered at $T=0$ K, except for the monochalcogenides of Sc, Y, La$(4f^0)$ and Lu$(4f^{14})$, which are superconductors at low temperatures. The case of antiferromagnetic GdX is the most thoroughly studied [5.111, 112]. The magnetic behavior of the intermediate-valence REX is quite anomalous [5.2, 4, 5a,b, 84, 113, 115]. The mixed-valent SmX do not order magnetically [5.2, 4, 84], despite the superposition of $Sm^{2+}\,4f^6({}^7F_{J=0})$ and $Sm^{3+}\,4f^5({}^6H_{5/2})$ at high temperatures. TmSe exhibits antiferromagnetism below $T_N \approx 3$ K [5.113], in spite of its mixed valence behavior. This transition is associated with strong transport anomalies [5.116–118].

Because of the small spatial extent of the $4f$ wave function, the exchange interactions in the REX are due to (indirect) superexchange and additionally, in the metallic compounds, are of the Ruderman-Kittel-Kasuya-Yoshida (RKKY) type [5.119, 120]. The ferromagnetic exchange coupling between nearest-neighbor (nn) RE ions (J_1) originates in the overlap of $4f$ and $5d$ states on one site and the "transmission" of the "spin information" onto a nn site via the extended $5d$ states (indirect cation-cation superexchange) [5.8a,b, 9, 11a, 119, 121]. The antiferromagnetic next-nearest-neighbor (nnn) exchange (J_2) involves the $5d$ states of the RE ion and the p orbitals of the intervening chalcogen ion. The exchange constants vary across a chalcogen series according to the changes in the lattice parameter and energy gaps and, in the case of the metals, also as a function of the concentration of conduction electrons [5.120]. As a result, a wide variety of magnetic phases is possible within one chalcogen series. This is, for instance, the case of the EuX. A summary of their magnetic properties will now be given as they are relevant to the discussion of Raman scattering. J_1 dominates over J_2 in EuO and EuS and these compounds become ferromagnetic at $T_C = 69.3$ K and 16.6 K, respectively. In EuSe, $J_1 \approx |J_2|$ results in a rather complex magnetic phase diagram (see inset of Fig. 5.7) [5.122, 123]. In the temperature range $2.8 < T < 4.6$ K, EuSe is in the AF-I phase, the spins are parallel to a [110] direction and oriented $\uparrow\uparrow\downarrow\downarrow$ along successive (111) planes. A small magnetic field induces a transition into a ferrimagnetic phase where the spins are oriented $\uparrow\uparrow\downarrow$. For $T < 1.8$ K and internal magnetic field $H_i \lesssim 0.4$ kOe, the AF-II phase ($\uparrow\downarrow$) is stable. Finally, a field-induced ferromagnetic alignment occurs for $H_i \gtrsim 1$–2 kOe. In EuTe, J_2 dominates. In the absence of an applied magnetic field, EuTe is in the AF-II phase below $T_N = 9.6$ K, while it becomes a "canted" antiferromagnet for $H_i \gtrsim 0.5$ kOe [5.124]. The total ferromagnetic alignment of the spins occurs for $H_i > 70$ kOe at $T \approx 0$ K.

We now turn to the discussion of the optical properties of the semiconducting REX. The metals and mixed-valence REX will not be considered here since their optical properties have only a marginal importance for the purpose of Raman scattering[4]. In the $RE^{2+}X$, the absorption edge is due to $4f^{n+1} \rightarrow 4f^n 5d^1$ transitions. The energy gaps of the known semiconductors are indicated in Fig. 5.1 with the zero of the energy scale at $\Delta = 2.1$ eV. This particular scale has been chosen to emphasize the relationship between δ, i.e., the metal-semiconductor free energy difference and the energy gap E_g [5.90]. In the single-site approximation, these quantities are exactly equal [5.88–90]. The actual difference $E_g - \delta$ ($\lesssim 1$ eV) accounts for the energy of a RE^{3+} impurity in the semiconducting host, if a completely relaxed final state is assumed [5.90].

The few available calculations of the band structure of semiconducting REX [5.126, 127] have not been very successful in interpreting the optical data. Better results have been achieved by using the methods of ligand field theory, which account for the many-electron nature of the $4f$ states. The Hamiltonian describing the excited $4f^n 5d^1$ configuration is given by [5.8a]

$$H = H_{so}^{f} + H_{cf}^{d} + H_{ex}^{f,d} + H_{so}^{d} . \tag{5.1}$$

H_{so}^{f} is the spin-orbit coupling in the $4f$ configuration; its energy spectrum is that of the excited hole state left behind by removing one $4f$ electron. This term leads to a splitting of the $4f$ levels characterized by $\lambda_f \approx 0.2$ eV [5.8a]. The crystal field Hamiltonian H_{cf}^{d} has a negligible effect on the $4f$ core but splits the $5d$ levels into

4 Considered relative to the energy range normally accessible to resonant Raman studies, the resonant transitions involving the $4f$ states in the mixed-valence and the metallic REX occur, respectively, too low and too high in energy. For the metals, the onset of the $4f^n \rightarrow 4f^{n-1}5d^1$ ($RE^{3+} \rightarrow RE^{4+}$) optical transition is typically 4–7 eV [5.9, 162a] as compared to larger $4f$ binding energies observed in x-ray photoemission [5.125].

states of t_{2g} and e_g symmetry by an amount of typically 1.5–3 eV [5.9]. The coupling constants for the remaining terms, i.e., the *f–d* exchange interaction and the spin-orbit coupling for the $5d$ electron, have nearly the same value of ≈ 0.1 eV [5.8a].

The room temperature optical absorption spectra of the EuX show two major, broad (FWHM ≈ 0.7 eV), and almost structureless features: E_1 and E_2, which correspond to transitions between $4f^7$ and the crystal-field split $4f^6(^7F_J)5d^1$:t_{2g} and e_g levels [5.8b, 5.9–11a, 128]. The E_1–E_2 splitting decreases from EuO (≈ 3.2 eV) to EuTe (≈ 1.3 eV) following the increase in lattice parameter. In most cases the accessible Raman data of the EuX involve the resonant excitation near the E_1 transitions. Only a few experiments have been reported using laser frequencies in the region between the E_1 and E_2 peaks, and below the band edge. The onset of *ferromagnetic* ordering (induced by an external magnetic field in the cases of EuSe and EuTe) leads to a splitting of the E_1 structure into three narrow components (FWHM ≈ 0.1 eV): E_1', E_1, and E_1'' [5.129, 130]. Experiments using modulation techniques have shown the presence of a fourth component E_1''' at higher energies [5.9, 131]. The exchange interaction between the excited $5d$ electron and the neighboring Eu ions in the $4f^7$ configuration is responsible for the red shift of the absorption edge observed in the ferromagnetic EuX [5.8a,b, 11a, 132].

Optical absorption data of YbX show a number of relatively narrow features (FWHM ~ 0.1 eV at $T \sim 10$ K) which correspond to $4f^{14} \rightarrow 4f^{13}5d^1$ transitions [5.133]. H_{so}^f splits the $4f^{13}$ states into $^2F_{7/2}$ and $^2F_{5/2}$ levels, separated by ≈ 1.2 eV. This splitting is comparable to the t_{2g}–e_g separation. The multiple LO(Γ) scattering observed in the YbX occurs for laser frequencies in resonance with $4f^{14} \rightarrow 4f^{13}(^2F_{7/2})5d^1$ transitions. In the case of the semiconducting SmX, the H_{so}^f term gives rise to a splitting of the excited $4f^5$ configuration into ^{6}H, ^{6}F and ^{6}P multiplet levels. Although the $4f^6 \rightarrow 4f^5(^6F)5d^1(t_{2g})$ transitions occur in a very convenient region (≈ 1.5–2 eV [5.9, 134, 135, 162a]) of the spectrum for the purpose of light scattering, there have not yet been any reports of resonant Raman studies.

5.2 Semiconductors

5.2.1 Magnetic-Phase Dependent Scattering by Phonons in EuX

Light scattering by phonons in EuX shows a remarkable correspondence with the degree of magnetic ordering in such a way that the *translational* symmetry of the Raman tensor is determined, not only by the symmetry of the lattice, but also of the spin system. This property of the scattering is not the result of a particularly large spin-phonon interaction in the ground state, but originates indirectly via the spin-orbit coupling in the excited intermediate states of the scattering process. It is important to point out the resonant nature of this form of scattering. For excitation energies ω_i far below the fundamental gap, the Raman spectra show only 'normal' features, i.e., second-order scattering by phonons, which does not correlate with the magnetic properties [5.30, 31, 42].

The $4f^7 \to 4f^6 5d^1$ optical transitions dominate the cross section for photon energies in almost the whole visible range. The sensitivity of the scattering on the state of magnetic ordering relies on the coupling of spin and orbital motions in the excited $4f^6$ hole state [5.36, 37, 40–44, 53]. This interaction, together with the electron-phonon coupling, leads to electronic polarizabilities which depend on *both* the phonon coordinates and the spins of the Eu^{2+} ions [5.36–45a, b, 51, 53]. The electron-phonon interaction couples the intermediate $4f^6 5d^1$ states mostly with displacements which are fully symmetric with respect to an Eu site, accounting for the large volume contraction associated with the promotion of the $4f$ electron [5.37–44, 53]. This symmetry constraint largely favors the scattering by phonons belonging to the longitudinal-optical (LO) branch.

Most of the available experimental data can be qualitatively understood by considering the *one phonon–one spin* polarizability $\alpha^{(1,1)}$, which is the leading term in an expansion of the polarizability tensor in powers of the $4f$ spin-orbit coupling constant λ_f (Chap. 4) [5.35–45a, b, 49–51, 53]. The corresponding scattering cross section is proportional to a two-spin correlation function. This gives rise to spin-disorder induced scattering by phonons in the paramagnetic phase [5.35–38, 40–45a, b, 50, 52, 53] and to scattering by folded phonons in the magnetically-ordered (nonferromagnetic) phases [5.27, 28, 35, 39, 44–51, 53]. The processes associated with spin disorder evolve into scattering due to the simultaneous excitation of a phonon-magnon pair in the ordered phases [5.41, 42, 44].

The presence of large spin fluctuations in the region close to the transition temperatures leads to an enhancement of the scattering due to phonons with wave vectors matching those of the critical fluctuations [5.36, 45a, b, 46, 50, 51]. This effect has been observed in EuSe for scattering involving, other than $\alpha^{(1,1)}$, the spin-polarizability arising from the phonon-modulation of exchange constants [5.45, 46, 50, 51].

Higher-order terms in the expansion of α in powers of λ_f become important closer to the resonance maximum [5.37, 38, 40–44, 53]. Scattering which is derived from the *one phonon–two spin* polarizability $\alpha^{(1,2)}$ has been identified in EuS [5.37, 38, 42], EuSe [5.42] and EuTe [5.42, 44]. The contributions due to $\alpha^{(1,1)}$ and $\alpha^{(1,2)}$ can be separated owing to their different symmetry properties and resonance enhancement [5.37, 40, 42, 43, 53].

In the ferromagnetic phase, scattering due to the LO(Γ) phonon and its overtones dominates the spectra [5.14, 18–21, 25, 27, 35, 44, 47, 49, 50]. The occurrence of a phonon-overtone sequence is a characteristic feature of Raman scattering in EuX involving *all* the processes mentioned above. What is different in the case of the ferromagnetic phase is the resonant behavior of the scattering, characterized by peaks at roughly the positions of the narrow lines (E_1, E_1'') of the magnetoreflectance spectra and a several orders of magnitude enhancement [5.27, 35, 50]. The LO(Γ) phonon scattering in EuX is closely related to the 'forbidden' Fröhlich LO(Γ) scattering found in many polar (nonmagnetic)

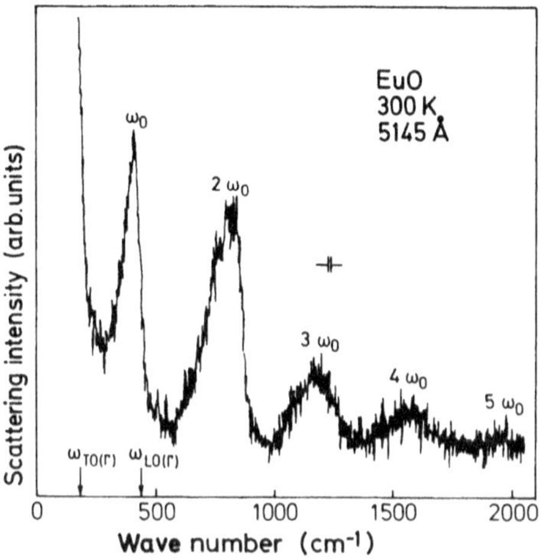

Fig. 5.2. Raman spectrum of EuO at 300 K [5.31]

materials [5.20, 32, 44, 49, 52]. However, a total identification with this form of scattering cannot explain the anomalous selection rules and the large suppression of the scattering intensity above T_C, observed in the EuX [5.18–21, 25, 27, 35, 44–50]. A microscopic model which considers some of the differences between the 'normal' and the 'magnetic' (Fröhlich) LO(Γ) scattering was given by *Abiko* [5.52]. This work identified the quenching above T_C with the large spread in the energy of the excited states arising from the spin disorder. The origin of the anomalous scattering with polarizations of incident and scattered electric field vectors perpendicular to each other [5.35, 50] has not been discussed in the literature. Nevertheless, this configuration contains the antisymmetric Γ_{15}^+ component, which is an indication that the scattering most likely involves the excitation of a $q \sim 0$ acoustic magnon.

The symmetry properties and the frequency dependence of the scattering associated with the spin-dependent polarizabilities have been discussed in detail in Chap. 4. This will not be repeated here. We will present instead an exhaustive review of the characteristic experimental and theoretical results for each particular member of the EuX family, in as far as to contrast the latter with the other families of the REX series.

EuO has been investigated in a broad range of temperatures above and below T_C [5.30,31,34,41,42,44]. Studies of the frequency dependence of the cross section have not been too extensive because of the relatively low energy position of the E_1 absorption band (< 1.5 eV). Figure 5.2 shows the Raman spectrum of EuO in the paramagnetic phase with its fundamental maximum at $\omega_0 = \omega_{LO(L)}$ and the $n \cdot \omega_0$ overtone sequence [5.31]. A theoretical calculation of the line shape was reported by *Zeyher* and *Kress* [5.43]. The maximum at LO(L) originates from weighting factors arising from the full symmetric (Γ_1^+) coupling of the $4f^6$ hole and the phonons [5.37,40–44,53]. The scattering in Fig. 5.2 is induced by spin disorder and associated with $\alpha^{(1,1)}$. It is seen mainly in the antisymmetric Γ_{15}^{+} scattering component. Upon cooling below T_C, the intensity is quenched while the peaks shift to higher energies [5.30, 31, 34, 41, 44] (see also Fig. 4.7 in the previous chapter). This shift is due to a change in the excitation spectrum of the spin system from essentially zero-energy (quasielastic) excitations in the paramagnetic phase to magnons of the ordered state [5.41, 42, 44].

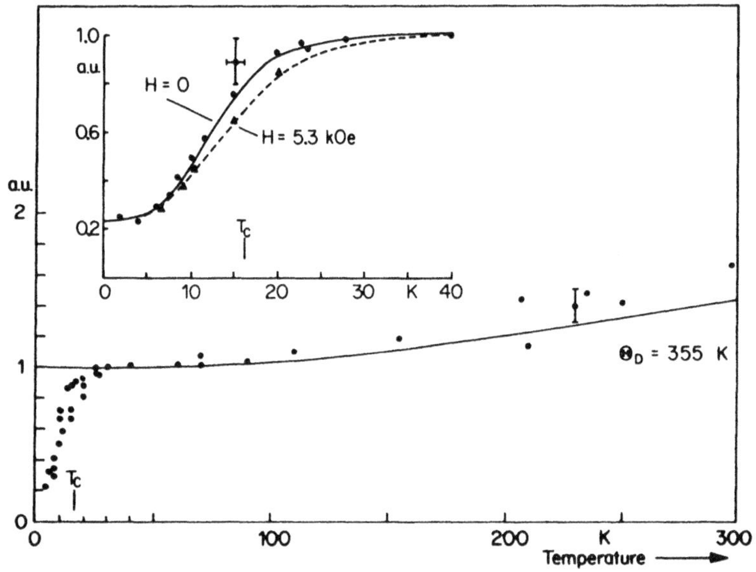

▲

Fig. 5.3. Temperature dependence of the integrated intensity of the fundamental maximum in EuS ($\omega_0 \approx 240\ cm^{-1}$) with excitation at $\lambda = 4416$ Å. The solid line is a theoretical fit using the Bose population factor. The inset shows in detail the data near T_C for $H = 0$, and $H = 5.3$ kOe fitted to the two-spin correlation function [5.16]

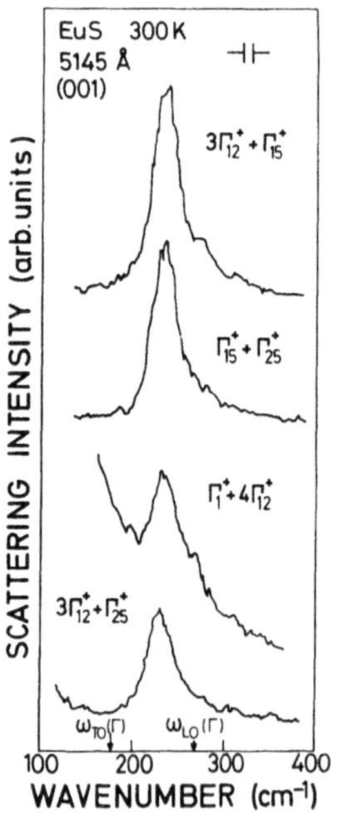

Fig. 5.4. Raman spectrum of EuS under 5145 Å laser excitation in four different scattering configurations for a (001) crystal plane [5.37]

The behavior of the spin-disorder induced scattering in EuS is similar to that described above. The data of *Schlegel* and *Wachter* [5.16] on the temperature dependence of the scattering intensity, reproduced in Fig. 5.3, show an almost perfect fit to a two-spin correlation function in the vicinity of T_C, indicating the $\alpha^{(1,1)}$ origin of the scattering. Consistent with these results, a symmetry analysis of the spectra reveals the predominance of the Γ_{15}^{+} symmetry component [5.37, 42] (see also Fig. 4.2 in Chap. 4). This situation corresponds to a value ω_i that falls in between the main absorption maxima E_1 and E_2 of EuS. Spectra obtained with ω_i close to the maximum of the E_1-band (Fig. 5.4) show also scattering from other components, associated with $\alpha^{(1,2)}$ [5.37]. Notorious from the data of Fig. 5.4 is the absence of Γ_1^{+} scattering [5.37]. This is a result of the transformation properties of the electron-phonon coupling (transforming as Γ_1^{+}) plus the fact that the full symmetric component of the spin operators does not induce transitions in the spin system [5.37, 38, 40–44, 53].

The temperature- and magnetic-field dependence of the spin-disorder scattering in EuS have been investigated by a number of workers [5.16, 19, 30, 31, 34, 42]. Theoretical calculations of these effects have been reported in [5.29, 36, 40, 45b, 51, 53] (the results of *Safran* [5.51] are shown in Fig. 4.8 of Chap. 4). The frequency dependence of the cross section for the different symmetry components, measured by *Merlin* et al. [5.37], has been considered in [5.37, 43] (see also Fig. 4.5) and by *Ousaka* et al. [5.53]. Lineshape analyses of the spectra have been given in [5.37, 43, 53] (Fig. 4.3).

Using an excitation energy ~0.1 eV below the fundamental gap, *Güntherodt* et al. [5.30, 31, 42] obtained *allowed* second-order phonon spectra in EuS. Their results are shown in Fig. 5.5, together with the two-phonon (overtone) density of states calculated in [5.43]. The similarities between the quantitative theoretical interpretation of this spectrum and that of the second-order Raman spectrum of SrO have been emphasized in [5.30, 31, 42].

LO(Γ) phonon scattering in EuS was first reported by *Tekippe* et al. [5.21]. It is only observable in the ferromagnetic phase and exhibits a strong dependence on excitation energy [5.21, 35, 50]. Figure 5.6 shows the transition from the spin-disorder induced spectra, characteristic of the paramagnetic phase, to the LO(Γ) scattering associated with the spin alignment [5.35]. The selection rules for this scattering were determined experimentally by *Silberstein* et al. [5.35, 50]. Theories that partly explain these results are presented in [5.28, 32, 52].

The properties of the scattering in the paramagnetic and (magnetic-field induced) ferromagnetic phases of EuSe follow the same trends exhibited by the other chalcogenides. The unique behavior of the scattering due to LO(Γ) phonons in EuX was first recognized by *Tsang* et al. [5.14, 20] in their experiments on EuSe. The temperature, magnetic-field and excitation-energy dependences and the selection rules for the LO(Γ) scattering in EuSe have been studied by several authors [5.14, 18, 20, 25, 27, 35, 50]. Data on the temperature and magnetic-field dependence of the spin-disorder induced scattering have been reported in [5.18–20, 22, 27, 34, 35, 45a, 46, 50]. A theoretical analysis of these results is given in [5.40, 43, 53]. The resonant enhancement of the spin-disorder

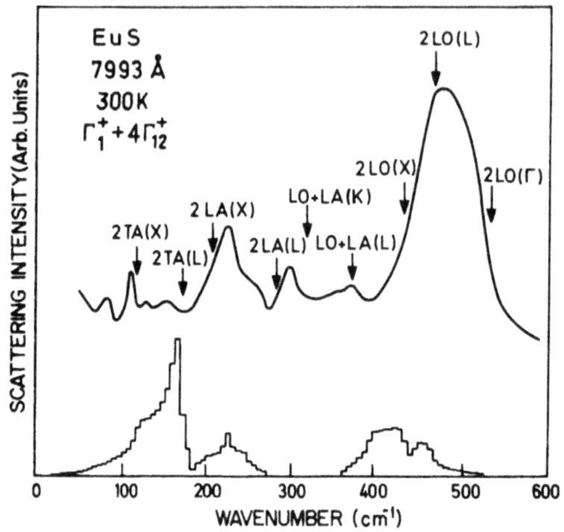

Fig. 5.5. Room temperature Raman spectrum of EuS in the $\Gamma_1^+ + 4\Gamma_{12}^+$ configuration for an excitation frequency below the absorption edge [5.30, 42]. The histogram represents the two-phonon (overtone) density of states from [5.43]. The assignment in terms of two-phonon critical points is based on the dispersion curves calculated in [5.43]

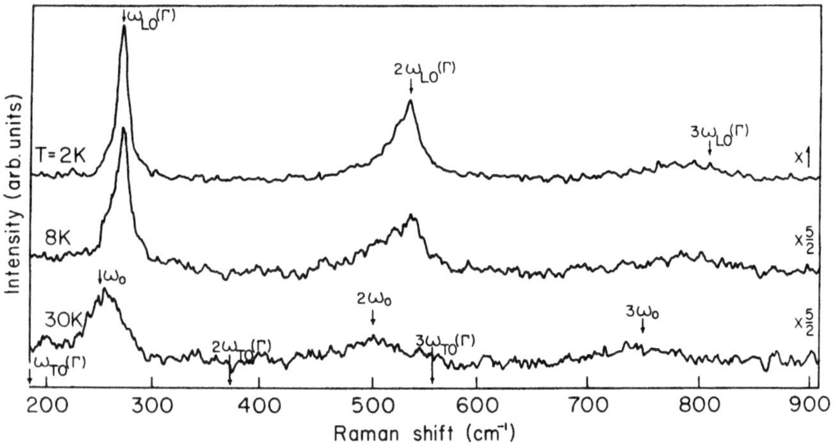

Fig. 5.6. Raman spectra of EuS at different temperatures above and below $T_C = 16.5$ K for a laser excitation $\omega_i = 2.16$ eV in the vicinity of the E_1'' peak; from [5.35]

scattering for the different symmetry components was measured by *Güntherodt* et al. [5.42]. Scattering due to allowed second-order phonon processes has been obtained in [5.42, 50] and compared with calculated two-phonon densities of states in [5.43].

Raman scattering by "*folded*" phonons (i.e., brought to the Γ point by the folding of the Brillouin zone) was originally observed by *Silberstein* et al. in EuSe [5.27, 35, 50]. Their results, shown in Fig. 5.7, represent some of the most convincing evidence of the fundamental role played by the spin system in the scattering processes discussed in this section. The traces for $T = 4.2$, 2 and 1.7 K show, respectively, features due to phonons with wave vectors $\boldsymbol{q} = \boldsymbol{q}_L/2$, $2\boldsymbol{q}_L/3$ and $\boldsymbol{q}_L$ [5.35, 50]. These wave vectors determine the folding of the fcc Brillouin

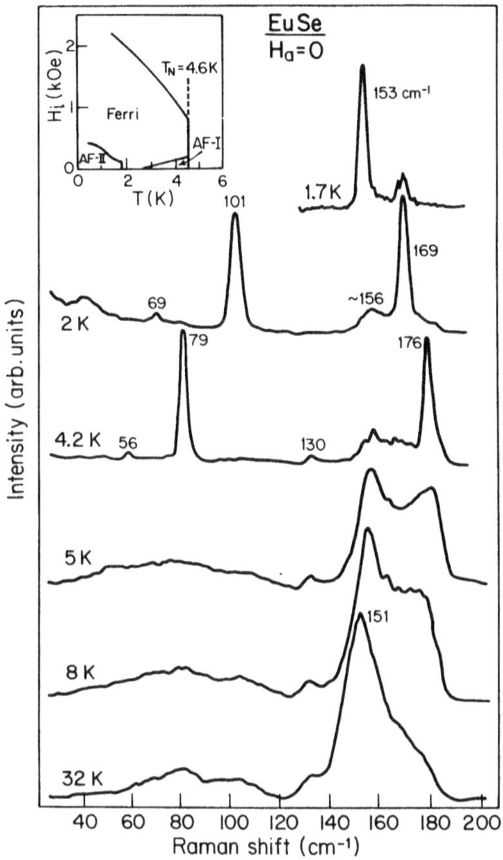

Fig. 5.7. Raman spectra of EuSe in the $y(zx+zy)z$ scattering configuration for $\omega_i = 1.91$ eV (for the spectrum at $T = 1.7$ K, $\omega_i = 1.83$ eV was used); from [5.50]. The Raman frequencies for the principal spectral features are indicated for the paramagnetic phase (32, 8 and 5 K), the AF-I phase (4.2 K), the ferrimagnetic phase (2 K) and the AF-II phase (1.7 K) [5.123]. (The absorption edge of EuSe is at ≈ 1.96 eV for $T = 27$ K). The inset shows the magnetic phase diagram of EuSe [5.123] as a function of internal magnetic field H_i

zone, yielding the 'magnetic' Brillouin zone of the different phases. The phonon frequencies associated with these wave vectors are folded into the zone center and become Raman-active. The theory of this often-called magnetic 'Bragg' scattering was given by *Ousaka* et al. [5.39, 53] and *Safran* et al. [5.36, 51]. The 'Bragg' scattering, like the spin-disorder induced scattering, is associated with $\alpha^{(1,1)}$ but, unlike the latter, it originates in the *coherent* term of the two-spin correlation function [5.36, 39, 51, 53] (see first term of Eq. (4.17)). From a microscopic point of view, it stems from the fact that the spin-orbit coupling leads to *elastic* scattering of the excited $4f^6 5d^1$ states by the spin system.

Another interesting aspect of the data in Fig. 5.7 is the increase in the intensity of the structure at $\approx 180\ \text{cm}^{-1}$ for temperatures close to, but above the AF-I transition (see also Fig. 4.14). This effect, and also the presence of the additional feature near $118\ \text{cm}^{-1}$ (see also Fig. 4.15) in a narrow range of temperatures above the transition have been identified with spin fluctuations [5.36, 45a, b, 46, 50, 51].

EuTe was long considered to be an anomalous case within the EuX family [5.22, 34] because of results showing that the scattering in the paramagnetic phase, instead of a sharp decrease, exhibited an increase in intensity upon cooling below T_N [5.22, 34, 41, 44] (see also Fig. 4.11). These results, obtained with $\omega_i \approx E_1''$, led to some controversy in the literature as to what extent the scattering in EuX should be ascribed to magnetic effects [5.22–24, 26, 34]. The fact is that these particular data correspond to a resonant situation in which the contributions due to $\alpha^{(1,2)}$ and higher-order terms become important [5.42, 44]. The scattering cross section due to, e.g., $\alpha^{(1,2)}$ is proportional to a four-spin correlation function which, for some of its components, shows a temperature behavior in qualitative agreement with the data [5.40, 41].

Off the resonance maximum, EuTe behaves in a 'normal' way in the sense that the scattering is mostly determined by the $\alpha^{(1,1)}$ term. Raman spectra obtained under these conditions show the characteristic quenching of the spin-disorder scattering at low temperatures and, in addition, the appearance of the magnetic-'Bragg' scattered LO(L) phonon line below T_N [5.44, 47–49] (see Fig. 4.10). Theoretical work on scattering in the paramagnetic phase is reported in [5.40, 43, 44] and for the AF-II phase in [5.39, 45, 51]. Other references for EuTe are [5.30, 43, 44] (allowed second-order phonon scattering) and [5.15, 17] (scattering due to screened LO-phonon modes in n-type EuTe). Data on LO(Γ) phonon scattering are presented in [5.25, 27, 44, 47, 49]. The latter results follow the same pattern exhibited by the other members of the EuX family (Figs. 4.12, 13).

5.2.2 Multiple Scattering by LO(Γ) Phonons in YbX

The YbX (X = S, Se, Te) compounds show multiphonon scattering in a narrow range ($\approx$0.15 eV) of excitation energies, resonant with the E_1, E_2: $4f^{14} \rightarrow 4f^{13}(^2F_{7/2})5d^1(t_{2g}, e_g)$ transitions [5.33, 34, 54, 55]. The LO(Γ) line (nominally forbidden in the fcc structure) and its overtones occur for polarizations of the incident and scattered photons parallel to each other, irrespective of the crystal orientation [5.26, 33–34, 54, 55]. This is the only intrinsic form of scattering known in these materials, as allowed second-order phonon spectra have not yet been obtained. Moreover, results showing evidence of first-order scattering by $q \neq 0$ phonons in YbTe [5.22, 33, 34] could not be reproduced in [5.54] and are unlikely to represent an intrinsic feature of the scattering.

Results of Raman experiments on YbS are given in [5.23, 34, 55]. Figure 5.8 shows a series of multiphonon spectra of this compound in the vicinity of the E_2 resonance [5.54]. Resonant data on YbSe ($\omega_i \approx E_2$) and YbTe ($\omega_i \approx E_1$) are presented, respectively, in [5.26, 34] and [5.22, 34]. The temperature dependence of the scattering is considered in [5.26, 34, 54].

Various models have been proposed to account for the properties of LO(Γ) phonon scattering in YbX. The apparent relationship with the Eu compounds, i.e., the fact that both YbX and EuX exhibit nominally forbidden first-order

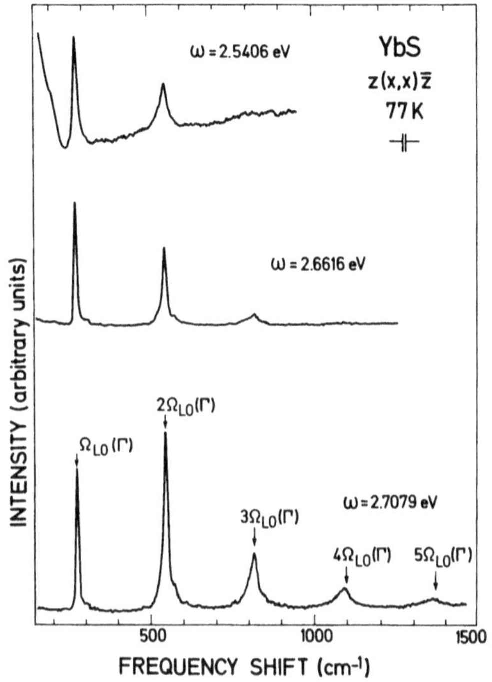

Fig. 5.8. Raman spectra of YbS at 77 K for different excitation energies. The frequencies of the harmonics of the LO(Γ) phonon are indicated [5.54]

scattering and also strong multiphonon processes, is largely emphasized in the works of *Vitins* and *Wachter* [5.22–24, 26, 33, 34] who considered the analogy as evidence favoring a hot-recombination model of the scattering. Irrespective of the question of the validity of such a model, a problem with this approach is that it overestimates the importance of the multiple scattering (merely a result of the large electron-phonon interaction associated with the excitation of the $4f$ shell) at the expense of ignoring the different nature of the symmetry-breaking mechanisms in the two families. The role played by the spin system in modifying the *translational* symmetry of the Raman tensor in the Eu compounds has now been well established. The origin of the scattering in YbX is addressed in the works of *Merlin* et al. [5.54, 55] and *Abiko* [5.56]. These authors pointed out the similarities with the "forbidden" LO(Γ) phonon scattering observed in many polar materials and concluded that, like in those cases, the symmetry breakdown is an effect of the *spatial* dispersion of the scattering tensor which relies on the coupling of the electronic states and the LO phonons through the Fröhlich interaction [5.54–56]. As indicated previously (Sect. 5.2.1), this mechanism is also responsible for the "forbidden" LO(Γ) scattering in the ferromagnetic phase of EuX.

The theories of *Merlin* et al. [5.45, 55] and *Abiko* [5.56], although markedly different in many respects, recognize the importance of the localized nature of the $4f^{13}$ hole in determining the properties of the scattering. In *Abiko*'s model, the resonantly excited intermediate state is considered to be a $4f^{13}5d^1$ exciton

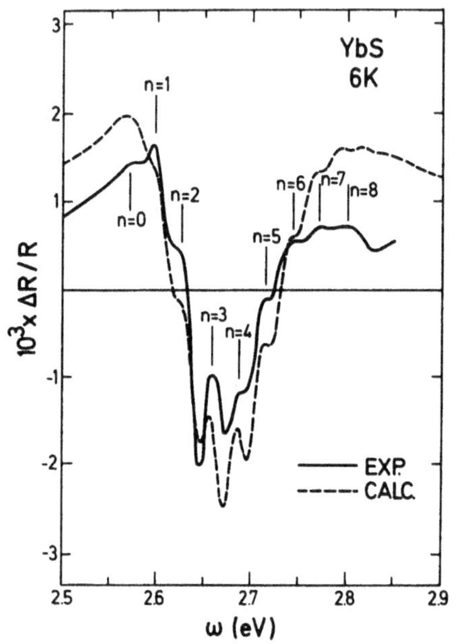

Fig. 5.9. Wavelength-modulated reflectance spectrum of YbS [5.54]. The dashed curve is a plot of the theoretical expression described in the text. The n-phonon structures are indicated by bars

immersed in the $6s$ band. This situation leads to a double resonance in the scattering tensor which, under certain conditions, results in an enhancement of the $q \approx 0$ LO scattering [5.56]. A simple explanation of the large multiple scattering is provided by *Merlin* et al. [5.54, 55] who considered the problem of a $4f^{13}5d^1$ dispersionless exciton interacting with LO phonons. What occurs in this case is a "displacement" of the phonon coordinates in the excited electronic state which (as in other configuration-coordinate models) results in a scattering cross section totally determined by Franck-Condon overlap integrals [5.54, 55]. The same values of the parameters of the model, obtained from a fit to the Raman data of YbS, were used by *Merlin* et al. [5.54] to calculate the wavelength-modulated reflectance spectrum in the region of resonant excitation. Their results, together with the experimental curve, are shown in Fig. 5.9. The good agreement between the two supports the claim that the fine structure in the reflectance spectrum is due to multiphonon processes involving LO(Γ) phonons. The fact that the frequency of this mode as obtained from Fig. 5.9 is 15 % lower than the value determined in the Raman spectra has been considered as evidence against the interpretation of the inelastic light-scattering data in terms of hot luminescence [5.54, 55].

5.3 Metals

The trivalent rare-earth monochalcogenides $RE^{3+}X$ are fcc metals. Cation and/or anion vacancies (up to 30 %) are easily obtained without significantly changing the lattice parameters or the crystal structure [5.136]. Studies of the phonon spectrum by means of infrared spectroscopy are prevented because of the high metallic reflectivity. Moreover, neutron experiments meet with the difficulty of strong absorption cross sections of the rare earths. On the other hand, the defect structure of the $RE^{3+}X$ has made possible phonon investigations through defect-induced phonon Raman scattering [5.57–59]. These studies are facilitated by the fact that the plasma reflection edges occur within the visible frequency range, yielding to a large penetration depth of the laser light and relatively small reflection losses. Shifts in the plasma edges over more than 1 eV may occur with varying defect concentration.

5.3.1 Defect-Induced Scattering

In Fig. 5.10 we show Raman spectra of Gd_xS at 300 K for $x=1.0$, 0.8 and 0.7 which have been discussed in detail by *Güntherodt* et al. [5.58a].

The assignment of first-order scattering by acoustic and optical phonons of $Gd_{1.0}S_{1.0}$ denoted by A and O, respectively, is based on the strong increase in intensity with increasing defect concentration as shown by the data on $Gd_{0.8}S$ and $Gd_{0.7}S$. In the latter case a further intensity increase (compared to $Gd_{0.8}S$) is suppressed due to increased defect-defect interactions. Second-order scattering (2A, O+A, 2O) has been assigned by the temperature dependence of the (2A) intensity according to the Bose-Einstein factor and by corresponding (O+A, 2O) frequency combinations. A symmetry analysis of the scattering intensity, particularly for the first-order optic phonon scattering, has shown almost equal

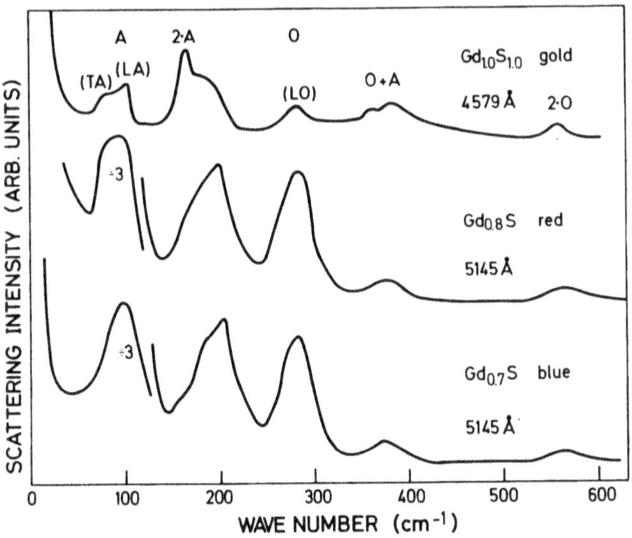

Fig. 5.10. Raman spectra of single crystals of $Gd_{1.0}S_{1.0}$ and $Gd_{0.8}S$, and of an evaporated thin film of $Gd_{0.7}S$ at 300 K [5.58a]; first-order scattering from acoustic (A) and optic (O) phonons and second-order scattering (2A, O+A, 2O) are indicated

contributions from the Γ_{12}^{+} and Γ_{25}^{+} Raman tensor components and a negligible Γ_1^{+} component. The latter fact, which has been attributed to the lack of resonant optical excitation from the $(Gd^{3+})4f^7$ level, is of importance for the $4f$ electron-phonon coupling (discussed in Sect. 5.2) and the following discussion on mixed valence materials (Sect. 5.5). For the same reason, i.e., no resonant optical excitation from the $4f$ level, there have been no observations of magnetic-order effects on the Raman spectra of GdS for temperatures below the Néel temperature of about 50 K [5.58a].

Of general importance for light scattering in metals is also the observation of phonon broadening effects and the smearing of structures in the Raman spectra of polished GdS samples compared to cleaved single crystals [5.58a]. This could be further corroborated by most recent, systematic studies of Raman scattering in rare-earth intermetallics [5.137]. Freshly fractured, preferably single crystalline surface areas of intermetallics, are in any case superior to polished surfaces.

Raman spectra of polished samples of GdS, GdSe, GdTe and LaTe have been reported by *Treindl* and *Wachter* [5.59]. No differences between polished and cleaved samples were indicated.

Defect-induced scattering of the metallic, intermediate-valence compounds, such as $Sm_{0.75}Y_{0.25}S$, SmS and TmSe, will be discussed in Sect. 5.5.

5.3.2 Superconductors. The Model of Local Cluster Deformabilities

The strong electron-phonon coupling in superconductors, showing up in pronounced phonon anomalies, is expected to yield strong Raman scattering intensities as well, through a strong phonon modulation of the electric susceptibility [5.141b]. The interrelationship between phonon anomalies and Raman intensities in the superconductor YS has been investigated by *Güntherodt* et al. [5.70]. The symmetry-analyzed Raman spectra of YS at 300 K are shown in the upper part of Fig. 5.11. First-order scattering from acoustic and optic phonons is seen near 100 cm^{-1} and 300 cm^{-1}, respectively, whereas second-order scattering has been identified near 200 cm^{-1}. Practically all the scattering intensity appears in the Γ_1^{+} component. If this spectrum was simply due to defect-induced Raman scattering, the dominance of the Γ_1^{+} scattering component would be difficult to comprehend in view of its absence in the Raman spectrum of GdS (Sect. 5.3.1). Moreover, no $4f$ states are involved in the optical excitation spectrum of YS. The key to the understanding is the fact that the scattering intensity from acoustic phonons is much larger than that of optical phonons, pointing to a connection with the LA(L) phonon anomaly [5.138]. In a lattice dynamical model calculation [5.107], the latter has been attributed to a local charge deformability of (Γ_1^{+}) breathing symmetry on the S ion due to (S)$3d_{xy} \rightarrow 3d_{xy}$ excitations near the Fermi energy [5.140].

Assuming radial coupling between nearest neighbors in a NaCl lattice, the electron-lattice interaction can be described in terms of three local cluster deformabilities of breathing (Γ_1^{+}), quadrupolar (Γ_{12}^{+}) and dipolar (Γ_{15}^{-}) sym-

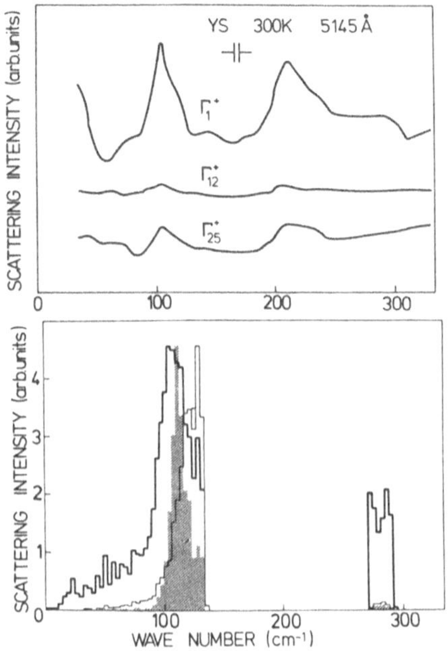

Fig. 5.11. *Upper part:* polarized Raman spectra of YS at 300 K; intensity near 200 cm^{-1} due to second order scattering (see also Fig. 5.17 for YSe). *Lower part:* calculated one-phonon density of states weighted by $[n(\boldsymbol{q},j)+1]/\omega(\boldsymbol{q},j)$ (*thick solid line*); *shaded area:* $\Gamma_1^+(S)$ contribution; *thin solid line:* $\Gamma_{12}^+(S)$ contribution [5.70]

metry [5.139]. The consideration of these displacement-induced local charge deformabilities in the dynamical matrix, in addition to the rigid ion-part, has been shown by *Bilz* et al. [5.107] to yield a good lattice dynamical description of the measured phonon dispersion of YS [5.138].

In a simple model, the Raman scattering cross section has been derived in terms of the same local cluster deformabilities which describe the strong anomalies in the phonon dispersion curves [5.70]. The time-dependent polarization operators $P_{\alpha\beta}(t)$ determining the Raman intensity [5.141a] have been expanded into normal coordinates $A(\boldsymbol{q},j)$ and impurity-induced distortions $B(\boldsymbol{q},\kappa)$ at a lattice site κ; j labels the phonon branches. The first-order terms read as follows:

$$\sum_{\kappa} P_{\alpha\beta}(\boldsymbol{q},j\kappa)B(-\boldsymbol{q},\kappa)A(\boldsymbol{q},j)=\tilde{P}_{\alpha\beta}(\boldsymbol{q},j)A(\boldsymbol{q},j). \qquad (5.2)$$

Equation (5.2) describes the symmetry breaking effect of impurities with respect to the q selection rule so that $\tilde{P}_{\alpha\beta}(\boldsymbol{q},j)\neq 0$ for $\boldsymbol{q}\neq 0$. Instead of using a many-parameter fit for the expansion coefficients of $\tilde{P}_{\alpha\beta}(\boldsymbol{q},j)$ in ordinary space, cluster-related projection operators of symmetry Γ_1^+ and Γ_{12}^+ are used to describe the site symmetry of the cluster ion at site κ for radial displacements of the nearest neighbors. (The Γ_{15}^- symmetry is not considered in the present context). Thus the symmetry-projected components of the Raman spectra are calculated with the $\tilde{P}_{\alpha\beta}(\Gamma_i,\kappa)$ as constant parameters, depending only on the cluster-deformation symmetry Γ_i^+. In Fig. 5.11 we compare the measured Raman spectra of YS (upper part) with the calculated one-phonon density of states (bold lines in lower part), derived from the fit of the measured phonon dispersion curves [5.107]. The density of states has been weighted by $[n(\boldsymbol{q},j)+1]/\omega(\boldsymbol{q},j)$. The rather poor description of the measured spectra by the density of states emphasizes the importance of particular matrix elements in the electron-phonon coupling. By taking into account the Γ_1^+ cluster deformability around the S ion, we obtain the hatched, symmetry-projected one-phonon density of states. The full symmetric (Γ_1^+) monopolar or breathing cluster deformability is proportional to $\left|\sum_{\gamma=1}^{3} \sin(k_\gamma a) e_\gamma(\boldsymbol{k}j)\right|^2$, where γ is an index of Cartesian coordinates, a is half the cubic lattice constant and e_γ is the eigenvector of the ligand ions around the central cluster ion in the γ-direction due to the phonon with wave vector $\boldsymbol{k}$ and branch j. This Γ_1^+ local cluster of deformability is analogous to the Γ_1^+ electron-lattice coupling described in (4.6) of Chap. 4. This weighting factor mainly enhances contributions from the zone boundary. The hatched area in Fig. 5.11 reproduces clearly the dominant scattering intensity from acoustic phonons and the almost negligible scattering intensity from optic phonons. The difference between the measured and calculated Γ_1^+ peak positions near 110 cm^{-1} is attributed to slightly different stoichiometries in the bulk as seen by neutron scattering and at the surface of the same sample as seen by Raman scattering.

Considering a quadrupolar (Γ_{12}^+) cluster deformability at the S ion, we obtain the Γ_{12}^+ projected one-phonon density of states (thin line in the lower part of Fig. 5.11) with a maximum clearly separated from that of the hatched (Γ_1^+) area. The absence of any scattering intensity in the measured Γ_{12}^+ spectrum supports the assumption of only a breathing charge deformability of the S ion as the origin of both the LA(L) phonon anomaly and the strong (Γ_1^+) Raman intensity of YS near 110 cm^{-1}. A similar situation is found in YSe as seen in Figs. 5.14, 5.17.

The above analysis has also been applied to superconducting transition metal compounds such as $TiN_{0.95}$ [5.70]. The two maxima in the unpolarized, first-order, defect-induced Raman spectrum of *Spengler* et al. [5.142] near 220 cm^{-1} and 320 cm^{-1}, respectively, have been attributed to nearly equal quadrupolar and breathing deformabilities at the N ion. These Γ_{12}^+ and Γ_1^+ charge deformabilities also describe the TA(L) and LA(L) phonon anomalies, respectively [5.143].

5.4 Semiconductor-Metal Transitions

The divalent semiconducting rare-earth monochalcogenides undergo semiconductor-metal transitions under pressure or as a function of cation or anion substitution (see Sect. 5.1). The interest in Raman studies of these transitions arose mainly from the intriguing possibility of following phononic and electronic properties upon approaching the concurrent configuration crossover. The latter is attributed to a reduction of the $4f^{n+1} \to 4f^n 5d^1$ excitation gap with decreasing lattice parameter due to the increased $5d(t_{2g} - e_g)$ crystal field splitting [5.144]. Hence, the spin-orbit split (J) multiplet levels of the $4f^{n+1}$ configuration should be strongly affected due to increased configuration interactions [5.145]. Anomalous phonon behavior had been indicated by the softer bulk modulus of SmS in its semiconducting phase ($B=475$ kbar) compared to the pressure-transformed intermediate-valence phase ($B=498$ kbar) [5.61, 96]. The latter result is rather unexpected and in contrast to the soft bulk modulus ($B \approx 0$) of $Sm_{1-x}Y_xS$ for $x \gtrsim 0.15$ [5.146] and $B=250$ kbar of $Tm_{0.99}Se$ [5.97].

Unfortunately, Raman studies of these semiconductor-metal transitions under hydrostatic pressure have so far failed because of the small scattering intensities (average about 50 counts/s) and a possible chemical reaction of the sample surface with the pressure-transmitting liquid. Thus, only solid solution systems of $Sm_{1-x}RE_xS$ and $Sm_{1-x}RE_xSe$ have been investigated, despite complications in the interpretation of the data due to the presence of a highly concentrated third component.

5.4.1 Phonon Anomalies

Using Raman scattering, the impending valence instability of Sm in semiconducting SmS has been identified by *Güntherodt* et al. [5.60, 62] in terms of a 17% softening of the LO(L) phonon frequency. As a reference divalent EuS, which has the same lattice constant, has been chosen. In Fig. 5.12 (upper part) we show

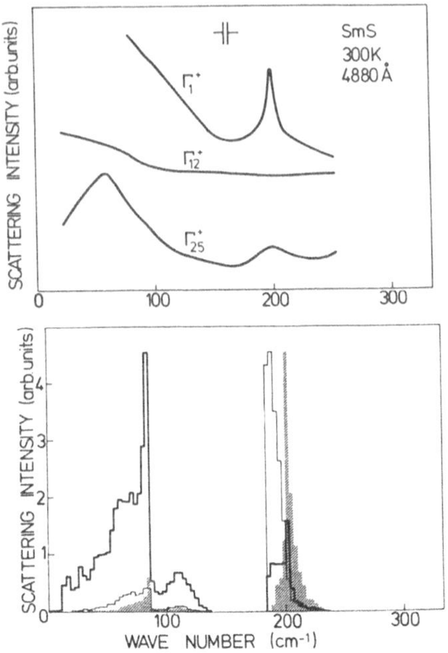

Fig. 5.12. *Upper part:* polarized Raman spectra of semiconducting SmS at 300 K. *Lower part:* calculated one-phonon density of states weighted by $[n(\mathbf{q}j)+1]/\omega(\mathbf{q}j)$ (*thick solid line*); *shaded area*: Γ_1^+ (Sm) contribution; *thin solid line*: Γ_{12}^+(Sm) contribution [5.70]

the symmetry-analyzed Raman spectrum of semiconducting SmS. First-order scattering from optic phonons shows up near 200 cm^{-1} predominantly in the Γ_1^+ component. The Raman spectrum has been analyzed within the same framework of local cluster deformabilities described in Sect. 5.3.2. In the lower part of Fig. 5.12 we show the one-phonon density of states (bold line) derived from a fit of the measured phonon dispersion of SmS [5.110] using the model of local intraionic charge deformabilities [5.107]. This model fit has also predicted the dispersion of the LO phonon branch, which so far has not been measured by neutron scattering [5.106, 110]. For the one-phonon density of states weighted by the Γ_1^+ breathing cluster deformability at the Sm site (hatched area) we find good agreement with the measured Γ_1^+ spectrum near 200 cm^{-1} (see lower part of Fig. 5.12, bold line). The assumption of a quadrupolar (Γ_{12}^+) cluster deformability at the Sm site (thin line in lower part of Fig. 5.12) leads to a maximum which is clearly separated from that of the hatched area near 200 cm^{-1}. This contribution is clearly not important as there is no scattering intensity in the measured Γ_{12}^+ spectrum. Hence, Raman scattering provides a crucial check for testing the relative weight of local cluster deformabilities of different symmetry types. (The analysis of the scattering from acoustic phonons in SmS is obscured by the strong background seen in Fig. 5.12).

The large (Γ_1^+) breathing response of the charge density around the Sm ion has been attributed to virtual $4f^6 \rightarrow 4f^5 5d^1$ excitations across the 0.15 eV *f–d* excitation gap, which is much smaller than the 1.65 eV gap of EuS. This large breathing charge deformability of the Sm ion in SmS also relates to the LO phonon anomaly near the *L* point within the lattice dynamical model [5.107]. It should be pointed out that this anomaly gives strong evidence for *f–d* hybridization and the precursor of the valence instability of semiconducting SmS.

In a model calculation by *Baba* et al. [5.147], the renormalization of the phonon frequencies of SmS has been expressed as a function of the energy gap between the $4f^6$ level and the bottom of the conduction band. A larger phonon softening has been obtained for the semiconducting phase because of the smaller energy gap, as compared to the metallic phase with a larger (although negative) gap. The microscopic origin of the renormalization has been attributed to the phonon-induced on-site *f–d* hybridization interaction which is enhanced for the smaller energy gap of the semiconducting phase of SmS.

5.4.2 Electronic Raman Scattering near Configuration Crossover

Spin-orbit levels and crystal-field split levels of $4f$ states of rare-earth ions in insulating hosts have been extensively studied by *Koningstein* et al. [5.148] using electronic Raman scattering. No measurements have so far been reported for metals, except $CeAl_2$ [5.137] and most recently CeB_6, $EuPd_2Si_2$ [5.149]. In this section we discuss electronic Raman scattering from $4f$ spin-orbit levels in $Sm_{1-x}RE_xSe$ and $Sm_{1-x}RE_xS$ solid solutions in the vicinity of the configuration

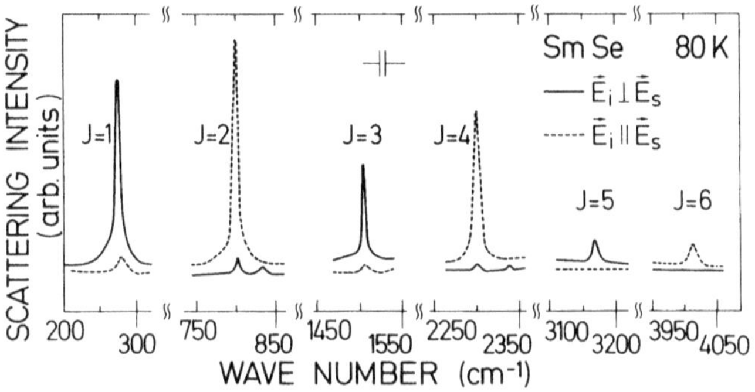

Fig. 5.13. Electronic Raman scattering from the $(Sm^{2+})4f^6(^7F_J)$ configuration of SmSe at 80 K under 5145 Å laser excitation using backscattering from a cleaved (100) face; $\boldsymbol{E}_i \| \boldsymbol{E}_s$: $\Gamma_1^+ + 4\Gamma_{12}^+$, $\boldsymbol{E}_i \perp \boldsymbol{E}_s$: $\Gamma_{25}^+ + \Gamma_{15}^+$, with $\boldsymbol{E}_{i(s)}$ the electric-field vector of the incident (scattered) photon [5.66]

crossover. Such investigations have a direct bearing not only on the valence fluctuation problem, but also on the fundamentals of Raman scattering.

The ground state of Sm^{2+} in the Sm monochalcogenides is $4f^6$ $(^7F_{J=0,1,\ldots 6})$, with a 0.6 eV wide spin-orbit split ^7F_J multiplet. Electronic Raman scattering from the different J multiplet levels of cleaved (100) SmSe [5.66] is shown in Fig. 5.13. The odd J levels show up for perpendicular incident ($\boldsymbol{E}_i$) and scattered ($\boldsymbol{E}_s$) polarization vectors, whereas the opposite is true for the even J levels. It has been shown for the case of SmS that the $J=1$ peak (and in principle the $J=3,5$ peaks) appears only in the antisymmetric Γ_{15}^+ component [5.62]. The scattering intensity from either odd or even J levels decreases monotonically with increasing J values as seen in Fig. 5.13. This has been attributed to the approaching of the higher J levels to the $4f$–$5d$ excitation gap (the excitation to the $J=6$ level near 4010 cm^{-1} coincides with the 0.5 eV gap of SmSe). This conclusion is further supported by the fact that the $J \geq 3$ levels are not observed in SmS, which shows a $4f$–$5d$ gap of about 1200 cm^{-1} (0.15 eV).

Electronic Raman scattering has been observed in SmS, SmSe and SmTe by *Nathan* et al. [5.63, 64], with particular emphasis on the temperature dependence of the singlet-triplet ($J=0\to1$) excitation. In all three compounds the $J=0\to1$ transition has been found to shift to a lower frequency upon cooling below room temperature. This temperature dependence of the singlet-triplet excitation could be fitted by results obtained in the random phase approximation [5.150], using the free-ion spin-orbit coupling constant ($\lambda = 293.5$ cm^{-1}) and by introducing an exchange interaction energy θ. The latter was found to decrease from 44 cm^{-1} in SmS to 8 cm^{-1} in SmTe. This seems to parallel a decreasing $5d$ admixture into the $4f^6$ ground state with increasing f–d excitation gap from SmS to SmTe.

Substitution of the cation in the solid solution system $Sm_{1-x}RE_xSe$ by, e.g., Y or La reduces the lattice parameter and the $4f$–$5d$ excitation gap, without yielding the transition into the metallic, mixed valence phase [5.151]. In Fig. 5.14 we show the polarized Raman spectra of $Sm_{1-x}Y_xSe$ at 80 K for $x=0, 0.25, 0.50, 0.75$ and 1.0 obtained by *Güntherodt* et al. [5.66]. For the sake of completeness,

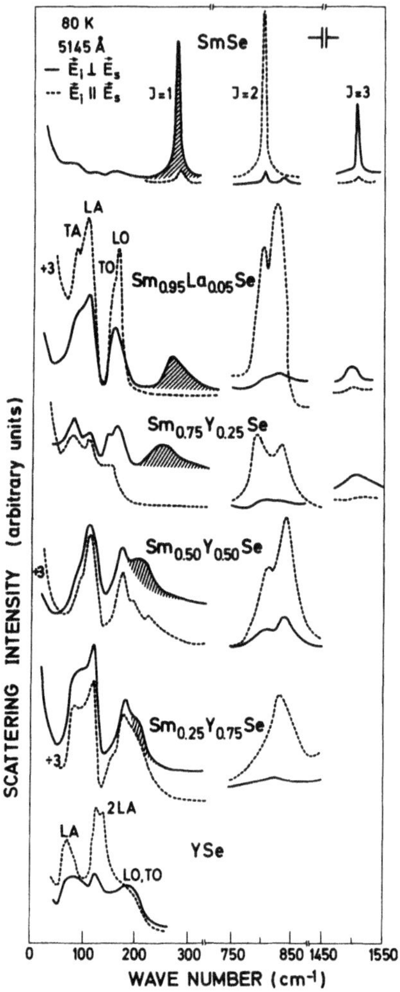

Fig. 5.14. Electronic Raman scattering from the J multiplet levels of $Sm_{1-x}Y_xSe$ at 80 K for $0 \leq x \leq 1.0$; for $x=0.05$ the La-substituted $Sm_{0.95}La_{0.05}Se$ is shown [5.66]. The scattering configuration is the same as in Fig. 5.13. The hatched area indicates the electronic scattering from the $J=0 \rightarrow 1$ excitation. Phonon scattering is seen below 200 cm^{-1}

$Sm_{0.95}La_{0.05}Se$ has also been included. For the latter sample one observes below 200 cm^{-1} first-order defect-induced Raman scattering from acoustic and optic phonons which is absent in pure SmSe. The $J=1$ peak of $Sm_{0.95}La_{0.05}Se$ has drastically broadened compared to that of pure SmSe at 275 cm^{-1} and has shifted to 266 cm^{-1}. The $J=1$ peak is broadened further with increasing x and merges with the optical phonon density of states for $x \geq 0.50$. For $x=0.75$, the $J=1$ peak has shifted to about 210 cm^{-1} and is barely seen. (The Raman spectrum of YSe is similar to that of YS discussed in Sect. 5.3.2). The $J=3$ peak of SmSe in Fig. 5.14 does not shift with increasing x, but becomes strongly broadened and finally can no longer be resolved for $x \geq 0.50$. On the other hand, the persistence of the peak related to the $J=2$ level up to $x=0.75$ indicates that the $4f$–$5d$ gap is still finite, i.e., of the order of 0.1 eV (800 cm^{-1}). This is consistent with the fact

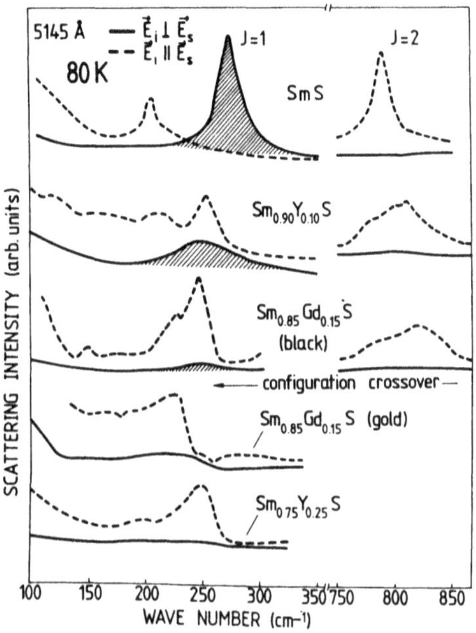

Fig. 5.15. Polarized Raman spectra of $Sm_{1-x}Y_xS$ ($x=0$, 0.10, 0.25) and of $Sm_{0.85}Gd_{0.15}S$ in the "black" and pressure-transformed ($p>4$ kbar) "gold" phase at 80 K [5.66]. The scattering configuration is the same as in Fig. 5.13. Dashed line below 300 cm^{-1}: phonon scattering. Hatched area: electronic scattering from the $J=0\rightarrow1$ excitation

that $Sm_{1-x}Y_xSe$ for all values of x does not undergo a transition into the homogeneously mixed-valent phase [5.151]. Otherwise, the intensity changes and splittings of the $J=2$ level with increasing x are unexplained.

The solid solution system $Sm_{1-x}Y_xS$ was first investigated using Raman scattering by *Smith* et al. [5.64] and *Tsang* [5.65] for concentrations near ($x \lesssim 0.15$) and beyond ($x>0.15$) configuration crossover (CC). Polarized Raman spectra by *Güntherodt* et al. [5.66] have shown a clear separation into phonon and electronic ("magnetic") Raman scattering and the evolution with CC. Figure 5.15 shows the polarized Raman spectra of cleaved (100) faces of $Sm_{1-x}Y_xS$ for $x=0$, 0.10 and 0.25 at 80 K. The case of $Sm_{0.85}Gd_{0.15}S$ has also been included, since the very same sample can be investigated in its nontransformed black phase (near CC) and in its pressure-transformed ($p>4$ kbar) gold phase (beyond CC). For SmS ($x=0$) in Fig. 5.15, phonon scattering is shown by the dashed line below 300 cm^{-1} (see Sect. 5.4.1 and Fig. 5.12), whereas electronic scattering from the $J=1$ level is represented by the solid line (hatched area) near 275 cm^{-1}. Electronic scattering from the $J=2$ level is shown by the dashed line near 780 cm^{-1}. The $J\geq3$ levels are not observed. With increasing x ($x\leq0.15$) the $J=1$ level is reduced in intensity, strongly broadened and shifts to 250 cm^{-1} for $x=0.15$. In the same sense the $J=2$ level is subject to strong broadening. Beyond CC ($x\geq0.15$) no contributions from the $J=0\rightarrow1$ and $J=0\rightarrow2$ excitations could be identified, contrary to previous unpolarized Raman measurements [5.65]. In the latter no distinction could be made between electronic and phonon contributions. The dominant phonon scattering intensity of, e.g., $Sm_{0.75}Y_{0.25}S$ near 250 cm^{-1}, will be discussed in more detail in Sect. 5.5.1.

From the data in Fig. 5.15 it can be concluded that the broadening of the $J=1$ peak must have exceeded at least 200 cm^{-1} (25 meV) FWHM in order to be no longer resolved. This result is consistent with neutron scattering studies on metallic SmS under pressure [5.152] showing no evidence of $J=0\to1$ excitations. On the other hand, a peak near 31 meV (250 cm^{-1}) found in neutron scattering on powdered $Sm_{0.75}Y_{0.25}S$ has been attributed to scattering from the $J=0\to1$ excitation [5.153]. A similar peak is found in single crystals near 25 meV (200 cm^{-1}) [5.154]. This point will be discussed further in Sect. 5.5.3 in the context of Raman scattering by a localized mode within the acoustic-optic phonon gap ("gap mode").

The lifetime broadening of excited $4f$ multiplet levels due to mixing with conduction-electron states has been predicted theoretically by *Hirst* [5.145] for SmS near and beyond CC. For a particular case near CC, which might be realized by the samples $Sm_{0.25}Y_{0.75}Se$ or $Sm_{0.90}Y_{0.10}S$, sharp levels are expected for $J=0$, 1, 2, 5 and 6, whereas some finite broadening is indicated for $J=3$, 4. However, in the Raman data of the above two samples, the $J\geq3$ multiplet levels are not observable. Moreover, the $J=1,2$ levels are found to be subject to strong broadening. The minimum level broadening deduced from the Raman data is of the same order of magnitude as the level width obtained from a discussion of the energy balance of mixed valence ions based on the ionic configuration model [5.155].

5.5 Intermediate Valence Materials

The primary interest in investigating mixed valence materials using Raman scattering arose from the estimate [5.2, 4] that the fluctuation rate may be on the same time scale as the lattice vibrations. Hence, Raman scattering experiments in intermediate valence materials by *Güntherodt* et al. [5.61, 62, 67, 70], *Treindl* and *Wachter* [5.68, 69], and *Stüsser* et al. [5.73, 74] have so far been concerned with the investigation of phonon anomalies and their relation with the electron-phonon interaction. In particular, polarized Raman scattering [5.67, 70, 72] has provided an experimental test of the relative importance of the different charge deformabilities introduced in the lattice dynamical model calculations [5.67, 72, 107].

Spin relaxations of the valence-fluctuating rare-earth ions have been observed in quasielastic neutron scattering [5.156]. On the other hand, charge relaxations are expected to show up in quasielastic light scattering [5.157]. Such measurements are presently awaiting their experimental realization using multipass tandem Fabry-Perot interferometry (for technical details see [5.158]). The use of an iodine absorption cell together with 5145 Å laser excitation [5.159] has proven to be inadequate to discriminate against surface roughness scattering of these opaque materials [5.160].

Brillouin scattering in intermediate valence compounds was first carried out by *Barth* and *Güntherodt* [5.161] using a high-contrast multipass Fabry-Perot interferometer. The elastic constants have been derived from the measured sound velocities of bulk or surface acoustic waves in rare-earth intermetallics such as $REAl_2$, $REPd_3$ and $RECu_2Si_2$, and in TmSe. In the latter case the sound velocities of bulk acoustic waves could be measured because of the large penetration depth of visible laser light of the order of 1000 Å. The determined elastic constants have been shown to be in good agreement with ultrasonic measurements [5.97] and neutron scattering data [5.105]. In particular, the negative c_{12} found by *Boppart* et al. [5.97] has been confirmed.

Here we want to describe Raman scattering in intermediate valence rare-earth chalcogenides, analyzed within the same framework of electron-lattice interactions introduced for superconducting compounds in Sect. 5.3.2 and for semiconducting SmS in Sect. 5.4.1. We also discuss the coupling between localized charge-density fluctuations and phonons of the same local symmetry to form a bound state (Sect. 5.5.3).

5.5.1 Phonon Anomalies and Raman Intensities

The intermediate valence phase of the solid solution system $Sm_{1-x}Y_xS$ with $x \gtrsim 0.15$ was the first striking example of anomalous electron-lattice interactions associated with valence fluctuations. The bulk modulus is soft for $x \gtrsim 0.15$ due to the elastic constant $c_{12} < 0$ [5.146]. Strong phonon anomalies have been identified in the [111] direction of $Sm_{0.75}Y_{0.25}S$ [5.70, 104–108]. Besides the LA phonon anomaly for $0 < k < 3/4\ k_L$ associated with the soft bulk modulus, a corresponding anomaly was found for the LO phonon breathing mode near the L point. In addition, there is an anomaly near the Γ point of similar magnitude to that near the L point [5.107]. Similar, but less pronounced anomalies have been found in TmSe [5.72, 97, 105].

The polarized Raman spectrum of a cleaved single crystal of $Sm_{0.75}Y_{0.25}S$ [5.70] is shown in the upper part of Fig. 5.16. As discussed in Sect. 5.4.2, the contribution from electronic Raman scattering is negligible. The Raman spectrum in Fig. 5.16 is dominated by the Γ_1^+ scattering intensity near 245 cm^{-1} and the weaker one near 85 cm^{-1}, superimposed on the rising background. The one-phonon density of states (bold lines) derived from the fit [5.107] of the measured phonon dispersion [5.104] is shown in the lower part of Fig. 5.16. The hatched area results from weighting the density of states with the Γ_1^+ breathing cluster deformability, which represents the dominant electron-phonon matrix element in the Raman scattering process. Obviously, the TO phonon density of states near 270 cm^{-1} does not contribute to the hatched area. The largest contribution is due to the LO phonon breathing mode near the L point. The small hatched area in the acoustic phonon region near 90 cm^{-1} is due to the Γ_1^+ deformability of the Sm ion induced by the longitudinal breathing motion of the surrounding Sm ions. The hatched areas in Fig. 5.16 describe the measured Raman spectrum very well, except for the shoulder in the Γ_1^+ spectrum near

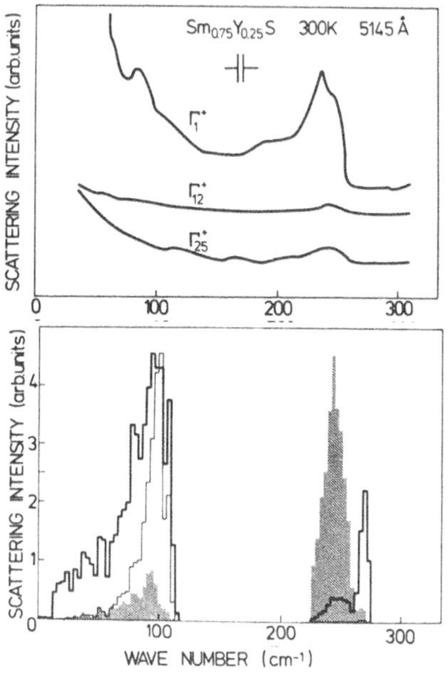

Fig. 5.16. *Upper part:* polarized Raman spectra of $Sm_{0.75}Y_{0.25}S$ at 300 K. *Lower part:* calculated one-phonon density of states weighted by $[n(\boldsymbol{q}j)+1]/\omega(\boldsymbol{q}j)$ (*thick solid line*); *shaded area:* $\Gamma_1^+(\mathrm{Sm})$ contribution; *thin solid line:* $\Gamma_{12}^+(\mathrm{S})$ contribution [5.70]

190 cm^{-1}, which will be discussed in Sect. 5.5.3. The assumption of a quadrupolar (Γ_{12}^+) deformability of the S ion, previously used in the lattice dynamical model [5.107], yields the thin-line acoustic-phonon contribution near 100 cm^{-1} in the lower part of Fig. 5.16. However, since the measured Γ_{12}^+ spectrum does not show any intensity near 100 cm^{-1}, such a local Γ_{12}^+ charge-density distortion is unrealistic. Thus, it had to be attributed to residual effects of screened Coulomb interactions [5.70].

The absence of scattering intensity in the Γ_{25}^+ spectrum in Fig. 5.16 indicates that intersite Sm-Sm or S-S interactions are negligible. Hence, the assumption of a local electron-lattice interaction, i.e., of Sm-S interactions only, in the lattice dynamical model for $Sm_{0.75}Y_{0.25}S$ [5.107] is corroborated experimentally.

The very same Γ_1^+ charge deformability of the mixed-valent Sm ion due to $4f^{n+1}\leftrightarrow 4f^n5d^1$ excitations used for the description of the Raman intensities in Fig. 5.16 has been used to describe the phonon anomalies [5.107]. Therefore we can conclude that the dominant Γ_1^+ scattering intensities of $Sm_{0.75}Y_{0.25}S$ near 245 cm^{-1} and 85 cm^{-1}, respectively, arise mainly from the LO and LA phonon anomalies in the [111] direction.

The connection between strong Raman intensities and phonon anomalies (see also [5.141b]) is further illustrated in Fig. 5.17 by the Raman spectra of cleaved (100) faces of the superconductor YSe and the intermediate valence compound TmSe as compared to the semiconductor SmSe [5.73]. The unpolarized Raman spectrum of SmSe does not show any appreciable scattering intensity in the range of acoustic and optical phonon frequencies. Besides a presumably low defect concentration, this result is consistent with the absence of

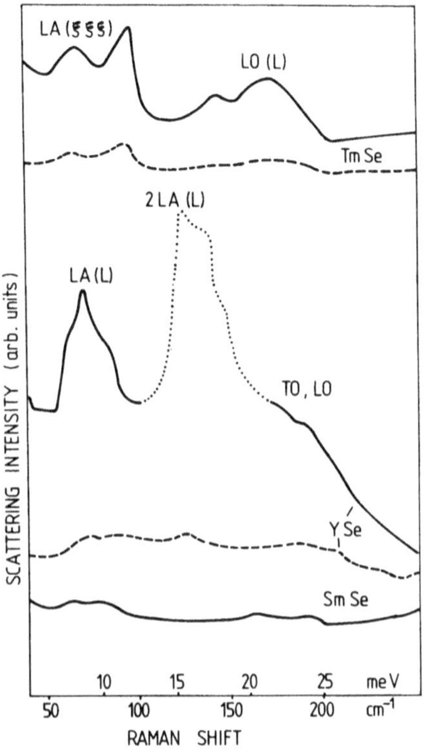

Fig. 5.17. Raman spectra of cleaved (100) faces of TmSe, YSe and SmSe at 300 K using 5309 Å (TmSe) and 5145 Å (YSe, SmSe) laser excitation. Unpolarized spectrum for SmSe; polarized spectra for TmSe and YSe with $\boldsymbol{E}_i \| \boldsymbol{E}_s (\Gamma_1^+ + 4\Gamma_{12}^+)$ and $\boldsymbol{E}_i \perp \boldsymbol{E}_s$ $(\Gamma_{25}^+ + \Gamma_{15}^+)$, where $\boldsymbol{E}_{i(s)}$ is the incident (scattered) electric field vector [5.73]

significant phonon anomalies of this 0.5 eV *f–d* gap semiconductor. On the other hand, the one-phonon scattering intensity of YSe for $\boldsymbol{E}_i \| \boldsymbol{E}_s$ (solid line) is enhanced for acoustic phonons near 70 cm^{-1} as compared to the weak intensity from optical phonons near 180 cm^{-1} (the dotted line indicates second-order scattering). This result is analogous to the case of YS (Sect. 5.3.2) and emphasizes the strong Raman intensity associated with the LA(L) phonon anomaly. On the other hand, for TmSe the $\boldsymbol{E}_i \| \boldsymbol{E}_s$ spectrum (solid line) shows about equal intensities from acoustic and optical phonons. The similarity with the spectrum of $Sm_{0.75}Y_{0.25}S$ in Fig. 5.16 becomes evident by comparing the maxima of the latter near 85 cm^{-1} and 245 cm^{-1}, respectively, with those near 70 cm^{-1} and 175 cm^{-1} in Fig. 5.17. The latter two frequencies coincide with the LA and LO phonon anomalies of TmSe in the [111] direction [5.105]. Hence it can be concluded from Fig. 5.17 that in superconducting or intermediate-valent NaCl-type compounds, the Raman intensity consists of a one-phonon density of states weighted by specific matrix elements of the electron-phonon coupling which are enhanced near phonon anomalies.

Raman scattering in TmSe was first investigated by *Treindl* and *Wachter* [5.68, 69]. Analogous to previous observations of LO(L) phonon softening in semiconducting and metallic SmS, and in $Sm_{0.75}Y_{0.25}S$ [5.62], LO(L) phonon

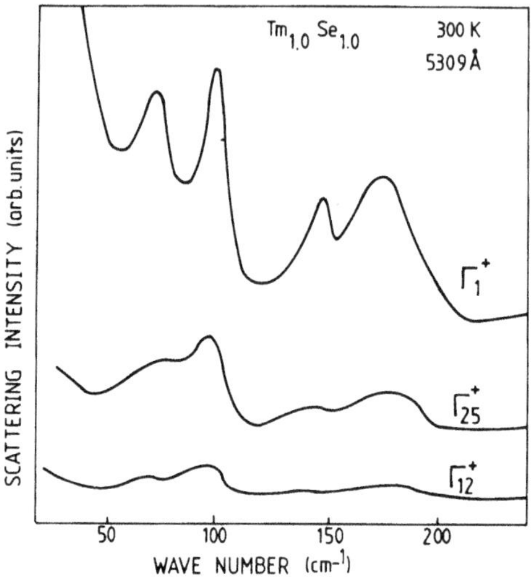

Fig. 5.18. Polarized Raman spectra of a (100) cleaved TmSe single crystal at 300 K [5.73]

softening has been found in going from $Tm^{3+}_{0.87}Se$ to $Tm^{2.72+}_{1.05}Se$ [5.68]. This softening increases linearly with increasing valence mixing. On polished samples of Tm_xSe ($x=0.97$, 1.0, 1.05), a peak is found near 60 cm^{-1} which increases in intensity with increasing valence mixing from $x=0.97$ to $x=1.05$ and which is absent in $Tm_{0.87}Se$. This peak has also been identified for cleaved samples of $Tm^{2.95+}Se_{0.91}Te_{0.09}$ and $Tm^{2.75+}Se_{0.83}Te_{0.17}$ [5.69]. It has been concluded that this peak is not created by mechanical polishing, which only reduces the scattering background. The 60 cm^{-1} peak, which appears to be connected with the intermediate valence state, has been attributed to either an anomaly in the LA phonon branch or, even more likely, to a localized low-energy electronic excitation near 60 cm^{-1} which may interact with the LA phonon branch. This assignment has been questioned in [5.73] on the basis that no such distinct localized electronic excitation was found in neutron scattering experiments on TmSe for temperatures above 100 K [5.105, 114].

A different point of view has been taken by *Stüsser* et al. [5.73] who interpreted their polarized Raman spectra of cleaved Tm_xSe single crystals in terms of phonon-induced local, intraionic charge deformabilities. In Fig. 5.18 the polarized Raman spectrum of a cleaved (100) face of nominally stoichiometric TmSe is shown. Similar to $Sm_{0.75}Y_{0.25}S$ in Fig. 5.16, the one-phonon scattering intensity appears predominantly in the Γ_1^+ symmetry component. In addition, however, there appears also an appreciable contribution of the Γ_{25}^+ component near 97 cm^{-1}. In analogy to the above discussion of $Sm_{0.75}Y_{0.25}S$, the strong Γ_1^+ scattering intensities near 70 cm^{-1} and 180 cm^{-1}, respectively, have been attributed to phonon-induced monopolar (breathing) charge deformabilities of the mixed-valent Tm ions which also describe, in a lattice dynamical model [5.72], the LA and LO phonon anomalies in the [111] direction of TmSe [5.105]. The microscopic origin of these charge deformabilities is due to

$4f^{13} \leftrightarrow 4f^{12}5d^1$ excitations near the Fermi energy. The LO phonon assignment for the 180 cm^{-1} peak in Fig. 5.18 is consistent with that in [5.68], concluded from the softening of the LO(L) phonon with varying stoichiometry of Tm_xSe. On the other hand, the difference in frequency between the peak near 70 cm^{-1} in Fig. 5.18 and that observed by *Treindl* and *Wachter* [5.68] near 60 cm^{-1} has been attributed for the latter case to a surface-bound softening of the bulk modulus, and hence a softening of the LA(ξ, ξ, ξ) branch ($0 < \xi < 4/5\, q_L$), by mechanical polishing [5.73].

The scattering intensity near 97 cm^{-1} in Fig. 5.18 has no analogy in the spectrum of $Sm_{0.75}Y_{0.25}S$ (Fig. 5.16). Hence, it has been assigned to a monopolar (Γ_1^+) and quadrupolar (Γ_{25}^+) charge deformability of the highly polarizable Se ions. The Γ_{25}^+ deformability involves Se-Se interactions. Moreover, the Γ_{25}^+ intensity near 70 and 180 cm^{-1} in Fig. 5.18 points to (quadrupolar) Tm-Tm interactions, which have also been included in the fit of the phonon dispersion of TmSe [5.72]. Consequently, the interpretation of the Raman intensities is not as straightforward as in the case of $Sm_{0.75}Y_{0.25}S$.

The Γ_1^+ first-order scattering intensity near 145 cm^{-1} in Fig. 5.18 appears within the gap of acoustic and optical-phonon branches of TmSe [5.105]. The symmetry analysis rules out any involvement of TO phonons [5.68] since these modes should show up mainly in the Γ_{12}^+ symmetry component [5.67, 107]. This anomalous scattering intensity ("gap mode"), which is similar to the shoulder near 190 cm^{-1} in the case of $Sm_{0.75}Y_{0.25}S$ (Fig. 5.16), will be discussed in Sect. 5.5.3.

Concluding this section, we present in Table 5.1 a list of local charge deformabilities which have so far been identified theoretically and/or experimentally in various materials of the NaCl-type crystal structure. The first

Table 5.1. Phonon-induced local charge deformabilities at cation or anion sites in compounds with NaCl structure: theoretical predictions from lattice dynamical model calculations (○) and experimental verification (●) [5.67, 107]

Symmetry of deformability	Effect on	○ calculated (Lattice dyn.) ● measured						Microscopic origin
		$Sm^{2+}S$	$Sm_{0.75}Y_{0.25}S$	TmSe	YS	YSe	TiN	
Cation (cat)								
Γ_1^+ (Sm)	{LO(L) / LA($\xi\xi\xi$)	○●	○●					$4f^6 \to 4f^5 + e^-$
Γ_1^+ (Tm)	LA($\xi\xi\xi$)			○●				$4f^{13} \to 4f^{12} + e^-$
Γ_{15}^- (Sm)	LO(Γ)	○●[a]	○					$4f^6 \to 5d^1(4f^5)$
Γ_{12}^+ (cat)	TO(L)							
Anion (an)								
Γ_1^+ (S)	LA(L)				○●			$3d_{xy} \to 3d_{xy}$
Γ_1^+ (Se)	LA(L)			●		●		$4d_{xy} \to 4d_{xy}$
Γ_1^+ (N)	LA(L)						○●	$d_{xy} \to d_{xy}$
Γ_{15}^- (an)	LO(Γ)							
Γ_{12}^+ (N)	TA(L)						○●	$3d_{xy} \to 2p$

[a] See [5.162b]

column shows the symmetries of the deformabilities at cation or anion sites. Their dominant effect on the different phonon branches near symmetry points or along symmetry lines is indicated in column two. Columns 3–8 show the examples investigated so far, consisting of unstable-valence (SmS), fluctuating-valence ($Sm_{0.75}Y_{0.25}S$, TmSe) and superconducting (YS, YSe, TiN) materials. The last column of Table 5.1 gives the microscopic origin of the charge deformabilities. (The first evidence for a dipolar charge deformability of Sm^{2+} in SmS has been found recently by infrared spectroscopy [5.162b]).

5.5.2 Metallic SmS

In a first attempt to overcome experimental difficulties with Raman measurements of SmS under hydrostatic pressure, *Güntherodt* et al. [5.62] have investigated the mechanically transformed metallic surface of SmS. The semiconductor-metal transition of SmS under hydrostatic pressure above 6.5 kbar [5.83] can be simulated near the surface by mechanical polishing [5.163]. The mechanically collapsed metallic surface layer of about 2000 Å thickness has been reported to be single crystalline [5.164, 165] with a lattice constant of $a_0 = 5.68$ Å. The Raman spectrum of a mechanically collapsed, cleaved (100) face of SmS at 300 K is shown as the top curve in Fig. 5.19. For comparison, the Raman spectra of $Gd_{1.0}S_{1.0}$, $Gd_{0.8}S$ and YS are also shown in Fig. 5.19. Quite analogous to the assignments for GdS and YS in Sect. 5.3, the Raman spectrum of metallic SmS exhibits first-order scattering from acoustic (A) and optic (O) phonons as well as second-order scattering (2A, O+A, 2O). The vertical dashed line in Fig. 5.19 marks the average position of the maximum of the optic-phonon density of states of GdS and YS. With respect to this $RE^{3+}S$

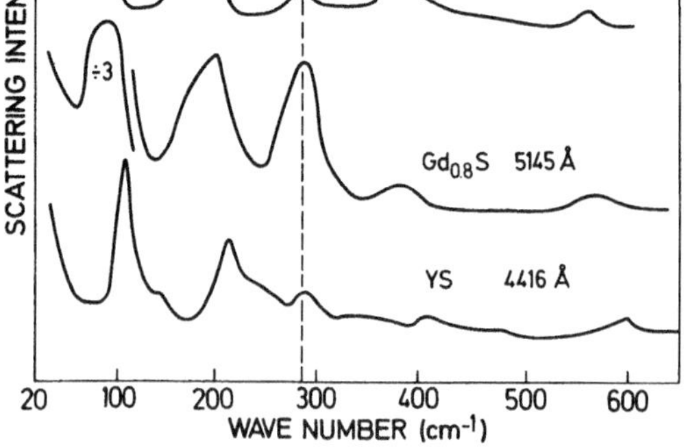

Fig. 5.19. Unpolarized Raman spectra of (100) cleaved single crystals of YS, $Gd_{1.0}S_{1.0}$ and $Gd_{0.8}S$ at 300 K [5.58a, 70] and polarized ($\Gamma_1^+ + 4\Gamma_{12}^+$) Raman spectrum of mechanically polished (100)-cleaved, metallic SmS at 300 K [5.62]

"reference line", metallic SmS shows a softening of the average optic-phonon, i.e., LO(L) phonon frequency by about 2%. This softening is rather small compared to the 15% LO(L) phonon softening of $Sm_{0.75}Y_{0.25}S$ [5.62]. However, it is in qualitative agreement with the much larger bulk modulus (498 kbar) of SmS under hydrostatic pressure [5.61, 96] as compared to that of $Sm_{0.75}Y_{0.25}S$ (71 kbar) [5.146]. Recent neutron scattering experiments of SmS under hydrostatic pressure [5.106] have given a qualitative indication of a much less pronounced LA phonon anomaly near the zone center in the [111] direction, as compared to that of $Sm_{0.75}Y_{0.25}S$ [5.104]. The LA phonon anomaly of metallic SmS, about halfway to the zone boundary in the [111] direction, does not appear to be connected with an anomaly of the bulk modulus. Hence, by analogy with $Sm_{0.75}Y_{0.25}S$, one concludes that the concurrent LO(L) phonon anomaly in metallic SmS is also much less pronounced in agreement with the Raman data [5.62].

No anomalous temperature dependence of the Raman spectrum of metallic SmS has been observed. The maximum of the scattering intensity from optic phonons at 300 K near 270 cm^{-1} (Fig. 5.19) shifts by 9 cm^{-1} towards higher frequencies upon coooling to 100 K, with no further change down to 5 K. The absence of strong temperature-dependent changes of the scattering intensity or phonon frequencies is consistent with the behavior of $Sm_{0.75}Y_{0.25}S$. Neutron scattering experiments in this material have shown a strong temperature dependence of the LA phonon anomaly in the [111] direction [5.104].

5.5.3 Bound Polaronic Charge Fluctuation Mode

As described above in Sects. 5.5.1, 5.5.2, Raman scattering in intermediate valence materials has been concerned mainly with investigations of phonon anomalies. On the other hand, charge relaxations of the valence-fluctuating ions are expected to show up in quasielastic light scattering [5.157]. A coupling of such charge density fluctuations to phonons may shift spectral weight from quasielastic to inelastic scattering, on the phonon frequency scale. An interesting question in the field of intermediate valence materials concerns the possible existence of a *bound state* of localized charge density fluctuations coupled to phonon modes of the same local symmetry.

Raman scattering has been performed by *Stüsser* et al. [5.74, 166] for the intermediate-valence phases of the solid solution systems $Sm_{1-x}R_xS$ (R=Y, La, Pr, Gd, Tb, Dy, Tm; $0.15 < x < 1.0$). A cation-mass independent mode has been observed in between the gap of acoustic and optical phonon branches for all Sm concentrated ($x < 0.5$) intermediate-valence phases. The Raman spectra of $Sm_{1-x}R_xS$ with R=Y, Pr, Gd, Dy in the upper part of Fig. 5.20 show at 300 K a maximum of the scattering intensity near 200 cm^{-1}. This "gap mode" for $x < 0.50$ is due to first-order scattering as demonstrated by its temperature dependence (Fig. 5.20 for $Sm_{0.78}Gd_{0.22}S$). For $x > 0.5$ the intensity observed in, e.g., $Sm_{0.25}Dy_{0.75}S$ at 300 K near 180 cm^{-1} (dashed line), is due to second-order scattering by phonons as proven by its quenching at 80 K. The

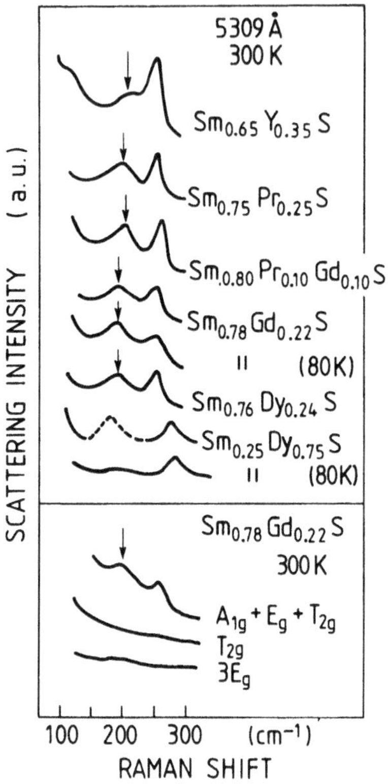

Fig. 5.20. Raman spectra of (100) cleaved intermediate-valence $Sm_{1-x}R_xS$ with R=Y, Pr, Gd, Dy at 300 K. *Upper part:* unpolarized spectra; spectra at 80 K for $x=0.22$ Gd and $x=0.75$ Dy, respectively, prove first and second-order scattering (*dashed line*) at 300 K. The "gap mode" is seen near 200 cm^{-1}. *Lower part:* polarized Raman spectra of $Sm_{0.78}Gd_{0.22}S$ at 300 K [5.74]

absence of a significant cation-mass defect in $Sm_{1-x}R_xS$ with R=Pr, Gd, Dy rules out an interpretation of the "gap mode" as a local vibrational mode of the substituted cations.

The symmetry analysis of the "gap mode" intensity of, e.g., $Sm_{0.78}Gd_{0.22}S$ in the lower part of Fig. 5.20 shows the dominance of the Γ_1^+ (A_{1g}) component. The latter is consistent with the scattering by electron density fluctuations (Chap. 2, Sect. 2.2). Moreover, the "gap mode" frequency appears to follow shifts of the LO(L) phonon frequency. This has been considered as evidence that the "gap mode" is due to a coupling of the incoherently fluctuating localized $4f$ charge density of the valence-fluctuating rare-earth ion to lattice vibrations of the same local symmetry. The latter are dominantly determined by the LO(L) phonon breathing mode. Hence this local, full symmetric (Γ_1^+) coupling is believed to yield a *bound state* ("bound fluctuon" or "bound polaronic exciton") split off from the LO phonon branch.

These conclusions have been corroborated further by investigations of the intermediate valence phases of $SmS_{1-y}As_y$, $TmSe_{1-y}Te_y$, and $TmSe_{1-y}S_y$ [5.74, 166]. In particular, the peak near 145 cm^{-1} in the Raman spectrum of TmSe in Fig. 5.18 has also been identified as a "gap mode". Its quantitative theoretical description should be a challenging task in the field of nonlinear lattice dynamics [5.167].

In the phonon dispersion of $Sm_{0.75}Y_{0.25}S$ measured by neutron scattering, a dispersionless mode has been found near 175 cm^{-1} within the acoustic-optical phonon gap [5.104]. This mode has been assigned to a mass-defect type local vibrational mode due to the Y ion substituting for the heavier Sm ion. In the Raman spectrum of $Sm_{0.75}Y_{0.25}S$ (Fig. 5.16), the maximum near 185 cm^{-1} ("gap mode") is most likely due to the density of states of this dispersionless mode. However, the absence of a cation-mass scaling of the "gap mode" in Fig. 5.20 calls for a reinterpretation of the dispersionless mode of $Sm_{0.75}Y_{0.25}S$, possibly in terms of the "bound fluctuon" mode discussed above.

5.6 Higher Rare-Earth Chalcogenides

In this section we want to discuss Raman scattering in rare-earth chalcogenides other than the monochalcogenides. Among the very few materials to be dealt with here is the rare-earth counterpart of magnetite (Fe_3O_4), i.e., the inhomogeneous mixed valence compound Eu_3S_4 or $Eu^{2+}Eu_2^{3+}S_4^{2-}$. The particular interest in this material is due to the expected strong electron-phonon coupling associated with the thermally activated $Eu^{2+} \rightarrow Eu^{3+}$ electron hopping. This interaction is, in principle, of similar origin to that discussed for EuS in Sect. 5.2.1 and for intermediate valence compounds in Sect. 5.5.1. The intriguing feature of Eu_3S_4 is that the electron hopping rate can be tuned as a function of temperature. Hence, Raman measurements of Eu_3S_4 by *Vitins* and *Wachter* [5.75–77] and *Güntherodt* and *Wichelhaus* [5.78] have aimed at investigating the role of optic phonons in the electron hopping process and the relationship to the homogeneously mixed-valent materials.

Other compounds to be discussed here have been investigated with respect to their Raman-active modes and the electronic Raman scattering from $(4f)$ J multiplet levels.

5.6.1 Inhomogeneous Intermediate-Valence Materials

Mössbauer experiments in Eu_3S_4 [5.168] and Sm_3S_4 [5.169] have given evidence for a thermally activated hopping of electrons between di- and trivalent cations on equivalent lattice sites. Eu_3S_4 has been shown to undergo a transition into a charge-ordered state below 186 K [5.170]. Theoretical work concerning the $Eu^{2+} \rightarrow Eu^{3+}$ charge transfer in Eu_3S_4 at high temperatures has stressed the important role of vibrations of the S^{2-} ions in the transfer mechanism [5.171]. Because of the large volume changes between di- and trivalent rare-earth ions, the $Eu^{2+} \rightarrow Eu^{3+}$ electron transfer is expected to couple strongly to lattice vibrations (see also Sects. 5.2.1, 5.5.1). The breathing mode of the S^{2-} ions with respect to the valence-fluctuating $Eu^{2+/3+}$ ion has been searched for by Raman scattering.

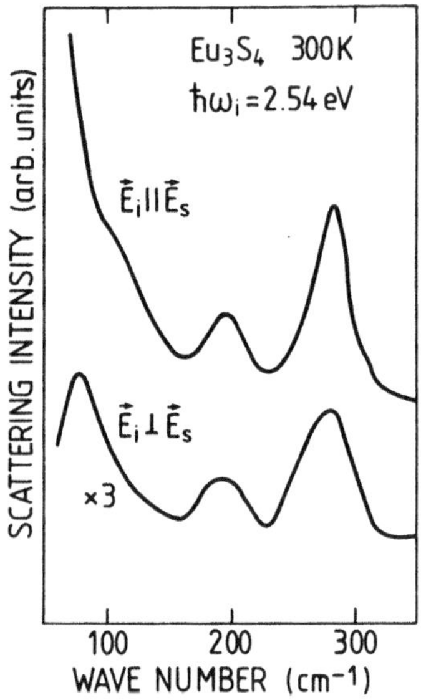

Fig. 5.21. Raman spectrum of an unoriented as-grown single crystal face of Eu_3S_4 at 300 K for $\boldsymbol{E}_i \| \boldsymbol{E}_s$ and $\boldsymbol{E}_i \perp \boldsymbol{E}_s$, where $\boldsymbol{E}_{i(s)}$ is the incident (scattered) electric-field vector [5.78]

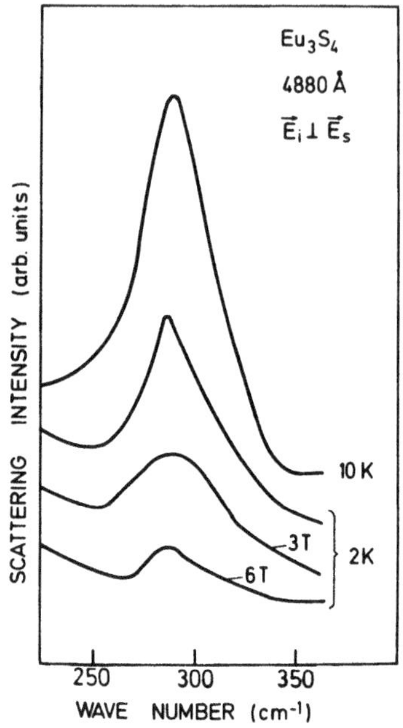

Fig. 5.22. Temperature and magnetic-field induced quenching of the scattering intensity of the 280 cm^{-1} peak of Eu_3S_4 (Fig. 5.21) in the vicinity of the Curie temperature $T_C = 3.8$ K [5.78]

Eu_3S_4 crystallizes in the cubic Th_3P_4 structure with 4 formula units per cell. There are 9 Raman-active $q=0$ vibrational modes expected [5.172]: $A_1+3E+5T_2$. The Raman spectrum for an as-grown face of an unoriented single crystal of Eu_3S_4 at 300 K is shown in Fig. 5.21 [5.78]. The three maxima due to first-order scattering are rather broad and apparently are not related to the expected ($q=0$) Raman-active modes. The scattering intensity near 280 cm^{-1} has been found to become strongly quenched upon cooling below the Curie temperature $T_C = 3.8$ K and by additionally applying a magnetic field up to 6 T. Figure 5.22 shows the 280 cm^{-1} maximum at 10 K and its partial quenching at 2 K. The intensity is further decreased in fields of 3 and 6 T. This observation is rather similar to the quenching of the spin-disorder induced phonon scattering in EuS (see Sect. 5.2.1). The maximum near 280 cm^{-1} in Figs. 5.21, 5.22 has been attributed to the spin-disorder induced one-phonon density of states from the LO branch weighted by the electron-phonon matrix element of Γ_1^+ symmetry [5.78]. The frequency position of this maximum is thus identified as due to the LO(L) phonon ("breathing") mode of S^{2-} ions with respect to a "magnetic" Eu^{2+} ion.

The temperature dependence of the Eu-S breathing mode from above to below the charge order-disorder phase transition near $T_t = 186$ K is shown in

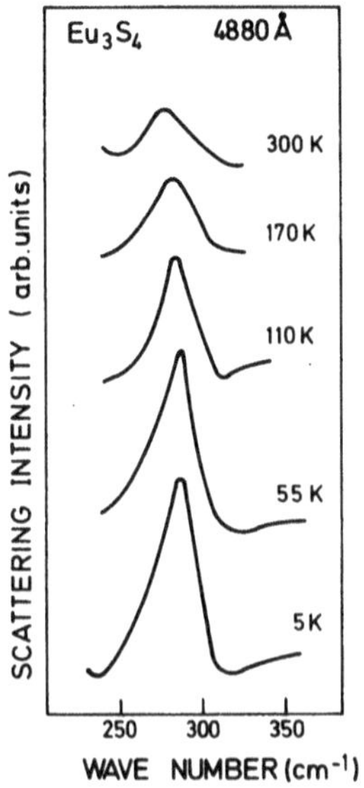

Fig. 5.23. Temperature dependence of the 280 cm^{-1} peak ("optic phonon breathing mode") of Eu_3S_4 (Fig. 5.21) above and below the order-disorder phase transition near $T_t = 186$ K [5.78]

Fig. 5.23. No abrupt change either in frequency or in width of the peak is observed near 186 K. The shift of the peak position from 280 cm^{-1} at 300 K to 288 cm^{-1} at 5 K corresponds to the effect of thermal expansion [5.170]. The discontinuous change in thermal expansion $\Delta l/l = 3 \times 10^{-4}$ at the first-order phase transition near $T_t = 186$ K gives rise to a small phonon frequency change within the experimental resolution [5.78]. The absence of anomalous behavior of the LO(L) phonon breathing mode for temperatures up to 300 K may be due to the fact that the electron hopping frequency $\nu = \nu_0 \exp(-\Delta E/kT)$, with $\nu_0 = \nu_{LO(L)} = 8.4 \times 10^{12}$ s^{-1} and $\Delta E = 0.16$ eV [5.170], amounts to 2.2×10^{10} s^{-1} at 300 K, which is off resonance from the breathing mode frequency ν_0. Contrary to the behavior of EuS (Fig. 5.3), the integrated intensity of the 280 cm^{-1} peak in Fig. 5.23 increases upon cooling. This has been attributed to a reduction in the spin-fluctuation rate of the Eu^{2+} ions going in parallel with the decreasing $Eu^{2+} \rightarrow Eu^{3+}$ electron hopping rate.

The transition of Eu_3S_4 below $T_t = 186$ K into a charge-ordered state is not predominantly a structural one, i.e., there is no significant displacive effect on the Eu and S positions [5.170]. This is indicated by x-ray [5.173] and recent neutron-diffraction measurements [5.174] which show that the change in the atomic positions is extremely small, probably orthorhombic [5.174]. This may be the reason why no additional Raman-active modes have been observed in the charge-ordered phase ($T < T_t = 186$ K) of Eu_3S_4.

In previous work on Raman scattering from pressed-powder Eu_3S_4 samples by *Vitins* and *Wachter* [5.75–77], an "anomalous vibrational mode" of the S^{2-} ions had been identified near 425 cm^{-1}. It was assumed that the electron hopping is activated by this mode. Raman investigations by *Güntherodt* and *Wichelhaus* [5.78] on Eu_3S_4 single crystals have identified the 425 cm^{-1} peak and its overtones as a S–S stretching vibration. This molecular vibration has been found to be due to laser-induced changes of the chemical composition near the sample surface. Similar S–S stretching vibrations have been observed between 400–450 cm^{-1} in a number of rare-earth disulfides (Sect. 5.6.2).

Previous Raman experiments in Sm_3S_4 [5.75, 76] have shown only electronic Raman scattering from the $J=0\rightarrow1$ and $J=0\rightarrow2$ excitations of the $4f^6(^7F_J)$ configuration of Sm^{2+}. Electronic Raman scattering of Sm_3S_4 up to $J=4$ and crystal field splittings of the different J levels have been identified in recent measurements by *Mörke* et al. [5.79b]. In addition, out of the 9 Raman-active modes only 3 rather broad bands have been observed between 150 cm^{-1} and 350 cm^{-1}. The phonon mode near 290 cm^{-1} appears to coincide with the electronic $J=0\rightarrow1$ excitation. The possible interaction between the phonon and the $J=1$ level still remains to be investigated.

5.6.2 Miscellaneous Materials

The rather few examples listed here can certainly not cover all the activity in the field and therefore we apologize to all those whose work has been omitted.

Raman investigations of Th_3P_4-structure compounds such as Sm_2S_3, Ce_3S_4, La_3Se_4 and Th_3P_4 have revealed out of the expected 9 Raman-active modes only three rather broad bands between 150 and 400 cm^{-1}, and two, in some cases very weak structures, below 150 cm^{-1} [5.79b]. This is consistent with observations in $Sr_{1-x}\square_{x/3}Nd_{2+2/3x}S_4$ solid solutions for which three strong bands between 185 and 275 cm^{-1} were also found [5.172]. Moreover, changes in the width of these bands have been observed as a function of vacancy concentration ($\square_{x/3}$). On the other hand, all 9 Raman-active modes have been identified in $SrLa_2S_4$ [5.172].

Most surprisingly, no Raman investigations of the cubic-to-tetragonal structural phase transition of La_3S_4 near 103 K or of La_3Se_4 near 70 K [5.170, 175] have so far been reported. Detailed Raman measurements of the lanthanide sesquioxides, such as La_2O_3, Pr_2O_3 and Nd_2O_3, have been carried out [5.79a]. These compounds crystallize in the hexagonal A-type structure and exhibit $2A_{1g}+2E_g$ Raman-active modes. These investigations could confirm, among others, the crystal structure with D^3_{3d} space group proposed by Pauling.

The Raman spectra of rare-earth disulfides show a pronounced peak between 400–450 cm^{-1} [5.78]. An example is shown in Fig. 5.24 for LaS_2, SmS_2 and EuS_2 at 300 K. The scattering intensity appears predominantly for parallel polarization vectors of incident and scattered photons ($\boldsymbol{E}_i \| \boldsymbol{E}_s$) as demonstrated for EuS_2. All these rare-earth disulfides contain as a common structural feature $(S–S)^{2-}$ ions according to the fictitious formula $RE^{3+}S^{2-}(S_2^{2-})_{1/2}$. Each peak in Fig. 5.24 is attributed to the S–S stretching vibration of the $(S–S)^{2-}$ ions. Such S–S stretching vibrations have been found in a variety of other materials near similar frequencies [5.176]. Another typical feature of these molecular stretching modes is shown for EuS_2 in Fig. 5.25. Weak overtones of the fundamental near 425 cm^{-1} are observed at 300 K which become more intense at 5 K [5.78]. As discussed in Sect. 5.6.1, the peaks in the Raman spectrum of Eu_3S_4 near 425 and

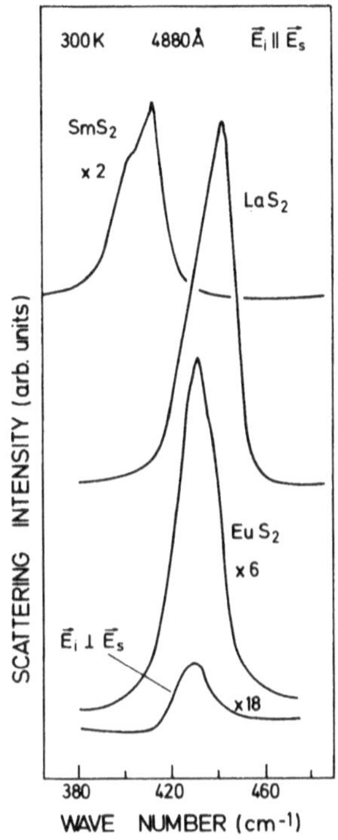

Fig. 5.24

Fig. 5.24. Raman spectra of single crystals of LaS_2, SmS_2 and EuS_2 at 300 K. The intensities are all referred to that of LaS_2

Fig. 5.25. Raman spectra of a single crystal of EuS_2 at 300 K and 5 K [5.78]
▼

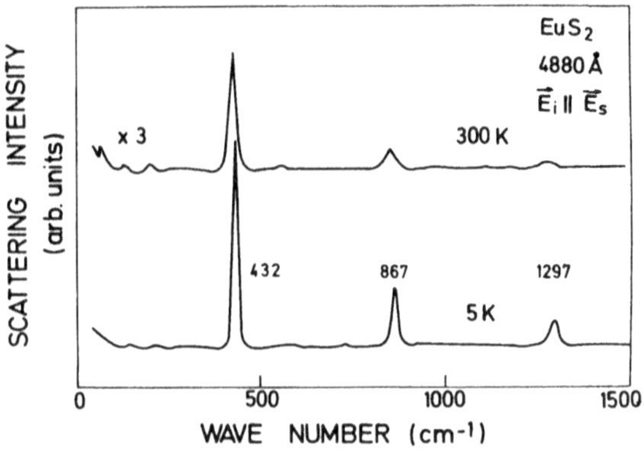

Fig. 5.25

850 cm^{-1} reported by *Vitins* and *Wachter* [5.75–77] are due to the S–S stretching vibration and are not intrinsic features. Raman investigations of Eu_3S_4 by *Güntherodt* and *Wichelhaus* [5.78] have shown a strong increase in intensity of the extrinsic 425 cm^{-1} peak with increasing laser power. An enhanced laser-induced change in surface chemistry, i.e., the formation of S_2 molecules, has been concluded.

5.7 Conclusions

In this chapter we have presented an overview of the rather diverse features observed by Raman scattering in rare-earth chalcogenides. Common to the various compounds is the dominant full symmetric (Γ_1^+) $4f$ electron-phonon coupling under excitation from the localized $4f$ states. Moreover, the model of local cluster deformabilities or phonon-induced intraionic charge deformabilities has been shown to give a consistent description of the electron-lattice interaction and of Raman intensities throughout the family of rare-earth chalcogenides. The investigations have not only revealed new scattering mechanisms, such as simultaneous spin-phonon excitations, but have also provided experimental verification of the concept of local cluster deformabilities applied to lattice dynamics. Hence, Raman scattering has given insight into the symmetry and strength of the electron-phonon interaction in rare-earth chalcogenides and has stimulated microscopic studies. From the materials' point of view, Raman scattering has proved to be a useful tool to get started on rare-earth materials whose purity, perfection, size or isotope composition often preclude investigations by neutron scattering. Thanks to defect-induced, spin-disorder induced or magnetic "Bragg" scattering, the information gained from Raman measurements has not been restricted to $q=0$ excitations only. Thus, Raman investigations in rare-earth compounds have served to advance the understanding both of new scattering mechanisms and new physical phenomena, such as valence fluctuations of rare-earth ions in solids.

The field of Raman scattering in rare-earth chalcogenides has now reached a certain degree of maturity. First steps towards Raman scattering in the vast, untouched field of the intermetallics of the rare earths such as $REAl_2$ [5.137], $REBe_{13}$ or $RECu_2Si_2$, have recently been undertaken [5.177]. For instance, in $CeAl_2$ the predicted "bound state" [5.178] due to the interaction of the Γ_5^+ optical phonon with $\Gamma_7-\Gamma_8$ crystal-field excitations within the $4f^1$ configuration has been identified [5.137], coinciding with the lower sublevel of the split Γ_8 crystal-field level. The observation of $4f$ crystal-field levels in metallic materials, in general, remains a challenging task for future light scattering experiments.

On the other hand, Brillouin scattering has been shown to be a powerful tool for studying both phonons and magnons in metals [5.158]. First experiments in rare-earth intermetallics have been carried out [5.161] to study the elastic properties, particularly of intermediate-valence materials. With respect to the latter, future emphasis will be on studies of quasielastic light scattering due to valence or charge fluctuations.

Acknowledgements. The authors would like to express their gratitude to G. Abstreiter, E. Anastassakis, M. Barth, W. Bauhofer, F. Canal, P. Grünberg, R. Humphreys, A. Jayaraman, and N. Stüsser for their cooperation and participation in various stages of the experimental work. We would like to thank G. Benedek, H. Bilz, M. Cardona, P. Entel, A. Frey, N. Grewe, W. Kress, and R. Zeyher for many stimulating discussions and for efficient support in different theoretical aspects of our work. The preparation, characterization and supply of single crystals and thin films by H. Bach, E. Bucher, K. Fischer, F. Holtzberg, R. Suryanarayanan, and W. Wichelhaus are gratefully acknowledged. We are indebted to S. Wood for help with the manuscript.

The preparation of this article was supported in part by the U.S. Army Research Office under Contract No. DAAG-29-82-K0057.

References

5.1 See Z.B.Goldschmidt: In *Handbook on the Physics and Chemistry of Rare Earths*, Vol. 1, ed. by K.A.Gschneidner, Jr., L.Eyring (North-Holland, Amsterdam 1978) p. 1

5.2 C.M.Varma: Rev. Mod. Phys. **48**, 219 (1976)

5.3a *Proc. of Intern. Conf. on Magnetic Semiconductors*, Jülich, FR. Germany (1975), ed. by W.Zinn (North-Holland, Amsterdam 1976)

5.3b *Proc. of Intern. Conf. on Magnetic Semiconductors*, Montpellier, France (1979), J. Physique **41**, C5 (1980)

5.4 *Valence Instabilities and Related Narrow Band Phenomena*, ed. by R.D.Parks (Plenum Press, New York 1977)
5.5a *Valence Fluctuations in Solids*, ed. by L.M.Falicov, W.Hanke, M.B.Maple (North-Holland, Amsterdam 1981)
5.5b *Valence Instabilities*, ed. by P.Wachter, H.Boppart (North-Holland, Amsterdam 1982)
5.6 S.Methfessel, D.C.Mattis: In *Handbuch der Physik*, Vol. *XVIII*, ed. by S.Flügge (Springer, Berlin, Göttingen, Heidelberg 1968) p. 1
5.7 C.Haas: Crit. Rev. Solid State Sci. **1**, 47 (1970)
5.8a T.Kasuya: Crit. Rev. Solid State Sci. **3**, 131 (1972)
5.8b P.Wachter: Crit. Rev. Solid State Sci. **3**, 189 (1972)
5.8c E.L.Nagaev: Sov. Phys. -Usp. **18**, 863 (1976)
5.9 G.Güntherodt: In *Festkörperprobleme* XVI, ed. by J.Treusch (Vieweg, Braunschweig 1976) p. 95
5.10 R.Suryanarayanan: Phys. Stat. Sol. **B85**, 9 (1978)
5.11a P.Wachter: In *Handbook on the Physics and Chemistry of Rare Earths*, Vol. 2, ed. by K.A.Gschneidner, Jr.,L.Eyring (North-Holland, Amsterdam 1978) p. 507
5.11b A.Jayaraman: In *Handbook on the Physics and Chemistry of Rare Earths*, Vol. 2, ed. by K.A.Gschneidner, Jr., L.Eyring (North-Holland, Amsterdam 1978) p. 575
5.12 F.Holtzberg, S. von Molnar, J.M.D.Coey: In *Handbook on Semiconductors*, Vol. 3, ed. by S.P.Keller (North-Holland, Amsterdam 1982) p. 803
5.13 R.K.Ray, J.C.Tsang, M.S.Dresselhaus, R.L.Aggarwal, T.B.Reed: Phys. Lett. **37**A, 129 (1971)
5.14 J.C.Tsang, M.S.Dresselhaus, R.L.Aggarwal, T.B.Reed: In *Proc. 11th Intern. Conf. on the Physics of Semiconductors* (PWN – Polish Scientific Publishers, Warsaw 1972) p. 1273
5.15 C.R.Pidgeon, G.D.Holah, R.B.Dennis, J.S.Webb: *Proc. of the 11th Intern. Conf. on the Physics of Semiconductors* (PWN – Polish Scientific Publishers, Warsaw 1972) p. 1280
5.16 A.Schlegel, P.Wachter: Solid State Commun. **13**, 1865 (1973)
5.17 G.D.Holah, J.S.Webb, R.B.Dennis, C.R.Pidgeon: Solid State Commun. **13**, 209 (1973)
5.18 V.J.Tekippe, R.P.Silberstein, M.S.Dresselhaus, R.L.Aggarwal: In *Physics of Semiconductors*, ed. by M.H.Pilkuhn (Teubner, Stuttgart 1974) p. 904
5.19 J.C.Tsang, M.S.Dresselhaus, R.L.Aggarwal, T.B.Reed: Phys. Rev. **B9**, 984 (1974)
5.20 J.C.Tsang, M.S.Dresselhaus, R.L.Aggarwal, T.B.Reed: Phys. Rev. **B9**, 997 (1974)
5.21 V.J.Tekippe, R.P.Silberstein, M.S.Dresselhaus, R.L.Aggarwal: Phys. Lett. **49**A, 295 (1974)
5.22 J.Vitins, P.Wachter: Solid State Commun. **17**, 911 (1975)
5.23 J.Vitins, P.Wachter: AIP Conf. Proc. **29**, 662 (1976)
5.24 J.Vitins, P.Wachter: Helv. Phys. Acta **48**, 435 (1975)
5.25 V.J.Tekippe, R.P.Silberstein, L.E.Schmutz, M.S.Dresselhaus, R.L.Aggarwal: In *Light Scattering in Solids*, ed. by M.Balkanski, R.C.C.Leite, S.P.S.Porto (Flammarion, Paris 1976) p. 362
5.26 J.Vitins, P.Wachter: J. Magnetism Magn. Mat. **3**, 161 (1976)
5.27 R.P.Silberstein, L.E.Schmutz, V.J.Tekippe, M.S.Dresselhaus, R.L.Aggarwal: Solid State Commun. **18**, 1173 (1976)
5.28 S.A.Safran, B.Lax, G.Dresselhaus: Solid State Commun. **19**, 1217 (1976)
5.29 N.Suzuki: J. Phys. Soc. Jpn. **40**, 1223 (1976)
5.30 G.Güntherodt: *Proc. of the 13th Intern. Conf. on the Physics of Semiconductors*, ed. by F.G.Fumi (Tipografia Marves, Rome 1976) p. 291
5.31 P.Grünberg, G.Güntherodt, A.Frey, W.Kress: Physica (Utrecht) **B89**, 225 (1977)
5.32 S.A.Safran, G.Dresselhaus, M.S.Dresselhaus, B.Lax: Physica **B89**, 229 (1977)
5.33 J.Vitins, P.Wachter: Physica **B86–88**, 213 (1977)
5.34 J.Vitins: J. Magnetism Magn. Mat. **5**, 212 (1977)
5.35 R.P.Silberstein, V.J.Tekippe, M.S.Dresselhaus: Phys. Rev. **B16**, 2728 (1977)
5.36 S.A.Safran, G.Dresselhaus, B.Lax: Phys. Rev. **B16**, 2749 (1977)
5.37 R.Merlin, R.Zeyher, G.Güntherodt: Phys. Rev. Lett. **39**, 1215 (1977)
5.38 R.Merlin, G.Güntherodt, R.Zeyher, W.Kress, P.Grünberg, F.Canal: In *Lattice Dynamics*, ed. by M.Balkanski (Flammarion, Paris 1977) p. 87
5.39 Y.Ousaka, O.Sakai, M.Tachiki: Solid State Commun. **23**, 589 (1977)

5.40 O.Sakai, M.Tachiki: J. Phys. Chem. Solids **39**, 269 (1978)
5.41 R.Merlin, R.Zeyher, G.Güntherodt: In *Physics of Semiconductors 1978*, ed. by B.L.H. Wilson (The Institute of Physics, London 1979) p. 145
5.42 G.Güntherodt, R.Merlin, P.Grünberg: Phys. Rev. B**20**, 2834 (1979)
5.43 R.Zeyher, W.Kress: Phys. Rev. B**20**, 2850 (1979)
5.44 G.Güntherodt: J. Magnetism Magn. Mat. **11**, 394 (1979)
5.45a S.A.Safran, G.Dresselhaus, R.P.Silberstein, B.Lax: J. Magnetism Magn. Mat. **11**, 403 (1979)
5.45b S.A.Safran, R.P.Silberstein, G.Dresselhaus, B.Lax: Solid State Commun. **29**, 339 (1979)
5.46 R.P.Silberstein, S.A.Safran, M.S.Dresselhaus: J. Magnetism Magn. Mat. **11**, 408 (1979)
5.47 L.E.Schmutz, G.Dresselhaus, M.S.Dresselhaus: J. Magnetism Magn. Mat. **11**, 412 (1979)
5.48 G.Güntherodt, G.Abstreiter, W.Bauhofer, G.Benedek, E.Anastassakis: J. Magnetism Magn. Mat. **15–18**, 777 (1980)
5.49 G.Güntherodt, R.Merlin, G.Abstreiter: J. Magnetism Magn. Mat. **15–18**, 821 (1980)
5.50 R.P.Silberstein: Phys. Rev. B**22**, 4791 (1980)
5.51 S.A.Safran: J. Physique **41**, C5–223 (1980)
5.52 S.Abiko: Z. Physik B**39**, 53 (1980)
5.53 Y.Ousaka, O.Sakai, M.Tachiki: J. Phys. Soc. Jpn. **48**, 1269 (1980)
5.54 R.Merlin, G.Güntherodt, R.Humphreys, M.Cardona, R.Suryanarayanan, F.Holtzberg: Phys. Rev. B**17**, 4951 (1978)
5.55 R.Merlin, G.Güntherodt, R.Humphreys: In *Physics of Semiconductors 1978*, ed. by B.L.H. Wilson (The Institute of Physics, London 1979) p. 875
5.56 S.Abiko: J. Phys. Soc. Jpn. **48**, 1245 (1980)
5.57 E.Anastassakis, H.Bilz, M.Cardona, P.Grünberg, W.Zinn: In *Proc. 3rd Intern. Conf. on Light Scattering in Solids*, ed. by M.Balkanski, R.C.C.Leite, S.P.S.Porto (Flammarion, Paris 1976) p. 367
5.58a G.Güntherodt, P.Grünberg, E.Anastassakis, M.Cardona, H.Hackfort, W.Zinn: Phys. Rev. B**16**, 3504 (1977)
5.58b G.Güntherodt, P.Grünberg, E.Anastassakis: In *Lattice Dynamics*, ed. by M.Balkanski (Flammarion, Paris 1977) p. 90
5.59 A.Treindl, P.Wachter: Phys. Lett **64**A, 147 (1977)
5.60 G.Güntherodt, R.Keller, P.Grünberg, A.Frey, W.Kress, R.Merlin, W.B.Holzapfel, F.Holtzberg: In [Ref. 5.4, p. 321]
5.61 G.Güntherodt, R.Merlin, A.Frey, F.Holtzberg: In *Lattice Dynamics*, ed. by M.Balkanski (Flammarion, Paris 1977) p. 130
5.62 G.Güntherodt, R.Merlin, A.Frey, M.Cardona: Solid State Commun. **27**, 551 (1978)
5.63 M.I.Nathan, F.Holtzberg, J.E.Smith, Jr., J.B.Torrance, J.C.Tsang: Phys. Rev. Lett. **34**, 467 (1975)
5.64 J.E.Smith, Jr., F.Holtzberg, M.I.Nathan, J.C.Tsang: In *Light Scattering in Solids*, ed. by M.Balkanski, R.C.C.Leite, S.P.S.Porto (Flammarion, Paris 1976) p. 313
5.65 J.C.Tsang: Solid State Commun. **18**, 57 (1976)
5.66 G.Güntherodt, A.Jayaraman, E.Anastassakis, E.Bucher, H.Bach: Phys. Rev. Lett. **46**, 855 (1981)
5.67 G.Güntherodt, A.Jayaraman, H.Bilz, W.Kress: In [Ref. 5.5a, p. 121]
5.68 A.Treindl, P.Wachter: Solid State Commun. **32**, 573 (1979)
5.69 A.Treindl, P.Wachter: Solid State Commun. **36**, 901 (1980)
5.70 G.Güntherodt, A.Jayaraman, W.Kress, H.Bilz: Phys. Lett. **82**A, 26 (1981)
5.71 H.Boppart, A.Treindl, P.Wachter: In [Ref. 5.5a, p. 103]
5.72 W.Kress, H.Bilz, G.Güntherodt, A.Jayaraman: J. de Physique **42**, Coll. C6, 3 (1981)
5.73 N.Stüsser, M.Barth, G.Güntherodt, A.Jayaraman: Solid State Commun. **39**, 965 (1981)
5.74 N.Stüsser, G.Güntherodt, A.Jayaraman, K.Fischer, F.Holtzberg: In [Ref. 5.5b, p. 69]
5.75 J.Vitins: J. Magnetism Magn. Mat. **5**, 234 (1977)
5.76 J.Vitins, P.Wachter: Physica **89**B, 234 (1977)
5.77 J.Vitins, P.Wachter: Phys. Rev. B**15**, 3225 (1977)
5.78 G.Güntherodt, W.Wichelhaus: Solid State Commun. **39**, 1147 (1981)
5.79a J.Zarembowitch, J.Gouteron, A.M.Lejus: Phys. Stat. Sol. B**94**, 249 (1979)

5.79b I.Mörke, G.Travaglini, P.Wachter: In [Ref. 5.5b, p. 573]
5.80 E.Bucher, K.Andres, F.J.DiSalvo, J.P.Maita, A.C.Gossard, A.S.Cooper, G.W.Hull,Jr.: Phys. Rev. **B11**, 500 (1975)
5.81 M.Campagna, E.Bucher, G.K.Wertheim, D.N.E.Buchanan, L.D.Longinotti: Phys. Rev. Lett. **33**, 885 (1974)
5.82 R.Suryanaranan, G.Güntherodt, J.L.Freeouf, F.Holtzberg: Phys. Rev. **B12**, 4215 (1975)
5.83 A.Jayaraman, V.Narayanamurti, E.Bucher, R.G.Maines: Phys. Rev. Lett. **25**, 1430 (1970)
5.84 M.B.Maple, D.Wohlleben: Phys. Rev. Lett. **27**, 511 (1971)
5.85 A.Chatterjee, A.K.Singh, A.Jayaraman: Phys. Rev. **B6**, 2285 (1972)
5.86 A.Jayaraman, A.K.Singh, A.Chatterjee, S.Usha Devi: Phys. Rev. **B9**, 2513 (1974)
5.87 A.Jayaraman, P.D.Dernier, L.D.Longinotti: High Temp. – High Press. **7**, 1 (1975)
5.88 J.F.Herbst, R.E.Watson, J.W.Wilkins: Phys. Rev. **B17**, 3089 (1978), and references therein
5.89 B.Johansson, A.Rosengren: Phys. Rev. **B11**, 1367 (1975)
5.90 B.Johansson: Phys. Rev. **B20**, 1315 (1979); J. Phys. **F4**, L169 (1974)
5.91 L.Brewer: J. Opt. Soc. Am. **61**, 1101 (1971)
5.92 L.J.Nugent, J.L.Burnett, L.R.Morss: J. Chem. Thermodyn. **5**, 665 (1973)
5.93 J.A.Freis, E.D.Cater: J. Chem. Phys. **68**, 3978 (1978)
5.94 S.I.Nagai, M.Shinmei, T.Yokokawa: J. Inorg. Nucl. Chem. **36**, 1904 (1974)
5.95 C.Bergman, P.Coppens, J.Drowart, S.Smoes: Trans. Faraday Soc. **66**, 800 (1970)
5.96 R.Keller, G.Güntherodt, W.B.Holzapfel, M.Dietrich, F.Holtzberg: Solid State Commun. **29**, 753 (1979)
5.97 H.Boppart, A.Treindl, P.Wachter, S.Roth: Solid State Commun. **35**, 489 (1980)
5.98 A.Werner, H.D.Hochheimer, A.Jayaraman, J.M.Leger: Solid State Commun. **38**, 325 (1981)
5.99 A.Jayaraman, E.Bucher, P.D.Dernier, L.D.Longinotti: Phys. Rev. Lett. **31**, 700 (1973)
5.100 F.Holtzberg: AIP Conf. Proc. **18**, 478 (1974)
5.101 L.J.Tao, F.Holtzberg: Phys. Rev. **B11**, 3842 (1975)
5.102 A.Jayaraman, P.D.Dernier, L.D.Longinotti: Phys. Rev. **B11**, 2783 (1975)
5.103 S. von Molnar, T.Penney, F.Holtzberg: J. Physique **37**, C4, 241 (1976)
5.104 H.A.Mook, R.M.Niklow, T.Penney, F.Holtzberg, M.W.Shafer: Phys. Rev. **B18**, 2925 (1978)
5.105 H.A.Mook, F.Holtzberg: In [Ref. 5.5a, p. 113]
5.106 H.A.Mook, D.B.McWhan, F.Holtzberg: Phys. Rev. **B25**, 4321 (1982)
5.107 H.Bilz, G.Güntherodt, W.Kleppmann, W.Kress: Phys. Rev. Lett. **43**, 1998 (1979) and references therein
5.108 P.Entel, N.Grewe, M.Sietz, K.Kowalski: Phys. Rev. Lett. **43**, 2002 (1979)
5.109 R.J.Birgeneau, E.Bucher, L.W.Rupp,Jr., W.M.Walsh,Jr.: Phys. Rev. **B5**, 3412 (1972)
5.110 R.J.Birgeneau, S.M.Shapiro: In [Ref. 5.4, p. 49]
5.111 W.Beckenbaugh, G.Güntherodt, R.Hauger, E.Kaldis, J.P.Kopp, P.Wachter: AIP Conf. Proc. **18**, 540 (1974)
5.112 F.Hulliger, T.Siegrist: Z. Phys. **B35**, 81 (1979) and references therein
5.113 H.Bjerrum-Møller, S.M.Shapiro, R.J.Birgeneau: Phys. Rev. Lett. **39**, 1021 (1977) and references therein
5.114 M.Loewenhaupt, E.Holland-Moritz: J. Magnetism Magn. Mat. **9**, 50 (1978)
5.115 B.Batlogg, H.R.Ott, P.Wachter: Phys. Rev. Lett. **42**, 278 (1979)
5.116 P.Haen, F.Holtzberg, F.Lapierre, T.Penney, R.Tournier: In [Ref. 5.4, p. 495]
5.117 F.Holtzberg, T.Penney, R.Tournier: J. Physique **40**, C5, 314 (1979)
5.118 M.Ribault, J.Flouquet, P.Haen, F.Lapierre, J.M.Mignot, F.Holtzberg: Phys. Rev. Lett. **45**, 1295 (1980)
5.119 P.Wachter: Phys. Repts. **44**, 161 (1978)
5.120 P.Wachter, E.Kaldis, R.Hauger: Phys. Rev. Lett. **40**, 1404 (1978)
5.121 T.Kasuya: IBM J. Res. Develop. **14**, 214 (1970)
5.122 P.Fischer, W.Hälg, W. von Wartburg, P.Schwob, O.Vogt: Phys. Kondens. Mater. **9**, 249 (1969)
5.123 R.Griessen, M.Landolt, H.R.Ott: Solid State Commun. **9**, 2219 (1971)
5.124 N.F.Olivera Jr., S.Foner, Y.Shapira, T.B.Reed: Phys. Rev. **B5**, 2634 (1972)

5.125 M.Campagna, G.K.Wertheim, Y.Baer: In *Photoemission in Solids II*, ed. by M.Cardona, L.Ley, Topics Appl. Phys., Vol. 27 (Springer, Berlin, Heidelberg, New York 1979) p. 217
5.126 S.J.Cho: Phys. Rev. B**1**, 4589 (1970)
5.127 K.Lendi: Phys. Kondens. Mater. **17**, 189 and 215 (1974)
5.128 G.Güntherodt: Phys. Kondens. Mater. **18**, 37 (1974)
5.129 C.R.Pidgeon, J.Feinleib, W.J.Scouler, J.Hanus, J.O.Dimmock, T.B.Reed: Solid State Commun. **7**, 1323 (1969)
5.130 C.R.Pidgeon, J.Feinleib, T.B.Reed: Solid State Commun. **8**, 1711 (1970)
5.131 J.Schoenes: Z. Physik B**20**, 345 (1975)
5.132 O.Sakai, A.Yanase, T.Kasuya: J. Phys. Soc. Jpn. **42**, 596 (1977)
5.133 R.Suryanarayanan, J. Ferré, B. Briat: Phys. Rev. B**9**, 554 (1974)
5.134 F.Holtzberg, J.B.Torrance: AIP Conf. Proc. **5**, 860 (1972)
5.135 B.Batlogg, E.Kaldis, P.Wachter: Phys. Rev. B**14**, 5503 (1976)
5.136 W.Beckenbaugh, J.Evers, G.Güntherodt, E.Kaldis, P.Wachter: J. Phys. Chem. Solids **36**, 239 (1975) and references therein
5.137 G.Güntherodt, A.Jayaraman, B.Batlogg, M.Croft, E.Melczer: Phys. Rev. Lett. **51**, 2330 (1983)
5.138 P.Roedhammer, W.Reichardt, F.Holtzberg: Phys. Rev. Lett. **40**, 465 (1978)
5.139 K.Fischer, H.Bilz, R.Haberkorn, W.Weber: Phys. Stat. Sol. (b) **54**, 285 (1972)
5.140 J.A.Appelbaum, D.R.Hamann: In *Physics of Transition Metals 1977*, ed. by M.J.Lee, J.M.Perz, E.Fawcett, AIP Conf. Proc. **39**, 111 (1978)
5.141a See M.Cardona, G.Güntherodt (eds.): *Light Scattering in Solids II*, Topics Appl. Phys., Vol. 50 (Springer, Berlin, Heidelberg, New York 1982) p. 41
5.141b For a review see M.V.Klein: In *Light Scattering in Solids III*, ed. by M.Cardona, G.Güntherodt, Topics Appl. Phys., Vol. 51 (Springer, Berlin, Heidelberg, New York 1982) p. 121
5.142 W.Spengler, R.Kaiser, A.N.Christensen, G.Müller-Vogt: Phys. Rev. B**17**, 1055 (1978)
5.143 W.Kress, P.Roedhammer, H.Bilz, W.D.Teuchert, A.N.Christensen: Phys. Rev. B**17**, 111 (1978)
5.144 B.Batlogg, J.Schoenes, P.Wachter: Phys. Lett. **49**A, 13 (1974)
5.145 L.L.Hirst: Phys. Rev. Lett. **35**, 1394 (1975)
5.146 T.Penney, R.L.Melcher, F.Holtzberg, G.Güntherodt: AIP Conf. Proc. **29**, 392 (1975)
5.147 K.Baba, M.Kobayashi, H.Kaga, I.Yokota: Solid State Commun. **35**, 175 (1980)
5.148 See, for example; J.A.Koningstein: J. Chem. Phys. **46**, 2811 (1967)
J.A.Koningstein, P.Grünberg: Canad. J. Chem. **49**, 2336 (1971) and references therein
5.149 E.Zirngiebl: Private communication
5.150 Y.L.Wang, B.R.Cooper: Phys. Rev. **172**, 539 (1968); ibid. **185**, 696 (1969)
5.151 M.Gronau: Ph. D. thesis, Ruhr-Universität Bochum (1979) unpublished
5.152 D.B.McWhan, S.M.Shapiro, J.Eckert, H.A.Mook, R.J.Birgeneau: Phys. Rev. B**18**, 3623 (1978)
5.153 H.A.Mook, T.Penney, F.Holtzberg, M.W.Shafer: J. Physique **39**, C-6, Suppl. 8, 837 (1978)
5.154 H.A.Mook: Private communication
5.155 J.Röhler, D.Wohlleben, G.Kaindl, H.Balster: Phys. Rev. Lett. **49**, 65 (1982)
5.156 M.Loewenhaupt, E.Holland-Moritz: J. Appl. Phys. **50**, 7456 (1979)
5.157 A.Lopez, C.Balseiro: Phys. Rev. B**17**, 99 (1979)
5.158 J.R.Sandercock: In *Light Scattering in Solids III*, ed. by M.Cardona, G.Güntherodt, Topics Appl. Phys., Vol. 51 (Springer, Berlin, Heidelberg, New York 1982) p. 173
5.159 G.E.Devlin, J.L.Davis, L.Chase, S.Geschwind: Appl. Phys. Lett. **19**, 138 (1971)
5.160 E.Zirngiebl: Diplomarbeit, Universität Köln (1983) unpublished
5.161 M.Barth, G.Güntherodt: In [Ref. 5.5b, p. 99]
5.162a G.Güntherodt, J.L.Freeouf, F.Holtzberg: Solid State Commun. **47**, 677 (1983)
5.162b B.Hillebrands, G.Güntherodt: Solid State Commun. **47**, 681 (1983)
5.163 E.Kaldis, P.Wachter: Solid State Commun. **11**, 907 (1972)
5.164 T.L.Bzhalava, T.B.Zhukova, I.A.Smirnov, S.G.Shulman, N.A.Yakovleva: Sov. Phys. Solid State **16**, 2428 (1975)
5.165 B.Batlogg, E.Kaldis, A.Schlegel, P.Wachter: Phys. Rev. B**14**, 5503 (1976)

5.166 N.Stüsser: Diplomarbeit, Universität Köln (1982) unpublished
5.167 H.Bilz, H.Büttner, G.Güntherodt, W.Kress, M.Miura: In [Ref: 5.5b, p. 67]
5.168 O.Berkooz, M.Malamud, S.Shtrikman: Solid State Commun. **6**, 185 (1968)
5.169 M.Eibschütz, R.L.Cohen, E.Buehler, J.H.Wernick: Phys. Rev. B**6**, 18 (1972)
5.170 R.Pott, G.Güntherodt, W.Wichelhaus, M.Ohl, H.Bach: Phys. Rev: B**27**, 359 (1983)
5.171 J.Mulak, K.W.H.Stevens: Z. Physik B**20**, 21 (1975)
5.172 P.L.Provenzano, W.B.White: *Proc. 12th Rare-Earth Res. Conf.*, Vail (1976) p. 522
5.173 H.H.Davis, I.Bransky, N.M.Tallan: J. Less-Common Met. **22**, 193 (1970)
5.174 W.Wichelhaus, A.Simon, K.W.H.Stevens, P.J.Brown, K.R.A.Ziebeck: Phil. Mag. B**46**, 115 (1982)
5.175 K.Westerholt, H.Bach, S.Methfessel: Solid State Commun. **36**, 431 (1980)
5.176 See, for example, M.C.Tobin: *Laser Raman Spectroscopy* (Wiley, New York 1971) p. 81
5.177 G.Güntherodt, A.Jayaraman, B.Batlogg, M.Croft, C.U.Segre: Verhandl. DPG (VI) **18**, 720, 725 (1983)
5.178 P.Thalmeier, P.Fulde: Phys. Rev. Lett. **49**, 1588 (1982)

6. Surface-Enhanced Raman Scattering: "Classical" and "Chemical" Origins

Andreas Otto

With 91 Figures

Surface science is very much in need of in situ methods to investigate solid-gas interfaces (for instance, in industrial heterogeneous catalysis) or solid-liquid interfaces (for instance, in electrochemical processes at battery electrodes). Spectroscopies with electrons or other particles are, with some minor exceptions, only possible at the solid-vacuum interface. Spectroscopy at solid-gas and solid-liquid interfaces is limited to optical spectroscopy within the spectral range of optical transparency of the gas or the liquid. Of particular interest is vibrational surface spectroscopy [6.1] because it may provide information on the chemical nature of adsorbates, chemical surface reactions and intermediate species in chemical surface reactions. Infrared vibrational spectroscopy is very difficult at a solid-water (e.g., an aqueous electrolyte) interface because of the high infrared absorption of water. However, there is progress also in this domain [6.2]. At the solid-vacuum and solid-gas interface, there is a wealth of infrared spectroscopy data (e.g. [6.3]).

6.1 Background

Raman spectroscopy in an aqueous environment [6.4] is possible due to the relatively weak inelastic light scattering by water outside the OH stretch and HOH bending bands of the water molecule. Besides this advantage, Raman spectroscopy does not suffer from problems of infrared spectroscopy below about 600 cm^{-1}, i.e., low intensity sources, low sensitivity detectors and problems of finding transparent windows. Nevertheless, up until 9 years ago, Raman spectroscopy was not considered as a useful method of surface vibrational spectroscopy (with the exception of adsorbates on transparent powders with a high surface-to-mass ratio [m^2/gr] (e.g., alumina and zeolite powders) [6.4b]: The Raman cross section for particular vibrations of free molecules hardly exceeds 10^{-13} Å^2 [6.5], whereas the average area per adsorbed molecule is of the order of 10 Å^2. This leads to a very low Raman intensity from a monolayer, in most cases submerged in the noise. (Nevertheless, a Raman spectrum of nitrobenzene on Ni(111) without enhancement was recently obtained by *Campion* et al. [6.6] with the help of multichannel detection [see also Sects. 6.9, 6.10]).

This situation has changed in the last 6 years. In 1974, *Fleischmann* et al. [6.7] observed a Raman band from pyridine adsorbed on a silver electrode. It was clear that the signal originated from adsorbed pyridine because vibrational frequencies were found which were different from the values for pyridine in the aqueous electrolyte. In 1977 it became evident thanks to the work of *Jeanmaire* and *Van Duyne* [6.8], and *Albrecht* and *Creighton* [6.9] that the occurrence of the Raman signal from adsorbed pyridine involved an enhancement by 6 orders of magnitude of the average Raman cross section per adsorbed molecule with respect to the molecule in an aqueous environment. This could no be explained by an increase in surface area related to the roughening of the silver electrode. (The surface area of an electrode "activated" by a charge transfer of 15 mC/cm^2 equivalent to the redeposition of about 70 monolayers of silver was recently measured by *Moerl* and *Pettinger* [6.10] by underpotential deposition of a monolayer of lead; it is about a factor of 5 bigger than that of a corresponding ideally flat surface).

The genuine need for in situ techniques of surface vibrational spectroscopy and the scientific challenge raised by the factor 10^6 just mentioned explains the great active and passive interest which "Surface-Enhanced Raman Spectroscopy" (SERS) has found in the scientific community. Early hypothetical models developed in 1978 to explain the experimental observations (compiled by *Van Duyne* [6.11] and *Creighton* [Ref. 6.1, Chap. 9]) have been reviewed by *Furtak* and *Reyes* [6.12]. Only in some of these models did "roughness" play a role at all [6.13–16]. In 1980, after the impact of 3 papers [6.17–19], the prevalent opinion was that SERS is mainly caused by the enhancement of the electromagnetic field at rough surfaces by electromagnetic resonances. This opinion dominated the 7th International Conference on Raman Spectroscopy [6.20]. It is best expressed in [6.27].

A review [6.22] containing experimental information available up to the end of 1980 was written with the intention of discussing the experimental evidence for an enhancement mechanism beyond the classical electromagnetic field enhancement. A more recent state of the field of SERS is presented in a collection of about twenty reviews in a volume edited by *Chang* and *Furtak* [6.23]. Reading [6.23] one will find that these authors have been far from a common consensus and that the collection of possible interpretations presented there is by no means complete. For instance, *Cooney* et al. [6.24] ascribed the enhancement of pyridine on silver to pyridine trapped in a carbon layer of high internal surface area formed by electroreduction from trace quantities of CO_2, formate or carbonate.

In view of the survey in [6.23], a comprehensive review of the present state of SERS including all theoretical models makes little sense. The intention of this chapter is first a critical evaluation of the "classical" electromagnetic resonance enhancement model to determine whether SERS is completely explained by it, and second, a presentation of current ideas on the additional "chemical enhancement".

6.2 The Phenomenon of Surface-Enhanced Raman Scattering, "Roughness" and Electromagnetic Resonance Effects

Surface-enhanced Raman scattering was first detected for pyridine C_5NH_5 (a benzene molecule with one CH group replaced by N) adsorbed at silver electrodes [6.7–9]. If one starts with a smooth silver electrode, e.g., a silver film deposited at room temperature, a strong Raman signal from the vibrational lines of pyridine appears only after an electrochemical oxidation-reduction cycle, a so-called "activation cycle". Such a cycle consists of the transformation of the topmost metallic silver layers into Ag^+ ions (in aqueous Cl^- electrolytes, a solid AgCl film is formed on the electrode) and the reduction of Ag^+ ions to redeposited metallic silver in the subsequent step. There are many experimental indications that this second step is the important part of the "activation". Examples of "SERS activation" are presented in Figs. 6.1,2. Figure 6.1 shows at the top the low-intensity Raman spectrum of water, in the middle the additional weak signal from pyridine dissolved in water, and at the bottom the signal from the electrode surface after activation. In this case, the lines from the monolayer of adsorbed pyridine are about 50 times stronger than the lines observed for the

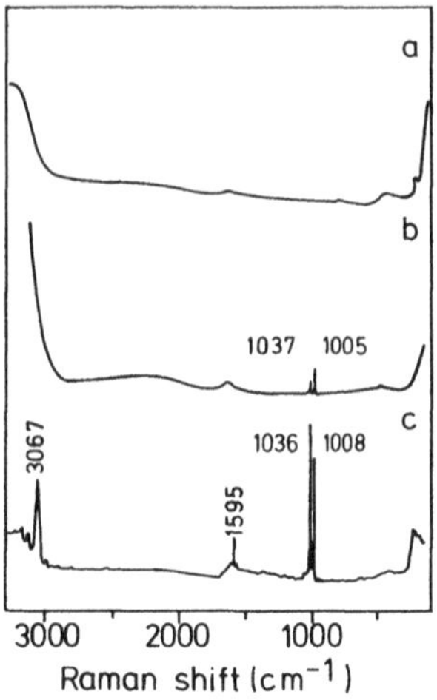

Fig. 6.1

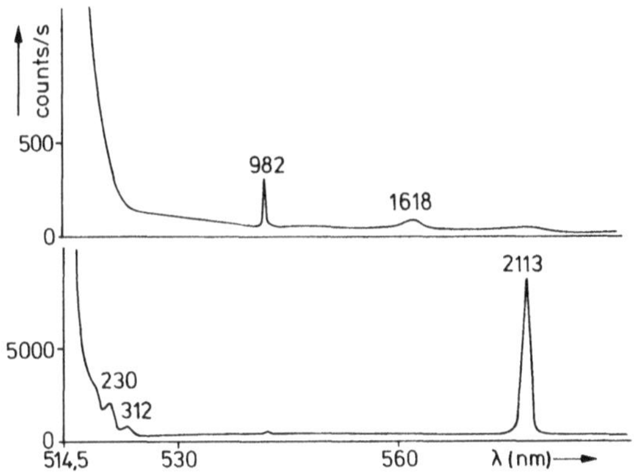

Fig. 6.2

Fig. 6.1. (**a**) Raman spectrum of a 0.1 M KCl electrolyte; (**b**) Raman spectrum of a 0.1 M KCl, 0.05 M pyridine electrolyte, 450 mW laser power. Lines at 1037 and 1005 cm^{-1} are due to pyridine. (**c**) SERS from pyridine in 0.1 M KCl, 0.05 M pyridine electrolyte after a dissolution and redeposition of a layer of about 250 Å thickness on the silver electrode; power = 50 mW. Intensities of (**c**) are divided by 10 with respect to (**b**). From [6.8.]

Fig. 6.2. Raman spectra from a polycrystalline silver electrode at $-1.0\ V_{SCE}$ in 0.1 M Na_2SO_4, 0.01 M KCN aqueous electrolyte, before any anodic sweep (*above*) and after switching for 5 s to +0.5 V (*below*). Laser wavelength 5145 Å. Stokes shifts at peaks are given in cm^{-1}. From [6.26] dissolved pyridine, although the laser power has been reduced by a factor of 9. The overall average enhancement, that is, the ratio of the intensities from a molecule adsorbed at the activated electrode versus a dissolved molecule, is estimated to be about 10^6. Figure 6.2 shows the analogous behavior for a silver electrode in a cyanide aqueous electrolyte. It has been shown by radioactive tracer experiments with CN^- that both the "unactivated" and the "activated surface" are covered with about one monolayer of cyanide [6.26]. SERS from cyanide is only observed after "activation". Laser illumination of a silver electrode during "activation" in halide solutions results in a further magnification of the SERS of pyridine, as reported by *Macomber* et al. [6.27] and others [6.28]. From now on, a metal surface for which one observes SERS will be referred to as a "SERS-active" surface. Of course, this is not a precise definition. Whether an experimentalist claims to have observed SERS or not also depends on the sensitivity of his experimental setup. An enhancement of say 100 could be unobservable [6.17]. Even so, it might be due to the same mechanisms which lead, under different conditions, to an enhancement of 10^6. The final definition of "surface-enhanced Raman scattering" will only be possible after the underlying mechanisms are understood. A "SERS-active" surface seems to imply that the occurrence or nonoccurrence of SERS is merely governed by the properties of the surface, and not by the adsorbate. As will become evident from Sect. 6.5.9, this is not the case; a silver sample may show SERS for some particular adsorbates, but not for all adsorbates. Therefore, the term "SERS-active" will always be given in quotation marks.

There are other ways of producing a "SERS-active" silver surface than by electrochemical activation. Mechanical polishing often does it [6.29, 30]. A silver film deposited in ultrahigh vacuum onto a cooled substrate (≈ 120 K) and kept at this temperature is "SERS-active" [6.31–32]. A silver film deposited at room temperature or warmed up from 120 K to room temperature is not, or no longer [6.32, 33], "SERS-active". Figure 6.3 shows a comparison of Raman spectra of pyridine adsorbed on a "SERS-active" silver film deposited at about 130 K ("cold deposited") and kept at 130 K, and on a silver single-crystal (110) surface cleaned by Ar^+ bombardment and annealed (well-defined LEED pattern), also at 130 K. The Ag(110) surface has been exposed to 10^4 L ($1\,L = 1$ Torr $\cdot$ s) of pyridine, so that a rather thick film ($\sim 5 \times 10^3$ monolayers of pyridine) has formed. The silver film has only been exposed to 1 L of pyridine. Even so, the Raman signal from the film is stronger than from the crystal, indicating 4 to 5 orders of magnitude enhancement for pyridine adsorbed on the film compared to that on the (110) surface. It was possible to obtain a pyridine Raman signal after exposure of the cold silver film to only 3×10^{-2} L, thus demonstrating the surprisingly high sensitivity of SERS under favorable conditions. In no case did I. Pockrand et al. (unpublished) observe any structure in the Raman spectra from cold-deposited silver films which could be assigned to carbon. This rules out the "intercalation-mechanism" proposed by *Cooney* et al. [6.24] (Sect. 6.1), at least in this case. What makes a silver surface "SERS-active"? The presence of the adsorbate during activation is not necessary. When pyridine or cyanide is

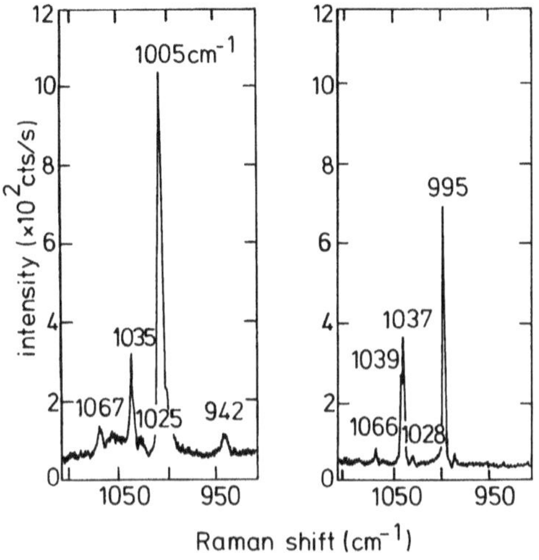

Fig. 6.3. SERS spectrum of pyridine on a cold "SERS-active" silver film exposed to 1 L of pyridine (*left*) compared to normal Raman scattering from a thick film of pyridine on an Ag(110) surface exposed to 10^4 L of pyridine at 120 K. From [6.34]

added after activation, the respective SERS spectrum is found as well. This indicates that it was a change in the topographical structure of the surface, or a change in the structure of the redeposited silver with respect to bulk silver, which made an electrode "active" [6.26]. The experiments on silver films deposited at 120 K or at room temperature under clean conditions in ultrahigh vacuum, without apparent coadsorbates [6.34], leave no other choice than to ascribe the "SERS activity" to a special surface topology, or to a change in the structure in the selvedge or in the bulk.

Scanning-electron-microscope pictures of heavily activated silver electrodes showed a rather coarse roughness on the 1000 Å scale (see figures in [6.30]). Silver-island films of average thickness 50 Å were found to be "SERS-active" [6.35] (for SEM pictures of "SERS-active" silver-island films of average thickness 90 Å, see [6.36] and Fig. 6.44). Silver colloids were found to be "SERS-active" [6.16], as were silver electrodes in tunnel junctions deposited on rough CaF_2 films [6.37]. For the 500 Å roughness scale of silver films evaporated on a CaF_2 film, see the electronmicrograph in [6.38]. *Rowe* et al. [6.17] reported SERS from pyridine on silver surfaces which were roughened to a scale of 500–2000 Å by exposing a UHV cleaned Ag(100) surface to I_2 vapour and 4880 Å laser light. Because of all these results, it has been widely accepted that "surface roughness" is very important for SERS (under "surface roughness" one should also include possible subtle changes of the surface or selvedge topography, the crystallographic orientation of microcrystalline surfaces, etc.). The scale of the roughness which induces the "activity" is under debate. The scale may be of 500 to 2000 Å [6.17], it may be "submicroscopic" ($\lesssim$100 Å) as pointed out by *Burstein* and *Chen* [6.39], or "atomic" [6.26]. There is an intrinsic difficulty in separating scales of roughness. "Submicroscopic" roughness may be present as well as "microscopic" roughness (i.e. roughness which can be seen in a commercial scanning electron microscope). When there is roughness, there is always atomic-scale roughness present (steps, kinks, high index "open" planes, adatoms, clusters, surface voids, grain boundaries) but not necessarily vice versa(!) because of the atomic nature of solids (Sect. 6.9 and [6.40–42]). "SERS-activity" was found for silver electrodes where the scale of the roughness was below the detection limit: *Pettinger* et al. [6.43] observed SERS after recycling only 1 to 5 monolayers of epitaxial (111) silver films. It is unlikely that this results in a rms value of the roughness which is much larger than the thickness of the dissolved silver film (one to five monolayers). *Schultz* et al. [6.30] carefully prepared mechanically polished electrodes. These were "SERS-active" without the oxidation reduction cycle described above. Within the resolution of their SEM (about 250 Å), no roughness could be detected. No direct experimental value exists as yet for the roughness of "SERS-active" silver films evaporated in UHV at low temperatures, although one may perhaps expect it to be of the order of 50 Å (Sect. 6.5.4). *Udagawa* et al. [6.44] reported a weak enhancement of about 250 on a Ag(100) surface. This may be an indication of an enhancement mechanism on a smooth surface or an indication that there is still residual atomic-scale roughness (Sect. 6.9). Because a silver surface is usually not better oriented than about 0.25 degrees with respect to the ideal crystallographic direction, a monocrystalline surface will always contain a surface concentration of atoms at steps with respect to atoms on terraces of 5×10^{-3} or greater, provided the steps are monoatomic.

Besides the relationship between roughness and "SERS-activity", there are other effects connected with SERS. In a SERS spectrum of silver, the vibrational lines of an adsorbate are always superimposed on a nearly structureless inelastic background. This is apparent in Fig. 6.2. In the top spectrum there is a very low background of about 50 Hz due to inelastic scattering from liquid water. After activation, the background has grown to about 300 Hz. Compared to this new background, the SO_4^{2-} stretch mode at 982 cm^{-1} and the water bending mode at 1618 cm^{-1} are weak. This background is an intrinsic effect of "SERS-active" silver and it can be enormously enhanced by drastic mechanical polishing [6.45]. It is present for silver films evaporated in ultrahigh vacuum on substrates at 120 K, before any deliberate exposure to an adsorbate [4.32]. Well-annealed silver (which is not "SERS-active") does not display the background, whereas gold and copper display a luminescence background in all cases [6.46]. The silver background has been asigned to luminescence [6.47] due to a continuum of electronic excitations of the metal [6.14, 39]. It shows a peculiar time dependence. *Heritage* et al. [6.47] observed in a picosecond Raman-gain experiment on an activated silver electrode that the reflectivity of the probe beam (Stokes-shifted with respect to the pump beam) was increased by the pump beam up to a delay time of 200 ps after switching off the pump beam. *Chen* et al. [6.48] found an "exceptionally strong" ani-Stokes background in second harmonic generation from "SERS-active" silver. For SERS active silver films, the temporal behaviour of this background showed a clear tail, several times the pulse width (of 10 ns).

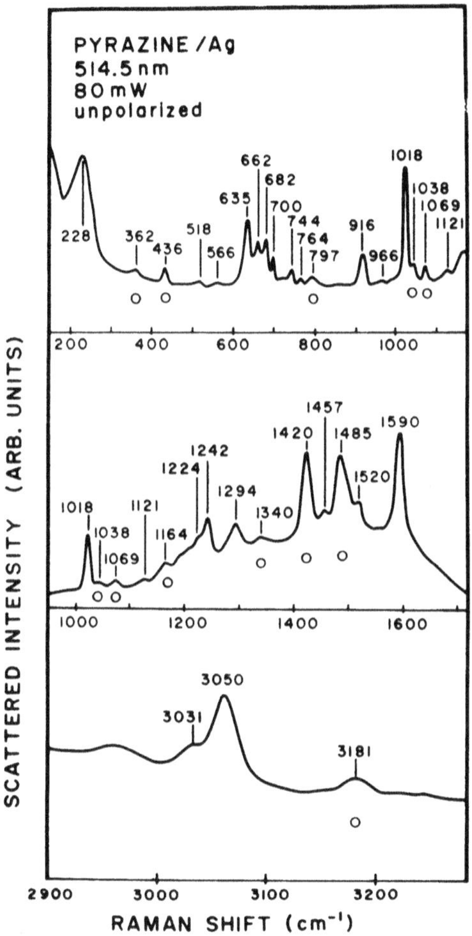

◀ **Fig. 6.4.** SERS spectrum of pyrazine adsorbed on an activated silver electrode at $-0.4\ V_{SCE}$ [6.49]. The circles mark lines which are ir active in free pyrazine

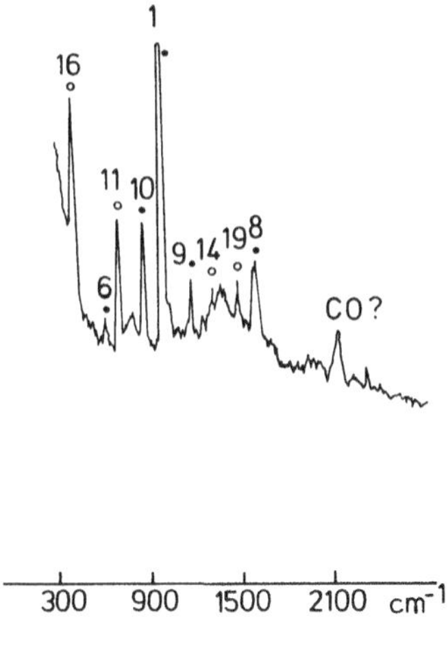

Fig. 6.5. SERS of benzene (C_6H_6) adsorbed on silver, excited with a 514.5 nm laser. Circles mark modes which in the free benzene molecule are ir active or silent, dots the Raman-active modes. Numbers refer to the Wilson mode numbers in Table 6.2. The feature labeled CO? has not been assigned in [6.51]

In SERS the Raman selection rules are relaxed, as detected for pyrazine on silver by *Dornhaus* et al. [6.49] and *Erdheim* et al. [6.50], and for benzene on silver by *Moskovits* and *DiLella* [6.51]. The SERS of pyrazine ($C_4N_2H_4$) on a "SERS-active" silver electrode is shown in Fig. 6.4. The assignment and the comparison with the vibrational frequencies of the liquid phase of pyrazine is given in Table 6.1. Free pyrazine has inversion symmetry and hence there is a Raman exclusion rule of infrared active vibrations [6.52]. The SERS, however, also shows the modes which are Raman inactive in the free molecule (open dots in Fig. 6.4), some of them with surprisingly high intensity. The SERS of benzene on a silver film deposited at 11 K [6.51] is shown in Fig. 6.5. In contrast to

Table 6.1. Comparison of vibrational frequencies of pyrazine in aqueous solution and of liquid (l) pyrazine with SERS frequencies on a silver electrode at $-0.4\ V_{SCE}$. Assignment and activity in the D_{2h}-symmetry of the free molecule. From [6.49]

Mode	Aqueous, (l): liquid	SERS	D_{2h}	Activity (D_{2h})
6a	615 vw	635 ?s	a_g	r
1	1017 vs	1018 vs	a_g	r
9a	1241 s	1242 s	a_g	r
8a	1594 s	1590 vs	a_g	r
2	3060 m	3050 m	a_g	r
16a	363 vw (393)	362 w	a_u	—
17a	(950) (1)	966 vw	a_u	—
6b	677 vw	635 ?s 662 s 682 s	b_{1g}	r
3	1120 vw	1121 w	b_{1g}	r
8b	1529 m	1520 w	b_{1g}	r
7b	3041–3060 (1)	3031 vw	b_{1g}	r
10a	758 vw	744 m (764) w	b_{2g}	r
4	703 s	700 m	b_{3g}	r
5	(922) vw	916 m-s	b_{3g}	r
16b	416– 417 (1)	436 m	b_{1u}	ir
11	786– 804 (1)	797 w	b_{1u}	ir
15	1063–1067 (1)	1069 w	b_{2u}/b_{3u}	ir
14	1342–1346 (1)	1340 vw	b_{2u}	ir
19b	1413–1418 (1)	1420 vs	b_{2u}	ir
20b	3066–3070 (1)	3060 ?	b_{2u}	ir
12	1021–1022 (1)	1038 vw	b_{3u}	ir
18a	1135–1148 (1)	1164 w	b_{3u}/b_{2u}	ir
19a	1484–1490 (1)	1485 s	b_{3u}	ir
13	3066, (3090) (1)	3183 ?	b_{3u}	ir

pyrazine, the free benzene molecule has so-called silent modes, neither Raman nor infrared active in first order: for instance, mode number 16 in Table 6.2. This mode is seen with good intensity in Fig. 6.5. It is shifted by only 7 cm^{-1} from its position in the spectrum of the liquid. The relaxation of the selection rules is a short-range effect (Sect. 6.5.7). *Moskovits* and *DiLella* [6.51] and *Sass* et al. [6.53] ascribed this to the strong gradient of the normal component of the electric field of the light at a metal surface. An alternative possibility will be discussed in Sect. 6.8. SERS is strongly depolarized, even for free-molecule modes which are polarized. Overtone or combination bands are only very weak in SERS [6.54], or not seen at all. The excitation spectra of SERS (intensity of SERS versus exciting laser frequencies) do not show narrow resonances. An example is given in Fig. 6.6. Surface-enhanced hyper-Raman scattering from SO_3^{2-} adsorbed on Ag powder was reported by *Murphy* et al. [6.56].

Table 6.2. Comparison of vibrational frequencies of liquid benzene C_6H_6 and of C_6H_6 on a "SERS-active" silver film. Assignment and activity in the D_{6h} symmetry of the free molecule. From [6.51] with a misprint corrected. (a) Wilson's mode number; (b) Herzberg's mode number; wavenumbers in brackets are for C_6D_6 [6.52]

a	b	Liquid	SERS	D_{6h}	Activity (D_{6h})
1	2	992 (943)	982	a_{1g}	r
2	1	3059	3062	a_{1g}	r
3	3	1346	—	a_{2g}	—
4	8	703	—	b_{2g}	—
5	7	989	?	b_{2g}	—
6	18	606	605	e_{2g}	r
7	15	3046	3047	e_{2g}	r
8	16	1596 (1552)	1587	e_{2g}	r
9	17	1178	1177	e_{2g}	r
10	11	849	864	e_{1g}	r
11	4	670	697	a_{2u}^{+}	ir
12	6	1008	—	b_{1u}	—
13	5	3062	?	b_{1u}	—
14	9	1309	1311	b_{2u}	—
15	10	1149	1149	b_{2u}	—
16	20	404 (352)	397	e_{2u}	—
17	19	966	?	e_{2u}	—
18	14	1036	1032	e_{1u}	ir
19	13	1479	1473	e_{1u}	ir
20	12	3073	?	e_{1u}	ir

SERS activity has been found so far on silver, copper and gold ([6.57] and references therein), on Li [6.58], K [6.59] and Na [6.60] and maybe on aluminum [6.61, 62]; there are also reports of SERS on Cd, Pt, Hg and Ni (Sect. 6.5.8). There is a large number of adsorbates for which SERS has been observed on silver. However, there seems to be chemical specificity, in particular far water, and much weaker enhancement for saturated hydrocarbons (Sect. 6.5.9).

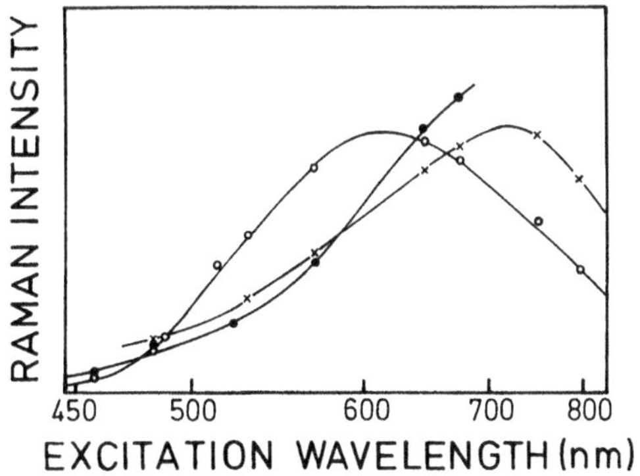

Fig. 6.6. Raman excitation profiles for adsorbates at electrochemically activated silver electrodes for the 1008 cm^{-1} line of pyridine (×) for the 2114 cm^{-1} line of cyanide (●) and the 1000 cm^{-1} line of triphenylphosphine (○). From [6.55]

The fairly general acceptance of the importance of "roughness" for SERS and the fact that the noble metals "enhance" well in those spectral ranges where they can support electromagnetic resonances of high quality-factor, lead to an explanation of SERS which shall be referred to as "classical enhancement" following *Jha* et al. [6.63]. The basic idea is the enhancement of the Raman scattered intensity by electromagnetic resonances of the local field, both for the incident field and the frequency-shifted emitted field. The solid surface is characterized by the boundary of a continuum with a dielectric constant $\varepsilon(\omega)$. In this way, one preempts the possibility that the electronic properties of a "SERS-active" sample are different at or near the surface compared to the bulk. Any specific adsorbate-metal interaction is neglected; the molecule "feels" the metal only through the enhancement of both the incident and emitted electromagnetic fields (Sect. 6.3). The response of the metal is not changed by the presence of the molecule. The self-interaction of the adsorbate via the metal is neglected. Without any further assumption, the classical model cannot explain the inelastic background which is caused by electron-hole excitations (Sect. 6.6) and the relaxation of Raman selection rules. In the sense of *Jha* et al., the definition here of "classical enhancement" does not include possible mechanisms like the modulation of the surface potential barrier by the adsorbate [6.63] or the relaxation of selection rules in the strong field gradient at the metal surface [6.53].

The discussion of the origin of SERS is not concerned with the question of whether classical enhancement exists or not. The predicate "classical" implies that electromagnetic field resonances do exist and there are many clear experimental verifications of this fact. The problem is whether the neglected mechanisms mentioned above do contribute to the enhancement and how much the electromagnetic field resonances are damped by electronic properties not described by the bulk value of $\varepsilon(\omega)$. The questions to be answered are [6.22]:

(i) Does "classical enhancement" explain all the observed enhancements?
(ii) If not, to what extent are other than 'classical" enhancement mechanisms involved?
(iii) Is SERS possible without "classical enhancement", for instance, on transition metals, and how strong is the enhancement in this case?

In this respect there are statements in the literature emphasizing different aspects:

"... the dominant contribution to SERS is electromagnetic rather than chemical in origin" [6.17];

"... The overall enhancement is the result of a combination of contributions, some of which may be quite specific to the particular metal-adsorbed molecule system" [6.39];

"We conclude that the active state of the molecule which demonstrates the surface-enhanced Raman effect involves a chemical bond with the metal of moderate strength. We are unable to unambiguously distinguish between purely chemical effects and those involving microroughness. In fact, the two may be intimately associated with each other through bonding site availability" [6.64];

"A strong Raman enhancement for an adsorbate on a silver surface is only possible when the adsorbate is bound to an atomic scale surface roughness" [6.26].

The hypothetical "nonclassical" contributions to the observed enhancement may be divided in "physical" and "chemical" effects (Sect. 6.7). Many possible mechanisms including charge-transfer excitation have been proposed by *Burstein* et al. [6.15] and others, see Sect. 6.8. It has also been proposed that this "chemical effect" is particularly strong at special surface sites ("active sites") of atomic scale roughness ([6.22, 26, 34, 65] and references therein; see also Sect. 6.9). One should also note the enhancement in Raman scattering from molecular oxygen adsorbed on polydiacetylene single crystals which has been recently reported by *Batchelder* et al. [6.66]. In a narrow frequency range, the Raman signal displays a resonance of about a factor of 25. In the field of SERS, this would correspond to a "nonclassical effect".

6.3 Classical Enhancement

In the majority of theoretical papers devoted to the enhancement by electromagnetic resonances, rather high enhancement factors have been calculated. They sometimes even exceeded the experimentally observed enhancement. This seems to have left no room and no need for additional or alternative hypothetical enhancement mechanisms and explains a statement made in 1981: "The most prevalent concept is that SERS is of purely electromagnetic origin" [6.21]. Apart from reviewing the relevant work, the intention here is to point out approximations made in the electromagnetic model calculations which lead to overestimated electromagnetic enhancement values. The experimental evidence for or against the assumptions of the theoretical papers, namely, the shape of the roughness and the concept of a dielectric constant for the material in the case of small-scale roughness and its value compared to that for bulk material, will be discussed in Sect. 6.5.3–6.

The relevant theoretical papers (see the compilation by Metiu in [6.23]) will not be presented chronologically. Calculations for single-particle or single-roughness protrusions will actually be discussed first. Thereafter, calculations for collective resonances on an ensemble of particles or surface bumps will be presented.

The basic idea of the classical field enhancement hypothesis is first the excitation of electromagnetic surface-plasmon polariton-type resonances on a nonflat metal surface (e.g., small colloidal particles, rough surfaces) by the incident laser light. On resonance, the electric field strength at the surface is increased with respect to that in the incident light wave. This leads to increased excitation of the Raman scatterer, either adsorbed at the surface or at a distance from the surface, which is small compared to the dimensions of the resonant colloid or to the bumps on a rough surface. Secondly, the Raman emission of the scatterer is additionally increased by resonance of the Stokes or anti-Stokes emission with electromagnetic surface-plasmon polariton-type excitations.

6.3.1 Single-Particle Resonances

Classical enhancement calculations for single spherical sol particles have been presented by *Kerker* and coworkers [6.67, 68]. The Rayleigh small-particle approximation, including only the dipole-mode excitation of the particles, was discussed in [6.67]; rigorous calculations including all modes are presented in [6.68]. They show that the Rayleigh approximation is very good for particle diameters smaller than $\lambda_L/15$. In the small-particle limit, the primary exciting field $\boldsymbol{E}_p(\omega_L, \boldsymbol{r}')$ for the Raman scatterer at $\boldsymbol{r}'$ near a sphere of radius a and dielectric constant $\varepsilon(\omega)$ in a surrounding medium of dielectric constant ε_0 is given by the sum of the incident field $\boldsymbol{E}_i(\omega_L \boldsymbol{r}')$ and the field of a dipole $\boldsymbol{\mu}(\omega_L)$ located at the center of the sphere (Fig. 6.7):

$$\boldsymbol{\mu}(\omega_L) = a^3 g_0 \boldsymbol{E}_i(\omega_L, \boldsymbol{r}')$$

$$g_0 = \frac{\varepsilon(\omega_L)/\varepsilon_0(\omega_L) - 1}{\varepsilon(\omega_L)/\varepsilon_0(\omega_L) + 2}. \tag{6.1}$$

$\boldsymbol{E}_p$ will force the molecule at $\boldsymbol{r}'$ to oscillate at the shifted frequency ω with dipole moment $\boldsymbol{p}(\omega) = \alpha \boldsymbol{E}_p(\omega_L, \boldsymbol{r}')$, where $\alpha\mathbf{1}$ is the Raman tensor of the free molecule (assumed isotropic). One should note that, classically, the Raman oscillator $\boldsymbol{p}(\omega)$ is a driven oscillator [6.69]. Channels of energy dissipation not available to the Raman oscillator in free space, for instance, energy dissipation directly to the electronic excitations in the metal (Sect. 6.4), will not change $\boldsymbol{p}(\omega)$. Therefore, the emitted Raman intensity, as calculated below, is not changed by additional dissipation of $\boldsymbol{p}(\omega)$ to the metal.

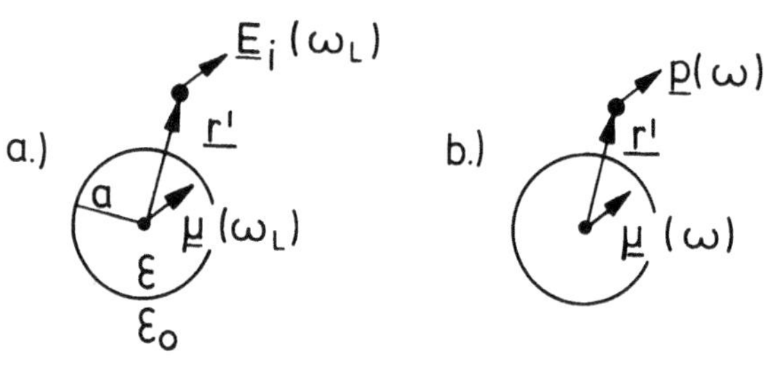

Fig. 6.7a, b. Small particle approximation for the primary field at the molecule position $\boldsymbol{r}'$ (**a**) and the dipolar sources of the Raman scattered field, if the molecule is isotropically polarizable, and an isotropic Raman scatterer (**b**)

This is different for luminescence from the molecule. In this case, $\boldsymbol{p}(\omega)$ is a free-running oscillator and energy dissipation to the metal decreases the luminescent light emission. The combined study of Raman scattering and luminescence from dyes coating metal spheres [6.70–73] is beyond the scope of this chapter.

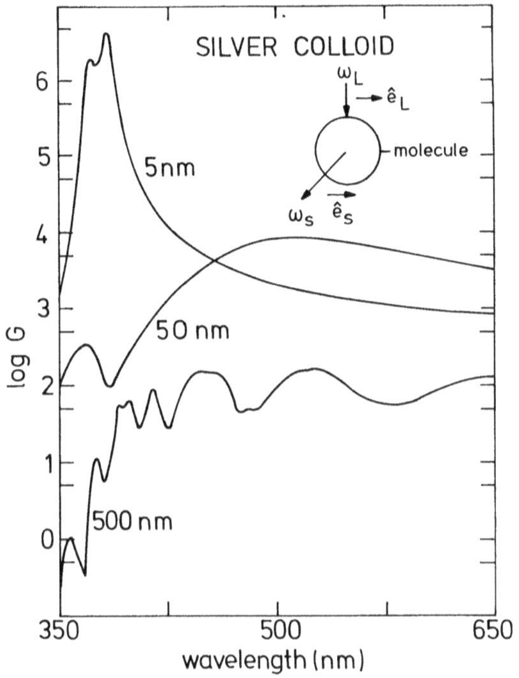

Fig. 6.8. Enhancement G versus exciting wavelength for a Raman band at 1010 cm^{-1} in the scattering configuration given in the insert for three radii of silver particles in water. $\hat{e}_L$ and $\hat{e}_s$ are the directions of the incident and scattered field. From [6.68]

The total Raman emission is given by the coherent addition of the field of $\boldsymbol{p}(\omega)$ and a dipole $\boldsymbol{\mu}(\omega)$ located in the center of the sphere:

$$\boldsymbol{\mu}(\omega) = a^3 g E_d(\omega, 0) \tag{6.2}$$

$$g = \frac{\varepsilon(\omega)/\varepsilon_0(\omega) - 1}{\varepsilon(\omega)/\varepsilon_0(\omega) + 2},$$

where $\boldsymbol{E}_d$ is the field of $\boldsymbol{p}(\omega)$ at the origin $\boldsymbol{r}' = 0$ in the absence of the sphere and $\boldsymbol{\mu}(\omega)$ represents the "antenna effect" of the sphere. Resonance for the incident and emission channel occurs at those frequencies ω_L and ω for which the denominators in (6.1, 2) are small [for $\varepsilon_0 = 1$ and $\varepsilon(\omega) = 1 - \omega_p^2/\omega^2$ at $\omega_p/\sqrt{3}$].

The Raman enhancement G for an adsorbed molecule ($r' = a$) and for the most favorable geometry ($\boldsymbol{E}_i \| \boldsymbol{r}'$) and $\boldsymbol{E}_i$ normal to the scattering plane (see insert of Fig. 6.8) is given by [6.67]

$$G = 5|(1 + 2g_0)\,(1 + 2g)|^2. \tag{6.3}$$

For a monolayer of molecules on the sphere described by oscillating electric dipoles normal to the surface and 90° scattering configuration, the average enhancement is

$$G = |(1+2g_0)\,(1+2g)|^2 \tag{6.4}$$

for any choice of polarization of the incident and scattered fields.

Equations (6.3, 4) are invariant with respect to an interchange of g_0 and g. This means that the contribution at the overall enhancement from the "excitation" and "emission" channels have the same analytical form. This correspondence for spherical particles in the small particle limit has lead many workers to calculate only the field enhancement in the "excitation" channel and to assume the correspondence quite generally. However, this is not warranted when retardation has to be taken into account. In the semiclassical treatment of radiation, the Stokes scattered intensity (frequency ω) is proportional to the incident intensity at frequency ω_L and the quantity

$$\left|\sum_i \frac{\langle f|1/2(\boldsymbol{p}\hat{\boldsymbol{A}}_{\mathrm{loc}}(\omega)+\hat{\boldsymbol{A}}_{\mathrm{loc}}(\omega)\boldsymbol{p})|i\rangle\;\langle i|1/2(\boldsymbol{p}\hat{\boldsymbol{A}}_{\mathrm{loc}}(\omega_L)+\hat{\boldsymbol{A}}_{\mathrm{loc}}(\omega_L)\boldsymbol{p})|g\rangle}{\omega_L-\omega_i}\right|^2 .$$

In this expression, according to the classical enhancement model, $(\boldsymbol{p}\hat{\boldsymbol{A}}_{\mathrm{loc}} + \hat{\boldsymbol{A}}_{\mathrm{loc}}\boldsymbol{p})/2$ is set equal to $\hat{\boldsymbol{A}}_{\mathrm{loc}}\boldsymbol{p}$, $|g\rangle$, $|i\rangle$ and $|f\rangle$ are ground, intermediate and final states of the scattering molecule, unchanged by adsorption to the metal. $\hat{\boldsymbol{A}}_{\mathrm{loc}}$ are the local vector-potentials in the absence of the molecule at the site of the adsorbed molecule normalized to the incident (ω_L) and scattered (ω) intensity. The classical enhancement is due to the increase of $\hat{A}_{\mathrm{loc}}$ (the absolute value of $\hat{\boldsymbol{A}}_{\mathrm{loc}}$) by the presence of the metal surface with respect to $\hat{A}_{\mathrm{free}}$ for the molecule in free space.

Thus, the classical enhancement G is, if one neglects polarization effects,

$$G = \left|\frac{\hat{A}_{\mathrm{loc}}}{\hat{A}_{\mathrm{free}}}(\omega_L)\right|^2 \left|\frac{\hat{A}_{\mathrm{loc}}}{\hat{A}_{\mathrm{free}}}(\omega)\right|^2 .$$

Both factors depend on the choice of the classical electromagnetic fields. For the excitation channel one always assumes an incident plane wave, a "reflected" ("scattered") wave and a field inside the metal, both unambiguously determined by the boundary conditions, from which $\hat{A}_{\mathrm{loc}}/\hat{A}_{\mathrm{free}}$ at ω_L follows.

One could formally reverse the time and the directions of currents and the $\boldsymbol{B}$-field of this solution so that the plane wave now corresponds to an emission channel. If one asked for the Stokes emission probability into this channel, the contribution from this emission channel to classical enhancement would be equal to the corresponding excitation channel. So far, there is "reciprocity". However, experimentally, the Stokes emission is collected by a high aperture lens, corresponding to an emitted spherical wave, of which a large solid angle is collected by the lens. Of course, one might decompose this conical spherical wave

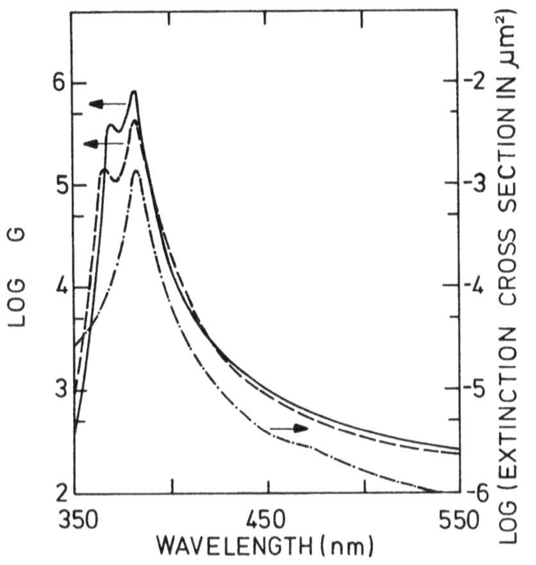

Fig. 6.9. Raman enhancement of vibrations at 1010 cm^{-1} (——) and 1400 cm^{-1} (---) for a monolayer adsorbed on a silver sphere in water versus the wavelength of the exciting light (in the Rayleigh limit). Optical extinction cross section of a silver sphere (10^{-6} μm^3 volume), dash-dotline. From [6.75]

into plane waves with different directions of propagation. But one is not allowed to average over the enhancements $|\hat{A}_{\text{loc}}/\hat{A}_{\text{free}}|^2$ for the corresponding plane waves. Instead, one has to square the average $\hat{A}_{\text{loc}}$. This is because the plane waves composing a spherical wave have fixed phase relations. Therefore, without doing any calculation, we do not expect exact "reciprocity" between enhancements in the "excitation" and "emission" channel. Indeed, in any of the cases where the emission was calculated explicitly, for instance, for spherical particles with retardation [6.68] (see below) and for the ATR configuration (see Sect. 6.3.3, especially Fig. 6.23), there is no "reciprocity".

In resonance, the dominant term in (6.4) is $g \cdot g_0$. For small Raman shifts, g is about equal to g_0 and $G \approx g_0^4$. This means that the Raman scattered intensity is proportional to the fourth power of the field enhancement when the laser frequency is tuned into the dipolar surface plasmon resonance of a small metal particle. For $\varepsilon_0 = 1$ and with $\varepsilon = \varepsilon_1 + i\varepsilon_2$, resonance occurs if $\varepsilon_2 \ll 1$ for $\varepsilon_1 \approx -2$. At resonance $|g_0| \approx 3/\varepsilon_2$, $G \approx \varepsilon_2^{-4}$; hence, the smaller the damping ε_2, the larger the classical enhancement. For larger Raman shifts, g_0 and g are not both in resonance and the enhancement becomes smaller (see also Fig. 6.9). As negative ε_1 and small ε_2 mean high optical reflectivity, the classical-enhancement mechanism should be strong for noble metals in the frequency range below the interband transitions but weak for transition metals (see discussion connected with Fig. 6.18). The exact calculations [6.68] of $\boldsymbol{E}_\text{p}$ include the Lorentz-Mie field (which means all multipole electromagnetic modes of the sphere) but only the dipolar spherical mode in the Raman emission channel, thus breaking the correspondence between "excitation" and "emission". Figure 6.8 depicts the calculated enhancement over the excitation wavelength range 350 to 650 nm for the scattering configuration in the insert, with water as the surrounding medium and a Raman shift of 1010 cm^{-1} (a pyridine vibration). The optical constants of bulk silver from [6.74] have been used. The sharp peaks ($>10^6$) near 370 and 382 nm for $a=5$ nm reflect the dipolar resonances in g and g_0. This is obvious from Fig. 6.9, which shows the average G for a monolayer adsorbed on a silver sphere [6.75] of small diameter (well within the Rayleigh limit). The resonance in g shifts with the wave number of the adsorbate vibration; the resonance in g_0 does not and coincides with the maximum of extinction.

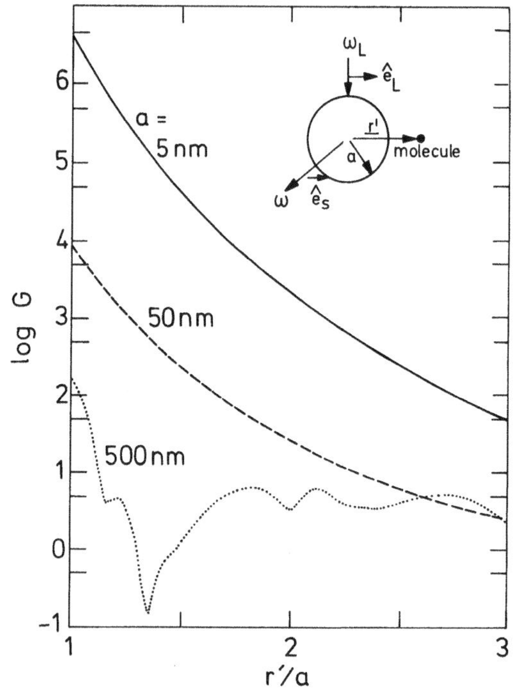

Fig. 6.10. Enhancement G for the scattering configuration in the inset versus distance r' of the molecule from the surface of silver particles of radius $a=5$ nm and laser wavelength $\lambda_0=382$ nm, of $a=50$ nm and $\lambda_0=511$ nm and of $a=500$ nm and $\lambda_0=528$ nm. From [6.68]

Note that the extinction cross section is the sum of absorption and scattering cross sections. With increasing wave number of the vibration, the maximum enhancement decreases fairly fast because the resonances in g and g_0 are further apart. This effect should lead to a characteristic intensity decay of the SERS intensity with increasing wave number of the vibrations. For a larger radius a, the maximum of G decreases and is shifted to higher wavelengths. The oscillations for $a=500$ nm originate from oscillations in the Lorentz-Mie field. Figure 6.10 shows the dependence of the enhancement G on the distance of the molecule from the sphere for three combinations of radius a and exciting wavelength λ_0.

Calculations similar to those of Kerker et al. have been presented by *McCall* et al. [6.18]. They assumed implicitly that the Raman shift is very small so that

$g_0 \approx g$ (6.1, 2). For $\varepsilon_0 = 1$ they obtained the maximum enhancement G, for $\varepsilon_1(\omega) = -2$,

$$G_{\max} = 6[(\varepsilon_1 - 1)/\varepsilon_2]^4 (a/r')^{12}. \tag{6.5}$$

The r'^{-12} dependence reflects the fourth power of the inverse cube dependence of the dipole field. One should note that the variation of $G_{\max}$ with distance depends on the size of a, as evident from Fig. 6.10. The smaller the a, the weaker the long-range character of the enhancement. For a rough film, McCall et al. assumed a distribution $P(\alpha)$ of resonance frequencies given by $\varepsilon_1(\omega) = -\alpha$ (for a spherical resonance $\alpha = 2$). The net enhancement for molecules on the film surface ($a \approx r'$) is given by

$$G_{\text{film}} \sim \int P(\alpha) \left| \frac{\varepsilon - 1}{\varepsilon + \alpha} \right|^4 d\alpha. \tag{6.6}$$

For the frequency ω, where $\varepsilon_1(\omega) = -2$, $P(2) \sim 0.1$, $\varepsilon_2 \sim 0{,}2$, one has an enhancement of about 10^4 [6.18].

Fields at the surface of noble-metal microsphere in incident parallel light have been calculated by *Messinger* et al. [6.76]. Examples for silver spheres in water with radii $a = 22$ and 100 nm are presented in Fig. 6.11. Q_s is the ratio of the power elastically scattered by the sphere to the power incident on its geometrical cross section. Q_s measures the ability of a sphere to extract power from an incident plane wave and scatter it over all solid angles. Q_{abs} is the ratio of the power absorbed by the sphere to the power incident on its geometrical cross section. Q_{NF} is the near-field form of Q_s, which becomes clear from a comparison of the expressions for Q_s and Q_{NF}:

$$Q_s = \lim_{R \to \infty} \frac{R^2}{\pi a^2} \int_R \frac{|E_s|^2}{|E_i|^2} d\Omega$$
$$Q_{NF} = \lim_{\substack{R > a \\ R \to a}} \frac{R^2}{\pi a^2} \int_R \frac{|E_s|^2}{|E_i|^2} d\Omega, \tag{6.7}$$

where E_s is the scattered field, E_i the incident field and Ω is the solid angle. The integral is over the surface of a concentric sphere of radius R. The contribution of the radial components of E_s to Q_{NF} is called Q_R.

For small silver spheres ($a \sim 22$ nm) in the dipolar resonance, the local field strength averaged over the surface is about 1000 times the incident field strength. For $a = 100$ nm, however, this factor is less than 100. This reflects the fact that the energy of the "photon trapped by the resonance" is confined to a smaller volume for a 22 nm sphere than for one of 100 nm. It is more likely for a small sphere than a large one that the photon is absorbed rather than scattered. In the resonances above 400 nm wavelength, the ratio Q_s/Q_{NF} is larger than in the small sphere dipolar resonance. About 3/4 of the near-field is radial.

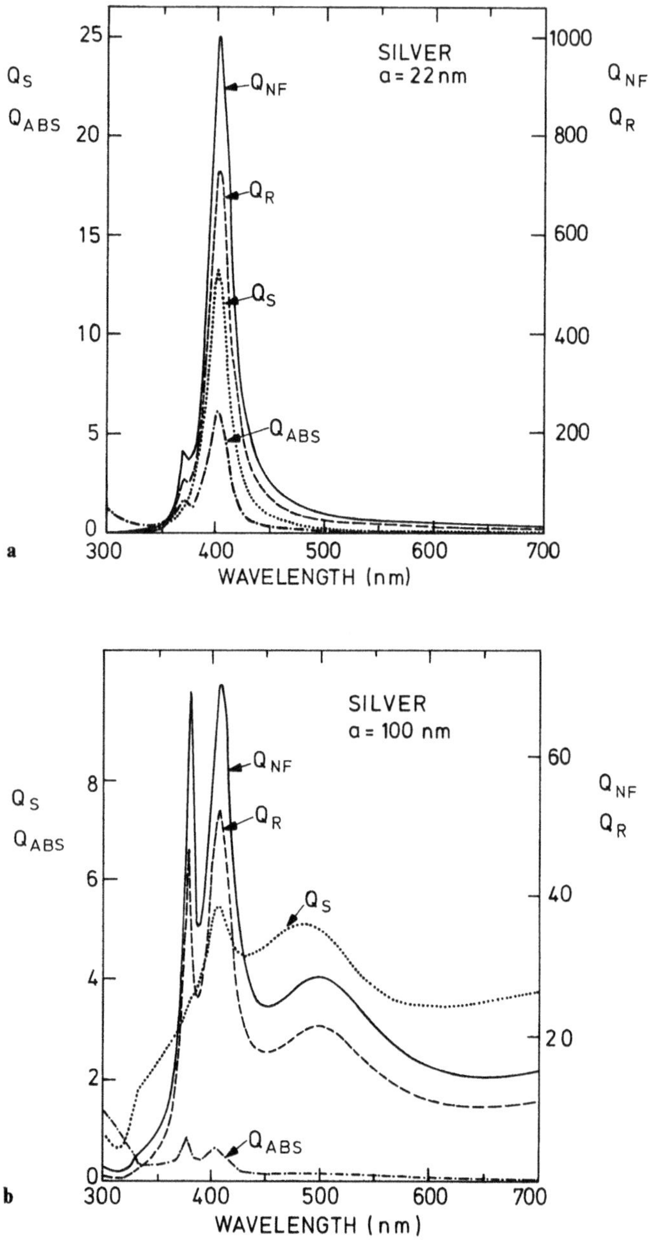

Fig. 6.11a, b. Relative intensity of the scattered light Q_s and the adsorbed light Q_{abs}, average enhancement Q_{NF} of the electric field at the surface and of its normal component Q_R versus wavelength of the incident light, for silver spheres of radius $a = 22$ (**a**) and 100 (**b**) nm. From [6.76]

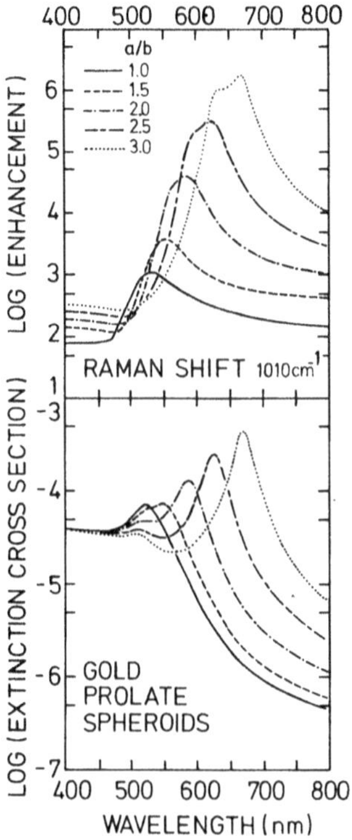

Fig. 6.12. *Top:* enhancement of 1010 cm^{-1} Raman line versus excitation wavelength for a monolayer adsorbed on randomly oriented gold prolate spheroids in water for various axial ratios a/b. *Bottom:* extinction cross section for small gold prolate spheroids in water for various axial ratios. From [6.75]

Fig. 6.13. Spheroidal protrusion with adsorbed molecule on a flat plane

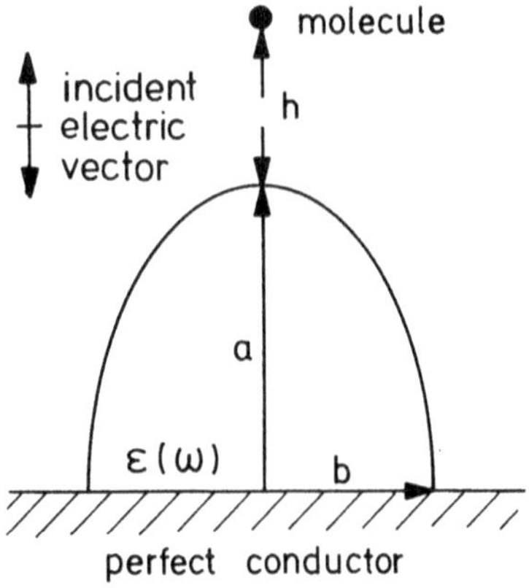

The trends with particle size will probably also hold for the scale of roughness (the "size of the bumps") of a continuous film, although collective resonances (see below) will modify this conclusion. One may thus expect higher field enhancement with decreasing scale of the roughness. There is, however, an opposite trend: an increase of damping in the metal with decreasing size of colloids or bumps (Sect. 6.5.6).

Wang and *Kerker* [6.75] calculated the enhanced Raman scattering for monolayers on prolate and oblate spheroids in the Rayleigh small-particle approximation. The results are averages over the random orientation of the spheroid with respect to the incident field direction. The molecules are described by an electric dipole moment oriented normal to the surface. Wang and Kerker presented examples for molecules with Raman shifts of 1100 and 1400 cm^{-1} on prolate and oblate spheroids of silver, gold and copper. With increasing aspect ratios (long-to-short axis), the wavelengths of maximum enhancement, maximum absorption and the maximum value of the enhancement increase. The wavelength of maximum extinction and the long-wavelength maximum of the enhancement coincide for the small-particle limit. In Fig. 6.12, the results for gold prolate spheroids are presented. They are compared with experimental results in Sect. 6.5.6.

Classical-enhancement calculations for prolate spheroids on an ideally conducting plane ($\varepsilon \to \infty$) (Fig. 6.13) have been presented by *Gersten* and *Nitzan* [6.77, 78]. All dimensions are assumed to be small with respect to the laser wavelength, so that the problem can be solved electrostatically. The relevant electric fields are the applied external field, the field from the polarized molecule, and the field due to the spheroid of dielectric constant $\varepsilon(\omega)$ at the surface plane of the conductor. Because the latter is taken to be infinitely conductive, one may assume zero electric potential at the plane and consider images of the molecule and the hemispheroid with respect to the plane. The direction of the incident field vector is taken to be along the symmetry axis of the spheroid. Because of the high symmetry of the problem, the resonance of the system should correspond to the resonance of the dipolar mode of the full spheroid. However, this configuration can only be approximately achieved experimentally for grazing incidence of the exciting light. Intuitively, one expects rather substantial changes in the response for normal incidence of the exciting light. In this case, the electric field at the surface of the ideally conducting plane will be zero, in contrast to the calculated configuration. This must change the resonance of the spheroid considerably (see the discussion of Ruppin's calculations below).

Gersten and Nitzan included the feedback interaction of the molecule with itself (the so-called "image effect"). However, in the numerical evaluations they usually did not include the "image effect" contribution to the overall enhancement (for a discussion of the relevance of the "image effect", see [6.26, 79, 80–82] and Chap. 7). They presented formulae for the elastic light scattering cross section σ which consists of 3 contributions, the pure Mie scattering σ_M from the "naked" spheroid, the Rayleigh scattering cross section σ_R resulting from the presence of the molecule and a cross term between Rayleigh and Mie scattering. σ_R scales with the Raman enhancement factor G:

$$G = |1 + (1-\varepsilon)\xi_0 Q_1'(\xi_1)/[\varepsilon Q_1(\xi_0) - \xi_0 Q_1'(\xi_0)]|^4$$

$$\sigma_R = \frac{32\pi}{3}\left(\frac{\omega}{c}\right)^4 \alpha^2 G \tag{6.8}$$

$$Q_1(\xi) = \frac{1}{2}\xi \log\frac{1+\xi}{\xi-1}$$

$$\xi_0 = a/f \quad \xi_1 = (a+h)/f \quad f = (a^2 - b^2)^{1/2},$$

where $Q_1(\xi)$, is the Legendre function of the second kind and ξ_1 is a spheroidal coordinate defined in such a way that the surface of the hemispheroid in Fig. 6.4 is given by $\xi \equiv \xi_0$ and the position of the molecule by $\xi = \xi_1$.

There are two sources for enhancement: a surface-plasmon-type enhancement where ω_L fulfills the surface-plasmon-frequency condition

$$\varepsilon(\omega_L) Q_1(\xi_0) - \xi_0 Q_1'(\xi_0) = 0 \tag{6.9}$$

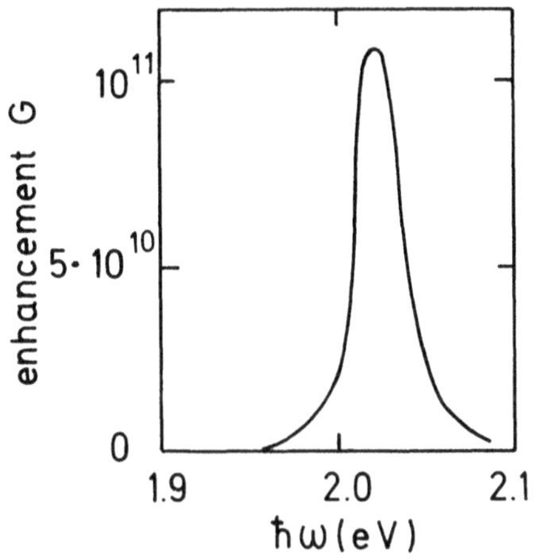

Fig. 6.14. Enhancement ratio versus laser photon energy for a molecule in the configuration of Fig. 6.13 (silver protrusion) with $a = 500$ Å, $b = 100$ Å, $h = 5$ Å and molecular polarizability of the molecule 10 Å^3. From [6.78]

which simplifies for the case where the spheroid becomes a sphere to $\varepsilon(\omega) + 2 = 0$. Since σ_M, σ_R and σ_{MR} all contain the surface-plasmon-type resonance like G, it should be possible to check the theory by measuring Rayleigh and Raman intensities versus ω_L during formation of the film roughness. When the spheroid becomes more needle-like ($a \gg b$), the resonant frequency is shifted to lower values. For extreme ratios a/b, the resonant field is concentrated near the tip of the needle (the lightning rod effect [6.77]). The enhancement in this case is approximately

$$G = \frac{1}{16} \frac{1}{\left(b^2/2a^2 + \frac{h}{a}\right)^4} \frac{1}{\ln(2a/b)^4}. \tag{6.10}$$

By combining the surface-plasmon-typė resonance and the field concentration at the tip of the spheroid, enormous enhancements may be achieved for molecules adsorbed at the tip, as given for the example in Fig. 6.14. *Gersten* and *Nitzan* [6.78] envisaged that a realistic surface contains many protrusions as well as many flat regions. Only a small fraction of the adsorbed molecules will be located on the protrusions which are resonant at a given ω_L. Hence, "it may be that while average enhancement ratios of 10^4–10^6 are observed, the dominant process is produced by only a small fraction of the molecules which, individually, would give rise to much larger enhancement" [6.78].

Following the calculations of the elastic cross section, *Gersten* et al. [6.83] calculated the inelastic scattering caused by shape oscillation of the protrusion ($a = a_0 + A \cos \Omega t$, $b = b_0 - \gamma A \cos \Omega t$, γ being the Poisson ratio). Ω is of the order of $2\pi \times$ sound velocity$/a_0$, corresponding to a Raman shift of some cm^{-1} for $\simeq 50$ Å. According to (6.9) for a given b, the resonant frequency of the spheroid increases with decreasing a. Because the Raman signal Ω originates from the spheroids which are in resonance at ω_L, Ω will increase with ω_L. Low frequency

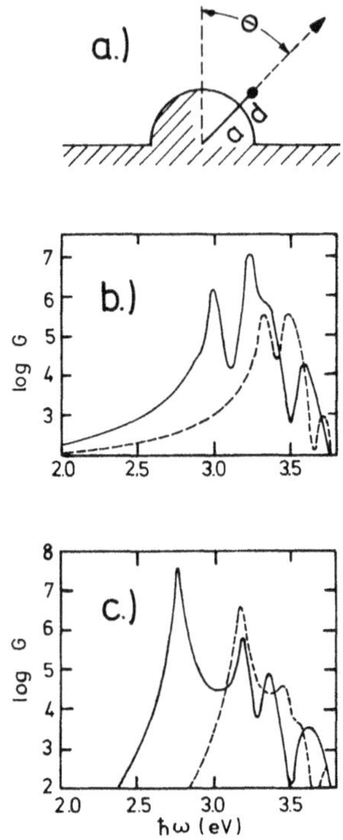

Fig. 6.15. (**a**) Hemispherical bump geometry; the molecule is located at a distance d from the bump of radius a. (**b**) Frequency dependence of the enhancement $G=|E/E_0|^4$ for a molecule located at $d/a=0.05$, $\theta=87{,}5°$ near a silver surface. *Solid curve:* molecule in water; *dashed curve:* molecule in vacuum. The incident field is parallel to the average surface. (**c**) Like (**b**), but with incident field vertical to the average surface. From [6.86]

modes, shifting to higher Ω with increasing ω_L, have indeed been observed [6.84] and assigned to shape oscillations of protrusions. *Fleischmann* et al. [6.85] observed that these low frequency modes at about $8.5\ \mathrm{cm}^{-1}$ on an activated silver electrode in 1 M aqueous KCl electrolyte were very strong at an electrode potential of -0.2 V. However, this band was virtually removed at -0.6 V. Following the interpretation of *Weitz* et al. [6.84], this would mean that bumps on a scale of 50 Å are flattened out by a potential change of 0.4 V. This should be checked further.

As mentioned above, one expects the resonances for spheroids on a metal surface to be different for the polarization of the exciting field parallel or perpendicular to the average surface. This is borne out by the calculations of *Ruppin* [6.86] for hemispherical surface bumps. Ruppin employed a hemispherical bump which had the same dielectric constant as the plane from which it protruded. In this way the collective modes were not artificially excluded from the plane, but were surface plasmons of the plane-bump system as a whole. The enhancement of the field of the incident channel is calculated at arbitrary locations around the bump in the electrostatic approximation. Ruppin assumed that for the emission channel, the enhancement is equal to that of the incident channel. Also, he neglected the Raman shift. (Optical constants for silver from *Johnson* and *Christy* [6.74] have been used. Ruppin's results are reproduced in Fig. 6.15 (note that, due to the electrostatic approximation, the results depend only on the ratio d/a). The positions of the peaks are characteristic of the hemispherical bump geometry; for other bump shapes the peak position will differ. There are considerable differences in the position of the resonance at lowest energy for the polarization of the incident field. Ruppin has only published results for a particular position of the scatterer; it would be interesting to have the average enhancement as a function of the distance from the surface (see Sect. 5.6). One should note that in Ruppin's and in Gersten and Nitzan's electrostatic approximation, the "applied field" E_0 is some superposition of the incident and reflected laser field. It is not clear whether one is allowed to use for E_0 the electric field at a smooth surface, as one may obtain it from ordinary metal optics. Also, the simple assumption $G=|E/E_0|^4$ only applies when the polarization direction of the scattered light is the same as that of the incident light, i.e., when the Raman tensor α of the molecules is of the form $\alpha \cdot \mathbf{1}$. If this is not the case, the excitation and the emission channel will not be simultaneously in resonance and the overall enhancement will be smaller.

A frequently-used scattered geometry is oblique incidence of the laser light at about 60° angle of incidence while the Raman scattered light is collected in a near-normal direction. In this configuration, the relevant resonance in the emission channel is the one for E_0 parallel to the macroscopic surface, whereas the resonance in the excitation channel would depend on the polarization of the incident light. It would be interesting to have the theoretical predictions of the

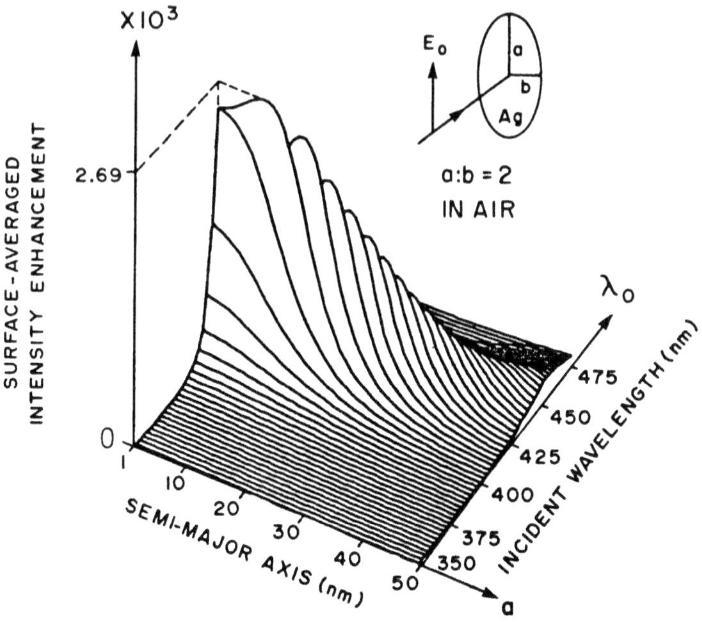

Fig. 6.16. Surface-averaged intensity enhancement calculated with electrodynamic retardation for a 2:1 Ag prolate spheroid in air as a function of a semi-major axis and incident wavelength. Incident field polarized along the spheroid. From [6.87]

relative intensities and of the depolarization of the Raman and Rayleigh scattered light in this case in order to compare it with experiments.

As pointed out by *Kerker* and coworkers [6.68], the electrostatic Rayleigh approximation is only valid for particles with dimensions smaller than $\lambda_L/15$. *Barber* et al. [6.87] calculated the average of the surface field squared (equivalent to Q_{NF} in (6.7), termed intensity enhancement in Fig. 6.16) electrodynamically for rotational spheroids of silver with bulk optical constants [6.74]. For spheroids with an aspect ratio of 2:1, the surface-averaged field enhancement reaches a maximum of 2690 at $\lambda_L = 407.5$ nm and a major axis length a of 16 nm. The electrostatic approximation yields almost the same result but, of course, independent of the spheroid size (for more details see [6.87]).

In the electrodynamic calculation it is evident from Fig. 6.16 that, as the size is increased, (i) the major peak shifts to longer wavelength (had $\varepsilon(\omega_L)$ been a constant, this size-dependent resonance would have occurred at a fixed a/λ_L ratio [6.87]), (ii) the intensity-enhancement decreases rapidly from the electrostatic value and (iii) the resonance becomes broader. These last two results are associated with the phase retardation between various portions of the spheroid which tends to decrease the enhancement and broaden the linewidth [6.87].

Both *McCall* et al. [6.18] and *Gersten* and *Nitzan* [6.78] applied the results for single spheres and single prolate spheroids to a statistical distribution of bumps or protrusions on a surface. In this way they neglected the coupling between the roughness hillocks. This coupling must be important, however, because the distance between hillocks may be of the order of the decay length of the resonating fields.

6.3.2 Collective Resonances

This subsection is devoted to classical surface-plasmon-type enhancement calculations which consider explicitly collective interactions between the resonating structures. *Moskovits* [6.13] considered collective resonances of bumps on a metal surface and the dependence of the resonance frequency on the bump density. To facilitate the calculations, the bumpy surface was replaced by a transition layer of metal spheres in a dielectric matrix, which in turn is treated by the Maxwell-Garnett theory. Accordingly, the composite material [dielectric constants of the matrix ε_0 and of the metal spheres $\varepsilon(\omega)$] is characterized by an effective dielectric constant $\bar{\varepsilon}(\omega)$ obtained by solving

$$\frac{\bar{\varepsilon}(\omega)-\varepsilon_0}{\bar{\varepsilon}(\omega)+2\varepsilon_0}=q\,\frac{\varepsilon(\omega)-\varepsilon_0}{\varepsilon(\omega)+2\varepsilon_0}, \tag{6.11}$$

where q is the volume fraction of the composite material filled with metal spheres. There is a singularity in $\bar{\varepsilon}(\omega)$ called by *Marton* and *Lemon* [6.88] a "conduction resonance". This is a coupled dipole mode of all the spheres for

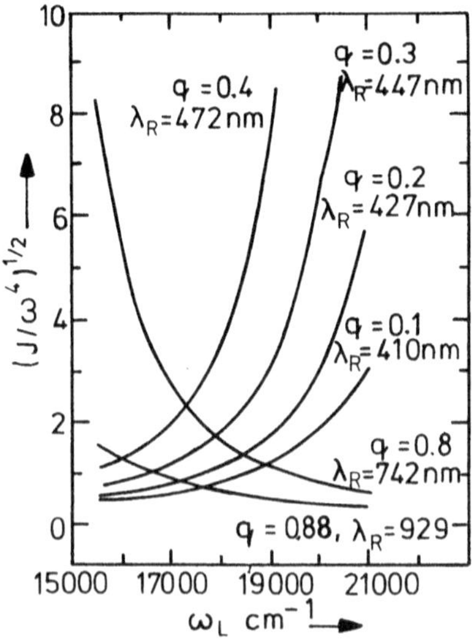

Fig. 6.17. Dependence of the Stokes intensity J [arb. units] on the exciting frequency ω_L for various values of the resonance wavelength λ_R. The corresponding values of the packing ratios q are also shown. From [6.14]

$\omega = \omega_R$, where ω_R fulfills

$$\varepsilon(\omega_R) = -\varepsilon_0 \frac{2+q}{1-q} \tag{6.12}$$

which leads again to $\varepsilon(\omega_R) + 2\varepsilon_0 = 0$ for $q \to 0$.

Assuming $\varepsilon(\omega) = \varepsilon_b - \omega_p^2/\omega^2$, where ε_b is the interband and $-\omega_p^2/\omega^2$ the free-electron contribution, *Moskovits* [6.13] derived for the resonance wavelength λ_R

$$\lambda_R = \lambda_p \left(1 + \varepsilon_b + \frac{2+q}{1-q}\,\varepsilon_0\right)^{1/2} \quad \lambda_p = \frac{2\pi c}{\omega_p}. \tag{6.13}$$

Then λ_R shifts from $\lambda_p(1+\varepsilon_b+2\varepsilon_0)^{1/2}$ to infinity with increasing q. Rather than calculating the emission of an adsorbed molecule, *Moskovits* assumed that "the wave function of the metal-bump-adsorbate system should include an explicit contribution from the conduction resonance" [6.89]. Therefore he inserted the resonance frequency ω_R into the formula for the intensity J of preresonantly-enhanced Raman scattering (ω: Stokes frequency):

$$J \sim [\omega^2(\omega_L^2 + \omega_R^2)/(\omega_R^2 - \omega_L^2]^2. \tag{6.14}$$

The results are presented in Fig. 6.17. The choice of q gives some freedom to fit J to experimental excitation spectra of SERS. For high density-of-surface bumps, the resonance shifts to the red.

Corresponding calculations of the change in reflectivity from a bumpy surface when varying the coverage of adsorbed molecules and including

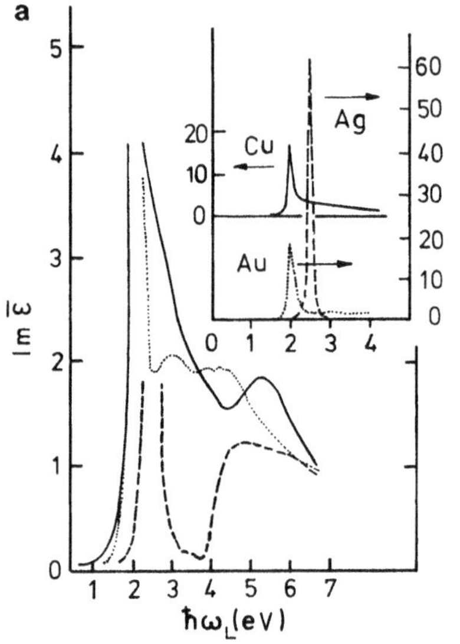

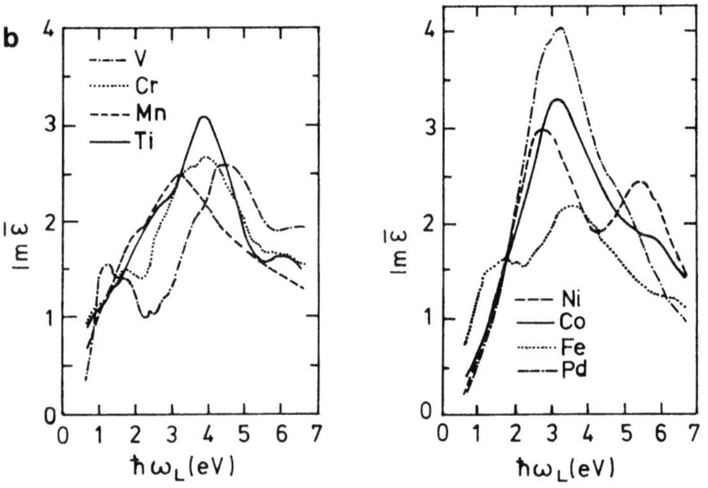

Fig. 6.18a, b. Imaginary part of the effective dielectric constant $\bar{\varepsilon}(\omega)$ of composite films (matrix with $\varepsilon_0 = 1$, filling factor $q = 0.5$) vs laser frequency for spheres of noble (**a**) and transition metals (**b**). From [6.89]

"coherent charge transfer" from the molecules to the bumps are presented in [6.90]. In [6.91] *Moskovits* fit this result to changes ΔR in the experimental reflectivity R on Ag(111) electrodes caused by electrochemical activation [6.43]. According to Moskovit's ideas, the electrode is smooth before and bumpy after the activation. Thus, ΔR should be caused by the creation of the bumps. However, in [6.90] ΔR is fitted by assuming a modulation $\Delta(n\alpha)$ (pyridine coverage n with polarizability α). Nevertheless, R was measured before and after activation at the same electrode potential and the total coverage with pyridine can only change by the increase of the surface area due to the formation of bumps. Thus one may ask how meaningful is the excellent fit in [6.91].

As a measure of the possible classical enhancement of bumpy films, *DiLella* et al. [6.89] have published the frequency dependence of Im $\{\bar{\varepsilon}(\omega)\}$ for $q=0.5$ and $\varepsilon_0=1$, see (6.11), for noble and transition metals, which is reproduced in Fig. 6.18. The results show the strong resonances which are possible for the noble metals. Moskovits's approach neglects the difference in the collective resonance for the polarization direction parallel and perpendicular to the macroscopic surface. Such differences have to be expected on the basis of *Ruppin*'s calculations [6.86]. For aggregated silver films deposited at 85 °C with an average thickness below 90 Å, *Yamaguchi* et al. [6.92] found optical anisotropy. The computation of the optical constants of these films from experimental ellipsometric data only became possible by using anisotropic optical parameters.

Burstein and *Chen* [6.39, 93] considered collective resonances in metal-island films on a substrate by treating the dipole moments induced in the metal islands as point dipoles and taking into account the contributions to the local field at a given island from the dipoles at neighbouring islands and from the image dipoles in the dielectric substrate. Retardation effects were neglected. The net effect of these contributions to the local field was to shift the collective resonances to longer wavelengths with respect to the resonance of the individual islands in the case of collective resonances parallel to the substrate, and to shorter wavelengths in the case of collective resonances perpendicular to it. The overall enhancement

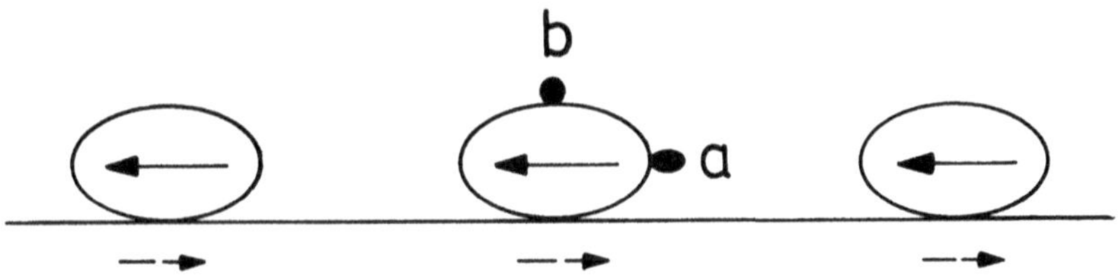

Fig. 6.19. Ordered array of ellipsoids of revolution on a flat surface with minor axes normal to the surface. Arrows characterize dipole moments of the ellipsoids and their image dipoles parallel to the substrate surface, a and b depict adsorption sites

factor comes again from the product of two response functions, described by g_0^2 and g^2 above and called $\lambda(\omega_L)$ and g_s by Burstein and Chen, characterizing the resonant enhancement of the incident and the emitted fields. To obtain theoretical expressions for these response functions, Burstein and Chen took an idealized Ag-island film consisting of ellipsoids of revolution which were uniform in size and distributed in an ordered array on a flat dielectric substrate, with their symmetry axes oriented normal to the surface (Fig. 6.19). For a molecule in position a in Fig. 6.19, the incident channel response parallel to the film is given by

$$\gamma(\omega_L)=\frac{|\varepsilon_1(\omega_L)|^2}{|1+[\varepsilon(\omega_L)-1]\,(f_\parallel+\beta_{\mathrm{id}_\parallel}+\beta_{\mathrm{nd}_\parallel})|^2}\,. \tag{6.15}$$

The maximum value of $\gamma(\omega_L)$ at the parallel collective resonance frequency $\omega_{\|}$ is

$$\gamma(\omega_{\|}) = \frac{|\varepsilon_1(\omega_{\|})|^2 \, |\varepsilon_1(\omega_{\|}) - 1|^2}{|\varepsilon_2(\omega_{\|})|^2}, \tag{6.16}$$

where $f_{\|}$ is the geometrical depolarization factor of the ellipsoids and $\beta_{\mathrm{id}_{\|}}$ and $\beta_{\mathrm{nd}_{\|}}$ represent the contributions to the local field acting on the ellipsoid from its own image dipoles in the substrate and from dipoles and image dipoles at neighbouring ellipsoids. β_{id} and β_{nd} have been calculated [6.94] for square lattices of ellipsoids. The "in-phase" contributions to g_s^2 from excited molecules on different ellipsoids in position a in Fig. 6.19 was given by Burstein and Chen as

$$g_s(\omega) \approx |1 - f_{\|}|^2 \, \frac{|\varepsilon(\omega) - 1|^2}{|1 + [\varepsilon(\omega) - 1] \, (f_{\|} + \beta_{\mathrm{id}_{\|}} - \beta_{\mathrm{nd}_{\|}})|^2} \tag{6.17}$$

which is nearly the same as $\gamma(\omega_L)$. When $\omega \approx \omega_L$, the resonances in γ and g_s may occur together and consequently one may calculate enormous enhancement [6.39]. This is correct in the sense of the "reciprocity" discussed in Sect. 6.3.1, irrespective of whether one considers just one single adsorbed molecule on a particular island or the "in-phase" contribution [6.39] of an adsorbate layer covering all islands.

However, from an experimental point of view, one is interested in the large space-angle Raman emission from one molecule or from an adsorbate layer covering all islands. As vibrations of molecules on different islands are, to first order, independent of each other, Raman emission from adsorbates on different islands is an incoherent process. Thus, one may discuss the overall emission with the picture of just one molecule adsorbed on one particular island, finally integrating over the emitted intensity of all different molecules. That this reasoning is correct is demonstrated by experiments in regular microlithographic silver particle surfaces (see detailed discussion in Sect. 6.5.2): in second harmonic generation (SHG), *Wokaun* et al. [6.95] observed a well-defined diffraction order, proving the predominantly coherent emission of the second harmonic from different silver particles. However, for Raman experiments on the same silver particle surfaces [6.96], no grating diffraction effects were reported.

Let us regard the calculations of Burstein and Chen discussed above from the viewpoint of overall Raman emission from the vibration of a molecule adsorbed on a particular island. The factor β_{nd} is calculated by *Yamaguchi* et al. [6.92] for the incident channel. The dipole moments of all islands are driven coherently by the same incident wave; the distance between neighbour islands is assumed to be small compared to the wavelength of the incident light. Therefore, the dipole moments of neighbour islands are of the same strength and phase as the dipole moment of the given island. The dipolar interaction between islands will scale like (distance)$^{-3}$. However, this does not hold for the Raman channel. The

Raman dipole moment of an island, driven by the molecules adsorbed on it, will induce a corresponding moment on neighbour islands like (distance)$^{-3}$ so that the overall influence from neighbour islands on the given island scales like (distance)$^{-6}$. Formally, this will lead to a different value for β_{nd} in $g_s(\omega)$ compared to $\gamma(\omega_L)$, so that, in general, $\gamma(\omega_L)$ and $g_s(\omega)$ will not have the resonance at the same frequency. This will lead to a weaker enhancement. The splitting between a "collective resonance" in the excitation channel and an "individual resonance" in the emission channel should be quite a general phenomenon. We will come back to the calculation of *Burstein* et al. [6.39, 93] in Sect. 6.5.3 in order to estimate the classical Raman enhancement for silver island films.

A different approximation of the electromagnetic enhancement G for adsorbates on island films was proposed by *Weitz* et al. [6.97]. The absorption $A(\omega)$ of the island film is expressed as

$$A(\omega)=qd\varepsilon_2(\omega)\,\frac{\omega}{c}\,\frac{\overline{|E_{\mathrm{in}}|^2}}{|E_0|^2}\,, \tag{6.18}$$

where q is the volume fraction of silver in the film of effective thickness d, E_0 is the incident field strength and $\overline{|E_{\mathrm{in}}|^2}$ represents an effective average field intensity inside the islands.

In the spirit of the electromagnetic model [6.97], the electromagnetic enhancement is expressed as the product of the local field enhancements at the excitation frequency ω_L and emission frequency ω. The local field at the surface of the island outside the metal is taken as $\varepsilon E_{\mathrm{in}}$, thus neglecting field components parallel to the surface. This leads to

$$\begin{aligned} G &\approx \frac{c^2}{q^2d^2}\quad A(\omega_L)A(\omega)\cdot F \\ F &= \frac{|\varepsilon(\omega_L)|^2\;|\varepsilon(\omega)|^2}{\varepsilon_2(\omega_L)\,\varepsilon_2(\omega)\,\hbar\omega_L\cdot\hbar\omega}\,. \end{aligned} \tag{6.19}$$

The implications of this approximation will be discussed in Sect. 6.5.6.

6.3.3 Resonances on Gratings, Rough Surfaces, and by Attenuated Total Reflection

For a metal-vacuum interface in the form of a grating with sinusoidal modulation ξ_g and grating vector $\boldsymbol{g}$, the enhancement of the driving field outside the metal surface has been given by *Jha* et al. [6.63] to first order in ξ_g. For p-polarized incident light, an additional field strength E_g^p is induced at the surface proportional to the incident field E_i, to $(\xi_g \cdot k_g)$, and to the surface response function Δ^{-1} given by

$$\Delta = \varepsilon(\omega_L)\left(k_g^2 - \frac{\omega^2}{c^2}\right)^{1/2} + \left[k_g^2 - \varepsilon(\omega_L)\frac{\omega^2}{c^2}\right]^{1/2}$$

$$\boldsymbol{k}_g = \boldsymbol{g} + \boldsymbol{k}_t \qquad k_t = \frac{\omega_L}{c}\sin\theta, \tag{6.20}$$

where θ is the angle of incidence. Resonance occurs for $\Delta \approx 0$, a condition which also yields the dispersion relation of surface-plasmon polaritons on a smooth surface [6.98]. For p-polarized light and a proper choice of θ, the maximal enhancement of the electric-field component normal to the surface is proportional to $\varepsilon^2/\varepsilon_2$. Jha et al. have not explicitly calculated the emission from a molecule near to or adsorbed on a grating. They assumed that the additional tensorial enhancement factor due to the conversion of the near-zone Stokes field into outgoing radiation depends strongly on the angle of scattering and, except for those angles where the surface-plasmon field can couple out resonantly from the grating, it is still of order unity. Accordingly, in [6.63] the classical Raman enhancement is only given by the square of the incident field enhancement (see discussion in Sect. 6.5.8).

Aravind et al. [6.99] have calculated in first-order perturbation theory, the emission from a dipole above a sinusoidal silver grating. Typical results are presented in Fig. 6.20 which compares the emission for a dipole at some

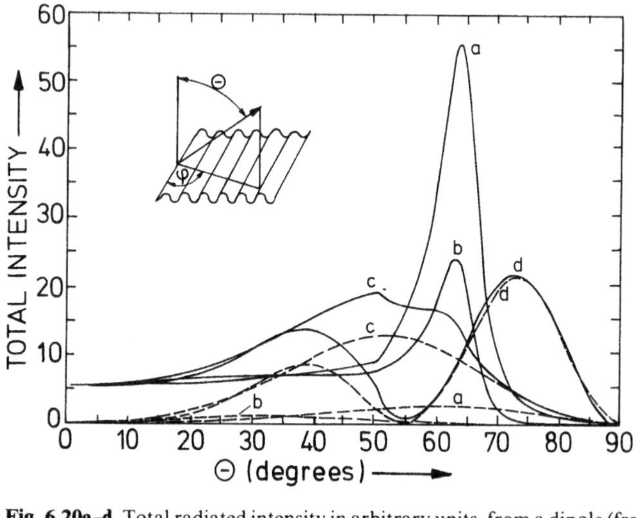

Fig. 6.20a–d. Total radiated intensity in arbitrary units, from a dipole (frequency $\hbar\omega = 3$ eV) in front of a sinusoidal silver grating, grating wavelength 8200 Å, amplitude 1148 Å (*full lines*), and in front of a smooth silver surface (*dashed lines*), for distances of the dipole from the mean plane 580 Å (**a**), 1160 Å (**b**), 2320 Å (**c**) and 4640 Å (**d**). Azimuthal angle φ and polar angle θ of the emitted light are defined by the inset. For all curves, $\varphi = 80°$. From [6.99]

distances above a grating and above a smooth surface. The extra emission near the polar angle $\theta=60°$ is due to the excitation of a surface-plasmon polariton by the dipole (frequency $\hbar\omega=3.0$ eV) and its subsequent transformation into a radiative mode by grating outcoupling. These authors point out that their method is not accurate for the response of the grating to the Fourier component of the dipole field for wave vectors larger than the reciprocal grating height. Their results, therefore, become less and less reliable as the dipole-surface distance diminishes.

The increase $I_{\text{rough}}-I_{\text{flat}}$ of the radiation from an induced dipole by roughening the surface has been calculated by *Aravind* and *Metiu* [6.100] for small-scale roughness (rms value ≈ 30 Å). The strength of the scattering process by roughness is determined, in the Born approximation, by the correlation function $\langle \Delta\varepsilon(\boldsymbol{r})\Delta\varepsilon(\boldsymbol{r}')\rangle$ which is proportional to $\langle\xi(x,y),\,\xi(x'y')\rangle$, where $\xi=\xi(xy)$ describes the surface roughness. The Fourier transform of this quantity is taken to be

$$\langle\xi(\boldsymbol{k}_{\|})\xi^*(\boldsymbol{k}'_{\|})\rangle=(2\pi)^2\delta(\boldsymbol{k}_{\|}+\boldsymbol{k}'_{\|})\delta^2a^2\mathrm{e}^{-(ak_{\|}/2)^2}, \tag{6.21}$$

where δ^2 is the mean square amplitude of roughness and a its correlation length. Figure 6.21 shows the quantity $(I_{\text{rough}}-I_{\text{flat}})/I_{\text{flat}}$ for silver with $\delta=30$ Å, the angle of detection 45° and, different correlation lengths. In the visible range the corresponding increase hardly exceeds a factor of ten.

For an overview of the extensive work of *Metiu* and coworkers related to electromagnetic resonances, the reader is referred to [6.101].

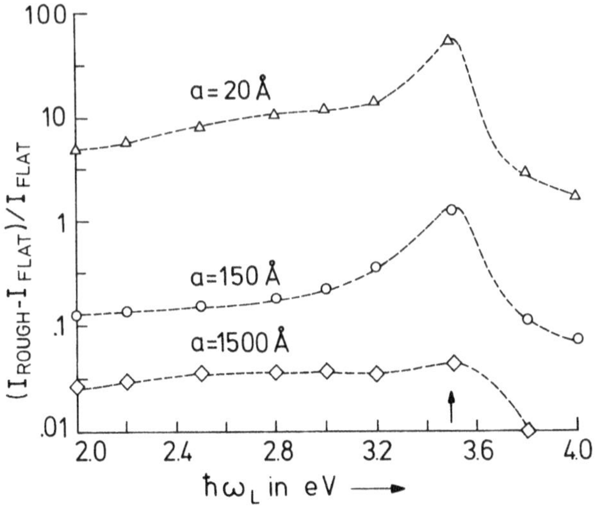

Fig. 6.21. I_{flat} and I_{rough} are the intensities radiated by a dipole in the presence of flat and rough silver surfaces. Mean amplitude of roughness 30 Å, angle of detection from normal 45°. The parameter a is the roughness correlation length. The arrow indicates the unretarded surface plasmon frequency. From [6.100]

A different approach to estimating the enhanced outcoupling via statistical surface roughness was presented by *Weber* and *Ford* [6.102]. The average total power $P_{tot}(\omega)$ dissipated at the Raman-shifted frequency ω from a dynamical Raman dipole $\mu(\omega)$ normal to the surface at a distance d from the surface is given by

$$P_{tot}(\omega)=\frac{1}{2}\,\omega|\mu(\omega)|^2\,\mathrm{Im}\left\{\int\limits_0^\infty \frac{k_\parallel^3}{q}\,(1+re^{-2qd})\,dk_\parallel\right\},$$
$$q=(k_\parallel^2-\varepsilon\omega^2/c^2)^{1/2}. \tag{6.22}$$

The second term under the integral corresponds to the field of μ after reflection at the surface. The reflection coefficient r is calculated following *Kliewer* and *Fuchs* [6.103, 104] and contains the nonlocal transverse and longitudinal response of the metal due to excitation of electron-hole pairs. The range of integration in (6.20) may be split into $0<k_\parallel<\frac{\omega}{c}$, $\frac{\omega}{c}<k_\parallel\lesssim 2k_{SPP}$ ($k_{\parallel SPP}$ is the $k_\parallel$ vector of the surface plasmon polariton at frequency ω) and $2k_{\parallel SPP}\lesssim k\lesssim d^{-1}$. Thus, the total emission is split into radiative emission, emission into surface-plasmon polaritons (SPP), and emission into the continuum of driven surface modes (DSM):

$$P_{tot}(\omega)=P_{photon}(\omega)+P_{SPP}(\omega)+P_{DSM}(\omega). \tag{6.23}$$

Table 6.3. Ratios of photon, surface-plasmon polariton and driven surface mode emission from a dipole, radiating with frequency ω at a distance d from a silver surface (see text). From [6.102]

$\hbar\omega$ [eV]	d	P_{SPP}/P_{photon}	P_{DSM}/P_{photon}
1.96	1.23	1.7	3.4×10^6
2.41	1.24	2.7	2.5×10^6
2.60	1.24	3.3	2.3×10^6

After modelling the optical properties of silver within the Fuchs-Kliewer nonlocal approach (free electron contribution to $\hbar\omega_p=9.33$ eV, $\hbar\tau^{-1}=0.058$ eV, $k_F=1.2\times10^8\ \mathrm{cm}^{-1}$, bound electron contribution to dielectric constant $\varepsilon_b=3.6$), Weber and Ford obtained for the ratios P_{SPP}/P_{photon} and P_{DSM}/P_{photon} for $d=1.23$–1.24 Å the values given in Table 6.3. A very large emission channel is opened by the driven surface modes, which are essentially electron-hole pair excitations. In the local approximation, with the response of the metal described by $\varepsilon(\omega)$, P_{DSM} is at least one order of magnitude smaller. This point will be discussed further in Sect. 6.4. The DSM channel will contribute to Raman radiation by conversion into radiative modes via the statistical surface roughness. The outcoupling efficiency is calculated as emission from a normal current $j(r)$:

$$j(r) = -\mathrm{i}\omega E_z(\boldsymbol{\mu})\delta(z)\hat{z}\left(\frac{\varepsilon-\varepsilon_0}{\varepsilon+2\varepsilon_0}\right)\xi(x,y), \tag{6.24}$$

where $E_z(\boldsymbol{\mu})$ is the normal component of the electric field of the Raman dipole. The total emission is assumed to be composed of P_{photon}, P_{SPP} (because surface plasmon polaritons, being high-Q modes, will ultimately be scattered into observable radiation by roughness [6.102] and of P_{rough}, the emission from $j(\boldsymbol{r})$. The overall enhancement of emission G_{rough} is given by

$$G_{\text{rough}} = (P_{\text{photon}} + P_{\text{SPP}} + P_{\text{rough}})/P_{\text{photon}}. \tag{6.25}$$

$\xi(x,y)$ describes a boss-like surface structure with

$$\langle\xi(x',y')\xi(x,y)\rangle = a^2 \exp\{-[(x-x')^2+(y-y')^2]/a^2\}$$
$$d \ll a \ll \frac{2\mu c}{\omega}. \tag{6.26}$$

$P_{\text{rough}}/P_{\text{photon}}$ is for $\hbar\omega = 1.96$ eV about 2×10^3. This means that about 10^{-3} of the electron-hole pair excitation P_{DSM} is converted into radiation by the surface roughness, and that $G_{\text{rough}} \approx P_{\text{rough}}/P_{\text{photon}} \approx 2\times10^3$.

Weber and *Ford* [6.105] estimated the enhancement of the electric field at the surface on the basis of energy conservation. The incident beam may couple to surface-plasmon polaritons on a plane surface via a prism or an inscribed grating structure. The dissipated power is proportional to the square of the electric field at the surface $|\boldsymbol{E}_{\text{sp}}(0^+)|^2$ (taken, like in all other calculations described here, outside the metal) and to the imaginary part of the k-vector of the surface-plasmon polariton, which describes lateral damping. This energy dissipation per unit area must be equal to

$$I = I_0(1-R)\cos\theta$$
$$I_0 \sim |E_0|^2, \tag{6.27}$$

where I_0 and E_0 are the incident intensity and electric field, R the power reflectance (irrespective of prism or grating coupling), and θ the angle of incidence. This leads to the enhancement

$$\frac{|E_{\text{sp}}(0^+)|^2}{|E_0|^2} = \frac{2\cos\theta\,\varepsilon_1^2(1-R)}{\varepsilon_0^{1/2}\varepsilon_2(|\varepsilon_1|-\varepsilon_0)^{1/2}} \tag{6.28}$$

[following [6.106], the ratio $2\cos\theta(|\varepsilon_1|-1)^{3/2}\varepsilon_2^{-1}$ was derived for $R=0$ and $\varepsilon_0=1$, which for $|\varepsilon_1|\gg1$, is nearly identical to (6.28)].

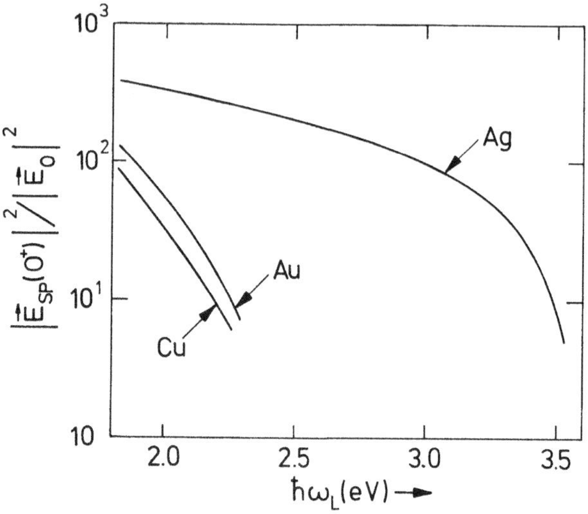

Fig. 6.22. Calculated upper limit for the electrical-field enhancement achievable with a perfect grating coupler of spacing $d = 800$ nm for Ag, Cu and Au. From [6.105]

Weber and Ford assumed that the above enhancement ratio is also an upper limit for a randomly roughened surface if one replaces $(1-R)$ by the roughness-induced power absorbance A. For $A=1$, the ratio of $|E_{sp}(0^+)|^2/|E_0|^2$ is plotted for Ag, Au and Cu in Fig. 6.22. For silver in the range 2–3 eV and typical values of A of 10–20 %, the field enhancement would be of order ten. If one follows *Aravind* and *Metiu* [6.100], the enhancement in the emission channel in the range 2–3 eV can reach a value of 10; according to Weber and Ford it is about 2×10^3. Hence, one would expect 1 to 3 orders of magnitude overall classical Raman enhancement.

One may raise one argument against the conclusion of [6.105]. The application of (2.28) to a randomly rough surface implies that the ratio of the average components of $E_{sp}(0^-)$ in and out of phase with the dielectric polarization $P_{sp}(0^-)$, is the same for a surface-plasmon polariton-type resonance on a smooth and on a nonsmooth surface. This point needs further consideration. It implies that the ratio of the average normal component to the average tangential compoment of $E_{sp}(0^+)$ is the same for all types of electromagnetic resonances.

An exact calculation for classical Raman enhancement has been presented for the excitation of a surface-plasmon polariton on a smooth surface by attenuated total reflection (ATR) [6.98] by *Chen* et al. [6.107]. Figure 6.23 compares the Raman signal expected for low frequency shifts for a thin overlayer f in the external reflectance (ER) with the ATR configuration. Γ_{zz} is the ratio of $E_z(\text{film})/E_z(\text{air})$ for the incident light at ω_L and G_{zz} is the ratio $E_z(\text{air})/E_z(\text{film})$ for the scattered field at ω. In the special case of Fig. 6.23 for $\varepsilon_f = 1$ (i.e., scattering molecules on top of the silver film), an ATR enhancement of 340 with

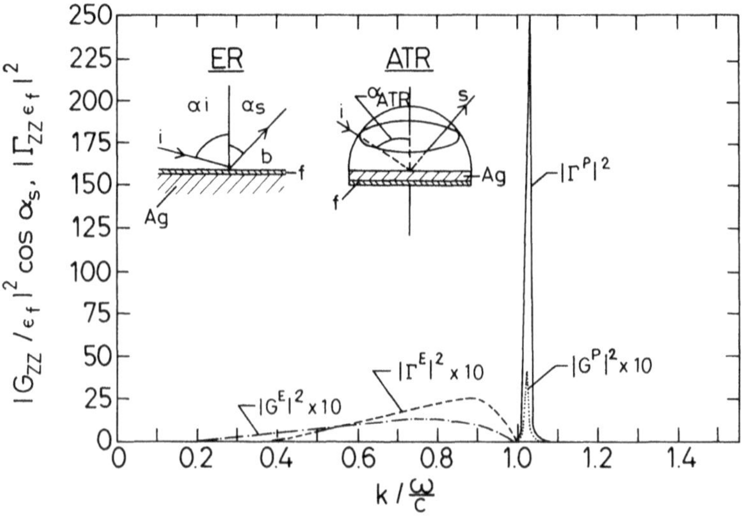

Fig. 6.23. Transfer functions $|\Gamma_{zz}\varepsilon_f|^2$ and $|G_{zz}/\varepsilon_f|^2 \cos\alpha_s$ for the ER configuration (superscript E) and the ATR configuration (superscript P) for a thin overlayer f on silver; λ(incident) $\simeq \lambda$(scattered) ≈ 6471 Å and the thickness of silver films $\simeq 550$ Å. ε(silver) $= -19.6 + \mathrm{i}\,0.59$ and ε(prism) $= 2.25$. From [6.107]

respect to the optimum external reflectance case is obtained, whereby a factor of about 100 is due to the incident field enhancement by tuning into the surface-plasmon polariton resonance and a factor of about 3 is due to enhanced emission to the prism side under an angle corresponding to the surface plasmon-polariton resonance.

For the same geometry, *Sakoda* et al. [6.108] calculated a factor of 150 for the incident field enhancement and a factor of 290 for the emission channel, leading to an overall enhancement of 4×10^4. Whereas the calculations of Chen et al. apply to the Raman scattering from an "adsorbate vibrational surface phonon" of a 2 dimensional wave vector $\boldsymbol{q}$ connecting the wave vectors $\boldsymbol{k}_\mathrm{L}$ and $\boldsymbol{k}_\mathrm{f}$ of the incident and emitted field ($\boldsymbol{k}_\mathrm{f} = \boldsymbol{k}_\mathrm{L} - \boldsymbol{q}$), the enhancement calculations of Sakoda et al. refer to a single adsorbed molecule.

6.3.4 Comments

In conclusion, rigorous calculations including retardation of the classical enhancement have so far only been presented for two ideal cases, namely, a spherical particle embedded in an infinite medium [6.68] and the ATR configuration with an ideal smooth surface [6.107]. If the silver film has the optical constants of perfectly crystalline silver [6.74], the enhancement for silver spheres in the range of the visible Ar^+ and Kr^+ laser lines is at most 4 orders of magnitude. For other "single colloid particle" or "single bump" geometries, only the enhancement of the exciting field is calculated, in most cases without retardation. Enhancement of the emitted field is assumed to be of the same analytical form, an unwarranted assumption if retardation effects become important. Only crude approximations exist for the classical enhancement on a surface with random roughness where the electromagnetic excitations of individual protrusions are coupled. The surface is either modelled as a regular array of ellipsoids [6.93] or as a transition layer with the dielectric constant of a composite material [6.89].

These calculations do not go beyond the unretarded dipole approximation. In some cases, the interaction between protrusions is neglected and the average enhancement is assumed to be given by the average concentration of "protrusions in resonance" [6.18, 78]. The difficult problem of angle-integrated emission from a point source on a grating surface or on a surface of truly statistical roughness has not yet been tackled. In this case the collective resonances for the excitation channel and the emission channel will be different, as discussed above. A first approach to electromagnetic resonances of statistically rough surfaces in the context of SERS has been made by *Laor* and *Schatz* [6.109]. They present calculations for the dipolar coupling of clusters of hemispheroids on a perfectly conducting surface and for random distributions of these hemispheroids.

They also concerned themselves with the interaction of the protrusions which had been considered as isolated by *Gersten* and *Nitzan* [6.78]. In detail, a total of 68 hemispheroids were chosen with axis *a* and *c* randomly distributed between 0 and 20 nm and spread randomly on a square region of 100 × 100 nm. Periodic boundary conditions were used to extend this to an infinite planar array. Enhancements of the square of the incident field averaged over the surface are shown for Ag and Cu in Fig. 6.24 (for bulk optical constants). The interaction

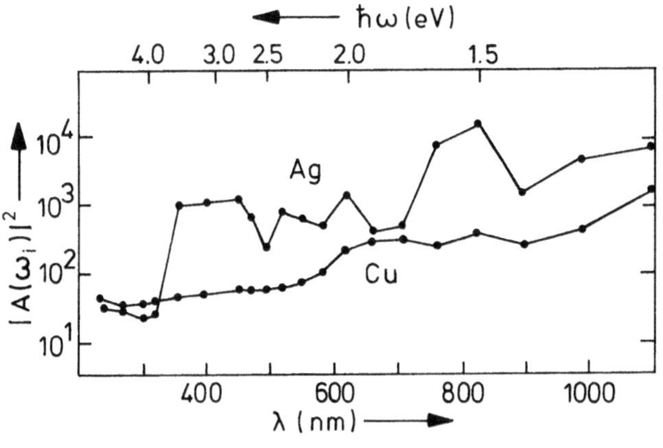

Fig. 6.24. $|A(\omega_i)|^2$ is the average of the square incident field enhancement for a random distribution of hemispheroids on a flat surface for Ag and Cu. The calculated points are connected by straight-line segments. From [6.109]

between the protrusions smears out the resonance and lowers the overall field enhancement. A remark made by *Aravind* et al. [6.110] is quoted here:

"Some very crude argument can be made to suggest that the effect of these interactions may be dramatic. An excited resonance turns on such a high local field because the energy of the absorbed photon is packed into a small spatial region. If many such resonance sustaining centers are brought together, the interaction between resonances spreads the photon energy among all these centers; therefore, energy per unit volume (hence the local field) tends to be diminished".

There is an experimental result of *Kreibig* et al. [6.111] on collective resonances of silver colloids which underscores this point. These authors prepared silver particles as aqueous colloidal systems by chemical reduction. The particles were embedded in a dielectric substance by adding dissolved gelatine and afterwards extracting the water. The filling factor was varied between 0.001 and 0.4 while all other properties of the particle system remained unchanged. The gelatine is assumed to form spherical shells around each particle already in the highly diluted aqueous solution so there was no direct contact between particles after the drying process and a rather uniform distance distribution was found by electron microscopic observation. Examples of the corresponding absorption spectra are given in Fig. 6.25.

With increasing filling factor q, the photon energy of maximum absorption decreases, in good agreement with the theory outlined in connection with (6.13). The resonance in liquid solution is much broader than calculated (Fig. 6.11b), probably because of increased damping in small spheres (see discussion in Sect. 6.5.6). Even this resonance becomes much broader with increasing q. At

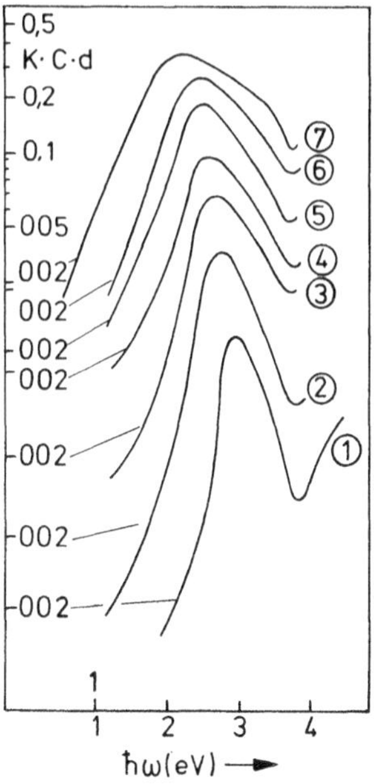

Fig. 6.25. Extinction spectra of noncoagulated Ag particles with various filling factors q (K: extinction constant, C: particle concentration, d: sample thickness, mean particle diameter $2a = 10$ nm). *Curve 1:* liquid solution, $q < 10^{-5}$. *Curves 2 to 7:* solid samples, $q = 0.003$, 0.013, 0.025, 0.12, 0.21, 0.40. The vertical scales are displaced as indicated by the 0.02 marks. From [6.111]

$q=0.21$, the halfwidth is about 1.25 eV and the overall resonant extinction between 1 and 4 eV does not exceed a factor of 10. The halfwidth has to be considered as the halfwidth of the resonant response of the collection of interacting spheres to a plane wave.

The absorption spectra for silver particle samples with partial coagulation show even broader resonances [6.111]. The same deterioration of electromagnetic resonances is observed for "SERS-active" silver island films, as will be discussed in Sect. 6.5.3.

6.4 Adsorbate-Surface Plasmon Polariton Interaction Compared to Adsorbate-Metal Electron Interaction

As outlined in Sect. 6.2, the classical-enhancement model neglects any metal-molecule interaction apart from the change of the local field induced by the metal surface. In this sense, the coupling of the electronic excitations of the adsorbate to the photons is mediated only by local electromagnetic resonances which correspond to surface-plasmon polaritons (SPP) at plane surfaces [6.98]. The self-interaction of the molecule via excitations of the metal (corresponding to the "classical image dipole model" is discussed in a self-consistent way in Chap. 7 and in [6.82]. Here a hypothetical Raman process will be discussed in a perturbative way [6.112] where the coupling of the photons to the electronic excitations of the adsorbate is mediated by electronic excitations of the metal. One may characterize the processes by the Feynman diagrams in Fig. 6.26a, b.

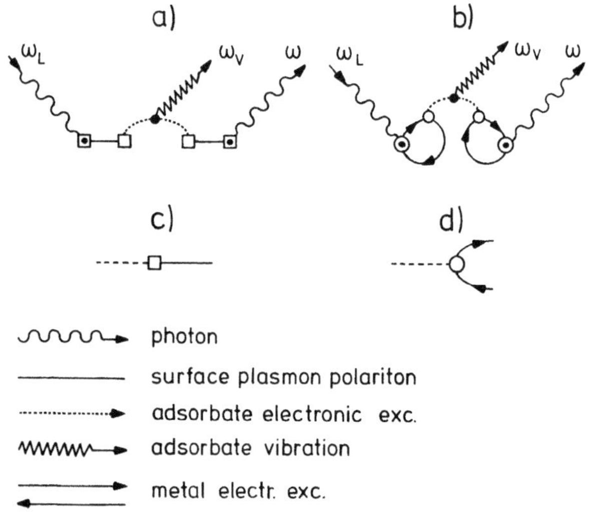

Fig. 6.26a–d. Feynman diagrams of two Raman scattering processes from an adsorbate involving (**a**) SPP's and (**b**) electronic excitations of the metal. Energy transfer from an adsorbate-excited electronic state to SPP's (**c**) and to electronic excitations of the metal (**d**)

The points in the vertices for photon-SPP coupling and photon-eh coupling symbolize the coupling mediated by surface roughness or by defects. The interaction with the vibration of the adsorbate is assumed to be indirect via an electronic excitation of the adsorbate. The cross section of both processes will depend on the matrix elements depicted in Fig. 6.26c,d, as well as the roughness- or defect-mediated coupling to photons. These matrix elements correspond to the de-excitation of an excited adsorbed molecule by energy transfer to SPP or electronic excitation of the metal. An abundance of theoretical and experimental literature has been written on this problem from which one may estimate the relative strength of the interaction of adsorbate with surface-plasmons and adsorbate with metal electrons. Although this is only half the answer to the question of the relative strength of processes *a* and *b*, the comparison mentioned above is worthwhile.

The physics of the interaction of an electronically excited molecule at a distance *d* from a metal surface has been reviewed in [6.113]. Experimental information is obtained by time-resolved luminescence spectroscopy [6.114]. The competing decay channels of the electronic excitation of the molecule (or adsorbate) at a free-electron metal surface are photon emission, plasmon-polariton emission and direct energy transfer to the metal. The direct energy transfer to the free-electron metal can be due to:

a) resistance damping of free-electron currents which are described by a complex, *k*-independent dielectric function $\varepsilon(\omega)$ of a Drude-type [6.113];

b) excitation of intraband electron-hole pairs (Landau damping, nonvertical intraband transitions) which are described by a wave-number dependent dielectric function $\varepsilon(\omega, k)$, for instance, the Lindhard function for a free-electron gas [6.115], or the Mermin modification of this dielectric function [6.116] (Chap. 2).

Decay channels are usually calculated for a dynamical dipole vibrating with frequency ω_D at a distance *d* from a smooth surface. *Morawitz* and *Philpott* [6.117] calculated the fluorescence efficiency of dipole transitions near a metal with a Drude dielectric constant without damping. The two competing channels are the decay into photons and surface-plasmon polaritons; the latter propagate undamped along the surface thus transporting the energy away from the radiating dipole. Figure 6.27 gives the ratio of luminescence (photon emission) to surface-plasmon polariton excitation for a dipole transition polarized normal to the surface. At small distances the luminescence is quenched near $\omega_p/\sqrt{2}$, the surface-plasmon polariton limiting frequency.

Chance et al. [6.113] have calculated the normalized decay rate for a complex $\varepsilon(\omega)$, thus allowing for direct energy transfer to the metal. This decay rate contains contributions from luminescence, surface-plasmon polariton emission and direct energy transfer to the metal. The ratio of the surface-plasmon polariton decay to the total decay rate has been calculated within the framework of this theory by *Weber* and *Eagen* [6.118] (Fig. 6.28). While approaching the metal, surface-plasmon polariton emission becomes more probable than

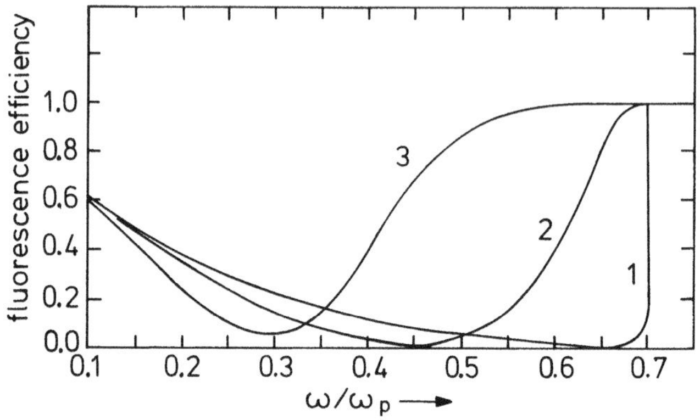

Fig. 6.27. Fluorescence efficiency of a dipole transition perpendicular to the metal surface as a function of the frequency ω, normalized to ω_p for distances $d/\lambda_p = 0.05$ (1), 1.0 (2) and 2.0 (3); λ_p: plasma wavelength. From [6.117]

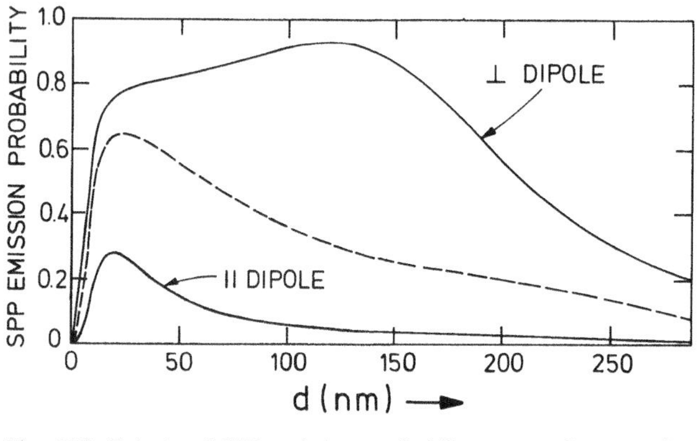

Fig. 6.28. Calculated SPP emission probability versus distance of a dipole on an Ag/methanol interface. The *dashed curve* is for an orientationally averaged dipole, $\lambda = 6000$ Å, ε(silver) $= -15.5 + 10.5\,i$, ε(methanol) $= 1.766$. From [6.118]

luminescence, in agreement with Fig. 6.27. For very small distances, however, the direct energy transfer rate increases like d^{-3}, thus making the surface-plasmon polariton emission unlikely for an adsorbate. This has been verified experimentally by *Pockrand* et al. [6.119].

A dye monolayer (structure formula of Fig. 6.29) was placed, by the Langmuir-Blodgett technique, at various distances d in front of a silver film, using intermediate fatty-acid layers as spacers. The sample was illuminated at normal incidence. The laser frequencies in the spectral range of the absorption band are at photon energies above the fluorescence band (Fig. 6.29). The intensity of the emitted SPP's at the fluorescence wavelength λ_{fl} was detected in an ATR configuration. The SPP intensity was coupled out through the thin silver film and a half-cylinder (Fig. 6.29). Figure 6.30 shows the dependence of dye-

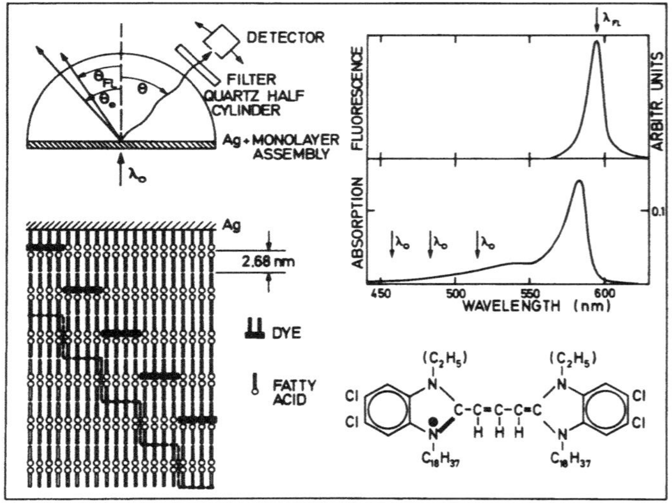

Fig. 6.29

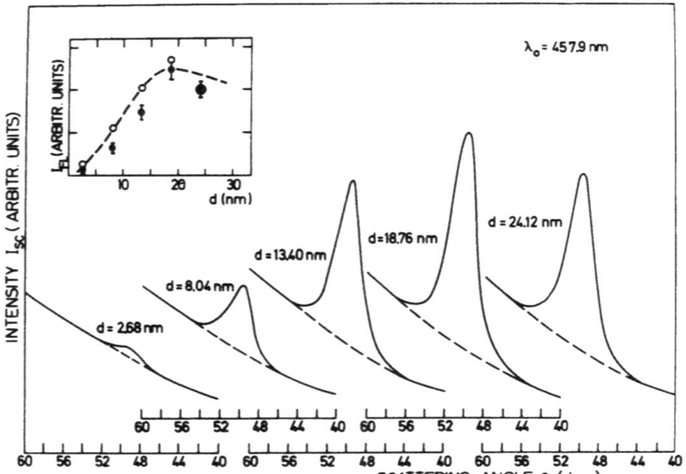

Fig. 6.30

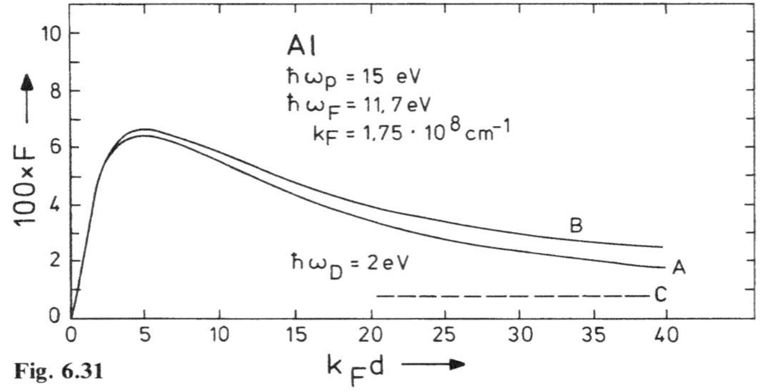

Fig. 6.31

Fig. 6.29. Layer system and molecular structure, absorption and fluorescence spectrum of the dye depicted in the figure; ATR outcoupling configuration. θ_0 and $\theta_{\rm fl}$ are the angles under which light is emitted from the SPP at wavelengths $\lambda_{\rm o}$ and $\lambda_{\rm FC}$. For clarity, the scattering angles are given only at one side of the film, although all phenomena are symmetrical in the observation plane with respect to the normal. From [6.119]

Fig. 6.30. Distance dependence of dye-SPP coupling for laser wavelength 457.9 nm, as monitored by the outcoupled intensity above 500 nm wavelength with a cutoff filter. The inset (*full circles*) shows the efficiency of the nonradiative energy transfer process by plotting the area under the peak versus the distance *d*. See [6.119]

Fig. 6.31. Damping function *F* (defined by energy transfer $d\bar{E}/dt = \mu^2 \omega_{\rm p} F/16 d^3$ as a function of distance *d* for a nearly free electron model of Al, and $\hbar\omega_{\rm D} = 2$ eV. Collision frequency $\hbar\tau^{-1} = 0$ for curve A and $\hbar\tau^{-1} = 0.234$ eV for curves B and C. From [6.116]

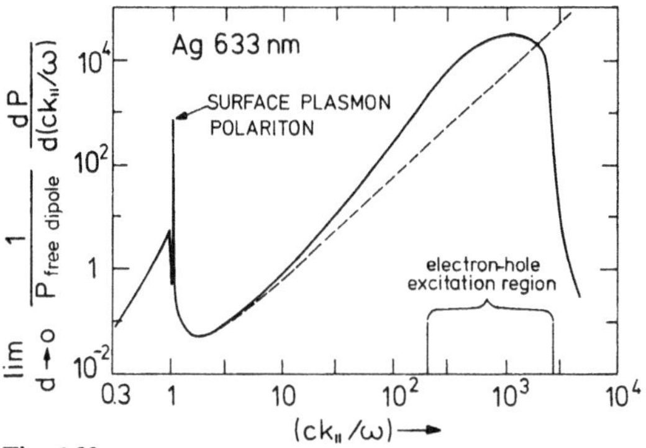

Fig. 6.32

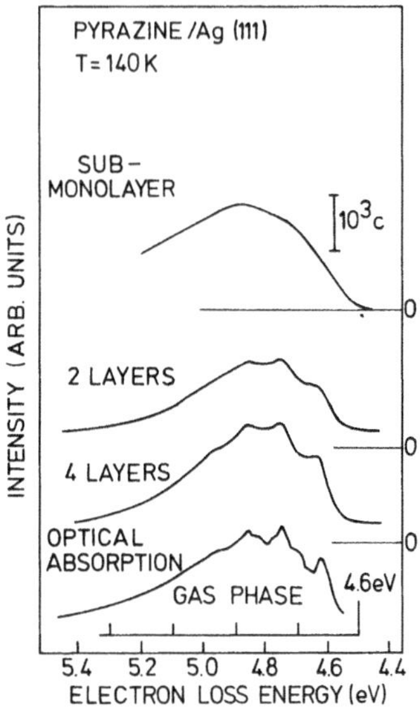

Fig. 6.33

Fig. 6.32. Dissipated power P from an oscillating dipole (corresponding to 633 nm wavelength) on a silver surface, with (——) and without (---) excitation of electron-hole pairs in the silver metal. From [6.120]

Fig. 6.33. Electronic energy loss spectra for the $^1B_{2u}$ excitation of a monolayer and subsequent adsorbed layers of pyrazine on Ag(111), compared to the optical absorption spectrum of [6.123]. From [6.121]

SPP coupling on distance; it tends to zero for direct contact of the dye with the silver surface.

According to [6.115, 116], direct energy transfer to the metal by the creation of electron-hole pairs (Landau damping) is, at small distances, a stronger decay channel than resistance damping. Figure 6.31 shows the prefactor F of the d^{-3} dependence for electron-hole excitation compared with the prefactor for resistance damping, as calculated with the Mermin dielectric constant [6.116]. The difference between curve B and A reflects the resistance damping (average collision time τ) which clearly becomes negligible compared to damping by electron-hole pair excitation for $d \to 0$. (Note that the decay rate $F \cdot \mu^2\omega_D/16d^3$ reaches a maximum at $d=0$.) Curve C is the approximate resistance damping given by $F=2\,\mathrm{Im}\,\{[\varepsilon(\omega)-1]/[\varepsilon(\omega)+1]\}$, $\varepsilon(\omega)=1-\omega_p^2[\omega(\omega+i/\tau)]^{-1}$. This approximation breaks down when $d<v_F\tau$ (v_F = Fermi velocity). For larger values of $k_F d$, curve C is roughly equal to the difference between curves A and B, which means that Landau damping and resistive damping are approximately additive. Results similar to those of *Persson* and *Schaich* [6.116] have been obtained by *Ford* and *Weber* [6.120]. Figure 6.32 shows the $k_{\parallel}$-dependent spectrum of power

dissipated from a point dipole oscillating normal to a silver surface. The solid curve corresponds to the nonlocal model described in Sect. 6.3.3, the dashed curve to the local result ("conduction current" interaction). There is a strong coupling to e–h excitations compared to the coupling to surface-plasmon polaritons (see also Table 6.1).

Demuth and *Avouris* [6.121, 122] observed strong adsorbate-metal coupling at short distances for pyrazine on silver by electron energy loss spectroscopy (EELS). Figure 6.33 shows the dependence of the EELS spectra on the coverage of the Ag(111) surface with pyrazine, within the range of the electronic $^1B_{2u}$ $(\pi\rightarrow\pi^*)$ excitation. As follows from the losses in the vibrational range in connection with the surface selection rule of EELS, the pyrazine molecules lie parallel to the silver surface independent of the amount of adsorbed pyrazine. For 4 adsorbed layers, the spectrum displays the fine structure found in the optical absorption in the gas phase. For submonolayer coverage, this fine structure disappears due to lifetime broadening, corresponding to about 100 meV for the first layer and 20 meV for the second. These numbers are in reasonable agreement with "resistance damping" [6.113], although the relatively short lifetime for the second layer "suggests the presence of additional mechanisms such as electron-hole pair excitations" [6.121] (according to Fig. 6.32, the decrease of lifetime broadening with d is slower than d^{-3}).

From the theoretical work presented up to this point, it follows that adsorbate electron-hole-pair interaction is stronger than adsorbate "conduction current" interaction which, in turn, is stronger than the adsorbate surface-plasmon-polariton interaction. However, when the dynamical frequency ω_p of the dipole approaches the surface-plasmon frequency (the limit of the SPP dispersion for wave vectors much larger than ω_p/c), the distinction between e–h excitations and surface-plasmon excitations becomes difficult because the coupling shows a resonance at the surface-plasmon frequency. This follows from the calculations of *Persson* and *Schaich* [6.116] presented in Fig. 6.34 for Al. With $k_F d=100$, F tends to a maximum at the surface-plasmon frequency $\omega_p/\sqrt{2}$, while the curve for $k_F d=10$ has its maximum slightly above this value because of dispersion of the surface plasmon at k-vectors of the order of $10\,d^{-1}$.

The work described above is concerned with coupling mechanisms at a plane metal surface. Qualitatively, the relative strength of the coupling constants of an adsorbate to surface-plasmon polaritons, surface plasmons, "conduction current" and electronic excitations should also hold for a rough surface, provided the local radius of curvature at the site of adsorption is large compared with the lattice constant ("macroscopic roughness"). The excitations analogous to the surface-plasmon polaritons at a plane surface are the electromagnetic resonant modes of the surface protrusions. The conclusion for a "macroscopically" rough surface is the following: provided the roughness-mediated coupling of photons to SPP and to electronic excitations is of comparable strength, the Raman process in Fig. 6.26b should be stronger than that in Fig. 6.26a. The process in Fig. 6.26b is subject to the same classical-enhancement mechanism as the one in Fig. 6.26a. The coupling of photons to electronic excitations will be enhanced by

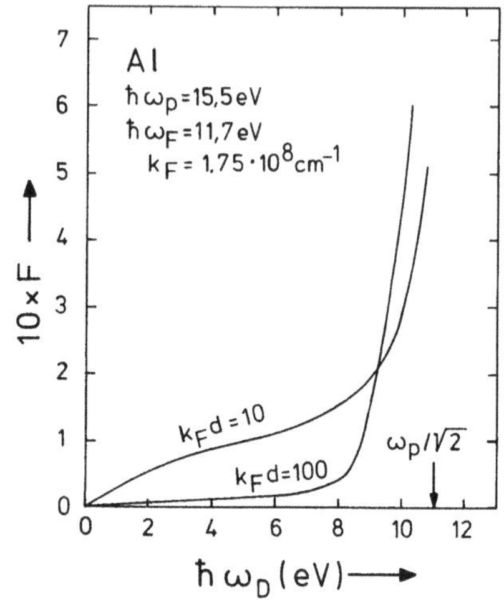

Fig. 6.34. Damping function F of a dipole near an Al surface versus dipole frequency ω_D for two values of d and zero resistance damping. From [6.116]

resonances due to intermediate surface-plasmon polaritons. If ω_L or ω coincide with the surface plasmon frequency (3.6 eV at a silver-vacuum interface and 10.6 eV at an aluminium-vacuum interface), the process depicted in Fig. 6.26a could become important. The strong coupling at a plane surface when ω approaches the surface-plasmon frequency $\omega_p/\sqrt{2}$ is due to the confinement of the electromagnetic field to the immediate neighborhood of the surface. Such a confinement also takes place for particles of very small dimensions (see the discussion in relation to Fig. 6.11). In this case it becomes difficult to differentiate between coupling to pure e–h excitations and electromagnetic excitations. In fact, below a certain minimum particle dimension, the conceptual difference between the two mechanisms breaks down.

6.5 Is SERS only an Electromagnetic Resonance Effect? Selected Relevant Experiments

In this section, a limited number of experimental results will be selected to discuss this question. First, SERS results which indicate a pure electromagnetic field enhancement will be presented. Second, data which have been published as evidence for pure field enhancement or in which only the field aspect has been analyzed, but where the discussions are not completely convincing, will be evaluated. Finally, results which indicate an enhancement mechanism beyond electromagnetic resonances will be discussed.

6.5.1 Spacer Experiments

All proposed mechanisms beyond the classical enhancement require chemisorption or at least a distance between molecule and metal surface which is smaller

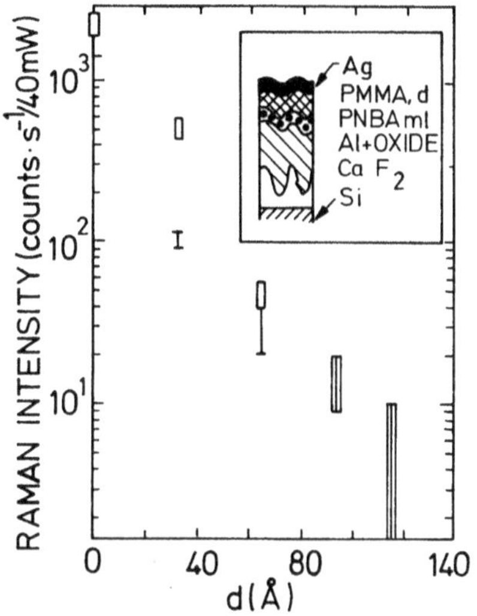

Fig. 6.35. Spacer experiment (see inset). Mass thicknesses are: CaF_2 800 Å, Al 2500 Å, oxide ≈ 30 Å, PNBA 1 monolayer, PMMA d [Å], Ag 200 Å. Raman intensity of the 1595 cm^{-1} (*open symbole*) and 1425 cm^{-1} line (*closed symbols*) versus d. From [6.19]

than 3 Å. (The only exception is the hypothetical "Raman reflectivity" [6.14, 124] which is not applicable to the "spacer experiments" because the van der Waals interaction between molecule and metal is screened by the spacer layer). Therefore, the demonstration of a long-range enhancement mechanism is unambiguous evidence for long-range electromagnetic, classical enhancement without any other contributions. In one of the experiments reported by *Murray* et al. [6.19], a spacer layer of a weak Raman scatterer, polymethylmethacrylate (PMMA), of variable thickness d (see insert in Fig. 6.35) was placed between a strongly Raman-active monolayer (ml) of chemisorbed para-nitrobenzoic acid (PNBA) and a rough silver film. The fabrication of the multilayer structure in Fig. 6.35 is by no means trivial so it will be discussed here in detail.

The Si substrate is fairly smooth; it shows only a 100 Å ripple under a scanning electron microscope. A CaF_2 layer on top of this substrate, typically 200–800 Å average thickness, produces a uniform-size roughness. An aluminum layer of about 2500 Å thickness, covered with a natural oxide layer of 20–30 Å, is evaporated on top of the CaF_2 layer. The PNBA is deposited as a 0.01M ethanol solution onto the native Al oxide, spun dried, then washed with tetrahydrofuran or chlorobenzene to remove all but the strongly chemisorbed first layer (as confirmed by ir spectroscopy [6.19]). The PMMA polymer is deposited by spin-coating from a chlorobenzene solution. The thickness d of PMMA was measured by a combination of ellipsometry and reflection ir spectroscopy. It is not mentioned in [6.19] as to what extent the original rough surface of the CaF_2 film is contoured by the uppper surface of the PMMA film. (We would expect a smoothing effect by the spin-coating procedure and therefore a varying thickness of the spacer layer.) For PMMA layers deposited on a smooth oxidized Al film with an intermediate PNBA monolayer, the thickness was found uniform to ±2 Å [6.19]. A silver film of about 200 Å thickness is evaporated on top of the PMMA film at a base pressure of 10^{-7} to 10^{-8} Torr. According to [6.19] the silver layer is modulated by crevices, somewhat shallower than 500 and 100 Å wide, separating interconnected 500–1000 Å diameter ellipsoids. The Ag deposition is the crucial step in the sample preparation because a 200 Å silver film, evaporated on a smooth warm substrate (above a temperature of 80 K) forms a nonuniform surface coverage of spheroidal bumps of 200–2000 Å in diameter.

Murray et al. [6.19] also state that these films often produced a large enhancement, even though the substrate was smooth. The samples which did not contour the substrate were discarded. In any case, the silver film was rough on a scale of 500–1000 Å. *Murray* et al. discuss whether there was any direct contact of the scatterer with the silver layer for $d=20$ to 50 Å, either caused by dissolution of the monolayer of PNBA in the PMMA spacer or by pinholes in the spacer. The first point could be discarded by various tests. To check for pinholes in the PMMA layer, the authors ran a series of experiments in which water was used as a solvent to remove PNBA monolayers on samples with and without covering PMMA spacer layers (in these experiments, the CaF_2 layer was absent). The effect of the water soaking on the concentration of PNBA was determined by subsequent Raman scattering experiments in which, after soaking, a CaF_2 and a silver layer were deposited on top of the covering PMMA layer. Water soaking for 5 min in the complete absence of the PMMA layer reduced the SERS intensity by 60 ± 8 %. However, on samples with 50 Å of PMMA on top of the monolayer, the PNBA Raman intensity was only reduced after the 5 min soak by 30 ± 20 % on one sample and 4 ± 5 % on another. From these values, Murray et al. calculated an upper limit for the diffusion constant of PNBA through the PMMA film of 10^{-16} cm^2/s, which is interpreted as molecular diffusion through the PMMA network rather than removal of PNBA molecules through >100 Å scale pinholes. They concluded that within the time of sample preparation and Raman measurements (1–2 h), diffusion of PNBA into the PMMA spacer layer is negligible and that SERS does not occur to any significant extent via direct contact of silver and scatterer at the bottom of pinholes. The Raman measurements were performed with the incident and scattered light penetrating through the silver film. In the calibration of the enhancement, the transmission through the rough ~ 200 Å silver layer must be taken into account twice. The enhancement of about 10^5 quoted by Murray et al. is with respect to the local-field intensity at the scatterer expected for smooth films. A uniform smooth 200 Å thick Ag film will transmit roughly 10 % of the incident or scattered light (however, a rough silver film may transmit more than a smooth one [6.38]). Figure 6.35 shows the dependence of the signal from the PNBA spacer layer for two Raman peaks above the background. There is clear evidence of long-range enhancement for molecule-silver separation d as large as 100 Å. The signal for direct contact ($d=0$) fits into the general decay of the signal with d. This may indicate the absence of a significant short-range enhancement mechanism, though due to the scatter of the data, a short-range mechanism may be compatible with the results (C. Murray: private communication). In no case did Murray et al. report an enhanced signal from the material directly in contact with the silver film (PMMA or CaF_2). There is another relevant preliminary result which should be mentioned. *Allara* et al. [6.125] coated rough silver films (which were evaporated "with linear mass thickness gradient from 0–300 Å") with a 15–30 Å aluminium layer which was subsequently completely oxidized. Organic monolayers are strongly chemisorbed to the oxide and exhibit Raman enhancements of about 10^4 for an oxide thickness of 30 Å. This technique appears to be very promising for Raman scattering studies of catalytic reactions on supported catalysts.

6.5.2 SERS from Regular Arrays of Silver Particles

Liao et al. [6.96] examined SERS from a regular array of silver particles. The rationale for this work was the following. As shown in [6.78] (Sect. 6.3.1), the resonant frequency of the metal protrusions depends on their shape and size. Therefore, on a statistically rough surface, no narrow resonance like that in Fig. 6.14 can be expected. To obtain a comparison with the classical plasmon-resonance theories, microlithographic techniques were used to produce a regular array of isolated, uniformly-sized silver particles of 100 nm dimensions and variable shapes. The substrate for this array consisted of SiO_2 posts on a silicon wafer, 500 nm high and 100 nm in diameter. These posts were arranged in a square lattice of 300 nm dimensions. By evaporating silver at grazing incidence onto the substrate along a nonchanneling direction, so that the posts shadowed each other, isolated silver particles were obtained on the tops of the posts. They were quite uniform in size and, as a first approximation, could be considered as ellipsoids with a 3:1 aspect ratio. Other aspect ratios could be obtained by varying the angle of evaporation. The diameter of the SiO_2 posts dictated the minor axis to be about 100 nm. For photographs of the samples see [6.126]. A monolayer of CN radicals is adsorbed onto the silver: SERS from the CN stretch mode at 2144 cm^{-1} is then observed. Figure 6.36, trace $\varepsilon_0 = 1$, shows the relative frequency dependence of the normalized (ω^4 factor eliminated) Raman enhancement versus the incident photon energy for the sample in nitrogen atmosphere (all spectra in Fig. 6.36 are for a sample with nominally 3:1 ellipsoids; samples with nominally 2:1 ellipsoids display the same enhancement as in Fig. 6.36 but the peak of the curve is shifted to 500 nm [6.126]).

By calibration of the absolute detection sensitivity of the apparatus and by assuming a Raman cross-section of the free CN^- stretch mode of 3×10^{-30} cm^2 and one monolayer coverage, an enhancement of about 10^7 was estimated [6.96].

Using *Gersten* and *Nitzan*'s electrostatic theory [6.78] and assuming a 12% distribution in aspect ratios and the dielectric function of bulk silver, an enhancement of 1.3×10^8 was calculated by Liao et al., in reasonable agreement with the experimental value. In the same way the width and the location of the resonance can be fitted to experimental data. When the sample is immersed into water or cyclohexane, the resonance is shifted to lower photon energy (traces

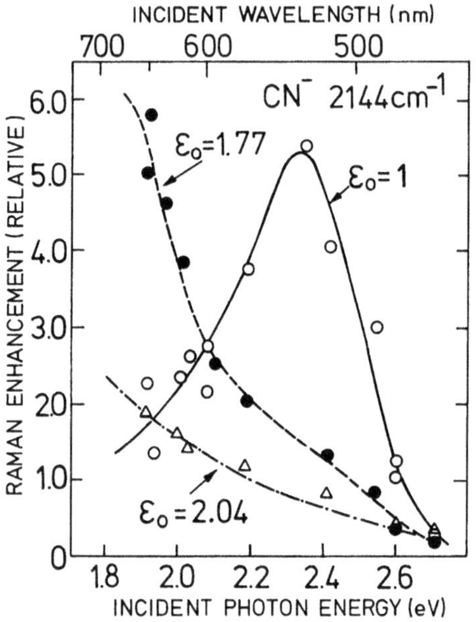

Fig. 6.36. Dependence of the Raman enhancement for cyanide (vibrational band at 2144 cm^{-1}) adsorbed on a regular array of silver particles approximated by ellipsoids with 3:1 aspect ratio. Surrounding medium is nitrogen ($\varepsilon_0 = 1$), water ($\varepsilon_0 = 1.77$), or cyclohexane ($\varepsilon_0 = 2.04$). From [6.96]

$\varepsilon_0 = 1.77$ and $\varepsilon_0 = 2.04$ in Fig. 6.36). This is in qualitative agreement with the plasmon resonance theory. If the medium adjacent to the spheroid is not a vacuum but a medium of dielectric constant ε_0, the resonance condition (6.9) has to be changed to

$$\varepsilon(\omega_L)Q_1(\xi_0) - \varepsilon_0 \xi_0 Q_1'(\xi_0) = 0. \tag{6.29}$$

The resonance for 3.9:1 aspect ellipsoids with a 12 % distribution around this ratio would be at 1.9 eV in water and 1.5 eV in cyclohexane. So far, everything looks convincing. However, there is one surprising experimental detail: the plasmon resonance is shifted from 2.4 eV to below 1.8 eV by immersion in liquids, nevertheless a considerable Raman intensity between 2.2 and 2.7 eV remains. The shift of the plasmon resonance diminishes the Raman intensity at 2.4 eV only by a factor of 4 to 6. Is this indicative of a second nonclassical enhancement mechanism of about 10^4–10^6, not influenced by immersion, on which a classical enhancement factor of 10 to 10^3, due to plasmon resonances, is superimposed? Note that the highest ratio between on and off-resonance signals in Fig. 6.36 is only 25. The samples of Liao et al. were mounted at 60° in the p-polarized configuration but "sample orientation was found not to be a particularly sensitive parameter" [6.96]. For prolate ellipsoids one would expect considerable intensity variations when changing the polarization of the exciting field and analyzing the polarization of Rayleight and Raman scattered light (Sect. 6.3.1). This might provide a test of the classical theory.

The results above led *Barber* et al. [6.87] to an electrodynamic rather than an electrostatic calculation of the surface-averaged intensity enhancement. As is

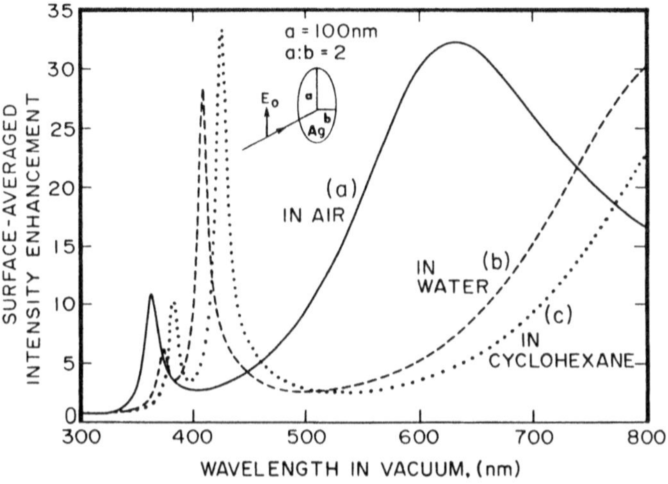

Fig. 6.37. Wavelength dependence of the surface-averaged intensity enhancement for a 2:1 Ag prolate spheroid with semi-major axis $a = 100$ nm in air, water and cyclohexane. Incident field polarized parallel to the long axis of the spheroid. From [6.87]

evident from Fig. 6.16, one must expect lower values than those given by the electrostatic approximations for particles of about 100 nm in size. For silver spheroids of 2:1 aspect ratio and 100 nm length of the larger axis (Fig. 6.37), the calculated wavelength dependence for $\varepsilon_0 = 1$ (in air) consisted of a broad peak at 630 nm and a considerably narrower one at 365 nm. The broad peak is a continuation of the dipolar surface-plasmon resonance observed in Fig. 6.16. The narrow peak represents multipole surface plasmon resonances [6.87]. The broad peak shifts to longer wavelength upon changing the surrounding medium, in a way qualitatively similar to the experimentally observed SERS dependence on wavelength (Fig. 6.36). On the other hand, no narrow SERS peaks were observed near 400–450 nm. If one squares the intensity enhancement to obtain the Raman enhancement, one finds, according to Fig. 6.37, a "classical enhancement" of about three orders of magnitude, thus leaving another 4 orders of magnitude yet to be explained.

6.5.3 Optical Properties and "Classical Enhancement" of Silver-Island Films

SERS from adsorbates on silver-island has been found by *Burstein* and collaborators [6.35, 127], *Seki* and *Philpott* [6.128, 129], *Weitz* et al. [6.97] and *Lyon* and *Worlock* [6.130]. SERS on an evaporated silver film which has a continuously varying average thickness and hence varying states of aggregation was investigated by *Bergman* et al. [6.21]. In all cases distinct Raman spectra of the adsorbates with good signal-to-noise ratios have been found. This implies an overall enhancement of at least 10^4 since in most experiments, enhancements of

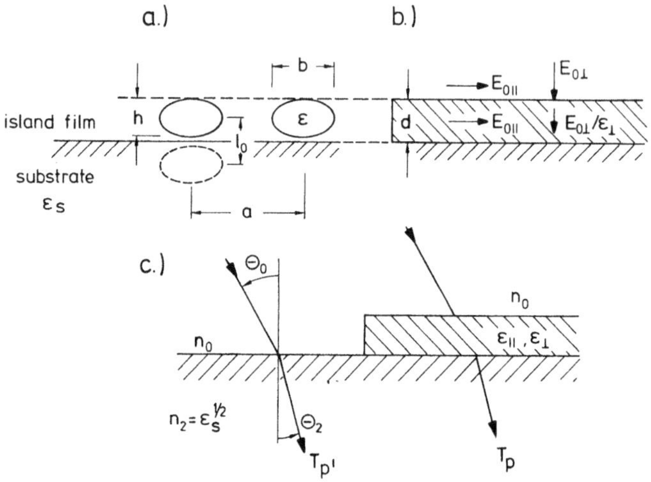

Fig. 6.38. (**a**) Island film composed of rotational ellipsoids with axis of rotation perpendicular to the substrate surface. Square-lattice constant a. Dielectric constant of the islands (substrate) is $\varepsilon(\varepsilon_s)$. (**b**) Approximation of the island film in (**a**) by an effective medium of thickness d with anisotropic dielectric constant $\varepsilon_{\|}, \varepsilon_{\perp}$, and relationship between external field E_0 and internal field in the film. (**c**) Transmission T_p' and T_p for p-polarized light at oblique incidence for the bare substrate and in the presence of the effective film. n_2: index of refraction of the transparent subtrate

the order of 10^2 cannot be discriminated from noise [6.17]. Only in one case [6.97] was the optical absorption spectrum of these silver-island films simultaneously investigated. On the other hand, there are detailed results for the optical properties of silver-island films available thanks to *Yamaguchi* et al. [6.94, 131, 132].

In the following we will see that island films with the optical parameters reported by Yamaguchi et al. would only yield a "classical enhancement" of about 3 orders of magnitude. *Yamaguchi* et al. [6.94, 131] calculate the effective anisotropic dielectric constant $\varepsilon_{\mathrm{eff}\,\|}(\omega)$ and $\varepsilon_{\mathrm{eff}\perp}(\omega)$ of an island film on a substrate. The island film is assumed to consist of rotational ellipsoids, the rotational axis being normal to the film. The aspect ratio is defined as $\gamma=b/h$ (Fig. 6.38). The center of the ellipsoids is at a vertical distance of $l_0/2$ from the substrate. The frequency-dependent dielectric constants of the island and the substrate are ε and ε_s, respectively. The thickness of a homogeneous film with the same average weight per area as the island is called d_w. The optical properties of the island film, especially its optical transmittance, are assumed to be equal to the optical properties of a film with a homogeneous thickness $d=l_0/2+h/2$ (Fig. 6.38) and an anisotropic dielectric constant $\varepsilon_{\|}$, $\varepsilon_{\perp}$. If the island film is composed of small ellipsoids of identical size, shape and orientation, and if the islands form a regular two-dimensional lattice with one island per 2-dim Wigner-Seitz cell, all islands will have the dipole moment P, provided that the lattice constant is much smaller than the wavelength of light. In this case, one equalizes the electric polarization in the effective film of thickness d to the sum over the moments P of the N islands per cm^2 of the substrate surface:

$$(\varepsilon_{\parallel}-1)E_{0\parallel}d=NP_{\parallel}=NV\alpha_{\parallel}E_{1\parallel}, \tag{6.30}$$

$$(\varepsilon_{\perp}-1)\frac{E_{0\perp}}{\varepsilon_{\perp}}d=NP_{\perp}=NV\alpha_{\perp}E_{1\perp}. \tag{6.31}$$

V is the volume of the ellipsoids, $\alpha_{\parallel}$ and $\alpha_{\perp}$ the polarizability of the individual ellipsoid parallel and perpendicular to the substrate, and $E_{1\parallel}$ and $E_{1\perp}$ the local fields giving rise to $P_{\parallel}$ and $P_{\perp}$. NV/d is the filling factor q. Equations (6.30, 31] lead to

$$(\varepsilon_{\parallel}-1)=q\alpha_{\parallel}\frac{E_{1\parallel}}{E_{0\parallel}}, \tag{6.32}$$

$$\left(1-\frac{1}{\varepsilon_{\perp}}\right)=q\alpha_{\perp}\frac{E_{1\perp}}{E_{0\perp}}. \tag{6.33}$$

The local field E_1 at a given ellipsoid is the sum of the external field E_0 and the dipolar fields of all other ellipsoids and their image dipoles, including that of the given ellipsoid and those of all other ellipsoids due to the presence of the substrate [6.94]. The calculation of E_1 has been presented in detail in [6.94] for a square lattice of islands with lattice constant a. The result is

$$E_{1\parallel,\perp}=\frac{E_{0\parallel,\perp}}{1+\alpha_{\parallel,\perp}\beta_{\parallel,\perp}}, \tag{6.34}$$

$$\alpha_{\parallel,\perp}=\frac{(\varepsilon-1)}{1+f_{\parallel,\perp}(\varepsilon-1)}, \tag{6.35}$$

$$\beta_{\parallel}=-\frac{\gamma^2}{24\eta^3}\frac{\varepsilon_s-1}{\varepsilon_s+1}-0.716\frac{2}{\varepsilon_s+1}\frac{d_w}{2a}, \tag{6.36}$$

$$\beta_{\perp}=-\frac{2\gamma^2}{24\eta^3}\frac{\varepsilon_s-1}{\varepsilon_s+1}+0.716\frac{2\varepsilon_s}{\varepsilon_s+1}\frac{d_w}{2a}. \tag{6.37}$$

$f_{\parallel,\perp}$ are depolarization factors of the ellipsoids and $\eta=l_0/h$. The first and second terms in (6.36, 37) correspond to β_{id} and β_{nd} in (6.15). The first terms describe the contribution of the dipole moment of the image of a given ellipsoid (see dotted ellipsoid in Fig. 6.38), the second terms describe the contributions of the dipolar moment of all other ellipsoids and their images. One should note that the second terms are at least of the same order as the first ones.

By defining $F_{\|,\perp} = f_{\|,\perp} + \beta_{\|,\perp}$ and $g = \mathrm{Re}\{(\varepsilon-1)^{-1}\}$ $\Delta g_{\mathrm{bulk}} = \mathrm{Im}\{(\varepsilon-1)^{-1}\}$, *Yamaguchi* et al. [6.94] obtained

$$\varepsilon_{\|} - 1 = q\,\frac{1}{F_{\|} + g + \mathrm{i}\Delta g_{\mathrm{bulk}}}, \tag{6.38}$$

$$1 - \varepsilon_{\perp}^{-1} = q\,\frac{1}{F_{\perp} + g + \mathrm{i}\Delta g_{\mathrm{bulk}}}. \tag{6.39}$$

Here it is understood that the bulk optical constants ε are used. This will lead to resonances, when $F_{\|,\perp} = -g$, the quality of the resonances depending on Δg_{bulk}.

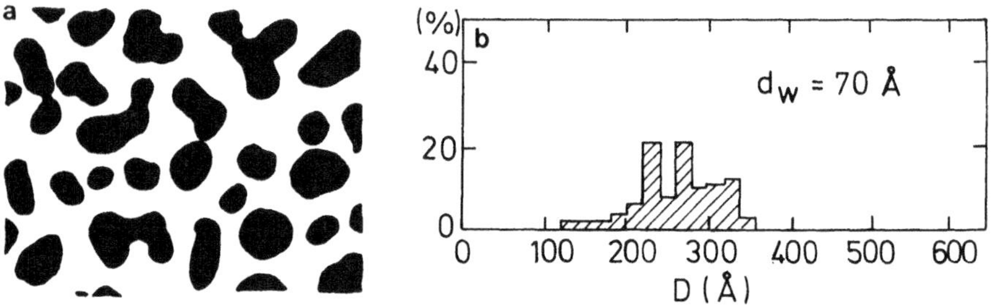

Fig. 6.39. (**a**) Electron micrograph of a silver film deposited onto PVA, with mass thickness $d_w = 70$ Å. (**b**) Particle size histogram of this film. D is the "diameter" of the particles. From Fig. 8 in [6.94]

This quality will be reduced if the idealisations of the model are removed. First, the island film will have a non δ-shaped distribution of sizes V, shapes and orientations, and of inter-island distances (see electron micrographs and size histograms, Fig. 6.39 and [Ref. 6.94, Fig. 8]). Second, the optical constants of the islands might be different from the bulk values (Sect. 6.5.6). Third, retardation effects will change the island-island interaction [6.131]. All these effects are taken into account by replacing Δg_{bulk} in (6.38, 39) by Δg:

$$\Delta g = \Delta g_{\mathrm{bulk}} + \Delta g_{\mathrm{size}} + \Delta g_{\mathrm{interaction}} + \Delta g_{\mathrm{shape}}. \tag{6.40}$$

In this sense, F and Δg become parameters to be fitted to the experiments, whereas g is calculated from the bulk optical constants ε. $F_{\|}$, $F_{\perp}$, and Δg are derived from transmission experiments (Fig. 6.38c). The transmission for p-polarized light is measured for the clean substrate and for the substrate covered with the island film. This yields the transmissions T_{p}' and T_{p}, respectively. The quantity A_{p} defined by $A_{\mathrm{p}} = (T_{\mathrm{p}}'/T_{\mathrm{p}}) - 1$ is for a small value of $t = 2\pi d/\lambda$ given approximately by [6.94]:

$$A_p \approx -k_{p1}\,\mathrm{Im}\,\{\varepsilon_\parallel\}t + k_{p2}\,\mathrm{Im}\,\{1/\varepsilon_\perp\}t = A_{p\parallel} + A_{p\perp}$$

$$k_{p1} = \frac{\cos\theta_0\cos\theta_2}{n_0\cos\theta_2 + n_2\cos\theta_0} \tag{6.41}$$

$$k_{p2} = \frac{n_0^2 n_2 \sin^2\theta_0}{n_0\cos\theta_2 + n_2\cos\theta_0},$$

where θ_0 and θ_2 are the angles of incidence and refraction (Fig. 6.38c). A_p will have two peaks corresponding to peaks of $-\mathrm{Im}\,\{\varepsilon_\parallel\}$ and $\mathrm{Im}\,\{1/\varepsilon_\perp\}$ (time dependence: $\exp(\mathrm{i}\omega t)$). At normal incidence, $\theta_0 = \theta_2 = 0$ and

$$A_p = A_0 = \frac{-1}{n_0 + n_2}\,\mathrm{Im}\,\{\varepsilon_\parallel\}t. \tag{6.42}$$

For s-polarized light at oblique incidence, only the resonance due to $-\mathrm{Im}\,\{\varepsilon_\parallel\}$ is observed [6.132]. $A_{p\perp}$ is evaluated from A_p and A_0 with

$$A_{p\perp} = A_p - A_0 k_{p\perp}(n_0 + n_2). \tag{6.43}$$

In this way it is possible to separate the resonances due to $\varepsilon_\parallel$ and $\varepsilon_\perp$. *Yamaguchi* et al. [6.94] measured the ratio T_p/T'_p directly with a dual-beam spectrophotometer, using a bare substrate as a reference sample. Their results are presented in

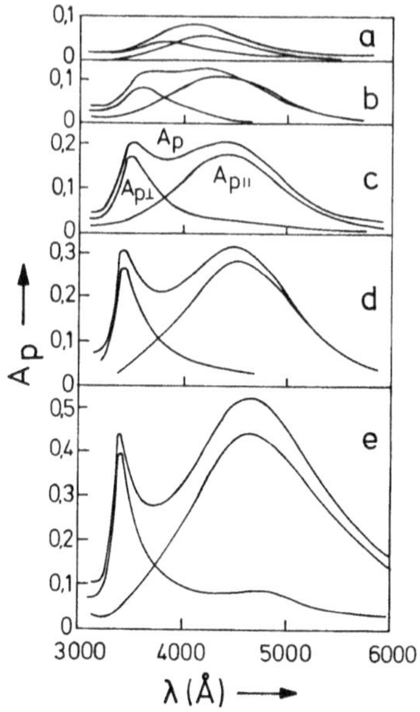

Fig. 6.40a–e. Observed spectra of A_p for silver-island samples of Fig. 6.38; p-polarized light and angle of incidence 60° for different samples, $d_w = 10$ Å (**a**), 20 Å (**b**), 30 Å (**c**) 50 Å (**d**), 70 Å (**e**). Decomposition of A_p in $A_{p\perp}$ and $A_{p\parallel}$. From [6.94]

Fig. 6.40 for five samples of silver-island films on a sheet of polyvinyl alcohol (PVA) of 20 μ thickness. Silver was deposited onto the PVA sheet at a deposition rate of about 0.5 Å/s at room temperature in a vacuum of 10^{-5} Torr. The weight thickness d_w was monitored with a quartz-crystal oscillator. As soon as the deposition ended, the film was annealed at 100 °C for one hour. After annealing it was cooled to room temperature and exposed to air. The optical measurements were carried out 1 h after exposing the film to air. One observes an increase in splitting between the resonances in $A_{p\perp}$ and $A_{p\parallel}$ with d_w. $F_\perp$ and $F_\parallel$ are evaluated from the peak positions of $A_{p\perp}$ and $A_{p\parallel}$ by using the bulk optical constants [6.133–135] to calculate $g = \mathrm{Re}\,\{(\varepsilon-1)^{-1}\}$. F is simply the negative value of g at the peak wavelength. The results for $F_\perp$ and $F_\parallel$ are presented in Fig. 6.41.

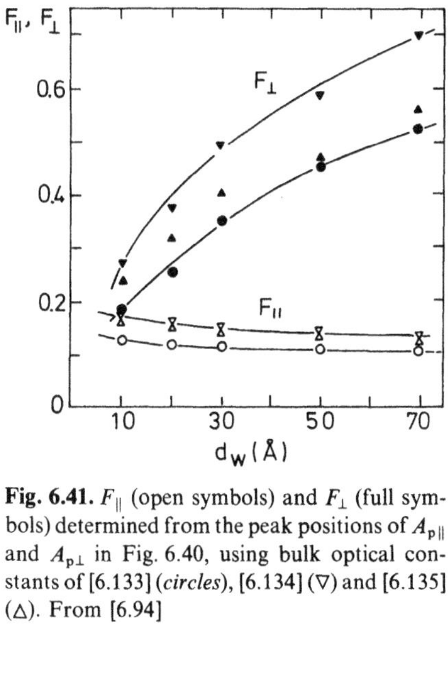

Fig. 6.41. $F_\parallel$ (open symbols) and $F_\perp$ (full symbols) determined from the peak positions of $A_{p\parallel}$ and $A_{p\perp}$ in Fig. 6.40, using bulk optical constants of [6.133] (*circles*), [6.134] (∇) and [6.135] (Δ). From [6.94]

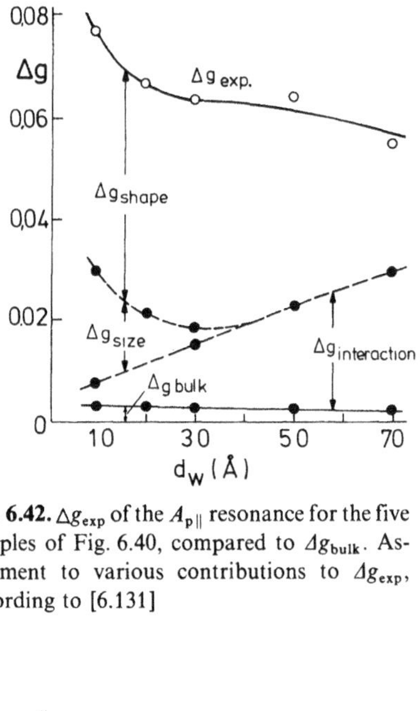

Fig. 6.42. Δg_{exp} of the $A_{p\parallel}$ resonance for the five samples of Fig. 6.40, compared to Δg_{bulk}. Assignment to various contributions to Δg_{exp}, according to [6.131]

The experimental value of Δg, see (6.38), is evaluated from the maximum value of A_p, called A_p(peak) and the corresponding wavelength λ(peak):

$$\Delta g = \frac{2\pi}{\lambda(\text{peak})}\,\frac{k_{pi}d_w}{A_p(\text{peak})} \tag{6.44}$$

The results of *Yamaguchi* et al. [6.94, 131] for the experimental value of $\Delta g_\parallel$ are presented in Fig. 6.42, as well as the value of Δg_{bulk}. The quality of the $\varepsilon_\parallel$ resonance is twenty times lower than for the idealized model described by (6.38). Figure 6.42 also shows the different contributions to Δg_{exp}, according to [6.131].

With the experimental data of Figs. 6.41, 42, one may now try to calculate the local field contribution to SERS in the $\varepsilon_\parallel$-resonance. In the idealized model, one may calculate the electric field E_i in the interior of an ellipsoid using

$$E_i = \frac{1}{1+(\varepsilon-1)f} E_{loc} \tag{6.45}$$

where f are the depolarizing factors. This yields with (6.34, 36),

$$E_{i\|} = \frac{E_{0\|}}{1+(\varepsilon-1)\,(f_\| + \beta_\|)} = \frac{E_{0\|}(g_\| + i\Delta g_\|)}{g + i\Delta g + F_\|}. \tag{6.46}$$

The field at point a just outside the ellipsoid in Fig. 6.19 is $\varepsilon E_{i\|}$, whereas at point b it is equal to $E_{i\|}$. For point a, this corresponds to (6.15) [6.39] (note that $\beta = \beta_{id} + \beta_{nd}$). For the average quadratic field enhancement $|E_{\|\,surface}(0^+)/E_{0\|}|^2 = G_F$ integrated over the surface of the ellipsoid, one obtains

$$G_F \approx \frac{1}{3}\,(2+|\varepsilon|^2)\,\frac{g_\|^2 + \Delta g_\|^2}{(g+F_\|)^2 + \Delta g_\|^2}. \tag{6.47}$$

If one chooses, for example, the values in the parallel resonance of sample e in Fig. 6.40 ($\lambda_{peak} \approx 4600$ Å, $|\varepsilon| \sim 7.2$, $|g| \sim 1.22$, $g_{exp} \sim 0.055$), one obtains about 18 for the first bracket in (6.47) and for the second bracket about 6, yielding $G_F \approx 110$.

For the emission channel one may use Burstein and Chen's equation (6.17), (for molecule a in Fig. 6.19) which yields the average field enhancement G_A in the "antenna channel":

$$G_A \approx \left|\frac{1}{3}\,(2+1|1-f_\||^2\,|\varepsilon-1|^2)\right| \cdot \left|\frac{g_\| + i\Delta g_\|}{g_\| + i\Delta g + F_\|}\right|^2. \tag{6.48}$$

However, because of the different interaction of bumps in the antenna channel, $F_\|$ is different from the values in the excitation channel (Sect. 6.3). If interactions between neighbouring islands are completely neglected, $F_\| = f_\|$. A rough estimate may again be achieved from the data of *Yamaguchi* et al. [6.94]. Apparently, the resonance of interacting islands is at $\lambda \approx 4200$ Å. At 4200 Å, $|\varepsilon_1| \approx 5.1$, and $F_\| = |\varepsilon_1(4200\,\text{Å}) - 1|^{-1}$. Δg is again taken as 0.06 (shape and size contributions, Fig. 6.42). This yields about 2.4 for the second bracket in G_A for an emission wavelength at about 4800 Å (laser wavelength 4600 Å, Raman shift 1000 cm^{-1}, $g \approx 0.11$). The first factor yields for $f_\| \approx 1/3$ and $\varepsilon \approx -9$ the value of 15, leading to $G_A \approx 37$.

A different way to estimate G_A would be to set it equal to $|1+2(\varepsilon-1)/(\varepsilon+2)|^2$ following (6.4). This would yield $G_A \simeq 12$. The overall enhancement $G_\| = G_F G_A$ would be about 4×10^3 or 1.4×10^3. One should note that $G_\|$ is calculated with respect to $E_{0\|}$ (Fig. 6.38). Nevertheless, $E_{0\|}$ is changed with respect to the bare surface by the presence of the "effective film". $\boldsymbol{E}_0$ is the vector sum of the incident

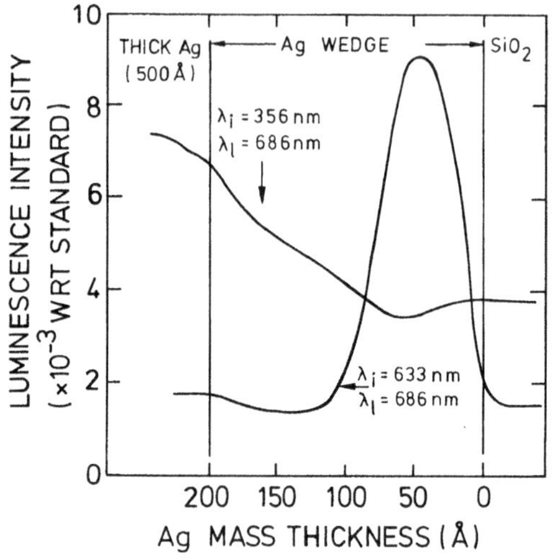

Fig. 6.43. Luminescence intensity at $\lambda_l = 686$ nm of nile blue adsorbed on a wedge-shaped silver film on a quartz substrate, excited at $\lambda_i = 346$ and $\lambda_i = 633$ nm versus mass thickness of the silver film. The WRT standard is the luminescence obtained from a 10^{-5} M solution of Rhodamine B in ethane in a 1 mm cell, measured in the same experiment. From [6.136]

and reflected field, the latter being increased by the presence of the island film. Hence, G has to be corrected by a factor $|E_{0\parallel}$ (with film)$/E_{0\parallel}$ (without film)$|^4$ in order to compare it to the experimental enhancement. If $E_{0\parallel}$ is reduced by 30 % by the presence of the film, the experimental G would be reduced by a factor of 4. In Sect. 6.5.6 (colloids), the possibility that the chemisorbed molecules are not exposed to the field outside the surface $E_{\text{surface}}(\text{O}^+)$ of the ellipsoids, but to a field of a value between $E_{\text{surface}}(\text{O}^+)$ and E_i will be discussed. This would reduce the contribution from the first factors of G_F and G_A, see (2.47, 48).

In conclusion, enhancements of more than 5 orders of magnitude from silver-island films are unlikely to arise exclusively from electromagnetic resonances. This point is corroborated by a fluorescence experiment by *Glass* et al. [6.136]. These authors adsorbed one monolayer of nile blue on a wedge-shaped silver film whose mass thickness varied from 0 to 500 Å (Fig. 6.43). In this way one produces a silver-island film of increasing aggregation, which eventually goes over into a continuous thick silver film [6.21]. The luminescence intensity of nile blue at a wavelength $\lambda_l = 686$ nm was measured as a function of the position of the wedge-shaped film. For the exciting light, two wavelengths were chosen: one at 633 nm which falls into an absorption band of nile blue, and another at 356 nm which falls into a second weaker absorption band of that dye. The wavelength of 633 nm lies within the bandwidth of the electromagnetic resonances of the wedge-shaped film, whereas the wavelength of 356 nm lies in the region where the silver film is most transparent.

A pronounced maximum of luminescence is observed for a silver mass-thickness near 60 Å when excited with 632.8 nm radiation (Fig. 6.43). However, when the same film is excited with 360 nm radiation, no luminescence enhancement is observed. Since only the wavelength of the exciting light was changed, this result suggests [6.136] that the observed enhancement is due to local-field amplification of the incident intensity and not to the local-field enhancement of the luminescence. One should note that the square of the local-field enhancement of the incident radiation of 633 nm wavelength at the optimum mass thickness is only about five times stronger than for the continuous thick part of the silver film. In the sense of the interpretation of *Glass* et al., the local-field enhancement of the thick part of the film must be low, because no SERS was detected there [Ref. 6.21, Fig. 1, Curve C]. A consistent interpretation of Fig. 6.43 may be obtained with the help of the expression for the fluorescence intensity I_{fl} [6.137]:

$$I_{\mathrm{fl}} \sim \sigma_{\mathrm{F}} |A(\omega_{\mathrm{i}})|^2 \frac{\gamma_{\mathrm{F}} |A(\omega_{\mathrm{l}})|^2}{\gamma_{\mathrm{F}} |A(\omega_{\mathrm{l}})|^2 + \Gamma_{\mathrm{nr}}} . \tag{6.49}$$

Here ω_{i} and ω_{l} are the laser and luminescence frequencies, σ_{F} and γ_{F} the absorption cross section and the radiative decay rate of the free molecule, and Γ_{nr} the nonradiative decay rate of the molecule adsorbed onto silver, including both the molecular relaxation rate and the energy transfer rate to the metal. $A(\omega)$ is the local-field enhancement.

At $\lambda_{\mathrm{L}} = 2\pi c/\omega_{\mathrm{L}} = 356$ nm $A(\omega_{\mathrm{i}}) \approx 1$ and for $\Gamma_{\mathrm{nr}} < \gamma_{\mathrm{F}} |A(\omega_{\mathrm{l}})|^2$, no luminescence enhancement in the emission channel is expected, in agreement with the experiments. Accordingly, the five-fold enhancement of I_{fl} observed for $\lambda_{\mathrm{i}} = 633$ nm may be assigned to the resonance in $|A(\omega_{\mathrm{i}})|^2$. Thus, the averaged fourth power of field enhancement on the silver-island films produced by Glass et al. is probably less than 3 orders of magnitude, in agreement with the estimate made above.

Seki [6.129] obtained interesting results for SERS of pyridine on different silver surfaces. He compared a silver-island film (average thickness 75 Å), a silver-island film overcoated with a cold-deposited film at 150 K of about 300 Å thickness, and a cold-deposited film directly deposited on a fused quartz substrate (Fig. 6.44a). Other electron micrographs of an island film and of an overcoated island film are reproduced in Fig. 6.44b [6.138]. All samples were prepared in situ and then exposed to either pyridine, CO, or both. The SERS-intensity of pyridine adsorbed on the cold-deposited film with the underlying island structure is 10–20 times stronger than for the cold-deposited film without island structure [6.129]. This factor of 10–20 is ascribed by Seki to the electromagnetic resonance enhancement caused by the "bumpy structure" of the underlying island film.

Both "cold films" on the quartz substrate and on the silver-island film lose their "SERS-activity" irreversibly by warming up to room temperature [6.129]. This is in contrast to the island films. These films are formed by evaporation (about 1 Å/s) on the quartz substrate at about 360 K (the irreversible annealing of the SERS-activity of cold-deposited silver has been found before [6.33, 32] and has been assigned to the annealing of atomic scale roughness [6.32]; see Sect. 6.5.4. The pictures in Figs. 6.44a,b were obtained after exposing them to adsorbates (CO and pyridine), warming them up to room temperature and then exposing them to the atmosphere. In spite of possible annealing of the surface

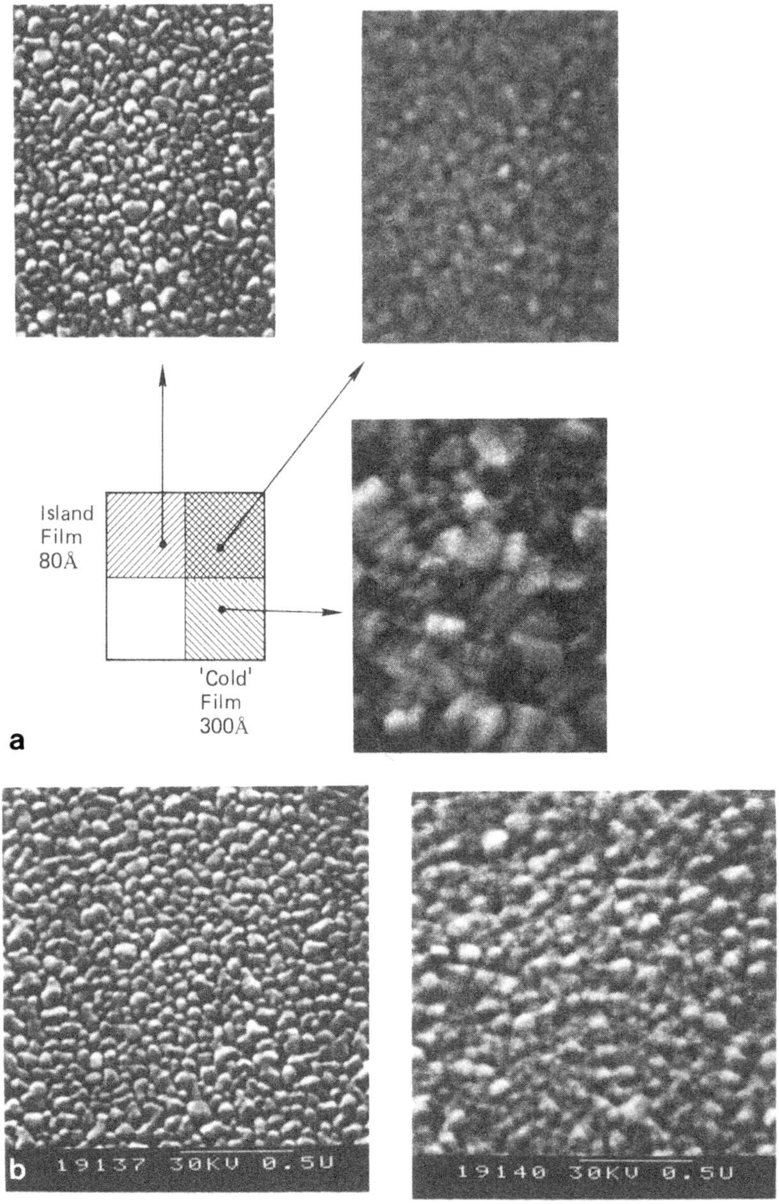

Fig. 6.44. (**a**) Deposition pattern of the different silver films on the 1/2″ × 1/2″ fused quartz substrate is given in the lower left corner. SEM micrographs of the surface, taken after the experiment, are indicated by the arrows: *upper left*, island film; *upper right*, island plus "cold film"; *lower right*, "cold film". The width of the micrographs corresponds to 1.6 μm. Courtesy of *Seki* [6.129] (**b**) Electron micrographs of the (85 Å thick) silver-island film (*left*) and the (35 Å thick) cold silver film deposited over the island film (*right*). Calibration bar at bottom is 5000 Å. The contrast was increased to bring out features for the cold film. Courtesy of *Seki* [6.138]

roughness of the cold-deposited film, the bumpy structure of the underlying island film was still visible. Seki wrote in [6.138]: "The cold film shows the profile of the islands in relief to some degree but it can be seen that the surface also has additional features of its own. At this scale they appear similar to those which have not been intentionally exposed to CO. It is possible that changes due to atmospheric exposure during the transfer to the scanning electron microscope have masked more subtle differences that existed in the UHV". Hence, according to the experiment described above, the samples shown in Fig. 6.44a (righ-hand corner) and Fig. 6.44b are "SERS-inactive"! So, the considerable roughness of the sample in Fig. 6.44b (right-hand side) and the electromagnetic resonances caused by this roughness are not enough to show the SERS phenomenon, given the experimental sensitivity of Seki's apparatus. For even more surprising effects see Sect. 6.5.9.

6.5.4 Optical Properties of "Cold-Deposited" Silver Films

As described in Sect. 6.2, a silver film evaporated in UHV on a substrate at low temperature is "SERS-active" (for instance, the silver film displaying the pyridine SERS spectrum in Fig. 6.3 was evaporated at $\approx 10^{-10}$ Torr and 120 K on a copper substrate with a deposition rate of 1 nm/s and a total thickness of 2000 Å).

The optical properties of silver films deposited at 120 to 140 K are compiled in Figs. 6.45b–d. In Fig. 6.45b, the quantity P determined by ellipsometry at oblique incidence is a measure of absorption [6.139]. The subscript of P denotes the temperature T at which P was measured after deposition at 140 K and warming up to T. P_{300} is the value of P of a well-annealed silver film evaporated at room temperature (not of a film warmed up to room temperature!). The difference spectrum P_{140}–P_{300} (Fig. 6.45b) shows a certain similarity to the spectrum of A_p of a silver-island film of $d_w = 70$ Å (see Fig. 6.45a, Sect. 6.5.3 and Fig. 6.40e).

The quantity $[R(295) - R(T)]/R(295)$ (Fig. 6.45c) was evaluated from measurements of the near-normal incidence specular reflection R [6.140]. $R(295)$ denotes R after the film, deposited at 125 K, was warmed up to 295 K. The difference $R(T) - R(295)$ is caused by absorption; the intensity of the diffuse scattered light is too low to account for the reduced specular reflectivity. The broad resonance-like structure of $[R(295) - R(T)]/R(295)$ shows similarities to the lower-wavelength structure in P_{140}–P_{300} and A_p. There is no evidence for resonance below 400 nm. *Hunderi* and *Myers* [6.141] measured the optical constants of opaque silver films evaporated onto a sapphire substrate held at 140 K. The optical conductivity (Fig. 6.45d) of these films shows a broad resonance peaking near 500 nm and a shoulder at lower wavelengths. The resonant absorption between 450 and 650 nm disappears during warming up to room temperature whereas the structure in $P_T - P_{300}$ near 350 nm remains.

The absorption spectra of silver films deposited at 140 K are very different from those of rough silver films evaporated at room temperature on rough CaF_2

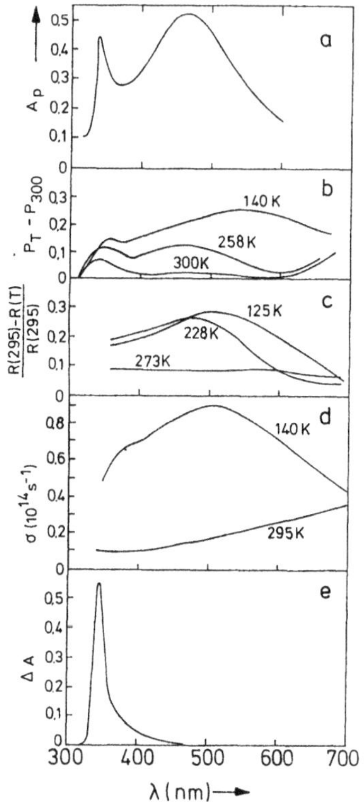

Fig. 6.45. Optical properties of cold-deposited silver films compared to those of silver-island and "rough" silver films, (**a**) A_p spectrum of a silver-island film of 70 Å mass thickness (Fig. 6.40), (**b**) Ellipsometric data [6.139] for a silver film deposited at 140 K. P is a measure of absorption (see text). Spectra denoted 258 K and 300 K are for the same film after warming up to 258 and 300 K, respectively. For P_{300} see text. (**c**) Near-normal incidence reflectivity R for a silver film deposited at 125 K, measured at 125 K and after warming up to 273 K and 295 K. From [6.140]. (**d**) Apparent optical conductivity of a silver film, deposited at 140 K, measured at 140 K and after warming up to 295 K. From ellipsometric measurements in [6.141]. (**e**) Difference ΔA in absorption of a "rough" film deposited at room temperature and a "smooth" film (see text). From [6.144]

films. A CaF_2 film investigated by scanning electron microscopy shows roughness of a lateral scale $\gtrsim$100 Å [Ref. 6.142, Fig. 4]. Scanning electron micrographs of the silver films obtained in this way ([Ref. 6.38, Fig. 1.] and [Ref. 6.143, Fig. 3]) show surface-roughness correlation lengths in the range of 800 to 1000 Å. These films show a decrease in specular reflectivity at a wavelenght of 5000 Å of only 5% with respect to a smooth film [6.38, 143]. Figure 6.45e shows the difference ΔA in absorption A between a "rough" and a "smooth" film [6.144]. The rough film was produced by evaporation of 3 μm silver on mica in UHV at room temperature. Scanning electron microscope pictures displayed a roughness correlation length of the same size as that for the films evaporated on CaF_2. The rms roughness is estimated as 60–70 Å. The "smooth" film was produced by annealing a film of 3000 Å thickness on mica for 1000 s at 140 °C in UHV. The absorption A was measured directly by a calorimetric method [6.144]. The extra absorption ΔA is caused by roughness-mediated coupling to surface plasmons [6.145].

DiLella et al. [6.139] assigned the enormous absorption of the cold-deposited silver films centered near 550 nm at low temperatures to electromagnetic resonances:

"On annealing, this feature decreases in both height and breadth and shifts to higher energies, eventually almost disappearing when the film is annealed to 300 K. This behaviour is exactly that predicted by the rough film model. When deposited on a cold substrate the film is covered with very small, closely packed irregularities. Their small size results in a broad band while their high density causes the collective electron oscillation to occur at low photon energies. On annealing, the small features coalesce into larger ones, causing the bandwidth to decrease; at the same time the density of surface bumps is reduced, producing the band center shift to lower wavelengths. Eventually the degree of roughness is sufficiently reduced to make the conduction electron resonance feature almost imperceptible".

Hunderi and *Myers* [6.141] ascribe the anomalous properties of the cold film to structural disorder, e.g., grain boundaries (see below).

There is a similarity between the optical properties of silver-island films (with 250 Å islands, see Fig. 6.45a) and of cold-deposited silver films, not only for the absorption centered at 550 nm, but also for the absorption peak (or shoulder) below 400 nm observed ellipsometrically at oblique incidence. Note that this feature was absent for normal incidence onto the silver-island films. It is also absent in the absorption spectra of cold-deposited silver films at normal incidence (see discussion of spectra displayed in Fig. 6.45c). Thus, the roughness of the cold-deposited silver films probably has a correlation length below 250 Å and a high filling factor q for "bumps on the surface". On the other hand, rough films with a correlation length near 800–1000 Å do not show similarities to silver-island films (compare Fig. 6.45a,e). From comparison of Fig. 6.45b,e, one may perhaps conclude that the resonance observed for the cold-deposited silver film near 350 nm transforms during annealing into a structure attributed to the surface plasmon on plane surfaces. If this is true, the changes of the optical spectra in Fig. 6.45b,e reflect changes in the roughness correlation length from values below 250 Å to larger values above 800 Å during annealing. This would be in qualitative agreement with the well-understood annealing of sinusoidal surface roughness with spatial wavelength λ larger than $\sim$1000 Å. In this case, the annealing rate at constant temperature is proportional to λ^{-4} [6.146]. The transformation of the dominant correlation length to larger values during annealing may also perhaps explain the observations of *Pettenkofer* [6.140] displayed in Fig. 6.46a,b. Figure 6.46a shows the change in transmission at a wavelength of 568.2 nm for a thin silver film evaporated in UHV on glass. The transmission of the thin film increases in the same way as the reflectivity of the thick opaque film. The Rayleigh scattered light (in Fig. 6.46b, measured for $\lambda = 568.2$ nm) shows different variations for the polarized and depolarized components $J_{\parallel}$ and $J_{\perp}$.

At low temperature, the ratio of $J_{\perp}/J_{\parallel}$ is below 1%, near 270 K it reaches about 7%. These observations are in contrast to those on silver films deposited at room temperature on a rough CaF_2 film. The ratio $J_{\perp}/J_{\parallel}$ for these films is about 25% [6.38, 143, 149]; the transmission of such a silver film of 700 Å thickness is increased by 50% with respect to a smooth film of the same thickness [6.38]. Accordingly, one may tentatively assign the increase in transmission and Rayleigh scattering to the transformation of short correlation roughness into roughness with larger correlation length during annealing to room temperature.

Note that the cold-deposited silver films have not become "smooth" at room temperature. They still are "rough", as indicated by Rayleigh scattering (Fig. 6.46b).

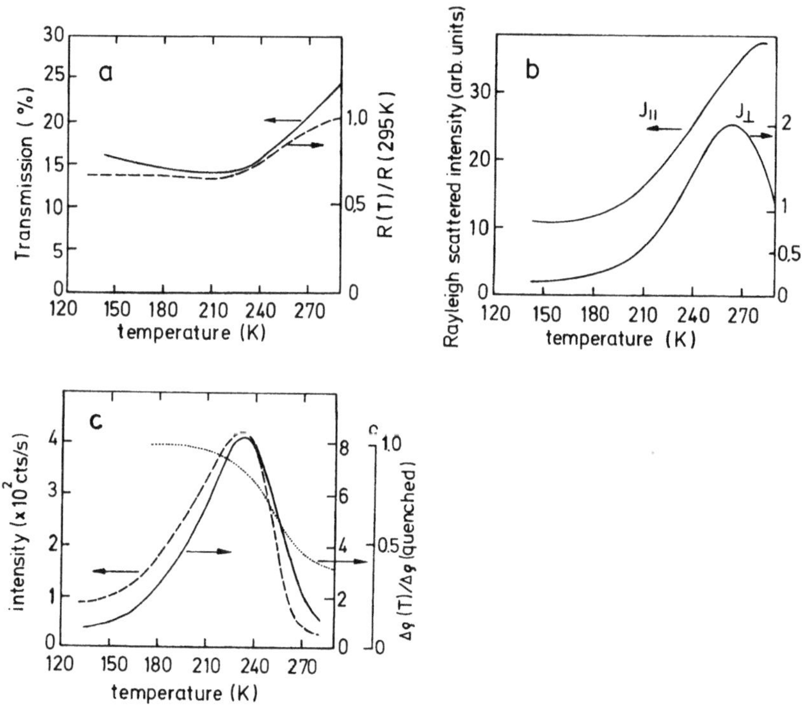

Fig. 6.46. Optical properties of silver films evaporated on substrates at 120 K during annealing to room temperature with average $dT/dt \approx 1$ K/min. (**a**) Temperature dependence of the specular reflectivity R at 5145 Å of an opaque "SERS-active" silver film normalized to R after reaching room temperature. Temperature dependence of the transmission at 5682 Å of a thin silver film on glass. From [6.140]. (**b**) Temperature dependence of Rayleigh scattered intensity J at 5682 Å. Angle of incidence. 10°, angle of emission 15° around the normal. $J_{\parallel}$, $J_{\perp}$: scattered light parallel and perpendicular to the incident polarization. Note different scales for $J_{\parallel}$ and $J_{\perp}$. [6.140]. (**c**) Temperature dependence of the background (——) and the disorder-induced phonon peak at 161 cm^{-1} (---) ($\lambda_{\text{Laser}} = 5145$ Å) [6.147] and the annealing of divacancies (···) [6.148]

A "bumpy" silver film is equivalent to a "silver-island film" on a "silver substrate". One might thus hesitate to invoke the theory of the optical properties of metallic island films on a dielectric substrate (Sect. 6.5.3). However, similar theoretical results for metal films of short-correlation-length roughness have been obtained by *Kretschmann* et al. [6.150]. These authors showed that the surface plasmon splits into two excitations under the influence of surface roughness. The dispersion relation for the surface-plasmon polariton on a smooth surface is implicitely contained in [6.150]

$$\begin{aligned} &n_s(\omega,k)=0 \\ &n_s(\omega,k)=\varepsilon(\omega)K_2+K_1 \\ &K_2=[K^2-\varepsilon(\omega)]^{1/2}\ K_1=(K^2-1)^{1/2} \\ &K=kc/\omega. \end{aligned} \tag{6.50}$$

The dispersion relation on a rough surface is given by $n_r(\omega,k)=0$ with $n_r(\omega,k) = n_s(\omega,k)-\omega^2/c^2(\varepsilon-1)^2\langle s^2\rangle I(K)$. $\langle s^2\rangle$ is the rms roughness of the surface, $I(K)$ is an integral involving all possible first-order scattering events of an excitation at ω,k on a rough surface [6.150].

For $K \gg 1$, n_r is approximately given by

$$\begin{aligned} &\frac{n_r}{K}=\varepsilon+1-\frac{(\varepsilon-1)^2a^2}{\varepsilon+1} \\ &a=\frac{\omega^2}{c^2}\langle s^2\rangle I_1(K) \\ &I_1(K)=\frac{\omega^2}{c^2}\int\limits_{K'>2} f\left[\frac{\omega}{c}(\boldsymbol{K}-\boldsymbol{K}')\right]KK'(1-\cos\theta)^2\,d^2K', \end{aligned} \tag{6.51}$$

where f is the Fourier transform of the roughness autocorrelation function. $I_1(K)$ describes all possible first-order scattering processes of a surface plasmon with normalized wave vector $\boldsymbol{K}$ into surface plasmons with normalized wave vector $\boldsymbol{K}'$ ($K'>2$). The angle between $\boldsymbol{K}$ and $\boldsymbol{K}'$ is ϕ. n_r/K is easily transformed into

$$\frac{n_r}{K}=\left(\varepsilon+\frac{1-a}{1+a}\right)\left(\varepsilon+\frac{1+a}{1-a}\right)\frac{1}{(1-a^2)(\varepsilon+1)}. \tag{6.52}$$

For $a>0$, there are two solutions of this dispersion relation reflecting the splitting of the surface plasmon compared to the case of a smooth surface given by

$$\frac{n_s}{K}=(\varepsilon+1).$$

The square of the surface response function $(n_R/K)^{-1}$ obtained from (6.52) for $K\gg 1$ is plotted in Fig. 6.47 for $\mathrm{Im}\{\varepsilon\}=0.1$ (independent of ω) versus $\mathrm{Re}\{\varepsilon\}$ for various values of a^2. At $a^2=0$ (smooth surface) there is the usual surface plasmon response for $\mathrm{Re}\{\varepsilon(\omega)\}\simeq -1$. With increasing a^2, this splits into two excitations, one for $\varepsilon(\omega)<-1$ and one for $0>\varepsilon(\omega)>-1$. This is an approximate result for $K\gg 1$. *Kretschmann* et al. [6.150] mentioned similar results for $K<1$ which are appropriate for the response to incident light. In Fig. 6.47, the corresponding

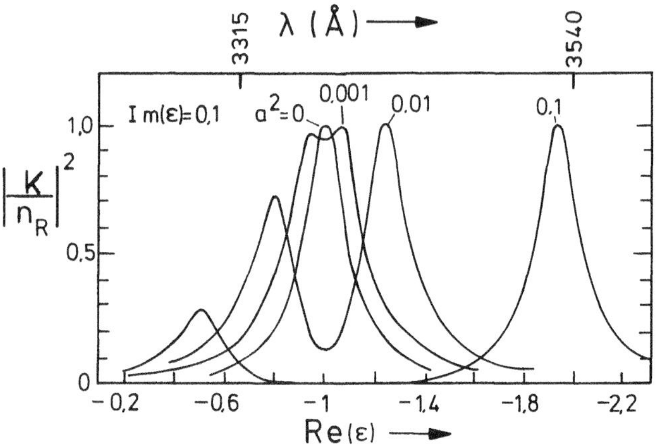

Fig. 6.47. Square of the surface plasmon response function K/n_r (see text) for $K \gg 1$ versus Re $\{\varepsilon\}$ for Im $\{\varepsilon\} \equiv 0.1$ for various roughness parameters a^2. The wavelength where the bulk optical constants of silver [6.74] have the values on the abscissa are given on the upper horizontal scale. From [6.150]

wavelength of light for silver is also given using the bulk optical constants of [6.74]. Clearly, much higher values of a^2 and enhanced "damping" is needed to fit this theoretical result to the observed spectra of Figs. 6.45b–d. The apparent decrease of the splitting between the two resonances in Figs. 6.45b,c could be explained by a decrease of the parameter a^2 during annealing. Note that a is a complicated function of the rms roughness $\langle s^2 \rangle^{1/2}$ and the roughness correlation length.

So far, only the optical properties of cold-deposited evaporated silver films due to roughness-induced electromagnetic resonances have been discussed. However, there are indications that the dielectric constant $\varepsilon(\omega)$ of cold-deposited silver is different from that of well-annealed bulk silver due to defects. This will lead to increased damping of the electromagnetic resonances and may explain why the quality of these resonances in Figs. 6.45b–d at 140 K is even lower than in the silver-island films (Fig. 6.45a).

In the following, experiments relevant to defects in cold-deposited silver films will be discussed. Figure 6.48 displays unpublished resistance measurements of *Chauvineau* [6.151] for silver films deposited at 120 K in UHV. The electrical resistance $R(T)$ (curve *I* in Fig. 6.48) of a 20 nm thin Ag film deposited on glass shows in the first warming up cycle ($dT/dt = 3$ K/min) a large and irreversible decrease with a pronounced step at 260 K. Subsequent heat treatment gives a linear and reversible $R(T)$ curve characteristic of a metallic film (lower curve I in Fig. 6.48). When a thin silver film of 8 nm thickness was deposited at 120 K on a well-annealed silver film, the electrical resistance variation $\Delta R(T)$ between the first irreversible annealing and subsequent reversible resistance behaviour shows one broad step from 170 to 280 K. Without further information one might assign the extra resistance of the unannealed films to scattering of the electrons in the case of sample I by (a) surface roughness with correlation length below 250 Å;

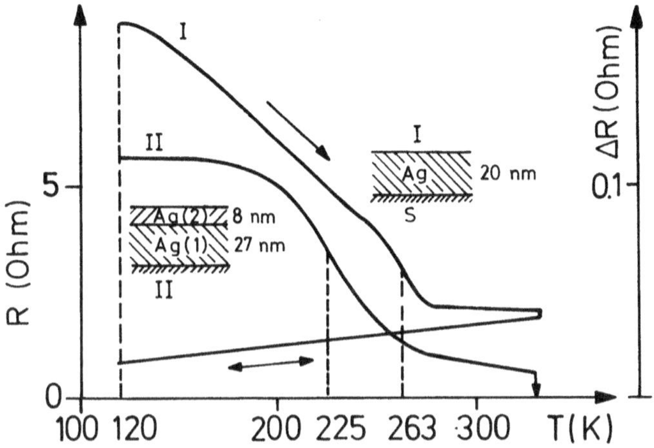

Fig. 6.48. Resistivity (R) measurements on silver films, evaporated at 120 K, after J. P. Chauvineau (unpublished 1981). I: R of a thin (20 nm) Ag film deposited on a glass substrate at 120 K during annealing at a constant rate ($dT/dt = 3$ K/min) (see text); II: $\Delta R(T)$ of a sandwich film between the first irreversible annealing ($dT/dt = 3$ K/min) and subsequent reversible resistance variations; Ag(1): "smooth" Ag film, annealed at 330 K for one hour; Ag(2): silver film, condensed on Ag(1) film at 120 K

(b) a high density of grain boundaries of unoriented microcrystalline grains. Thermal annealing will allow the growth of crystals with a (111) plane nearly parallel to the substrate plane at the expense of the smaller misoriented crystals. In this sense, the disappearance of grain boundaries would be mainly responsible for the irreversible decrease of the electrical resistance [6.151]; (c) point defects or dislocation in the bulk of the film.

Figure 6.49 shows relevant resistance measurements of *Schumacher* and *Stark* [6.152]. First, silver films of 20 nm thickness were evaporated at 300 K, annealed at 350 K and their square resistance $R(T = 350\ \mathrm{K})$ measured at 350 K. Subsequently, the films were cooled to various temperatures between 10 and 350 K and additional silver with average thickness d_2 was evaporated on top of them with a deposition rate of about 0.01 monolayer/s. The resistance change ΔR was measured during the increase of d_2. At temperatures below 150 K there was a strong increase of ΔR for the first evaporated monolayers. This reflects the increase of electronic surface scattering by the atomic-scale surface roughness produced by depositing a small quantity of silver on silver at low temperatures. Similar studies of gold on gold have been reported by *Chauvineau* [6.153]. The first step-like increase in resistance is not cancelled by a further increase in d_2 (Fig. 6.49). This reflects frozen defects due to imperfect auto-epitaxy of silver on silver. Atomic-scale surface defects could lead to defects like dislocations and twin boundaries in the bulk of the growing crystals in the additionally evaporated films [6.151]. For example, an adatom adsorbed on a (111) terrace can occupy a regular *fcc* site or an irregular *hcp* site. Accordingly, the defect structure in the Ag (2) films and in sample I in Fig. 6.48 will be different, leading

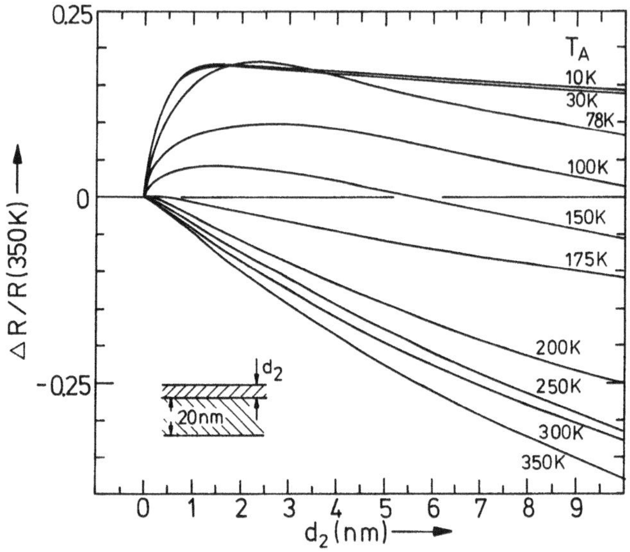

Fig. 6.49. Relative variation of square dc-resistance of silver films of 20 nm thickness (evaporated at 300 K, annealed at 350 K, cooled to temperatures T_A between 10 K and 350 K) versus thickness d_2 of a layer of silver, evaporated on top of the original film at temperature T_A. From [6.152]

to different annealing curves [6.151]. With increasing d_2 in Fig. 6.49, there is little decrease of ΔR below 100 K. This reflects the low bulk conductivity of the cold-deposited films. Apart from this experimental evidence for the effects of grain boundaries, atomic-scale surface roughness and dislocations, there is also experimental evidence for point defects in silver from resistivity measurements. Hot silver wires were quickly cooled to liquid nitrogen temperature and the annealing of the frozen-in lattice defects upon warming was monitored by dc resistivity measurements. A big step-like decrease in resistivity (see the dotted curve in Fig.(6.46c) occurred around 250 K [6.148, 154, 155]. This decrease is unanimously assigned to the disappearance of the concentration of divacancies by diffusion to the surface or internal voids [6.148, 155, 156].

There is evidence for the influence of point defects in cold-deposited "SERS-active" silver films. *Pockrand* and *Otto* [6.147] observed disorder-induced Raman scattering from phonons on these films (Fig. 6.50). This assignment is corroborated by analogous results from cold-deposited copper and gold films (Table 6.4) [6.57]. The appearance of a spectrum similar to the phonon density of states can be explained by the existence of point-like defects in the cold-deposited films.

The irreversible variation of the intensity of the phonon signal and the inelastic background during the warming-up cycle of a cold-deposited silver film is presented in Fig. 6.46c. The similar variation of both intensities indicates a common defect-related origin. The strong decrease in intensity between 230 and 260 K coincides with the temperature range where divacancies are annealed.

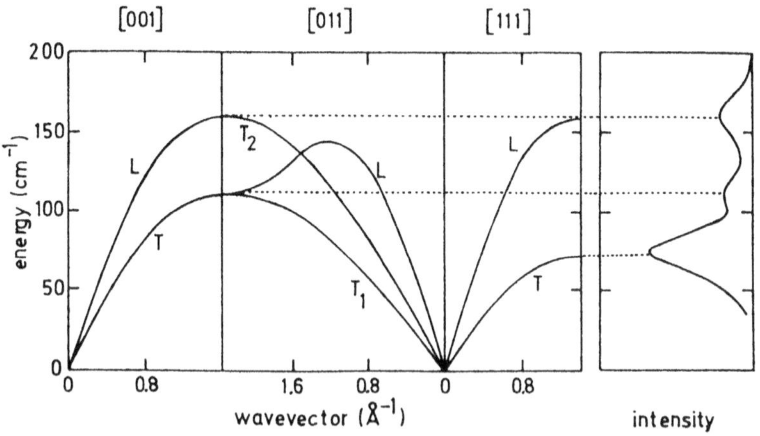

Fig. 6.50. *Left:* dispersion of phonons in crystalline silver from neutron data [6.157, 158]. *Right:* Raman scattering intensity from cold-deposited silver films. From [6.147]

Table 6.4. Disorder-induced Raman scattering from acoustical phonons. Wave number of Raman structures (first column) and of critical points in the phonon spectra from neutron scattering (second column) for Ag [6.157, 158], Cu [6.159] and Au [6.160]. After [6.57]

Ag		Cu		Au	
[cm^{-1}]					
73	73	(102)	112	59	61
112	113	167	172	97	95
—	145	211	213	—	132
161	160	252	277	152	159

Therefore, the phonon signal was assigned to photon-electron and electron-phonon coupling at divacancies, the background to photon-electron coupling at divacancies [6.147]. Consequently, there must also be optical absorption in the range of the used laser frequencies due to photon-electron coupling at divacancies. This may hold also for grain boundaries and dislocations. As outlined above, these bulk absorption mechanisms will lead to increased damping of the electromagnetic resonances.

Apart from bulk-defect induced absorption one has to consider another as yet hypothetical absorption mechanism, namely, absorption in very small clusters of silver atoms at the surface. *Schulze* et al. [6.161] reported the absorption spectra of monomeric and dimeric silver particles in a xenon matrix (Fig. 6.51). The absorption bands between 3000 and 3500 Å are due to $^2S_{1/2} \rightarrow {}^2P_{1/2,3/2}$ atomic transitions [6.162]. The two absorption bands near 4000 Å and below 3000 Å are caused by Ag_2 molecules [6.161]. When the gas-to-

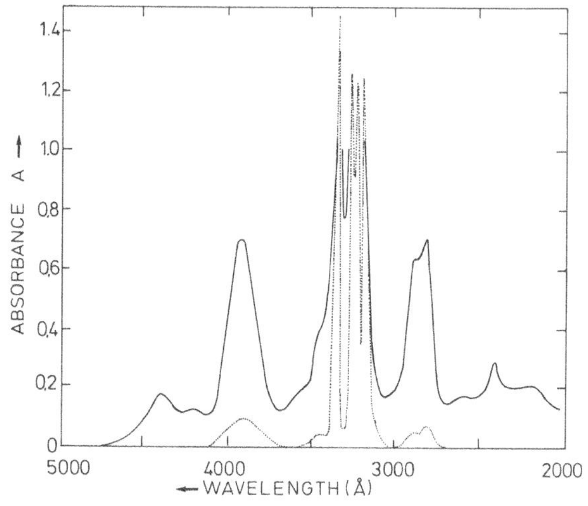

Fig. 6.51. *Dotted line:* absorbance spectrum of monomeric and dimeric silver in a xenon matrix formed at 7 K. Xe-to-Ag ratio 670:1. Mass density of silver is 1.5×10^{16} cm^{-2} (~one monolayer). *Solid line:* absorbance spectrum for Xe-to-Ag ratio 257:1, mass density of silver 5.9×10^{16} cm^{-2}. From [6.161]

metal ratio is decreased, an increase in the intensity of the bands of dimeric silver is observed (Fig. 6.51). In addition, two new bands of the same relative intensity and symmetric with respect to the dimer bands appear, together with a broad absorption over the whole region 2000–5000 Å. An absorption structure at even longer wavelengths has been reported in [6.163]. *Schulze* et al. [6.161] attributed the additional bands to trimeric silver molecules and the broad background to silver aggregates with more than 3 atoms. The absorption in Fig. 6.51 is definitely not an electromagnetic resonance but reflects the electronic excitation spectrum of very small clusters when crystal momentum is not yet a good quantum number. Only eventually, when the silver-to-rare-gas ratio of the samples is increased, does one observe dipolar electromagnetic resonance in silver microcrystals larger than 30 Å diameter [6.164, 165]. The Raman spectrum of Ag_2 and Ag_3 clusters [6.166] shows vibrational bands in the same frequency range as the spectrum in Fig. 6.50. As yet it remains an open question as to whether such small silver clusters exist at the surface of cold-deposited silver and whether they contribute to the optical absorption.

A further unanswered question is why one observes an increase of the phonon and inelastic background spectrum between 120 and 220 K (Fig. 6.46c). This increase resembles the increase of the Rayleigh scattering seen in Fig. 6.46b. If one multiplies the Rayleigh intensity $I_{\|}$ with the concentration of divacancies, one reproduces roughly the observed inelastic intensity variations. The increase in Rayleigh scattering may reflect an increase in long-correlation at the expense of short-correlation roughness (following the trend of elastically scattered light with the size of isolated spherical particles discussed in Sect. 6.3.1a) or just an annealing of bulk defects leading to a decrease in $\varepsilon_2(\omega)$ and therefore an increase in the strength of electromagnetic resonances in the supra-atomic roughness. Hence, phonon and background signals are governed both by the concentration of defects and by electromagnetic resonance in the supra-atomic surface roughness. Likewise, the irreversible development of the SERS intensity on cold-deposited silver films during warming up to room temperature [6.32] may be tentatively explained by the concentration of surface (and maybe bulk) defects and by the variation of electromagnetic resonances. Obviously, more work is needed to understand the optical properties of cold-deposited silver films.

In conclusion, the classical contribution to SERS in the case of cold-deposited silver films is certainly not greater than in the case of silver-island films. This contribution is probably smaller than 10^3. In this case one has to invoke a "nonclassical effect" of some kind to explain the overall enhancement. An interesting observation indicates a "nonclassical" contribution to SERS from cold-deposited silver films: the 1006 cm^{-1} line of pyridine adsorbed at 120 K (Fig. 6.3) is almost completely depolarized [6.167]. For free pyridine, the Raman scattering from the corresponding vibration is polarized. The observed SERS depolarization has been assigned by proponents of exclusive classical enhancement to the depolarization of the incident and emitted light by surface roughness. Of course, this argument should also apply to the Rayleigh scattered light. However, the Rayleigh scattered light from cold-deposited films at 120 K is nearly completely polarized (Fig. 6.46b).

6.5.5 Second Harmonic Generation from "SERS-Active" Surfaces

Chen et al. [6.48] observed that second harmonic generation (SHG) from a "SERS-activated" silver electrode exposed to air is enhanced by a factor of 10^4 with respect to a smooth silver-air surface. Although they did not exclude a "chemical effect", they only discussed the enhancement by electromagnetic resonances. They compared SHG from smooth, evaporated silver films and bulk silver samples roughened (activated) by electrolytic cycling. When pyridine was added to the electrolyte, these activated samples yielded the "normal" surface enhancement of the pyridine Raman signal ($\sim 10^6$). The power density at the sample surface during the 10 ns pulses from a Q-switched Nd-YAG laser ($\lambda = 1.06$ μm) was at such a low level that no surface damage occurred. The SHG from the smooth film was well collimated in the specular reflection direction, whereas the signal from the rough, activated surface was nearly isotropic in angular distribution and independent of both input and output polarizations. In addition to the SHG intensity at 0.53 μm, a broad background, highly diffuse and extending from ~ 3500 Å to the infrared, also appeared. The background was weak for the smooth film, but exceptionally strong for the rough surface (Fig. 6.52). The SHG signal from the "SERS-active" bulk silver surface, integrated over all emission angles (2π sr), was found to be 10^4 times larger than the collimated SHG from the smooth, "SERS-inactive" film (Fig. 6.53). The SHG intensity shows the quadratic dependence on input energy, whereas the anti-Stokes background shows a cubic dependence.

According to Chen et al., the nonlinearity originates from the first one or two layers of metal atoms at the surface. The resonant field enhancement is expected to be strongest in these layers. The second harmonic of ω_L results from the

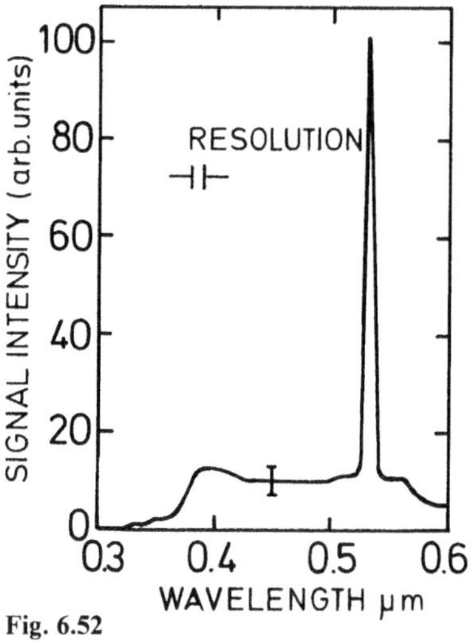

Fig. 6.52

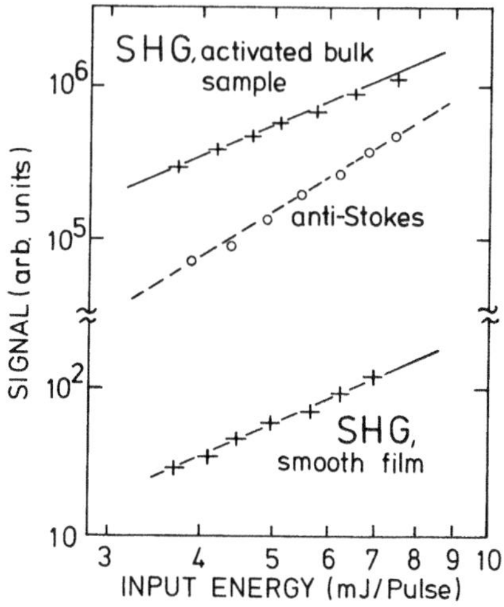

Fig. 6.53

Fig. 6.52. Spectral distribution of the nonlinear signal from a rough "SERS-active" bulk silver sample. Pump beam at 1.06 μm, pulses of 10 ns duration, 7 mJ energy, 45° angle of incidence on a 5 mm diameter spot. From [6.48]

Fig. 6.53. Dependence of the second-harmonic-generation (SHG) of silver on laser energy. The upper and lower curves show the quadratic dependence of the diffuse SHG signal from the rough activated bulk sample and of the collimated SHG signal from the smooth film, respectively. The dashed curve shows the cubic dependence of the diffuse anti-Stokes signal from the rough activated bulk sample. From [6.48]

nonlinear polarization $\boldsymbol{P}(2\omega_L)$ induced by the laser field $\boldsymbol{E}$ near the metal surface:

$$\boldsymbol{P}(2\omega_L)=\alpha(\nabla\cdot\boldsymbol{E})\boldsymbol{E}+\beta(\boldsymbol{E}\cdot\nabla)\boldsymbol{E}+\gamma\boldsymbol{E}\times(\nabla\times\boldsymbol{E}), \tag{6.53}$$

where α, β, γ are coefficients characteristic of the metal [6.168].

Because of the large dielectric constant of the metal, and hence the strong variation of the normal component of $\boldsymbol{E}$ at the surface, the first two terms in (6.53) will dominate. They are mainly due to the first one or two layers of metal atoms at the surface. $\boldsymbol{P}(2\omega_L)$ should vary over a rough surface because of the local fields due to electromagnetic resonances. If for the normal components $\boldsymbol{E}_{loc}=g\boldsymbol{E}$ (where $\boldsymbol{E}$ is the field strength of the laser at a smooth surface), then

$$\boldsymbol{P}_L(2\omega)\approx\alpha_{eff}(\nabla\cdot\boldsymbol{E})\boldsymbol{E}+\beta_{eff}(\boldsymbol{E}\cdot\nabla)\boldsymbol{E}, \tag{6.54}$$

with

$$\begin{aligned}\alpha_{eff}&=g^2(\omega_L)\,g(2\omega_L)\,\alpha\\ \beta_{eff}&=g^2(\omega_L)\,g(2\omega_L)\,\beta.\end{aligned} \tag{6.55}$$

Chen et al. [6.48] have assumed that α and β are approximately the same at a smooth silver surface as at an "activated" one. The local-field enhancement η of SHG from a rough surface over that from a smooth surface is approximated by

$$\eta \simeq |g^2(\omega_L)\, g(2\omega_L)|^2_{\max} \cdot x, \tag{6.56}$$

where x is the fractional area of those regions with maximum local field enhancement. Chen et al. estimated that $x \approx 0.05$, $g(2\omega_L) \approx 1$ and hence derived $g(\omega_L) \approx 20$. It is not clear why they attributed the enhanced signal to $g(\omega_L)$ (at $2\pi c/\omega_L = 1.06\ \mu m$) rather than to $g(2\omega_L)$.

Wokaun et al. [6.95] performed SHG experiments with 1.06 μm laser light at a "wedge-shaped" silver-island film (Fig. 6.43) which gradually became a thick continuous film. They compared the intensity ratio between the maximum signal from the island part of the film to that for the continuous film. This ratio was only about 8, rather than 10^4 (for an analogous gold sample, this ratio was even smaller than one!). They attributed this low ratio to increased SHG from the thick film due to increased residual surface roughness.

A weak point of the interpretation of Chen et al. is the assumption that α and β are "constants of the material", also valid at the surface. Equation (6.53) is valid in the bulk, where contributions to SHG from terms of the form $E_i E_j$ $(i,j = xyz)$ are forbidden by inversion symmetry. In the bulk, α and β are constants determined by the electronic structure, but the main contribution to SHG comes from the surface layer where the derivatives of E are large. At a silver surface, the electronic structure is different from the bulk (see [6.169, 170] and references therein). A smooth surface will have a different electronic structure than a surface with atomic-scale roughness [6.171]. Therefore, α and β will depend on the surface structure. Because the inversion symmetry and, in the case of a rough surface, the translational symmetry parallel to the surface are broken, the quadratic terms $E_i\, E_j$ will also contribute. Therefore, it might even be better to drop the macroscopic concept of (6.53) altogether. SHG is a three-photon process; it will become resonant by electronic surface excitations at $\hbar\omega_L$ and $2\hbar\omega_L$. Experimental evidence for surface excitations of clean and adsorbate-covered silver-electrode surfaces is presented in Sects. 6.8, 6.9. The observations by Chen et al. of the "unusually broad" and "exceptionally strong" luminescence background (Fig. 6.52) for the "SERS-active" sample may even prove the existence of intermediate electronic excitations for an "activated" sample. This background must be due to a continuum of electronic excitations serving as intermediate and final states for nonlinear optical processes.

The sensitivity of SHG to minute changes at the surface is also demonstrated by the detection of molecular monolayers by SHG [6.172]. The specific adsorption of a monolayer of chloride and pyridine at a surface enhances SHG by more than one order of magnitude. On the other hand, *Wokaun* et al. [6.95] were unable to detect a significant change in the SHG intensity from silver-island films after chemisorption of one monolayer of CN. A more recent experiment on SHG and SERS from activated silver electrodes in an electrolyte with

5×10^{-2} M pyridine [6.173] showed a decay of both signals within minutes of the activating oxidation-reduction cycle. Chen et al. found this transient behaviour difficult to understand (a flattening-out of roughness of the order of 50 Å within minutes is unlikely). *Timper* et al. [6.174], who observed the same effect in SERS, proposed as an explanation the migration of adatoms freshly redeposited after activation, to kinks, steps and microclusters. *Glass* et al. [6.136], when using 514.5 nm rather than 1.06 μm excitation, did not observe the SHG due to the postulated resonance in $g(\omega_L)$. They attributed this to the damping of the second harmonic (at 268 nm) by carrier excitation in the silver particles. In summary, the interpretation of the enhancement in SHG from "SERS-active" surfaces is subject to the same debate as SERS itself.

6.5.6 SERS from Colloids

As discussed in Sect. 6.3, rigorous calculations of the classical enhancement exist for isolated spheres. SERS experiments on isolated spherical metal particles (colloids) suspended in a liquid or a solid matrix should be a good test of these theories. The difficulty is keeping the fraction of particle clusters low. Such clusters will have peculiar absorption and scattering spectra.

The early results of *Creighton* et al. [6.16] are still very often presented as proof of an exclusive "classical enhancement", although *Creighton* has later [6.175] shown electron micrographs proving that this samples were chain-like particle aggregates. A review of the excitation spectra of aggregated sols was given in [6.23].

Qualitative agreement with the electromagnetic enhancement model was found for adsorbates on small silver crystallites in argon matrices. Silver was vaporized in an argon atmosphere, where cluster growth occurs through thermomolecular processes. The small silver particles were swept into a separate chamber where the argon gas was removed and the particles were co-condensed with CO [6.176], ethylene, acetylene or ethane [6.177]. In the cases of CO, ethylene and acetylene, the excitation profile of the Raman lines increased in going from the red to the violet, in agreement with optical extinction. No SERS from ethane was found (Sect. 6.5.9).

Stable silver sols were prepared by *Wetzel* and *Gerischer* [6.178]. The excitation profile of adsorbed pyridine increased from red to blue, in qualitative agreement with the extinction spectrum (Fig. 6.58). The excitation spectrum was not followed up to and beyond the spectral position of the extinction maximum because of the lack of suitable laser lines.

On the other hand, data exist [6.180, 181] which show definite discrepancies between excitation and extinction spectra. *Kerker* et al. [6.180] reported Raman signals, enhanced more than 10^5-fold from citrate ions adsorbed on colloidal silver. The measured wavelength dependence of Raman intensity differed from that predicted by the theoretical model [6.68] (Sect. 6.3.1). Figure 6.54 shows the absorbance spectrum of 500:1 diluted silver sol with adsorbed citrate. The

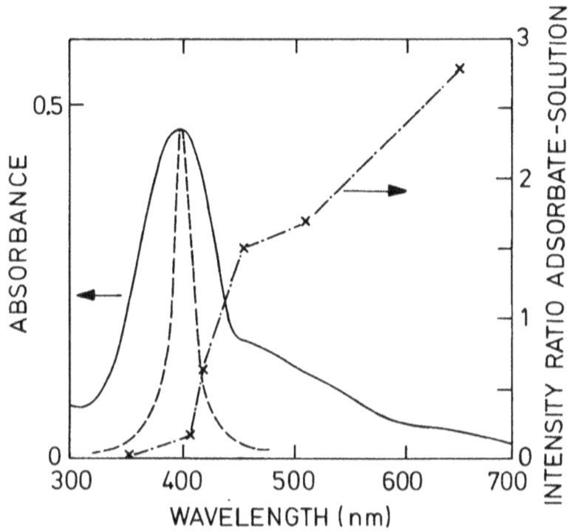

Fig. 6.54. Absorption spectrum of 500:1 diluted silver sol with adsorbed citrate (1 cm cell) ——, theoretical extinction efficiency normalized at peak absorbance ---, integrated and normalized (see text) band intensity of the $1400\,\mathrm{cm}^{-1}$ Raman band of adsorbed citrate as a function of the exciting laser wavelength (crosses). From [6.180]

absorbance is log (I_0/I), where I_0 is the incident and I the transmitted intensity. Absorbance and extinction are used synonymously here. Extinction is actually light attenuation in the forward direction by *absorption* and *scattering*. The particles are covered by a monolayer of citrate. This experimental spectrum is compared with a calculation using the Lorenz-Mie theory [6.183] with the optical constants of bulk silver [6.74]. The best fit is shown in Fig. 6.54 for silver particles of 21 nm radius. The experimental width of the resonance is about 3 times the theoretical one (see below).

Although the relative intensities of the Raman bands of adsorbed citrate showed no dependence on excitation wavelength, the absolute intensities, and hence the enhancements, were strongly dependent on wavelength. Kerker et al. compared the integrated strength of the $1400\,\mathrm{cm}^{-1}$ band for citrate adsorbed on the colloidal particles (in the same solution as for the absorbance measurement) to the signal from a concentrated citrate solution (1.37 M). The relative enhancement by adsorption to the colloidal particles is shown in Fig. 6.54 versus the exciting laser wavelength [Ref. 6.180, Table 1]. Whereas the theoretical curve for a 21 nm radius particle shows a sharp peak ($>10^5$) at 400 nm with a drop-off at higher and lower wavelengths, the measured enhancements do not display this peak but increase monotonically with increasing wavelength. Kerker et al. proposed as possible sources of the discrepancy "the effects of(1) small particle size upon the optical constants of silver, (2) polydispersivity in particle size, (3) nonsphericity, (4) aggregation of particles into clusters and (5) mechanisms for the enhancement other than, or in addition to, our theoretical model". There may indeed by some concentration of aggregated particles in the Ag colloids used by Kerker et al., as becomes apparent by a comparison with [6.181].

Von Raben et al. [6.181] investigated the extinction spectrum of gold colloids. With a comparable concentration of aggregated and nonaggregated gold particles, a shoulder on the long-wavelength side of the main extinction peak was

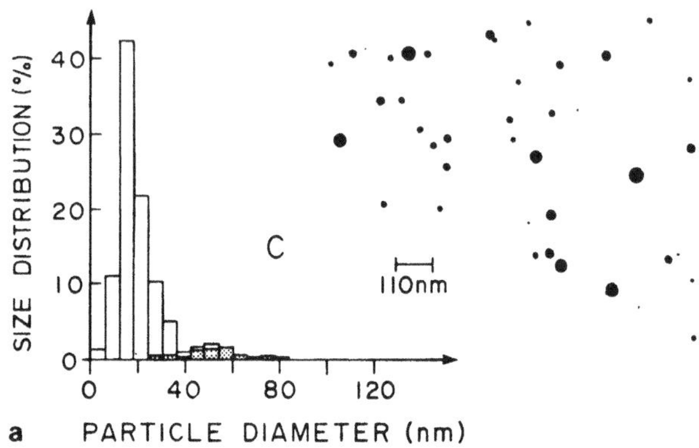

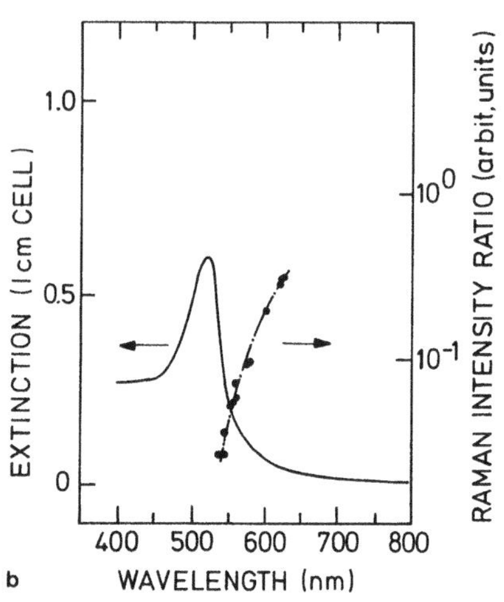

Fig. 6.55. (**a**) Histogram of size distribution from a TEM micrograph of a special gold colloid. For nonaggregated particles (*unshaded bars*), their diameters are used. For aggregated particles (*shaded bars*), the longest dimension of the clusters is used. From [6.181]. (**b**) Extinction spectrum for the same colloid and wavelength dependence of the $[Au(CN)_2]^-$ normalized Raman intensity at 2138 cm^{-1} (see text). From [6.181]

visible, as for the absorbance curve in Fig. 6.54. This shoulder was absent for a very small fraction of aggregated gold particles (Fig. 6.55a,b). These figures are from a sol with only 7.2% of the aggregated particles; the mean diameter of the nonaggregated particles was 16.8 nm, of those aggregated 45 nm. All particles were well within the Rayleigh limit: about 50 nm for the spheres (Sect. 6.3.1). The SERS and extinction spectra remained stable for months. The main extinction peak in Fig. 6.55b at 530 nm is due to the dipolar surface-plasmon resonance of Au spheres with diameters small enough to be in the Rayleigh limit.

As the gold colloid was prepared by reduction of K $Au(CN)_2$, the particles were covered by adsorbed $Au(CN)_2$. The SERS from the colloid yielded for $\lambda_L > 550$ nm a strong $[Au(CN)_2]^-$ peak at 2138 cm^{-1}. However, the peak of the aqueous $[Au(CN)_2]^-$ ion at 2164 cm^{-1} in solutions containing no Au-colloids prior to the precipitation of the particles was barely detectable. For the exciting laser wavelength $\lambda_L < 514.5$ nm, no $[Au(CN)_2]^-$ peak at 2138 cm^{-1} was

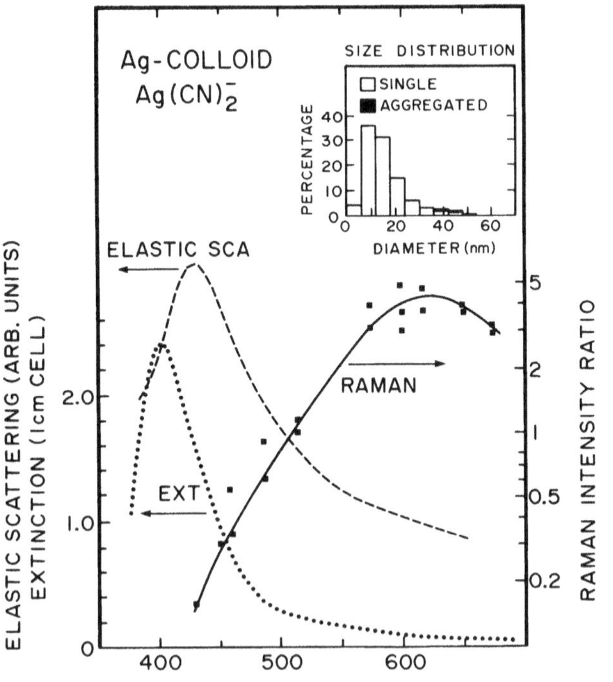

Fig. 6.56

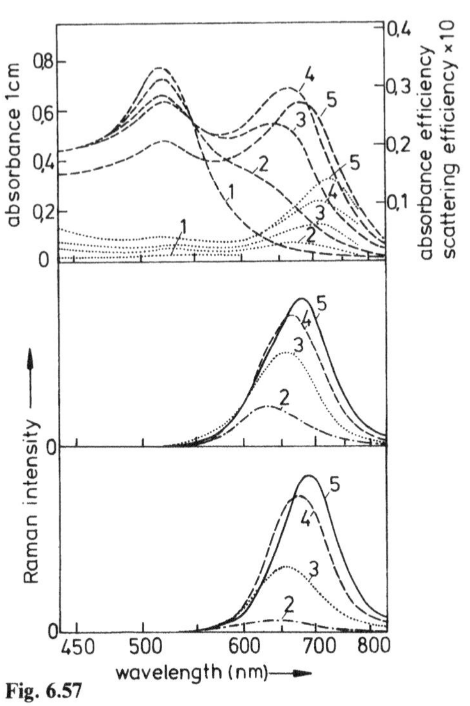

Fig. 6.57

Fig. 6.56. Extinction and elastic scattering spectra at $90 \pm 30°$ for silver colloids (size distribution in insert) compared to the Raman excitation spectrum of the CN stretch mode from adsorbed $[Ag(CN)_2]^-$. From [6.182]

Fig. 6.57. *Top:* extinction (---) and elastic scattering (···) spectra of a gold sol; *middle:* excitation profile for the $1014\ cm^{-1}$ Raman band of pyridine adsorbed on the colloid particles. Measurements before addition of pyridine [curve (1)], 3 hours (2), 8 hours (3), 24 hours (4) and 58 hours (5) after making the colloid 8×10^{-5} M in pyridine; *bottom:* Raman excitation profiles for gold sols calculated by means of (6.19) from the experimental extinction data. From [6.179] detected, although the extinction maximum almost falls into this range. Surprisingly, the difference in Raman spectra between the colloidal solution and pure water in the range of the O–H modes of water near 3425 cm^{-1} only show a small residual peak compared to the SERS signal of $[Au(CN)_2]^-$ (see also Sects. 6.5.7, 9). This residual peak shows no resonance in the range of the colloidal extinction maximum either. From the classical theory, especially from the long-range effect (6.5), one would expect a considerable enhancement of the Raman spectrum of water near 500 nm. The intensity of the $[Au(CN)_2]^-$ SERS line at 2138 cm^{-1} measured with respect to the benzene mode at 1584 cm^{-1} is plotted in arbitrary units in Fig. 6.55. It starts to grow monotonically at the long-wavelength side of the extinction maximum.

Although the colloid particles are not monodispersed, the fact that they are all within the Rayleigh limit should yield classical enhancement peaks near $\lambda_L = 530$ nm and $\lambda_{Stokes} = 530$ nm (corresponding for the 2138 cm^{-1} Stokes line to $\lambda_L = 476$ nm). Nonsphericity of the colloids cannot account for the discrepancy between extinction spectrum and Raman excitation spectrum. Nonsphericity could shift peak positions (Fig. 6.12) with respect to spherical particles, but the positions of the peak in extinction and the longer wavelength peak of the Raman excitation spectrum should roughly coincide.

Von Raben et al. [6.182] found similar results for silver colloids, covered by $[Ag(CN)_2]^-$ (Raman line at 2143 cm^{-1} [6.26]) (Fig. 6.56). Whereas the extinction and the elastic scattering (at $90 \pm 30°$) agree approximately with the predictions of the electromagnetic theory (Fig. 6.11), the Raman excitation spectrum does not. Only 7% of the colloids are aggregated (see insert in Fig. 6.56), the single colloids are well within the small-particle limit.

Are the results of [6.180–182] in disagreement with classical theory? *Blatchford* et al. [6.179] discussed the problem as if they were in agreement:

"It appears to us to be more likely that the observed Raman scattering was almost entirely contributed by the small fraction of particle aggregates".

They presented the excitation profile of the 1014 cm^{-1} Raman band of pyridine adsorbed on colloidal gold and the extinction and elastic scattering spectra of these colloids measured as they slowly aggregate in the presence of pyridine (Fig. 6.57). With aggregation to strings of particles [6.179], a longitudinal electromagnetic resonance developed around 700 nm. Simultaneously, the excitation spectrum, similar to the one in Fig. 6.55b, became stronger. No resonance was observed near the single-particle resonance [6.184]. Similar results were obtained by *Mabuchi* et al. [6.185]. This behaviour is qualitatively in agreement with the spectra at the bottom of Fig. 6.57 calculated with the help of (6.19) [6.97], the absorbance spectra in Fig. 6.57 (top) and the optical constants of silver [6.74]. The author has applied the theory to the results of *Wetzel* and *Gerischer* [6.178]. Figure 6.58 shows the strong increase of the factor F in (6.19) for $\hbar(\omega_L - \omega) = 0.1$ eV, 0.2 eV toward the red which yields, with the experimental absorption and (6.19), an excitation profile peaking near 3 eV and increasing again below 2.2 eV. For this second rise there is no indication in the Raman

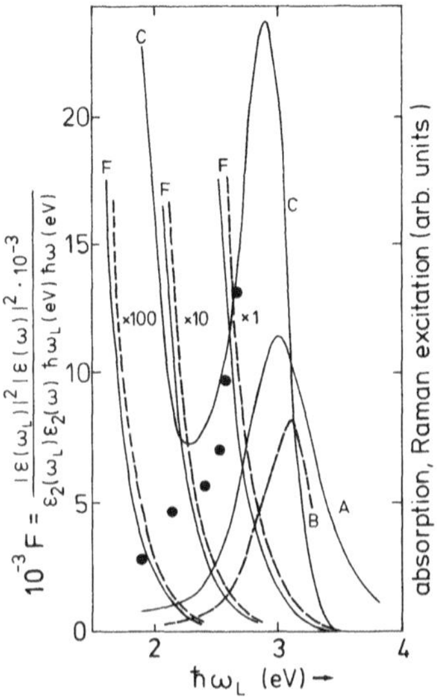

Fig. 6.58. Factor F (6.19) calculated with the optical constants of silver of [6.74] for $\hbar(\omega_L - \omega) = 0.1$ eV (*solid line*) and 0.2 eV (*dashed line*). A, B: optical absorption (extinction) of silver colloids [6.178, 182]. C: Raman excitation spectrum calculated with (6.19) and spectrum A. Dots: experimental excitation spectrum [6.178] of pyridine on the colloid with absorption spectrum A

excitation profile of the 1010 cm^{-1} mode of pyridine adsorbed on the silver colloids [6.178].

From Fig. 6.58 it is obvious that the spectral behaviour of F and the low frequency tail of the absorption is very crucial for the calculated excitation profiles. Therefore it would be helpful to test the approximation of (6.19) by performing exact calculations with the help of the theory of Kerker and coworkers (Sect. 6.3). We note that in some cases [6.176–178] the excitation spectra are qualitatively in agreement with the extinction profiles and in others [6.180–182] they are not. Instead of a strong contribution of aggregated particles, one may also invoke an additional "nonclassical" enhancement mechanism and assume different weights between these meachnisms or a variable excitation profile for the nonclassical excitation (see Sect. 6.8) in the two groups of experiments. One should also consider the recent results of *Bachackashvilli* et al. [6.186]. Dye molecules and pyridine or citrate were simultaneously adsorbed on silver colloid. Pyridine and citrate revealed the usually large SERS while the *coadsorbed* dye showed no enhancement (within a factor ~10 to 100) [6.186].

It is surprising to find no sign of the classical resonance in the results of *von Raben* et al. (Fig. 6.56). Possible reasons could be (a) an increase in $\varepsilon_2(\omega_L)$ for small particles (see below), (b) an extra surface absorption, for instance, by charge transfer between metal and adsorbate (see Sect. 6.8), and (c) that in all theories in Sect. 6.3, the adsorbates are exposed to the field outside the surface. This is not necessarily so. In the dielectric continuum model, there is a step for the normal component E_n of the electric vector at the surface. In reality, there will be a steady transition to $E_n(0^-) = [1/\varepsilon(\omega_L)]E_n(0^+)$ from $E_n(0^+)$ above and below the surface. The adsorbate may sit in this zone especially for the "atomically open" surface which must exist for spherical particles. One should note that the classical enhancement is mainly caused by the resonance of the field E_n normal to the surface (Fig. 6.11). For the case where the adsorbate is exposed to $E_n(0^-)$ in the ATR experiment (Fig. 6.23), the possible weakening of the surface-plasmon polariton resonance has been estimated to be 5.9 for $E^2(0^-)$ versus $E^2(0^+)$ [6.187], which would lead to a reduction in the Raman enhancement G by a factor of ≈ 35. Any effect due to the strong gradient of E_n at the surface (as proposed in [6.53, 139]) should display nearly the same resonant behaviour as $E_n(0^+)$.

As pointed out in Sect. 6.3, the enhancements are usually calculated from $\varepsilon_2(\omega)$ of crystalline bulk materials. However, one always observes increased damping of the dipolar mode in small particles. For instance, for silver colloids in glass (stained glass), the same increase [Ref. 6.188, Fig. 6.5] in the halfwidth with decreasing particle diameter has been found by *Doremus* [6.189], *Kreibig* and *Fragstein* [6.190], and *Smithard* [6.191]. Between 100 and 20 Å diameter of the colloids, the halfwidth of the resonance grows from 200 to about 1000 Å. For spherical silver microcrystallites prepared by gas aggregation, *Abe* et al. [6.164] found an increase in the halfwidth which was only half as strong as that observed for colloids in glasses. The shift of the resonance is controversial. Whereas *Smithard* [6.191] found a strong red shift with decreasing colloid diameter (between 100 and 10 Å diameter from 4150 to 4500 Å), *Genzel* et al. [6.192] observed a slight blue shift for Ag colloids in glass with decreasing diameter. *Abe* et al. [6.164] observed a blue shift for spherical microcrystallites prepared by gas aggregation (3800 Å at 100 Å diameter, 3600 Å at 25 Å diameter).

Theoretical models involve diffuse surface scattering of electrons [6.190] or the quantum size effect [6.193]. In the diffuse surface scattering model, the effective collision frequency ω_c of the free electron is given by $\omega_c = \omega_{c\,\mathrm{bulk}} + \alpha v_f/\phi$, where α is of order 1 and ϕ is the particle diameter. Both models predict an increase in damping with decreasing ϕ. The scattering model predicts a slight shift towards the red with decreasing ϕ, the quantum size model a shift towards the blue [Ref. 6.194, Figs. 1, 2]. It is obvious that any increase in $\varepsilon_2(\omega)$ near the surface due to diffuse scattering or quantum size effects will change the classical enhancement, due to its proportionality to $\varepsilon_2(\omega)^{-4}$. A calculation of the photoabsorption of small metal spheres [6.195] included the excitation of electron-hole pairs. Below the frequency of the dipolar mode the photoabsorption is raised by a factor of 10–100 with respect to the photoabsorption calculated from the classical theory employing only $\varepsilon(\omega)$.

As pointed out at the end of Sect. 6.3, exact calculations of the "classical enhancement" also exist for the ATR resonant excitation of surface plasmon polaritons (SPP) [6.107, 108] (Fig. 6.23). In the relevant Raman experiments from monolayer adsorbates on silver films, the SPP resonance was only investigated for the incident channel, whereas the Raman scattered light emitted into the space "below the prism", approximately normal to the prism base-plane, was collected [6.196–198]. For this case, enhancements of about 2 orders of magnitude are expected by tuning into the SPP resonance. *Dornhaus* et al. [6.197] were able to detect a Raman signal from a monolayer of the strong Raman scatterer isonicotinic acid, adsorbed on a room-temperature-evaporated silver film, when tuning into the SPP resonance. When the silver films were in contact with a pyridine-containing electrolyte, *Pettinger* et al. [6.196] and *Tom* et al. [6.198] were unable to obtain a Raman signal from pyridine without the "activation" described in Sect. 6.2. After activation, the quality of the SPP resonance was reduced, leaving only a residual enhancement of less than 10. The lowered quality after activation was attributed to the formation of a strongly absorbing film of Ag-pyridine surface complexes [6.196] (after redeposition of less than a monolayer of Ag) or to the formation of roughness [6.198] (after redeposition of about 20 atomic layers of Ag).

6.5.7 Short-Range Effects in SERS

On cold-deposited silver films, a short-range SERS effect has been found [6.199, 200, 129]. The SERS from pyridine was essentially saturated after the exposure to one monolayer of pyridine (for details see [6.199]). An effect confined to the first monolayer of pyridine was also observed on a mechanically polished and Ar^+-sputtered polycrystalline silver foil [6.201, 202].

Short-range effects in SERS are *per se* not in contradiction with the classical model. As described in Sect. 6.3.1 the enhancement G decays roughly like $(a/r')^{12}$ for a molecule at distance r' from the center of an isolated sphere of radius a. Therefore, the range of the effect depends very much on a. For instance, G is 10 times smaller for a molecule sitting 10 Å above a sphere of 50 Å radius, as compared to G for a physisorbed molecule ($r'=a$). For a sphere radius of 500 Å, the decrease in G at a distance of 10 Å would only be 22 %. Various apparent short and long-range effects [6.17] have been explained in this way [6.200]. However, only in SERS from colloids can one invoke the distance dependence for isolated spherical particles. For a statistically rough surface, one has necessarily collective resonances (Sect. 6.3.2) which involve strong fields between neighbouring protrusions. *Aravind* et al. [6.110] calculated the local-field enhancement for a wave incident upon 2 spheres at a distance equal to 0.2 of the radii of the spheres. The enhancement of the square of the field on the connecting line between the centers of the spheres varies only by a factor of 1.5 when one proceeds from the surface of one sphere to the middle between the spheres. Therefore, there should be comparable enhancement of the exciting field for molecules between neighbouring protrusions and for physisorbed molecules. One would expect a slow decay of the exciting field enhancement with increasing separation between the molecule and the nearest protrusion, slow over distances of the order of the radius of the protrusion or their average separation. On a surface with a statistical distribution of bumps of about 50 Å diameter and a "filling factor" of 20 %, the exciting-field enhancement at point P in Fig. 6.59 (distance to the next bump about 25 Å) should not be lowered by as much as 1.6×10^{-2} compared to point P′, as would be predicted by the $(a/r')^6$ law. Also, the

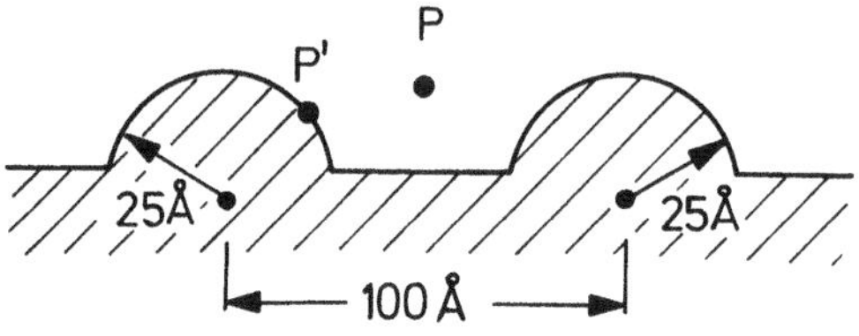

Fig. 6.59. Two neighbouring bumps on a rough surface with filling factor of 20 % (see text)

calculated enhancement of the exciting field cannot be used for the enhancement within the emission channel because the phase relations of the resonant fields in the emission channel are different from those in the excitation channel (Sects. 6.3, 6.5.3). Molecules near a particular bump (or protrusion) will mainly have electromagnetic interaction with this bump. Therefore, the distance dependence will be more short ranged than for the excitation channel; even so, the $(a/r')^6$ law most likely also overestimates this short-range behaviour.

Wood et al. [6.200] explained the long-range enhancement found by *Rowe* et al. [6.17] and the short-range enhancement on "cold-deposited" silver films [6.200] as being due to different kinds of roughness. Their experimental results are given in Fig. 6.60. The surface labelled "iodine rough" consists of about 500 Å radius Ag particles (approximately spherical in shape) separated by ~1500 to 3000 Å [6.17]. The roughness of the cold-deposited silver film is estimated by *Wood* [6.203] to be ~50 Å in size (Sect. 6.5.4). Both surfaces are covered by 5 ± 1 monolayers of pyridine. The vibrational line at 991 cm^{-1} is assigned to multilayer pyridine which is not in physical contact with the surface. (See Sect. 6.9 for a different interpretation. Results similar to those of *Rowe* et al. [6.17] in Fig. 6.60 for an "iodine rough" silver sample have been found by *Seki* [6.129] for silver-island films). The line at 1003 cm^{-1} is assigned to pyridine in contact with the surface, in good agreement with the line at 1006 cm^{-1} for less than a monolayer pyridine coverage (Fig. 6.3, see also Sect. 6.9). The evaporated film shows a shorter-range enhancement than the "iodine rough" surface. Wood

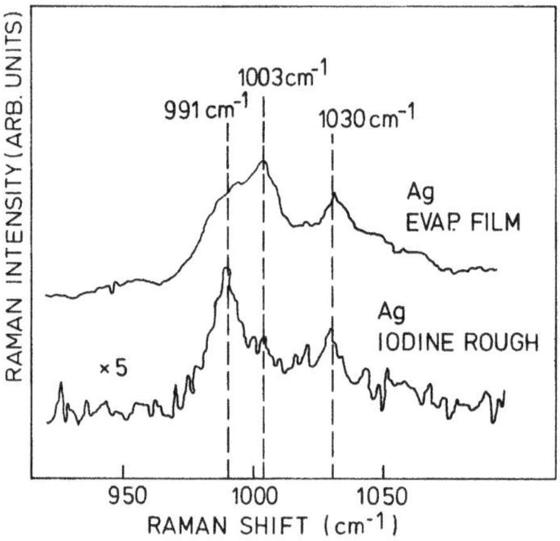

Fig. 6.60. Raman spectra of 5 ± 1 monolayers pyridine adsorbed on two silver surfaces. Upper curve is from evaporated silver film, lower is from photochemically roughened silver. Vertical dotted lines indicate Raman peaks at 991 cm^{-1} (multilayer pyridine), 1003 cm^{-1} (first layer pyridine) and 1030 cm^{-1} (both types of pyridine). From [6.200]

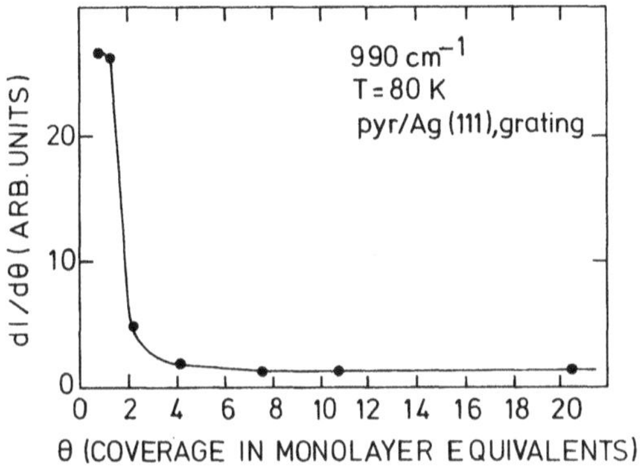

Fig. 6.61. Incremental enhancement factor $dI/d\theta$ as a function of coverage for pyridine on a modulated Ag(111) surface. From [6.204]

et al. attributed this to the different electromagnetic enhancement range for surfaces with 100 – 1000 Å roughness and smaller than 100 Å roughness.

On the other hand, *Eesley* [6.201] found dominant short-range enhancement for mechanically polished and Ar^+-sputtered surfaces for which SEM revealed "elevated" Ag plateaus of approximately 1000 – 3000 Å in diameter, separated by 1 – 3 µm. The Ag-areas between plateaus were smooth surfaces covered with ≈200 Å diameter Ag balls separated by 200 – 400 Å (between centers) and lower density (≈400 Å) balls separated by 1000 – 5000 Å. A dose of 330L was needed to observe the bulk pyridine Raman scattering above the signal noise. From this multilayer signal level, Eesley estimated the surface layer enhancement with respect to the multilayer case to be larger than 10^3. This is not the long-range effect predicted by the electromagnetic model for structures of the order of 200 – 3000 Å.

There is further experimental evidence for a short-range nonclassical enhancement mechanism. *Sanda* et al. [6.204] observed SERS from a (111) surface of a silver crystal. This surface had a sinusoidal grating structure with 10,000 Å periodicity and about 1000 Å modulation, inscribed into it by X-ray lithographic techniques [6.205]. A surface-plasmon polariton resonance was excited by choosing the appropriate angle of incidence and a p-polarized laser beam. The sample was cooled to 80K and exposed to increasing quantities of pyridine. A larger enhancement ($\sim 10^4$) of the pyridine vibrational line at 990 cm^{-1} was found for the first adsorbed layer and a comparatively small enhancement ($\sim 10^2$) for further condensed layers (see also Sect. 6.9).

The incremental enhancement factor (per increase in pyridine coverage θ) is presented in Fig. 6.61. The first monolayer in direct contact with the silver surface contributes more to the Raman signal at 990 cm^{-1} than the 20 consecutive layers. The field of the excited surface-plasmon polariton decays comparatively slowly with distance from the surface over several thousand Å. This cannot account for the strong difference between the first and the consecutive layers. In the sense of C. A. Murray's paper in [6.23] one might invoke a very small-scale statistical roughness of bumps of ≈35 Å radius. This is

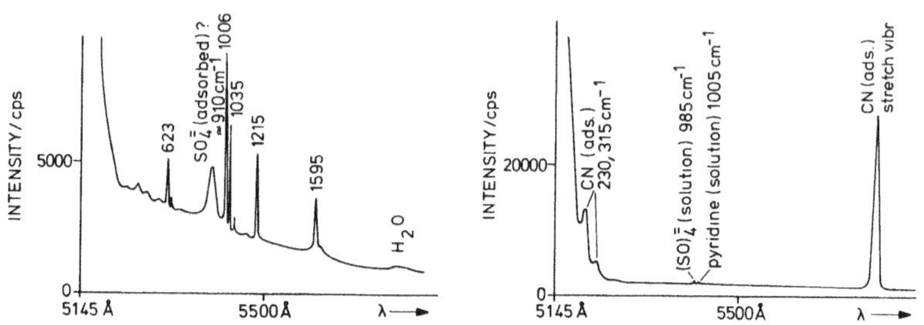

Fig. 6.62. Spectra of a polycrystalline Ag-electrode in an electrolyte of about 0.1 M Na_2SO_4 and 0.05 M pyridine at $U = -0.8$ V_{SCE}. Left spectrum is obtained after "activation" of the electrode. The main lines at 623, 1006, 1035, 1215 and 1595 cm^{-1} originate from pyridine. Right spectrum is obtained after adding 0.01 M concentration of CN^- to electrolyte without further activation. From [6.206]

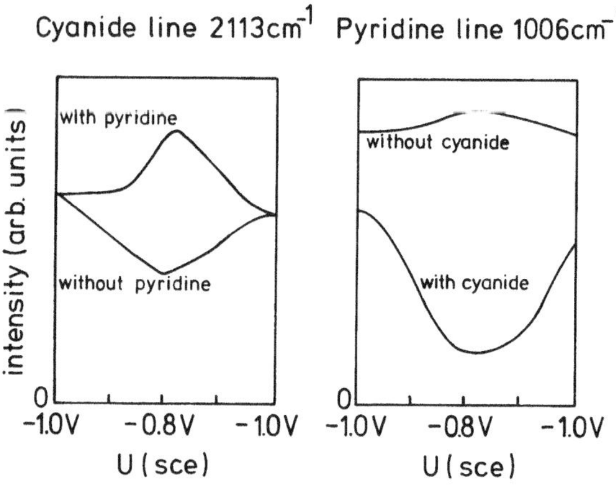

Fig. 6.63. Intensity variation at 1006 cm^{-1} (pyridine line plus background, *right*) and at 2113 cm^{-1} (cyanide line plus background, *left*) at sweeps from -1.0 V to -0.8 V_{SCE} and back with 10 mV/s. Spectra were taken from the silver electrode in electrolytes with either 0.1 M Na_2SO_4 and 0.05 M pyridine only (*upper trace right*), or 0.01 M KCN (*lower trace left*), or both 0.05 M pyridine and 0.01 M KCN (*upper trace left, lower trace right*). From [6.206]

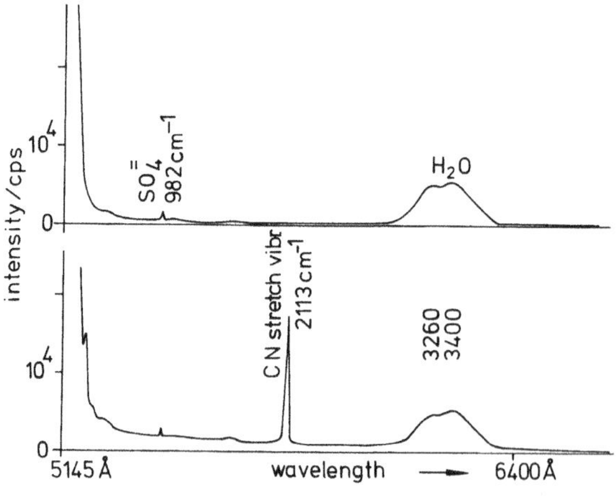

Fig. 6.64. Spectrum of a polycrystalline Ag electrode at -1.0 V_{SCE} in a 0.1 M Na_2SO_4, 0.01 M KCN electrolyte before (*top*) and after activation (*bottom*). The weak structure near 1618 cm^{-1} is caused by the water bending mode (Fig. 6.2). From [6.206]

very unlikely for the following reason. The sample of Sanda et al. showed a well-defined, low background LEED pattern, comparable to a flat unmodulated (111) region. Satellites were observed in the beam profiles. They indicated a distribution of steps and terraces parallel to the direction of the grooves of the surface-grating. The peak in this distribution corresponded to a terrace width-to-step height ratio of about 10:1. The intensity of the main peak relative to the side lobes indicated that roughly 90% of the surface is of (111) orientation. Therefore, in the author's opinion, these results show the existence of a short-range, nonclassical enhancement mechanism.

Billmann and *Otto* [6.206] argued that the strong enhancement ($\sim 10^6$) for cyanide and pyridine on silver electrodes was caused by a short-range mechanism for distances from the surface smaller than 3 Å. The main results and their interpretation are as follows. A silver electrode is activated by an oxidation-reduction cycle in a Na_2SO_4 electrolyte containing pyridine. The spectrum in the left part of Fig. 6.62 is obtained at an electrode potential of $U = -0.8\ V_{SCE}$, which corresponds roughly to a vanishing potential-step between silver electrode and electrolyte. One observes the characteristic vibrational line of pyridine. When CN^- is added to the electrolyte (without any further change of the electrode), the spectrum changes dramatically (right part of Fig. 6.62). If the potential is changed to -1.0 V, the pyridine lines reappear and the CN^- stretch line decreases a reversible way (Fig. 6.63). There is only one explanation for this result [6.206]: the SERS signals are caused by adsorbed pyridine and cyanide because the properties of the bulk of the electrolyte are not changed with potential. In the same experiment, there is no indication of an enhancement of the Raman spectrum of water as depicted in Fig. 6.64. One might argue that the first layer in direct contact with the silver surface is composed of pyridine and cyanide, but at least the next layers will contain plenty of water molecules. Therefore, according to the classical models, one should observe SERS from water when one observes SERS from cyanide or pyridine. This is not the case. This fact indicates that the enhancement of 10^6 contains a short-range effect of at least 10^2. The failure to observe the electromagnetic resonance from well-characterized colloids in water by means of SERS of water has already been described (Sect. 6.5.6).

Rzepka et al. observed first-order Raman scattering from NaCl, NaBr and NaI induced by embedded Ag colloids [6.207] and sodium colloids [6.208]. These silver particles were formed in Ag-doped crystals by injection of electrons at high temperature. The crystals were then suitably heat-treated in order to eliminate F-centers and their quasimolecular aggregates. The size of the silver particles was estimated to be 50–400 Å. The absorption spectra showed the characteristic bands due to the dipolar small-particle resonances (insets of Fig. 6.65). The Raman spectra of the samples in Fig. 6.65, with embedded silver particles, are definitely different from the low-intensity second-order Raman spectrum of the pure crystals and from the spectra induced by F-centers [6.209]. No overtones from the host phonons were detected. According to *Taurel* [6.210], these spectra are related to the first-order phonon spectrum of the host lattice,

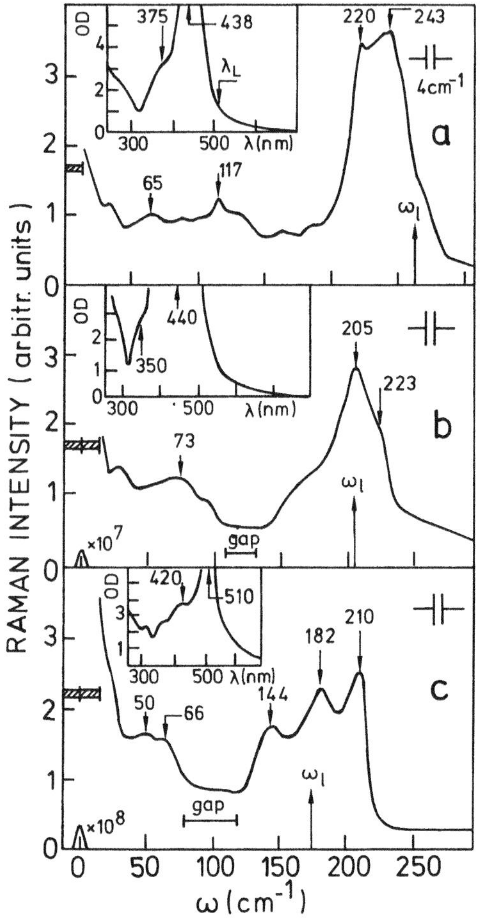

Fig. 6.65a–c. Raman spectra of NaCl (**a**), NaBr (**b**) and NaI (**c**) crystals containing Ag particles at 78 K with laser wavelength 514.5 nm. The corresponding absorption spectra (OD: optical density) are shown in the insets. Hatched horizontal bars characterize spectral regions where density filters were used to scan over the Rayleigh line. From [6.207]

the best indication being the gap between acoustic and optic phonons (Fig. 6.65b,c). Taurel did not exclude seeing some structure due to the vibrational modes of the metallic particles (Sect. 6.5.4). Though not displayed in Fig. 6.65, broadening of the Rayleigh line was observed, suggesting a single electron-hole excitation in the colloidal particles [6.208]. Resonances were observed when the laser lines lay within the electromagnetic-resonance band of the silver particles. This resonance, for the case of NaI, contributed about a factor of three to the enhancement, this factor being somewhat different for the different modes observed. Modifications of the spectra also occurred when the colloid size was changed. "Classical enhancement" of the second-order Raman spectrum of the clean crystal was not observed. Thus, a short-range interaction at the interface must be responsible for the occurrence of the first-order Raman scattering.

A further indication of a short-range effect is the relaxation of Raman selection rules described in Sect. 6.2 (Figs. 6.4, 5). This relaxation may be caused by strong field gradients at the metallic surface layer [6.53, 139] or by charge-transfer excitations (Sect. 6.8).

In conclusion, the well-documented short-range effects cannot be explained on the basis of electromagnetic resonances with roughness features greater than 50 Å in size. On the other hand, saturation of the Raman intensity after approximately two to three monolayers (see, for instance, Fig. 6.61 and the discussion of Pockrand's results [6.211] in Sect. 6.8) may still be attributed to electromagnetic resonances in protrusions considerably smaller than 50 Å, comprising 100 – 1000 silver atoms. Aggregates of such small dimensions will probably have different properties than the bulk, especially a different "effective" dielectric constant (Sect. 6.5.6). With decreasing size of the protrusions, the simple macroscopic electromagnetic model must eventually break down. The optical properties of silver microcrystals and colloids of size below 50 Å are considerably different from those of bigger silver colloids (Sect. 6.5.6). The optical absorption of very small silver clusters isolated in an argon matrix shows structure in the visible range [6.161, 164] which cannot be explained by electromagnetic resonances (Sect. 6.5.4).

6.5.8 SERS on Metals of Low Reflectivity

As discussed in Sect. 6.3, the classical enhancement scales roughly like $[\varepsilon_2(\omega)]^{-4}$. By alloying Pd to Ag, one may steadily increase $\varepsilon_2(\omega)$. *Furtak* and *Kester* [6.212] reported the SERS of pyridine adsorbed onto $Ag_{1-x}Pd_x$ electrodes for $x \leq 0.04$. Four samples of 0, 2, 4 and 5% Pd received similar mechanical polishing. Scanning electron microscopy with 200 Å resolution revealed similar surface topography (polishing grooves ~500 – 1000 Å across). In order to avoid surface segregation, no oxidation-reduction cycle was employed. A test for the surface composition by cathodic hydrogen development confirmed that the surface stoichiometry was the same as for the bulk. Because of the localization of the interfacial voltage drop between electrode and electrolyte within 25 Å of the metal, any voltage-induced intensity changes must be attributed to pyridine within a distance of 0 – 25 Å from the metal surface. The results are shown in Fig. 6.66. For $x > 0.04$, the signals at -0.7 and $-0.35\ V_{SCE}$ are the same and, consequently, Furtak and Kester assigned these signals to pyridine in the electrolyte. This assumption implies that the intensity-potential relation for the 1215 cm^{-1} line (Fig. 6.82) is not changed by alloying.

Furtak and Kester estimated that the Raman-scattering enhancement factor must be lower than about 2000 to escape detection in their system. They compared the trend in intensity with the relaxation time τ for surface-plasmon decay calculated from the optical data of Ag-Pd alloys of *Schmidt* and *Lynch* [6.213]. Between $x=0$ and $x=0.042$, τ drops by a factor of 4, corresponding roughly to a drop in "classical enhancement" by a factor of about 250 which is stronger than observed. Furtak and Kester's data were analyzed by *Tsang* et al. [6.214]. They found the best agreement with theory by considering only the

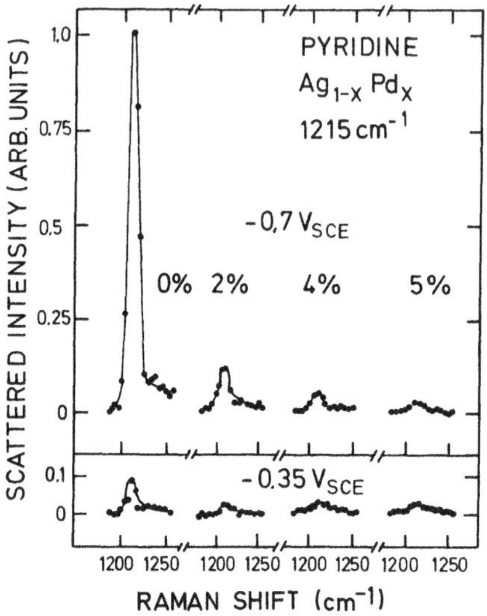

Fig. 6.66. Raman signals from the 1215 cm^{-1} mode of pyridine on mechanically abraded $Ag_{1-x}Pd_x$ electrodes at two applied potentials. From [6.212]

enhancement in the incident channel, which is proportional to ε_2^{-2}. Furtak and Kester did not exclude theories which involve the direct excitation of electron-hole pairs as the primary step in SERS. In alloys, additional electronic damping channels could then compete with this process.

For $\varepsilon_2(\omega) \ll |\varepsilon_1(\omega)|$ and $\varepsilon_1(\omega) < 0$, the optical reflectivity at normal incidence is given approximately by

$$R(\omega) \approx 1 - \frac{2\varepsilon_2(\omega)}{[1+|\varepsilon_1(\omega)|]\,|\varepsilon_1(\omega)|^{1/2}}\,. \tag{6.57}$$

For small $\varepsilon_2(\omega)$, the loss of reflectivity is directly proportional to $\varepsilon_2(\omega)$. In the silver-palladium alloys, the reflectivity at 514.5 nm decreases from 97.8 % for pure Ag to about 92.5 % after alloying it with 4.2 % Pd (calculated with the $\varepsilon(\omega)$ of [6.213]). The intrinsic reflectivities in the visible range of silver, gold, aluminum, cadmium, platinum and liquid mercury in Fig. 6.67 show $R < 0.9$ for Cd and $R < 0.8$ for Pt and Hg, in contrast to the noble metals and Al. Hence, the classical resonances of Cd, Pt, and Hg should be of much lower quality than for the noble metals (compare with Fig. 6.18). The quality should even be lower than for the Ag alloy with 4 % Pd. Therefore, from classical considerations, one would exclude strong SERS from adsorbates on Cd, Pt, and liquid Hg. There are several reports of enhanced Raman scattering from these metals, many of them, however, not confirmed by independent research.

A SERS spectrum for pyridine on a Cd-electrode, comparable in intensity to SERS of pyridine on a Ag-electrode, has been published by *Loo* [6.219] (Fig. 6.68). The Cd-electrode was "activated" by anodizing at -0.6 V_{SCE}. The spectrum exhibits a broad background; the intensity of the Raman band is

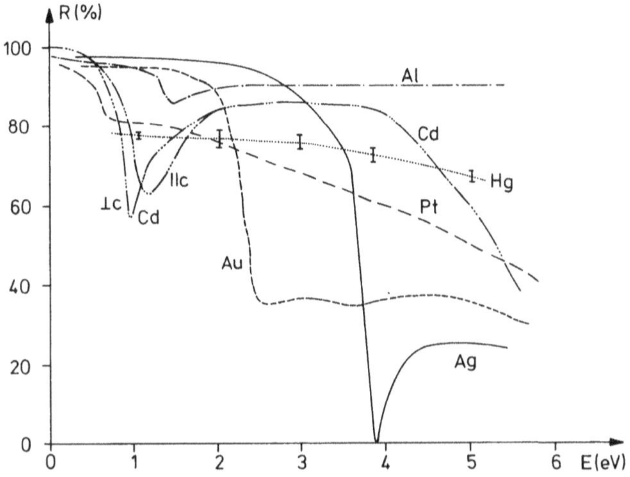

Fig. 6.67. Normal incidence reflectivity versus photon energy for Ag and Au [6.215], Al, Cd [6.216], Pt [6.217] and liquid Hg [6.218]

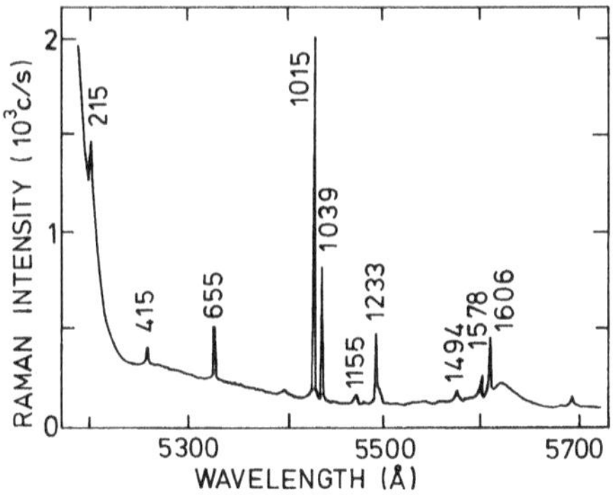

▲

Fig. 6.68. SERS for pyridine on an "activated" Cd-electrode at $-0.8\ V_{SCE}$ in 0.1 M KCl + 0.05 M pyridine. Laser wavelength 514.5 nm, power 150 mW, angle of incidence 60 to 70°. From [6.219]

Fig. 6.69. Raman spectra of a Pt-electrode in 1 M H_2SO_4 and 10^{-2} M phenylhydrazine. The peak at 985 cm^{-1} is from the breathing vibration of SO_4^{2-} in solution [6.222]. After [6.221] ▶

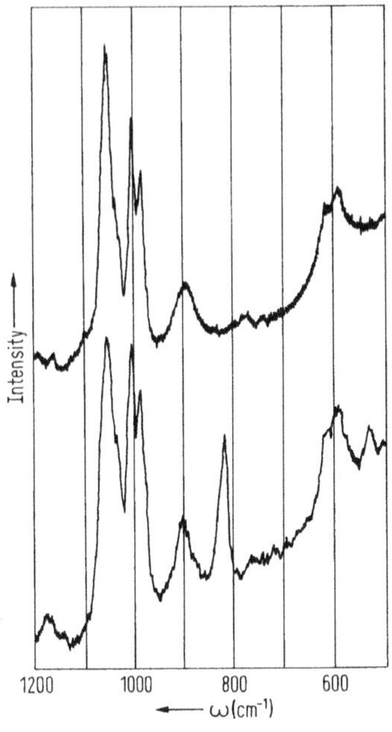

voltage dependent. *Pettinger* and *Moerl* have not been able to reproduce any SERS signal of pyridine on a Cd-electrode [6.220]. The first Raman spectrum of an adsorbate on a Pt-electrode was presented by *Heitbaum* [6.221]. Both traces in Fig. 6.69 are Raman spectra of phenylhydrazine $(C_6H_5NHNH_3)^+$ on the Pt-electrode, the upper one before, the lower one after electrochemical oxidation to

S_6H_5OH. The concentration of phenylhydrazine is 10^{-2} M in 1 M H_2SO_4. This low concentration, and the fact that no signal was seen before the "activation" of the Pt-electrode [6.222], makes it very unlikely that the signal in Fig. 6.65 is an ordinary Raman effect from the solution in front of the electrode. The "activation" consisted of repetitive high-speed triangular potential sweeps applied to the electrode. In every sweep, probably 1–2 monolayers of platinum are subject to site exchange due to the formation of a superficial platinum oxide. The activation was finished when the electrode surface looked "yellowish gray", in other words, "slightly platinized".

Krasser and *Renouprez* [6.223] investigated the Raman scattering of benzene chemisorbed on small platinum clusters (clusters of irregular shape with diameter ~100 Å and silica-supported platinum clusters, diameter ~100 Å). Strong Raman intensities were found for short-wavelength excitation. Weak signals of CO on Pt-electrodes [6.224], iodine on Pt-electrodes [6.225] and pyridine adsorbed on sputtered platinum films [6.226] have been reported as well.

Krasser and *Renouprez* [6.227] observed a Raman enhancement of 10^2–10^4 for coadsorbed CO and hydrogen on silica-supported nickel particles. The enhancement depended strongly on the excitation wavelength, the maximum lying between 450 to 520 nm. *Yamada* et al. [6.226] evaporated films of Ni, Pd, and Pt on glass at room temperature in a vacuum of 2×10^{-5} Torr. Subsequently, the films were exposed to saturated pyridine vapour, which was then pumped off. Signals at 1008, 1004 and 1015 cm^{-1} were found. Under the same conditions, Yamada et al. also observed a weak signal of pyridine on a cleaved NiO crystal. *Stenzel* and *Bradley* [6.228] saw weak Raman spectra for CO adsorbed at high pressures (390 Torr, 1 Torr) on Ni (100), (110) and (111) surfaces at about 200 K. In this case there is no evidence for an enhancement mechanism.

Great interest arose from the report of *Naaman* et al. [6.229] of a surface enhancement of about 10^4–10^5 from liquid mercury. In this case the electromagnetic resonance must be of low quality and large scale supra-atomic roughness on the surface of liquid mercury seems unlikely. Thus, the usual coupling to the electromagnetic resonances by supra-atomic roughness should not be possible. The experiment was performed in the following way. A mercury drop hanging at the end of a glass capillary (so-called hanging-drop electrodes) was suspended in a pyrex cell. The cell was evacuated to a pressure of 10^{-3} Torr, then exposed to the equilibrium vapour pressure of pyridine, benzene, or cyclohexane at 22 °C. Several minutes were required for the SERS signal to build up to full intensity. Figure 6.70 compares these spectra from the Hg surface with the vapour phase spectra, which are about 20 times weaker. (A trivial explanation, that the spectra with the Hg drop are stronger because of the reflection of the laser beam from the surface could only account for a factor of two, not twenty.) The vibrational lines from the surface signal are superimposed on a background. The laser beam in the experiment of Naaman et al. is only focussed on the Hg drop to a diameter of 2 mm. The light scattered at 90° is collected from an area of about 0.33 mm^2. The volume from which the gas phase

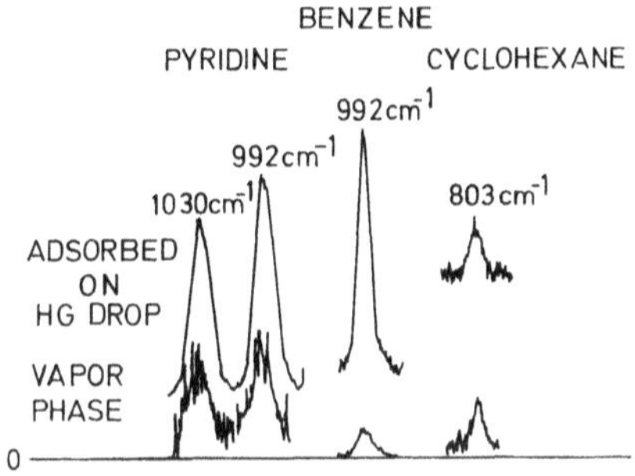

Fig. 6.70. Comparison of Raman scattering from pyridine, benzene, and cyclohexane adsorbed on a mercury drop (*upper traces*) with vapour phase spectra (*lower traces*, ordinates for pyridine and cyclohexane are multiplied by 10). Vapour pressures were respectively 200, 120, and 100 Torr. The spectra are plotted with a common baseline to indicate the intensity of the broad continuum which appears underneath the Raman lines. From [6.229]

spectra are monitored is 2 mm^3. A monolayer on the Hg surface is assumed to have one molecule per 30 $Å^2$. With these data, those authors deduced an enhancement of about 10^5. Even allowing for an error of one order of magnitude in the above estimate, this is a strong enhancement. Naaman et al. have also published spectra from a Hg drop in an aqueous solution of pyridine. In this case the signal is only doubled by the presence of the Hg. From private communications with L. A. Sanches et al., B. Pettinger, R. R. Van Duyne and M. Moskovits it follows that the situation in the case of Raman scattering from liquid Hg is still controversial.

In summary, the reported Raman enhancement from transition metals, Cd and liquid Hg surfaces cannot be understood on the basis of a pure classical electromagnetic resonance effect. The experimental situation, however, is not as clear as for SERS on Ag surfaces. For many of the mentioned metals, there are private communications of unsuccessful attempts to reproduce the data. More experimental work is needed, especially in the case of Hg and Cd.

6.5.9 Chemical Specificity of SERS

The enhancement due to electromagnetic resonances – in the classical sense – must clearly be the same for any molecule at a given site. Therefore, the classical enhancement of Raman-active vibrations does not depend on the chemical nature of the molecules. The only restriction to this statement could originate from image effects by the presence of the "ideal" metal surface. According to *Hexter* and *Albrecht* [6.230] only those components of the Raman tensors which are invariant under the simultaneous action of reflection by the surface plane and charge conjugation contribute to the scattering intensity. Therefore, different orientations of a molecule with respect to the surface normal could lead to different Raman intensities. Of course, this rule cannot explain why non Raman active modes are seen in SERS. According to *Moskovits* and *DiLella* [6.231] there is evidence that SERS is better described without taking the image of the scatterer into account. It is questionable whether the concept of the classical "image" and the classical "image potential" is a good one very near to the surface. According to [6.232 and 233] the local exchange correlation hole of the

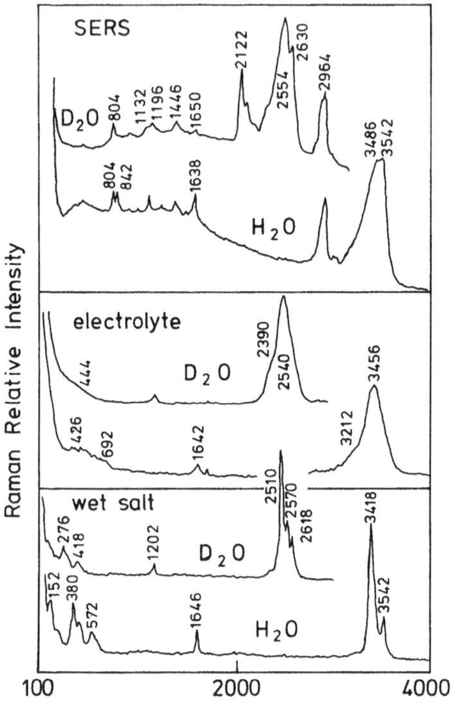

Fig. 6.71

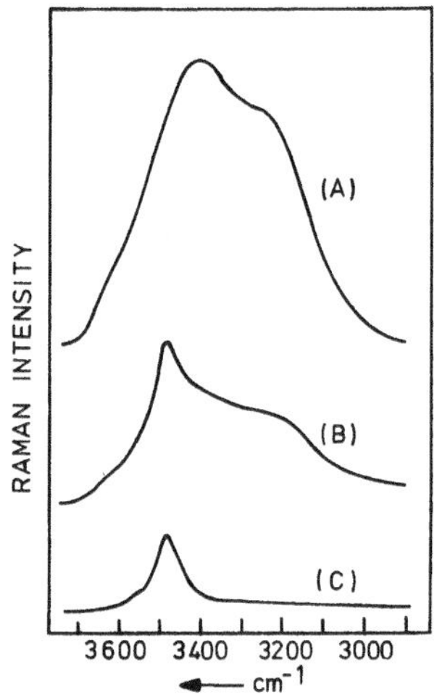

Fig. 6.72

Fig. 6.71. SERS of water at the Ag interface with concentrated electrolyte (10 M NaBr) after an oxidation reduction cycle from -0.3 V to 0.0 V and back to -0.3 V versus Ag/AgCl with 20 mV s^{-1} compared to Raman scattering from the same electrolyte and from water in wet NaBr. After [6.238]

Fig. 6.72. Raman spectra in the OH stretching frequency region of molar KCl/H_2O with the laser beam focussed at the silver electrode. All spectra were measured at a potential of -0.2 V_{SCE}. (A) Polished electrode; (B) "activated" electrode; (C) "activated" electrode pushed against the cell window. From [6.85]

Lang-Kohn model [6.234, 235] transforms steadily to the "classical" image charge only at some distance from the surface. Any strong differences in enhancement for different adsorbates, i.e., a pronounced chemical specificity of SERS, cannot be understood on a classical basis. In a forthcoming paper [6.236] the lack of dependence of the enhancement on the identity and bonding state of the molecule is slated to be discussed. However, there is evidence against such independence, e.g., the chemical specificity of water. For electrolytes containing SO_4^{2-} or ClO_3^-, there is no surface enhancement for water at a silver electrode as discussed in the last section. However, SERS from water was observed for silver [6.85, 237] and copper electrodes [6.237] in aqueous electrolytes with halide ions after activation of the electrodes by an oxidation-reduction cycle. The results of *Pettinger* et al. for a silver electrode are given in Fig. 71 [6.238]. There are sharp bands in the O – D or OH stretching and in the DOD and HOH bending spectral regions. At lower wave numbers, vibrational modes of water are evident. The spectra are considerably different from those of the electrolyte without the electrode, indicating that many vibrational bands are due to "surface water". For Cl^- concentrations below 2 M only broad (and comparatively weak) bands are observed. The enhancement in the range of the stretch modes is about 10^5 [6.238]. The frequencies of the "surface water" stretch vibrations are shifted up with respect to liquid water. This indicates a weakening of the hydrogen bond between molecules of the "surface water". There is a clear correlation between SERS from water and the presence of chloride ions.

A somewhat different result for SERS of water on silver was reported by *Fleischmann* et al. [6.85]. Their spectra for a molar KCl electrolyte in the range of the OH stretch vibrations are displayed in Fig. 6.72. Spectrum (A) was obtained after focussing the laser at a silver electrode, polished to a mirror using alumina powder of 0.05 μm average diameter. Spectrum (B) was obtained at a potential of −0.2 V after an additional "activation" with a cyclic potential ramp to +0.3 V at a speed of 5 mV s^{-1}. A sharp new feature is seen at 3498 cm^{-1}. In order to differentiate the surface spectrum from that of bulk water, the electrode was pushed against the cell window [spectrum (C)]. Only the SERS signal from "surface water" at 3498 cm^{-1} is seen in this case. A signal at 3500 cm^{-1} from a silver electrode exposed to air has already been reported in [6.14]. The position of the OH stretch band observed by Fleischmann et al. depends on the halide species: whereas this band is at 3498 cm^{-1} for Cl^-, it is at 3523 for a Br^- containing electrolyte. In an I^- electrolyte, this band splits into two intense ones at 3493 and 3533 cm^{-1} [6.85].

The specific adsorption of halide ions on silver is different. According to electrochemical knowledge at a potential of −0.9 V no Cl^- ions are adsorbed, in contrast to iodine ions. Fleischmann et al. reported:

> "The fact that the Raman bands of adsorbed water were present at −0.9 V in iodine solution confirmed that the water and halide are not adsorbed independently because the Raman spectrum of the chloride system at −0.9 V shows no trace of adsorbed water". "The analysis of the entire surface Raman spectrum for a roughened silver electrode in 1 M KCl at potentials varying from −0.2 to −0.6 V showed that the Raman intensities began to fall rapidly at −0.5 V and that at −0.6 V only about 3 % of the intensity at −0.2 V remained" [6.85].

From the SERS spectra in H_2O, D_2O and HDO electrolytes, one can exclude the possibility that the SERS signals in the OH stretch region are due to adsorbed OH groups rather than to adsorbed water [6.239].

Moskovits and *DiLella* [6.58] compared the Raman signals of an equimolar mixture of CO and N_2 several monolayers thick on silver films evaporated at ~20 K. The enhancement for nitrogen was at least 100 times smaller than for CO, although the Raman cross sections of the free molecules are nearly identical. This has been corroborated in [6.59] for silver islands in CO and N_2 matrices.

DiLella and *Moskovits* [6.240] also observed strong Raman scattering from the first monolayer of cis- and trans- 2 butene, isobutylene, 1 – butene and cis – 2 butene adsorbed on a silver film deposited at 11 K. Similar experiments using methane, ethane, or propane as an adsorbate did not produce SERS spectra. This result was corroborated by *Manzel* et al. [6.177] for colloidal siver particles of ≈ 100 Å in size formed by gas aggregation in an argon matrix: they obtained SERS spectra from the first layer of adsorbed ethylene and acetylene but not from ethane. This chemical specificity also holds for silver films evaporated at 120 K: Pockrand obtained strong SERS spectra after exposing the films to 1 L of ethylene, but no specific signal after exposure to ethane and methane [6.25]. These experiments are very strong evidence against the exclusive electromagnetic model.

Lyon and *Worlock* [6.130] reported that silver-island films evaporated directly onto a silicon substrate did not enhance the Raman signal from the zone-center phonon nor from any other phonons of Si. They do, however, enhance scattering from PNBA added on top of the silver film. It is still an open question as to how this result is compatible with the opinion that the PNBA signal in this experiment is enhanced entirely classically.

Chemical specificity has also been found by *Chen* et al. [6.241]. Benzoic acid in contact with silver-island films displays SERS only when its carboxyl group is in direct contact with the silver surface but not when the benzene ring opposite the carboxyl group makes contact with such a surface. *Loo* and *Furtak* [6.242] observed a kind of "chemical activation" of a gold electrode. A gold electrode (in a pyridine-containing electrolyte) with a roughness factor (actual over nominal surface area) of 5–6 when irradiated with the 514.5 nm Ar^+ laser line did not show a Raman signal from pyridine, in agreement with other observations. However, when only 0.15 monolayers of silver were deposited on the gold electrode, a SERS signal from pyridine was detected.

Seki [6.36, 243] reported that the physical proximity of pyridine and CO to a silver surface is not enough to yield SERS from these molecules. He evaporated silver-island films on saphire and quartz substrates [6.128], cooled the sample to liquid He temperature and deposited 10^{15} pyridine molecules/cm^2 [6.36]. The results are given in Fig. 6.73. A SERS spectrum of pyridine was not observed initially. Only after warming the substrate to temperatures near 100 K did the peaks appear. The peaks remained after cooling back to 10 K. Even after warming to $T > 300$ K and recooling, some signal from pyridine remained. Similar results were found for cold-deposited (at 150 K) silver films which were cooled to 10 K [6.243]. These results are corroborated by measurements using CO [6.243]. After deposition at 10 K of 4×10^{14} molecules/cm^2 of CO on the cold-deposited (at 150 K) silver film, the SERS intensity of the CO stretch vibration at 2142 cm^{-1} increased when warming the sample from 10 to 20 K. As for pyridine, lowering the temperature did not cause the intensity to decrease. In contrast to pyridine, warming up to 60 K caused the CO to desorb and the 2142 cm^{-1} peak disappeared. (The 2115 cm^{-1} peak reported, for instance, in [6.34] at 120 K, is most probably caused by an unknown reaction of CO [6.259]). Cooling the substrate, of course, did not restore the peak at 2142 cm^{-1} but the CO could be redeposited again and the "thermal activation" process could be repeated, clearly indicating that this process did not involve an irreversible

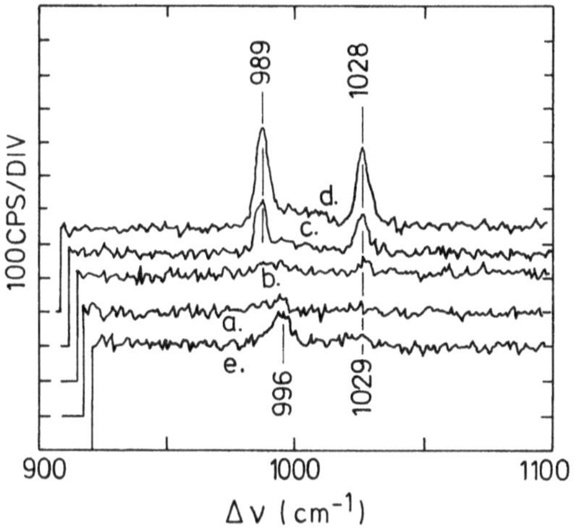

Fig. 6.73. A series of Raman spectra of pyridine on an Ag island film of 90 Å thickness. The baselines are shifted as indicated. *Curve* (*a*): $T = 10$ K after deposition of 10^{15} pyridine molecules/cm^2 (*b*): $T = 77$ K; (*c*): $T = 90$ K; (*d*): $T = 10$ K after warming to 100 K; (*e*): $T = 80$ K after warming the substrate to $T > 300$ K. From [6.36]

annealing of the surface topography of the silver film. Another surprising observation of *Seki* [6.243] is that it was not possible to obtain the SERS spectrum of CO on the silver-island film even though such films exhibit SERS for pyridine, although, as described above, he obtained SERS spectra of both pyridine alone or CO alone on cold-deposited (at 150 K) silver films. When a silver-island film and a silver film evaporated at 150 K were simultaneously exposed to a CO pressure of 10^{-3} Torr at 90 K, the silver-island film did not show the SERS spectrum of CO, in contrast to the "cold-deposited" film. It is, however, most likely that the islands also are covered with CO. *Seki* [6.243] summarized his observations as follows:

"Molecules deposited at liquid He temperature on silver surfaces known to exhibit SERS do not show full enhancement initially. Full enhancement is achieved by thermal activation. In cases that the deposited molecules can be desorbed at temperatures lower than the substrate temperature at which the silver film was deposited, this process can be repeated.

The Raman spectrum of pyridine is easily seen on the silver island film but it has not been possible to observe the Raman spectrum of CO even though it was quite certain that the CO molecules were at the silver surface."

With regards to the observed "thermal activation" (so-called low temperature anneal [6.243] one might think about an activated surface migration of molecules to "SERS active sites" [6.243] or on activated transition from a so-called precursor state (physisorbed) into a more chemisorbed state or an activated transition from one bonding state into another. Such temperature-dependent transitions in the state of adsorption have been observed for oxygen on copper by *Schmeisser* et al. [6.244].

Seki also investigated the effect of CO on the SERS spectrum of pyridine on silver [6.138]. This SERS spectrum on a cold-deposited (at 150 K) silver film (for details see [6.138]) at a temperature of 90 K showed line shifts and intensity variations, when the film was exposed at 90 K to a static CO pressure of 10^{-3} Torr. No signal from CO in the range of 1800–2200 cm^{-1} was observed. On the other hand, coldly evaporated silver, without adsorbed pyridine, shows strong SERS from CO when exposed at 105 K to static CO-pressures of 10^{-6}–10^{-3} Torr [6.245] (Sect. 6.6). In *Seki's* experiment [6.138] CO is present: one observed its influence on the SERS spectrum of pyridine. For exclusive classical enhancement models one should observe a SERS signal from CO, but this it not the case.

From the publications by *Demuth* et al. [6.121, 122, 204, 246, 247, 249–251], only those results pertinent to the topic of this section will be discussed. On smooth Ag(111) surface, a compressional phase transformation of chemisorbed pyridine was observed by means of vibrational spectroscopy with high-resolution electron energy loss and by uv photoemission spectroscopy [6.246]. Below a coverage of about 0.4 monolayers, pyridine was π-bonded to the surface, its molecular plane parallel to the surface. Above 0.4 monolayers the plane of the pyridine molecules was inclined to the surface, the bonding to the surface now involving the nitrogen lone-pair electrons.

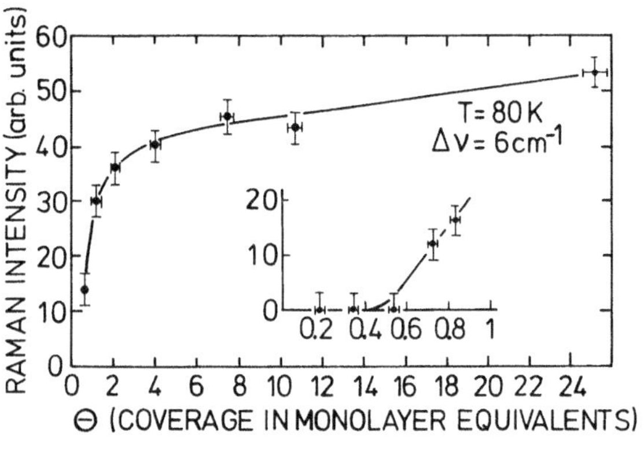

Fig. 6.74. Raman intensity for the 990 cm^{-1} peak of pyridine on a silver (111) grating structure as a function of coverage in monolayers. The insert shows the detailed low coverage behaviour. Laser wavelength 5145 Å. From [6.246, 249]

SERS investigations on Ag(111) was performed after inscribing a quasisinusoidal grating with 10000 Å periodicity and 1000 Å modulation into this surface [6.204] (Sect. 6.5.7). After tuning to the surface-plasmon polariton resonance (Sect. 6.5.7), the intensity variation of the 990 cm^{-1} line of adsorbed pyridine was measured as a function of coverage [6.247, 249] (Fig. 6.74). A signal was only seen for more than 0.4 monolayers when the compressed phase was observed on smooth Ag(111). Thus, an electromagnetic resonance and physical proximity to the surface is not enough to induce the SERS effect. The short-range effect at higher coverages was discussed in Sect. 6.5.7.

An analogous experiment with benzene rather than pyridine did not yield any signal unit about 8 monolayers of benzene [6.247]. This was attributed to the fact that adsorbed benzene on flat Ag(111) always lies flat on the surface (π-bonded) [6.247] because of the lack of nitrogen lone-pair electrons.

However, SERS experiments with adsorbed pyridine and benzene on cold-deposited silver films yield different results. A SERS spectrum of pyridine on silver films evaporated at 120 K has already been seen for a few percent of a monolayer [6.199]. However, the spectral positions of the vibrations are slightly different (Fig. 6.3) and more vibrational lines from pyridine are observed than on "modulated" Ag(111). Likewise, a rich SERS spectrum of benzene on silver films evaporated at 120 K at submonolayer coverage has been obtained by *Pettenkofer* and *Otto* [6.252]. This corroborated the results of [6.231] (Sect. 6.2, Fig. 6.5). The different results on "modulated" Ag(111) and on cold-deposited silver films must be attributed to different bonding configurations on different surfaces. After an unsuccessful attempt to remove the carbon contamination of a "modulated Ag(111) surface" by sputtering, *Demuth* et al. [6.247] observed a vibrational line at 1005 cm^{-1}, in agreement with the result of Fig. 6.3. Much work is needed to understand these results in detail (Sect. 6.9).

6.6 Indications for the Interaction of Metal Electrons with Adsorbates in SERS

As described in Sect. 6.2, SERS always coexists with an inelastic background [6.14, 26, 29, 253]. In Sect. 6.5.4 we have described its intensity variation during the annealing of a "cold-deposited" silver film and in Sect. 6.5.5 its surprisingly strong intensity in second harmonic generation experiments. This background is intrinsic: it exists without any adsorbate [6.32,45]. It seems to be generally accepted that the background is caused by a continuum of electronic excitations of the metal [6.14,254,241]. However, there is no consensus as to the relationship between this background and SERS.

One way to probe for interactions of an adsorbate with metal electrons would be to vary the metallic surface charge. In the case of SERS at metal electrodes, this charge can be easily modulated by varying the potential of the electrode with respect to the electrolyte. For single-crystal silver electrodes, the differential capacitance of the double layer is of the order of 50–100 μF per nominal cm^2 of surface. Potential variations of 0.1 V lead to changes in the surface charge of about 4 $\mu C/cm^2$ [6.255], equivalent to about 0.05 electron charges per surface atom. Simultaneously, one changes the local electrostatic potential at the site of an adsorbate with respect to the electrostatic potential of the metal. This point will be important in the context of Sect. 6.8.

Relative intensity variations of various pyridine SERS lines as a function of the potential U_{SCE} of the silver electrode were first reported by *Jeanmaire* and *Van Duyne* [6.8] (Fig. 6.82). The intensities followed quasi-instantaneously and reversibly the changes in U_{SCE} (within 1 ms) [6.11]. *Marinyuk et al.* [6.256] observed that the intensity of the background in a 0.1 M KCl with 0.05 M

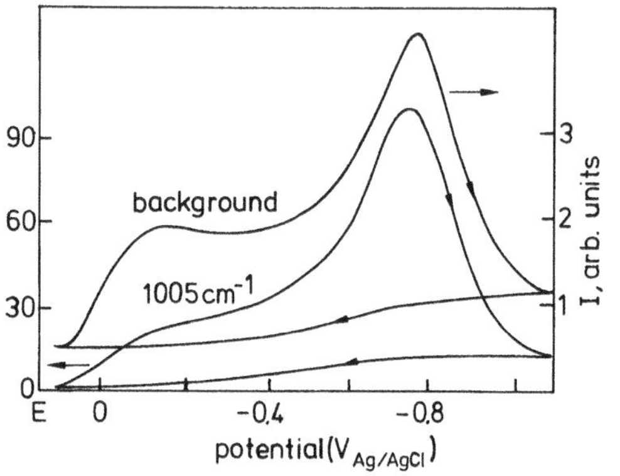

Fig. 6.75. Dependence on potential of the intensity of the line at 1005 cm^{-1} in the SERS spectrum of pyridine on a silver electrode in 0.1 M KCl + 0.05 M pyridine and for the background near this line. Linear scan from about +0.1 to −1.1 and back, scan rate 25 mV/s, $\lambda_L = 632.8$ nm. Potential scale has to be shifted 8 mV to the left in order to convert to U_{SCE}. From [6.256]

pyridine electrolyte closely followed the potential variation of the pyridine SERS line at 1005 cm^{-1} (Fig. 6.75), whereas in an electrolyte, containing less than 5×10^{-4} M pyridine, the potential variation of the background was much weaker. According to these authors both SERS and background intensity are linked to the surface concentration of silver adatoms. The results (Sect. 6.8) show that it is more likely that the variation of the SERS with potential is due to "tuning in and out" of a charge transfer resonance. Only the irreversible loss of intensity for potentials below −0.8 V is probably caused by an irreversible loss of "active sites" (Sect. 6.9). In any case, the results in Fig. 6.71 show a close connection between the processes responsible for SERS and the background.

Burstein et al. [6.127] raised the point of possible Fano-type interferences [6.257] between Raman scattering by the adsorbed molecule and the background continuum. The absence of these interferences would indicate that the inelastic background is due to "incoherent" luminescence. In the context of Raman scattering, Fano line shapes have been observed for the zone-center phonon in heavily doped Si (Chap. 2). In SERS, pronounced asymmetric line shapes for ethylene on cold-deposited silver films were found [6.259]. They were fitted to the Fano profile intensity:

$$I \sim \frac{(q+\varepsilon)^2}{1+\varepsilon^2}, \qquad \varepsilon = (\omega - \omega_A)/\Gamma. \tag{6.58}$$

For the meaning of the parameter ω_A, q, and Γ, see Sect. 2.7 and [2.257, 258].

Ethylene on silver displays, among other features, two strong vibrational lines at about 1324 cm,$^{-1}$ (symmetric in-plane CH bending) and at about 1589 cm^{-1} (C−C stretch). The line shapes of the 1324 cm^{-1} vibrations for the laser excitation wavelength at 514.5 nm and 647.1 nm are given in Fig. 6.76. These line shapes are fitted with (6.58) for $\Gamma = 3$ cm^{-1} and $q = -4.6$ (for 514.5 nm excitation) and $q = -3$ (for 647.1 cm excitation). A noninterfering background compatible with Fano's theory for the case of complex matrix elements (Sect. 2.7) has been used for the fit. The fit is not convincing since the

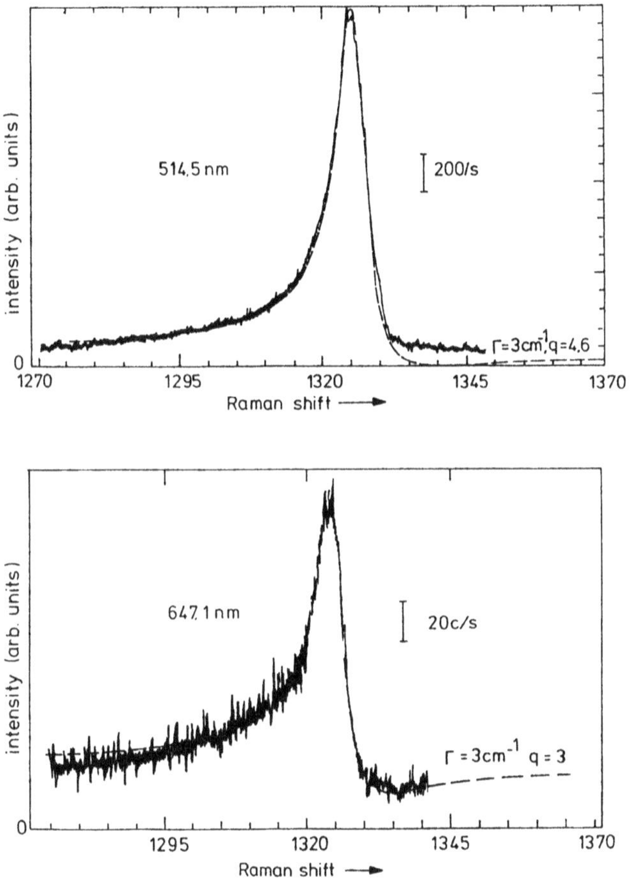

Fig. 6.76. SERS spectra of ethylene (100 L exposure) on a cold-deposited silver film [in the range of the symmetric inplane CH bending vibration, excitation wavelength 514.5 nm (160 mW) and 647.1 nm (190 mW)]. *Dotted line* is the fit to the Fano profile. Note that q changes with excitation wavelength. Accuracy of wave number scale: $\pm 1\ \mathrm{cm}^{-1}$

antiresonance characteristic of the Fano theory does not appear clearly in the experiments. The origin of the asymmetric line shapes could also be translational disorder in an adsorbed film [6.260].

When the cold-deposited film is exposed to less than 100 L of ethylene, the line shapes indeed become even more asymmetric. But if the cold-deposited (at 120 K) silver film is first annealed to 210 K, subsequently cooled to 120 K and exposed to 3 L of ethylene, no asymmetry is apparent [6.259]. In summary, any definite conclusion from these experiments would be premature.

Kötz and *Yeager* [6.261] found that the frequencies of vibrational bands of pyridine, pyrazine, p-nitroso dimethylaniline and cyanide adsorbed on a silver electrode decrease linearly with cathodic potential. This observation is independent of the anions in the electrolyte (e.g., Cl^- or OH^-). For example, the

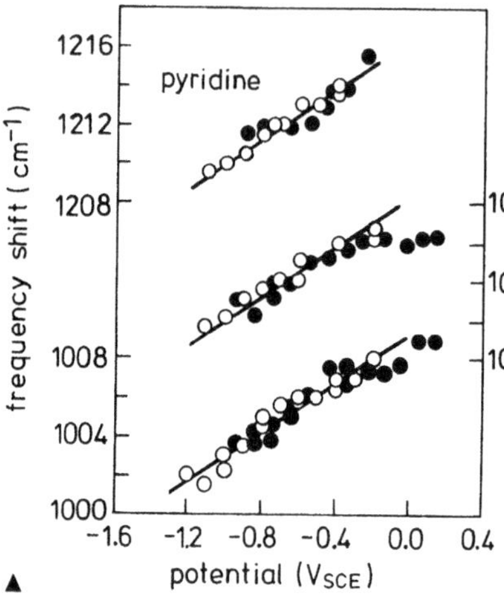

▲ **Fig. 6.77.** Potential dependence of three ring vibration modes of pyridine adsorbed on a polycrystalline Ag-electrode. Electrolytes: 2×10^{-2} M pyridine in 1 M NaOH (○) or 0.1 M KCl (●). Laser wavelength 5145 Å. From [6.261]

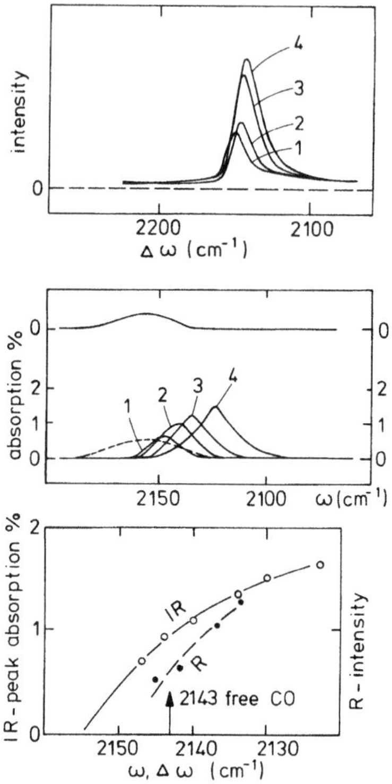

Fig. 6.78. *Top:* pressure dependence of Raman intensity (arbitrary units) from CO on a cold-deposited silver film at 105 K ($\lambda_L = 4880$ Å, 300 mW). Static CO pressure [Torr]: $1:10^{-6}$; $2:10^{-5}$; $3:10^{-4}$; $4:10^{-3}$. From [6.245] *Middle:* infrared absorption spectra of CO adsorbed on silver films at 77 K. *Upper trace:* film deposited at room temperature, CO pressure $\approx 10^{-3}$ Torr. *Lower trace:* film deposited and maintained at 77 K (solid lines), static CO pressure [Torr]: $1:<5\times10^{-6}$; $2:5\times10^{-5}$; $3:2\times10^{-4}$; 4:1,5 Torr and at 10^{-3} Torr after annealing (*broken line*). From [6.263] *Bottom:* ir peak absorption [%] and Raman [R −] intensity [arbitrary units] versus CO stretch frequency (ω, $\Delta\omega$, respectively). Arrow denotes stretch frequency of the free CO molecule. From [6.245, 263]

frequency shifts for 3 pyridine lines are shown in Fig. 6.77. *Kötz*, *Yeager*, and *Anderson* et al. [6.262] proposed as an explanation for the case of adsorbed CN^- a potential dependent charge transfer from the CN antibonding 5σ orbital to the metal.

Wood et al. [6.245] investigated SERS of CO on cold-deposited silver films in UHV at 105 K under a CO pressure between 10^{-6} and 10^{-3} Torr. The line at ≈ 2140 cm^{-1} (CO stretch) shifts to lower frequency with increasing pressure (Fig. 6.78, top). *Chesters* et al. [6.263] observed for the same system a similar shift by infrared reflection-absorption spectroscopy (Fig. 6.78, middle). After annealing the cold silver film to room temperature, the "SERS activity" is lost, as already discussed in Sect. 6.5.4. Likewise, the infrared absorption of CO adsorbed at 77 K on an annelared silver film and on a silver film evaporated at room temperature is comparatively low (Fig. 6.78, middle). The agreement on frequency shift versus signal intensity between Raman and infrared spectroscopy is good (Fig. 6.78, bottom). These results raise the following problems. Provided the surface selection rule of infrared spectroscopy [6.264] is applicable in this case, the ir signal must be caused by CO molecules with their CO axis perpendicular to the local surface. In such a case, dipole-dipole coupling predicts a shift to higher frequency with increasing coverage ([6.265] and references therein) which is indeed often observed experimentally, for instance, for CO on Pd (100) [6.266] and on Pt low index faces ([6.267] and references therein). The fact that a shift is sometimes not observed, as for CO on single-crystal copper surfaces [6.264], is assigned to a counteracting "chemical shift" with increasing coverage [6.264, 268] which, in the case of silver, would be stronger than the dipolar shift. Two-dimensional disorder for incomplete coverage leads to an asymmetric line shape [6.260]. The broader side of the line is expected to be the side opposite the direction of the line shift upon increasing coverage [6.260], which was indeed observed for CO on Pd (100) [6.266]. This is opposite to the asymmetry of the SERS lines of CO on silver (top of Fig. 6.78). *Wood* et al. mention that the asymmetry may be due to a weak vibrational line at 2118 cm^{-1} [6.245] or 2115 cm^{-1} [6.34] observed after pumping-off the gaseous CO. It is surprising that the infrared signal for CO pressure below 5×10^{-6} Torr for the cold-deposited film is stronger than that for the annealed film exposed to 10^{-3} Torr CO. In the author's opinion, the Raman and infrared signals of CO on cold-deposited silver films originate from CO adsorbed at atomic-scale surface defects and also indicate an infrared enhancement mechanism, for instance, due to local-field effects on an atomic scale [6.269] (Sect. 6.9).

6.7 Conclusion on "Classical Enhancement"

Electromagnetic resonances do certainly contribute to the enhancement of the Raman signal of adsorbates on metals. Nevertheless, there are many experimental observations which show that the electromagnetic resonances are not the complete answer to the SERS problem. The most noteworthy will be listed again (Sect. 6.5):

a) The excitation spectra of SERS from colloids are, in some cases (according to the interpretation given by the authors), in disagreement with the classical resonance enhancement. One should note that, apart from the ATR configuration, the classical enhancement for adsorbates on a metal sphere is the only case which has been calculated rigorously.

b) The strong short-range effect, observed under many different conditions, can only be explained on the basis of electromagnetic resonances if one assumes isolated roughness features smaller than 50 Å in size (Sect. 6.5.7). It is questionable whether predictions of classical theories based on the bulk dielectric constants and on models for isolated particles or protrusions are of much use in this case.

c) SERS has been reported on metals of relatively low smooth-surface reflectivity, where classical electromagnetic resonances are of low quality. However, many of these results have not been independently reconfirmed.

d) A "SERS-active" silver or copper surface does not display SERS for every molecular species even when the Raman cross sections of the free molecules are comparable. There are several independent indications suggesting that the physical proximity of pyridine, CO, or benzene to a "SERS-active" silver or copper surface is not enough to induce the SERS effect. Only weak signals have been found for N_2 and saturated hydrocarbons.

Concerning the calculations of "classical enhancement" with values $\approx 10^5$ for silver-island films [6.39], it was shown in Sect. 6.5.3 that a random silver-island film may have much lower "classical enhancement" than anticipated. This may also hold for "cold-deposited" silver films (Sect. 6.5.4). There is as yet no theory which accurately calculates the electromagnetic resonances on surfaces of statistical roughness and the contribution to the Raman enhancement from these resonances in the excitation and emission channel, although some progress has been made [6.270, 271] (Chapt. 7). Likewise, there is as yet no evidence which shows convincingly that the observed experimental enhancement (often larger than 10^5) is of purely electromagnetic origin.

For all these reasons, especially those in (d), the author concludes that SERS is not of purely electromagnetic origin in the sense of electromagnetic resonances induced by supra-atomic surface roughness. The statement of *Burstein* and *Chen* [6.39] that

> "The overall enhancement is the result of a combination of contributions some of which may be quite specific to the particular metal-adsorbed molecule system",

is certainly correct.

As yet, all the mechanisms contributing to the SERS phenomenon beyond "classical enhancement" are poorly understood or not positively identified by experiment. There are further "physical" enhancement models like the electronic "self-energy" of the adsorbate, corresponding to the classical "image dipole" (Chap. 7), or models elaborating on the displacement of metallic electrons at the surface by vibrating molecules (the classical "Raman reflectivity" [6.15] or "potential barrier modulation" [6.63]; see also [6.272–274]). In general, these "physical" enhancement models predict a short-range enhancement at atomically smooth surfaces. However, in some recent experiments on smooth surfaces, such a short-range mechanism was not observed (Sect. 6.9).

"Chemical" enhancement models involve dynamic charge transfer between metal and adsorbate. Because of some recent developments, a discussion of the "chemical effect" and the related concept of "SERS-active" sites seems worthwhile. The restriction of this discussion to the photon-driven charge-transfer resonant Raman effect does not imply that there are no other "nonclassical" enhancement mechanisms. It would be surprising if the "physical enhancement models" mentioned above or local-field effects on an atomic scale [6.269] did not contribute to the enhancement. On the other hand, there is as yet no positive experimental evidence for these mechanisms.

6.8 Charge-Transfer Excitations and SERS

The proposals for a contribution of dynamical charge transfer (CT) between metal and adsorbate to Raman scattering may be roughly divided into two groups. The first group concerns "charge injection into or withdrawal from" [6.176] the metal accompanying by the adsorbate vibration. The second group considers a photon-driven resonant CT [6.15, 273, 276–279] (Chap. 7).

In the first group, *Aussenegg* and *Lippitsch* [6.280] envisaged a modulation of the polarizability during the vibration of the adsorbate because the transferred charge is much more polarizable when it is on the metal side than when it is on the adsorbate side. No frequency dependence of the polarizability was discussed: the process is nonresonant. This process and the concept of the modulation of the metallic reflectivity by the adsorbate [6.14] were connected as a model of Raman scattering from chemisorbed molecules by *McCall* and *Platzman* [6.281]. *Abe* et al. [6.176] discussed a model in which the effective polarizability of small colloids is modulated by a modulation of the electron density, characterized by the plasma frequency ω_p. An adsorbate vibration with normal coordinate Q may change ω_p by changing the effective volume of the colloid characterized by a sphere radius R and by CT:

$$\frac{\partial\omega_p}{\partial Q}=\frac{\partial\omega_p}{\partial R}\,\frac{\partial R}{\partial Q}+\left(\frac{\partial\omega_p}{\partial Q}\right)_{CT}. \tag{6.59}$$

Though not explicitly stated in [6.176], a coherent action of all adsorbed molecules on one sphere is assumed in this model.

Through several new experiments, a picture of the "chemical effect" based on the second group of models seems to be emerging. *Persson's* model [6.277] will be reviewed here. He considered Newns-Anderson resonances of an adsorbate at a silver surface. The electronic density of states of the metal and of the adsorbate, ϱ_a are depicted in Fig. 6.79. The discrete energy level of the lowest unoccupied orbital of the adsorbate is shifted and broadened into a resonance by short time excursions of electrons from the metal into this orbital $|a\rangle$. This is equivalent to a partial filling of this broadened level by electrons (Fig. 6.79). CT involves the excitation of an electron from a metal state below the Fermi energy to about the maximum of ϱ_a. For this model, Persson calculated the ratio of Raman scattering intensity of the adsorbed molecule with respect to the free molecule. The photon-electron interaction is simply given by

$$H_{\text{photon-electron}}=-edE_z\hat{n}_a, \tag{6.60}$$

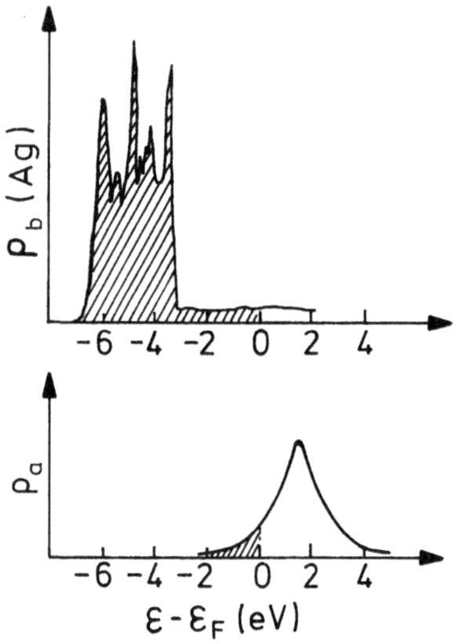

Fig. 6.79. Schematic density of states ϱ_a associated with a molecule adsorbed on a silver surface. ϱ_b(Ag) is the bulk density of states of silver. From [6.277]

where d is the distance of the "center of charge" of the orbital $|a\rangle$ to the metal image plane, E_z the incident field normal to the surface, and $\hat{n}_a$ the operator describing the number of electrons in orbital $|a\rangle$.

The electron-molecule vibration interaction is given by

$$H_{\text{electron-vibr}} = \varepsilon_a'(0) Q \hat{n}_a, \tag{6.61}$$

where $\varepsilon_a'(0)$ stems from the expansion of the energy ε_a of the orbital $|a\rangle$ of the free molecule with respect to the normal coordinate Q of the vibration:

$$\varepsilon_a(Q) = \tilde{\varepsilon}_a + \varepsilon_a'(0) Q + \dots . \tag{6.62}$$

Note that for $Q=0$, the energy of the neutral molecule, not of the "negative ion state", is at a minimum [6.282]. The "chemical enhancement" ratio G_{chem} thus becomes:

$$G_{\text{chem}} = |(ed)^2 \varepsilon_a'(0)\, G(\omega_L, \omega)/\alpha'(0)/2|^2 . \tag{6.63}$$

$\alpha'(0)$ is the derivative of the electronic polarizability of the free molecule with respect to Q which governs the ordinary nonresonant Raman scattering, ω_L is the laser frequency, ω the Stokes frequency and $G(\omega_L, \omega)$ contains the resonant Raman scattering caused by the charge-transfer excitations [note that $d^2 G(\omega_L, \omega)$ goes to zero for $d \to \infty$].

For a Lorentzian $\varrho_a(\varepsilon)$:

$$\varrho_a(\varepsilon) = \frac{\Gamma/\pi}{\pi(\varepsilon - \tilde{\varepsilon}_a)^2 + \Gamma^2} \tag{6.64}$$

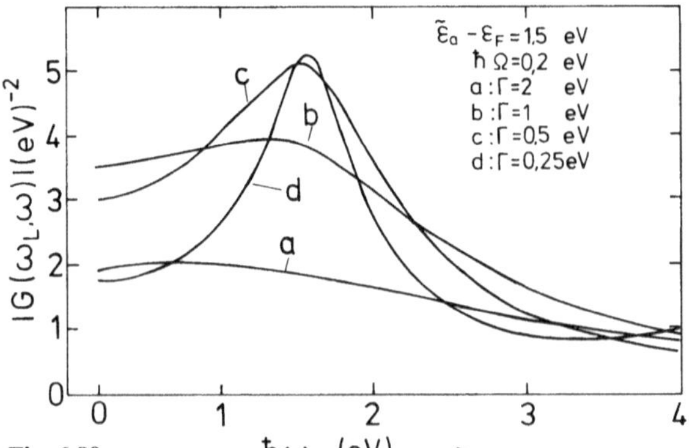

Fig. 6.80

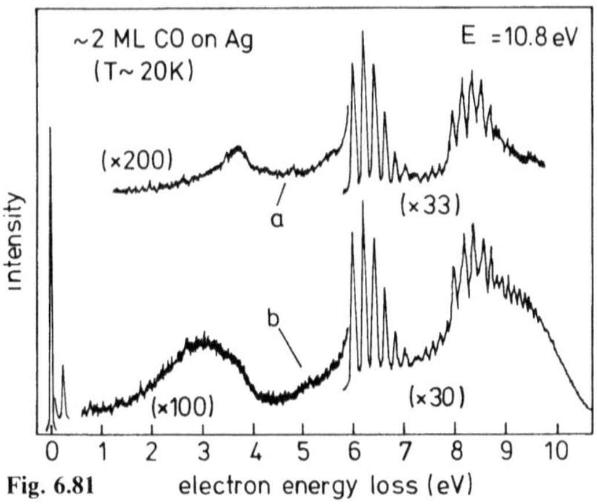

Fig. 6.81

Fig. 6.80. $\left|G\left(\omega_L, \omega_L - \frac{0.2\,\text{eV}}{\hbar}\right)\right|$ (see text) as a function of the energy $\hbar\omega_L$ of the incident photon. $\tilde{\varepsilon}_a$ is the energy where $\varrho_a(\varepsilon)$ has its maximum. Γ is the halfwidth of the Lorentzian $\varrho_a(\varepsilon)$. From [6.277]

Fig. 6.81. Electron energy loss spectra for 2 monolayers of CO at 20 K on Ag films prepared by deposition at room temperature (a) and at 20 K (b). Initial electron energy 10.8 eV. From [6.251]

and for $\hbar(\omega_L - \omega) = 0.2$ eV, G is given as function of ω_L for $\tilde{\varepsilon}_a - \varepsilon_F = 1.5$ eV and various parameters Γ in Fig. 6.80. For $\hbar\omega_L = 2.4$ eV, $\hbar(\omega_L - \omega) = 0.12$ eV, $2\Gamma = 0.6$ eV, $\varepsilon_a'(0) = 10$ eV/Å and $\alpha'(0) \approx d^2 \approx 1$ Å^2, Persson obtained

$$\eta_{\text{chem}} \approx 30.$$

The first observation of charge-transfer excitations from silver adsorbates on silver was made by Demuth and collaborators. Whereas on "smooth" Ag(111) only relatively weak charge-transfer structures were found in the electron energy loss spectra of adsorbed pyridine and pyrazine [6.122, 250], surprisingly strong structures did appear for cold-deposited films [6.251]. Figure 6.81 displays the electron energy loss spectra for 2 monolayers of CO adsorbed at 20 K on differently prepared silver films. In one case the silver film was vapour-deposited at room temperature and subsequently cooled to 20 K; in the other case it was vapour-deposited onto a substrate of 20 K and kept at this temperature. The electronic $5\sigma - 2\pi^*$ excitations of the adsorbate at 6 and 8 eV do not depend on the film preparation. However, striking differences were observed at lower energies. For CO on the room-temperature film, only a peak at 3.7 eV was seen (ascribed to a surface plasmon [6.251]), whereas the low-temperature film displayed a broad band between 1.8 and 4 eV with a maximum at about 3 eV. This band was not observed on the clean low-temperature film. Similar excitations were observed for O_2 at 3.5 eV, for C_2H_4 at 2.8 eV, and for pyridine at 2.5 eV, but only for the silver film evaporated at 20 K. For adsorbed N_2 no such energy loss band was observed, neither on the room-temperature nor on the low-temperature deposited film. This absence rules out a significant contribution of surface plasmons to the intensity of the observed bands. Therefore, these losses are assigned to charge-transfer excitations. According to *Schmeisser* et al. [6.251], these excitations must be localized at sites of "microscopic" roughness (atoms, clusters of atoms or other defects [6.251] termed "atomic scale roughness" in [6.34]) for 2 reasons:

a) For CO and C_2H_4, the results can be compared to optical absorption measurements on Ag atoms isolated in CO and C_2H_4 matrices, where additional absorption bands at 3.1 eV (CO) and at 2.25 (and 4.16 eV) (C_2H_4) were observed [6.283, 284]. The low-energy excitations in both systems were assigned from SCF-Xα-SW calculations [6.284] to 5s(Ag)$\rightarrow 2\pi^*$ (CO, C_2H_4) charge-transfer excitations.

b) The charge-transfer excitation losses are only observed for the cold-deposited Ag films. Apparently only those have enough sites of "microscopic" or "atomic" scale roughness. This suggests that the irreversible loss of the "SERS activity" of a cold-deposited film by warming up to room temperature is caused by annealing of atomic-scale surface roughness [6.32, 34, 147]. Although the reason for the absence of the charge-transfer transition for N_2 is unknown, we note that both *Moskovits* and *DiLella* [6.58] and *Schulze* and coworkers [6.59] found SERS of nitrogen on silver about 100 times weaker than SERS of CO on silver while the Raman cross sections of free CO and N_2 were comparable. This indicates that the CT excitations at sites of atomic scale roughness do contribute about two orders of magnitude to the Raman enhancement.

Other unassigned electronic excitations which may be relevant to SERS have been found by *Krasser* et al. [6.285]. Matrix-isolated silver clusters in solid argon showed a very strong luminescence peaking at about 2.8 eV. When pyridine was co-condensed with Ag in the argon matrix, this luminescence was absent but a new relatively sharp luminescence band near 2.05 eV was observed. The contribution of electronic excitations to the Raman cross-section of the adsorbates should be observable as a resonance in the Raman excitation spectra (see, e.g., Fig. 6.6). However, also electromagnetic resonances will contribute to the excitation spectra. This makes the assignment of resonances in the excitation spectra of adsorbates on cold-deposited silver films [6.211, 286, 287] to electronic surface excitations or to electromagnetic resonances difficult. One point is particularly important: the excitation spectra of pyridine [6.211, 286] and O_2,

C_2H_4 and CO [6.287] on cold-deposited Ag films display resonances with maxima between 2.1 and 2.3 eV (comparable to those in Fig. 6.6) which do not agree with the charge-transfer excitation energies reported by *Schmeisser* et al. [6.259]. Because of this, *Pockrand* [6.211, 287] has assigned the resonances in the excitation spectra to electromagnetic resonances. In the context of the short-range effect discussed in Sect. 6.5.7, Pockrand investigated the coverage dependence of the Raman signal. The excitation resonance shifted to lower photon energies with coverage; this shift was nearly complete after a coverage of 3 monolayers of pyridine. He assigned this effect to the shift of an electromagnetic resonance and determined the size of the relevant roughness features to be only 10–20 Å [6.211]. Accordingly, the decrease of the hypothetical "classical enhancement" would be of the order of 10–100 for the second layer of adsorbed molecules [6.211], in agreement with the experiment [6.199] (about the same distance dependence as observed by *Sanda* et al. [6.204] on a modulated Ag(111) surface; see Fig. 6.61).

The extremely small size of the relevant roughness features raises all the questions asked at the end of Sect. 6.5.7. It cannot be excluded that the resonances observed in the excitation spectra on cold-deposited silver films [6.211, 286, 287] are partly due to charge-transfer excitations whose energy is shifted by changes of the surface potential of the cold-deposited films (see below). They could also be related to the electronic state observed by *Krasser* et al. [6.285]. *Moskovits* and *DiLella* [6.58] reported very different excitation spectra of the CO-stretching SERS band of CO adsorbed on cold-deposited silver films (maximum below 1.9 eV) and on silver colloidal particles (above 2.5 eV). This was presented [6.58] as a strong argument against "atomic-scale roughness" and it would also serve as an argument against CT–SERS mechanisms. On the other hand, it may simply indicate different ratios between the electromagnetic enhancement and the "chemical effect at sites of atomic-scale roughness" for differently prepared samples (Sect. 6.5.6). *Pettinger* et al. [6.43] and *Girlando* et al. [6.288] measured the excitation profile of a 1005–1008 cm^{-1} band of pyridine on very weakly activated silver electrodes in Cl^- electrolytes (redeposited silver corresponding to a thickness of 8 monolayers in the case of [6.43]). The profile increased from 2.8 to 1.9 eV, reaching a maximum near 1.6 eV [6.288]. *Pettinger* et al. [6.43] also measured the relative change ΔR of the reflectance R where ΔR is R(smooth) – R(activated). The frequency dependence of $\Delta R/R$ matches approximately the excitation spectrum discussed above [6.288]. The $\Delta R/R$ spectrum and the SERS excitation spectrum were assigned to electronic absorption and to the corresponding resonance Raman effect in a silver-pyridine-halide complex which is formed during the activation cycle [6.43]. *Marinyuk* et al. [6.289] determined an excitation spectrum of pyridine and tetraamylammonium on silver, copper and gold electrodes from the intensity ratio of Stokes and anti-Stokes lines. The energy of the maxima of the excitation spectra (1.92 eV on silver, 1.55 eV on gold and 1.77 eV on copper) did not depend on the electrode potential. These energies were assigned to electron-transfer excitations from metal adatoms that form a complex with the adsorbate molecule to the Fermi level of the metal. More work is needed to understand SERS excitation spectra (see, for instance, the proposals in [6.290]).

In this respect, SERS on silver electrode surfaces offers an advantage to experiments on silver surfaces in vacuum. As discussed in Sect. 6.6, the potential step U between electrode and electrolyte is variable by potentiostatic control. In this way, the energy of electronic adsorbate states and of surface states may be shifted with respect to bulk metal states. In other words, the energy scales for the density of electronic states of the metal and of the adsorbate in Fig. 6.79 may be shifted with respect to each other. Hence, the charge-transfer excitation energy will vary with the difference in the potentials of the electrode and the reference electrode. On the other hand, supra-atomic roughness and hence electromagnetic resonances will be little influenced by U [6.291].

One may assume that the overall Raman enhancement G is given by

$$G = G_{\mathrm{EM}}(\omega_{\mathrm{L}}, \omega) \cdot G_{\mathrm{chem}}(\omega_{\mathrm{L}}, \omega, U) \cdot G_{?}(\omega_{\mathrm{L}}, \omega) \cdot \theta(U). \tag{6.65}$$

G_{EM} is the SERS enhancement by electromagnetic resonances in supra-atomic roughness discussed at great length in this article, G_{chem} is the resonance enhancement by charge-transfer excitations, $G_{?}(\omega_{\mathrm{L}}, \omega)$ may account for any additional enhancement mechanisms and $\theta(U)$ contains the surface coverage and possible adsorbate and atomic-scale rearrangement effects. For $\omega_{\mathrm{L}} = \mathrm{const}$ (and hence for a given adsorbate vibration Ω, also $\omega = \omega_{\mathrm{L}} - \Omega$, is constant, the maximum of G vs U is given by the maximum of $G_{\mathrm{chem}}(\omega_{\mathrm{L}}, \omega, U)$ vs U (provided that $d\theta/dU = 0$!), because in this case $\partial G/\partial U = \partial G_{\mathrm{chem}}/\partial U = 0$. *Jeanmaire* and *Van Duyne* [6.8] have indeed observed maxima of G with respect to U (similar to Fig. 6.82), albeit maxima at different U for different vibrations. In the sense of our discussion, these maxima may reflect a coverage or configuration dependence [maxima in $\theta(U)$] or they may reflect the maxima of G_{chem} by shifting the negative-ion resonance into the right position so as to have the optimum resonance Raman effect.

The decision between these two possibilities comes from an experiment of *Billmann* and *Otto* [6.294]. The intensity of several vibrational bands of pyridine on an "activated" silver electrode is given as function of U for two laser wavelengths λ_{L} in Fig. 6.82. There is a shift in the maxima which cannot be assigned to changes in $\theta(U)$ (however, irreversible changes of θ do occur for $U_{\mathrm{SCE}} < -0.8$ V [6.256, 294], see Fig. 6.75, probably due to a loss of "active sites", see Sect. 6.9. For the 1006 cm^{-1} line, results similar to those in Fig. 6.82 were recently obtained by *Furtak* and *Macomber* [6.299] when scanning U in a positive direction. This proves that the main trends in Fig. 6.82 are unaffected by irreversible electrode surface-reconstruction. Therefore, the observed dependence on ω_{L} of the values U of U_{SCE}, where the intensity reaches a maximum, is attributed to the implicit dependence of U on ω_{L} contained in

$$\frac{\partial G_{\mathrm{chem}}(\omega_{\mathrm{L}}, \omega_{\mathrm{L}} - \Omega, U_{\mathrm{SCE}})}{\partial U_{\mathrm{SCE}}} = 0. \tag{6.66}$$

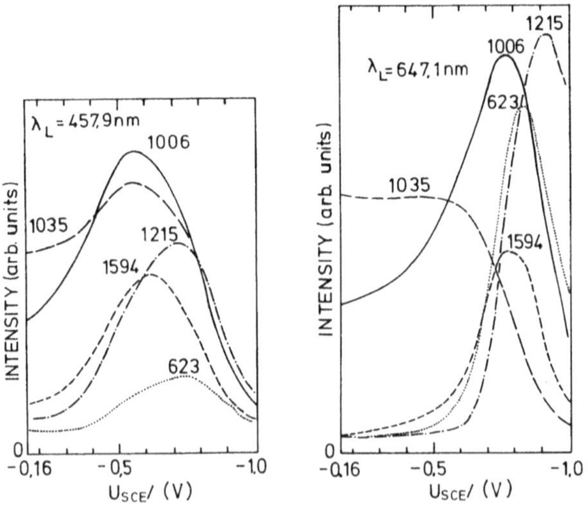

Fig. 6.82. Intensity of adsorbed pyridine vibrational bands at 623 cm^{-1}, 1006 cm^{-1}, 1035 cm^{-1}, 1215 cm^{-1} and 1594 cm^{-1} versus potential of the polycrystalline activated silver electrodes in 0.1 M KCl, 0.05 M pyridine electrolyte measured with respect to a saturated calomel electrode (SCE). For laser wavelength $\lambda_L = 457.9$ and 647.1 nm, U_{SCE} was scanned from -0.16 to -1.0 V with 50 mV/s. From [6.294]

This dependence is just the energy shift of the charge-transfer resonance with respect to the bulk metal states.

The observed $\hbar\omega_L$-U dependence for 4 vibrational bands of pyridine on silver electrodes is given in Fig. 6.83 (for a discussion of the 1035 cm^{-1} band see [6.294]). This dependence most probably reflects the energy shift of the charge-transfer resonance [6.296]. Note that the direction of the shift rules out the transfer of an electron from the adsorbate-to-bulk metal states or from a silver adatom to empty bulk silver states [6.289].

Going back to Fig. 6.82, the intensity variation with U_{SCE} is assigned to "tuning in" or "tuning out" of the resonance. From this interpretation one may infer a chemical enhancement by dynamical charge transfer of 1 to 2 orders of magnitude [6.294], in line with the theoretical estimates of *Persson* [6.277]. Now one may also try an explanation of the breakdown of Raman selection rules described in Sect. 6.2 (see also Figs 6.4, 5) based on the comparison with electron energy loss spectra at shape resonances.

The vibrational excitation of free benzene molecules (C_6H_6 and C_6D_6) by electron impact was studied by *Wong* and *Schulz* [6.300]. Figure 6.84 shows the electron energy loss spectrum in the lowest shape resonance (at 1.8 eV energy of the incident electrons) which consists of the neutral molecule plus the electron in the e_{2u} orbital in the ground vibrational level. Only the vibrations ν_{16}, ν_1 and ν_8 (Wilson notation, see Table 6.2) are strongly excited. Apart from the presence of overtones, there is a striking similarity to the SERS spectrum of benzene on silver

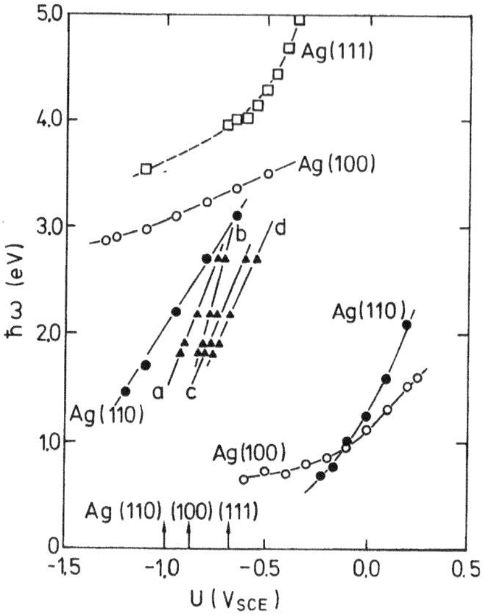

Fig. 6.83. Peak positions U (*full triangles*) of the scans intensity versus voltage U_{SCE} (Fig. 6.82) as a function of the exciting laser photon energy $\hbar\omega_L$ for the pyridine vibrational bands near 623 cm^{-1} (b), 1006 cm^{-1} (d), 1215 cm^{-1} (a) and 1594 cm^{-1} (c) for a 0.1 M KCl, 0.05 M pyridine electrolyte. All other signs denote transition energies for excitation into surface bands observed in electroreflectance in a 0.5 M NaF electrolyte [6.295]. Arrows at the bottom line mark the "potentials of zero charge" for the different silver surfaces [6.295]. From [6.294]

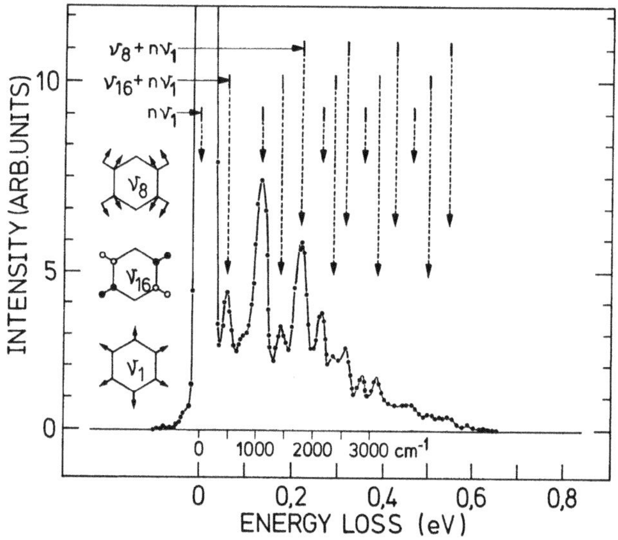

Fig. 6.84. Energy loss spectrum of electrons crossing a C_6D_6 molecular beam; incident electron energy 1.18 eV, scattering angle of electron 70°. Only the modes ν_1, ν_8 and ν_{16} (Wilson's notation) become appreciably excited. The geometry of these modes is indicated (only one each of the doubly-degenerate modes ν_8 and ν_{16} are shown). From [6.300]

(Fig. 6.5), given the limited resolution (≈ 130 cm^{-1}) of the electron energy loss spectrum.

There is also a very good correspondence between SERS spectra of ethylene on cold-deposited silver [6.301, 302] and energy loss spectra of ethylene in the shape resonance at 1.95 eV electron energy [6.303]. In both spectra the ν_2, ν_3 and

ν_7 modes of a_g, a_g and b_{3u} symmetry are strong and have comparable intensity ratios, whereas the C–H stretch modes are very weak, if at all observable. In the case of the energy loss spectra, this is caused by the low coupling [corresponding to ε_a' in (6.61)] of the electron in the ethylene b_{2g} orbital to the C–H stretch modes [6.303]. One has difficulties in understanding the mode selectivity in the SERS spectra of benzene and ethylene with the models proposed so far.

The model of *Sass* et al. [6.53], in which vibrational modes are "Raman-activated" by electric field gradients, can explain the benzene SERS spectrum [6.51] but not the fact that the signal of the C–H stretch mode in the ethylene spectrum is so low. According to *Adrian* [6.279], the main resonant Raman scattering by charge transfer is of a Frank-Condon type involving totally symmetric vibrational modes. (This corresponds to (6.61), because $\varepsilon_a'(0)$ is zero for non totally-symmetric modes, provided the electronic ground state of the molecule is not degenerate). The resonance for non totally-symmetric modes through Herzberg-Teller mechanisms is much weaker [6.279]. With the model of *Persson* [6.277] and *Adrian* [6.279] (which corresponds to the process depicted in Fig. 6.85a), one can understand why the C–H stretch modes of ethylene are very weak in SERS, but one would expect weak intensity for the non totally-symmetric modes of ethylene and benzene, in contrast to the experimental result. Another proposal is a reduction of the ground state symmetry of the molecule by adsorption, by which modes may become Raman active [6.51, 53]. This cannot explain the low intensity of the ethylene C–H stretch modes either.

However, if the electron-photon coupling (steps 1 and 4 in Fig. 6.85b) is separated from the CT process (steps 2 and 3 in Fig. 6.85b), as proposed by *Burstein* et al. [6.15], mechanism c, and in [6.22], the vibrational selection rules are the same as those for inelastic electron scattering [6.15]. This is easily

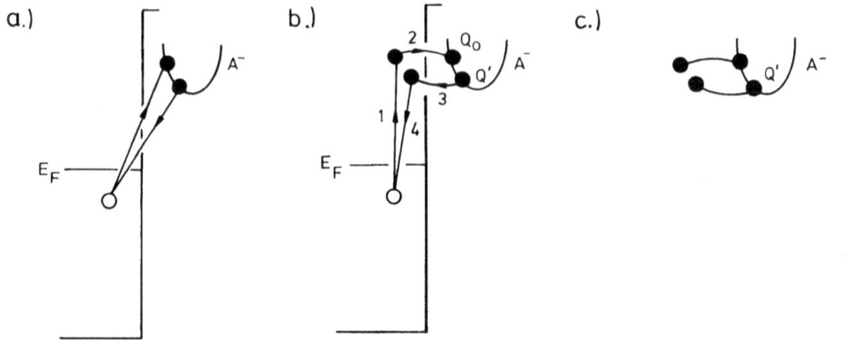

Fig. 6.85. (**a**) Resonant Raman charge-transfer process discussed in [6.277, 279]. (**b**) Resonant Raman CT process discussed in [6.15]. Steps 1 and 4 correspond to the radiative excitation and recombination, respectively, of an e–h pair. Steps 2 and 3 correspond to the hopping of the excited electron onto and back from the "virtual bound state" A^-. An energy versus configuration curve for A^- and a total symmetric mode is shown. Q_0 is the normal coordinate Q at the potential minimum of the neutral adsorbate A; Q' is the value of Q when the electron is transferring back to the metal. (**c**) Shape resonance as discussed in [6.300]

understood by comparing the Raman process in Fig. 6.85b with the shape resonance process in Fig. 6.85c. The selection rules in the latter case have been discussed by *Wong* and *Schulz* [6.300]. The representation of the allowed vibrations are given by the direct product of the representation of the intermediate molecular state and the scattered partial wave (in analogy to elastic scattering, it is usually assumed to belong to the same representation as the incident partial wave, which has again the same representation as the intermediate molecular state). Additionally, one has to consider constraints given by the nature of the interatomic bonds changed by the filling of the unoccupied molecular orbital [6.300, 303]. For degenerate molecular orbitals (e.g., the e_{2u} benzene orbital), non totally-symmetric modes become allowed [6.300]. Contributions to other scattered partial waves also allow coupling to non totally-symmetric modes [6.300, 303]. Because the Raman scattering process in Fig. 6.85b involves the continuum of free-electron states above E_F, the same partial waves are available as in free space. It has been pointed out in [6.22] that the electron-photon coupling (steps 1 and 4 in Fig. 6.85b) is stronger than normal at sites of atomic scale roughness (Sect. 6.9). The Raman scattering process (Fig. 6.85b) will be resonant when the photon energy equals the energy difference between the initial electron state and the intermediate molecular state (A^- in Fig. 6.85b).

The strength of the resonance will be stronger when the initial states are confined to a narrow energy range, see Sect. 6.9. One must not expect a coincidence of the photon energy for maximum charge transfer and the electron energy in the shape resonance of the free molecule. For an adsorbate the charge-transfer energy will depend on the ground state charge transfer between the metal and the adsorbate, on the work function of the metal and the ionization energy of the adsorbate, on the energy of the metal states from which the electron starts and on relaxation effects due to the hole on the metal side after charge transfer. As is evident from Fig. 6.80, the lifetime of the charge-transfer state determines the enhancement. Hence, subtle changes in the adsorption configuration of weakly adsorbed molecules, also the presence of neighbouring molecules could lead to drastic changes in the enhancement. The author ascribes the absence of overtones in the benzene SERS spectra (Fig. 6.5) compared to the EELS spectra (Fig. 6.84) to the shorter lifetime of the e_{2u} orbital of benzene in the adsorbed state. One may now understand why the careful search for a pyridine-substrate vibration on cold-deposited silver films was unsuccessful [6.147]. This vibration is unlikely to couple to the π^* orbitals concentrated on the carbon skeleton of pyridine in case of a bond between the nitrogen atom in pyridine and the silver surface: according to [6.247, 249], SERS is only observed for N-bonded pyridine (Sect. 6.5.9). On the other hand, it is difficult to understand the absence of a pyridine-substrate SERS signal within the concept of the first group of CT models [6.176, 280, 281].

The comparison of the ordinary and surface-enhanced Raman scattering by the C–H stretch modes of ethylene allows for an estimate of the lower limit of required "chemical enhancement" on cold-deposited silver films. In ordinary

Raman scattering, the intensity ratio of the ν_5-B_{1g} C–H stretch vibration ($\sim$3000 cm^{-1}) to the ν_3-Ag-CH_2 scissor mode ($\sim$1350 cm^{-1}) is about 0.3–0.4 [6.304]; in SERS from cold-deposited silver smaller than $\approx$1:150 [6.301, 302]. If the C–H stretch vibration only experiences classical enhancement, then the "chemical enhancement" for the ν_3-mode must be $\approx$50, in agreement with the estimates above. For an overall enhancement of about 10^4–10^5 from cold-deposited silver films (Sect. 6.2), this yields an upper limit of classical enhancement of 3 orders of magnitude, in agreement with the conclusion of Sect. 6.5.4. The breakdown of Raman selection rules has also been observed by *Lund* et al. [6.60] for benzene on vapour-deposited *sodium* surfaces at 15 K, indicating the presence of a CT mechanism also in this case.

The discussion in this section was only concerned with the photon-driven transfer of an electron from a metal state to an unoccupied molecular orbital of the adsorbate. In principle, Raman enhancement and the relaxation of selection rules is also possible by a photon-driven transfer of an electron from an occupied molecular orbital to a metal state above the Fermi level.

6.9 Evidence for "SERS-Active Sites"

As outlined in Sect. 6.2, the Raman scattering from adsorbates at sites of atomic scale roughness (ASR) maybe stronger than for an adsorbate on "atomically smooth" surface terraces [6.26], due to a charge-transfer mechanism [6.34].

Sites of atomic scale roughness for an elemental single crystalline fcc lattice may be defined as follows (changes in lattice parameters by surface reconstruction are neglected): surface atoms are those atoms whose coordination number (number of nearest-neighbour atoms), less than 12, may be increased by adding further atoms to the crystal from outside without changing the position of atoms already present. Sites of ASR are given by those surface atoms whose geometrical configuration with first-neighbour surface atoms is different from surface atoms at (111) (100) and (110) surfaces. For a nonperfect lattice dislocations and grain boundaries ending at the surface, stacking fault boundaries in the surface layer provide sites of ASR. Further sites of ASR are related to special relaxations of lattice parameters by surface reconstruction. Supra-atomic roughness is defined by structure in the surface roughness correlation function $S(\boldsymbol{x})$ for $|\boldsymbol{x}|$ greater than or equal to twice the lattice constant ($\boldsymbol{x}$ is a distance vector parallel to the average surface). There is no supra-atomic roughness without atomic scale roughness [6.42].

The results of *Schmeisser* et al. [6.251], supporting the notion of CT mechanisms at sites of ASR have already been discussed in Sect. 6.8. As further support of this suggestion, some experimental work will be reviewed here which was obtained after the articles [6.34,65] were written. As already outlined in Sect. 6.2, the Raman scattering results of *Udagawa* et al. [6.44] for pyridine on silver (100) (Fig. 6.86) at ca. 150 K left the question open as to whether the

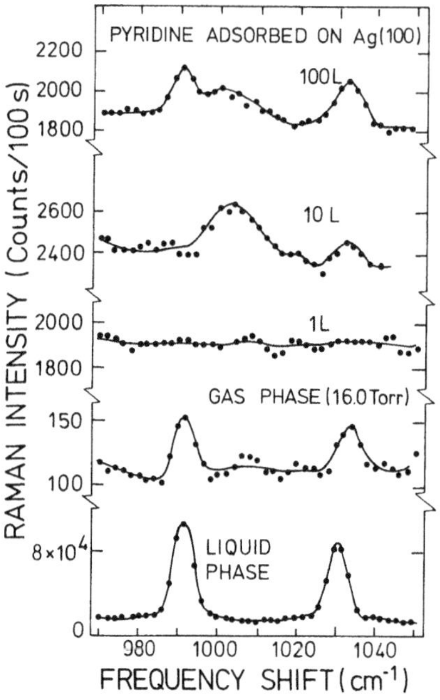

Fig. 6.86. Raman spectra of pyridine molecules on a "smooth" Ag(100) surface for three exposures, and in the gaseous and liquid phases. Incident laser power at 5145 Å is 100 mW, resolution 5.5 cm^{-1}. 5L corresponds to one monolayer. After [6.44]

pyridine vibration at about 1003–1008 cm^{-1} ("line E") was due to the first layer of adsorbed pyiridine, the vibration at about 990–995 cm^{-1} ("line N") due to physisorbed pyridine in subsequent layers, or whether the E-line reflected pyridine adsorbed at residual sites of ASR and the N-line pyridine adsorbed at atomically smooth parts of the surface and in subsequent layers. One should note that (a) SERS spectra from electrodes and cold films at low pyridine coverage feature only the E-line (Figs. 6.1, 3), (b) for this line a strong short-range effect is always found (Sect. 6.5.7), and (c) an electrode "activation" consisting of the redeposition of less than one monolayer of silver is enough to make the E-line appear [6.43]. *Campion* reported unenhanced Raman scattering from a monolayer of pyridine adsorbed on silver (111) [6.305] (Fig. 6.87) and silver (100) [6.306]. The results are those expected from classical metal optics and Raman scattering of the free molecule:

a) The relative intensity of the two totally symmetric modes of the free molecule (at 992 and 1030 cm^{-1}) and of the other vibrations near 1220, 1580 and 3050 cm^{-1} is similar to the case of liquid pyridine. This is in contrast to SERS.
b) The intensity of the Raman bands is linear in exposure from 0.5 L to 25 L. There is no short-range enhancement like in SERS (Sect. 6.5.7).
c) The Raman intensity is about 10 times stronger for p-polarized oblique-incident light than for s-polarized light. The Raman scattered signal is polarized, in contrast to the SERS signal.

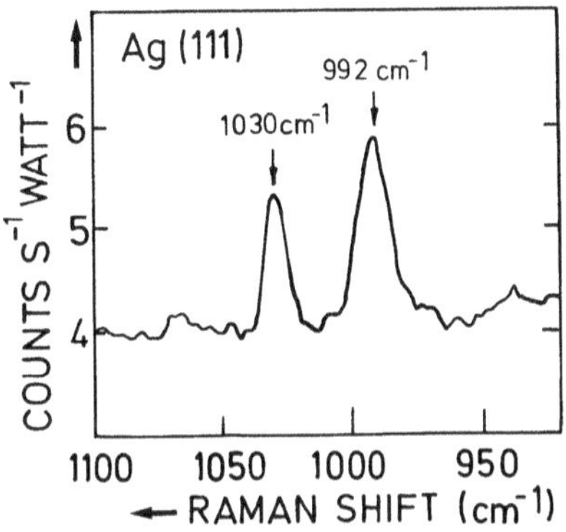

Fig. 6.87. Raman spectrum of pyridine on Ag(111) after 5L exposure at 110 K. The count rate shown includes background. From [6.305]

Campion compared these results with the data of *Sanda* et al. [6.247, 249] on "modulated" Ag(111), which yielded a short-range effect (Fig. 6.74). He suggested that the latter result may be due to chemisorption to "sites that are not available on the (111) surface" (steps were suggested in [6.22]).

It has been argued that the difference between the spectra of [6.44, 305] might be due to different substrate temperatures during pyridine exposure (150 and 110 K, respectively). Pyridine would overcome a potential barrier only above 110 K to settle into a state with the E-vibration. This is unlikely, as the E-line is already observed at 18 K [6.243] and at 120 K (Fig. 6.3). Another explanation could be a higher concentration of sites of residual roughness on the sample used by *Udagawa* et al. [6.44] compared to the sample used by *Campion* [6.305]. This explanation becomes more likely by considering Fig. 6.88 reproduced from *Tsang* et al. [6.307]. These authors prepared a silver grating by evaporating 400 Å of silver onto a oxidized silicon wafer with corrugated grating profile, kept at 300 K (remember that a silver film evaporated on a smooth substrate at 300 K is "SERS inactive"). Subsequently, the silver grating was cooled to 100 K and exposed to pyridine. The intensity of the N-line when tuning into the

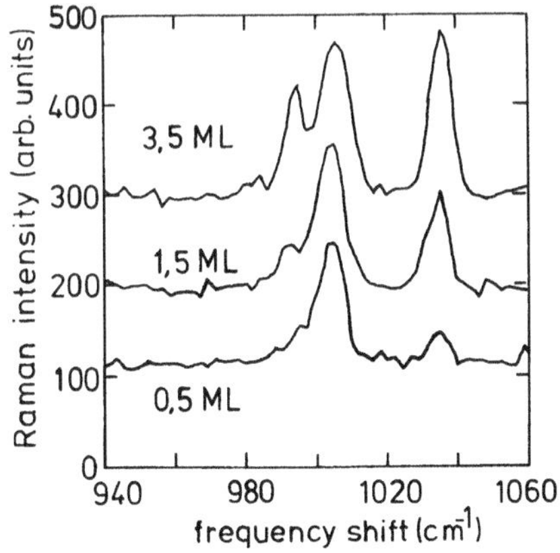

Fig. 6.88. Raman spectrum of three different coverages of pyridine adsorbed onto 20 Å of Ag deposited onto an Ag grating held at 100 K during the measurement. The Ag-grating was deposited at 300 K. From [6.307]

surface plasmon-polariton resonance grew approximately linearly with exposure, but was not yet discernible for 1 monolayer. No E-line was observed. This behaviour probably reflects "classical" enhancement without the short-range "nonclassical effect". When 10 or 20 Å of silver were additionally evaporated on top of the silver grating described above, after cooling it to 100 K (cold "SERS-active" silver), the E-line was obtained for a submonolayer coverage of pyridine. This submonolayer signal is more enhanced than the Raman scattering from subsequent monolayers. The change in frequency from the N-line to the E-line indicates that new bonding sites for pyridine become available after deposition of cold silver.

How large are these "SERS-active sites"? As second-nearest-neighbour metal atoms already have little influence on the frequency of an adsorbate bound to metal atoms [6.26] they could be rather small, possibly of atomic scale. On silver electrode surfaces, *Pettinger* and *Moerl* [6.308] demonstrated that active sites are stabilized by Cu coverages of only 0.01 monolayers. This demonstrates the rather low concentration of active sites and the atomic extension of these sites. The absence of a short-range effect on the silver film, evaporated onto a grating at room temperature [6.307], and the presence of a short-range effect on a grating inscribed on a (111) single crystal surface (Figs. 6.61, 74) is noteworthy. In the author's opinion, this different behaviour can only be understood if one assumes, contrary to [6.249], that the step-site atoms on the modulated Ag(111) surface (about 10 % of the surface atoms) are special "active sites", as proposed in [6.23, 309].

Tevault and *Smardzweski* [6.310] reported on an interesting relationship between infrared spectra of pyridine plus silver-atom (and silver cluster) complexes and SERS. The complexes were produced by Ar matrix isolation at 15 K. In the absence of any silver, the pyridine breathing vibration was found at 992 cm^{-1} (Fig. 6.89, top). When a low concentration of Ag atoms was

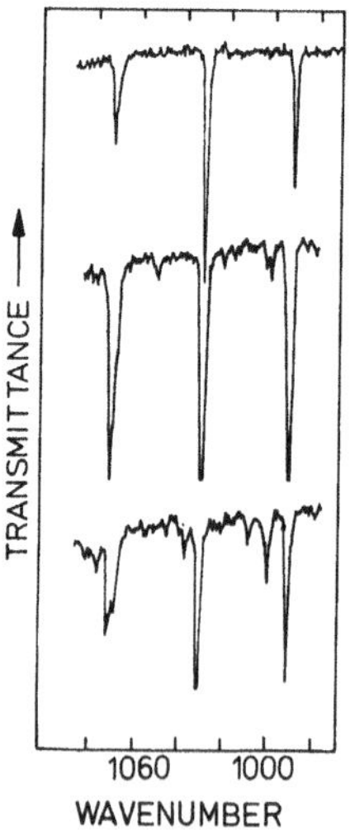

Fig. 6.89. Infrared spectra observed in the 980–1090 cm^{-1} region following deposition of Ar/pyridine = 300:1 samples. *Top:* blank. *Middle:* with a relatively low concentration of Ag atoms (pyridine/Ag ≈ 10:1). *Bottom:* with a higher Ag atom concentration (pyridine/Ag ≈ 3:1). From [6.309]

Table 6.5. Wave numbers of selected pyridine vibrations under various conditions

SERS cold silver	SERS electrode −0.6 V	"Smooth" silver no short range effect	Pyr.-Ag complexes ir-absorption	Ar-matrix ir-absorption	Liquid
623[a]	623[c]		614, 623[g]	602[b]	601[c]
624[b]	623[d]				
1006[j]	1008[c]	992[e]	1000, 1010[g]	992[b]	991[c]
1006[b]	1006[d]	991[f]			
1215[i]	1215[c]		1212[g]	1221[b]	1215[c]
1218[b]	1215[d]				
1596[h]	1594[c]		1592[g]	1599[b]	1581[c]
1595[b]	1594[d]				

[a]: [6.147] [b]: [6.311] [c]: [6.8] [d]: [6.49] [e]: [6.305]
[f]: [6.307] [g]: [6.310] [h]: [6.286] [i]: [6.167] [j]: [6.199]

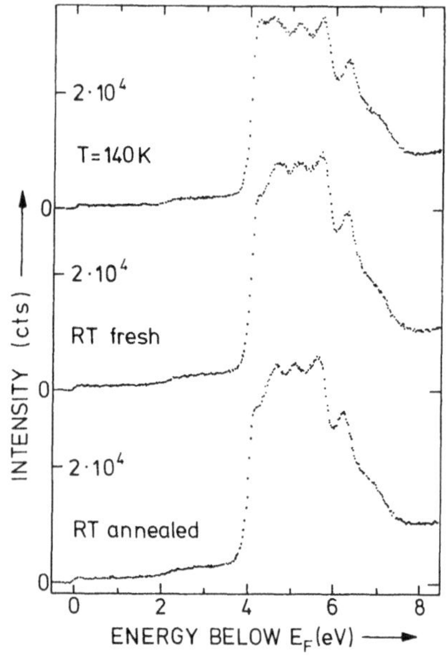

Fig. 6.90. UPS – He I ($\hbar\omega = 21.2$ eV) spectra of silver films deposited at 140 K and kept at this temperature (*top*) immediately after deposition at room temperature (RT) (*middle*) and after deposition at RT and 18 h annealing at RT (*bottom*). The weak structure between 2 and 4 eV below the Fermi energy E_F is caused by the weak intensity (4 %) HeI satellite line at 23.09 eV photon energy. From [6.170]

codeposited, a new absorption was found at 1000.2 cm^{-1} (Fig. 6.89, middle). When the concentration of Ag atoms was raised, an additional absorption band appeared at 1009.8 cm^{-1} (Fig. 6.89, bottom). Equivalent data were observed for deuterated pyridine. Tevault and Smardzewski assigned the 1000.2 and 1009.8 cm^{-1} bands to pyridine complexed to a single silver atom and to small Ag clusters (most likely Ag_2). The SERS-E line lies between the frequencies 1000.2 and 1009.8 cm^{-1}. Similar analogies exist for the other pyridine SERS bands (Table 6.5). These results suggest a localized chemical enhancement [6.310] and atomic dimensions for the active sites. If the author's interpretation of the experimental results in [6.44, 204, 305–307] is correct, there is as yet no convincing experimental evidence for a short-range enhancement mechanism for pyridine on atomically smooth silver surfaces. This is surprising in view of the many reasonable models for such a short-range mechanism (Sect. 6.7). As this problem is of fundamental theoretical importance, further careful experimental work is needed to clarify it.

As a further test of the concept of "SERS-active" sites of atomic scale roughness, simultaneous SERS [6.312] and ultraviolet photoemission spectroscopy (UPS) experiments [6.170] were performed for oxygen on cold-deposited silver films. Figure 6.90 shows the He I spectra of a silver film evaporated at 140 K ($\approx$2000 Å thick), and of a silver film evaporated at 300 K immediately after deposition and after annealing at 300 K for about 18 h, all unexposed. The work function of the unexposed silver films (obtained from the secondary electron cutoff) was 4.4 ± 0.1 eV and depended neither on the deposition temperature nor on annealing. No drastic differences in the He I spectra exist between RT annealed and cold-deposited films. The differences are comparable to those observed after evaporating ~3 Å of silver at 120 K on a RT-deposited silver film [6.169]. This may indicate that the main difference between the films deposited at 140 K and the RT-annealed films is a greater density of surface defects for the cold-deposited films. Also, the unchanged work function of 4.4 ± 0.1 eV for all the films speaks against a drastic restructuring of the surface by annealing (the work function of Ag(111), (100) and (110) is 4.46, 4.22 and 4.14 eV, respectively [6.313]).

This point is corroborated by the He II spectra. The spectrum of cold-deposited silver films is almost identical to the one taken with 40 eV synchrotron radiation [6.169] and He II radiation [6.314] for a RT deposited film. The He I and He II spectra are very different from UPS [6.315] and XPS [6.316] spectra from small silver clusters. For clusters with less than 150 silver atoms, the full crystal field splitting of bulk silver *d*-states as compared with the spin-orbit splitting of *d*-states of silver atoms has not yet been developed [6.315]. This indicates the surface of the cold-deposited silver films does not contain in any significant amount little bumps of less than 150 atoms (corresponding to hemispherical bumps of about 20 Å diameter).

Koch et al. [6.169] observed an extra photoemission [6.317] at the top of the *d*-band ≈4.2 eV below E_F after cold deposition of 3 Å of silver on RT deposited silver. This extra structure was assigned to localized *d*-electron states of surface defects. The intensity ratio of the defect structure near 4.2 eV below E_F to the one at 6.25 eV (which is probably of purely bulk origin because it is least affected by oxygen adsorption and no surface states are expected at this energy [6.318] is 1.25 for the cold-deposited film, 1.08 for the freshly RT deposited film and 1.01 for the RT annealed film. In the work of *Koch* et al. [6.169], the change in intensity at 4.2 eV below E_F after annealing of 3 Å cold-deposited silver was only 4% (measured at 40 eV photon energy), compared to 25% in our case (Fig. 6.90). A crude estimate of the surface defect concentration may be made as follows: the energy-integrated *d*-band emission from surface defects as inferred from Fig. 6.87 relative to the overall *d*-band emission is about 4%. If one assumes that the overall *d*-band emission corresponds to the unattenuated signal from about 3 layers of silver, then the surface defect concentration would be ~12%. Apparently the defect concentration increases with the amount of cold-deposited silver.

Though one might expect a different local work function at defect sites [6.319], the work function should be dominated by terraces. Hence, the same work function of about 4.4 eV is obtained for films evaporated at RT and at 140 K. Changes in the UPS spectra under oxygen exposure are observed for the silver films evaporated at 140 K but not for those deposited at RT and annealed. It is generally accepted that a perfect Ag(111) surface exposed to molecular oxygen at room temperature below 10^{-3} Torr is inert: adsorption occurs only at defect sites [6.320, 21]. Thus probably holds also for a Ag(100) surface [6.321]. The lack of change in the UPS spectra after exposing a RT-annealed film to oxygen at temperatures of 140–150 K indicates that these films are also almost

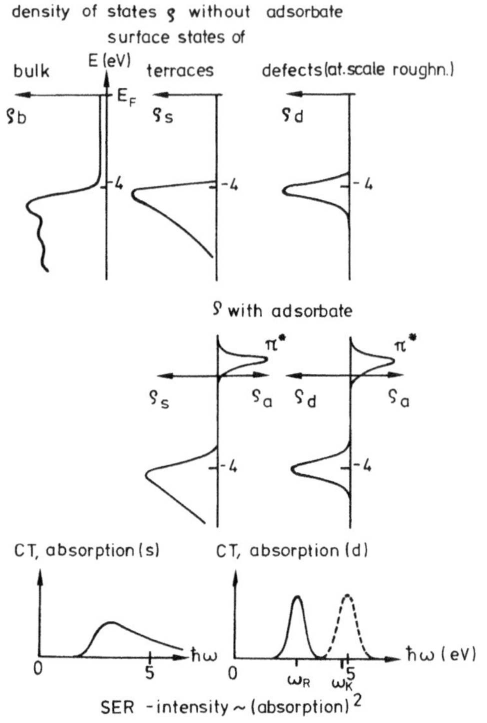

Fig. 6.91. *Top:* densities $\varrho_b, \varrho_s, \varrho_d$ of electronic states of bulk silver, of the surface states at smooth parts of the silver surface and at defects (atomic scale roughness). *Middle:* lowest unoccupied orbital (Newns Anderson resonance) of adsorbate π^*. *Bottom:* schematic optical absorption cross section of CT excitations at smooth surfaces (s) and at defects (d). ω_K is the excitation energy in Koopman's "frozen energy level" scheme, ω_R is the relaxed excitation energy

inert to molecular oxygen at lower temperatures. Therefore, oxygen must be adsorbed at surface-defect sites of the cold-deposited silver, but not at the atomically smooth terraces. From the change in work function after oxygen exposure, an independent estimate of the surface defect concentration was possible, yielding the same value of 12 % given above.

The Raman spectrum of the "SERS-active" cold-deposited silver film, exposed to oxygen under similar conditions to the UPS experiment, yielded two lines [6.312]. The vibration at 1053 cm^{-1} was assigned to the superoxide species O_2^- adsorbed via one oxygen atom to silver defect sites and the vibration at 697 cm^{-1} to a peroxide species O_2^{2-} adsorbed symmetrically to silver defect sites [6.170]. Hence the SERS was due to dioxygen adsorbed at sites of ASR, amounting to a surface coverage of about 10 %. This result does not yet prove CT excitation at sites of ASR nor that adsorbates at smooth parts of the surface are not subject to a CT Raman enhancement mechanism. On the other hand, the defect-related Ag 4*d* surface state may be the initial state of the CT excitation. One might speculate about why the resonant Raman enhancement by CT is more efficient at sites of ASR than on atomically-smooth parts of the surface (Fig. 6.91). In the sense of Koopman's theorem, excitation energies from the surface states about 4 eV below the Fermi level to the lowest unoccupied molecular orbit (LUMO) of the adsorbate above E_F would be in the ultraviolet range [6.169]. However, relaxation effects due to the interaction of the electron in the LUMO and the hole in the localized surface defect states would decrease the excitation energy. According to the UPS data presented here, the surface defect states have a narrow energy distribution compared to the width of the *d*-band. The surface states of a smooth Ag(100) surface start approximately at the top of the bulk bands and extend about 2/3 over the range of the bulk *d*-band [6.314, 318]. At grazing take-off of the He II-excited photoelectrons, the upper part of the silver *d*-band structure was enhanced with respect to the lower part of the *d*-band structure, indicating surface states in the upper part of the *d*-band [6.314]. When silver films evaporated at RT were covered with about 10 Å of Al [6.169], the energy distribution curve of the photoelectrons changed in the upper two-thirds of the *d*-band. This was interpreted as suppression of the silver surface states. Due to the smaller halfwidth of the surface state at defects compared to the width of bulk states or surface state bands, CT at defect sites would lead to a narrower resonance and hence higher Raman enhancement than CT at smooth parts of the surfaces (Fig. 6.88). The experimental results of *Demuth* et al. (see Sect. 6.8) are in agreement with this conjecture. As already described above, on Ag(111) only relatively weak and broad CT structures were found for adsorbed pyridine [6.250], whereas the CT excitations of CO on cold-deposited silver films appear better defined [6.251]. In addition to a more efficient CT mechanism, at sites of atomic scale roughness there is the possibility of local field enhancements [6.269].

The results of [6.294] (Fig. 6.83) also suggest "SERS-active" sites of atomic scale. Here a comparison with recent electroreflectance measurements on silver (111), (100), and (110) surfaces of *Boeck* and *Kolb* [6.295] is helpful. These authors found structures in the electroreflectance which are the first derivative of rather broad absorption peaks (for a fixed potential ~0.5 eV wide, for a fixed wavelength about 0.3 V width in potential, comparable to the results in Fig. 6.82). The dips in this derivative showed characteristic shifts with potential of the silver electrodes (Fig. 6.83). The absorption structure is interpreted as electronic transitions from occupied bulk states to vacant surface states. Self-consistent pseudopotential calculations of the electronic structure [6.322, 323] for the silver (110) and (100) surfaces reveal surface bands in the gaps of the projected bulk density of states above the Fermi energy (for zero potential difference between metal and electrolyte. This so-called potential of zero charge is given in Fig. 6.83). These surface states decay within one lattice constant to the metal and the electrolyte (vacuum) side [6.323]. Reasonable agreement with the experimental results was reached for the Ag(100) surface [6.323] by means of transitions to two different surface bands. The shift of the surface bands with respect to the bulk due to charging was modelled by an external potential [6.323] and relatively good agreement was obtained with the shift observed for the Ag(100) surface. However, the larger slope of 3–4 eV/V observed for Ag(110) is not easy to understand [6.295]. Even if the full potential difference U between silver electrode and electrolyte is applied to bulk and surface electronic states, one would only expect a shift of 1 eV/V. The comparison of the shifts for the different faces suggests a local field effect on atomic scale in the following sense.

The electrostatic potential difference between the bulk of the metal and the electrode may be produced by a dipole layer involving the outermost metal layer and its surface states, adsorbates, the first layers of water (Helmholtz layers) and the counter-ions. In the electrochemical experiment, only the dipole layer moment averaged over the surface of the electrode is under potentiostatic control. This concept is analogous to the concept of a dipole layer at a metal-vacuum interface contributing to the work function [6.324]. Face and site-specific variations of this dipole moment are responsible for observed variations of the work function of a given metal. The concept of a local work function on an atomic scale, as it follows from measurements on stepped metal surfaces [6.325] and photoemission on xenon adlayers [6.319, 326], is well accepted [6.327]. This means there are varying dipole moments on an atomic scale. This will certainly also hold for the dipole layer at an electrode surface.

It follows from field-ion microscopy [6.328] that sites of atomic scale roughness (adatoms, adatoms at steps, kink-site and step-site atoms) are more polarizable than atoms on smooth terraces. Hence, sites of ASR are seen with good contrast to field-ion microscopy [6.328]. One may therefore expect that sites of ASR at an electrode surface are more polarizable than atomically smooth parts of the surface (the world "polarizable" is used here in the sense of a polarizability relating induced electric dipole moments to electric field, and not in the electrochemist's sense of a polarizable electrode).

If this is so, the change in the local static dipole moment at sites of ASR will be larger than the average one, only the latter being under potentiostatic control. This means that the shift discussed above between surface states and bulk states is larger at sites of ASR than on the average. In this sense, we tentatively assign the large shift observed in [6.295] for the (110) surface of Ag to surface states localized at the (110) atomic ridges and the shifts observed in [6.294] to CT excitations at sites of ASR. In the latter case, the same conclusion was independently reached in [6.299]. Further evidence for "SERS-active sites" has been presented by *Marinyuk* et al. ([6.329] and references therein), by *Pettinger* et al. ([6.308] and references therein), by *Seki* [6.243] and by *Macomber* and *Furtak* [6.330]. The results of *Chang* and coworkers will be presented in a forthcoming review [6.331], therefore they will not be dicussed here.

Silver adatoms as active sites were first proposed by *Lazorenko-Manevich* et al. [6.332] and independently by *Billmann* et al. [6.26], whereas a resonance Raman effect due to a silver-halide pyridine complex was first proposed by *Pettinger* et al. [6.43]. The importance of active sites has also been emphasized by *Dornhaus* [6.333]. There is still controversy about the relevance of "adatoms" and "atomic scale roughness" [6.203, 42]. The state of the art and the discussion may be characterized by a little poem by *McKee* [6.334]:

> "We seek it here, we seek it there, in defect, steps, most everywhere. Is it fact? – Or merely a sleight? That demmed elusive active site!"

As yet, experimental indications for active sites remains indirect. SERS intensity variations with temperature and electrode potential variation indicate variations of the concentration of active sites. These observations may provide new knowledge about the reconstruction and electrocrystallization of surfaces. Future work has to take into account that adsorbates may change the surface topography. *Chang* and coworkers [6.331] observed by SEM that apparently smooth surface planes of silver particles constituting an "island film" developed an orange-peel appearance after exposure to NO. Silver surfaces of (331) orientation develop facets in a reversible way upon oxygen exposure [6.335]. These points, however, are beyond the scope of this chapter.

6.10 Relevance for Catalysis

In the context of the "chemical effect at sites of atomic scale roughness", one should point out that atomic surface structure (different from low-index metal faces) is becoming more important in the discussion of heterogenous catalysis [6.336–339]. *Somorjai* [6.337] writes: "The relative concentration of various surface sites-steps, terrace atoms, kinks and appropriately placed additives determines the activity and selectivity of the catalyst. Once an understanding of the working catalyst on the atomic scale is reached, improvement of catalytic activity is almost automatic". If the picture of the chemical effect outlined here is basically correct, SERS may become a tool for research on non-smooth, polycrystalline surfaces concerning catalytic or electrocatalytic processes in silver and copper, but maybe also on some other metals (Sect. 6.5.8).

One may argue that silver films are only "SERS-active" when cold-deposited and that they are no longer "SERS-active" when warmed to room temperature due to annealing of atomic scale and supra-atomic roughness [6.34]. However, higher temperatures are used in catalytic reactions (for instance, 200 – 300 °C for ethylene oxydation on a silver catalyst). Nevertheless strong SERS was observed on mechanically polished silver exposed to air at room temperature [2.14, 29]. *Dorain* et al. [6.340] observed the catalytic transformation of SO_3^{2-} to SO_4^{2-} from SO_2 and O_2 on a silver powder at temperatures between 27 °C and 108 °C by SERS. According to them, the lack of subsequent formation of SO_3^{2-} with SO_2 exposure of the silver powder at 27 °C implies blockage or destruction of active sites. If one accepts that the "chemical effect at sites of atomic scale roughness" contribute two orders of magnitude to the SERS enhancement, one has to conclude that the "SERS-active" surface structure may also be stabilized above room temperature. More work concerning this point is needed. If one believes that the enhancement is mainly "classical", there are no difficulties at all, apart from the effect that $\varepsilon_2(\omega)$ of Ag increases with temperature. Another direction worth pursuing is the coating of silver islands by catalytically active oxides and the use of "classical enhancement" to study chemical reactions at the oxide surface [6.125].

Concerning the future of Raman research in surface science (Sect. 6.1), it becomes obvious that predictions about the enhancing mechanism only on the basis of the theory of electromagnetic resonances are too simplistic. On the other hand, we have no generally accepted models of the "nonclassical effect". Hence an experimental approach is advisable. For instance, in the work of *Krasser* and *Renouprez* [6.227] on silica-supported Ni-particle catalysts (Sect. 6.5.8)

"a lot of catalysts never showed an enhancement"

[6.341], but some did! There may also be a future in unenhanced Raman scattering from adsorbates on well-characterized surfaces as demonstrated by *Campion* and coworkers [6.5, 306].

Note Added in Proof

This contribution presents the author's state of information at the end of 1982. More recent work on SERS may be found in [6.342–344]. Only some new results pertaining to the discussion in this contribution are mentioned below.

Section 6.5.3. The estimate of the maximum "classical enhancement" of 10^3 of silver island films is corroborated experimentally in [6.345].

Section 6.5.4. Cold deposited silver films are highly porous, as observed by combined photoelectron spectroscopy and thermodesorption spectroscopy of adsorbed Xe [6.346, 348] confirming older data for cold deposited Cu [6.349].

Section 6.5.8. On surfaces of PbTe, covered by a 20 Å thick layer consisting of carbon, Pb, and Te oxides, a band extending from 700 to 2200 cm^{-1} on top of a background extending beyond 3000 cm^{-1} was found. It increased by a factor of 100 when the surface was roughened. The continuum was assigned to electronic excitations and/or vibrational modes intrinsic to adsorbed carbon. The excitation spectrum showed a factor of ~200 difference between the red and green-blue regions. It was assignment to interband transitions at maxima of the joint density of states. An electromagnetic enhancement mechanism could only account for a factor of 2. More details on the Raman spectra of pyridine on Ag, Au, Ni, Pd, Pt, Ti, Co, NiO and TiO_2 are found in [6.351].

Section 6.5.9. Results [6.243] similar to those of Fig. 6.73 have been explained in [6.352] and [6.346] by thermally activated migration of pyridine into the porosity of the cold deposited silver films (see below).

Section 6.6. Evidence for enhanced excitation of vibrations of CO and ethylene on cold deposited Cu films in infrared transmission spectroscopy by adsorbate-metal electron interaction will be presented in [6.353]. The mechanism is similar to the charge transfer mechanism of SERS [6.347].

Section 6.8. SERS of cyanide on silver electrodes involves a photon-driven charge transfer (CT) of an electron from cyanide to silver [6.354]. This process cannot involve localized Ag4d states (Fig. 6.91). Therefore, a revised model of SERS by CT is proposed in [6.347], involving localized Ag5s states near E_F. In [6.355] the unenhanced spectrum of perdeuterobenzene on smooth Ag(111) was reported. It was not necessary to invoke quadrupolar transitions due to strong field gradients [6.53].

Section 6.9. No SERS was observed for pyridine on smooth Ag(110) in [6.356], thus resolving the apparent contradiction between [6.305, 306] and [6.44]. An unambiguous assignment of the E-line (~1010 cm^{-1}) of pyridine adsorbed on Cu films deposited in UHV to pyridine adsorbed on surface defects has been given in [6.309]. This result implies a short range nonclassical enhancement of at least two orders of magnitude on top of any "classical enhancement" for pyridine at surface defects ("active sites"). Further evidence for "SERS active sites" on silver electrodes is presented in [6.357, 358, 354]. In [6.352] active sites as "cavity sites" in the cold deposited porous films is envisioned while in [6.346] "sites within pores" are suggested. Both papers attribute the enhancement at "SERS active sites" to local electromagnetic resonances within the "cavity sites" or "pores". This mechanism should not be chemically specific. In order to test the hypothesis of [6.352, 346] cold deposited Ag an Cu films were exposed at low temperatures to CO or O_2, such as to cover also the "internal surface" of the porous films by a monolayer [6.359]. Whereas SERS of CO was observed, the Raman signal of O_2 was about two orders of magnitude weaker. This proves the existence of a contribution to SERS of about two orders of magnitude which is not caused by electromagnetic resonances.

The author's present understanding of the "chemical contribution" to SERS is outlined in [6.347]. The interpretation of SERS is still a controversial matter, in spite of the enormous work in this field.

References

6.1 R.F.Willis (ed.): *Vibrational Spectroscopy of Adsorbates*, Springer Ser. Chem. Phys., Vol. 15 (Springer, Berlin, Heidelberg, New York 1980)
6.2 T.Davidson, B.S.Pons, A.Bewick, P.P.Schmidt: J. Electro-Anal. Chem. **125**, 237 (1981) and references therein
6.3 M.L.Hair: *Infrared Spectroscopy in Surface Chemistry* (A.Arnold, London 1967)
L.H.Little: *Infrared Spectra of Adsorbed Species* (Academic Press, New York 1966)
6.4a A.Weber (ed.): *Raman Spectroscopy of Gases and Liquids*, Topics Current Phys., Vol. 11 (Springer, Berlin, Heidelberg, New York 1980)
6.4b H.Yamada: Reviews Appl. Spectr. **17**, 227 (1981)
6.5 J.G.Skinner, W.G.Nilsen: J. Opt. Soc. Am. **58**, 113 (1968)
6.6 A.Campion, J.K.Brown, V.M.Grizzle: Surf. Sci. **115**, L153 (1982)
6.7 M.Fleischmann, P.J.Hendra, A.J.McQuillan: Chem. Phys. Lett. **26**, 123 (1974)
6.8 D.L.Jeanmaire, R.P.Van Duyne: J. Electroanal. Chem. **84**, 1 (1977)
6.9 M.G.Albrecht, J.A.Creighton: J. Am. Chem. Soc. **99**, 5215 (1977)
6.10 L.Moerl, B.Pettinger: Solid State Commun. **43**, 315 (1982)
6.11 R.P.Van Duyne: In *Chemical and Biological Applications of Lasers* **4**, ed. by C.B.Moore (Academic Press, New York 1979)
6.12 T.E.Furtak, J. Reyes: Surf. Sci. **93**, 351 (1980)
6.13 M.Moskovits: J. Chem. Phys. **69**, 4159 (1978)
6.14 A.Otto: In Proc. Intern. Conf. on Vibrations in Adsorbed Layers, Jülich, June'78; Surf. Sci. **92**, 145 (1980)
6.15 E.Burstein, Y.J.Chen, C.Y.Chen, S.Lundquist, E.Tossatti: Solid State Commun. **29**, 567 (1979)
6.16 J.A.Creighton, C.G.Blatchford, M.G.Albrecht: J. Chem. Soc. Faraday II, **75**, 790 (1979)
6.17 J.E.Rowe, C.V.Shank, D.A.Zwemer, C.A.Murray: Phys. Rev. Lett. **44**, 1770 (1980)
6.18 S.L.McCall, P.M.Platzman, P.A.Wolff: Phys. Lett. **77**A, 381 (1980), submitted on 12 March 1980
6.19 C.A.Murray, D.L.Allara, M.Rhinewine: Phys. Rev. Lett. **46**, 57 (1981) and private communication
6.20 Proc. 7th Intern. Conf. Raman Spectroscopy, ed. by W.F.Murphy, Ottawa (1980)
6.21 J.G.Bergman, D.S.Chemla, P.F.Liao, A.M.Glass, A.Pinczuk, R.M.Hart, D.H.Olson: Opt. Lett. **6**, 33 (1981)
6.22 A. Otto: Appl. Surf. Sci. **6**, 309 (1980)
6.23 *Surface-Enhanced Raman Scattering*, ed. by R.K.Chang, T.E.Furtak (Plenum, New York 1982)
6.24 R.P.Cooney, M.W.Howard, M.R.Mahoney, R.P.Mernagh: Chem. Phys. Lett. **79**, 459 (1981)
6.25 I.Pockrand: Vhdlg. der DPG **5**, 944 (1982)
6.26 J.Billmann, G.Kovacs, A.Otto: Surf. Sci. **92**, 153 (1980)
6.27 S.H.Macomber, T.E.Furtak, T.M.Devine: Chem. Phys. Lett. **90**, 439 (1982)
6.28 F.Barz, J.G.Gordon II, M.R.Philpott, M.J.Weaver: Chem. Phys. Lett. **91**, 291 (1982); T.T.Chen, K.U. von Raben, J.F.Owen, R.K.Chang, B.L.Laube: Chem. Phys. Lett. **91**, 494 (1982)
6.29 A.Otto: Surf. Sci. **75**, L392 (1978)
6.30 S.G.Schultz, M.Janik-Czachor, R.P.Van Duyne: Surf. Sci. **104**, 419 (1981)
6.31 T.H.Wood, M.V.Klein: J. Vac. Sci. Technol. **16**, 459 (1979)
6.32 I.Pockrand, A.Otto: Solid State Commun. **38**, 1159 (1981)
6.33 T.H.Wood, M.V.Klein: Solid State Commun. **35**, 263 (1980)

6.34 A.Otto, I.Pockrand, J.Billmann, C.Pettenkofer: In [6.23]
6.35 C.Y.Chen, I.Davoli, G.Ritchie, E.Burstein: Surf. Sci. **101**, 363 (1980)
6.36 H.Seki: J.Vac. Sci. Technol. **18**, 633 (1981)
6.37 J.C.Tsang, J.R.Kirtley, J.A.Bradley: Phys. Rev. Lett. **43**, 772 (1979)
6.38 D.Beaglehole, O.Hunderi: Phys. Rev. B**2**, 309 (1970)
6.39 E.Burstein, C.Y.Chen: In [Ref. 6.20, p. 346]
6.40 J.M.Blakeley, M.Eizenberger: In *The Chemical Physics of Solid Surfaces and Heterogeneous Catalysis* Vol. 1, ed. by D.A.King (Elsevier, Amsterdam 1981)
6.41 J.M.Blakeley: *Introduction to the Properties of Crystal Surfaces* (Pergamon Press, Oxford 1973) Fig. 6, 8(b)
6.42 A.Otto: Phys. Rev. B**27**, 5132 (1983)
6.43 B.Pettinger, U.Wenning, D.M.Kolb: Ber. Bunsenges. Phys. Chem. **82**, 1326 (1978)
6.44 M.Udagawa, Chih-Cong Chou, J.C.Hemminger, S.Ushioda: Phys. Rev. B**23**, 6843 (1981)
6.45 A.Otto, J.Timper, J.Billmann, G.Kovacs, I.Pockrand: Surf. Sci. **92**, L55 (1980)
6.46 A.Mooradian: Phys. Rev. Lett. **22**, 185 (1969)
6.47 J.P.Heritage, J.G.Bergman, A.Pinczuk, J.M.Worlock: Chem. Phys. Lett. **67**, 229 (1979)
6.48 C.K.Chen, A.R.B. de Castro, Y.R.Shen: Rev. Lett. **46**, 145 (1981)
6.49 R.Dornhaus, M.B.Long, R.E.Benner, R.K.Chang: Surf. Sci. **93**, 240 (1980)
6.50 G.R.Erdheim, R.L.Birke, J.R.Lombardi: Chem. Phys. Lett. **69**, 495 (1980)
6.51 M.Moskovits, D.P.DiLella: J. Chem. Phys. **73**, 6068 (1980)
6.52 L.M.Sverdlov, M.A.Kovner, E.P.Krainov: *Vibrational Spectra of Polyatomic Molecules* (Wiley, New York 1974)
6.53 J.K.Sass, H.Neff, M.Moskovits, S.Holloway: J. Phys. Chem. **85**, 621 (1981)
6.54 B.Pettinger: Chem. Phys. Lett. **78**, 404 (1981)
6.55 C.G.Blatchford, J.R.Campbell, J.A.Creighton: Surf. Sci. **108**, 411 (1981)
6.56 D.V.Murphy, K.U. von Raben, R.K.Chang, P.B.Dorain: Chem. Phys. Lett. **85**, 43 (1982)
6.57 I.Pockrand: Chem. Phys. Lett. **85**, 37 (1982)
6.58 M.Moskovits, D.P.DiLella: In [6.23]
6.59 W.Schulze, B.Breithaupt, K.-P.Charlé, U.Kloss, Ber. Bunsenges. Phys. Chem. (in press)
6.60 P.A.Lund, R.R.Smardzewski, D.E.Tevault: Chem. Phys. Lett. **89**, 508 (1982)
6.61 T.H.Wood, M.V.Klein: *Enhanced Raman Scattering from Adsorbates on Metal Films in Ultrahigh Vacuum* (unpublished manuscript)
6.62 T.Lopez-Rios, C.Pettenkofer, I.Pockrand, A.Otto: Surf. Sci. Lett. **121**, L541 (1982)
6.63 S.S.Jha, J.R.Kirtley, J.C.Tsang: Phys. Rev. B**22**, 3973 (1980)
6.64 T.E.Furtak, G.Trott, B.H.Loo: Surf. Sci. **101**, 374 (1980)
6.65 B.Pettinger, H.Wetzel: In [6.23]
6.66 D.N.Batchelder, N.J.Poole, D.Bloor: Chem. Phys. Lett. **81**, 560 (1981)
6.67 D.S.Wang, H.Chew, M.Kerker: Appl. Opt. **19**, 2256 (1980)
6.68 M.Kerker, D.S.Wang, H.Chew: Appl. Opt. **19**, 3373 (1980) [Errata: Appl. Opt. **19**, 4159 (1980)], first submitted to J. Chem. Phys. on 18 March 1980, but rejected
6.69 I am grateful to J.Worlock for a discussion of this point
6.70 C.Y.Chen, G.Ritchie: In [6.23]
6.71 A.M.Glass, P.F.Liao, J.G.Bergman, D.H.Olson: Opt. Lett. **5**, 368 (1980)
6.72 S.Garoff, D.A.Weitz, T.J.Gramila, C.D.Hanson: Opt. Lett. **6**, 245 (1981)
6.73 C.F.Eagen: App. Opt. **20**, 3035 (1981)
6.74 P.B.Johnson, R.W.Christy: Phys. Rev. **6**B, 4370 (1972)
6.75 D.S.Wang, M.Kerker: Phys. Rev. B**24**, 1777 (1981)
6.76 B.J.Messinger, K.U. von Raben, R.K.Chang, P.W. Barber: Phys. Rev. B**24**, 649 (1981)
6.77 J.I.Gersten: J. Chem. Phys. **72**, 5779 (1980)
6.78 J.I.Gersten, A.Nitzan: J. Chem. Phys. **73**, 3023 (1980)
6.79 P.R.Hilton, D.W.Oxtoby: J. Chem. Phys. **72**, 6346 (1980)
6.80 P.J.Feibelmann: Phys. Rev. B**22**, 3654 (1980)
6.81 G.C.Schatz: In [6.23]
6.82 T.K.Lee, J.L. Birman: In [6.23]
6.83 J.I.Gersten, D.A.Weitz, T.J.Gramila, A.Z.Genack: Phys. Rev. B**22**, 4562 (1980)
6.84 D.A.Weitz, T.J.Gramila, A.Z.Genack: In [6.23]

6.85 M.Fleischmann, P.J.Hendra, I.R.Hill, M.E.Pemble: J. Electroanal. Chem. **117**, 243 (1981)
6.86 R.Ruppin: Solid State Commun. **39**, 903 (1981)
6.87 P.W.Barber, R.K.Chang, H.Massoudi: Phys. Rev. Lett. **50**, 997 (1983)
6.88 J.P.Marton, J.R.Lemon: Phys. Rev. B**4**, 271 (1971)
6.89 D.P.DiLella, A.Gohin, R.H.Lipson, P.McBreen, M.Moskovits: J. Chem. Phys. **73**, 4282 (1980)
6.90 M.J.Dignam, M.Moskovits: J. Chem. Soc., Faraday Trans II, **69**, 65 (1973)
6.91 M.Moskovits: Solid State Commun. **32**, 59 (1979)
6.92 T.Yamaguchi, S.Yoshida, A.Kinbara: J. Opt. Soc. Am. **62**, 634 (1972)
6.93 C.Y.Chen, E.Burstein: Phys. Rev. Lett. **45**, 1287 (1980)
6.94 T.Yamaguchi, S.Yoshida, A.Kinbara: Thin Solid Films **21**, 173 (1974)
6.95 A.Wokaun, J.G.Bergman, J.P.Heritage, A.M.Glass, P.F.Liao, D.H.Olson: Phys. Rev. B**24**, 849 (1981)
6.96 P.F.Liao, J.G.Bergman, D.S.Chemla, A.Wokaun, J.Melugailis, A.M.Hawryluk, N.P.Economou: Chem. Phys. Lett. **82**, 355 (1981)
6.97 D.A.Weitz, S.Garoff, T.J.Gramila: Opt. Lett. **7**, 168 (1982)
6.98 A.Otto: In *Optical Properties of Solids, New Developments*, ed. by B.O.Seraphin (North-Holland, Amsterdam 1976)
6.99 P.K.Aravind, E.Hood, H.Metiu: Surf. Sci. **109**, 95 (1981)
6.100 P.K.Aravind, H.Metiu, Chem. Phys. Lett. **74**, 301 (1980)
6.101 H. Metiu: "Surface Enhanced Spectroscopy", to appear in Progress in Surf. Science
6.102 W.H.Weber, G.W.Ford: Phys. Rev. Lett. **44**, 1774 (1980)
6.103 K.L.Kliewer, J.R.Fuchs: Phys. Rev. **172**, 607 (1968)
6.104 J.R.Fuchs, K.L.Kliewer: Phys. Rev. **185**, 905 (1969)
6.105 W.H.Weber, G.W.Ford: Opt. Lett. **6**, 122 (1981)
6.106 A.Otto: Optik **38**, 566 (1973)
6.107 Y.J.Chen, W.P.Chen, E.Burstein: Phys. Rev. Lett. **36**, 1207 (1976)
6.108 K.Sakoda, K.Oktaka, E.Hanamura: Solid State Commun. **41**, 393 (1982)
6.109 U.Laor, G.C.Schatz: Chem. Phys. Lett. **82**, 566 (1981)
6.110 P.K.Aravind, A.Nitzan, H.Metiu: Surf. Sci. **110**, 189 (1981)
6.111 U.Kreibig, A.Althoff, H.Pressmann: Surf. Sci. **106**, 308 (1981)
6.112 E.Burstein, S.Lundquist, D.L.Mills: In [6.23]
6.113 R.R.Chance, A.Prock, R.Silbey: Advances in Chem. Phys. **37**, 1 (Wiley, New York 1978)
6.114 K.H.Drexhage: *Progress in Optics* **12**, 165 (North-Holland, Amsterdam 1974)
6.115 B.N.J.Persson: Solid State Commun. **27**, 417 (1978)
6.116 B.N.J.Persson, W.L.Schaich: J. Phys. C, **14**, 5583 (1981)
6.117 H.Morawitz, M.R.Philpott: Phys. Rev. B**10**, 4863 (1974)
6.118 W.H.Weber, C.F.Eagen: Opt. Lett. **4**, 236 (1979)
6.119 I.Pockrand, A.Brillante, D.Möbius: Chem. Phys. Lett. **69**, 499 (1980)
6.120 G.W.Ford, W.H.Weber: Surf. Sci. **109**, 451 (1981)
6.121 J.E.Demuth, Ph.Avouris: Phys. Rev. Lett. **47**, 61 (1981)
6.122 Ph.Avouris, J.E.Demuth: J. Chem. Phys. **75**, 4783 (1981)
6.123 J.E.Parkins, K.K.Innes: J. Molec. Spectr. **15**, 407 (1965)
6.124 T.Maniv, H.Metiu: Chem. Phys. Lett. **79**, 79 (1981)
6.125 D.L.Allara, C.A.Murray, S.Bodoff: Bull. Am. Soc. **26**, 338 (1981)
6.126 P.F.Liao: In [6.23]
6.127 E.Burstein, C.Y.Chen, S.Lundquist: in *Proc. US-USSR Symp. Theory of Light Scattering on Condensed Matter* (Plenum Press, New York 1979)
6.128 H.Seki, M.R.Philpott: J. Chem. Phys. **73**, 5376 (1980)
6.129 H.Seki: J. Chem. Phys. **76**, 4412 (1982)
6.130 S.A.Lyon, J.Worlock: Bull. Am. Phys. Soc. **26**, 378 (1981)
6.131 T.Yamaguchi, S.Yoshida, A.Kinbara: J. Opt. Soc. Am. **64**, 1563 (1974)
6.132 S.Yoshida, T.Yamaguchi, A.Kinbara: J. Opt. Soc. Am. **62**, 1415 (1972)
6.133 D.C.Skillmann: J. Opt. Soc. Am. **61**, 1264 (1971)
6.134 H.Ehrenreich, H.R.Philipp: Phys. Rev. **128**, 1622 (1962)
6.135 G.B.Irani, T.Huen, F.Wooten: J. Opt. Soc. Am. **61**, 128 (1971)

6.136 A.M.Glass, A.Wokaun, J.P.Heritage, J.G.Bergman, P.F.Liao, D.H.Olson: Phys. Rev. B**24**, 4906 (1981)
6.137 D.A.Weitz, S.Garoff, C.D.Hanson, T.J.Gramila, J.I.Gersten: J. Lumines. **24/25**, 83 (1981)
6.138 H.Seki: J. Vac. Sci. Technol. **20**, 584 (1982)
6.139 D.P.DiLella, R.H.Lipson, P.McBreen, M.Moskovits: J. Vac. Sci. Technol. **18**, 453 (1981)
6.140 C.Pettenkofer: Diploma thesis, Universität Düsseldorf (1981)
6.141 O.Hunderi, H.P.Myers: J. Phys. F. **3**, 683 (1973)
6.142 H.E.Bennett, J.M.Bennett, E.J.Ashley, R.J.Motyka: Phys. Rev. **163**, 755 (1968)
6.143 S.O.Sari, D.K.Cohen, K.D.Scherkoske: Phys. Rev. B**21**, 2162 (1980)
6.144 W.Kaspar, U.Kreibig: Surf. Sci. **69**, 619 (1977)
6.145 E.Kretschmann, E.Kröger: J. Opt. Soc. Am. **65**, 150 (1975)
6.146 H.P.Bonzel: CRC Critical Reviews in Solid State Sci. **6**, 171 (1976)
6.147 I.Pockrand, A.Otto: Solid State Commun. **37**, 109 (1981)
6.148 L.J.Cuddy, E.S.Machlin: Phil. Mag. **7**, 745 (1962)
6.149 J.Bodesheim, A.Otto: Surf. Sci. **45**, 441 (1974)
6.150 E.Kretschmann, T.L.Ferrell, J.C.Ashley: Phys. Rev. Lett. **42**, 1312 (1979)
6.151 J.P.Chauvineau: Private communication
6.152 D.Schumacher, D.Stark: Surf. Sci. **123**, 384 (1982)
6.153 J.P.Chauvineau: Surf. Sci. **93**, 471 (1980)
6.154 L.M.Clarebrough, R.L.Segall, M.H.Loretto, M.E.Hargreaves: Phys. Mag. **9**, 377 (1964)
6.155 M.Doyama, J.S.Koehler: Phys. Rev. **127**, 21 (1962)
6.156 H.Mehrer, A.Seeger: Phys. Stat. Sol. **39**, 647 (1970)
6.157 W.Drexel, W.Gläser, F.Gompf: Phys. Lett. **28**A, 531 (1969)
6.158 W.A.Kamitakahara, B.N.Brookhouse, Phys. Lett. **29**A, 531 (1969)
6.159 E.C.Svensson, B.N.Brookhouse, J.M.Rowe: Phys. Rev. **155**, 619 (1967)
6.160 J.W.Lynn, H.G.Smith, R.M.Nicklow: Phys. Rev. B**8**, 3493 (1973)
6.161 W.Schulze, H.U.Becker, D.Leutloff: J. Phys. C**2**, 7 (1977)
6.162 W.Schulze, D.M.Kolb, H.Gerischer: J. Chem. Soc. Faraday 2, **71**, 1763 (1975)
6.163 W.Schulze, H.U.Becker, H.Abe: Ber. Bunsenges. Phys. Chem. **82**, 138 (1978)
6.164 H.Abe, W.Schulze, B.Tesche: Chem. Phys. **47**, 95 (1980)
6.165 T.Welker, T.P.Martin: J. Chem. Phys. **70**, 5683 (1979)
6.166 W.Schulze, H.U.Becker, R.Minkwitz, K.Manzel: Chem. Phys. Lett. **55**, 59 (1978)
6.167 I.Pockrand: Unpublished
6.168 N.Bloembergen, R.K.Chang, S.S.Jha, C.H.Lee: Phys. Rev. **174**, 813 (1968)
6.169 E.E.Koch, J.Barth, J.H.Fock, A.Goldmann, A.Otto: Solid State Commun. **42**, 897 (1982)
6.170 J.Eickmans, A.Goldmann, A.Otto: Surf. Sci. **127**, 153 (1983)
6.171 J.Tersoff, L.M.Falicov: Phys. Rev. B**24**, 754 (1981)
6.172 C.K.Chen, T.F.Heinz, D.Ricard, Y.R.Shen: Phys. Rev. Lett. **46**, 1010 (1981)
6.173 C.K.Chen, T.F.Heinz, D.Ricard, Y.R.Shen: Chem. Phys. Lett. **83**, 455 (1981)
6.174 J. Timper, J.Billmann, A.Otto, I.Pockrand: Surf. Sci. **101**, 348 (1980)
6.175 J.A.Creighton: Oral Commun. at the Intern. Conf. on Vibrations at Surfaces, Namur (Sept. 1980), see also [6.179]
6.176 H.Abe, K.Manzel, W.Schulze, M.Moskovits, D.P.DiLella: J. Chem. Phys. **74**, 792 (1981)
6.177 K.Manzel, W.Schulze, M.Moskovits: Chem. Phys. Lett. **85**, 183 (1982)
6.178 H.Wetzel, H.Gerischer: Chem. Phys. Lett. **76**, 460 (1980)
6.179 C.G.Blatchford, J.R.Campbell, J.A.Creighton: Surf. Sci. **120**, 435 (1982)
6.180 M.Kerker, O.Sijman, L.A.Bumm, D.S.Wang: Appl. Opt. **19**, 3253 (1980)
6.181 K.U. von Raben, R.K.Chang, R.L.Laube: Chem. Phys. Lett. **79**, 465 (1981)
6.182 K.U. von Raben, J.F.Owen, R.K.Chang, B.L.Laube: Submitted to J. Phys. Chem.
6.183 M.Kerker: *The Scattering of Light and other Electromagnetic Radiation* (Academic Press, New York 1969)
6.184 The result in Fig. 6.57 raise the question of why no enhancement in the range of the single-particle dipolar resonances was observed before the apparent onset of aggregation. According to theory (Fig. 6.12), there should be an enhancement of about 3 orders of magnitude. Due to the onset of interband transitions below 500 nm, the maximum of "classical enhancement"

increases considerably the more the electromagnetic resonances are shifted to the red (Fig. 6.12). This could be significant in the case of gold but less so for silver [6.75]. Therefore a small fraction of particle strings would contribute less to the overall enhancement in the case of silver than in the case of gold. Hence, Creighton's explanation of the results of von Raben et al. is less likely in the case of silver (Fig. 6.56) than in the case of gold colloids (Fig. 6.55).
6.185 M.Mabuchi, T.Takenaka, Y.Fujiyoshi, N.Uyeda: Surf. Sci. **119**, 150 (1982)
6.186 A.Bachackashvilli, S.Efrima, B.Katz, Z.Priel: Chem. Phys. Lett. **94**, 571 (1983)
6.187 A.Otto: Surf. Sci. **101**, 99 (1980)
6.188 U.Kreibig: J. Phys. **F4**, 999 (1974)
6.189 R.H.Doremus: J. Chem. Phys. **42**, 414 (1965)
6.190 U.Kreibig, C.V.Fragstein: J. Phys. **224**, 307 (1969)
6.191 M.A.Smithard: Solid State Commun. **13**, 153 (1973)
6.192 L.Genzel, T.P.Martin, U.Kreibig: Z. Phys. B**21**, 339 (1980)
6.193 A.Kawabata, R.Kubo: J. Phys. Soc. Japan **21**, 1765 (1966)
6.194 J.D.Ganiere, R.Rechsteiner, M.A.Smithard: Solid State Commun. **16**, 113 (1975)
6.195 D.R.Penn, R.W.Rendell: Phys. Rev. Lett. **47**, 1067 (1981)
6.196 B.Pettinger, A.Tadjeddine, D.M.Kolb: Chem. Phys. Lett. **66**, 544 (1979)
6.197 R.Dornhaus, R.E.Benner, R.K.Chang, I.Chabay: Surf. Sci. **101**, 367 (1980)
6.198 H.W.K.Tom, C.K.Chen, A.R.B. de Castro, Y.R.Shen: Solid State Commun. **41**, 259 (1982)
6.199 I.Pockrand, A.Otto: Solid State Commun. **35**, 861 (1980)
6.200 T.H.Wood, D.A.Zwemer, C.V.Shank, J.E.Rowe: Chem. Phys. Lett. **82**, 5 (1981)
6.201 G.L.Eesley: Phys. Lett. **81**A, 193 (1981)
6.202 G.L.Eesley, J.M.Burkstrand: Phys. Rev. B**24**, 582 (1981)
6.203 T.H.Wood: Phys. Rev. B**24**, 2289(1981)
6.204 N.Sanda, J.Warlaumont, J.E.Demuth, J.C.Tsang, K.Christmann, J.A.Bradley: Phys. Rev. Lett. **45**, 1519 (1980)
6.205 D.F.Barbe (ed.): *Very Large Scale Integration (VLSI)*, Springer Ser. Electrophys., Vol. 5 (Springer, Berlin, Heidelberg, New York 1982)
6.206 J.Billmann, A.Otto: Appl. Surf. Sci. **6**, 356 (1980)
6.207 E.Rzepka, L.Taurel, S.Lefrant: Surf. Sci. **106**, 345 (1981)
6.208 E.Rzepka, S.Lefrant, L.Taurel: Solid State Commun. **30**, 801 (1979)
6.209 M.Ghomi, E.Rzepka, L.Taurel: Phys. Stat. Sol. (b) **92**, 447 (1979)
6.210 L.Taurel: Private communication
6.211 I.Pockrand: Chem. Phys. Lett. **92**, 509 (1982)
6.212 T.E.Furtak, J.Kester: Phys. Rev. Lett. **45**, 1652 (1980)
6.213 B.F.Schmidt, D.W.Lynch: Phys. Rev. B**3**, 4015 (1971)
6.214 J.C.Tsang, S.S.Jha, J.R.Kirtley: Phys. Rev. Lett. **46**, 1044 (1981)
6.215 P.Winsemius: Thesis, Leiden (1973)
6.216 R.J.Bartlett, D.W.Lynch, R.Rosei: Phys. Rev. B**3**, 4074 (1971)
6.217 A.Y.-C.Yu, W.E.Spicer, G.Hass: Phys. Rev. **171**, 834 (1968)
6.218 T.Inagaki, E.T.Arakawa, M.W.Williams: Phys. Rev. **23**, 5246 (1981)
6.219 B.H.Loo: J. Chem. Phys. **75**, 5955 (1981)
6.220 B.Pettinger, L.Moerl: J. Electron. Spectrosc. Rel. Phen. **29**, 383 (1983)
6.221 J.Heitbaum: Z. physik. Chemie **105**, 307 (1977)
6.222 J.Heitbaum: Private communication
6.223 W.Krasser, A.J.Renouprez: Solid State Commun. **41**, 231 (1982)
6.224 R.P.Cooney, M.Fleischmann, P.J.Hendra: JCS, Chem. Comm. **235** (1977)
6.225 R.P.Cooney, E.S.Reid, P.J.Hendra, M.Fleischmann: J. Am. Chem. Soc. **99**, 2002 (1977)
6.226 H.Yamada, Y.Yamamoto, N.Tani: Chem. Phys. Lett. **86**, 397 (1982)
6.227 W.Krasser, A.J.Renouprez: J. Raman Spectrosc. **11**, 425 (1981)
6.228 J.M.Stenzel, E.B.Bradley: J. Raman Spectrosc. **8**, 203 (1979)
6.229 R.Naaman, S.J.Buelow, O.Cheshnovsky, D.R.Herschbach: J. Phys. Chem. **84**, 2692 (1980)
6.230 R.M.Hexter, M.G.Albrecht: Spectrochim. Acta **35**A, 233 (1979)
6.231 M.Moskovits, D.P.DiLella: Chem. Phys. Lett. **73**, 500 (1980)
6.232 J.A.Appelbaum, D.R.Hamann: Rev. Mod. Phys. **48**, 479 (1976)

6.233 J.Hölzl, F.K.Schulte: *Springer Tracts Mod. Phys.*, Vol. **85**, 1 (Springer, Berlin, Heidelberg, New York 1979)
6.234 N.D.Lang, W.Kohn: Phys. Rev. **B1**, 4555 (1970)
6.235 N.D.Lang, W.Kohn: Phys. Rev. **B3**, 1215 (1971)
6.236 D.A.Zwemer, J.E.Rowe, C.V.Shank: To be published (announced in [6.200])
6.237 B.Pettinger, M.R.Philpott, J.G.Gordon II: Surf. Sci. **105**, 469 (1981)
6.238 B.Pettinger, M.R.Philgott, J.G.Gordon II: J. Chem. Phys. **74**, 934 (1981)
B.Pettinger, H.Wetzel: Ber. Bunsenges. Phys. Chem. **85**, 473 (1981)
6.239 T.T.Chen, J.F.Owen, R.K.Chang, B.L.Laube: Chem. Phys. Lett. **89**, 356 (1982)
6.240 D.P.DiLella, M.Moskovits: J. Phys. Chem. **85**, 2042 (1981)
6.241 C.Y.Chem. I. Davoli, E.Burstein: In *Proc. US-USSR Symp. Theorey Light Scattering in Condensed Matter*, ed. by E. Burstein, C.Y. Chen, S. Lundquist (Plenum Press, New York 1979)
6.242 B.H.Loo, T.E.Furtak: Chem. Phys. Lett. **71**, 68 (1980)
6.243 H.Seki: Solid State Commun. **42**, 695 (1982)
6.244 D.Schmeisser, K.Jakobi, D.M.Kolb: Appl. Surf. Sci. **1/12**, 164 (1982)
6.245 T.H.Wood, M.V.Klein, D.S.Zwemer: Surf. Sci. **107**, 625 (1981)
6.246 J.E.Demuth, K.Christmann, P.N.Sanda: Chem. Phys. Lett. **76**, 201 (1980)
6.247 J.E.Demuth, P.N.Sanda, J.M.Warlaumont, J.C.Tsang, K.Christmann: In [6.248]
6.248 R.Caudano, J.M.Gilles, A.A.Lucas (eds.): *Vibrations at Surfaces* (Plenum Press, London 1982)
6.249 P.N.Sanda, J.E.Demuth, J.C.Tsang, J.M.Warlaumont: In [6.23]
6.250 J.E.Demuth, P.N.Sanda: Phys. Rev. Lett. **47**, 57 (1981)
6.251 D.Schmeisser, J.E.Demuth, Ph.Avouris: Chem. Phys. Lett. **87**, 324 (1982)
6.252 C.Pettenkofer, A.Otto: Unpublished
6.253 R.L.Birke, J.R.Lombardi, J.I.Gersten: Phys. Rev. Lett. **43**, 71 (1979)
6.254 C.Y.Chen, E.Burstein: Bull. Am. Phys. Soc. **24**, 341 (1979)
6.255 G.Valette, A.Hamelin: Electroanal. Chem. Interf. Electrochem. **45**, 301 (1973)
6.256 V.V.Marinyuk, R.M.Lazorenko-Manevich, Ya.M.Kolotyrkin: Sov. Electrochem. **17**, 527 (1981)
6.257 U.Fano: Phys. Rev. **124**, 1866 (1961)
6.259 I.Pockrand, A.Otto: To be published
6.260 B.N.J.Persson, R.Ryberg: Phys. Rev. **B24**, 6954 (1981)
6.261 R.Kötz, E.Yeager: J. Electroanal. Chem. **123**, 335 (1981)
6.262 A.B.Anderson, R.Kötz, E.Yeager: Chem. Phys. Lett. **82**, 130 (1981)
6.263 M.A.Chesters, J.Pritchard, M.L.Sims: In *Adsorption-Desorption Phenomena*, ed. by F.Ricca (Academic Press, New York 1972) p. 277
6.264 P.Hollin, J.Pritchard: In [6.1]
6.265 M.Scheffler: Surf. Sci. **81**, 562 (1979)
6.266 A.M.Bradshaw, F.M.Hoffmann: Surf. Sci. **72**, 513 (1978)
6.267 D.A.King: In [6.1]
6.268 B.N.J.Persson, A.Liebsch: Surf. Sci. **110**, 356 (1981)
6.269 A.Otto: Vhdl. der DPG **5**, 945 (1982)
6.270 K.Arya, R.Zeyher, A.A.Maradudin: Solid State Commun. **42**, 461 (1982)
6.271 B.N.J.Persson, A.Liebsch: Preprint: "Optical properties of inhomogeneous media"
6.272 S.S.Jha: In [6.23]
6.273 A.G.Mal'shukov: Solid State Commun. **38**, 907 (1981)
6.274 T.Maniv: Phys. Rev. **B26**, 2858 (1982)
6.275 H.Ueba: In [6.23]
6.276 K.Arya, R.Zeyher: Phys. Rev. **B24**, 1852 (1981)
6.277 B.N.J.Persson: Chem. Phys. Lett. **82**, 561 (1981)
6.278 H.Ueba, S.Ichimura, H.Yamada: Surf. Sci. **119**, 433 (1982)
6.279 F.J.Adrian: J. Chem. Phys. **77**, 5302 (1982)
6.280 F.R.Aussenegg, M.E.Lippitsch: Chem. Phys. Lett. **59**, 214 (1978)
6.281 S.L.McCall, P.M.Platzman: Phys. Rev. **B22**, 1660 (1980)

6.282 B.N.J.Persson: Private communication
6.283 M.Moskovits, G.A.Ozin (eds.): *Cryochemistry* (Wiley, New York 1976) p. 261
6.284 D.F.McIntosh, G.A.Ozin, R.P.Messmer: Inorg. Chem. **19**, 3321 (1980)
6.285 W.Krasser, U.Kettler, P.S.Bechthold: Chem. Phys. Lett. **86**, 223 (1982)
6.286 I.Pockrand, J.Billmann, A.Otto: J. Chem. Phys. **78**, 6384 (1983)
6.287 I.Pockrand: Chem. Phys. Lett. **92**, 514 (1982)
6.288 A.Girlando, J.G.Gordon, D.Heitbaum, M.R.Philpott, H.Seki, J.D.Swalen: Surf. Sci. **101**, 417 (1980)
6.289 V.V.Marinyuk, R.M.Lazorenko-Manevich, Ya.M.Kolotyrkin: J. Electroanal. Chem. **110**, 111 (1980)
6.290 A.Otto: In *Raman Spectroscopy*, ed. by J.Lascombe, P.V.Huong (Wiley, New York 1982) p. 49
6.291 Variation of U should change the metallic surface charge and therefore influence electromagnetic resonances in principle. However, the spectral position of electromagnetic resonances is little effected by U, for instance, the surface-plasmon-induced absorption on roughened electrodes only shifts from 3.52 at $U=-0.7$ V to 3.40 eV at $U=0$ V [6.292]. Rayleigh scattering from an "activated" silver electrode does not change when changing the potential [6.293]. In the case where the "surface bumps" are very small (10–20 Å) [6.211], electromagnetic resonances of the bumps may be more effected; however, this would lead to opposite shifts than those observed [6.294]
6.292 D.M.Kolb, R.Kötz: Surf. Sci. **64**, 96 (1977)
6.293 A.Otto, J.Timper, J.Billmann, I.Pockrand: Phys. Rev. Lett. **45**, 46 (1980)
6.294 J.Billmann, A.Otto: Solid State Commun. **44**, 105 (1982)
6.295 W.Boeck, D.M.Kolb: Surf. Sci. **118**, 613 (1982)
6.296 The gradient of the $\hbar\omega_L \sim U$ lines (Fig. 6.83) was found to be about the same in a Br^- electrolyte, but smaller for pyridine in an SO_4^{2-} electrolyte, and larger in an SCN^- electrolyte [6.297]. A possible explanation was given in [6.290] based on the local electrostatic potential variation caused by the anions. Intensity versus potential scans at different laser frequencies were also described independently from [6.294] by *Macomber* and *Furtak* [6.298] for thiourea adsorbed on silver electrodes in 0.1 M thiourea, 0.1 M K_2SO_4, 10^{-2} N H_2SO_4 electrolyte. No detectable differences in the intensity potential scans were detected for excitation energies between 1.92 and 2.55 eV. Macomber and Furtak concluded that either the molecular states involved were very broad or that the CT transition concept was not operative in this situation. In the meantime, *Furtak* and *Macomber* [6.299] have confirmed the results of Figs. 6.82 and 83
6.297 J.Billmann, Ph.D.Thesis: Universität Düsseldorf (1984) unpublished
6.298 S.H.Macomber, T.E.Furtak: Chem. Phys. Lett. **90**, 59 (1982)
6.299 T.E.Furtak, S.H.Macomber: Chem. Phys. Lett. **95**, 328 (1983)
6.300 S.F.Wong, G.J.Schulz: Phys. Rev. Lett. **35**, 1429 (1975)
6.301 T.H.Wood, D.A.Zwemer: J. Vac. Sci. Technol. **18**, 649 (1981)
6.302 I.Pockrand: Springer Tracts Mod. Phys. **104** (August 1984)
6.303 I.C.Walker, A.Stamatovic, S.F.Wong: J. Chem. Phys. **69**, 5532 (1978)
6.304 G.R.Elliot, G.E.Leroi: J. Chem. Phys. **59**, 1217 (1973)
6.305 A.Campion: J. Electron Spectrosc. Rel. Phen. **29**, 397 (1983)
6.306 A.Campion, D.R.Mullins: Chem. Phys. Lett. **94**, 576 (1983)
6.307 J.C.Tsang, J.R.Kirtley, T.N.Theis: J. Chem. Phys. **77**, 641 (1982)
6.308 L.Moerl, B.Pettinger: Solid State Commun. **43**, 315 (1982)
6.309 Ü.Ertürk, I.Pockrand, A.Otto: Surf. Sci. **131**, 367 (1983)
6.310 D.E.Tevault, R.R.Smardzewski: J. Chem. Phys. **77**, 2221 (1982)
6.311 H.D.Ladouceur, D.E.Tevault, R.R.Smardzewski: J. Chem. Phys. **78**, 980 (1983)
6.312 C.Pettenkofer, I.Pockrand, A.Otto: Surf. Sci. **135**, 52 (1983)
6.313 M.Chelvoyohan, C.H.B.Mee: J. Phys. C, Solid State Phys. **15**, 2305 (1982)
6.314 S.P.Kowalczyk: J. Vac. Sci. Technol. **16**, 520 (1979)
6.315 G.Apai, S.T.Lee, M.G.Mason: Solid State Commun. **37**, 213 (1981); R.C.Baetzold: J. Am. Chem. Soc. **103**, 6116 (1981)

6.316 W.F.Egelhoff Jr: J. Vac. Sci. Technol. **20**, 668 (1982)
6.317 M.Cardona, L.Ley (eds.): *Photoemission in Solids I*, Topics Appl. Phys., Vol. 26 (Springer, Berlin, Heidelberg, New York 1978) Chap. 1
6.318 J.R.Smith, F.J.Arlinghaus, J.G.Gay: Phys. Rev. B**22**, 4757 (1982)
6.319 J.Hulse, J.Küppers, K.Wandelt, G.Ertl: Appl. Surf. Sci. **6**, 453 (1980)
6.320 H.Albers, W.J.J. Van der Waal, O.L.J.Gijzeman, G.A.Bootsma: Surf. Sci. **77**, 1 (1978)
6.321 H.A.Engelhardt, D.Menzel: Surf. Sci. **57**, 591 (1976)
6.322 K.-M.Ho, B.N.Harmon, S.H.Liu: Phys. Rev. Lett. **44**, 1531 (1980)
6.323 D.M.Kolb, W.Boeck, K.-M.Ho, S.H.Liu: Phys. Rev. Lett. **47**, 1921 (1981)
6.324 N.D.Lang: *Solid State Physics* **28**, 225 (Academic Press, New York 1973)
6.325 H.Wagner: *Springer Tracts Mod. Phys.*, Vol. **85**, 151 (Springer, Berlin, Heidelberg, New York 1979)
6.326 J.Küppers, K.Wandelt, G.Ertl: Phys. Rev. Lett. **43**, 928 (1979)
6.327 N.D.Lang, A.R.Williams: Phys. Rev. B**25**, 2940 (1982)
6.328 T.T.Tsong: Progress in Surf. Sci. **10**, 165 (1980)
6.329 V.V.Marinyuk, R.M.Lazorenko-Manevich, Ya.M.Kolotyrkin: Solid State Commun. **43**, 721 (1982)
6.330 S.H.Macomber, T.E.Furtak: Sol. State Commun. **45**, 267 (1983)
6.331 R.K.Chang: CRC Critical Review in Solid State Science
6.332 R.M.Lazorenko-Manevich, V.V.Marinyuk, Ya.M.Kolotyrkin: Doklady Akad. Nauk. SSR, **244**, 664 (1979)
6.333 R.Dornhaus: *Springer Tracts Mod. Phys.* **22**, 201 (Springer, Berlin, Heidelberg, New York 1982)
6.334 C.S.McKee: In *Chemical Physics of Solids and their Surfaces*, Vol. 8 (Roy. Soc. of Chem. 1980)
6.335 R.A.Marbrow, R.M.Lambert: Surf. Sci. **71**, 107 (1978)
6.336 M.Bondart: In *Interactions on Metal Surfaces*, ed. by R.Gomer, Topics Appl. Phys., Vol. 4 (Springer, Berlin, Heidelberg, New York 1975) Chap. 7
6.337 G.A.Somorjai: Surf. Sci. **89**, 496 (1979)
6.338 D.L.Doering, H.Poppa, J.T.Dickinson: J. Vac. Sci. Technol. **18**, 460 (1981)
6.339 P.Gallezot: Surf. Sci. **106**, 459 (1981)
6.340 P.B.Dorain, K.U. von Raben, R.K.Chang, B.L.Laube: Chem. Phys. Lett. **84**, 405 (1981)
6.341 W.Krasser: In [6.248]
6.342 F.R.Aussenegg, A.Leitner, M.E.Lipitsch (eds.): *Surface Studies with Lasers*, Springer Ser. Chem. Phys., Vol. 33
6.343 Proc. of the Conf. Int. Ellipsometrie et Autres Methodes Optiques pour L'Analyse des Surfaces et Films Minces, Paris, France (1983) J. Physique **12**, C 10 (1983)
6.344 I.Pockrand: Springer Tracts Mod. Phys. **104** (August 1984)
6.345 S.A.Lyon, J.M.Worlock: Phys. Rev. Lett. **51**, 593 (1983)
6.346 E.V.Albano, S.Daiser, G.Ertl, R.Miranda, K.Wandelt, N.Garcia: Phys. Rev. Lett. **51**, 2314 (1983)
6.347 A.Otto, J.Billmann, J.Eickmans, Ü.Ertürk, C.Pettenkofer: Surf. Sci. **138**, 319 (1984)
6.348 J.Eickmans, A.Goldmann, A.Otto: accepted at ECOSS VI (to be published)
6.349 J.A.Allen, C.C.Evans, J.W.Mitchell: In *Structure and Properties of Thin Films*, ed. by C.A.Neugebauer, J.B.Newkirk, D.A.Vermilyea (Wiley, New York 1959)
6.350 J.E.Potts, R.Merlin, D.L.Partin: Phys. Rev. B**27**, 3905 (1983)
6.351 H.Yamada, Y.Yamamoto: Surf. Sci. **134**, 71 (1983)
6.352 H.Seki, T.J.Chuang: Chem. Phys. Lett. **100**, 393 (1983)
6.353 M.A.Chesters, Ü.Ertürk, A.Otto: accepted at ECOSS VI (to be published)
6.354 J.Billmann, A.Otto: Surf. Sci. **138**, 1 (1984)
6.355 A.Campion, V.M.Grizzle, D.R.Mullin, J.K.Brown: In [Ref. 6.343, p. 341]
6.356 C.Pettenkofer, A.Otto: In [Ref. 6.343, p. 337]
6.357 T.Watanabe, N.Yanagihara, K.Honda, B.Pettinger, L.Moerl: Chem. Phys. Lett. **96**, 649 (1983)
6.358 T.E.Furtak, D.Roy: Phys. Rev. Lett. **50**, 1301 (1983)
6.359 C.Pettenkofer, Ü.Ertürk, A.Otto: accepted at ECOSS VI (to be published)

7. Theory of Surface-Enhanced Raman Scattering

Karamjeet Arya and Roland Zeyher

With 15 Figures

In this chapter we first discuss a general theory of SERS which is based on the authors' earlier work. Using a polariton description for the interaction of the photon with the molecule and the bounded metal, a general expression for the cross section of SERS is derived in Sect. 7.3. Taking only one or two electronic states on the molecule into account, we show that the expression for the cross section can be written down in a factorized form corresponding to the contributions (a–c) of Sect. 7.1 to the enhancement discussed previously. These three contributions are discussed in Sect. 7.4, 7.5, and 7.6, respectively.

7.1 Background

Until recently, Raman scattering did not seem to be a useful tool for studying vibrational spectra of monolayers of adsorbed molecules on metal surfaces. The Raman signal was usually too weak to be detectable; there are only about 10^{14} molecules in a monolayer participating in the scattering process compared to about 10^{20} molecules in the case of a solid or a liquid. In 1974, *Fleischmann* et al. [7.1] observed Raman scattering from pyridine molecules adsorbed on a rough silver surface. They attributed the scattering to the large increase in the number of admolecules caused by a large increase in the effective surface area due to roughness. However, in 1977, *Jeanmaire* and *Van Duyne* [7.2] and *Albrecht* and *Creighton* [7.3] repeated these experiments. They simultaneously came to the conclusion that the observed Raman intensity was due to a large increase in the Raman cross section of the pyridine molecule when it became adsorbed on the Ag surface. Since then there have been extensive studies (see, e.g., the reviews [7.4–8] and Chap. 6) both theoretical [7.10–42] and experimental [7.43–62] on this surprising effect which is now generally known as surface-enhanced Raman scattering (SERS). Many more molecules, e.g., CO, CN, Cl, etc., have been found to give enhanced cross section. However, there are only a few substrates which do show SERS: Ag (enhancement $\sim 10^6$) and Cu, Au, Pt, Ni, and Hg (enhancements $\sim 10^2$–10^4, see Chap. 6). Another important point in SERS is that the surface roughness plays a crucial role in obtaining very large enhancements. Experiments on pyridine adsorbed on a smooth Ag surface give enhancements of only about 400 [7.62]. Somewhat surprising is, however, the recent observation of large enhancements (10^4–10^5) for molecules adsorbed on a liquid mercury surface which is certainly flat [7.61]; see also Chap. 6.

The precise origin of SERS is still not completely understood. However, it is now generally believed that one or more of the following three effects, together with surface roughness, play an important role.

a) The bounded metal has electromagnetic modes which are localized near the surface and have large field amplitudes there. If an incident or a scattered photon can excite these modes substantially, the molecule will be exposed to large effective fields varying with the frequency of the incident or scattered photon. As a result, the electron-photon (e-p) interaction and hence the cross section will be enhanced. A rough metal surface is especially effective in producing large electric fields. In the absence of admolecules, the incident or scattered s or p-photons cannot couple to the surface modes in the case of an ideal surface. The coupling becomes only nonzero because of the presence of the admolecules and, in a more direct and effective way, by deviations from the ideal plane surface. In addition, the local electric fields due to the localized electromagnetic modes may become very large near the rough structures of the surface such as needles, bumps, etc.

b) The incident or scattered photon fields can be substantially changed due to polarization effects associated with the presence of the molecule. For instance, all kinds of real or virtual dipolar transitions on the molecule (say, between the ground state of the molecule and unoccupied states in the molecule or in the metal, or because of infrared-active molecular vibration) induce image dipoles in the metal and thus change the electric field.

c) The electronic levels of the molecule may shift or broaden because of chemisorption effects. Thus the initial, final and intermediate electronic states for the scattering process may change compared with the free molecule. In particular, the continuum of unoccupied metal states can act as intermediate states.

From a more general point of view, (a) and (b) describe the local field effects. The total susceptibility function for the metal plus molecule system can be written as $\overleftrightarrow{\chi}(\boldsymbol{r}, \boldsymbol{r}', \omega)$. Without loss of generality, $\overleftrightarrow{\chi}$ can be split into $\overleftrightarrow{\chi}^{(0)}(\boldsymbol{r}, \boldsymbol{r}', \omega)$, describing the bounded metal in the absence of the admolecule, and the difference $\delta\overleftrightarrow{\chi}(\boldsymbol{r},\boldsymbol{r}', \omega) = \overleftrightarrow{\chi}(\boldsymbol{r},\boldsymbol{r}', \omega) - \overleftrightarrow{\chi}^{(0)}(\boldsymbol{r}, \boldsymbol{r}', \omega)$, which accounts for all effects associated with the admolecules. Modification (a) is taken into account if the photon propagators are calculated as polariton propagators in a medium with the susceptibility $\overleftrightarrow{\chi}^{(0)}$. Modification (b) is included if, in addition, $\delta\overleftrightarrow{\chi}(\boldsymbol{r},\boldsymbol{r}', \omega)$ is taken into account. In the case of a statistically rough surface, the final cross section must still be averaged over the ensemble of physically equivalent rough surfaces. Finally, modification (c) is taken into account if $\delta\overleftrightarrow{\chi}(\boldsymbol{r}, \boldsymbol{r}', \omega)$ is calculated using the true electronic states of the total system, i.e., molecule and metal, instead of the free molecular states.

Various theoretical models exist for SERS which take into account one or more of the above modifications approximately. Contribution (a) has been considered in [7.11–29]. In order to obtain the total field near the metal surface, one must solve Maxwell's equations in the presence of the bounded metal. Neglecting the $\boldsymbol{r}$-dependence of $\chi^{(0)}$, this can be done in a straightforward way for a few geometries of the metal. A semi-infinite metal with its plane surface superimposed with a weak grating [7.15, 19] and a metal sphere [7.14] can be done exactly. The cases of a general spheroid [7.27] and a hemispheroid sitting on a flat metal surface [7.23, 26] have been treated neglecting retardation effects. These calculations, roughly speaking, show that large total electric fields near the surface and hence large enhancements in the cross section can be expected if (i) the surface profile varies substantially on a scale given by the wavelength of the light and if (ii) the imaginary part of the dielectric function $\varepsilon(\omega)=1+4\pi\chi^{(0)}(\omega)$ is small for incident and scattered frequencies. For instance, in the case of a sphere with radius small compared with the light wavelength, the maximum enhancement near the Mie resonance is proportional to $(\mathrm{Im}\{\varepsilon\})^{-4}$.

The case of two metal spheres interacting via the electromagnetic field is discussed in [7.28] in the Rayleigh limit, i.e., without retardation. If $D \geqslant R$ (R is the radius of the sphere, D the distance between the spheres), the two spheres interact weakly and the resonance frequencies are those of the isolated spheres. For $D < R$, the resonances interact strongly; one large frequency is still given approximately by the noninteracting case whereas the second eigenfrequency shifts to lower values and loses intensity. A more general case has been treated by *Moskovits* [7.11]. He considered many spheres on a smooth mirror embedded in a dielectric medium. Due to the electrostatic interaction of the spheres (representing bumps), the total system possesses a new collective eigenfrequency which depends on the density of bumps. It is suggested that SERS is due to (pre)resonant Raman scattering with the above collective resonance as the dominant intermediate state. A similar theory given in [7.29] considers a film of silver-island consisting of symmetric ellipsoids distributed in a square array on a glass surface. The ellipsoids interact with one another. The frequency of the collective mode with infrared fields is determined and it is proposed that this (transverse) collective mode plays an important role for SERS.

A quite different approach has been given in [7.20, 21]. The rough surface is described by a statistical ensemble of equivalent surface profiles characterized by a transverse correlation length a and a root mean square deviation δ from the flat surface. A perturbation series in terms of averaged, translationally invariant photon propagators is set up and treated within certain approximations. Lowest-order perturbation theory was used in [7.20] and rather small enhancements for realistic values of a and δ were obtained. Large enhancements are obtained in the nonperturbative treatment of [7.21]. The approach of [7.20, 21] and that of *Moskovits* [7.11] are in some sense complementary. The former is more appropriate for small ratios δ/a corresponding to the situation where the fields parallel to the surface are rather homogeneous, whereas the latter is expected to work for large ratios δ/a and inhomogeneous field distributions. Correspondingly, the statistical approach yields rather frequency-independent enhancements in contrast to the results of Moskovits.

Finally, the case of a flat metal surface with a grating has been investigated both experimentally [7.43–46] and theoretically [7.15, 19]. Enhancements of 3 to 4 orders of magnitude have been found for Ag for those frequencies at which the incident or scattered photon can (for a given scattering configuration) resonantly excite the surface plasmon polariton.

The second contribution (b) accounts for all polarization effects associated with the presence of the molecule. It has been calculated within image potential models where the e-p interaction is assumed to be increased, largely due to the induced dipoles in the metal [7.30, 31]. From a microscopic point of view, these image potentials arise from the interaction of the molecular dipole transitions with the surface modes as discussed in [7.19, 32–36]. The resulting enhancements are quite sensitive to the microscopic details such as the position of the molecule with respect to the image plane and the magnitude of the various dipole matrix elements. Simple estimates show that contribution (b) can give enhancements between 1 and 2 orders of magnitude for optimal values of the microscopic parameters.

Contribution (c) has been considered in [7.5, 19, 37–42]. For instance, *Arya* and *Zeyher* [7.19] included chemisorption effects within the *Newns-Anderson* model [7.63, 64]. They considered electronic transitions between an occupied level of the molecule, broadened due to its interaction with the metal, and unoccupied states of the metal. If the molecule is close to the metal surface, such transitions are possible because of the overlap of the wave functions of the molecule and the neighbouring metal atoms. Recent experiments [7.54–55] on SERS support the presence of such transitions. Expected enhancements due to chemisorption are, in the most favourable cases, around 1 to 2 orders of magnitude.

7.2 Hamiltonian

The total Hamiltonian for the system consisting of the radiation field and a molecule adsorbed on the metal surface can be written as

$$\mathscr{H} = \mathscr{H}_0 + \mathscr{H}', \tag{7.1}$$

where

$$\mathscr{H}_0 = \sum_{Q\lambda} \hbar\omega_\lambda(Q) a^\dagger_{Q\lambda} a_{Q\lambda} + \sum_{s,t} T_{st} c^\dagger_s c_t + \hbar\Omega b^\dagger b, \tag{7.2}$$

$$\mathscr{H}' = \sum_{Q\lambda}\sum_{s,t} \alpha_{st}(Q\lambda) c^\dagger_s c_t A_{Q\lambda} + \sum_{Q\lambda, Q'\lambda'}\sum_{s,t} \beta_{st}(Q\lambda, Q'\lambda') c^\dagger_s c_t A_{Q\lambda} A^\dagger_{Q'\lambda'} + \sum_{s,t} \eta_{st} c^\dagger_s c_t (b + b^\dagger), \tag{7.3}$$

$$A_{Q\lambda} = a_{Q\lambda} + a^\dagger_{-Q\lambda}. \tag{7.4}$$

In (7.2), the first term corresponds to the radiation field in vacuum with $a^+_{Q\lambda}$ ($a_{Q\lambda}$) denoting the annihilation (creation) operator for a photon with wave vector $\boldsymbol{Q}$, polarization λ and frequency $\omega_\lambda(\boldsymbol{Q})$. The second term corresponds to the electronic part of the metal-molecule system. In writing down this term, we have used a local basis of orthonormal one-electron states ϕ_s. The index s denotes either a lattice site of the metal or the molecular site. It also contains implicitly the band (orbital) index for the metal (molecule). c_s ($c_s^\dagger$) is the annihilation (creation) operator for the electron at site s. The last term in (7.2) corresponds to a vibrational mode with energy $\hbar\Omega$. The operator b ($b^\dagger$) denotes the corresponding annihilation (creation) operator.

Equation (7.3) contains the various interaction terms. The first two terms, which are linear and bilinear in the vector potential, denote the electron-photon (e-p) interaction. The explicit expressions for α and β are given by

$$\alpha_{st}(\boldsymbol{Q}\lambda) = -\frac{e}{m}\frac{2\pi\hbar}{\omega_\lambda(\boldsymbol{Q})V}\langle\phi_s(\boldsymbol{r})|\mathrm{e}^{\mathrm{i}\boldsymbol{Q}\cdot\boldsymbol{r}}\hat{e}_\lambda(\boldsymbol{Q})\cdot\boldsymbol{p}|\phi_t(\boldsymbol{r})\rangle, \tag{7.5}$$

$$\beta_{st}(\boldsymbol{Q}\lambda, \boldsymbol{Q}'\lambda') = \frac{\pi e^2 h}{mV}\frac{\hat{e}_\lambda(\boldsymbol{Q})\cdot\hat{e}_{\lambda'}(\boldsymbol{Q}')}{[\omega_\lambda(\boldsymbol{Q})\omega_{\lambda'}(\boldsymbol{Q}')]^{1/2}}\langle\phi_s(\boldsymbol{r})|\mathrm{e}^{\mathrm{i}(\boldsymbol{Q}-\boldsymbol{Q}')\cdot\boldsymbol{r}}|\phi_t(\boldsymbol{r})\rangle, \tag{7.6}$$

where $\hat{e}_\lambda(\boldsymbol{Q})$ is the polarization vector of the photon, V is a normalization volume and $\boldsymbol{p}$ the momentum operator of the electron. The last term in (7.3) represents the electron-vibration interaction.

We also give here the definition of the photon Green's function which will be used in the next section in calculating the scattering cross section. The photon propagator at finite temperature [7.65–66] is defined as (in the following we use units with $\hbar = 1$)

$$D(\boldsymbol{Q}\lambda\tau, \boldsymbol{Q}'\lambda'\tau') = \langle T_\tau[A^\dagger_{\boldsymbol{Q}\lambda}(\tau)A_{\boldsymbol{Q}'\lambda'}(\tau')]\rangle, \tag{7.7}$$

where $A_{\boldsymbol{Q}\lambda}(\tau) = \mathrm{e}^{H\tau}A_{\boldsymbol{Q}\lambda}\mathrm{e}^{-H\tau}$, $\tau = \mathrm{i}t$ and T_τ is the time ordering operator. The Fourier transform of (7.7) can be written as

$$D(\boldsymbol{Q}\lambda, \boldsymbol{Q}'\lambda', \omega_m) = \frac{1}{2}\int_{-\beta}^{\beta} d\tau \mathrm{e}^{\mathrm{i}\omega_m\tau}D(\boldsymbol{Q}\lambda 0, \boldsymbol{Q}'\lambda'\tau'), \tag{7.8}$$

where $\beta^{-1} = k_\mathrm{B}T$; k_B is the Boltzmann constant, T the temperature and $\omega_m = (2\pi/\beta)m$, $m = 0, 1, 2\dots$. The free photon propagator $D^{(0)}(\boldsymbol{Q}\lambda, \boldsymbol{Q}'\lambda', \omega_m)$ is given by

$$D^{(0)}(\boldsymbol{Q}\lambda, \boldsymbol{Q}'\lambda', \omega_m) = \frac{2\omega_\lambda(\boldsymbol{Q})}{\omega_m^2 + \omega_\lambda^2(\boldsymbol{Q})}\Delta_{\boldsymbol{Q},\boldsymbol{Q}'}\Delta_{\lambda,\lambda'}, \tag{7.9}$$

where $\Delta_{\boldsymbol{Q},\boldsymbol{Q}'}$ denotes the Kronecker delta function. In a similar way one can also define the propagators for the vibrational mode and the electron *Green*'s functions [7.65].

7.3 Scattering Cross Section

7.3.1 General Expression

In this section we derive a general expression for the Raman cross section from a molecule adsorbed on a metal surface using the polariton approach [7.19, 66]. The incident photon interacts with the molecule-metal system and is scattered to other channels, for instance, an outgoing photon and molecular vibration. It thus gets a nonvanishing complex self-energy and a finite lifetime. Mathematically, the total scattering cross section of an incident photon with momentum $\boldsymbol{Q}$, polarization λ and frequency ω is related to the imaginary part of its self-energy Σ by

$$\sigma = (2V/c)\ \mathrm{Im}\ \{\Sigma(\boldsymbol{Q}\lambda, \omega + \mathrm{i}\eta)\}. \tag{7.10}$$

In (7.10) it is assumed that the analytic continuation from the discrete imaginary frequencies $\mathrm{i}\omega_m$ to the frequency $\omega + \mathrm{i}\eta$ near the real axis has been carried out.

The infinite set of Feynman diagrams considered for the calculation of Σ are shown in Fig. 7.1a–c. They wavy line describes the photon propagator defined in (7.7). The dotted line denotes the propagator of the vibration on the molecule. The thin full line in Fig. 7.1d is the electron propagator of the molecule-metal system. The double wavy line defined in Fig. 7.1b describes a photon propagator renormalized by polarization effects in the coupled molecule-metal system. Figure 7.1c describes the renormalized e–p interaction which appears at the two ends of Fig. 7.1a. The thick full line defined in Fig. 7.1d corresponds to the renormalized electron propagator. The lowest-order approximation for the cross section is given by Fig. 7.1a if (A) the vertex correction is neglected and the double wavy line is replaced by a single wavy line, and if (B) the thick solid lines

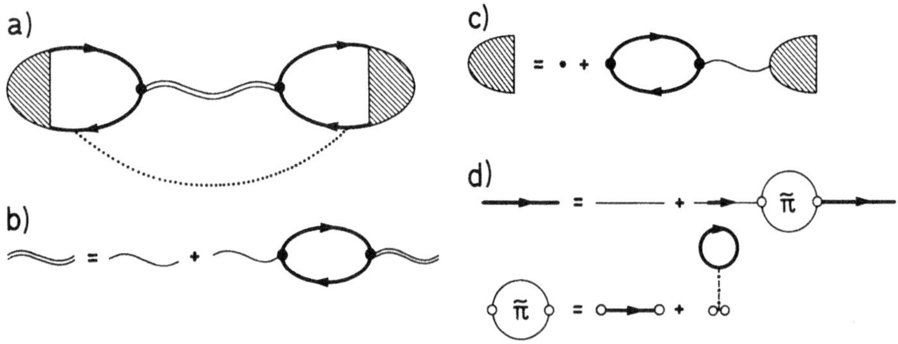

Fig. 7.1. (**a**) Self-energy of the photon propagator considered for Raman scattering; (**b**) Dyson equation for the photon propagator; (**c**) equation for the renormalized e-p vertex; (**d**) Dyson equation for the electronic propagator. The self-energy $\tilde{\Pi}$ consists of a hopping contribution due to one-particle potentials (*first term*) and a Hubbard contribution (*second term*) which is used only for the admolecule

are replaced by thin ones. Inclusion of correction (A) means that the scattering process is described in a polariton picture. Inclusion of correction (B) means that each single bubble is to be calculated with wave functions which contain chemisorption effects.

Thus, using the standard rules for evaluating Feynman diagrams, one can write down the analytic contribution from these diagrams to the self-energy $\Sigma(\boldsymbol{Q}\lambda,\omega_m)$. Carrying out the analytic continuation in $\Sigma(\boldsymbol{Q}\lambda,\omega_m)$ and using (7.10), the differential cross section for Stokes scattering is given by [7.19]

$$\frac{d\sigma^{\mathrm{St}}}{d\Omega_{\mathrm{s}}}=\frac{V^2\omega'^2}{4\pi^2c^4}|\tilde{g}(\boldsymbol{Q}\lambda,\boldsymbol{Q}'\lambda',\omega+\mathrm{i}\eta,-\Omega)|^2\,[1+f(\Omega)]. \tag{7.11}$$

In (7.11) $\boldsymbol{Q}'(\lambda')$ is the wave vector (polarization vector) for the scattered photon with $c|\boldsymbol{Q}'|=\omega'=\omega-\Omega$ and Ω_{s} is the solid angle. Similarly for anti-Stokes' scattering, we obtain

$$\frac{d\sigma^{\mathrm{ant-St}}}{d\Omega_{\mathrm{s}}}=\frac{V^2\omega''^2}{4\pi^2c^4}|\tilde{g}(\boldsymbol{Q}\lambda,\boldsymbol{Q}''\lambda'',\omega+\mathrm{i}\eta,\Omega)|^2f(\Omega) \tag{7.12}$$

with $c|\boldsymbol{Q}''|=\omega''=\omega+\Omega$. The other quantities in (7.11, 12) are defined as:

$$f(\Omega)=(\mathrm{e}^{\beta\Omega}-1)^{-1}, \tag{7.13}$$

$$\begin{aligned}\tilde{g}(\boldsymbol{Q}\lambda,\boldsymbol{Q}'\lambda',\omega,-\Omega)=&\sum_{\substack{s,t,u\\ s',t',u'}}V_{ts'}(\boldsymbol{Q}\lambda,\omega)\eta_{ut'}V^*_{u's}(\boldsymbol{Q},\omega-\Omega)\\ &\cdot[\mathscr{L}^{s't'u'}_{stu}(\omega,-\Omega)+\mathscr{L}^{stu}_{s't'u'}(-\omega,\Omega)],\end{aligned} \tag{7.14}$$

$$\mathscr{L}^{s't'u'}_{stu}(\omega_1,\omega_2)=\int\frac{d\omega}{2\pi}G_{s's}(\omega)G_{t't}(\omega+\omega_1)G_{u'u}(\omega+\omega_1+\omega_2), \tag{7.15}$$

where $G_{ss'}(\omega)$ is the electron propagator. $V_{st}(\boldsymbol{Q}\lambda,\omega)$ is the renormalized e–p interaction vertex (Fig. 7.1c) which can be expressed in terms of the renormalized photon propagator (Fig. 7.1b) as

$$V_{st}(\boldsymbol{Q}\lambda,\omega)=\sum_{\boldsymbol{Q}\lambda}D(\boldsymbol{Q}\lambda,\boldsymbol{Q}'\lambda',\omega)\alpha_{st}(\boldsymbol{Q}'\lambda')/D^{(0)}(\boldsymbol{Q}\lambda,\omega), \tag{7.16}$$

where

$$D^{(0)}(\boldsymbol{Q}\lambda,\omega)=2\omega_\lambda(\boldsymbol{Q})/[\omega^2_\lambda(\boldsymbol{Q})-\omega^2] \tag{7.17}$$

and $\alpha_{st}(\boldsymbol{Q})$ and η_{ut} are defined in (7.3).

The photon propagator $D(\boldsymbol{Q}\lambda, \boldsymbol{Q}'\lambda', \omega)$ satisfies the Dyson equation

$$D(\boldsymbol{Q}\lambda, \boldsymbol{Q}'\lambda', \omega) = [\Delta_{\boldsymbol{Q},\boldsymbol{Q}'}\delta_{\lambda,\lambda'} + \sum_{\boldsymbol{Q}''\lambda''} D(\boldsymbol{Q}\lambda, \boldsymbol{Q}''\lambda'', \omega)\Pi(\boldsymbol{Q}''\lambda'', \boldsymbol{Q}'\lambda', \omega)] \cdot D^{(0)}(\boldsymbol{Q}'\lambda', \omega), \tag{7.18}$$

where $\Pi(\boldsymbol{Q}\lambda, \boldsymbol{Q}'\lambda', \omega)$ is the electronic polarization due to the metal-molecule system. In [7.19] we have discussed the procedure to solve (7.18) and hence (7.16) by splitting Π into two parts. The first one contains all electronic transitions between metal states and the second the remaining transitions between molecular or between molecular and metal states. Following [7.19], one can write (7.16) as

$$V_{st}(\boldsymbol{Q}\lambda, \omega) = \sum_{s_1,t_1} \{[1 - X(\omega)]^{-1}\}_{st,s_1t_1} \tilde{\alpha}_{s_1t_1}(\boldsymbol{Q}\lambda, \omega) \tag{7.19}$$

with

$$X_{sts_1t_1}(\omega) = \sum_{s_2,t_2}{}' R^{t_1s_2}_{t_2s_1}(\omega) \sum_{\boldsymbol{Q}\lambda,\boldsymbol{Q}'\lambda'} \alpha^*_{s_2t_2}(\boldsymbol{Q}\lambda)\tilde{D}(\boldsymbol{Q}\lambda, \boldsymbol{Q}'\lambda', \omega)\alpha_{st}(\boldsymbol{Q}'\lambda'), \tag{7.20}$$

$$R^{s't'}_{st}(\omega) = -\int \frac{d\omega_1}{2\pi} G_{s's}(\omega_1)G_{t't}(\omega_1 + \omega), \tag{7.21}$$

$$\tilde{\alpha}_{st}(\boldsymbol{Q}\lambda, \omega) = \sum_{\boldsymbol{Q}'\lambda'} \tilde{D}(\boldsymbol{Q}\lambda, \boldsymbol{Q}'\lambda', \omega)\alpha_{st}(\boldsymbol{Q}'\lambda')/D^{(0)}(\boldsymbol{Q}\lambda, \omega). \tag{7.22}$$

The prime at the summation sign in (7.20) means that at least one of the four indices s_1, t_1, s_2, t_2 denotes a function localized at the molecular site. The meaning of $\tilde{D}$ in (7.20) becomes clear in $\boldsymbol{r}$-space:

$$\tilde{D}(\boldsymbol{Q}\lambda, \boldsymbol{Q}'\lambda', \omega) = \frac{[\omega_\lambda(\boldsymbol{Q})\omega_{\lambda'}(\boldsymbol{Q}')]^{1/2}}{2\pi c^2 V} \int d\boldsymbol{r} \int d\boldsymbol{r}' \, e^{i(\boldsymbol{Q}\cdot\boldsymbol{r}' - \boldsymbol{Q}\cdot\boldsymbol{r})} \cdot \hat{e}_\lambda(\boldsymbol{Q}) \cdot \overleftrightarrow{\mathscr{D}}(\boldsymbol{r}, \boldsymbol{r}'\omega) \cdot \hat{e}_{\lambda'}(\boldsymbol{Q}'). \tag{7.23}$$

The Green's function $\overleftrightarrow{\mathscr{D}}(\boldsymbol{r}, \boldsymbol{r}'\omega)$ satisfies the vector Helmholtz equation

$$\nabla^2_r \mathscr{D}_{ij}(\boldsymbol{r}, \boldsymbol{r}'\omega) + \frac{\omega^2}{c^2} \sum_k \int d\boldsymbol{r}'' \varepsilon_{ik}(\boldsymbol{r}, \boldsymbol{r}'', \omega)\mathscr{D}_{kj}(\boldsymbol{r}'', \boldsymbol{r}'\omega) = -4\pi\delta(\boldsymbol{r} - \boldsymbol{r}')\delta_{ij} \tag{7.24}$$

with $\overleftrightarrow{\varepsilon}(\boldsymbol{r}, \boldsymbol{r}', \omega)$ being the dielectric function of the bounded metal, excluding the molecule. $\overleftrightarrow{\mathscr{D}}$ is thus identical with the classical photon Green's function.

According to the above equations, the calculation of $V_{st}(\boldsymbol{Q}\lambda, \omega)$ requires (i) the calculation of the inverse of the matrix $[1 - X(\omega)]$ and (ii) the photon Green's function $\overleftrightarrow{\mathscr{D}}(\boldsymbol{r}, \boldsymbol{r}', \omega)$. (a) Since $\alpha_{st}(\boldsymbol{Q}\lambda, \omega)$ is the momentum matrix between two localized functions, it is of short range. Furthermore, one of the four indices s_1, t_1, s_2, t_2 under the summation sign in (7.20) denotes a function localized at the molecule. This means that at least one of the two short-ranged pairs $(s_1 t_1)$ and $(s_2 t_2)$ in R is localized near the molecule. Due to the short-range nature of chemisorption, the other pair must also be localized near the molecule. As a result, the matrix $X_{sts_1t_1}(\omega)$ is of short range in the sites s, t, s_1, t_1 and can be assumed to be nonzero only if all the sites are near the molecule. Thus, $[1-X(\omega)]^{-1}$ can be easily calculated because $X(\omega)$ is a small matrix of dimensions $l \times m \times n$ where l, n, m are the number of states in the molecule, the number of bands in the metal and the number of metal atoms which interact with the molecule, respectively. (b) Using a specific form for $\vec{\varepsilon}(\boldsymbol{r}, \boldsymbol{r}', \omega)$ (thus specifying in particular the properties of the metal surface), (7.24) can be solved by various methods [7.67–69]. Some of these methods will be discussed in the next section.

Inserting (7.19) in (7.14), the resulting expressions for the cross section are quite general. These include chemisorption effects as well as renormalization of the e–p interaction due to local field effects. These can be applied to an arbitrary electronic structure and to an arbitrary nonstatistical metal surface. Moreover, the cross sections have been brought into a form that makes their numerical evaluations for realistic systems possible.

In order to discuss these expressions in more physical terms, let us consider three different regions of distances z_0 between the molecule and the metal surface.

A) $z_0 \sim \infty$, i.e., Raman scattering from a free molecule. The polarization effects or vertex corrections are then small and can be neglected. As a result the cross section can be calculated by lowest-order perturbation theory.

B) $a \ll z_0 < \infty$, where a is the lattice constant of the metal. In this case the molecule and metal are still uncoupled with respect to the electronic matrix elements but are coupled with respect to the long-range Coulomb forces. The basic scattering process is still that of case (A); however, local field effects induced by the metal or the molecule have to be taken into account. There are two kinds of local field effects: (i) the incident (or scattered) photon excites surface polaritons because the periodicity of the surface is broken by the admolecule or surface roughness. The associated electric field is large near the metal surface and usually extends several hundred Å into the vacuum. Electronic transitions in the molecule are due to this enhanced field instead of the bare field of the incident (or scattered) photon. These effects are described in our theory by the difference between $\tilde{\alpha}_{st}(\boldsymbol{Q}\lambda, \omega)$ and $\alpha_{st}(\boldsymbol{Q}\lambda, \omega)$ and between $\tilde{D}(\boldsymbol{Q}\lambda, \boldsymbol{Q}'\lambda', \omega)$ and $D(\boldsymbol{Q}\lambda, \boldsymbol{Q}'\lambda', \omega)$. (ii) As discussed in [7.70, 71], the interactions of molecular transitions and the surface modes give rise to dynamic image dipoles in the metal which cause extra contributions to the incident and scattered radiation fields. These effects are described by the X-matrices in our theory. As will be discussed in Sect. 7.5, the microscopic expression for X yields the phenomenological image potentials of [7.30, 31] if the dielectric function of the metal is $\boldsymbol{Q}$ independent and if X is a scalar, i.e., if a two-level model for the molecule can be used.

(C) $z_0 \sim a$, corresponding to the case where the molecule is coupled to the metal also via the short-range tight-binding matrix elements. The scattering process discussed in (B) still applies but with one important modification. Even if X is a scalar, its z_0 dependence deviates strongly from its classical image potential form. For instance, the $1/z_0^3$ divergence at $z_0=0$ is removed and $X(z_0=0)$ is finite (see Sect. 7.5 for more details). These deviations from the classical form are due to two facts: (i) the extension of the molecular wave functions are comparable to z_0 so that an electron on the molecule can no longer be considered as a point charge; (ii) because of $z_0 \sim a$, the molecular transitions excite surface polaritons mainly at the edges of the two-dimensional Brillouin zone because of phase space reasons. This means, however, that the dispersion of the surface polaritons can no longer be neglected.

In addition to the above scattering process, there are now more scattering possibilities due to the nonvanishing tight-binding matrix elements between the molecule and the metal. For instance, the molecular vibration may be created by electronic transitions in the metal [7.41, 42] or between molecule and metal [7.19, 54, 55]. All these processes are contained, for instance, in (7.14) for a suitable choice for the localized functions $s, t, u \ldots$ Moreover, all the local field effects produced by these transitions are taken into account by the diagonal and nondiagonal elements of X.

7.3.2 Scattering Cross Section in the Case of a Molecule with Two Electronic States

The general expression (7.11 or 12) for the scattering cross section involves matrix elements $\alpha_{st}(\boldsymbol{Q}\lambda)$ between various localized functions. In order to be able to work out more details analytically, we consider in this section a simple two-state model for the molecule. The first state ϕ_a corresponds to an occupied level of the molecule which has, in general, a finite width due to chemisorption effects. The second state ϕ_b may be either an unoccupied broad state of the molecule or of a neighbouring metal atom. The second case may become important if the molecule is sufficiently close to the metal surface. In general, both cases should be considered at the same time.

Assuming a two-state model and G_{ss}' to be diagonal, (7.11) reduces to the simplified form

$$\left(\frac{d\sigma_{\lambda\to\lambda'}}{d\Omega_s}\right)^{\mathrm{St}} = \frac{V^2\omega'^2}{4\pi^2c^4}\left|\frac{\tilde{\alpha}_{ba}(\boldsymbol{Q}\lambda,\omega+\mathrm{i}\eta)}{1-X_{ba}(\omega+\mathrm{i}\eta)}\right|^2 \left|\frac{\tilde{\alpha}_{ba}(\boldsymbol{Q}'\lambda',\omega-\Omega+\mathrm{i}\eta)}{X_{ba}(\omega-\Omega+\mathrm{i}\eta)}\right|^2$$
$$\cdot\left|\mathscr{L}_{abb}(\omega+\mathrm{i}\eta,\,-\Omega)\eta_{bb}+\mathscr{L}_{baa}(-\omega-\mathrm{i}\eta,\Omega)\eta_{aa}\right|^2[1+f(\Omega)], \tag{7.25}$$

$$\mathscr{L}_{baa}(-\omega-\mathrm{i}\eta,\Omega) = -\int_{-\infty}^{\varepsilon_F} d\omega_1 \int_{-\infty}^{\varepsilon_F} d\omega_2 \int_{\varepsilon_F}^{\infty} d\omega_3$$
$$\cdot\frac{\varrho_a(\omega_1)\varrho_a(\omega_2)\varrho_b(\omega_3)}{(\omega+\omega_1-\omega_3+\mathrm{i}\eta)\,(\omega+\omega_2-\omega_3+\mathrm{i}\eta)}, \tag{7.26}$$

$$\mathcal{L}_{abb}(\omega+\mathrm{i}\eta,-\Omega)=\int_{-\infty}^{\varepsilon_F} d\omega_1 \int_{\varepsilon_F}^{\infty} d\omega_2 \int_{\varepsilon_F}^{\infty} d\omega_3 \cdot \frac{\varrho_a(\omega_1)\varrho_b(\omega_2)\varrho_b(\omega_3)}{(\omega-\omega_1-\omega_2+\mathrm{i}\eta)\,(\omega+\omega_1-\omega_3-\Omega+\mathrm{i}\eta)}. \tag{7.27}$$

$\varrho_{a,b}(\omega)$ are the spectral densities of the states ϕ_a, ϕ_b, respectively, and ε_F is the Fermi energy. Furthermore, $\tilde{\alpha}_{ba}$ and X_{ba} [(7.22) and (7.20)] can be expressed in terms of $\vec{\vec{\mathscr{D}}}(\boldsymbol{r},\boldsymbol{r}',\omega)$ by using (7.23):

$$\tilde{\alpha}_{ba}(\boldsymbol{Q}\lambda,\omega)=A_0(\boldsymbol{Q}\lambda)\,\frac{\omega_\lambda^2(\boldsymbol{Q})-\omega^2-\mathrm{i}\eta}{4\pi c^2}\sum_{\alpha,\beta}\hat{e}_{\lambda\alpha}(\boldsymbol{Q})\int d\boldsymbol{r}\,\mathrm{e}^{\mathrm{i}\boldsymbol{Q}\cdot\boldsymbol{r}}\int d\boldsymbol{r}'\,\mathscr{D}_{\alpha\beta}(\boldsymbol{r},\boldsymbol{r}'\omega) \cdot \phi_b^*(\boldsymbol{r}')p_\beta\phi_a(\boldsymbol{r}'), \tag{7.28}$$

$$X_{ba}(\omega)=R_{ba}(\omega)\,(e/mc)^2\sum_{\alpha,\beta}\int d\boldsymbol{r}\int d\boldsymbol{r}'\,\phi_b(\boldsymbol{r})p_\alpha\phi_a(\boldsymbol{r})\mathscr{D}_{\alpha\beta}(\boldsymbol{r},\boldsymbol{r}',\omega) \cdot \phi_b^*(\boldsymbol{r}')p_\beta\phi_a(\boldsymbol{r}'), \tag{7.29}$$

where

$$A_0(\boldsymbol{Q}\lambda)=-(e/m)\,(2\pi\hbar/V)^{1/2}, \tag{7.30}$$

$$R_{ba}(\omega)=\int_{-\infty}^{\varepsilon_F} d\omega_1\int_{\varepsilon_F}^{\infty} d\omega_2\,\frac{\varrho_a(\omega_1)\varrho_b(\omega_2)}{\omega_2-\omega_1-\omega}. \tag{7.31}$$

Similar expression holds for anti-Stokes scattering.

In SERS, one is usually interested in the relative cross section of an admolecule compared to that of an isolated molecule. The cross section for an isolated molecule is obtained from (7.25) by taking only the lowest-order contribution into account (assuming $\eta_{bb}=0$):

$$\left(\frac{d\sigma_{\lambda-\lambda'}}{d\Omega}\right)^{\mathrm{St}}=\frac{V^2\omega'^2}{4\pi^2c^4}\left|\frac{\alpha_{ba}(\boldsymbol{Q}\lambda)\eta_{aa}\alpha_{ba}^*(\boldsymbol{Q}'\lambda')}{(\omega-\omega_{ba})\,(\omega-\omega_{ba}-\Omega)}\right|^2[1+f(\Omega)]. \tag{7.32}$$

ω_{ba} is the energy difference of the unperturbed states b and a. Taking the ratio of (7.25) and (7.32), we obtain for the enhancement factor $\varrho(\omega)$:

$$\varrho(\omega)=\varrho_{\mathrm{I}}(\omega)\varrho_{\mathrm{II}}(\omega)\varrho_{\mathrm{III}}(\omega), \tag{7.33}$$

with

$$\varrho_{\mathrm{I}}(\omega)=\left|\frac{\tilde{\alpha}_{ba}(\boldsymbol{Q}\lambda,\omega+\mathrm{i}\eta)}{\tilde{\alpha}_{ba}(\boldsymbol{Q}\lambda)}\right|^2\left|\frac{\tilde{\alpha}_{ba}(\boldsymbol{Q}'\lambda',\omega-\Omega+\mathrm{i}\eta)}{\tilde{\alpha}_{ba}(\boldsymbol{Q}'\lambda')}\right|^2, \tag{7.34}$$

$$\varrho_{\mathrm{II}}(\omega)=\left|1-X_{ba}(\omega+\mathrm{i}\eta)\right|^{-2}\left|1-X_{ba}(\omega-\Omega+\mathrm{i}\eta)\right|^{-2}, \tag{7.35}$$

$$\varrho_{\mathrm{III}}(\omega)=\left|(\omega-\omega_{ba})\,(\omega-\omega_{ba}-\Omega)\,\mathscr{L}_{baa}(-\omega-\mathrm{i}\eta,\Omega)\right|^2. \tag{7.36}$$

The three factors $\varrho_{\mathrm{I}}, \varrho_{\mathrm{II}}, \varrho_{\mathrm{III}}$ describe the three effects (a), (b), (c), respectively, discussed in Sect. 7.1. ϱ_{I} and σ_{II} are local field effects. ϱ_{I} is caused by polarization effects of the bounded metal in the absence of the adsorbate. ϱ_{II} is due to polarization effects caused by the presence of the adsorbate. The distinction between the two may seem at first sight somewhat arbitrary; however, it arose quite naturally in the diagrammatic analysis. Also the relevant momenta in the two cases are quite different: ϱ_{I} involves small momenta determined by the photon momentum or, in case of a statistically rough surface, by the inverse of the transverse correlation length; ϱ_{II} involves large momenta of the order of $1/z_0$ or 1/(extension of the molecular wave functions). ϱ_{III} represents the enhancement due to chemisorption (shifting and broadening of the spectral functions ϱ_a, ϱ_b) with the possibility of charge transfer. Strictly speaking, the chemisorption effects are also present in the local field effects $\varrho_{\mathrm{II}}(\omega)$. This is because $X_{ba}(\omega)$ appearing in (7.35) involves $R_{ba}(\omega)$ which depends upon the spectral functions ϱ_a and ϱ_b (7.31).

In the next three sections, we will discuss these three contributions to SERS in detail.

7.4 Local Field Effects Caused by a Bounded Metal

In this section we discuss the quantity ϱ_{I} of (7.34) which describes the increase of the e-p interaction due to the presence of a bounded metal. The calculation of $\tilde{\alpha}(\boldsymbol{Q}\lambda,\omega)$ for a nonstatistical surface requires essentially the solution of the Helmholtz equation (7.24) for the photon Green's function $\overleftrightarrow{\mathscr{D}}(\boldsymbol{r},\boldsymbol{r}',\omega)$. Equation (7.24) is extremely difficult to solve for a realistic dielectric function $\varepsilon(\boldsymbol{r},\boldsymbol{r}',\omega)$ of the bounded metal [7.72–75]. Solutions have been obtained only for special cases, for instance, if only the collective response (plasmon pole approximation) is taken into account or for some approximation for the single-particle response. From (7.28, 34) it follows that the momentum transfers occurring in $\varrho_{\mathrm{I}}(\omega)$ are determined by the momentum of the incident and scattered photons and are thus small.

The detailed calculations of [7.74, 75] allow us to estimate the error being made if the exact dielectric function $\overleftrightarrow{\varepsilon}(\boldsymbol{r},\boldsymbol{r}',\omega)$ is approximated by the commonly used local and scalar form

$$\varepsilon_{ij}(\boldsymbol{r},\boldsymbol{r}',\omega)=\varepsilon(\boldsymbol{r},\omega)\delta(\boldsymbol{r}-\boldsymbol{r}')\delta_{ij}. \tag{7.37}$$

The $\boldsymbol{r}$-dependence, moreover, is assumed to be solely due to the geometry of the metal surface. Figures 7.2a,b show the results of a numerical calculation of the real and imaginary parts of the total electric field $E_z(z,\omega)$ near an ideal plane

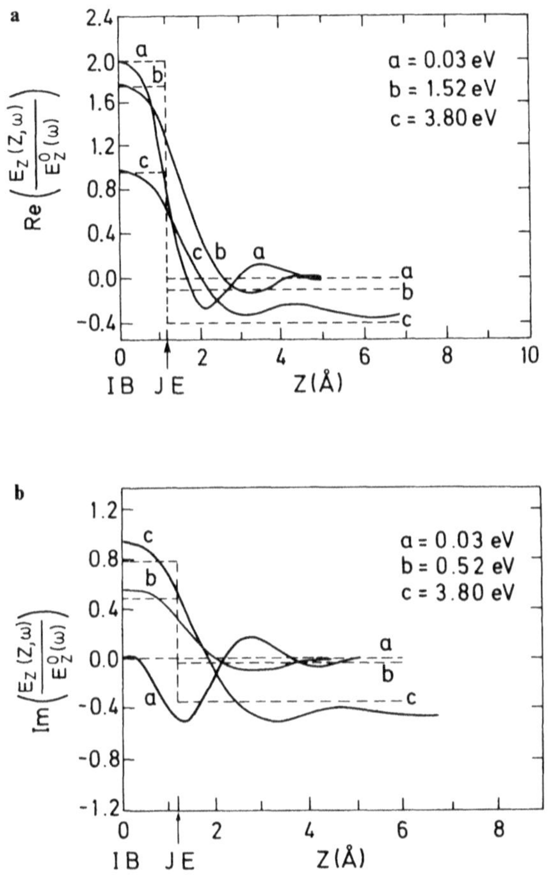

Fig. 7.2. (**a**) Real part of the total electric field $E_z(z, \omega)$ within the electron gas, normalized to the incident field $E_z^{(0)}$ for three frequencies: $\omega = 0.03$ eV, 1.52 eV and 3.8 eV. The full lines are the results of exact calculations within RPA of [7.74] and the dashed lines are those of scalar and local (within the plasmon pole approximation) ε of (7.37). For dashed lines the boundary of the metal is assumed at the jellium edge (JE). The infinite potential barrier at the interface is represented by IB. (**b**) Same as in Fig. 7.2a for the imaginary part of $E_z(z, \omega)$

electron gas-air interface for three incident photon energies $\omega = 0.03$ eV, 1.52 and 3.8 eV (solid lines). $E_z^0(\omega)$ is the incident electric field. The other parameters are: Fermi energy 3.8 eV, Fermi momentum 1 Å^{-1}, plasmon energy 6.82 eV, mean free path of the electrons 140 Å and the angle of incidence 45°. z measures the distance from the surface and $z = 0$ corresponds to the infinite high potential barrier (IB) which is assumed for the electrons to model the boundary. JE means jellium edge and denotes the position where the uniform background with the same but opposite charge density jumps to zero. The dashed lines represent the results of the local approximation (7.37). The nonlocality of ε affects the microscopic field in two ways: (a) for $z < 0$ the charge density is zero and the form of the electric fields is independent of ε whereas the numerical value for the reflection coefficient r still depends on ε. The calculations show that for realistic mean free paths < 1000 Å, r is not affected by the nonlocality and is correctly given by (7.37). Analogous results hold for the other asymptotic region deep inside the crystal. (b) In the local approximation, E_z jumps discontinuously between the asymptotic values whereas the microscopic electric fields join the two regions smoothly within a distance of a few Fermi wavelengths. In SERS one is interested in values of z which are somewhat smaller than zero (in the case of physisorption) or between zero and the jellium edge (in the case of chemisorption). Figure 7.2 suggests that in these regions the local approximation (7.37) yields either the correct or somewhat too large electric fields in the cases of physisorption or chemisorption, respectively. In view of the large effects discussed in SERS, it seems that (7.37) is an excellent approximation if only small momentum transfers are involved. Thus we will adopt (7.37) for ε throughout this section.

In Sect. 7.4.1 we discuss a systematic approach for solving (7.24) for a rough plane surface [7.68, 76]. This approach is then applied to two important special cases: Section 7.4.1a deals with a weak sinusoidal grating, Sect. 7.4.1b considers the case of a statistically rough surface. Finally, Sect. 7.4.2 presents results for $\varrho_I(\omega)$ in the case where the substrate is either an isolated metal sphere or has a more general geometry. In these cases (7.24) can be solved directly [7.67] (i.e., without perturbation approximation).

7.4.1 Plane Metal Surface with Roughness

Most of the SERS experiments are carried out on plane metal surfaces. Usually rough surfaces are used in order to achieve large enhancements. The roughness can be either well defined, e.g., in the form of a grating, or it can be random. The method described below is able to deal with both cases.

Let us consider a semi-infinite metal with a local dielectric function

$$\varepsilon(\boldsymbol{r},\omega)=\theta[z-\zeta(\boldsymbol{r}_\parallel)]+\varepsilon(\omega)\theta[\zeta(\boldsymbol{r}_\parallel)-z]. \tag{7.38}$$

Here $\zeta(\boldsymbol{r}_\parallel)$ describes the surface profile with respect to the plane surface ($z=0$). $\boldsymbol{r}_\parallel=(x,y)$ is the component of the position vector parallel to the (plane) surface. We assume, without loss of generality, that the average value of $\zeta(\boldsymbol{r}_\parallel)$ over the whole plane is zero. We split $\varepsilon(\boldsymbol{r},\omega)$ into two parts:

$$\varepsilon(\boldsymbol{r},\omega)=\varepsilon^{(0)}(z,\omega)+\Delta\varepsilon(\boldsymbol{r},\omega), \tag{7.39}$$

$$\varepsilon^{(0)}(z,\omega)=\theta(z)+\varepsilon(\omega)\theta(-z), \tag{7.40}$$

$$\Delta\varepsilon(\boldsymbol{r},\omega)=[\varepsilon(\omega)-1]\;\{\theta[\zeta(\boldsymbol{r}_\parallel)-z]-\theta(-z)\}, \tag{7.41}$$

where $\theta(z)$ is the step function. $\varepsilon^{(0)}(z,\omega)$ is the dielectric function for a plane dielectric vacuum interface and $\Delta\varepsilon$ denotes the change due to roughness.

Substituting (7.37, 39) into (7.24), $\overset{\leftrightarrow}{\mathscr{D}}(\boldsymbol{r},\boldsymbol{r}',\omega)$ obeys the following:

$$\nabla_r^2\overset{\leftrightarrow}{\mathscr{D}}(\boldsymbol{r},\boldsymbol{r}',\omega)+\frac{\omega^2}{c^2}\,\varepsilon^{(0)}(z,\omega)\overset{\leftrightarrow}{\mathscr{D}}(\boldsymbol{r},\boldsymbol{r}'\omega)\;=\;-4\pi\delta(\boldsymbol{r}-\boldsymbol{r}')\overset{\leftrightarrow}{I}-\frac{\omega^2}{c^2}\,\Delta\varepsilon(\boldsymbol{r},\omega)\overset{\leftrightarrow}{\mathscr{D}}(\boldsymbol{r},\boldsymbol{r}'\omega). \tag{7.42}$$

To solve this equation we follow the procedure due to *Maradudin* and *Mills* [7.68]. They first converted this differential equation into an integral equation and then solved it using a perturbation approach. We outline only briefly the procedure; the details can be found in [7.68, 76]. First we define the photon Green's function $\overleftrightarrow{\mathscr{D}}^{(0)}(\boldsymbol{r},\boldsymbol{r}',\omega)$ for a smooth surface ($z=0$) by

$$\nabla_r^2\overleftrightarrow{\mathscr{D}}^{(0)}(\boldsymbol{r},\boldsymbol{r}'\omega)+\frac{\omega^2}{c^2}\,\varepsilon^{(0)}(z,\omega)\overleftrightarrow{\mathscr{D}}^{(0)}(\boldsymbol{r},\boldsymbol{r}'\omega)=-4\pi\delta(\boldsymbol{r}-\boldsymbol{r}')\overleftrightarrow{I}. \tag{7.43}$$

Equations (7.42, 43) lead to the following integral equation for $\overleftrightarrow{\mathscr{D}}$:

$$\mathscr{D}_{ij}(\boldsymbol{r},\boldsymbol{r}'\omega)=\mathscr{D}_{ij}^{(0)}(\boldsymbol{r},\boldsymbol{r}'\omega)+\frac{\omega^2}{4\pi c^2}\sum_k\int d\boldsymbol{r}''\,\mathscr{D}_{ik}^{(0)}(\boldsymbol{r},\boldsymbol{r}''\omega)\,\Delta\varepsilon(\boldsymbol{r}'',\omega)$$
$$\cdot\,\mathscr{D}_{kj}(\boldsymbol{r}'',\boldsymbol{r}'\omega). \tag{7.44}$$

For small ζ one can expand $\theta(\zeta-z)$ in (7.41) retaining only first-order terms in ζ:

$$\Delta\varepsilon(\boldsymbol{r},\omega)=\zeta(\boldsymbol{r}_\|)\;[\varepsilon(\omega)-1]\delta(z). \tag{7.45}$$

Using this and the Fourier transformed quantities

$$\mathscr{D}_{ij}(\boldsymbol{r},\boldsymbol{r}',\omega)=\int\frac{d^2k_\|}{(2\pi)^2}\int\frac{d^2k'_\|}{(2\pi)^2}\,\mathrm{e}^{\mathrm{i}\boldsymbol{k}_\|\cdot\boldsymbol{r}_\|+\mathrm{i}\boldsymbol{k}'_\|\cdot\boldsymbol{r}'_\|}\hat{d}_{ij}(\boldsymbol{k}_\|\boldsymbol{k}'_\|\omega|zz'), \tag{7.46}$$

$$\mathscr{D}_{ij}^{(0)}(\boldsymbol{r},\boldsymbol{r}'\omega)=\int\frac{d^2k_\|}{(2\pi)^2}\,\mathrm{e}^{\mathrm{i}\boldsymbol{k}_\|\cdot(\boldsymbol{r}_\|-\boldsymbol{r}'_\|)}d_{ij}^{(0)}(\boldsymbol{k}_\|\omega|zz'), \tag{7.47}$$

$$\zeta(\boldsymbol{r}_\|)=\int\frac{d^2k_\|}{(2\pi)^2}\,\mathrm{e}^{\mathrm{i}\boldsymbol{k}_\|\cdot\boldsymbol{r}_\|}\zeta(\boldsymbol{k}_\|), \tag{7.48}$$

(7.44) becomes

$$\hat{d}_{ij}(\boldsymbol{k}_\|\boldsymbol{k}'_\|\omega|zz')=(2\pi)^2\delta(\boldsymbol{k}_\|+\boldsymbol{k}'_\|)d_{ij}^{(0)}(\boldsymbol{k}_\|\omega|zz')+\gamma\sum_k\int\frac{d^2k''_\|}{(2\pi)^2}\int dz''$$
$$\cdot\,d_{ik}^{(0)}(\boldsymbol{k}_\|\omega|zz'')\zeta(\boldsymbol{k}_\|-\boldsymbol{k}''_\|)\delta(z'')\hat{d}_{kj}(\boldsymbol{k}''_\|\boldsymbol{k}'_\|\omega|z''z') \tag{7.49}$$

with

$$\gamma=(\omega^2/4\pi c^2)\;[\varepsilon(\omega)-1]. \tag{7.50}$$

In (7.49) the integration over z'' can be carried over using [7.76]

$$\int f(z)\delta(z)g(z)dz=f(0^+)g(0^-), \tag{7.51}$$

where 0^+ enotes an infinitesimally small positive and negative quantity, respectively. This gives

$$\hat{d}_{ij}(\boldsymbol{k}_{\|}\boldsymbol{k}'_{\|}\omega|zz') = (2\pi)^2\delta(\boldsymbol{k}_{\|}+\boldsymbol{k}'_{\|})d^{(0)}_{ij}(\boldsymbol{k}_{\|}\omega|zz') + \gamma\sum_{k} d^{(0)}_{ik}(\boldsymbol{k}_{\|}\omega|z0^+)$$

$$\cdot\int\frac{d^2k''_{\|}}{(2\pi)^2}\,\zeta(\boldsymbol{k}_{\|}-\boldsymbol{k}''_{\|})\hat{d}_{kj}(\boldsymbol{k}''_{\|}\boldsymbol{k}'_{\|}\omega|0^-z'), \qquad (7.52)$$

where $\hat{d}_{ij}(\boldsymbol{k}_{\|}\boldsymbol{k}_{\|}\omega|0^-z')$ (appearing on the right-hand side) satisfies the following integral equation [substitute $z=0^-$ in (7.52)]:

$$\hat{d}_{ij}(\boldsymbol{k}_{\|}\boldsymbol{k}'_{\|}\omega|0^-z') = (2\pi)^2\delta(\boldsymbol{k}_{\|}+\boldsymbol{k}'_{\|})d^{(0)}_{ij}(\boldsymbol{k}_{\|}\omega|0^-z') + \gamma\sum_{k} d^{(0)}_{ik}(\boldsymbol{k}_{\|}\omega|0^-0^+)$$

$$\cdot\int\frac{d^2k''_{\|}}{(2\pi)^2}\,\zeta(\boldsymbol{k}_{\|}-\boldsymbol{k}''_{\|})\hat{d}_{kj}(\boldsymbol{k}''_{\|}\boldsymbol{k}'_{\|}\omega|0^-z'). \qquad (7.53)$$

Thus, (7.52) together with (7.53) gives a series solution for $\hat{d}_{ij}(\boldsymbol{k}_{\|}\boldsymbol{k}'_{\|}\omega|zz')$ in terms of $d^{(0)}_{ij}(\boldsymbol{k}_{\|}\omega|zz')$ which is given by [7.77]

$$\overleftrightarrow{d}^{(0)}(\boldsymbol{k}_{\|}\omega|zz') = -\begin{pmatrix} \hat{k}_x^2 g_{xx}+\hat{k}_y^2 g_{yy} & \hat{k}_x\hat{k}_y(g_{xx}-g_{yy}) & \hat{k}_x g_{xz} \\ \hat{k}_x\hat{k}_y(g_{xx}-g_{yy}) & \hat{k}_x^2 g_{yy}+\hat{k}_y^2 g_{xx} & \hat{k}_y g_{xz} \\ \hat{k}_x g_{zx} & \hat{k}_y g_{zx} & g_{zz} \end{pmatrix}, \qquad (7.54)$$

where $\hat{k}_x = k_x/k_{\|}$, $\hat{k}_y = k_y/k_{\|}$. The different elements of $\overleftrightarrow{g}(\boldsymbol{k}_{\|}\omega|zz')$ have lengthy expressions given in [7.68][1]. The poles of $d^{(0)}$ correspond to the three normal modes of the electromagnetic field in the presence of a plane metal surface. Two of them are extended s and p-polarized photons; the third one is the localized surface plasmon (sp) [7.78]. Since the (sp) contribution to $d^{(0)}$ will be needed later, we give its explicit expression here:

$$\overleftrightarrow{d}^{(0)}_{\text{sp}}(\boldsymbol{k}_{\|}\omega|zz') = -\frac{4\pi c^2 k_{\|}^2}{\omega^2(\nu+\varepsilon\nu_0)}\,[\theta(-z)\boldsymbol{e}^-(\boldsymbol{k}_{\|})\mathrm{e}^{\nu z} - \theta(z)\boldsymbol{e}^{+*}(\boldsymbol{k}_{\|})\exp(-\nu_0 z)]$$

$$\cdot[\theta(-z')\boldsymbol{e}^{-*}(\boldsymbol{k}_{\|})\mathrm{e}^{\nu z'} - \theta(z')\boldsymbol{e}^+(\boldsymbol{k}_{\|})\exp(-\nu_0 z')], \qquad (7.55)$$

where

$$\boldsymbol{e}^-(\boldsymbol{k}_{\|}) = \hat{z} + \mathrm{i}(\nu/k_{\|})\hat{k}_{\|}, \qquad (7.56)$$

$$\boldsymbol{e}^+(\boldsymbol{k}_{\|}) = \hat{z} + \mathrm{i}(\nu_0/k_{\|})\hat{k}_{\|}, \qquad (7.57)$$

---

1 Note that $\overleftrightarrow{d}^{(0)}$ in (7.54) has a different sign compared to that given in [7.68 or 7.76]. This is because of our different definition for $\overleftrightarrow{\mathscr{D}}$ and $\overleftrightarrow{\mathscr{D}}^{(0)}$, [see (7.42, 43)]. However, this does not make any difference in the final results.

$$\nu^2 = k_{\|}^2 - \varepsilon\omega^2/c^2, \tag{7.58}$$

$$\nu_0^2 = k_{\|}^2 - \omega^2/c^2. \tag{7.59}$$

The dispersion relation for (sp) is given by

$$k_{\|}^2 = (\omega^2/c^2)\,\{\varepsilon(\omega)/[\varepsilon(\omega)+1]\}. \tag{7.60}$$

Finally, the renormalized vertex $\tilde{\alpha}_{ba}(\boldsymbol{Q}\lambda, \omega)$ can be expressed in terms of $\hat{d}_{ij}(\boldsymbol{k}_{\|}\boldsymbol{k}'_{\|}\omega|zz')$:

$$\tilde{\alpha}_{ba}(\boldsymbol{Q}\lambda, \omega) = A_0(\boldsymbol{Q}\lambda)\,\frac{\omega_\lambda^2(\boldsymbol{Q}) - \omega^2 - \mathrm{i}\eta}{4\pi c^2} \sum_{\alpha,\beta} \hat{e}_{\lambda\alpha}(\boldsymbol{Q}) \int dz \exp(\mathrm{i}k_z z)$$
$$\cdot \int \frac{d^2k'_{\|}}{(2\pi)^2}\,\hat{d}_{\alpha\beta}(-\boldsymbol{k}_{\|}\boldsymbol{k}'_{\|}\omega|zz_0)\,\langle p_\beta\rangle, \tag{7.61}$$

where

$$\langle \boldsymbol{p}\rangle = \langle \phi_b(\boldsymbol{r})|\,\boldsymbol{p}\,|\phi_a(\boldsymbol{r})\rangle \tag{7.62}$$

and $\boldsymbol{Q} = (\boldsymbol{k}_{\|}, k_z)$. In writing (7.61) we have assumed the molecular wave functions $\phi_{a,b}$ to be localized at $z = z_0$ $(z_0 > 0)$.

a) Weak Sinusoidal Grating

A sinusoidal grating on a plane metal surface is described by one nonzero Fourier component for $\zeta(\boldsymbol{k}_{\|})$ in (7.48):

$$\zeta(\boldsymbol{k}_{\|}) = (2\pi)^2 \zeta_g \delta(\boldsymbol{k}_{\|} - \boldsymbol{g}). \tag{7.63}$$

Assuming that the amplitude of the grating is small compared to its wavelength, i.e., $\zeta_g g \ll 1$, (7.52) can be solved by perturbation theory. Inserting (7.63) into (7.52) and retaining terms only up to first order in ζ_g, one finds

$$\hat{d}_{ij}(-\boldsymbol{k}_{\|}\boldsymbol{k}'_{\|}\omega|zz') = (2\pi)^2\,[\delta(-\boldsymbol{k}_{\|} + \boldsymbol{k}'_{\|})\,\mathrm{d}_{ij}^{(0)}(-\boldsymbol{k}_{\|}\omega|zz') + \gamma\zeta_g\delta(\boldsymbol{k}'_{\|} - \boldsymbol{k}_g)$$
$$\cdot \sum_k d_{ik}^{(0)}(-\boldsymbol{k}_{\|}\omega|z0^+)\,d_{kj}^{(0)}(-\boldsymbol{k}_g\omega|0^- z')], \tag{7.64}$$

where $\boldsymbol{k}_g = \boldsymbol{k}_{\|} + \boldsymbol{g}$. Substituting this into (7.61), the renormalized vertex can be written as

$$\tilde{\alpha}_{ba}(\boldsymbol{Q}\lambda, \omega) = \tilde{\alpha}_{ba}^{(0)}(\boldsymbol{Q}\lambda, \omega) + \gamma\zeta_g\,\frac{\omega_\lambda^2(\boldsymbol{Q}) - \omega^2 - \mathrm{i}\eta}{4\pi c^2}\,A_0(\boldsymbol{Q}\lambda) \sum_{\alpha,\beta,\beta'} e_{\lambda\alpha}(\boldsymbol{Q})$$
$$\cdot \int dz\,\mathrm{e}^{\mathrm{i}k_z z} d_{\alpha\beta'}^{(0)}(-\boldsymbol{k}_{\|}\omega|z0^+)\,d_{\beta'\beta}^{(0)}(-\boldsymbol{k}_g\omega|0^- z_0)\,\langle p_\beta\rangle. \tag{7.65}$$

The first term on the right-hand side of (7.65),

$$\tilde{\alpha}_{ba}^{(0)}(Q\lambda,\omega)=\frac{\omega_\lambda^2(Q)-\omega^2-\mathrm{i}\eta}{4\pi c^2}A_0(Q\lambda)\sum_{\alpha,\beta}e_{\lambda\alpha}(Q)\int dz\exp(\mathrm{i}k_z z)$$

$$\cdot d_{\alpha\beta}^{(0)}(-\boldsymbol{k}_\parallel\omega|zz_0)\,\langle p_\beta\rangle, \tag{7.66}$$

is the renormalized vertex due to the presence of the smooth metal surface. The second term in (7.65) describes the effect of the grating.

In order to calculate $\varrho_\mathrm{I}(\omega)$, we assume the incident photon to be p-polarized with

$$\hat{e}_\lambda(Q)=(k_z\boldsymbol{k}_\parallel/k_\parallel-\boldsymbol{k}_\parallel\hat{z})/Q. \tag{7.67}$$

For a s-polarized photon there is no enhancement [i.e., $\varrho_\mathrm{I}(\omega)=1$] if $\boldsymbol{k}_\parallel$ is parallel to $\boldsymbol{g}$ [7.15]. For other angles $\varrho_\mathrm{I}(\omega)$ is also close to one. For simplicity we also assume that only the z-component of $\langle \boldsymbol{p}\rangle$ is nonzero. Then using (7.54) together with the explicit expressions for $\vec{g}$ of [7.68], one finds

$$\frac{|\tilde{\alpha}_{ba}(Q\lambda,\omega)|^2}{|\tilde{\alpha}_{ba}^{(0)}(Q\lambda,\omega)|^2}=1+\left|\zeta_g k_g\frac{\varepsilon(\omega)v_{0g}-v_g}{k_g^2-k_s^2}\frac{1}{1+\varepsilon(\omega)}\left(k_g-\frac{\mathrm{i}vv_g}{\varepsilon k_\parallel}\right)\exp(-v_{0g}z_0)\right|^2, \tag{7.68}$$

with $v_g^2=-\varepsilon(\omega)\omega^2/c^2+k_g^2$, $\quad v_{0g}^2=-\omega^2/c^2+k_g^2$, $v^2=k_\parallel^2-\varepsilon(\omega)\omega^2/c^2$, and

$$k_s^2(\omega)=(\omega^2/c^2)\,\{\varepsilon(\omega)/[1+\varepsilon(\omega)]\}. \tag{7.69}$$

For the following discussions it is convenient to decompose (7.69) into real and imaginary parts:

$$k_s^2(\omega)=k_{sr}^2(\omega)+\mathrm{i}k_{si}^2(\omega), \tag{7.70}$$

$$k_{sr}^2(\omega)=\frac{\omega^2}{c^2}\,\frac{[1+\varepsilon_1(\omega)]\varepsilon_1(\omega)+\varepsilon_2^2(\omega)}{|1+\varepsilon(\omega)|^2}, \tag{7.71}$$

$$k_{si}^2(\omega)=\frac{\omega^2}{c^2}\,\frac{\varepsilon_2(\omega)}{|1+\varepsilon(\omega)|^2}, \tag{7.72}$$

where ε_1 and ε_2 are the real and imaginary parts of ε.

If the wave vector $\boldsymbol{g}$ of the grating is such that $k_g^2\approx k_{sr}^2(\omega)$ and if $k_{si}^2(\omega)$ (i.e., ε_2) is small, then the right-hand side of (7.68) can become quite large. From (7.66) follows that $|\tilde{\alpha}_{ba}^{(0)}(Q\lambda,\omega)|\approx(2k_\parallel/Q|\alpha_{ba}(Q\lambda,\omega)|\approx|\alpha_{ba}(Q\lambda,\omega)|$. Thus, (7.68) can be approximately taken as the enhancement with respect to the free molecule. It is true for the incident and the scattered photon. However, because the grating contains one single Fourier component, only the incident or the scattered photon can come into resonance with (sp) polaritons. Near the resonance ($k_g^2 \sim k_{sr}^2$), (7.68) can therefore be further simplified yielding

$$\varrho_1(\omega) = 1 + \{2\zeta_g k_g[-\varepsilon_1(\omega)]^{3/2}/\varepsilon_2(\omega)\}^2 \; [1 - v^2/\varepsilon_1(\omega)k_{\|}^2]^2 \exp(-2v_{0g}z_0). \tag{7.73}$$

This result was first derived by *Jha* et al. [7.15, 16] by a somewhat different method [7.79]. The second term in (7.73) is due to the process where the photon picks up additional momentum $\boldsymbol{g}$ from the grating and thus is able to satisfy the conservation laws for the excitation of the (sp) polaritons. For a flat surface the second term in (7.73) is missing and $\varrho_1(\omega) \equiv 1$.

Figure 7.3 shows $\varrho_1(\omega)$ for Ag calculated from (7.68). The parameters are those of [7.15], i.e., $2\pi/g = 8000$ Å, $\zeta_g = 150$ Å, $z_0 \ll 1/g$, an angle of incidence of 24° and $\boldsymbol{k}_g$ parallel to $\boldsymbol{g}$. For $\varepsilon(\omega)$, we took the experimental values from [7.80]. The sharp peak at $\omega = 2.35$ eV corresponds to the resonant excitation of (sp) polaritons. Figure 7.4 shows curves for $\varrho_1(\omega)$ calculated for Ag, Cu, and Au as a function of the photon frequency. These calculations assume that the scattering configuration is chosen so that the incident photon is exactly in resonance with the (sp) polariton. $\varepsilon(\omega)$ is again taken from [7.80]; the other parameters are $2\pi/g = 8000$ Å, $\zeta_g = 150$ Å and $\boldsymbol{k}_g \| \boldsymbol{g}$. This figure clearly shows that the maximum

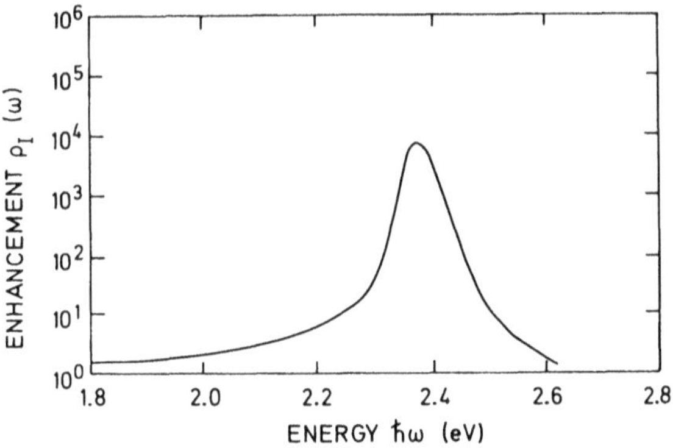

Fig. 7.3. $\varrho_1(\omega)$ calculated from (7.73) for a grating superimposed on a Ag smooth surface with $2\pi/g = 8000$ Å, $\zeta_g = 150$ Å, $\boldsymbol{k}_{g\|}\boldsymbol{g}$ and angle of incidence of 24°. $\varepsilon(\omega)$ is taken from [7.80]

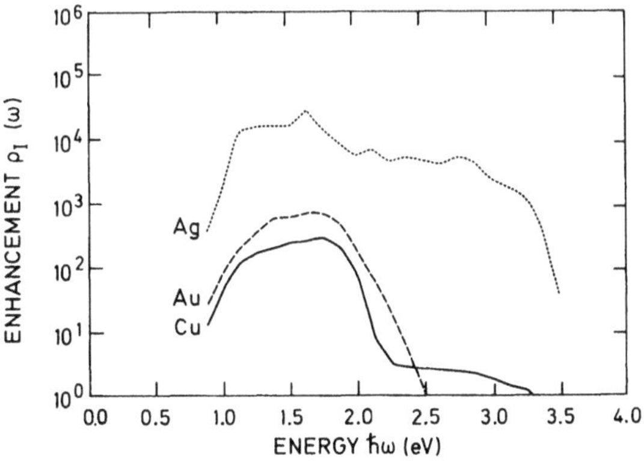

Fig. 7.4. Relative variation of $\varrho_1(\omega)$ for Ag, Cu and Au as a function of incident photon energy for an 8000 Å grating periodicity. $\varepsilon(\omega)$ is taken from [7.80]. Exact resonance condition with surface plasmon polariton has been assumed for each incident frequency. From [7.15]

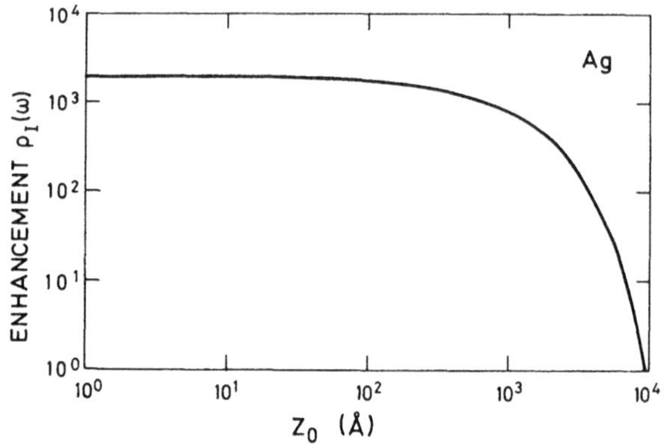

Fig. 7.5. $\varrho_1(\omega)$ as a function of the distance z_0 of the molecule from the image plane. Exact resonance condition for the (sp) polariton for Ag has been assumed with incident photon of 2.34 eV

enhancement in Cu and Au is smaller than in Ag. This is due to the fact that $\varepsilon_2(\omega)$ in Ag is about one order of magnitude smaller than in Cu and Au whereas $\varepsilon_1(\omega)$ is approximately the same in each case. In Fig. 7.5 we have plotted for Ag the dependence of $\varrho_1(\omega)$ on z_0, i.e., the distance of the admolecule from the plane surface. Again, exact resonance of the incident or scattered photon with (sp) polariton has been assumed. The parameters are $\omega = 2.34$ eV, $2\pi/g = 8000$ Å and $\zeta_g k_g = 0.2$. The figure shows that the enhancement extends to distances z_0 up to ~ 1000 Å and then decreases rapidly due to the finite range of the (sp) polaritons.

The above theoretical results are in good agreement with SERS experiments [7.43–46] for Ag films. The fact that large enhancements are only observed at angles for which the incident or the scattered photon can resonantly excite (sp) polaritons shows the important role played by (sp) polaritons. The experiments also show that strong SERS can be observed in scattering configurations where the light is incident on one side of the film and the molecule is on the other side for film with thicknesses well over 500 Å [7.46]. This observation can easily be explained in terms of (sp) polaritons because their fields extend up to ~ 1000 Å ($\sim \nu_{0g}^{-1}$ or ν_g^{-1}) into the vacuum on both sides of the film.

b) Randomly Rough Surface

In most SERS experiments, the roughness of the surface is of a random nature. Such a surface can be described by a distribution for the surface profile functions $\zeta(\boldsymbol{k}_{\|})$, see (7.48). The cross section for SERS is then obtained by calculating it first for a given surface profile $\zeta(\boldsymbol{k}_{\|})$ and then by averaging it over an ensemble of physical equivalent surface profiles. The averaging process is usually assumed to be Gaussian characterized by

$$\langle \zeta(\boldsymbol{k}_{\|}) \rangle = 0 \tag{7.74}$$

and

$$\langle \zeta(\boldsymbol{k}_{\|}) \zeta(\boldsymbol{k}'_{\|}) \rangle = (2\pi)^2 \delta(\boldsymbol{k}_{\|} + \boldsymbol{k}'_{\|}) \delta^2 g(|\boldsymbol{k}_{\|}/), \tag{7.75}$$

where $\langle\ \rangle$ denotes the ensemble average. δ^2 is the mean square height variation of the surface profile and $g(|\boldsymbol{k}_\parallel/)$ is given by

$$g(|\boldsymbol{k}_\parallel/) = \pi a^2 \exp(-a^2 k_\parallel^2/4), \tag{7.76}$$

where a is the transverse correlation length. Calculations for SERS have been carried out in [7.20] using perturbation theory and in [7.21] using a nonperturbative approach. We will mainly follow the latter.

The ensemble average of the cross section needs the average of the quantity $\langle|\tilde{\alpha}(\boldsymbol{Q}\lambda,\omega)|^2\,|\tilde{\alpha}(\boldsymbol{Q}'\lambda',\omega')|^2\rangle$, i.e., the average of four photon propagators, see (7.34). To a good approximation the incident and scattered propagators may, however, be averaged separately, i.e.,

$$\langle|\tilde{\alpha}(\boldsymbol{Q}\lambda,\omega)|^2\,|\tilde{\alpha}(\boldsymbol{Q}'\lambda',\omega')|^2\rangle \sim \langle|\tilde{\alpha}(\boldsymbol{Q}\lambda,\omega)|^2\rangle\,\langle|\tilde{\alpha}(\boldsymbol{Q}'\lambda',\omega')|^2\rangle. \tag{7.77}$$

This means that one has to calculate the average of one photon propagator $\langle\hat{d}_{ij}(\boldsymbol{k}_\parallel \boldsymbol{k}'_\parallel \omega|z_1 z_2)\rangle$ and that of the product of two photon propagators $\langle\hat{d}_{ij}(\boldsymbol{K}_\parallel \boldsymbol{K}'_\parallel \omega|z_1 z_2)\, d^*_{kl}(\boldsymbol{k}_\parallel \boldsymbol{k}'_\parallel \omega|z'_1 z'_2)\rangle$. Furthermore, it is sufficient to consider only the arguments $z_1 = z'_1 = 0^-$ and $z_2 = z'_2 = 0^+$ in all these calculations [7.51].

First we consider the average of the one photon propagator $\langle\hat{d}_{ij}(\boldsymbol{k}_\parallel \boldsymbol{k}'_\parallel \omega|0^-\,0^+)\rangle$. It can be calculated from (7.53) by making a series expansion and then averaging term by term using (7.74, 75). As shown in [7.76], one finds that

$$\langle\hat{d}_{ij}(\boldsymbol{k}_\parallel \boldsymbol{k}'_\parallel \omega|0^-\,0^+)\rangle = (2\pi)^2\,\delta(\boldsymbol{k}_\parallel + \boldsymbol{k}'_\parallel)\, d_{ij}(\boldsymbol{k}_\parallel \omega) \tag{7.78}$$

satisfies the Dyson equation

$$d_{ij}(\boldsymbol{k}_\parallel \omega) = d^{(0)}_{ij}(\boldsymbol{k}_\parallel \omega) + \sum_{k,l} d^{(0)}_{ik}(\boldsymbol{k}_\parallel \omega)\,\Sigma_{kl}(\boldsymbol{k}_\parallel \omega)\, d_{lj}(\boldsymbol{k}_\parallel \omega), \tag{7.79}$$

where $d^{(0)}_{ij}(\boldsymbol{k}_\parallel \omega) \equiv d^{(0)}_{ij}(\boldsymbol{k}_\parallel \omega|0^-\,0^+)$ is the unperturbed photon propagator corresponding to the plane surface (we suppress $0^-\,0^+$ arguments for simplicity). Σ_{ij} is the self-energy of the photon due to roughness and can be expressed as a power series of d_{ij}. Equation (7.79) is expressed diagramatically in Figs. 7.6a, b. Thick lines denote the averaged photon propagator $d_{ij}(\boldsymbol{k}_\parallel \omega)$ in the presence of a rough surface and thin lines correspond to $d^{(0)}_{ij}(\boldsymbol{k}_\parallel \omega)$. Figure 7.6b shows the lowest-order diagrams for $\Sigma_{ij}(\boldsymbol{k}_\parallel \omega)$ where the dotted line denotes the interaction $g(|\boldsymbol{k}_\parallel|)$ due to roughness. The leading term of the series for Σ has the analytic expression

$$\Sigma^{(1)}_{ij}(\boldsymbol{k}_\parallel \omega) = \delta^2 \lambda^2 \int \frac{d\boldsymbol{k}'_\parallel}{(2\pi)^2}\, g(|\boldsymbol{k}_\parallel - \boldsymbol{k}'_\parallel|)\, d_{ij}(\boldsymbol{k}'_\parallel \omega). \tag{7.80}$$

We now consider the average of two photon propagators which can be written as

$$\langle\hat{d}_{ij}(\boldsymbol{k}_\parallel \boldsymbol{k}'_\parallel \omega|0^-0^+)\,\hat{d}_{kl}(\boldsymbol{k}_\parallel \boldsymbol{k}''_\parallel \omega|0^-0^+)\rangle = (2\pi)^2\,\delta(\boldsymbol{k}'_\parallel - \boldsymbol{k}''_\parallel)\,\hat{T}_{ijkl}(\boldsymbol{k}_\parallel \boldsymbol{k}'_\parallel \omega). \tag{7.81}$$

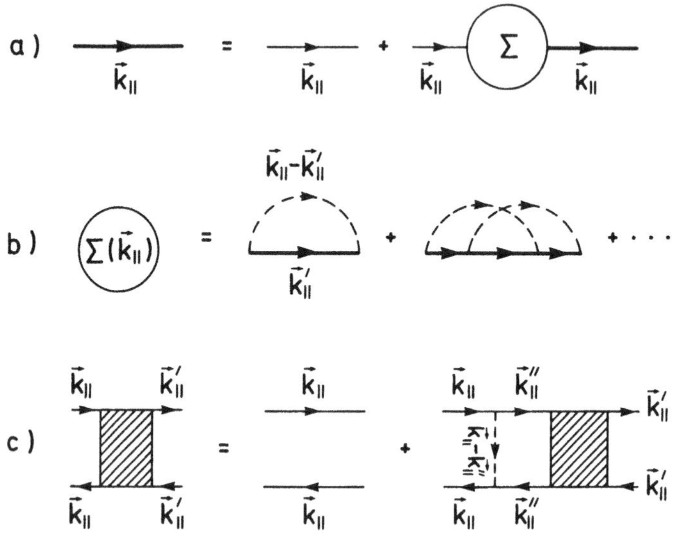

Fig. 7.6. (**a**) Dyson equation for the average one-photon propagator $d_{ij}(\boldsymbol{k}_{\|}\omega|0^-0^+)$; (**b**) self-energy diagrams contributing to $d_{ij}(\boldsymbol{k}_{\|}\omega|0^-0^+)$ due to roughness, the dashed line denotes the interaction $[\delta^2\gamma^2 g(|\boldsymbol{k}_{\|}-\boldsymbol{k}'_{\|}|)]$ due to roughness; (**c**) Bethe-Salpeter equation for the average two-photon propagator $\langle d_{ij}(\boldsymbol{k}_{\|}\boldsymbol{k}'_{\|}\omega|0^-0^+)d^*_{kl}(\boldsymbol{k}_{\|}\boldsymbol{k}'_{\|}\omega|0^-0^+)\rangle$

$\hat{T}$ can be calculated by making a series expansion using (7.53) and by averaging term by term. In this way we obtain a perturbation expansion in terms of averaged propagators. The resulting diagrams and the rules to calculate them are very similar to those of an electron moving in a random potential. Since our treatment takes only the term linear in ζ in (7.45) into account, we keep only the leading term (in ζ) for Σ, i.e., (7.80). This implies within the scheme of conserving approximations [7.77] that the kernel for the integral equation for $\hat{T}$ is given by g, i.e., by a dotted line diagramatically. This is illustrated in Fig. 7.6c. Analytically $\hat{T}$ satisfies the following Bethe-Salpeter equation

$$\hat{T}_{ijkl}(\boldsymbol{k}_{\|}\boldsymbol{k}'_{\|}\omega)=\sum_{m,n} d_{im}(\boldsymbol{k}_{\|}\omega)d^*_{kn}(k_{\|}\omega)\Bigg[(2\pi)^2\delta(\boldsymbol{k}_{\|}+\boldsymbol{k}'_{\|})\delta_{jm}\delta_{ln} + \delta^2\gamma^2\int\frac{d^2\boldsymbol{k}''_{\|}}{(2\pi)^2}g(|\boldsymbol{k}_{\|}-\boldsymbol{k}'_{\|}|)\hat{T}_{mjnl}(\boldsymbol{k}''_{\|}\boldsymbol{k}'_{\|}\omega)\Bigg], \tag{7.82}$$

which has to be solved self-consistently together with (7.79, 80). These equations can be simplified by retaining only the (sp) polariton part in $d^{(0)}_{ij}(\boldsymbol{k}_{\|}\omega)$ of (7.55). This is justified so long as δ and a are much smaller than the wavelength of the incident light. Within this approximation,

$$d^{(0)}_{ij}(\boldsymbol{k}_{\|}\omega)=-[4\pi c^2k^2_{\|}/\omega^2(\nu+\varepsilon\nu_0)]e^-_i(\boldsymbol{k}_{\|})e^+_j(\boldsymbol{k}_{\|}). \tag{7.83}$$

Using (7.79, 83), (7.82) becomes a scalar equation with

$$\hat{T}_{ijkl}(\boldsymbol{k}_{\|}\boldsymbol{k}'_{\|}\omega)=e_i^-(\boldsymbol{k}_{\|})e_k^{-*}(\boldsymbol{k}_{\|})e_j^+(-\boldsymbol{k}'_{\|})e_l^{+*}(-\boldsymbol{k}'_{\|})\tilde{T}(\boldsymbol{k}_{\|}\boldsymbol{k}'_{\|}\omega), \tag{7.84}$$

$$\tilde{T}(\boldsymbol{k}_{\|}\boldsymbol{k}'_{\|}\omega)=|\tilde{d}(\boldsymbol{k}_{\|}\omega)|^2\left[(2\pi)^2\delta(\boldsymbol{k}_{\|}+\boldsymbol{k}'_{\|}) + \delta^2\gamma^2\int\frac{d^2\boldsymbol{k}''_{\|}}{(2\pi)^2}B(\boldsymbol{k}_{\|},\boldsymbol{k}''_{\|})\tilde{T}(\boldsymbol{k}''_{\|}\boldsymbol{k}'_{\|}\omega)\right], \tag{7.85}$$

where

$$\tilde{d}(\boldsymbol{k}_{\|}\omega)= -(4\pi c^2k_{\|}^2/\omega^2)/[\nu+\varepsilon\nu_0+(4\pi c^2k_{\|}^2/\omega^2)\boldsymbol{e}^+(\boldsymbol{k}_{\|})\cdot\overleftrightarrow{\Sigma}(\boldsymbol{k}_{\|}\omega)\cdot\boldsymbol{e}^-(\boldsymbol{k}_{\|})], \tag{7.86}$$

$$B(\boldsymbol{k}_{\|},\boldsymbol{k}'_{\|})=g(|\boldsymbol{k}_{\|}-\boldsymbol{k}'_{\|}|)|\boldsymbol{e}^+(\boldsymbol{k}_{\|})\cdot\boldsymbol{e}^-(\boldsymbol{k}'_{\|})|^2. \tag{7.87}$$

The angular integration in (7.85) can be performed using the following partial wave expansion for B:

$$B(\boldsymbol{k}_{\|},\boldsymbol{k}'_{\|})=\sum_l B_l(k_{\|},k'_{\|})\,\mathrm{e}^{\mathrm{i}l(\phi_K-\phi_{K'})}. \tag{7.88}$$

In (7.88) $\boldsymbol{k}_{\|}$ has been decomposed into its radial component $k_{\|}$ and its angle ϕ_K. The index l runs over all integers. One thus finds that

$$\tilde{T}(\boldsymbol{k}_{\|}\boldsymbol{k}'_{\|}\omega)=\sum_l \tilde{T}_l(k_{\|}k'_{\|}\omega)\,\mathrm{e}^{\mathrm{i}l(\phi_K+\phi_{K'})} \tag{7.89}$$

with

$$\tilde{T}_l(k_{\|}k'\omega)=|\tilde{d}(k_{\|}\omega)|^2\left[\frac{2\pi}{k_{\|}}\delta(k_{\|}-k'_{\|}) + \delta^2\gamma^2\int\frac{dk''_{\|}}{2\pi}k''_{\|}B_l(k_{\|},k''_{\|})\tilde{T}_l(k''_{\|}k'_{\|}\omega)\right]. \tag{7.90}$$

Using a polynomial approximation for the modified Bessel functions occurring in $B_l(k_{\|},k''_{\|})$, the integral equation (7.90) can be reduced to a matrix equation which can be easily solved numerically.

The calculation of $\langle|\tilde{\alpha}_{ba}(\boldsymbol{Q}\lambda,\omega)|^2\rangle$ needs the quantity $\langle\hat{d}_{ij}(\boldsymbol{k}_{\|}-\boldsymbol{k}'_{\|}\omega|zz_0)\hat{d}_{kl}(\boldsymbol{k}_{\|}\boldsymbol{k}''_{\|}\omega|z'z_0)\rangle$. It can be expressed in terms of $\hat{T}$ using:

$$\begin{aligned}\langle\hat{d}_{ij}(\boldsymbol{k}_{\|}\boldsymbol{k}'_{\|}\omega|z0^+)\hat{d}^*_{kl}(\boldsymbol{k}_{\|}\boldsymbol{k}'''_{\|}\omega|z'0^+)\rangle &=(2\pi)^2\delta(\boldsymbol{k}'_{\|}-\boldsymbol{k}'''_{\|})\\ &\cdot\sum_{m,n}d^{(0)}_{im}(\boldsymbol{k}_{\|}\omega|z0^+)d^{(0)}_{kn}(\boldsymbol{k}'_{\|}\omega|z'0^+)\left[(2\pi)^2\delta(\boldsymbol{k}_{\|}+\boldsymbol{k}'_{\|})\delta_{mj}\delta_{nl}\right.\\ &\left.+\delta^2\gamma^2\int\frac{d^2\boldsymbol{k}''_{\|}}{(2\pi)^2}g(|\boldsymbol{k}_{\|}-\boldsymbol{k}''_{\|}|)\hat{T}_{mjnl}(\boldsymbol{k}''_{\|}\boldsymbol{k}'_{\|}\omega)\right],\end{aligned} \tag{7.91}$$

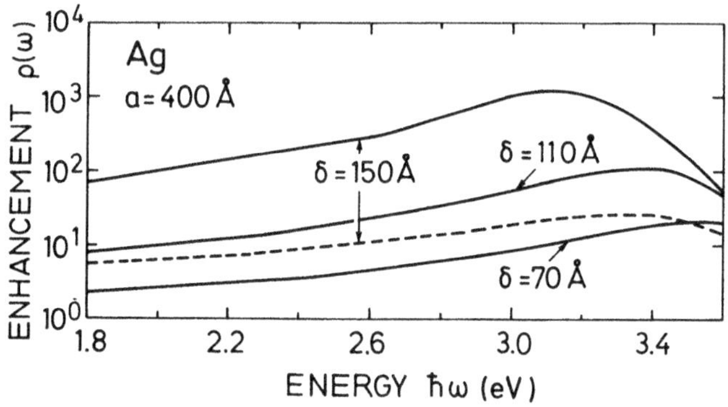

Fig. 7.7. Enhancement curves $\varrho_1(\omega)$ for p-polarized photons for a rough Ag surface using $\langle p_\parallel\rangle = \langle p_z\rangle$, an incident and scattered angle of 30° and a vibrational frequency $\Omega = 0.2$ eV. The broken line corresponds to lowest-order perturbation theory

where we have assumed $z_0 = 0^+$. Assuming p-polarized photons [polarization $\hat{e}_\lambda(\boldsymbol{Q})$ given by (7.67)] and using (7.54) for $d_{ij}^{(0)}(\boldsymbol{k}_\parallel \omega | z0^+)$, (7.61, 91) yield

$$\varrho_1^{\rm i}(\omega) = \langle|\tilde{\alpha}_{ba}(\boldsymbol{Q}\lambda,\omega)|^2\rangle / |\tilde{\alpha}_{ba}^{(0)}(\boldsymbol{Q}\lambda,\omega)|^2$$

$$= \Bigg[1 + \delta^2\gamma^2 \int \frac{d^2\boldsymbol{k}'_\parallel}{(2\pi)^2} \int \frac{d^2\boldsymbol{k}''_\parallel}{(2\pi)^2} \left|\frac{(\mathrm{i}\nu\hat{k}_\parallel + \varepsilon k_\parallel \hat{z})\cdot \boldsymbol{e}^-(\boldsymbol{k}''_\parallel)}{(\mathrm{i}\nu\hat{k}_\parallel + \varepsilon k_\parallel \hat{z})\cdot\langle \boldsymbol{p}\rangle}\right|^2$$

$$\cdot\, g(|\boldsymbol{k}_\parallel - \boldsymbol{k}''_\parallel|)\tilde{T}(\boldsymbol{k}''_\parallel \boldsymbol{k}'_\parallel \omega)|\boldsymbol{e}^+(-\boldsymbol{k}'_\parallel)\cdot\langle\boldsymbol{p}\rangle|^2\Bigg], \tag{7.92}$$

where $\tilde{\alpha}_{ba}^{(0)}(\boldsymbol{Q}\lambda,\omega)$ is given by (7.66). A similar relation holds for the scattered photon vertex $[\varrho_1^{\rm s}(\omega')]$. Thus, using (7.89, 90) for $\tilde{T}(\boldsymbol{k}''\boldsymbol{k}'\omega)$, the total enhancement $\varrho_1(\omega) = \varrho_1^{\rm i}(\omega)\varrho_1^{\rm s}(\omega')$ can be calculated.

Figure 7.7 shows calculated enhancement curves $\varrho_1(\omega)$ for a rough Ag surface using $a = 400$ Å and $\delta = 70$ Å, 110 Å, and 150 Å. In contrast to the case of grating, these curves are rather smooth functions of frequency as long as $\delta/a < 1/3$. This agrees with SERS experiments from rough surfaces which also yield rather structureless enhancement curves. Figure 7.8 shows calculated $\varrho_1(\omega)$ curves for a rough Cu surface with $\delta = 150$ Å and $a = 250$ Å, 300 Å and 400 Å. For both Ag and Cu, $\varepsilon(\omega)$ was taken from [7.80]. The calculations yield enhancements of up to 2 and 3 orders of magnitude in Cu and Ag, respectively.

We have also compared the results of lowest-order perturbation theory [i.e., approximating $\tilde{T}_l$ by only the first term in (7.90)] and those of our non-perturbative approach. We find that both results agree for $\delta/a < 0.1$. For $\delta/a > 0.1$, our theory predicts much larger enhancements compared to the lowest-order perturbation theory (compare the broken line corresponding to lowest-order perturbation theory and $\delta = 150$ Å and $a = 400$ Å with the corresponding full line in Fig. 7.7). Figures 7.7, 8 show that $\varrho_1(\omega)$ increases with increasing δ and for decreasing a. The calculation shows that $\varrho_1(\omega)$ diverges somewhat below $\omega_{\rm sp}$

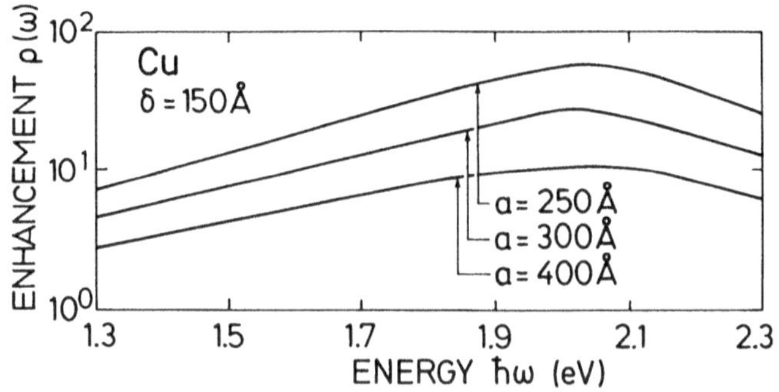

Fig. 7.8. Enhancement curves $\varrho_{\mathrm{I}}(\omega)$ for p-polarized photons for a rough Cu surface using $\langle p_{\|}\rangle = \langle p_z\rangle$, an incident and scattered angle of 30° and a vibrational frequency $\Omega = 0.2$ eV

$= 3.5$ eV for $\delta/a > 0.5$. At this point the lowest-order conserving approximation corresponding to uncrossed diagrams for the self-energy and the Bethe-Salpeter equation breaks down. As discussed in [7.81], it becomes important in this case to include also crossed diagrams (Fig. 7.6b) which correspond to rather localized field distributions parallel to the surface due to very rapidly varying or strongly fluctuating surface profiles.

The above calculations yield the following picture: For randomly rough surfaces with small δ and large a, the electric field is enhanced because (sp) polaritons are excited under resonant conditions. The resulting enhancement $\varrho_{\mathrm{I}}(\omega)$ for SERS is, however, rather small. For rather rough surfaces, resonant multiple scattering of (sp) polaritons becomes important (described diagramatically by the ladder summation) and can yield enhancements up to 2 and 3 orders of magnitude. For very rough surfaces (corresponding to large amplitudes δ or small transverse correlation lengths a), the imaginary part of the (sp) becomes important. As a result, energy is no longer conserved in the multiple scattering process, the momentum transfer becomes much larger because of phase space arguments and the crossed and uncrossed diagrams are equally important. This case has not yet been treated satisfactorily.

7.4.2 Sphere and Other Substrate Geometries

In Sect. 7.4.1 we considered a molecule adsorbed on a rough (plane) metal surface. Some SERS experiments [7.3, 49] have also been carried out for other geometries for the substrates. For instance, the case of pyridine molecule adsorbed on a Ag or Au particles of a colloidal solution has been investigated. The metal particles are approximately spherical in shape with radius ranging from ~50 Å to 1000 Å.

The theoretical calculation of the contribution ϱ_{I} for an isolated metal sphere as a substrate was first described by *McCall* et al. [7.12]. Their calculation neglected retardation effects assuming that the wavelength of the light is much larger than the radius of the sphere. *Kerker* and coworkers [7.13, 14] presented more elaborate calculations which also included retardation effects. In both cases the enhanced electric field was obtained by solving Maxwell's equations directly in the presence of a metal sphere. We derive these results using our more general approach which is based on the photon propagator $\overleftrightarrow{\mathscr{D}}(\boldsymbol{r},\boldsymbol{r}',\omega)$.

For the local dielectric function $\varepsilon(\boldsymbol{r},\omega)$ of (7.37), the Helmholz equation (7.24) can be written as

$$\nabla^2\overleftrightarrow{\mathscr{D}}(\boldsymbol{r},\boldsymbol{r}',\omega)+\frac{\omega^2}{c^2}\,\varepsilon(\boldsymbol{r},\omega)\overleftrightarrow{\mathscr{D}}(\boldsymbol{r},\boldsymbol{r}',\omega)=-4\pi c^2\delta(\boldsymbol{r}-\boldsymbol{r}')\overleftrightarrow{I}. \tag{7.93}$$

For a metal sphere of radius R, ε assumes the form

$$\varepsilon(\boldsymbol{r},\omega)=\theta(r-R)+\theta(R-r)\varepsilon(\omega). \tag{7.94}$$

Instead of (7.93), let us first solve the corresponding homogeneous Maxwell equation

$$\nabla^2\boldsymbol{A}(\boldsymbol{r})+\frac{\omega^2}{c^2}\,\varepsilon(\boldsymbol{r},\omega)\boldsymbol{A}(\boldsymbol{r},\omega)=0 \tag{7.95}$$

for the vector potential $\boldsymbol{A}(\boldsymbol{r})$. For $\varepsilon(\boldsymbol{r},\omega)$ given by (7.94), all the solutions of (7.95) are explicitly known [7.67, 82, 83]:

$$\boldsymbol{A}^{\mathrm{E}}_{klm}(\boldsymbol{r})=\frac{\mathrm{i}}{k}\,\nabla\times[f_l^{\mathrm{E}}(kr)\boldsymbol{X}_{lm}(\theta,\phi)] \tag{7.96}$$

$$=\left[-\hat{r}\,\frac{\sqrt{l(l+1)}}{kr}\,f_l^{\mathrm{E}}(kr)y_{lm}(\theta,\phi)+\frac{\mathrm{i}}{kr}\,\frac{\partial}{\partial r}\,(rf_l^{\mathrm{E}}(kr)\hat{r}\times\boldsymbol{X}_{lm}(\theta,\phi)\right],$$

$$\boldsymbol{A}^{\mathrm{M}}_{klm}(\boldsymbol{r})=f_l^{\mathrm{M}}(kr)\boldsymbol{X}_{lm}(\theta,\phi), \tag{7.97}$$

where (r,θ,ϕ) are the components of $\boldsymbol{r}$ in polar coordinates. X and f are given by

$$\boldsymbol{X}_{lm}(\theta,\phi)=\frac{1}{\sqrt{l(l+1)}}\,\boldsymbol{L}y_{lm}(\theta,\phi);\quad \boldsymbol{L}=-\mathrm{i}\boldsymbol{r}\times\nabla, \tag{7.98}$$

$$f_l^\sigma(kr)=\{[a^\sigma_{kl}j_l(kr)+c^\sigma_{kl}\eta_l(kr)]\theta(r-R)+b^\sigma_{kl}j_l(k_ir)\theta(R-r)\}\cdot(a^{\sigma 2}_{kl}+c^{\sigma 2}_{kl})^{-1/2} \tag{7.99}$$

where $\sigma=E$ or M. The coefficients c^σ_{kl} and b^σ_{kl} are obtained by using the boundary conditions for the electric and magnetic fields at the surface of sphere and are given by [7.82]

$$c^E_{kl}=\frac{-\varepsilon(\omega)\mathrm{j}_l(k_iR)\,[kR\mathrm{j}_l(kR)]'+\mathrm{j}_l(kR)\,[k_iR\mathrm{j}_l(k_iR)]'}{\varepsilon(\omega)\mathrm{j}_l(k_iR)\,[kR\eta_l(kR)]'-\eta_l(kR)\,[k_iR\mathrm{j}_l(k_iR)]'}\,a^E_{kl}, \tag{7.100}$$

$$b_{kl}^{\mathrm{E}}=\frac{-\sqrt{\varepsilon(\omega)}\eta_l(kR)\,[kR\mathrm{j}_l(kR)]'+\sqrt{\varepsilon(\omega)}\mathrm{j}_l(kR)\,[kR\eta_l(kR)]'}{\varepsilon(\omega)\mathrm{j}_l(k_iR)\,[kR\eta_l(kR)]'-\eta_l(kR)\,[k_iR\mathrm{j}_l(k_iR)]'}\,a_{kl}^{\mathrm{E}}, \quad (7.101)$$

$$c_{kl}^{M}=\frac{\mathrm{j}_l(k_iR)\,[kR\mathrm{j}_l(kR)]'-\mathrm{j}_l(kR)\,[k_iR\mathrm{j}_l(k_iR)]'}{-\mathrm{j}_l(k_iR)\,[kR\eta_l(kR)]'+\eta_l(kR)\,[k_iR\mathrm{j}_l(k_iR)]'}\,a_{kl}^{M}, \quad (7.102)$$

$$b_{kl}^{M}=\frac{\eta_l(kR)\,[kR\mathrm{j}_l(kR)]'-\mathrm{j}_l(kR)\,[kR\eta_l(kR)]'}{-\mathrm{j}_l(k_iR)\,[kR\eta_l(kR)]'+\eta_l(kR)\,[k_iR\mathrm{j}_l(k_iR)]'}\,a_{kl}^{M}, \quad (7.103)$$

where $k_i=\sqrt{\varepsilon(\omega)}k$. $\mathrm{j}_l(kr)$ and $\eta_l(kr)$ are spherical Bessel functions, $\mathrm{y}_{lm}(\theta,\phi)$ are spherical harmonics and the prime in (7.100–103) denotes differentiation of the quantity in brackets with respect to the argument of the spherical Bessel function. The eigenfrequencies corresponding to (7.96, 97) are degenerate and are given by

$$\begin{aligned}\omega_k&=ck \qquad r>R\\ &=ck_i \qquad r<R.\end{aligned} \quad (7.104)$$

$\boldsymbol{A}^{\mathrm{E}}$ and $\boldsymbol{A}^{\mathrm{M}}$ are usually called electric and magnetic modes, respectively.

There is a third type of eigensolution of (7.95) corresponding to the zero of the denominator of the right-hand sides in (7.100) or (7.101), i.e.,

$$\varepsilon(\omega)\mathrm{j}_l(k_iR)\,[kR\,\eta_l(kR)]'-\eta_l(kR)\,[k_iR\mathrm{j}_l(k_iR)]'=0 \quad (7.105)$$

and $a_{kl}^{\mathrm{E}}=0$. For $kR\ll l$ $(l\neq 0)$, (7.105) yields the eigenfrequencies

$$\varepsilon(\omega_l)=-(l+1)/l+\theta(kR)^2. \quad (7.106)$$

The corresponding eigensolutions are

$$\begin{aligned}A_{lm}^{\mathrm{s}}(\boldsymbol{r})=&\frac{1}{R^{3/2}}\,\frac{1}{(2l+1)^{1/2}}\,\{\theta(r-R)\,(R/r)^{l+2}\,[r(l+1)\mathrm{y}_{lm}(\theta,\phi)\\ &+\mathrm{i}\sqrt{l(l+1)}\,\hat{\boldsymbol{r}}\times\boldsymbol{X}_{lm}(\theta,\phi)]\\ &+\theta(R-r)\,(r/R)^{l-1}\,[\hat{r}l\mathrm{y}_{lm}(\theta,\phi)-\mathrm{i}\sqrt{l(l+1)}\,\hat{\boldsymbol{r}}\times\boldsymbol{X}_{lm}(\theta,\phi)]\}.\end{aligned} \quad (7.107)$$

This solution is localized near the surface of the metal sphere and decays as $(R/r)^{l+2}$ and $(r/R)^{l-1}$ outside and inside the sphere, respectively. The solution is similar to the (sp) polariton discussed previously for the plane metal surface though the dispersion relations are different. In the limit $l\to\infty$, both solutions become identical and satisfy $\varepsilon(\omega_l)+1=0$ [7.84].

Using the homogeneous solutions given above, $\overleftrightarrow{\mathscr{D}}(\boldsymbol{r},\boldsymbol{r}',\omega)$ can be written as

$$\overleftrightarrow{\mathscr{D}}(\boldsymbol{r},\boldsymbol{r}'\omega)=4\pi c^2\left[\frac{2}{\pi}\int dk\,k^2\sum_{l,m,\sigma}\frac{A^{\sigma *}_{klm}(\boldsymbol{r})A^{\sigma}_{klm}(\boldsymbol{r}')}{\omega_k^2-\omega^2-\mathrm{i}\eta}+\sum_{l,m}\frac{A^{s*}_{lm}(\boldsymbol{r})A^{s}_{lm}(\boldsymbol{r}')}{\omega_l^2-\omega^2-\mathrm{i}\eta}\right]. \tag{7.108}$$

The eigenfrequencies ω_k and ω_l are defined in (7.104, 105), respectively. Having obtained the photon propagator $\overleftrightarrow{\mathscr{D}}(\boldsymbol{r},\boldsymbol{r}'\omega)$, one can easily calculate $\tilde{\alpha}_{ba}(\boldsymbol{Q}\lambda,\omega)$ by substituting (7.108) into (7.28). We thus obtain [7.85]

$$\tilde{\alpha}_{ba}(\boldsymbol{Q}\lambda,\omega)=\sum_{l,m,\sigma}\frac{a^{\sigma}_{Ql}}{(a^{\sigma 2}_{Ql}+c^{\sigma 2}_{Ql})^{1/2}}\,\tilde{A}^{\sigma}_{lm}(\boldsymbol{Q}\lambda)\,\alpha_{ba}(Qlm\sigma) \tag{7.109}$$

with

$$\tilde{A}^{\mathrm{E}}_{lm}(\boldsymbol{Q}\lambda)=\frac{1}{\mathrm{j}_l(Qr)}\int \boldsymbol{X}^{*}_{lm}(\Omega)\cdot[\boldsymbol{Q}\times\hat{e}_\lambda(\boldsymbol{Q})]\exp(\mathrm{i}\boldsymbol{Q}\cdot\boldsymbol{r})\,d\Omega, \tag{7.110}$$

$$\tilde{A}^{\mathrm{M}}_{lm}(\boldsymbol{Q}\lambda)=\frac{1}{\mathrm{j}_l(Qr)}\int \boldsymbol{X}^{*}_{lm}(\Omega)\cdot\hat{e}_\lambda(\boldsymbol{Q})\exp(\mathrm{i}\boldsymbol{Q}\cdot\boldsymbol{r})\,d\Omega, \tag{7.111}$$

$$\alpha_{ba}(Qlm\sigma)=A_0(\boldsymbol{Q}\lambda)\,\langle\phi_b(\boldsymbol{r})|\boldsymbol{A}_{Qlm}(\boldsymbol{r})\cdot\boldsymbol{p}|\phi_a(\boldsymbol{r})\rangle. \tag{7.112}$$

If R is small compared to the wavelength of the light, the dominant contribution in (7.109) is obtained in the electric dipole approximation, i.e., from the term $\sigma=E$ and $l=1$. Assuming in addition that ϕ_a and ϕ_b are well localized at the molecular site $\boldsymbol{r}_0=(r_0,\theta_0,\phi_0)$ $(r_0>R)$ (Fig. 7.9), (7.109) simplified to

$$\tilde{\alpha}_{ba}(Q1mE)=A_0(\boldsymbol{Q}\lambda)\boldsymbol{A}^{\mathrm{E}}_{Q1m}(\boldsymbol{r}_0)\cdot\langle\boldsymbol{p}\rangle. \tag{7.113}$$

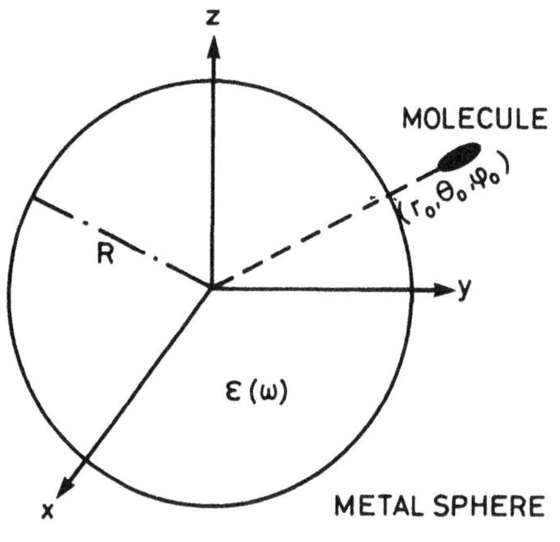

Fig. 7.9. Schematic diagram of the metal sphere and an adsorbed molecule

For the explicit calculation we consider the case of a 90° scattering geometry with $\hat{Q}\|\hat{z}, \hat{e}_\lambda(Q)\|\hat{x}$ and $\hat{Q}'\|\hat{y}, \hat{e}_\lambda(Q)\|\hat{x}$. Substituting these in (7.110) and using the spherical wave expansion for $\exp(\mathrm{i}Q\cdot r)$ and (7.98) for $X_{lm}(\theta, \phi)$, the angular integration in (7.110) can be easily performed. Taking $\langle p\rangle\|\hat{r}_0$, we obtain

$$\tilde{A}^{\mathrm{E}}_{1m}(Q\lambda)=\tilde{A}^{\mathrm{E}}_{1m}(Q'\lambda')=\sqrt{3\pi}\,(\delta_{m,-1}-\delta_{m,1}). \tag{7.114}$$

Thus from (7.109, 113, 114) and (7.96) we find

$$\tilde{\alpha}_{ba}(Q\lambda,\omega)=-A_0(Q\lambda)\langle|p|\rangle(1+c_Q'^2)^{-1/2}[f_1(Qr_0)/Qr_0]\,3\sin\theta_0\cos\phi_0, \tag{7.115}$$

$$c_Q'=c^{\mathrm{E}}_{Q1}/a^{\mathrm{E}}_{Q1}. \tag{7.116}$$

A similar relation holds for the scattered photon vertex. The total cross section for fixed distance r_0 is then obtained by averaging over all angles θ_0 and ϕ_0 (corresponding to the random position of the molecule), i.e.,

$$|\tilde{\alpha}_{ba}(Q\lambda,\omega)|^2|\tilde{\alpha}_{ba}(Q'\lambda',\omega)|^2=\int_0^{2\pi}d\phi_0\int_0^{\pi}d\theta_0\,\sin\theta_0|\alpha_{ba}(Q\lambda,\omega)|^2|\alpha_{ba}(Q'\lambda',\omega')|^2$$

$$=\frac{1}{5}|A_0(Q\lambda)A_0(Q'\lambda')|^2|\langle p\rangle|^4\left|\frac{3f_1(Qr_0)}{Qr_0(1+C_Q'^2)}\right|^2\left|\frac{3f_1(Q'r_0)}{Q'r_0(1+C_{Q'}'^2)}\right|^2. \tag{7.117}$$

The expression analogous to (7.117) for a free molecule is

$$|\alpha^{(0)}_{ba}(Q\lambda,\omega)|^2|\alpha^{(0)}_{ba}(Q'\lambda',\omega')|^2=|A_0(Q\lambda)|^2|A_0(Q'\lambda')|^2|\langle p_x\rangle|^4. \tag{7.118}$$

Averaging again over all angles corresponding to the random orientation of $\langle p\rangle$ of the free molecule and comparing the results with (7.117), we obtain for the enhancement

$$\varrho_1(\omega)=\left|\frac{\mathrm{j}_1(Qr_0)+c_Q'\eta_1(Qr_0)}{(1+c_Q'^2)Qr_0/3}\right|^2\left|\frac{\mathrm{j}_1(Q'r_0)+c_{Q'}'\eta_1(Q'r_0)}{(1+c_{Q'}'^2)Q'r_0/3}\right|^2. \tag{7.119}$$

For $Qr_0\ll 1$, one can simplify (7.119) by using $\mathrm{j}_1(Qr_0)\sim Qr_0/3$, $\eta_1(Qr_0)\sim 1/(Qr_0)^2$ and, using (7.116) and 7.110),

$$c_Q'\approx\frac{1-\varepsilon(\omega)}{1+\varepsilon(\omega)/2}\,(QR)^3/3. \tag{7.120}$$

Thus,

$$\varrho_1(\omega)\approx\left|1+2\,\frac{\varepsilon(\omega)-1}{\varepsilon(\omega)+2}\,(R/r_0)^3\right|^2\left|1+2\,\frac{\varepsilon(\omega')-1}{\varepsilon(\omega')+2}\,(R/r_0)^3\right|^2. \tag{7.121}$$

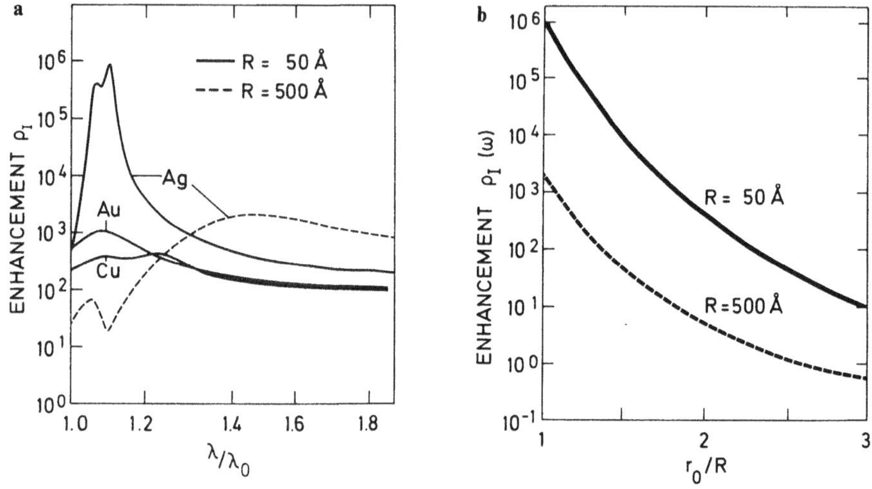

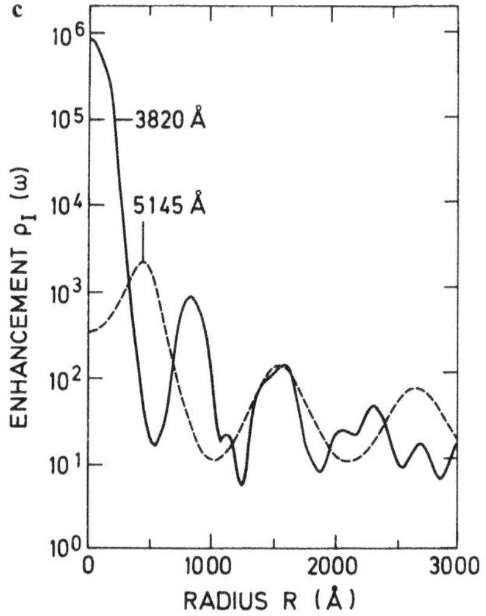

Fig. 7.10. (**a**) ϱ_{I} calculated from (7.119) as a function of incident wavelength for Ag, Cu and Au spheres surrounded with water. A molecule is assumed on the surface. Results for Ag are from [7.14]. To give results in the same figure, a different scale for the wavelength is chosen by using $\lambda_0 = 3500$ Å for Ag and $\lambda_0 = 5000$ Å for Cu and Au. In Figs. 7.10a–c $\varepsilon(\omega)$ for metals is taken from [7.80], the vibrational frequency Ω is $1010\ \mathrm{cm}^{-1}$ and a 90° scattering geometry is assumed (see text). (**b**) $\varrho_{\mathrm{I}}(\omega)$ calculated from (7.119) as a function of the distance r_0 of the molecule from the Ag sphere. From [7.14]. (**c**) $\varrho_{\mathrm{I}}(\omega)$ calculated from (7.119) as a function of the radius of the Ag sphere. The molecule is assumed on the surface of sphere. From [7.14]

It can be shown that the results (7.119, 121) are independent of the polarizations of the photons as long as 90° scattering is considered [7.14]. Results for other scattering geometries have been derived in [7.14]. If the position of the molecule is fixed parallel to the x-axis [i.e., $\theta_0 = 90°$, $\phi_0 = 0°$ in (7.115)], one obtains an additional factor of 5 in the expressions (7.119, 121).

Equation (7.121) shows that $\varrho_{\mathrm{I}}(\omega)$ depends on *three* parameters: the dielectric function of $\varepsilon(\omega)$ of the metal, the radius R of the metal sphere and the position r_0 of the molecule from the metal. In the following we discuss this dependence in more detail. Figure 7.10 gives the enhancement for Ag, Cu and Au as calculated from (7.119). The molecule is assumed to be on or very near the metal sphere (i.e., $r_0/R \sim 1$). The metal sphere is assumed to be surrounded by water so that $\varepsilon(\omega)$ denotes the relative dielectric function of the metal with respect to water (the value of ε for water is taken to be 1.76). The values of ε for Ag, Cu and Au are taken from [7.80] and the vibrational frequency of the pyridine molecule is taken to be $1010\,\mathrm{cm}^{-1}$ [7.14]. For $R=50$ Å, (7.119, 121) give almost the same results. In the case of Ag, the peak around 3820 Å corresponds to the excitation of the localized mode given by $\varepsilon_1(\omega_1)\approx -2$. The corresponding enhancement is quite large ($\sim 10^6$) because ε_2 for Ag is very small. The enhancement decreases rapidly for higher and lower frequencies but its absolute magnitude is still ≈ 100 for $\omega\approx 2\omega_1$. Similar enhancement curves (note the different frequency scale) are obtained for Cu and Au spheres of radii 50 Å. However, there is no sharp peak with a large absolute value as in Ag because for Cu and Au, ε_2 is almost 10 times larger than for Ag. Nevertheless, the value for the enhancement ranges between 2 to 3 orders of magnitude. The peak enhancement for Ag also decreases with increasing radius of the sphere (compare dotted line for $R=500$ Å with full line for $R=50$ Å for Ag). It becomes much broader and shifts to larger wavelengths (≈ 5000 Å).

Figure 7.10b shows $\varrho_1(\omega)$ for Ag as a function of r_0. The wavelength of the incident photon corresponds to the maxima of Fig. 7.10a, i.e., 3820 Å and 5110 Å for $R=50$ Å and 500 Å, respectively. In both cases the enhancement drops with increasing r_0; it is, however, still appreciable for distances $r_0 - R\approx 100$ Å. The decrease of the enhancement with increasing r_0-R slows down if R is increased. For very large R's, the enhancement curve is similar to that for a plane Ag surface shown in Fig. 7.5.

Figure 7.10c illustrates the dependence of $\varrho_1(\omega)$ on R (for Ag) for incident photon wavelengths of 3820 Å and 5145 Å. The former corresponds to the peak position for small R and the latter corresponds to the ion-argon laser line. The enhancement oscillates mainly in the 10^1–10^4 range except for radii <300 Å and an incident wavelength of 3820 Å. In this case $\varrho_1(\omega)$ increases sharply to $\sim 10^6$ for a radius of ≈ 50 Å. For still smaller R's, the enhancement remains constant at this value (which is determined by ε_1 and ε_2 of Ag).

So far we have considered three simple geometries for the substrate, i.e., a plane surface with a grating, a randomly rough surface and an isolated metal sphere. These cases can be experimentally realized; see for instance [7.43–46], [47, 48] and [7.49], respectively. However, in most SERS experiments, the rough surface may not be as simple. For example, in experiments which use an electrochemical procedure for roughening the metal electrode, it is believed that surface irregularities are in the form of bumps and the admolecules are in their neighbourhood [7.50, 51, 55]. In other experiments the metal may be deposited in the form of small islands [7.52, 53]. In colloidal solutions, the colloidal particles may be ellipsoidal or of any other shape.

Some attempts have been made to treat a few simple substrate geometries other than those discussed previously, e.g., an isolated spheroid [7.27], two spheres close to each other [7.28], a hemispheroid protruding from a flat surface [7.23, 26] and metal-island films [7.29]. All these treatments neglect retardation effects which in general would decrease the enhancement. These calculations, therefore, are valid only if the wavelength of the incident and scattered photons is much larger than the size of the spheroid, island, etc. The calculations are quite involved, so we will not give any details here. The results are similar to those in the case of a sphere: a localized mode exists for each geometry (formally the localized modes are given by $\varepsilon(\omega)+G=0$, where G is different for different geometries [7.11]; G is equal to 1 and 2 for a plane surface and a sphere, respectively; for a spheroid G depends on its three axis). Large enhancements are obtained if these localized modes are excited resonantly by the incident and/or scattered photon.

The case of an isolated metal spheroid has been discussed in detail by *Wang* and *Kerker* [7.27] considering a randomly oriented admolecule. For a prolate spheroid the enhancement was found to increase with increasing ratio a/b, where a and b are the large and small axes, respectively. Furthermore, the position of the peak of the enhancement curves shifted to longer wavelengths [7.27]. Similar results hold for oblate spheroids. *Aravind* et al. [7.28] considered two isolated spheres (each of radius R) and studied the interaction between the corresponding two localized modes (Mie resonances) when the spheres came closer. It was found that the interaction becomes important if the separation D (surface to surface) $\leqslant R/2$. As a result, the two degenerate resonances (corresponding to the isolated spheres) split into two separate resonances. The enhancement is especially large if the molecule lies in the space between the two spheres because of the large electric field there. If the molecule is outside that region, the maximum enhancement is rather small compared to that of an isolated sphere. These facts may be important, for instance, for the discussion of concentrated colloidal solutions.

A hemispheroid protruding form a plane surface is discussed in [7.23, 26]. For a hemiellipsoid (on a perfectly conducting plane) with the molecule lying on the top (along the major axis), the enhancement increases with increasing a/b [7.23] (a and b being the major and minor axes, respectively). For very large ratios a/b (i.e., highly prolate or needle-like ellipsoids), the field at the tip is very large (lighting rod effect) yielding large enhancements. For a hemisphere on a flat metal with the same ε, the enhancement is largest if the molecule is near both the sphere and the plane surface [7.26].

7.5 Local Field Effects Due to the Presence of a Molecule

The second contribution $\varrho_{\text{II}}(\omega)$ to SERS describes local field effects caused by the presence of a molecule. In our general formulation of SERS, these effects are contained in the factor $[1-X(\omega)]^{-1}$ in (7.19) where the matrix $X(\omega)$ is given by (7.20). Large enhancements may occur in frequency regions where the determinant of the matrix $[1-X(\omega)]$ is very small. Speaking in more physical terms, ϱ_{II} is due to the fact that a real or virtual dipole transition on the molecule or between the molecule and the metal excites an electron-hole pair in the metal. The resulting image potential acts back on the molecule and may increase the transition probability substantially. In the oversimplified form, these effects can be described within the classical theory of image point dipoles [7.30, 31] and were used for the first explanation of SERS [7.30]. More realistic treatments include quantum effects such as the extension of the molecule, charge transfer transitions or the $\boldsymbol{Q}$ dependence of ε [7.34–36].

Adopting the two-state model for the molecule of Sect. 7.3.2, $X(\omega)$ reduces to the scalar quantity of (7.29) and $\varrho_{\text{II}}(\omega)$ is given by (7.35). The basic quantity required for the calculation of $X(\omega)$ is the photon propagator $\overleftrightarrow{\mathscr{D}}(\boldsymbol{r},\boldsymbol{r}',\omega)$ in the presence of a metal. $\overleftrightarrow{\mathscr{D}}$ has been obtained in the previous section for a plane metal surface and a metal sphere within the local dielectric approximation. We will limit ourselves to the plane surface here. The details in the case of a sphere are essentially similar.

Inserting the Fourier transform of $\overleftrightarrow{\mathscr{D}}(\boldsymbol{r},\boldsymbol{r}',\omega)$ [(7.46)] into (7.29), $X_{ba}(\omega)$ can be written as

$$X_{ba}(\omega)=R_{ba}(\omega)\,(e/mc)^2\sum_{\alpha,\beta}\int d\boldsymbol{r}\int d\boldsymbol{r}'\int\frac{d^2\boldsymbol{k}_{\|}}{(2\pi)^2}\int\frac{d^2\boldsymbol{k}'_{\|}}{(2\pi)^2}\exp\left(\mathrm{i}\boldsymbol{k}_{\|}\cdot\boldsymbol{r}_{\|}+\mathrm{i}\boldsymbol{k}'_{\|}\cdot\boldsymbol{r}'_{\|}\right)$$

$$\cdot\,\phi_b(\boldsymbol{r})\,p_\alpha\phi_a(\boldsymbol{r})\,\hat{d}_{\alpha\beta}(\boldsymbol{k}_{\|}\boldsymbol{k}'_{\|}\omega|zz')\,\phi_b(\boldsymbol{r}')\,p_\beta\phi_a(\boldsymbol{r}'),\qquad(7.122)$$

where $\hat{d}_{\alpha\beta}(\boldsymbol{k}_{\|}\boldsymbol{k}'_{\|}\omega|zz')$ is given by (7.52). We simplify the following analysis by assuming a smooth surface and use $d^{(0)}_{\alpha\beta}$ instead of $\hat{d}_{\alpha\beta}$ in (7.122). $d^{(0)}_{\alpha\beta}(\boldsymbol{k}_{\|}\omega|zz')$ is given by (7.54) and contains two extended s and p-polarized photon modes and a localized (sp) mode. If the distance z_0 between the molecule and metal is much smaller than c/ω, it is sufficient to keep only the (sp) contribution to $d^{(0)}_{\alpha\beta}$. Using thus (7.55) for $d^{(0)}_{\alpha\beta}$, (7.122) can be written as

$$X_{ba}(\omega)=-R_{ba}(\omega)\,\frac{e^2}{\pi m^2\omega^2}\int d^2\boldsymbol{k}_{\|}\,\frac{k_{\|}^2}{(\nu+\varepsilon\nu_0)}\,|F_{ba}(\boldsymbol{k}_{\|})|^2\qquad(7.123)$$

with

$$F_{ba}(\boldsymbol{k}_{\|})=\int d\boldsymbol{r}\,\exp\left(\mathrm{i}\boldsymbol{k}_{\|}\cdot\boldsymbol{r}_{\|}\right)\phi_b^*(\boldsymbol{r})\,[\theta(z)\exp(-\nu_0 z)\boldsymbol{e}^{+*}(\boldsymbol{k}_{\|})\cdot\boldsymbol{p}$$

$$-\theta(-z)\,\mathrm{e}^{\nu z}\boldsymbol{e}(\boldsymbol{k}_{\|})\cdot\boldsymbol{p}]\,\phi_a(\boldsymbol{r}).\qquad(7.124)$$

Equation (7.123) can be simplified by neglecting the retardation effects which is justified because the main contribution in the $\boldsymbol{k}_{\|}$ integration in (7.123) comes from the region $|\boldsymbol{k}_{\|}|\gg\omega/c$. The integrand is a rather smooth but increasing function of $|\boldsymbol{k}_{\|}|$ until an effective cut-off $\sim 1/r_0$, where r_0 is the extension of the molecular wave function. For $|\boldsymbol{k}_{\|}|\gg\omega/c$, (7.56–59) give $\nu\sim\nu_0\sim|\boldsymbol{k}_{\|}|$ and $\boldsymbol{e}^-(\boldsymbol{k}_{\|})\sim\boldsymbol{e}^+(\boldsymbol{k}_{\|})\sim(\hat{z}+\mathrm{i}\hat{k}_{\|})$. Furthermore, the second term in (7.124) contributes only when the wave functions $\phi_{a,b}(\boldsymbol{r})$ extend inside the metal. However, inside the metal there is an additional contribution to (7.123) due to the virtual excitation of the bulk plasmon mode which we have not written down here explicitly. As discussed by *Evans* and *Mills* [7.71], this contribution cancels approximately the second term in (7.124). Equations (7.123, 124) can thus be written as

$$X_{ba}(\omega) = -R_{ba}(\omega)\,\frac{e^2}{\pi m^2\omega^2}\int d^2\boldsymbol{k}_{\parallel}\,\frac{k_{\parallel}}{[1+\varepsilon(\omega)]}\,|F_{ba}(\boldsymbol{k}_{\parallel})|^2 \tag{7.125}$$

$$F_{ba}(\boldsymbol{k}_{\parallel}) = \int d\boldsymbol{r}\,\exp(\mathrm{i}\boldsymbol{k}_{\parallel}\cdot\boldsymbol{r}_{\parallel})\,\mathrm{e}^{-k_{\parallel}z}\theta(z)\,\phi_b^*(\boldsymbol{r})\,(\hat{z}-\mathrm{i}\hat{k}_{\parallel})\cdot\boldsymbol{p}\phi_a(\boldsymbol{r}). \tag{7.126}$$

If $\phi_{a,b}$ correspond to the molecule lying at z_0 and $z_0 \gg r_0$, then $\phi_{a,b}$ can be taken approximately to be localized at z_0. The $\boldsymbol{k}_{\parallel}$ integration in (7.125) can then be performed and we obtain the result of classical image theory:

$$X_{ba}(\omega) = -R_{ba}(\omega)\,\frac{|\langle \boldsymbol{d}\rangle|^2}{8z_0^3}\,\frac{\omega_s}{\omega-\omega_s}, \tag{7.127}$$

where

$$\langle \boldsymbol{d}\rangle = \mathrm{i}e\,\frac{\langle \boldsymbol{p}\rangle}{m\omega_s} = \langle -e\boldsymbol{r}\rangle \tag{7.128}$$

is proportional to the dipole moment induced on the molecule. The second equality in (7.128) can be obtained by using a complete set of molecular states and a $\boldsymbol{p}$-sum rule as done in [7.86]. Also, in writing (7.127) we have taken the plasmon-pole approximation for $\varepsilon(\omega)$ so that $\omega^2[1+\varepsilon(\omega)] = 2(\omega^2-\omega_s^2) \sim 4\omega_s(\omega-\omega_s)$, where ω_s is the (sp) frequency.

If z_0 is comparable to r_0, one cannot assume the $\phi_{a,b}$ to be localized like a δ-function. The z_0 dependence of $X_{ba}(\omega)$ deviates then from $1/z_0^3$ in the region $z_0 \lesssim r_0$ and $X_{ba}(\omega)$ approaches a constant value for $z_0 \to 0$. To illustrate this, let us

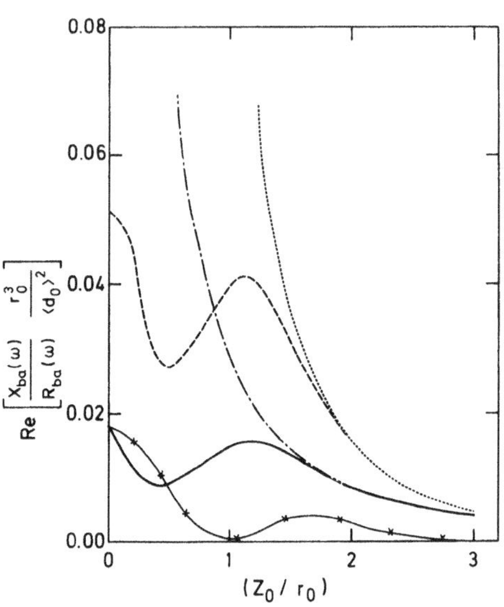

Fig. 7.11. Real part of $X_{ba}(\omega)$ calculated with different approximations from (7.125) as a function of the distance z_0 of the molecule from the plane surface. $\langle d_0\rangle \equiv e\hbar/m\omega_s r_0$ is roughly the dipole moment of a point dipole located at r_0. –·–· and ···· are the results for a point dipole with (7.134) and without spatial dispersion, respectively. —— and –––– are the results for an extended molecule (7.129, 130) with and without spatial dispersion, respectively. –×–×– is the result when wave functions $\phi_b(\boldsymbol{r})$ and $\phi_a(\boldsymbol{r})$ in (7.126) correspond to the metal atom (located at the surface) and the molecule, respectively, including dispersion. The other parameters used are $\omega = 2.3$ eV, $\omega_s = 3.6$ eV and in the case of dispersion, $\omega_{pf} = \sqrt{2}\omega_s$ and $c/\beta = 2.74 \times 10^2$

assume $\phi_{a,b}(\boldsymbol{r})$ are of Gaussian form:

$$\phi_a(\boldsymbol{r})=(2/\pi r_0^2)^{3/4}\exp[-(r_a/r_0)^2], \tag{7.129}$$

$$\phi_b(\boldsymbol{r})=2(2/\pi r_0^2)^{3/4}(\hat{z}\cdot \boldsymbol{r}_b/r_0)\exp[-(r_b/r)^2], \tag{7.130}$$

where $\boldsymbol{r}_a=\boldsymbol{r}-z_{0a}\hat{z}$ and $\boldsymbol{r}_b=\boldsymbol{r}-z_{0b}\hat{z}$. The integral in (7.126) can be carried out and we obtain

$$\begin{aligned}|F_{ba}(\boldsymbol{k}_{\parallel})|=(\hbar/2r_0)\exp(-A^2/2)\,&[(1+Ax-A^2)\,(1-\operatorname{erf}\{(x-A-2B)/\sqrt{2}\})\\ &\cdot\exp[-x(A+2B)]-\sqrt{\frac{2}{\pi}}(A+2B)\exp\{-[x^2+(A+2B)^2]/2\}],\end{aligned} \tag{7.131}$$

where we have defined the dimensionless quantities

$$\begin{aligned}B&=z_{0b}/r_0\\ A&=(z_{0a}-z_{0b})/r_0\\ x&=k_{\parallel}r_0\end{aligned} \tag{7.132}$$

and erf $\{x\}$ is the error function. The $|\boldsymbol{k}_{\parallel}|$ integration in (7.125) has to be done numerically. The resulting z_0 dependence of the real part of $X_{ba}(\omega)$ for $z_{0a}=z_{0b}=z_0$ (i.e., $\phi_{a,b}$ corresponding to molecule) is given by the dashed curve in Fig. 7.11. The z_0^{-3} dependence for a point dipole (7.127) is also given (dotted line) for comparison. It shows that the dashed curve starts deviating from a z_0^{-3} dependence for $z_0 \lesssim 2r_0$ and has a rather constant value for $0<z_0\lesssim 1.5r_0$ except for small oscillations.

As stated earlier, the main contribution to the integration in (7.125) comes from large $|\boldsymbol{k}_{\parallel}|$ values ($\sim 10^8$ cm^{-1}). So far in our treatment we have neglected spatial dependence of the dielectric function $\varepsilon(\boldsymbol{Q},\omega)$ which may, however, be important for large $|\boldsymbol{Q}|$. This has been considered in [7.34]. Let us assume

$$\varepsilon(\boldsymbol{Q},\omega)=1-\frac{\omega_{\mathrm{pf}}^2}{\omega^2-\beta^2Q^2}, \tag{7.133}$$

where ω_{pf} is the free-electron plasma frequency, $\beta^2=3v_{\mathrm{F}}^2/5$ and v_{F} is the Fermi velocity. In this case (7.125) changes to

$$X_{ba}(\omega)=-R_{ba}(\omega)\frac{e^2}{4\pi m^2}\int d^2\boldsymbol{k}_{\parallel}k_{\parallel}|F_{ba}(\boldsymbol{k}_{\parallel})|^2 f(\beta,k_{\parallel}), \tag{7.134}$$

$$f(\beta,k_{\parallel})=\frac{1}{[\omega-\omega_{\mathrm{s}}(k_{\parallel})]\omega_{\mathrm{s}}(k_{\parallel})}. \tag{7.135}$$

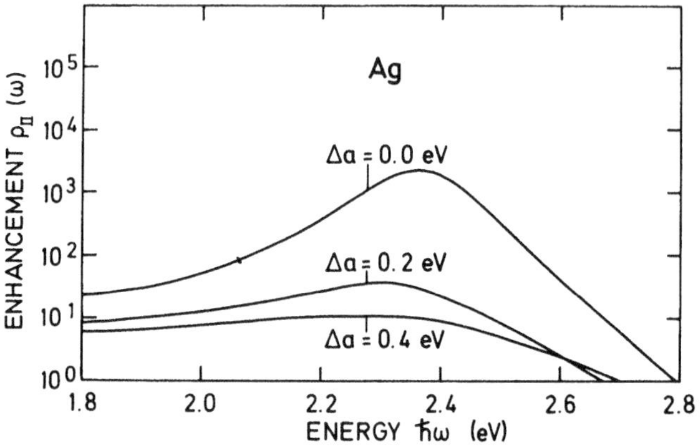

Fig. 7.12. Enhancement $\varrho_{\text{II}}(\omega)$ (7.35) due to image local field effects. Calculations assume that Re $\{X_{ba}(\omega)\}=1$ for $\omega=2.3$ eV. Δ_a denotes the width of the molecular level (see Sect. 7.6). $\Delta_a=0$ eV corresponds to the physisorbed case and $\Delta_a=0.2$ eV and $\Delta_a=0.4$ eV are for the chemisorbed case. $\varepsilon(\omega)$ for Ag is taken from [7.80] and the vibrational frequency $\Omega=0.2$ eV

The dispersion relation for surface plasmon frequency $\omega_s(k_{\|})$ is given by [7.34]

$$\omega_s^2(k_{\|})=\tfrac{1}{2}\,[\omega_{\text{pf}}^2+\beta^2k_{\|}^2+\beta^2k_{\|}^2(1+2\omega_{\text{pf}}^2/\beta^2k_{\|}^2)^{1/2}]. \tag{7.136}$$

Using (7.131) for $F_{ba}(\boldsymbol{k}_{\|})$, the real part of $X_{ba}(\omega)$ is calculated from (7.134) and its z_0 dependence is given in Fig. 7.11 (full line). The dash-dotted curve in Fig. 7.11 corresponds to a point dipole but with dispersion included (this curve goes as $1/z_0$ for very small z_0 [7.34]). These curves show that the extension of the molecular wave function strongly influences the z_0 dependence of $X_{ba}(\omega)$. We have also calculated $X_{ba}(\omega)$ when ϕ_a and ϕ_b correspond to a molecular and a metal state, respectively ($z_{ab}=0$). In this case X_{ba} decreases exponentially with z_0 (see full line with crosses in Fig. 7.11).

The above local field effects are only important if the real part of $X_{ba}(\omega)$ is ≈ 1 and, at the same time, the imaginary part of X is small. Note that X_{ba} is, in general, complex because $R_{ba}(\omega)$ and $\varepsilon(\omega)$ are complex. Unfortunately, there are no reliable matrix elements available to judge whether the above conditions are fulfilled for realistic systems. It is, however, possible to satisfy the above conditions for plausible values of z_0, $\langle \boldsymbol{d} \rangle$, etc. and for frequencies in the optical region [7.31, 34]. Choosing such a set of values we have plotted $\varrho_{\text{II}}(\omega)$ in Fig. 7.12 for Ag using $\varepsilon(\omega)$ from [7.80]. The calculation of $R_{ba}(\omega)$ is described in the next section. The maximum of ϱ_{II} corresponds to the resonance condition Re $\{X_{ba}(\omega)\}\approx 1$. The value for this maximum is especially large if $R_{ba}(\omega)$ is real ($\Delta_a=0$). For Ag it can reach 3 orders of magnitude and for Cu and Au it is less than 100 [7.19]. For a chemisorbed molecule, $R_{ba}(\omega)$ has a large imaginary part due to the finite width of the molecular levels, which decreases $\varrho_{\text{II}}(\omega)$ considerably. Furthermore, the condition that Re $\{X_{ba}(\omega)\}\approx 1$ may not be fulfilled in real situations so that the enhancements are actually smaller.

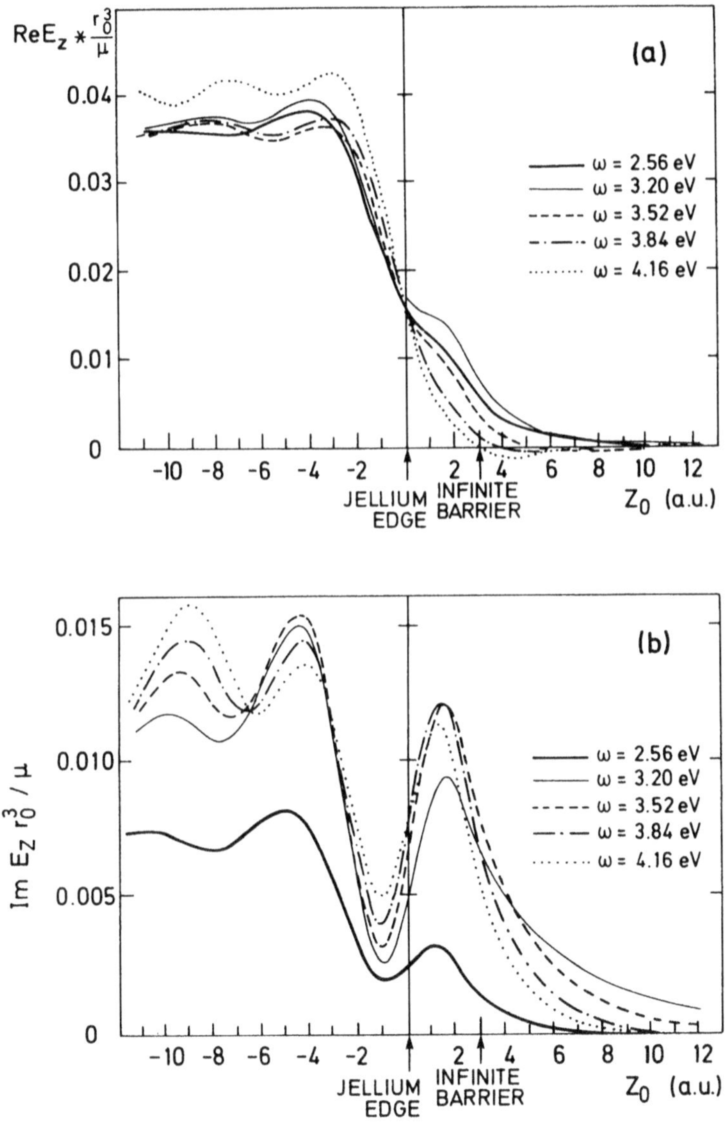

Fig. 7.13. The perpendicular component of the electric field, caused by the polarization of the metal by a dipole, determined at the dipole location. μ is the dipole moment, r_0 is the Bohr radius and z_0 is the distance between the jellium edge and the dipole location. The dipole is perpendicular to the surface and points towards the vacuum. The results are given for different frequencies above and below the surface plasmon frequency $\omega_{sp} = 2.976$ eV. (**a**) The real part of the field; (**b**) the imaginary part. From [7.75]. Results of the real part of this figure can be compared with those of Fig. 7.11 by choosing there $r_0 = 1$ a.u. and $\langle d_0 \rangle = \mu$

The above calculations disregard the single particle contribution to the polarization field and spatial dispersion effects connected with it. Recently, the polarization field of an oscillating dipole located near a metal surface has been calculated using a jellium model for the metal and an infinite high potential step for the boundary [7.75]. Both the single particle as well as the collective response were included. Figure 7.13a, b shows the calculated real and imaginary parts of the perpendicular component of the electric field induced by the metal at the location of the point dipole as a function of the distance z_0 from the jellium edge and for different frequencies of the oscillating dipole. The increase near the metal surface as well as its absolute value are similar to those of Fig. 7.11 if we identify the boundary with the jellium edge. This figure also indicates that field enhancements due to polarization effects should strongly depend on frequency in a way similar to the previous treatments where only the collective response was taken into account.

7.6 Chemisorption Effects

The third contribution ϱ_{III} to the enhancement of the cross section is due to the fact that the tight-binding matrix elements between localized functions on the molecule and on the metal are nonzero in the case of chemisorption. As a result the electronic levels of the admolecule become broadened and shifted. Because of the short-range coupling of molecule and metal, the electronic states of the molecule as well as those of the metal may serve as initial or intermediate states for the scattering process. In particular, the incident or scattered light or the molecular vibration may produce real or virtual electronic charge transfer transitions between molecule and metal.

Experimental evidence for chemisorption effects in SERS is rather indirect. For instance, recent experiments [7.54] of electron-energy loss spectra from pyridine chemisorbed on Ag(111) show charge transfer excitations between metal and molecule above ≈ 1.4 eV and also transitions between weakly perturbed molecular states. The broad onset of these transitions indicates that some levels of the chemisorbed pyridine are rather broad ($\Delta \approx 0.4$ eV). Another argument for the influence of chemisorption on SERS is based on an experiment [7.56] where several monolayers of pyridine were successively adsorbed on a clean smoothly modulated Ag metal surface. The molecules in the first monolayer show enhancements which are about 100 times larger than those in the following monolayers.

A theoretical model which takes chemisorption into account was first proposed by *Gersten* et al. [7.37]. They considered two electronic states of the molecule, one below and another above the Fermi energy, and calculated their self-energies due to charge transfer transitions. It was found that the real part of the self-energy depends on the frequency in such a way that the real part of the energy denominators in Raman cross section become zero if the incident frequency is somewhat smaller than the difference between the Fermi energy and the energy of the lower molecular level. Near that frequency the cross section should show a large enhancement if the imaginary part of the self-energy Δ is very small. Peak enhancement ratios of 10^4 to 10^6 are predicted if Δ is two or three orders of magnitude smaller than the energy difference of the two molecular levels. Such enhancements near the Fermi surface, however, have not been observed so far, probably because realistic dampings Δ are much larger than those used in the calculations.

The general theory of Sect. 7.3 contains chemisorption effects in terms of self-energy effects for the one-particle propagators (Fig. 7.1d). The self-consistent solution of the resulting equations is difficult to obtain without further approximations. Following [7.19] we use the approximations of the Newns-Anderson model of chemisorption [7.63, 64]. In this model one considers only one single discrete level for the molecule (say $|a\rangle$) which interacts with a set of continuum states $|\boldsymbol{k}\rangle$ of the metal. The metal states are assumed to be unaffected by the level $|a\rangle$. Thus, in Fig. 7.1d, only a thin unperturbed line is used for the metal electron propagator. The self-energy $\tilde{\Pi}$ for the molecular level contains two interactions. The first one is due to the hopping matrix element T_{ak}, the second is due to the Coulomb repulsion U which is important because of the possibility of charge transfer between molecule and metal. If $\varepsilon_a^{(0)} (= T_{aa})$ is the energy of the isolated molecule and $\langle n_a \rangle$ is the charge on it after chemisorption, then the shifted level will have energy ε_a:

$$\varepsilon_a = \varepsilon_a^{(0)} + U\langle n_a \rangle . \tag{7.137}$$

The corresponding propagator G_{aa} including the hopping term (first term in self-energy $\tilde{\Pi}$) is given by

$$G_{aa}(\varepsilon) = \left(\varepsilon - \varepsilon_a - \sum_{\boldsymbol{k}} \frac{|T_{a\boldsymbol{k}}|^2}{\varepsilon - \varepsilon_{\boldsymbol{k}} + \mathrm{i}\delta} \right)^{-1} . \tag{7.138}$$

The corresponding spectral density $\varrho_a(\varepsilon)$ is

$$\varrho_a(\varepsilon) = -\frac{1}{\pi} \operatorname{Im} \{G_{aa}(\varepsilon)\} = \frac{\Delta_a(\varepsilon)}{\pi \{[\varepsilon - \varepsilon_a - \Lambda_a(\varepsilon)]^2 + \Delta_a^2(\varepsilon)\}} , \tag{7.139}$$

where Λ_a and Δ_a are the real and imaginary parts of the last term in (7.138):

$$\Lambda_a(\varepsilon) - \mathrm{i}\Delta_a(\varepsilon) = \sum_{\boldsymbol{k}} \frac{|T_{a\boldsymbol{k}}|^2}{\varepsilon - \varepsilon_{\boldsymbol{k}} + \mathrm{i}\delta} . \tag{7.140}$$

Also

$$\langle n_a \rangle = \int_{-\infty}^{\varepsilon_F} d\varepsilon \varrho_a(\varepsilon) , \tag{7.141}$$

where ε_F is the Fermi energy. One has to solve (7.137–139) self-consistently.

The spectral function $\varrho_a(\varepsilon)$ of (7.139) resembles a Lorentzian. In the case where the energy dependence of Λ_a and Δ_a can be neglected, $\varrho_a(\varepsilon)$ becomes essentially a Lorentzian centred at the energy $\varepsilon = \varepsilon_a - \Lambda_a$ and with a width of $2\Delta_a$. In general, one has to retain the ε dependence of Λ_a and Δ_a. *Newns* [7.63] approximated (7.140) by considering the interaction of the molecule with only the nearest-neighbour atom(s) on the metal surface. The local density $\varrho_s^{(0)}(\omega)$ of metal states near the surface is assumed to be of the elliptical form

$$\varrho_s^{(0)}(\omega) = \frac{2}{\pi W}(1 - \omega^2/W^2)^{1/2}. \tag{7.142}$$

W is the halfwidth of the electronic band of the metal and ω is measured with respect to the center of the band. Thus, (7.140) reduces to

$$\Delta_a(\omega) = \frac{2}{W}|T_{as}|^2(1 - \omega^2/W^2)^{1/2}; \quad |\omega| < W, \tag{7.143}$$

$$\Lambda_a(\omega) = \frac{1}{\pi} P \int_P d\omega' \frac{\Delta_a(\omega')}{\omega - \omega'}, \tag{7.144}$$

where in (7.144), P denotes the principal part.

For a calculation of ϱ_{III} we consider electronic transitions between the occupied admolecular level $|a\rangle$ and the nearest-neighbour metal atom s. Thus we use, respectively, (7.139, 142) for the density of states $\varrho_a(\omega)$ and $\varrho_b(\omega)$ in (7.26, 27, 31). These expressions simplify if $\Delta_a(\omega)$ and $\Lambda_a(\omega)$ are small and if their energy dependence can be neglected. We then obtain for $\Lambda_a \sim 0$

$$R_{ba}(\omega + i\eta) \approx \int_{\varepsilon_F}^{\infty} d\omega' \frac{\varrho_s^{(0)}(\omega')}{\omega' - \omega - \varepsilon_a - i\Delta_a}, \tag{7.145}$$

$$\mathcal{L}_{baa}(-\omega - i\eta, \Omega) = \frac{1}{\Omega}[R_{ba}(\omega - \Omega + i\eta) - R_{ba}(\omega + i\eta)], \tag{7.146}$$

$$\mathcal{L}_{abb}(\omega + i\eta, -\Omega) = R_{ba}(\omega + i\eta) R_{ba}(\omega - \Omega + i\eta). \tag{7.147}$$

For a numerical estimate of ϱ_{III} we consider a pyridine molecule adsorbed on Ag. Silver has a *d*-band which is roughly 4 eV broad and which lies 4 eV below the Fermi level. The *sp*-bands are above the *d*-band and are taken to be 8 eV ($\hat{=}$ 2 W) wide (Fig. 7.14). For gaseous pyridine, the minimum energy gap is ≈ 5 eV.

Figure 7.15 shows the dependence of $\varrho_{\text{III}}(\omega)$ on the chemisorption parameters ε_a and Δ_a. ε_a is the position of the admolecular level $|a\rangle$ relative to the Fermi level, Δ_a denotes its width. The full and dashed lines in Fig. 7.15 correspond to ϱ_{III} enhancement curves with $\varepsilon_a = -2.5$ eV and $\varepsilon_a = -2.0$ eV, respectively, and for three different values of $\Delta_a = 0.1$ eV, 0.2 eV and 0.4 eV. These lines show that ϱ_{III} increases if ε_a approaches the incident or the scattered photon frequency as in

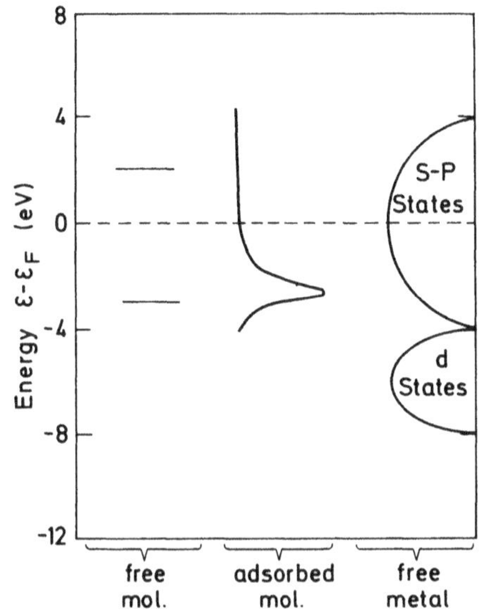

Fig. 7.14. Energy-level scheme used for the free molecule, admolecule and the substrate

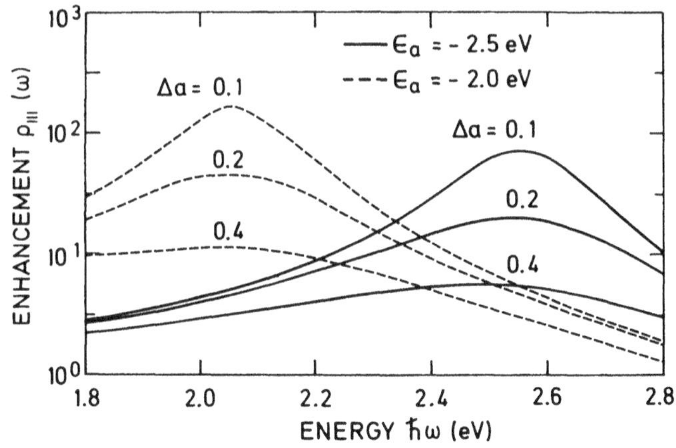

Fig. 7.15. Dependence of the enhancement $\varrho_{III}(\omega)$ due to chemisorption on Δ_a and ε_a (halfwidth and position of the admolecule level a, respectively). Width of the *sp*-band $2W = 8$ eV and vibrational frequency $\Omega = 0.2$ eV. Only the charge transfer excitations between the occupied molecular level and the unoccupied metal states are considered

usual resonant Raman scattering. The maximum around 2.5 eV in the full lines (2.0 eV in dashed lines) is caused by a threshold effect where transitions from the admolecule to the unoccupied metal states just above the Fermi level are near resonance with the incident photon [7.37]. ϱ_{III} decreases rapidly with increasing width Δ_a. The experiments of [7.54] on pyridine adsorbed on Ag show that $\varepsilon_a \approx -2.5$ eV and $\Delta_a \approx 0.2$–0.4 eV. Thus, the absolute value of ϱ_{III} is of the order of 5–50. $\varrho_{III} \approx 100$ was obtained [7.38] for a chemisorbed CO molecule using the same theory as above but different input parameters.

7.7 Conclusions

We have presented a general theory of SERS in Sects. 7.2, 7.3 using the RPA approximation. It was shown that the calculations of the cross section can be reduced to three subproblems:

a) the calculation of local field effects due to the bounded metal. This mean that for a non-statistical surface, Maxwell's equations must be solved for the bounded metal and for a statistical surface, the averaged two-photon propagator must be determined.

b) the calculation of local field effects due to the presence of the molecule. Using the localized nature of chemisorption, this problem can be reduced to the inversion of small matrices.

c) the calculation of the polarization bubbles using the eigenstates of the molecule-metal system, i.e., taking into account chemisorption effects.

Points (a), (b) and (c) account for classical field enhancements, polarization (image) potentials and chemisorption effects, respectively.

Accurate results have been obtained for (a) by various geometries of the substrate such as spheres, spheroids, etc. Somewhat surprisingly, these results often cannot explain the observed SERS from colloids or metal spheres. Better agreement between theory and experiment has been obtained in the case of a plane surface with a grating although the calculations use only lowest-order perturbation theory in the grating amplitude. The interesting case of a plane, rough surface has been attacked from two different sides. In the first approach, one starts from localized field distributions near the irregularities of the surface and then considers the interactions of these fields. In the second approach, one writes down a perturbation expansion for the averaged photon propagators and uses some approximation scheme. There is still a lot of work to be done before final conclusions can be reached. However, it seems that the electromagnetic field enhancement due to the roughness of the surface plays an important role in SERS.

Reliable calculations for the effect (b) are presently not available and are extremely difficult. The involved momenta are of the order of the Fermi momentum so that spatial dispersion effects, as well as single particle effects in the metallic response, cannot be neglected. Also the electronic wave functions of the moecule as well as the coupling of the molecule and metal states enter the theory in a subtle way. Thus, the importance of polarization fields for SERS has been so far neither proved nor disproved on any general basis. Instead, the available calculations suggest that many microscopic details play a role in this question. Similar conclusions hold also for the effect (c). Available model calculations based on the Newns-Anderson model of chemisorption suggest the possibility of enhancements up to two orders of magnitude because of the change in chemical structure of the adsorbed molecule compared to the free one and the availability of unoccupied metal states as possible intermediate states for the scattering process.

References

7.1 M.Fleischmann, P.J.Hendra, A.J.McQuillan: Chem. Phys. Lett. **26**, 163 (1974)
7.2 D.J.Jeanmaire, R.P.Van Duyne: J. Electroanal. Chem. **84**, 1 (1977)
7.3 M.G.Albrecht, A.Creighton: J. Am. Chem. Soc. **99**, 5215 (1977)
7.4 T.E.Furtak, J.Reyes: Surf. Sci. **93**, 351 (1980)
7.5 E.Burstein, C.Y.Chen, S.Lundquist: In *Light Scattering in Solids*, ed. by J.L.Birman, H.Z.Cummins, K.K.Rebane (Plenum Press, New York 1979) p. 479
7.6 E.Burstein, C.Y.Chen: In Proc. of the VII Intern. Conf. on Raman Spectroscopy, Ottawa, ed. by W.F.Murphy (North-Holland, Amsterdam 1980) p. 346
7.7 A.Otto: Appl. Surf. Sci. **6**, 309 (1980)
7.8 For example, see recent book on *Surface Enhanced Raman Scattering*, ed. by R.K.Chang, T.E.Furtak (Plenum Press, New York 1982)
7.9 R.Dornhaus: *Festkörperprobleme* XXII, ed. by P.Groose (Vieweg, Braunschweig 1982) p. 201
7.10 E.Burstein, Y.J.Chen, C.Y.Chen, S.Lundquist, E.Tosatti: Solid State Commun. **29**, 567 (1979)
7.11 M.Moskovits: Solid State Commun. **32**, 59 (1979)
7.12 S.L.McCall, P.M.Platzman, P.A.Wolff: Phys. Lett **77**A, 381 (1980)
7.13 D.-S.Wang, H.Chew, M.Kerker: Appl. Opt. **19**, 2256 (1980)
7.14 M.Kerker, D.-S.Wang, H.Chew: Appl. Opt. **19**, 4159 (1980)
7.15 S.S.Jha, J.R.Kirtley, J.C.Tsang: Phys. Rev. B**22**, 3973 (1980)
7.16 J.R.Kirtley, S.S.Jha, J.C.Tsang: Solid State Commun. **35**, 509 (1980)
7.17 J.C.Tsang, S.S.Jha, J.R.Kirtley: Phys. Rev. Lett. **46**, 1044 (1981)
7.18 G.S.Agarwal, S.S.Jha, J.C.Tsang: Phys. Rev. B**25**, 2089 (1982)
7.19 K.Arya, R.Zeyher: Phys. Rev. B**24**, 1852 (1981)
7.20 P.K.Aravind, H.Metiu: Chem. Phys. Lett. **74**, 301 (1980)
7.21 K.Arya, R.Zeyher, A.A.Maradudin: Solid State Commun. **42**, 461 (1982)
7.22 A.G.Mal Shukov: Solid State Commun. **38**, 907 (1981)
7.23 J.Gersten, A.Nitzan: J. Chem. Phys. **73**, 3023 (1980)
7.24 M.Moskovits: J. Chem. Phys. **69**, 4159 (1978)
7.25 J.I.Gersten, D.A.Weitz, T.J.Gramila, A.Z.Genack: Phys. Rev. B**22**, 4562 (1980)
7.26 R.Ruppin: Solid State Commun. **39**, 903 (1981)
7.27 D.-S.Wang, M.Kerker: Phys. Rev. B**24**, 1777 (1981)
7.28 P.K.Aravind, A.Nitzan, H.Metiu: Surf. Sci. **110**, 189 (1981)
7.29 C.Y.Chen, E.Burstein: Phys. Rev. Lett. **45**, 1287 (1980)
7.30 F.W.King, R.P.Van Duyne, G.C.Schatz: J. Chem. Phys. **69**, 4472 (1978)
7.31 S.Efrima, H.Metiu: Chem. Phys. Lett. **60**, 59 (1978); J. Chem. Phys. **70**, 1602 (1979); **70**, 2297 (1979)
7.32 G.L.Eesley, J.R.Smith: Solid State Commun. **31**, 815 (1979)
7.33 G.W.Ford, W.H.Weber: Surf. Sci. **109**, 451 (1981)
7.34 T.K.Lee, J.L.Birman: Phys. Rev. B**22**, 5953 (1980); **22**, 5961 (1980)
7.35 P.J.Feibelman: Phys. Rev. B**22**, 3654 (1980)
7.36 P.M.Echenique, R.H.Ritchie, N.Barberan, J.Inkson: Phys. Rev. B**23**, 6486 (1981)
7.37 J.I.Gersten, R.L.Birke, J.R.Lombardi: Phys. Rev. Lett. **43**, 147 (1979)
7.38 B.N.J.Person: Chem. Phys. Lett. **82**, 561 (1981)
7.39 S.L.McCall, P.M.Platzman: Phys. Rev. B**22**, 1660 (1980)
7.40 G.W.Robinson: Chem. Phys. Lett. **76**, 191 (1980)
7.41 A.Otto: Surf. Sci. **92**, 145 (1980)
7.42 A.Otto, J.Timper, J.Billmann, I.Pockrand: Phys. Rev. Lett. **45**, 46 (1980)
7.43 J.C.Tsang, J.R.Kirtley, J.A.Bradley: Phys. Rev. Lett. **43**, 772 (1979)
7.44 P.N.Sanda, J.W.Warlaumont, J.E.Demuth, J.C.Tsang, K.Christman, J.A.Bradley: Phys. Rev. Lett. **45**, 1519 (1980)
7.45 A.Girlando, M.R.Philpott, D.Heitmann, J.D.Swalen, R.Santo: J. Chem. Phys. **72**, 5187 (1980)
7.46 J.C.Tsang, J.R.Kirtley, T.N.Theis, S.S.Jha: Phys. Rev. B**25**, 5070 (1982)
7.47 A.Adams, J.C.Wyss, P.K.Hansma: Phys. Rev. Lett. **42**, 912 (1979)
7.48 J.C.Tsang, J.Kirtley: Solid State Commun. **30**, 617 (1979)

7.49 A.Creighton, C.B.Blatchford, M.G.Albrecht: J. Chem. Soc. Faraday Trans. II, **75**, 790 (1979)
7.50 R.Dornhaus, M.B.Long, R.E.Benner, R.K.Chang: Surf. Sci. **93**, 240 (1980)
7.51 B.Pettinger, M.R.Philpott, J.G.Gordon: J. Chem. Phys. **74**, 934 (1981)
7.52 C.Y.Chen, E.Burstein, S.Lundquist: Solid State Commun. **32**, 63 (1979)
7.53 A.Hartstein, J.R.Kirtley, J.C.Tsang: Phys. Rev. Lett. **45**, 201 (1980)
7.54 J.E.Demuth, P.N.Sanda: Phys. Rev. Lett. **47**, 57 (1981)
7.55 T.E.Furtak: Solid State Commun. **28**, 903 (1978)
7.56 C.A.Murray, D.L.Allara, M.Rhinewine: Phys. Rev. Lett. **46**, 57 (1981)
7.57 T.H.Wood, M.V.Klein: Solid State Commun. **35**, 263 (1980)
7.58 T.H.Wood: Phys. Rev. B**24**, 2289 (1981)
7.59 J.E.Rowe, C.V.Shank, D.A.Zwemer, C.A.Murray: Phys. rev. Lett. **44**, 1770 (1980)
7.60 T.E.Furtak, J.Kester: Phys. Rev. Lett. **45**, 1652 (1980)
7.61 R.Naaman, S.J.Bnelow, O.Cheshnovsky, D.R.Herschbach: J. Phys. Chem. **84**, 2694 (1980)
7.62 M.Udagawa, Chih-Cong Chou, J.C.Hemminger, S.Ushioda: Phys. Rev. B**23**, 6843 (1981)
7.63 D.M.Newns: Phys. Rev. **178**, 1123 (1969)
7.64 J.P.Muscat, D.M.Newns: Prog. Surf. Sci. **9**, 1 (1978)
7.65 A.A.Abrikosov, L.P.Gorkov, I.E.Dzyaloshinski: *Methods of Quantum Field Theory in Statistical Physics*, transl. and ed. by R.Silverman (Prentice-Hall, Englewood Cliffs, NJ 1963)
7.66 D.L.Mills, E.Burstein: Phys. Rev. **188**, 1465 (1969)
7.67 P.M.Morse, H.Feschbach: *Methods of Theoretical Physics* (McGraw-Hill, New York 1953)
7.68 A.A.Maradudin, D.L.Mills: Phys. Rev. B**11**, 1392 (1975)
7.69 G.S.Agarwal: Phys. Rev. B**14**, 846 (1976)
7.70 G.D.Mahan: Phys. Rev. B**5**, 739 (1972)
7.71 E.Evans, D.L.Mills: Phys. Rev. B**8**, 4004 (1973)
7.72 K.L.Kliewer, R.Fuchs: Phys. Rev. **172**, 607 (1968)
7.73 P.J.Feibelman: Phys. Rev. B**12**, 1319 (1975)
7.74 T.Maniv, H.Metiu: J. Chem. Phys. **76**, 2697 (1982)
7.75 G.E.Korzemewski, T.Maniv, H.Metiu: J. Chem. Phys. **76**, 1564 (1982)
7.76 A.A.Maradudin, W.Zierau: Phys. Rev. B**14**, 484 (1976)
7.77 L.P.Kadanoff, G.Baym: *Quantum Statistical Mechancs* (Benjamin, New York 1962)
7.78 J.M.Elson, R.H.Ritchie: Phys. Rev. B**4**, 4129 (1971)
7.79 A.Marvin, F.Toigo, V.Celli: Phys. Rev. B**11**, 2777 (1975)
7.80 P.B.Johnson, R.W.Christy: Phys. Rev. B**6**, 4370 (1972)
7.81 L.P.Gorkov, A.I.Larkin, D.E.Kumel'nitzkii: JETP Lett. **30**, 229 (1979)
7.82 R.Ruppin, R.Englman: J. Phys. C. **1**, 630 (1968)
7.83 J.D.Jackson: *Classical Electrodynamics* (Wiley, New York 1975) Chap. 16
7.84 Note that no surface polaritons of a magnetic type exist because for metals with $\varepsilon(\omega)$ in the frequency range of interest, the denominator in (7.102) or (7.103) never goes to zero to give any new solution [7.82]
7.85 For this, express $\hat{e}_\lambda(\boldsymbol{Q})\exp(\mathrm{i}\boldsymbol{Q}\cdot\boldsymbol{r})$ appearing in (7.28) in terms of the solutions of the Maxwell equation (7.95) which is possible because $\hat{e}_\lambda(\boldsymbol{Q})\exp(\mathrm{i}\boldsymbol{Q}\cdot\boldsymbol{r})$ is also a solution of (7.95). Then use the orthogonality relations to obtain (7.109–112). For more details see [Ref. 7.83, p. 768]. Also note that the surface mode $\boldsymbol{A}^{s}_{lm}(\boldsymbol{r})$ is orthogonal to the bulk modes $\boldsymbol{A}_{klm}(\boldsymbol{r})$ [i.e., also to $\hat{e}_\lambda(\boldsymbol{Q})\exp(\mathrm{i}\boldsymbol{Q}\cdot\boldsymbol{r})$] because of phase space
7.86 R.Zeyher, H.Bilz, M.Cardona: Solid State Commun. **19**, 57 (1976)

8. Pressure-Raman Effects in Covalent and Molecular Solids

Bernard A. Weinstein and Richard Zallen

With 45 Figures

The advantages of pressure for studying solids have been recognized since the pioneering work of Bridgman. Hydrostatic pressure greater than a few kbar (1 kbar $= 10^9$ dyne/cm^2 $= 0.1$ Gigapascal) is a stronger and, in principle, a much cleaner perturbation than the more common thermodynamic variable temperature. For example, in Si a pressure of 100 kbar (easily attainable with modern techniques) produces a volume decrease of 5%, whereas the total temperature induced volume change in Si from 0 K to melting is only 1.8%. For the softer molecular solids this contrast is even greater. Furthermore, temperature is complicated by the parallel action of thermal expansion and phonon population effects, but hydrostatic pressure is manifest solely through volume change.

The main drawback of pressure studies is their experimental complexity. Expensive bulky apparatus are sometimes required, seals blow out, windows crack, and the sample size is necessarily small to achieve high pressure. Advances during the last decade, such as modern diamond anvil-type presses, have greatly reduced the mechanical problems. However sample size still remains small. Consequently, the importance of modern laser sources, which have catalyzed the general resurgence of Raman scattering, cannot be understated for pressure measurements. It is the marriage of modern laser-Raman and high-pressure techniques which has brought about the blossoming of pressure-Raman studies during the last decade. Though young, the marriage has been fruitful, and its future holds considerable promise.

The present review deals with two classes of solids, the tetrahedral semiconductors and the molecular insulators. Citations to work on other materials have been included in the references where appropriate. We begin by discussing the role of pressure in Raman processes and the Grüneisen parameter concept (Sect. 8.1). A brief account of experimental techniques follows (Sect. 8.2). Pressure-Raman studies of tetrahedral semiconductors (including materials from groups IV, III-V, II-VI, and I-VII) are then treated. This includes discussions of changes in phonon frequencies (Sect. 8.3), changes in phonon line shapes (Sect. 8.4), phase transitions (Sect. 8.5) and pressure-tuned resonant Raman scattering (Sect. 8.6). Our account of pressure-Raman effects in molecular solids (Sect. 8.7) is divided into five parts. These deal with overall rationale (Sect. 8.7.1), line shifts in simple organic and inorganic molecular solids (Sect. 8.7.2), vibrational scaling (Sect. 8.7.3), the connection between pressure and temperature effects (Sect. 8.7.4) and finally, molecular-to-nonmolecular transitions at high pressure (Sect. 8.7.5). Throughout we have limited the discussion to symmetry-conserving hydrostatic pressure investigations. Symmetry-breaking uniaxial stress studies are outside both the scope and space of this chapter. Nevertheless, for completeness some hydrostatic Grüneisen parameters deduced from uniaxial measurements have been included in Table 8.2 with appropriate citations, and a bibliography of recent uniaxial stress Raman work follows the numbered references.

8.1 The Raman Effect

Raman scattering is a powerful probe of solid-state excitations in the frequency range 3–3000 cm^{-1}. Pressure can cause large changes in the energies and interactions of these excitations within a given solid phase, and/or it can cause transitions to new phases having quite different excitation spectra. Let us examine formally how these changes manifest themselves in Raman scattering measurements.

8.1.1 How Pressure Enters

A typical experimental configuration is depicted schematically in Fig. 8.1. Here ω_l, ω_s, and ω_i are the angular frequencies of the incident laser photon, the scattered photon, and the elementary excitation under study, and $\boldsymbol{k}_l$, $\boldsymbol{k}_s$, and $\boldsymbol{q}_i$ are the corresponding wave vectors *within the scattering medium.* As usual, kinematics requires

$$\omega_l = \omega_s \pm \omega_i, \qquad \boldsymbol{k}_l = \boldsymbol{k}_s \pm \boldsymbol{q}_i. \tag{8.1}$$

In relating $\boldsymbol{k}_l$ and $\boldsymbol{k}_s$ to their counterparts *outside* the pressure vessel, one must properly account for the refractive indices of the optical windows and the pressure transmitting medium, which are, *of necessity*, part of the apparatus.

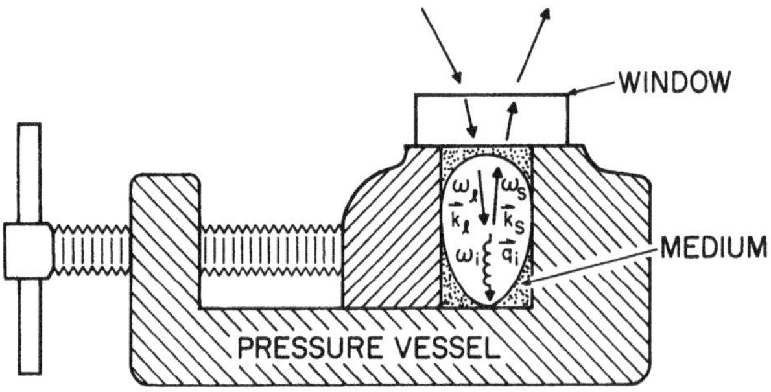

Fig. 8.1. Typical configuration for a hydrostatic pressure-Raman experiment in which a phonon ω_i, $\boldsymbol{q}_i$ is created (Stokes process). Of *necessity*, the window and medium are part of the sample chamber

Generally these indices will be pressure dependent and anisotropic so that the relationship will be nontrivial. In practice, this correction can often be neglected for Raman (but not Brillouin) studies because the dispersion relations $\omega_i(\boldsymbol{q}_i)$ are usually flat in the $\boldsymbol{q}_i \approx 0$ region of interest. However, if selection rules are to be tested closely, the correction should not be ignored. In that case the problem may still be circumvented for small energy excitations (typically valid to 1% for $\omega_i \leq 200\ \mathrm{cm}^{-1}$) by arranging that the set of surfaces (window, medium, sample) traversed by the light are all parallel to the wave vector $\boldsymbol{q}_i$. It is then correct to use in (8.1) the incident and scattered wave vectors outside the pressure vessel [8.1].

The mechanism for inelastic light scattering can be viewed as a modulation of the dielectric susceptibility by some elementary excitation of the solid [8.2a]. The standard derivation leads to [8.2b, 3]

$$\frac{d^2\sigma}{d\Omega\, d\omega_s} = vV \frac{\omega_s^4}{c^4} |\hat{\varepsilon}_s \cdot \chi' \cdot \hat{\varepsilon}_l|^2 \langle UU^\dagger \rangle_\omega \tag{8.2}$$

for the differential cross section for scattering into the solid angle and frequency increments $d\Omega$ at Ω and $d\omega_s$ at ω_s.

Here χ' is the higher-order susceptibility tensor appropriate to the elementary excitation of amplitude U, $\hat{\varepsilon}_l$ and $\hat{\varepsilon}_s$ are the unit polarization vectors of incident and scattered light, v is the interaction volume and V is the sample volume. The notation $\langle UU^\dagger \rangle_\omega$ denotes the power spectrum of $|U|^2$.

Equation (8.2) is useful for our present purposes because it separates the cross section into a shape factor $\langle UU^\dagger \rangle_\omega$ which describes the frequency spectrum of the excitation under study, and a strength factor $|\hat{\varepsilon}_s \cdot \chi' \cdot \hat{\varepsilon}_l|^2$ which contains the relevant interactions of light with the elementary excitation through other possible intermediate excitations of the solid. To be more specific, for a one-phonon Stokes process we have [8.2b]

$$\langle UU^\dagger \rangle_\omega = \frac{\hbar}{2N\omega_i} (n_i + 1) g_i(\omega), \tag{8.3}$$

where N is the number of oscillators in the solid (note that $d^2\sigma/d\Omega\, d\omega_s$ actually depends on V/N not V), $n_i = [\exp(\hbar\omega_i/kT) - 1]^{-1}$ is the Bose thermal population factor and $g_i(\omega)$ is a line-shape response function, often taken to be a Lorentzian:

$$g_i(\omega) = \frac{\Gamma_i/2\pi}{(\omega_i - \omega)^2 + (\Gamma_i/2)^2}. \tag{8.4}$$

For a multiphonon process we must replace $(n_i + 1)$ in (8.3) by the appropriate thermal factor [e.g., $(n_1 + 1)(n_2 + 1)$ for a summation process involving ω_1 and ω_2]; $g_i(\omega)$ becomes the multiphonon density of states which generally contains several sharp critical point features [8.2b, 4, 5].

The vast majority of high pressure Raman measurements have been concerned with changes in the frequencies of lattice vibrations ω_i. These changes show up as spectral shifts of the peak of the Lorentzian, or the structure in the density of states. If the lifetime of the phonons is affected by pressure (e.g., through anharmonic interactions) then, in addition to spectral shifts, the observed line shape will also be modified. This occurs through changes in the full width of the Lorentzian Γ_i, or other appropriate width parameters.

The selection rules for Raman scattering are contained within the inner product $\hat{\varepsilon}_s \cdot \chi' \cdot \hat{\varepsilon}_l$ which selects specific components of the tensor χ' (often called the Raman tensor). The form of this (usually symmetric) tensor is determined by crystal symmetry [8.6, 7]. For a cubic solid of point group O_h, tensor components transforming as the irreducible representations Γ_1 (finite trace diagonal, also A_{1g}), Γ_{12} (traceless diagonal, also T_{2g}) and $\Gamma_{25'}$ (symmetric off diagonal, also E_g) appear. Only elementary excitations having wave functions transforming as these representations will be Raman active; they will contribute only to the tensor components associated with their representations. If compression causes a phase transition involving a change in point group symmetry, the Raman selection rules will also change. In general this is manifest through the appearance of new features in the observed spectra as forbidden excitations become Raman active and as degeneracies are lifted.

The magnitude of allowed Raman tensor components can also be affected by pressure without a phase transition. Consider one-phonon scattering in semiconductors. A typical two-band term contributing to $\hat{\varepsilon}_s \cdot \chi' \cdot \hat{\varepsilon}_l$ has the form [8.8]

$$\hat{\varepsilon}_s \cdot \chi' \cdot \hat{\varepsilon}_l \sim \frac{\langle v|\hat{\varepsilon}_s \cdot \boldsymbol{p}|c\rangle \langle c|H_{ep}^{(1)}|c\rangle \langle c|\boldsymbol{p} \cdot \hat{\varepsilon}_l|v\rangle}{(\omega_g + \omega_i - \omega_l)(\omega_g - \omega_l)}. \tag{8.5}$$

Equation (8.5) represents the process depicted in Fig. 8.2. Here $\boldsymbol{p}$ is the electron momentum, $H_{ep}^{(1)}$ is the relevant electron-phonon interaction and $|v\rangle$, $|c\rangle$, and ω_g are defined in Fig. 8.2.

Compression can modify either the matrix elements or the energy denominators in (8.5). Matrix element changes have not been studied in great detail; a

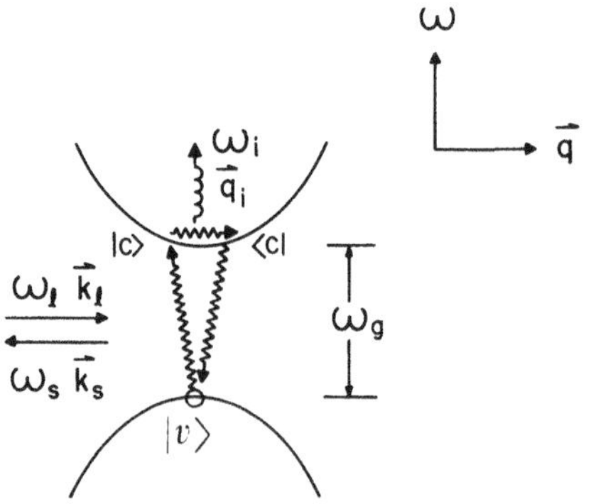

Fig. 8.2. A two-band process in which incident light ω_l, $\boldsymbol{k}_l$ excites an electron from the valence state $|v\rangle$ to the conduction state $|c\rangle$, the electron scatters to a different $\boldsymbol{q}$-state $\langle c|$ emitting a phonon ω_i, $\boldsymbol{q}_i$, and the electron recombines with the hole in $|v\rangle$ emitting scattered light ω_s, $\boldsymbol{k}_s$. The intermediate states $|c\rangle$ and $\langle c|$ should be summed over a complete set; ω_g is a $\boldsymbol{q} \approx 0$ direct bandgap

recent calculation estimates these effects to be too small for observation in Si [8.9]. In contrast, strong resonant Raman enhancement can be produced by pressure tuning the electronic gap ω_g through the zeros in the energy denominator at the incident and scattered frequencies.

To summarize, there are four principal pressure-Raman effects: these are frequency shifts of elementary excitations, line-shape changes, selection rule changes accompanying phase transitions and pressure-tuned resonant Raman scattering. We shall describe specific results that exemplify these categories, although space does not permit an exhaustive discussion in each case. Our main interest lies with covalent tetrahedral semiconductors and molecular solids. This is an important but partial subgroup of the extensive body of pressure-Raman studies. The reader is encouraged to refer to the several excellent reviews listed in the references for discussions of other solids [8.10–15].

8.1.2 Grüneisen Parameters and Scaling

Since much of our discussion will be concerned with pressure-induced phonon frequency shifts, a short discussion of Grüneisen parameters is appropriate.

The Grüneisen parameter γ_i for a quasi-harmonic mode i of frequency ω_i is defined by [8.16, 17]

$$\gamma_i = -\frac{\partial \ln \omega_i}{\partial \ln V} = \frac{1}{\beta}\frac{\partial \ln \omega_i}{\partial P}, \tag{8.6}$$

where β is the isothermal volume compressibility and the derivatives are evaluated at $P=0$. Essentially γ_i is an exponent that tells how ω_i *scales* with volume. Thus, if γ_i is independent of volume, (8.6) implies,

$$\frac{\omega_i(1)}{\omega_i(2)} = \left(\frac{V_2}{V_1}\right)^{\gamma_i}, \tag{8.7}$$

where $\omega_i(1)$ and $\omega_i(2)$ are the values of ω_i at volumes V_1 and V_2. The mode parameters γ_i are related to the macroscopic or average Grüneisen constant γ_{av} introduced by *Debye* [8.18] and *Grüneisen* [8.19] to discuss the equation of state of a solid. The Helmholtz free energy in the quasi-harmonic approximation is [8.20]

$$F = \Phi(V) + \Phi_D(T, V) - TS = \Phi(V) + \sum_i kT \ln[1 - \exp(-\hbar\omega_i/kT)], \tag{8.8}$$

where the $T=0$ internal energy $\Phi(V)$ is temperature independent and S is the entropy. In the Debye theory, $\Phi_D = Tf(\theta/T)$, where f is a universal function of θ/T with θ the Debye temperature. Using $-(\partial F/\partial V)_T = P$, we find the Debye equation of state

$$P = -\frac{\partial \Phi}{\partial V} + \gamma_{av}\frac{\Phi_D}{V}, \tag{8.9}$$

where $\gamma_{av} = -(\partial \ln \theta / \partial \ln V)_T$. Employing the relation $-\partial^2 F/\partial V \partial T = \alpha/\beta$, where α is the volume thermal expansion, yields

$$\frac{V\alpha}{C_v \beta} = \gamma_{av} = \frac{\sum_i \gamma_i c_v(i)}{C_v}. \tag{8.10}$$

Here C_v is the crystal heat capacity and $c_v(i) = \partial(n_i \hbar\omega_i)/\partial T$ is the Einstein heat capacity for mode i. Grüneisen's approximation, viz. $\gamma_{av} = \gamma_{av}(V)$ independent of T, is justified for two situations (assuming that γ_i is T-independent): for $T \gg \theta$, we have $c_v(i) \approx k$ and $\gamma_{av} \approx (1/N) \sum_i \gamma_i$; for mode independent γ_i, (8.10) reduces to $\gamma_{av} = \gamma_i$ at any temperature.

The latter case applies to the Debye continuum limit for which $\gamma_i = \gamma_{av} = -\partial \ln \theta / \partial \ln V$. Since $\theta = (\hbar/k)\omega_D$ and $\omega_D = (cV/N)^{1/2} q_D$, where c is a characteristic (adiabatic) elastic constant and $q_D = 2\pi N/V^{1/3}$, we expect $\theta \propto c^{1/2} V^{1/6}$. This led *Slater* to propose [8.20]

$$\gamma_{av} = -\frac{1}{2}\left(\frac{\partial \ln B}{\partial \ln V}\right)_T - \frac{1}{6}, \tag{8.11}$$

where B is the adiabatic bulk modulus. Equation (8.11) and subsequent refinements have been employed by many authors [8.17, 21, 22].

We note in passing that γ_{av} is an important geophysical parameter. In the high-T and adiabatic limits, one obtains $T \propto V^{-\gamma_{av}}$. This equation of state is thought to apply to the earth's core [8.23].

8.2 Experimental Aspects

In recent years static high pressure techniques have become more accessible to non-specialists. Most relevant to the present chapter are the combined developments of the diamond-anvil press (Fig. 8.3) and the ruby calibration scale (Fig. 8.4) [8.24–26]. These developments have made practical and common, quantitative Raman measurements to several hundred kbar [8.27, 28]. Pressure-Raman techniques actually differ little from standard light scattering and high pressure procedures. Since there have been abundant individual discussions of these, we shall not dwell at length on this subject (see, for example, [Ref. 8.3, Chap. 2], [Ref. 8.14, Chap. 4], and [8.29–8.33b]). Instead we wish to discuss several operational problems common to most pressure-Raman work.

A difficulty often encountered is a large background that overwhelms the Raman signal. This holds for most common apparatus, such as the diamond-anvil press (Fig. 8.3), the NaCl press [8.32], the oil bomb [8.29] and variations on these (the problem is less for a gas bomb [8.34]). The background can be due to direct and specular reflections, elastic and inelastic scattering, and fluorescence, from the cell windows, packing, walls, the pressure transmitting medium and the sample.

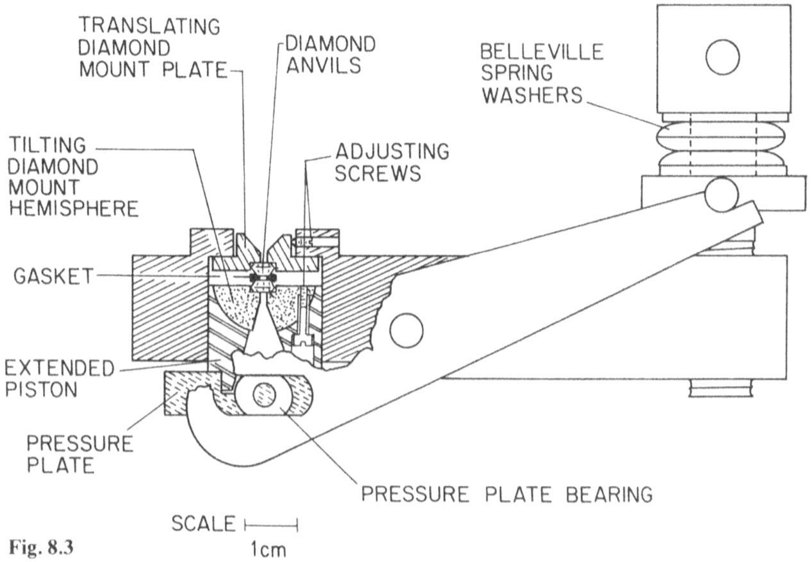

Fig. 8.3

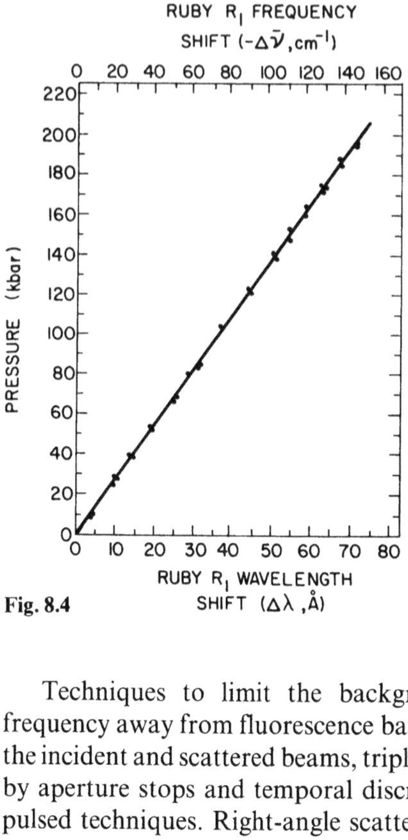

Fig. 8.4

Fig. 8.3. Cutaway cross-section drawing showing the essential parts of the modern diamond-anvil cell developed by the National Bureau of Standards (U.S.) group. Under proper operation, several hundred kbar can be routinely achieved without disturbing optical access. After [8.24]

Fig. 8.4. Original calibration for the ruby secondary pressure scale. The R_1 fluorescence line (6942 Å at $P=0$, $T=298$ K) shifts by 0.365 Å/kbar. After [8.26]

Techniques to limit the background include [8.27] varying the laser frequency away from fluorescence bands, sharp cut filters, crossed polarizers in the incident and scattered beams, triple monochromators, spatial discrimination by aperture stops and temporal discrimination (against slow fluorescence) by pulsed techniques. Right-angle scattering is always preferable, but not always

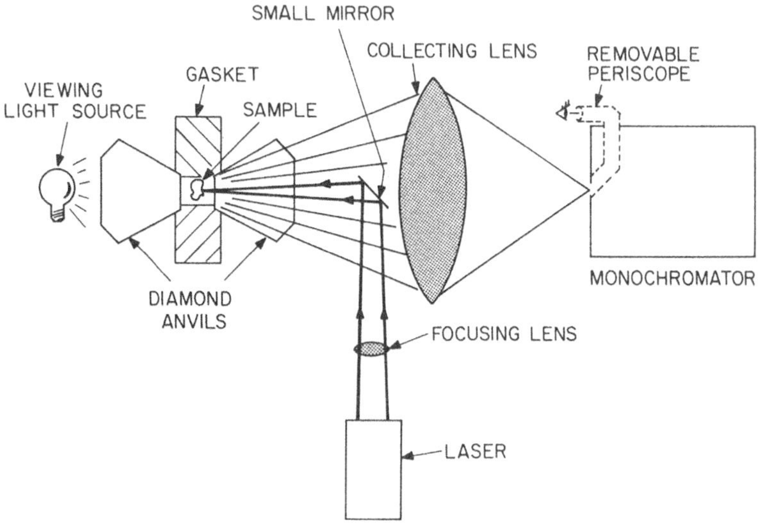

Fig. 8.5. A Raman experiment using the diamond cell in the back-scattering geometry. This schematic drawing is not to scale. After [8.27]

possible, as for opaque samples or the diamond-anvil press. In such cases the back-scattering geometry (generally superior to forward scattering) should be used; Fig. 8.5 shows a typical arrangement. As a rule of thumb for the diamond-cell, one should not expect to observe a signal of < 100 counts/s below a Raman shift of $50\ cm^{-1}$.

A secondary concern is a reduced signal due to the small (often microscopic) sample size required by many apparatus to reach high pressure. The ability to tightly focus ($\leq 30\ \mu m$ inside the pressure chamber) modern coherent sources generally minimizes this problem. However, one must take care not to exceed sample damage thresholds (the standard technique of focussing with a cylindrical lens cannot be used). For the diamond cell, it is typical to find signal reductions of only $1/2 - 2/3$ below levels measured outside the cell [8.27]. When working with small specimens, some means for visually observing the sample during alignment is a great advantage.

The ruby fluorescence (R_1 and R_2-lines) pressure scale (Fig. 8.4) developed by the U. S. National Bureau of Standards group [8.24, 26] has become a widely accepted secondary standard with $\sim 3\%$ accuracy. Subsequent work showed that the expression $P(\text{Mbar}) = 3.808\,[(\lambda/\lambda_0)^5 - 1]$, where $\lambda_0 = 6942\,\text{Å}$ and λ are the R_1 wavelengths at $P=0$ and $P>0$ ($T=300\text{K}$), respectively, could be used into the megabar regime [8.33a]. The R_1-scale was also extended to low temperature [8.35]. The ruby fluorescence technique is ideal for Raman experiments because the internal cell pressure can be quickly determined without changing the measuring equipment or geometry (the laser wavelength must be shorter than λ_0). Sometimes it is advantageous to calibrate the strongest Raman line, in an otherwise weak spectrum, against R_1, and then to measure the entire spectrum in a separate loading without ruby, thereby eliminating the ruby luminescence from the background [8.27].

8.3 Phonon Frequencies Under Pressure in Tetrahedral Semiconductors

Here we discuss the effects of compression on the entire phonon dispersion of several prototype materials. This has become possible through a combination of one and two-phonon Raman measurements, and our account is so divided. The reader should refer often to Tables 8.1 – 3 which extend the discussion to other solids and give detailed references.

8.3.1 One-Phonon Spectra and the Transverse Effective Charge

The diamond and zincblende structures can be viewed as two interpenetrating fcc sublattices, separated by 1/4 of a cube body diagonal along (111). The $\boldsymbol{q}=0$ (Γ-point) groups are O_h and T_d, respectively, with two identical atoms per primitive cell in the former and one cation-anion pair in the latter. This leads to three acoustic and three optic modes. Because the diamond structure has inversion symmetry the optic modes are degenerate (no LO-TO splitting); they transform as $\Gamma_{25'}$ (E_g) making them Raman active but infrared (ir) inactive. There will be just one first-order Raman peak, hereafter designated $O(\Gamma)$. The heteropolar zincblende materials do not possess inversion symmetry, so that longitudinal LO(Γ) and transverse TO(Γ) phonons are not degenerate. Instead we have [8.2b, 36]

$$\omega_{\mathrm{LO}}^2-\omega_{\mathrm{TO}}^2=\frac{4\pi N_{\mathrm{d}}e^{*2}}{\varepsilon_\infty \mu V}=\omega_{\mathrm{TO}}^2\,\frac{(\varepsilon_0-\varepsilon_\infty)}{\varepsilon_\infty}, \tag{8.12}$$

which defines Born's transverse dynamic effective charge e^*. Here N_{d}/V is the density of cation-anion dipole pairs, ε_∞ and ε_0 are the optical and static dielectric constants, and μ is the reduced mass. The middle expression defines the plasma frequency of the ions squared, and the far right-hand side makes explicit use of the Lyddane-Sachs-Teller relation $\omega_{\mathrm{LO}}^2/\omega_{\mathrm{TO}}^2=\varepsilon_0/\varepsilon_\infty$. The modes have Γ_{15} (E) symmetry; they are Raman and ir active. There will be two first-order Raman peaks corresponding to TO(Γ) and LO(Γ).

The effect of hydrostatic pressure to 125 kbar on $O(\Gamma)$ in Si is displayed in Fig. 8.6 after [8.27]. These high pressure results agree with earlier measurements by *Buchenauer* et al. [8.37] to 9 kbar. The Raman peak shifts to higher energy by about 25 cm^{-1} in 100 kbar without changing shape. This corresponds to $\gamma_{O(\Gamma)}=0.98\pm0.06$, nearly equal to unity. $\gamma_{O(\Gamma)}\approx 1.0$ is a general characteristic of covalent group IV semiconductors, and may be viewed as a norm for

Table 8.1. Glossary of Grüneisen parameters for tetrahedral semiconductors measured by pressure-Raman scattering (except for $\boldsymbol{q}=0$ acoustic modes[a]). Branches (in order of descending energy) and critical points are arranged to match usual phonon dispersion format [8.51]. Best values are listed; in most cases uncertainty occurs in last decimal place

Branch	Material	Critical points				Material	Critical points				Material	Critical points			
		X	K	Γ	L		X	K	Γ	L		X	K	Γ	L
LO	C[1-4]	1.3		1.19	1.27	GaAs[8,12,18,19,20]			1.23		ZnTe[3,22,23,27,28]	1.7		1.2	1.8
TO(2)		2.0		1.19	1.4		1.73		1.39	1.5		1.8		1.7	
TO(1)		2.0		1.19	1.4		1.73		1.39	1.5		1.8		1.7	
LA				1.06[a]					1.3[a]					1.55[a]	
TA(2)		0.4[b]		0.97[a]			−1.62	−0.4	0.53[a]	−1.7		−1.55	−0.4	0.20[a]	−1.0
TA(1)		0.4[b]	0.4[b]	0.51[a]			−1.62	−1.62	−0.05[a]	−1.7		−1.55	−1.55	−0.43[a]	−1.0
LO	Si[5-10]	0.9[c]	0.9[c]	0.98	0.9[c]	GaSb[8,12,21]			1.21		CuCl[29-33]			1.0[e]	
TO(2)		1.5		0.98	1.3				1.23					2.4[e]	
TO(1)		1.5	1.5	0.98	1.3				1.23					2.4[e]	
LA		0.9[c]	0.9[c]	1.1[a]	0.9[c]				1.4[a]					1.45[a]	
TA(2)		−1.4	−0.3	0.33[a]	−1.3				0.48[a]			−1.4[f]		−0.39[a]	−1.4[f]
TA(1)		−1.4	−1.0	−0.05[a]	−1.3				−0.004[a]			−1.4[f]	−1.4[f]	−1.87[a]	−1.4[f]
LO	Ge[6,8,9,11,12]			1.12		InP[18]			1.24		CuBr[30,31,33]			1.9[e]	
TO(2)				1.12			1.4		1.44	1.4				2.4[e]	
TO(1)				1.12			1.4		1.44	1.4				2.4[e]	
LA				1.3[a]											
TA(2)		−1.53		0.62[a]			−2.1	−0.3		−2.0		−0.4[f]			−0.4[f]
TA(1)		−1.53		0.17[a]			−2.1	−2.1		−2.0		−0.4[f]	−0.4[f]		−0.4[f]
LO	AlSb[8]			1.27		ZnS[3,22-26]	1.1		0.95	1.1	CuI[30,31,33,34]			1.6[e]	
TO(2)				1.23			1.0			1.0				2.2[e]	
TO(1)				1.23			1.0			1.0				2.2[e]	
LA									1.34[a]						
TA(2)							−1.2	−0.3	0.21[a]	−1.5		2.5			2.5
TA(1)							−1.2	−1.2	−0.90[a]	−1.5		2.5	2.5		2.5

Table 8.1 (continued)

Branch	Material	Critical points				Material	Critical points			
		X	K	Γ	L		X	K	Γ	L
LO	GaP[3,5,13–17]		1.27	0.95		ZnSe[3,12,22,23,27]	0.9		0.9	0.9
TO(2)		1.3		1.09	1.50		1.6		1.4	1.6
TO(1)		1.3		1.09	1.50		1.6		1.4	1.6
LA		1.0[d]	1.0[d]	1.4[a]			1.1		1.45[a]	1.1
TA(2)		−0.72	0.15	0.49[a]	−0.81		−1.3	−0.4	0.16[a]	−1.5
TA(1)		−0.72	−0.72	−0.2[a]	−0.81		−1.3	−1.3	−0.60[a]	−1.5

a From measurements of the elastic constants c_{11}, c_{44}, $(c_{11}\text{-}c_{12})/2$ corresponding, respectively, to LA, TA(2), TA(1)
b Calculated using optic mode data and a Born-von Karman model
c From fitting the temperature dependent thermal expansion
d From the line-shape study of [8.78]
e Measured at 40 K
f From measurements on $CuCl_{0.80}Br_{0.20}$ and $CuCl_{0.20}Br_{0.80}$
1 [8.39]
2 H. J. McSkimin, P. Andreatch, Jr.: J. Appl. Phys. **43**, 2944 (1972)
3 [8.38]
4 M. H.Grimsditch, E. Anastassakis, M. Cardona: Phys. Rev. B**18**, 901 (1978)
5 [8.27]
6 H. J. McSkimin, P. Andreatch, Jr.: J. Appl. Phys. **35**, 3312 (1964)
7 [8.52]
8 [8.37]
9 J. B. Renucci, M. A. Renucci, M. Cardona: Solid State Commun. **9**, 1651 (1971)
10 E. Anastassakis, A. Pinczuk, E. Burstein: Solid State Commun. **8**, 133 (1970)
11 [8.96]; see this reference also for results on 3*C*-SiC
12 F. Cerdeira, C. J. Buchenauer, F. H. Pollak, M. Cardona: Phys. Rev. B**5**, 580 (1972)
13 [8.41]
14 [8.53]
15 [8.60]
16 [8.78]
17 I. Balslev: Phys. Stat. Solidi (b) **61**, 207 (1974)
18 [8.42a]
19 H. J. McSkimin, A. Jayaraman, P. Andreatch, Jr.: J. Appl. Phys. **38**, 2362 (1967)
20 [8.97]
21 H. J. McSkimin, P. Andreatch, Jr., T. B. Bateman: J. Appl. Phys. **39**, 4127 (1968)
22 [8.54]
23 [8.43a]
24 [8.88]
25 E. Chang, G. R. Barsch: J. Phys. Chem. Sol. **34**, 1543 (1973)
26 [8.55]
27 B. H. Lee: J. Appl. Phys. **41**, 2988 (1970)
28 [8.61]
29 R. C. Hanson, K. Helliwell, C. Schwab: Phys. Rev. B**9**, 2649 (1974)
30 [8.59]
31 H. D. Hochheimer, M. L. Shand, J. E. Potts, R. C. Hanson, C. T. Walker: Phys. Rev. B**14**, 4630 (1976)
32 [8.83]
33 [8.84a]
34 [8.58]

Table 8.2. Supplementary listing of mode γ's for tetrahedral and related solids measured by various techniques

Material	Critical point (diamond, zb) mode γ's				
Ge(diam.)[a], tunneling 1.2 K, uniaxial	TA(L) −0.4	LA(L) 0.5	TO(L) 0.9	LO(L) 1.2	
ZnS(wurt.)[b,c], Raman hydrostatic	TA(X;K) (2nd order) −1.8	TO(Γ) (A_1, E_1) 1.8 2.0[c]	LO(Γ) (A_1, E_1) 1.0	TO(X;L) (2nd order) 1.3	LO(X;L) (2nd order) 1.4
ZnO(wurt.)[c,d], Raman hydrostatic	TA(L) (E_2) −1.8	TO(L) (E_2) 1.7	TO(Γ) (A_1) 2.1		
CdS(wurt.)[c–e], Raman hydrostatic, uniaxial 77 K	TA(L) (E_2) −2.7[c,d] −2.2	TO(L) (E_2) 1.6	TO(Γ) (E_1) (A_1) 1.0 1.8	LO(Γ) (E_1) (A_1) 1.1[d] 0.8* 1.4*	
CuI(rhom., tetrag.)[f] Raman hydrostatic	TA(L) (E) 3.2	LA(L) (A_1) 1.6	TO(Γ,L) (E) 1.5	[TA(X) LA(X) TO(Γ,X)] [(E_g) (A_{1g}) (E_g)] >0	
AgI(wurt.)[g] Raman hydrostatic	TA(L) (E_2) −7.5	InAs(zb)[h] Raman uniaxial		TO(Γ) 0.9	

* In e (below) the LO − TO splitting is assumed pressure independent
[a] R. T. Payne: Phys. Rev. Lett. **13**, 53 (1964) [b] [8.88] [c] [8.43a] [d] [8.38]
[e] R. J. Briggs, A. K. Ramdas: Phys. Rev. B**13**, 5518 (1976) [f] [8.58] [g] [8.89]
[h] F. Cerdeira, C. J. Buchenauer, F. H. Pollak, M. Cardona: Phys. Rev. B**5**, 580 (1972)

comparison. The measured $\gamma_{O(\Gamma)}$ values for diamond [8.38, 39] and Ge [8.37, 40] are within 20% of unity (see Table 8.1). Thus, for the strictly covalent materials $\omega_{O(\Gamma)}$ scales as V^{-1}, and the corresponding force constants scale quadratically with density.

This situation changes for the partially ionic zincblende materials. Typical data for the TO(Γ) and LO(Γ) Raman peaks of GaP [8.41] and InP [8.42a] are shown in Figs. 8.7, 8, and the measured results for other materials are summarized in Tables 8.1, 2. *Mitra* et al. [8.38, 43a] studied several of the tetrahedral solids in Tables 8.1, 2, as well as some alkali halides. They found that $\gamma_{LO(\Gamma)}$ maintains an approximately constant value close to one, but that $\gamma_{TO(\Gamma)}$ increases above unity with increasing effective charge. This correlation is demonstrated in Fig. 8.9, where $\gamma_{TO(\Gamma)}$ is plotted against the Szigetti effective charge per cation valence electron e_s^*/ze. Here $e_s^* = 3e^*/(\varepsilon_\infty + 2)$, where the factor $3/(\varepsilon_\infty + 2)$ corrects for the local field due to ion polarizability. *Mitra* and *Namioshi* [8.43b] considered several models to explain their data including a Born-Mayer treatment for the alkali halides, and the Phillips ionicity theory for

Table 8.3.† Effect of compression on Born's effective charge expressed as γ_{e^*}, and parameters necessary to compute γ_{e^*} from (8.13)

Material	ε_∞ [a]	$\gamma_{\varepsilon_\infty}$ [b]	ν_{TO} [c] [cm^{-1}]	ν_{LO} [c] [cm^{-1}]	γ_{TO}	γ_{LO}	γ_{e^*}
AlSb	10.2	−1.0	319	340	1.23	1.27	0.6
GaP	9.1	−0.6	367	403	1.09	0.95	−0.5
GaAs	10.9	−0.8	269	292	1.39	1.23	−0.5
GaSb	14.4	−1.3	231	241	1.23	1.21	−0.1
InP	9.6	−0.8	304	345	1.44	1.24	−0.3
ZnS	5.2	−0.2	271	352	1.9*	0.95	−1.0
ZnSe	5.9	−0.3	205	250	1.4	0.9	−0.8
ZnTe	7.3	−0.6	177	205	1.7	1.2	−1.0
CuCl	3.7	0.0	158[d]	210[d]	2.4	1.0	−1.3
CuBr	4.4	−0.2	135[d]	170[d]	2.4	1.9	0.5
CuI	5.5	−0.4	133[d]	151[d]	2.2	1.6	−1.1

† See text and [8.42b] for very recent results on 3C-SiC, AlN, BN, and BP

* The average of both values in Table 8.2 for the E_1/A_1 TO(Γ) modes in wurtzite ZnS

[a] [8.48]

[b] [8.45]; $\gamma_{\varepsilon_\infty}$ was calculated from the best fit line with slope −3.8 eV [Ref. 8.45, Fig. 4] using $E_g = [(C^2 + E_h^2)2/3]^{1/2}$ from [8.48]

[c] W. Richter: "*Resonant Raman Scattering in Semiconductors*", in Springer Tracts Mod. Phys. **78**, 174 (1976)

[d] H. D. Hochheimer, M. L. Shand, J. E. Potts, R. C. Hanson, C. T. Walker: Phys. Rev. B**14**, 4630 (1976); data recorded at 40 K

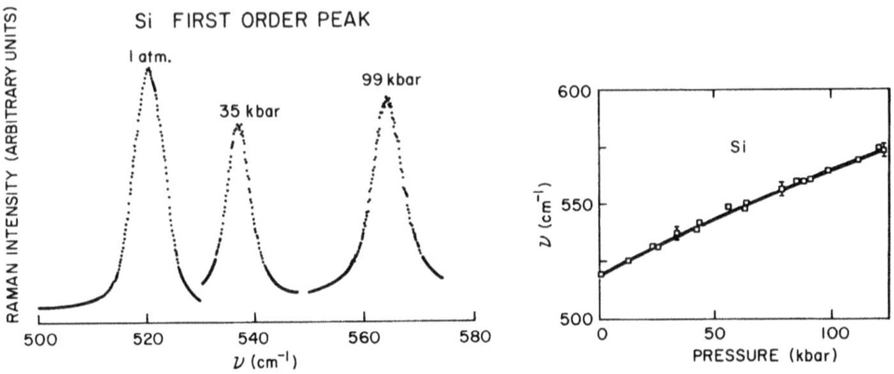

Fig. 8.6. Shift to higher frequency with pressure (300 K) of the first-order Raman peak of Si. In the right-hand plot the initial $P=0$ slope gives $\gamma_{O(\Gamma)} \approx 1$ for this prototypical homopolar semiconductor. After [8.27]

the more covalent semiconductors [8.48]. In the latter approach the homopolar gap was taken to scale as $V^{-2/3}$ and the bond ionicity to be proportional to e_s^{*2}. Both treatments predicted a decrease in e_s^* with pressure.

These experimental and theoretical findings show that compression tends to reduce the effective charge for a considerable sampling of different ionicity materials. However zincblende-type 3C−SiC is an exception (see below and

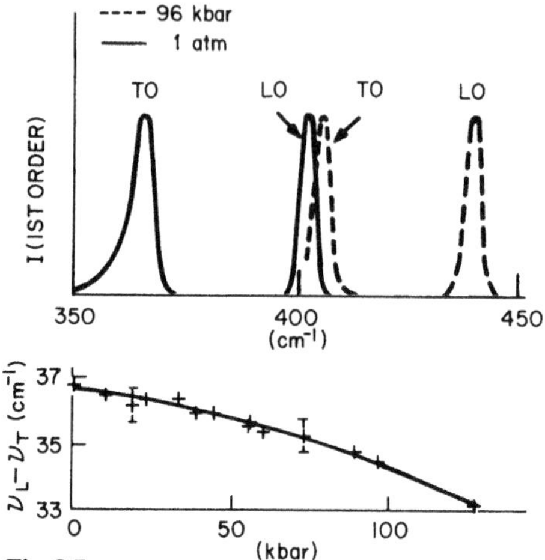

Fig. 8.7

Fig. 8.7. Shift with pressure (300 K) of the first-order TO and LO phonon peaks of GaP [8.41]. The LO−TO splitting decreases under compression as shown in the lower plot

Fig. 8.8. Shift with pressure (300 K) of the first-order TO and LO phonon peaks of InP. In the right-hand plot, phonon frequency is shown versus lattice constant (*linear lower scale*) and pressure (*upper scale*). After [8.42a]

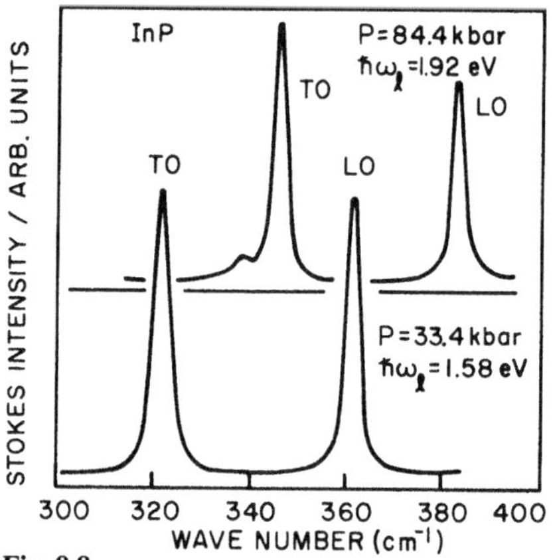

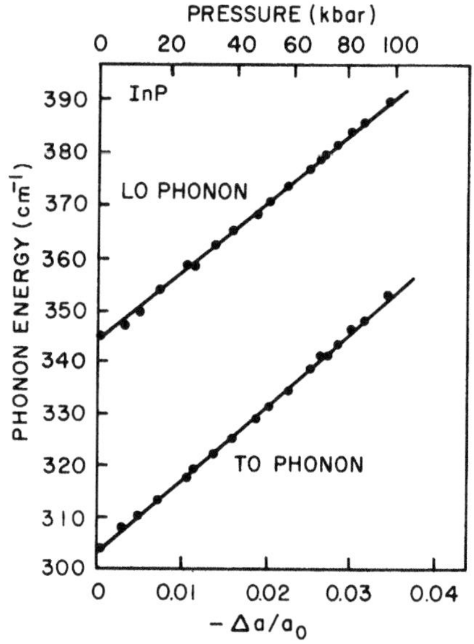

Fig. 8.8

[8.42b]). From (8.12), the Grüneisen parameter γ_{e*} for Born's dynamic charge is given by

$$\gamma_{e*} = -\frac{1}{2} + \frac{1}{2}\gamma_{\varepsilon_\infty} + \gamma_{\mathrm{LO}} + (\gamma_{\mathrm{LO}} - \gamma_{\mathrm{TO}})\frac{\varepsilon_\infty}{\varepsilon_0 - \varepsilon_\infty}. \qquad (8.13)$$

Values of γ_{e*} calculated using (8.13) are listed in Table 8.3 for several III-V, II-VI and I-VII solids.

Recently, *Trommer* et al. [8.42a, 44] studied in detail the volume dependence of e^* for GaAs and InP. They found $e^*/e = 2.18 + 4.4(\Delta r/r_0) - 88(\Delta r/r_0)^2$ and $e^*/e = 2.54 + 4.5(\Delta r/r_0) - 88(\Delta r/r_0)^2$ for GaAs and InP, respectively; here r is the nearest-neighbor separation. Thus, $\gamma_{e*} \approx -2/3$ for both materials with a non-negligible quadratic contribution. The variations of γ_{e*}, obtained in this way,

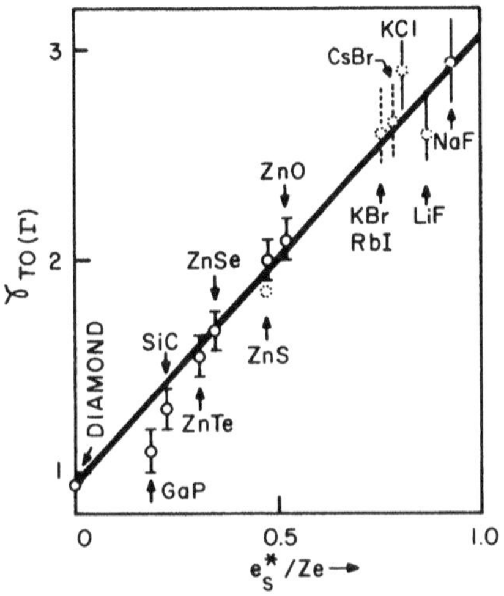

Fig. 8.9. Linear relationship between the Grüneisen parameter for TO(Γ) and the Szigetti effective charge per cation electron discovered by *Mitra* et al. [8.38, 43a]. Closed circles Raman data, open circles ir data

from the values listed in Table 8.3 for GaAs and InP, stem from using (8.13) in Table 8.3 and from the means of estimating $\gamma_{\varepsilon_\infty}$. In [8.42a, 44], $\gamma_{\varepsilon_\infty}=-1$, whereas in Table 8.3, $\gamma_{\varepsilon_\infty}$ was estimated according to the correlations established in [8.45].

To explain their pressure-Raman data on GaAs and InP, *Trommer* et al. applied two theoretical approaches appropriate for the e^* of partially covalent semiconductors. These are the bond orbital model (BOM) of *Harrison* and *Ciraci* [8.46] and the dielectric matrix pseudopotential (DMP) method of *Vogl* [8.47].

In the BOM e^* is given by

$$e^* = -\Delta ze + 4\alpha_p e + 4c\alpha_p(1-\alpha_p^2)e, \tag{8.14}$$

where $-\Delta ze = \frac{1}{2}(z_{\text{anion}} - z_{\text{cation}})e$ is the charge due to equal covalent sharing of the valence electrons, $c \approx 1$ is an adjustable parameter and α_p is the bond polarity giving the amount of asymmetric charge transfer. Note that both static $(4\alpha_p e)$ and dynamic $[4c\alpha_p(1-\alpha_p^2)e]$ terms must be included. α_p is defined by the covalent and polar energies W_2 and W_3 [8.46], which are similar to E_h and C in the theory of *Phillips* and *Van Vechten* [8.48]:

$$\alpha_p = \frac{W_3}{(W_2^2 + W_3^2)^{1/2}}. \tag{8.15}$$

W_2 is known to scale as $V^{-n/3}$ with $n \approx (2.0-2.5)$, and W_3 is thought to depend weakly on volume. Accordingly, under compression α_p decreases, and since for the materials of interest $8\alpha_p > \Delta z$, e^* also decreases. Neglecting the term in α_p^3 we have $\gamma_{e^*} \approx \gamma_{\alpha_p}$, and for small W_3 (strongly covalent solids) $\gamma_{e^*} \approx -2/3$, in agreement with the measurements of [8.42a, 44] for InP and GaAs. However, this simple approach cannot explain the nonlinear dependence of e^* on $\Delta r/r_0$.

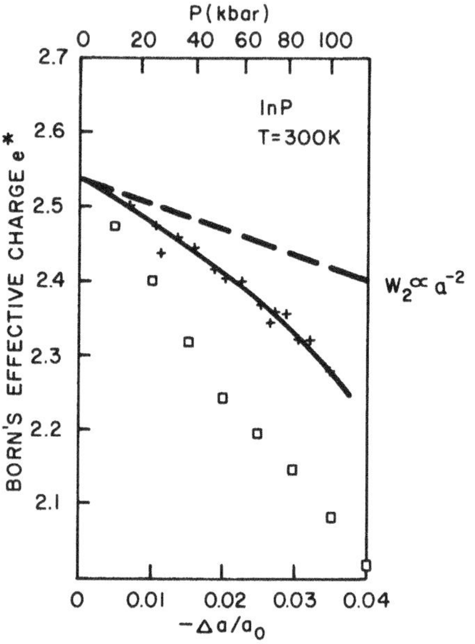

Fig. 8.10. Effective charge e^* versus lattice constant and pressure for InP. *Crosses* and *solid line* are data and fit; *dashed line* and *squares* are the results of BOM and full DMP calculations, respectively (see text). After [8.42a]

In the DMP method of *Vogl* an expression for e^* [Ref. 8.42a, Eq. (10)] is obtained by calculating the sublattice-displacement induced change in charge density using pseudopotential techniques. Care must be taken to maintain self-consistency. In the simplest Heine-Jones type band structure [8.49], an expression is obtained which is approximately equivalent to (8.15) for $\alpha_p^2 \leq 0.6$ (valid for groups III – V and II – VI). W_2 and W_3, determined only by the (111) form factors in this approximation, are expected to scale with volume as before [8.46,48]. Thus the DMP method can also explain $\gamma_{e^*} \approx -2/3$.

For InP in Fig 8.10, the result of an extensive DMP calculation (squares), including form factors with $|G| \leq (2\pi/a_0)\sqrt{11}$ (G and a_0 are a reciprocal lattice vector, and the lattice constant), is compared with experiment (crosses) and the BOM (dashed line). The extensive DMP result overestimates, and the BOM underestimates the effect by the same amount. Nevertheless, the sign and magnitude of the change are successfully reproduced, showing that both models contain the essential physics. Very recently the work of this group has been extended [8.42b]. Pressure-Raman measurements on 3*C*-SiC, AlN, BN, and BP gave for γ_{e^*} +0.67, −0.65, −0.17, −2.49, respectively, and DMP calculations yielded qualitative agreement for γ_{e^*}, reproducing the correct sign in each case. The increase in e^* with pressure for 3*C*-SiC was attributed to the lack of *p*-electrons in the carbon core.

A density functional pseudopotential (DFP) calculation of e^* in GaAs was also reported recently [8.50]. This method is well suited to pressure effects because it is not restricted to small displacements. It will be discussed further in Sects. 8.3.4, 5.2.

8.3.2 Phonon Dispersion at High Pressure – Two-Phonon Results

For the tetrahedral semiconductors, the interpretation of two-phonon pressure-Raman data is generally aided by the resemblance of the dominant Γ_1-component to the density of phonon overtone states [8.4,5]. This is just the single-phonon density on a doubled frequency scale. For example, compare the Raman spectra and density of states for Ge in Fig. 8.11. For III – V and II – VI materials, the similarity continues for the TA modes, but lessens for the optic modes (see Figs. 8.13, 14 and [8.51] for GaP and ZnS). This fortunate circumstance arises because the matrix elements in $\hat{\varepsilon}_s \cdot \chi' \cdot \hat{\varepsilon}_l$ [see (8.2,5)] do not strongly modulate the density obtained by integrating (8.4) over the kinematically allowed phonon states. Since, in addition, the crystal symmetry is invariant under hydrostatic conditions, it is possible to monitor the entire phonon dispersion under pressure. Examples will be given below. The prototypical materials Si, GaP, and ZnS will be discussed in detail; data for other solids are presented in Tables 8.1, 2. The known homologies at $P=0$ can often be used to extend these results to solids where less detailed pressure data exist.

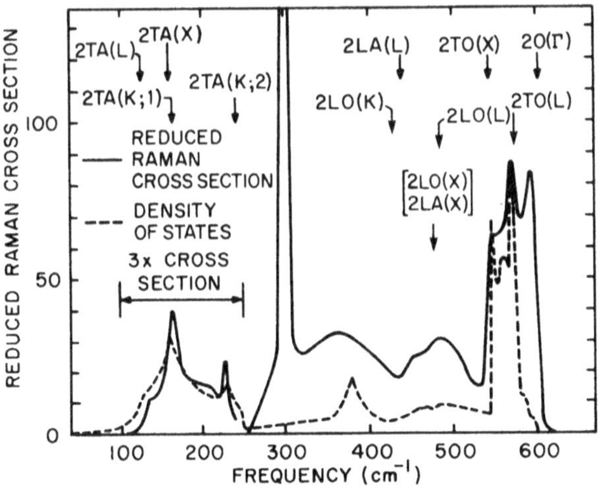

Fig. 8.11. Reduced [divided by $(n_i+1)^2$] $\Gamma_1+4\Gamma_{12}$ component of the two-phonon Raman spectrum of Ge ($P=0$, $T=300$ K) compared with the density of overtone states from neutron scattering. Note close similarity in 2TA region. Critical point designations match those in Tables 8.1, 2. After [8.4]

The two-phonon spectra of Si and GaP were studied by *Weinstein* and *Piermarini* [8.27, 41] in the first use of the modern (Ruby-calibrated, hydrostatic) diamond cell for Raman measurements. Lower pressure (≤ 10 kbar) measurements were reported for Si by *Richter* et al. [8.52] and for GaP by *Weinstein* et al. [8.53]. Second-order scattering in cubic ZnS was investigated to 150 kbar by *Weinstein* [8.54] and at lower pressure by *Brafman* and *Mitra* [8.43] and *Zigone* et al. [8.55]. Data from [8.27, 41, 54] are reproduced in Figs. 8.12 – 14. The critical point designations match those in Tables 8.1, 2 and in the phonon dispersion curves, Figs. 8.15 – 17. Note that the peaks and shoulders in the spectra are derived from a collection of modes in the vicinity of a critical point. For example, the strongest peak in the TA overtone spectrum is due to a large set of nearly degenerate phonons extending over the Brillouin zone (BZ) surface

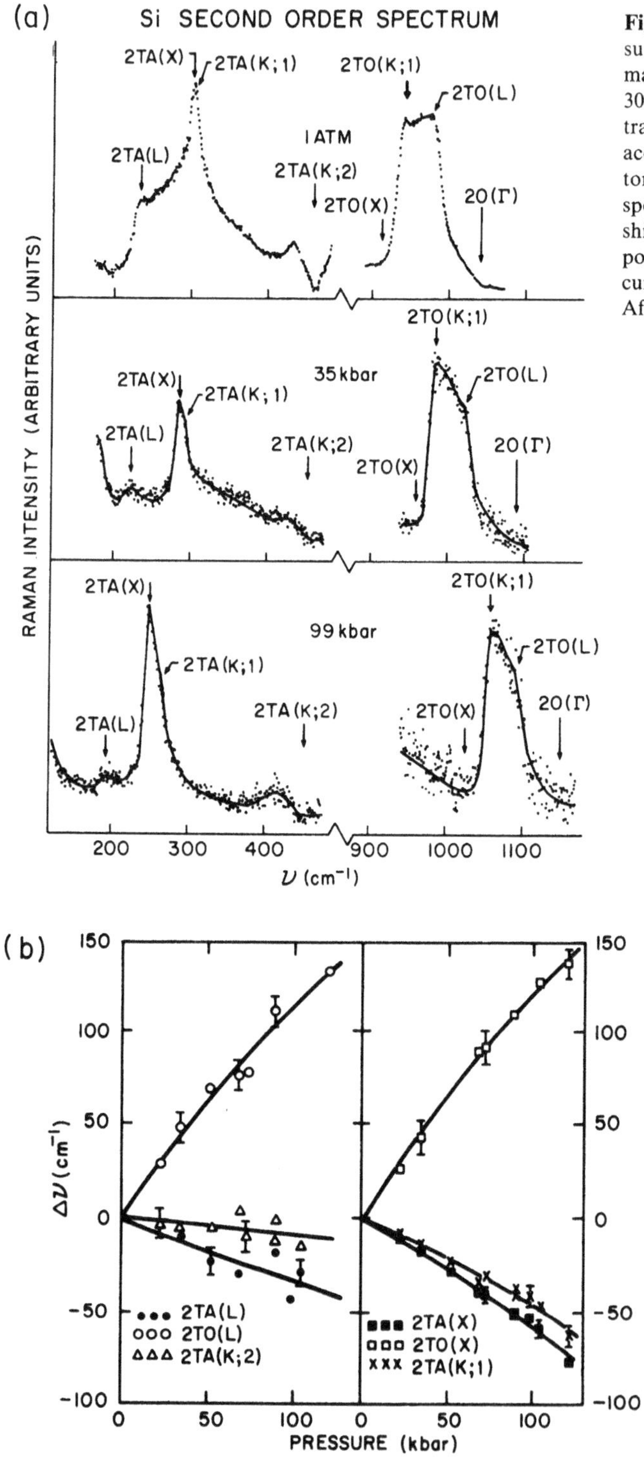

Fig. 8.12a,b. Effect of pressure on the two-phonon Raman spectrum of Si (at 300 K). (**a**) Measured spectra in the regions of strong acoustical and optical overtone scattering. (**b**) Corresponding plots of frequency shift versus pressure for important critical points. Solid curves are least-square fits. After [8.27]

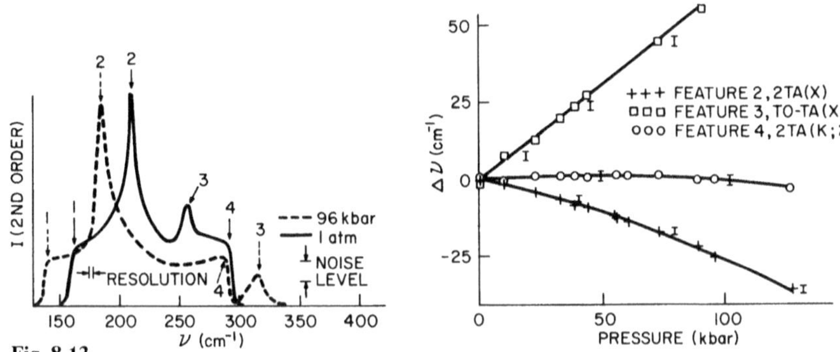

Fig. 8.13

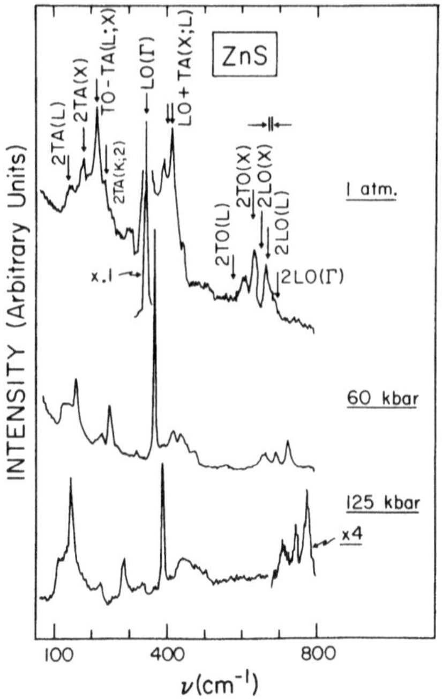

Fig. 8.14

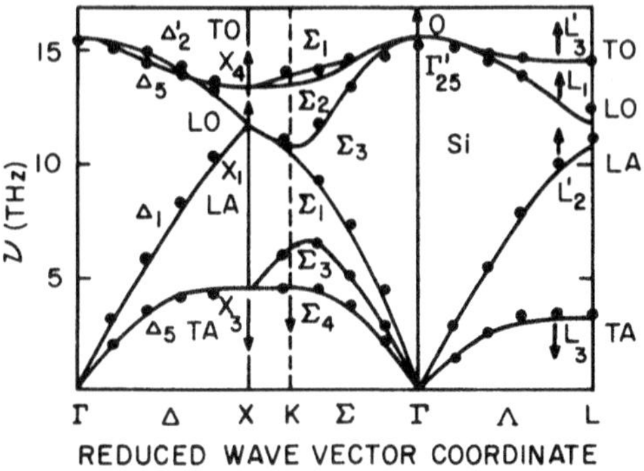

Fig. 8.15

Fig. 8.13. Raman spectrum of GaP (300 K) in the 2TA region as a function of pressure. The right-hand figure plots frequency shift versus pressure for important critical points; error bars are displaced from best fit curves. After [8.41]

Fig. 8.14. Effect of pressure on the Raman spectrum of cubic ZnS at 300 K. After [8.54]

Fig. 8.15. Phonon dispersion relations for Si ($P=0$, $T=300$ K) measured by neutron scattering (*points*) and fit to a model calculation (*curves*). Band labels and irreducible representations are indicated. *Arrows show direction of pressure response* (see [8.51] and references therein)

from X to K and into the zone by $\sim 1/3$ the distance to Γ along the Σ and Δ-directions.

The pressure results for Si, GaP, and ZnS are as follows. The BZ center optical phonons have already been discussed. Proceeding on the LO branch along the (00ζ) and $(\zeta\zeta\zeta)$ directions to the BZ boundary X and L-points, the mode Grüneisen parameters change little. There is only a $15-20\%$ increase in γ_i for GaP and ZnS (no BZ boundary LO data exist for Si). For the TO branches of

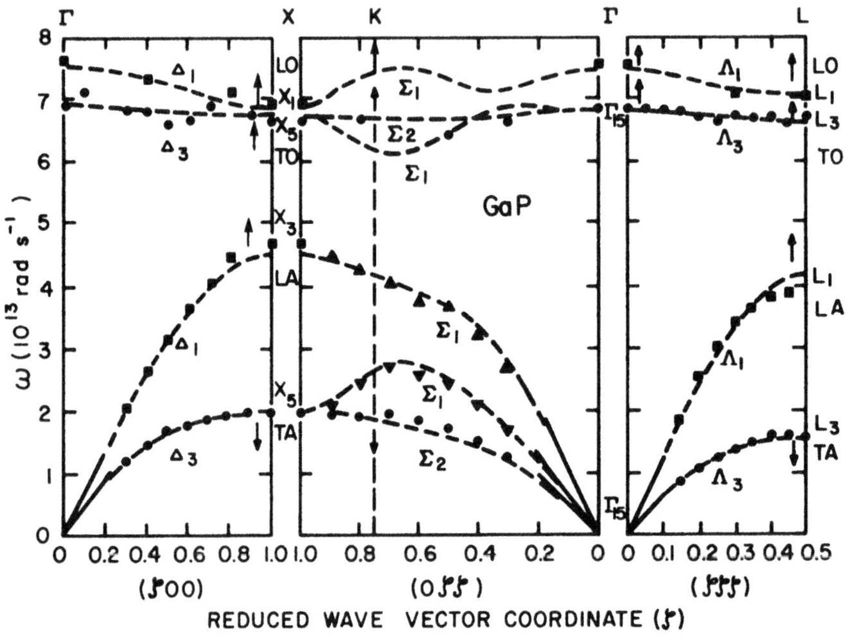

Fig. 8.16. Phonon dispersion relations for GaP. Key same as in Fig. 8.15

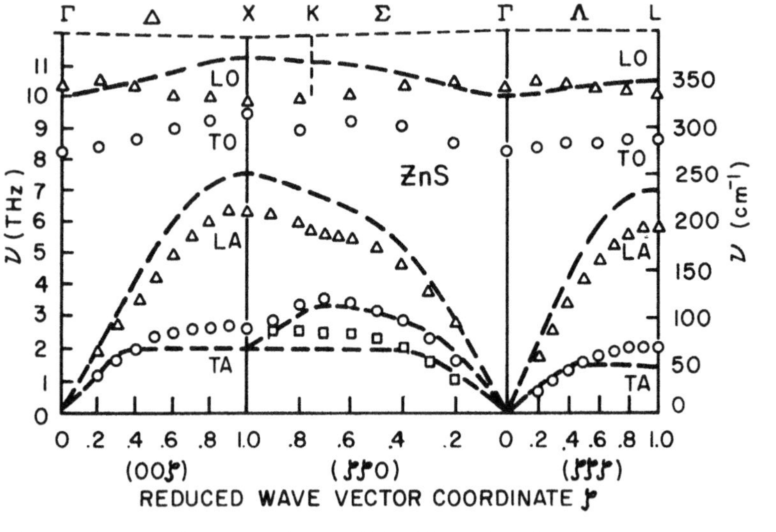

Fig. 8.17. Phonon dispersion relations (300 K) for ZnS at $P=0$ measured by neutron scattering (*points*), and at $P=150$ kbar deduced from pressure-Raman studies at critical points (*dashed curves*). 150 kbar marks onset of the metallization transition. See Sect. 8.5.2. After [8.54]

Si and GaP, γ_i also increases slightly away from $\boldsymbol{q}=0$ reaching ~ 1.4 at the X and L-points, but for ZnS the reverse is true. For each material the initial effect of pressure is to further flatten the already Einstein-like LO and TO branches [8.54]. However, one should be careful not to generalize this result to other solids until more data is collected. No pattern emerges as to whether pressure causes the LO and TO branches to cross or separate. For GaP, $\gamma_{\mathrm{TO}(L)} > \gamma_{\mathrm{LO}(L)}$ implying eventual crossing at the L-point, while for ZnS the situation is ambiguous (Tables 8.1, 2). The lack of a pattern is consistent with the $P=0$ dispersion curves for which LO and TO may (as in GaAs) or may not (as in GaP) cross [8.51]. No data exist to distinguish the pressure behavior of the two nondegenerate TO branches along low-symmetry directions.

Consider now the acoustic branches. There are no direct pressure data for BZ boundary LA modes, as these do not appear strongly in the two-phonon Raman spectra, and they are often dominated by nearby first-order peaks (see the densities of states in [8.51]). However, in group IV materials, LO(X) and LA(X) are degenerate by symmetry, and they remain close for many tetrahedral semiconductors at both the X and L-points (again see, for example, GaAs, InSb, ZnSe, and CdTe [8.51]). Consequently we take $\gamma_{\mathrm{LA}} \approx \gamma_{\mathrm{LO}}$ at the BZ boundary as a zeroth-order approximation. This is supported by recent calculations for ZnS, ZnSe, and ZnTe showing that γ_{LA}, γ_{LO} (and also γ_{TO}) change little across the BZ [8.56]. For Si, $\gamma_{\mathrm{LA}} \approx 0.9$ at the BZ boundary was obtained by fitting pressure-Raman data to the temperature-dependent thermal expansion [8.27].

The mode γ's for BZ boundary TA phonons in most tetrahedral semiconductors are anomalous in that they are negative. This means that the effective force constants for these modes weaken with pressure, i.e., the corresponding springs become softer rather than stiffer. Some modes in molecular solids have $\gamma_i < 0$ because a *macroscopic* compression actually causes a slight expansion in a particular direction, as for intrachain optical vibrations in trigonal Se [8.57]. This is not the case here; the mode softening corresponds to a real weakening of the effective repulsive forces determining short wavelength shear acoustic vibrations. $\gamma_{\mathrm{TA}(X)}$ is -1.4 for Si, -0.72 for GaP, -1.2 for ZnS, and $\gamma_{\mathrm{TA}(L)}$ exhibits similar large negative values (Table 8.1). In fact, the available data for these materials show that for the lowest energy acoustic branch negative γ_i of this magnitude apply to a rather large set of modes covering most of the BZ boundary and extending $\sim 1/3$ into the zone (Figs. 8.15–17). Tables 8.1, 2 show that a similar situation pertains to other tetrahedral semiconductors, with the notable exceptions of diamond [$\gamma_{\mathrm{TA}(X)} \approx 0.4$] [8.39] and CuI [$\gamma_{\mathrm{TA}(L)} \approx 3.0$] [8.58, 59]. The reason for this is not presently understood.

Along low-symmetry directions, where the shear branches are not doubly degenerate, data are more sparse. For the second (intermediate energy) acoustic branch near the K-point, viz. $\boldsymbol{q}=(\pi/a\sqrt{2})(110)$, $\gamma_i=-0.3$ for Si and ZnS and $\gamma_i=0.15$ for GaP. Thus we expect small negative or positive mode γ's for this branch near the BZ boundary.

Also listed for completeness in Table 8.1 are the γ_i for $\boldsymbol{q} \approx 0$ acoustic modes (from Brillouin scattering or pulse-echo measurements) computed by replacing

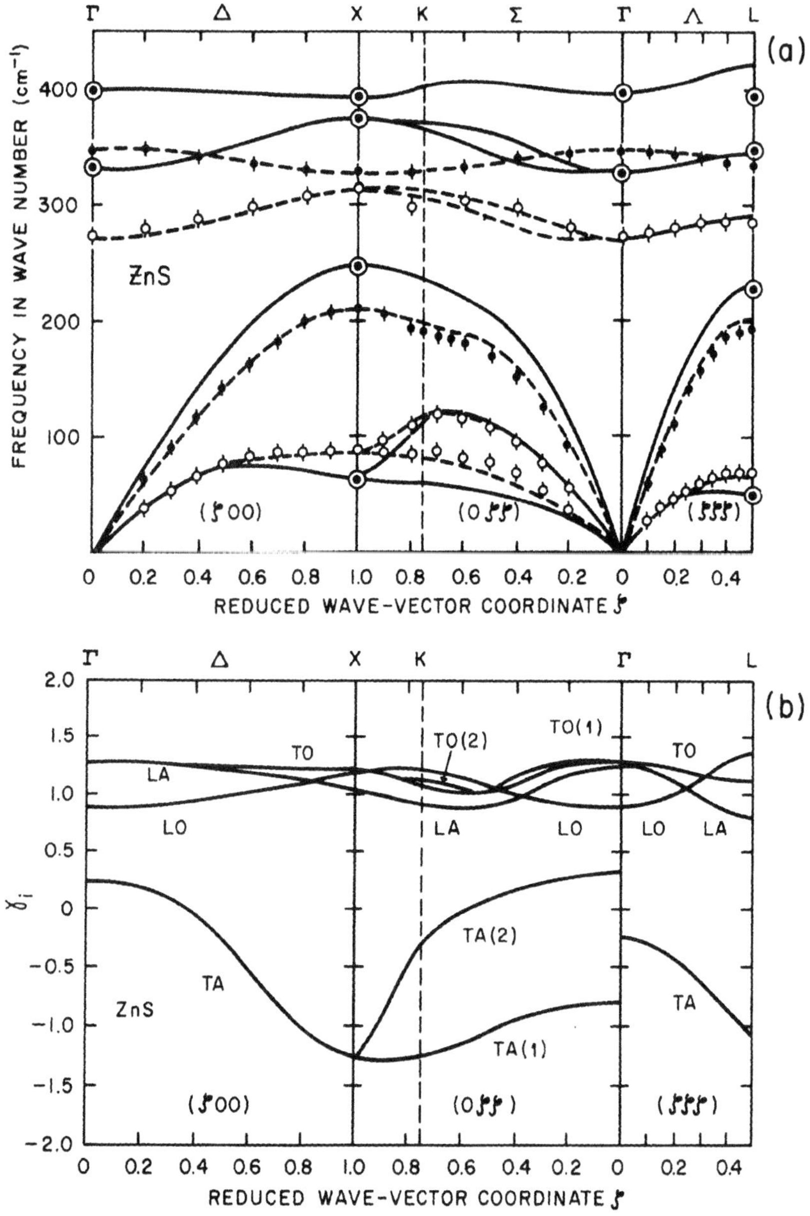

Fig. 8.18. (**a**) Calculated phonon dispersion (300 K) of ZnS at $P = 0$ (*dashed curves*) and $P = 150$ kbar using an 11 parameter rigid-ion model. $P = 0$ dispersion fit to neutron scattering data (*solid* and *open points*); $P = 150$ kbar dispersion fit to pressure-Raman data (*circled points*). Note close agreement with Fig. 8.17. (**b**) Corresponding mode Grüneisen parameter dispersion relations for ZnS. After [8.56]

B in (8.11) by the longitudinal and two shear elastic moduli c_{11}, c_{44} and $(c_{11}-c_{12})/2$, respectively. Only for the latter mode [corresponding to distortion parallel to $(1\bar{1}0)$, $\boldsymbol{q}$ parallel to (110)] is γ_i negative in some cases. However, the reported values $0.2 \geq \gamma_i \geq -0.5$ are generally small compared to the corresponding BZ boundary TA(K or X) mode parameters [8.60].

The above trends are summarized in the frequency dispersion curves of Figs. 8.15–17 where arrows (for Si and GaP) and dashed curves (for ZnS) indicate the response to pressure based solely on the Raman observations near critical points. For ZnS (Fig. 8.17), the dashed curves are a semiquantitative estimate of the frequency dispersion at 150 kbar (the transition pressure to an opaque phase). Similar estimates have been made for ZnSe and ZnTe [8.54, 61]. These totally empirical estimates compare favorably [see Fig. 8.18a for ZnS] with the more rigorous results of an 11 parameter semiempirical rigid-ion model for the zinc chalcogenides that will be discussed below [8.56]. Figure 8.18b shows the mode Grüneisen parameter dispersion curves for ZnS obtained from this calculation. From the data in Tables 8.1, 2, it would be straightforward to sketch similar high-pressure frequency dispersion curves and Grüneisen parameter dispersion curves which exhibit the correct behavior for other materials.

8.3.3 Thermal Expansion

Most tetrahedral semiconductors (except diamond and CuI) have a negative low temperature thermal expansion α [8.62]. [Of course, for $T \rightarrow 0$, $\alpha(T) \rightarrow 0$]. *Blackman* [8.16] suggested on the basis of (8.10) that this anomalous behavior stemmed from negative γ_i for low frequency BZ boundary TA modes.

Pressure-Raman measurements have made it possible to explicitly demonstrate Blackman's conjecture for several of the materials in Tables 8.1, 2. For Si, *Weinstein* and *Piermarini* [8.27] exploited the aforementioned similarity between the two-phonon Raman spectrum and the density of states to perform the summation over modes in (8.10). Interpolating γ_i linearly with frequency between critical points, they obtained the $\alpha(T)$ curves shown in Fig. 8.19. By fitting to α(300 K), $\gamma_{\mathrm{LA,LO}} \approx 0.9$ was deduced for the (unmeasured) BZ boundary longitudinal modes. A scheme for weighted averaging over high-symmetry directions was employed by *Soma* [8.22] in similar calculations. Also *Talwar* et al. [8.56] performed a more detailed calculation for ZnS, ZnSe, and ZnTe (Fig. 8.20) using the density of states and mode γ's derived from their rigid-ion model.

The mode γ's are microscopic quantities depending on interatomic forces; the thermal expansion [or by (8.10), γ_{av}] is an independently measurable macroscopic quantity. Thus, the agreement with experiment in Figs. 8.19, 20 is excellent corroboration of the pressure-Raman results. However, the macroscopic anomaly of negative thermal expansion has merely been traded for the microscopic anomaly of pressure-induced softening of BZ boundary TA phonons, which is not well understood at present. We shall comment further on this in relation to lattice dynamics theory.

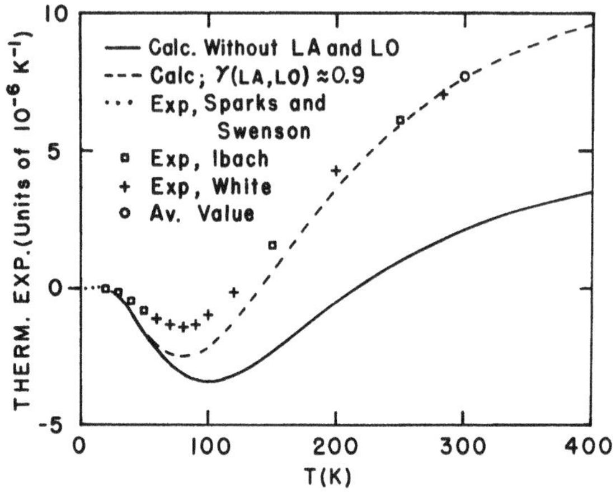

Fig. 8.19. Volume thermal expansion $\alpha(T)$ of Si calculated using pressure-Raman data compared with experiment (see text). After [8.27] and references therein

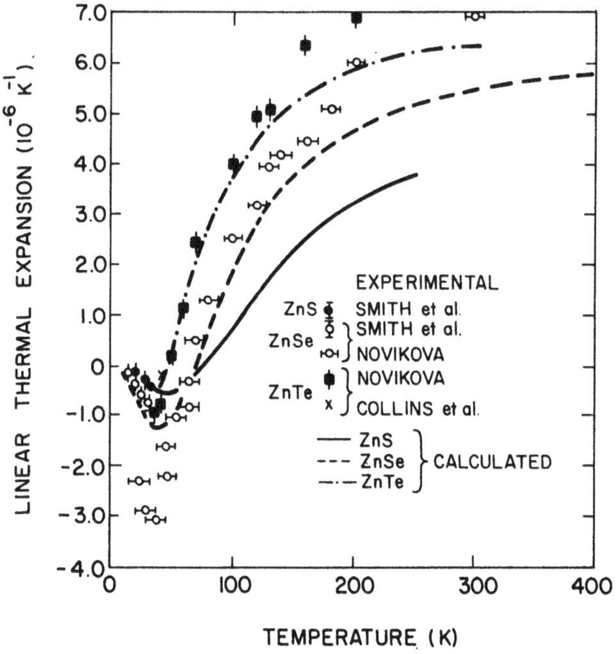

Fig. 8.20. Linear thermal expansion of ZnS, ZnSe, ZnTe calculated using the 11 parameter rigid-ion model of *Talwar* et al. compared with experiment. After [8.56] and references therein

8.3.4 Implications for Lattice Dynamics Theory

The pressure-Raman results just described provide important input, and also stringently test various lattice dynamics models. These divide into two broad classes: those employing semiempirical force constants such as the rigid-ion [8.56, 63] or shell [8.64] models and the microscopic theories based on the dielectric matrix formalism [8.65–67] or the total energy density functional [8.68, 69] approach. We shall discuss these classes in order.

Dolling and *Cowley* [8.64] and *Jex* [8.70] introduced anharmonic terms of third and/or fourth order (in the harmonic displacements) into shell model expansions of the interatomic potential. In [8.64], third-order 2-body forces between nearest neighbors were included, leading to two parameters, which were fit to the temperature-dependent thermal expansion. Mode γ's were calculated for $\boldsymbol{q}$ along the (001) and (111) directions for Ge. Considerable dispersion was found with negative but underestimated γ_{TA} at the BZ boundary. Calculated γ values ranged between 1.6 and -0.4 which is the correct magnitude (Tables 8.1, 2). *Jex* [8.70] followed a similar procedure, except that third-order terms to second neighbor, and for Ge, fourth-order terms to first neighbor were included. The model parameters were determined for C, Si, and Ge by fitting to $\gamma_{av}(T)$. Similar results were obtained ($-0.6 \leq \gamma_i \leq 1.4$) with $\gamma_{TA} < 0$, but γ_{TA} was again underestimated at the BZ boundary. *Bienenstock* [8.71] simulated a shell model for Ge by dividing a given phonon frequency into first neighbor rigid-ion and dipole-dipole contributions. The former were determined by the elastic constants for which the pressure dependence was taken from experiment. The latter were assumed to scale as $1/V^n$, leading to the single parameter n which was fit to $\gamma_{av}(T)$. Again, considerable dispersion was found for the mode γ's throughout the BZ ($-0.4 \leq \gamma_i \leq 1.8$), with $\gamma_{TA} < 0$ but too small at the BZ boundary.

These anharmonic potential parametrization models give the correct magnitude for the mode γ's. However, to reproduce accurately the measured dispersion of γ_i, higher order and more distant forces would have to be included. The proliferation of parameters would then have little physical meaning, so that a different approach is desirable. If an accurate harmonic model with a limited number of semiempirical parameters is known for $P=0$, such an option is available. One simply measures the pressure dependence of enough phonon frequencies to determine all the parameters, and then recalculates the entire phonon dispersion at each pressure. This fact was first applied by *Vetelino* et al. [8.63] to ZnTe using a 4 parameter rigid-ion model. The three elastic constants, $\omega_{TO(\Gamma)}$ and $\omega_{LO(\Gamma)}$ (all measured as a function of pressure) were used as input data. They found $-0.2 \leq \gamma_i \leq 1.8$ which is in reasonable agreement with experiment (Table 8.1), but again the negative γ_{TA} were too small. *Talwar* et al. [8.56] used an 11 parameter rigid-ion model in a similar calculation for ZnS, ZnSe, and ZnTe. The pressure-dependent input data were the elastic and lattice constants, and the Raman frequencies at Γ, X, L, and K. The predicted frequency and mode γ dispersion curves, shown in Fig. 8.18 for ZnS, are in excellent agreement with experiment [8.54], even for the BZ boundary TA modes. The computed Debye temperature, γ_{av}, and thermal expansion (Fig. 8.20) also agree with experiment at low T but deviate somewhat at high T. For all three materials the calculated BZ boundary TA branches do not retain their flatness at high pressure. This may be an artifact of the calculation, since at $P=0$ the flatness is maintained as the BZ boundary TA frequencies decrease through the series ZnS, ZnSe, ZnTe. The flatness for $P>0$ probably requires the inclusion of bond-charge effects, which are not treated in a rigid-ion model.

The bond-charge picture [8.72] is a convenient vehicle for including near range strictly covalent forces – viz., those arising from off-diagonal screening in the dielectric matrix formalism [8.65] (see below). Born-von Karman models may require forces to fifth neighbor to achieve equivalent accuracy [8.73]. For covalent semiconductors at $P=0$, *Weber's* [8.74] dynamic bond-charge model succesfully represents the BZ boundary TA phonons. Their low frequency and flat dispersion results because the effective ion bond-charge force is small compared to the force between bond charges. Consequently,for TA(X) and nearby modes, the ions vibrate like Einstein oscillators [at $\sim\omega_{TA(X)}$] within a nearly rigid lattice of *bond charges*. At least for C, Si, and Ge, there is no reason why the 4(or 5) parameter models of [8.74], in combination with available Raman and elastic constant pressure data, could not be employed to calculate high-pressure dispersion curves. However, this has not yet been done.

The pressure-Raman results severely test various microscopic lattice dynamics theories because they require these models to reproduce not only the $P=0$ dispersion curves but also the proper trends as the lattice constant is varied. Two microscopic pseudopotential-based approaches, the dielectric matrix (DMP) [8.65] and the density functional (DFP) [8.75] formalisms (P stands for pseudopotential in these abbreviations) have met with reasonable success in calculating pressure-dependent phonon properties for semiconductors.

Turner and *Inkson* [8.66] calculated the phonon dispersion at elevated pressure for C, Si, and Ge using the DMP formalism with a Shaw-type pseudopotential [8.76]. The reciprocal lattice Fourier components of the dielectric function $\varepsilon_{GG'}$ were computed in the Wannier representation. However, to simplify the inversion of $\varepsilon_{GG'}$, the extreme tight-binding approximation was used away from the BZ center, but with off-diagonal elements scaled to agree with a more exact $\boldsymbol{q}=0$ treatment. Using the measured pressure dependence of the electronic band structure and the lattice constant, these authors obtained good agreement with experiment for $\omega_{TA(X)}$, with the pressure-induced softening well reproduced (Fig. 8.21). They confirm that off-diagonal screening must be included to describe properly the bond-charge (noncentral) forces crucial to the BZ boundary TA modes. *Porod* et al. [8.67] also employed the DMP formalism, but in a plane-wave representation, to calculate the pressure dependence of $\omega_{TO(\Gamma)}$ and $\omega_{LO(\Gamma)}$. They simplified the inversion of $\varepsilon_{GG'}$ and simultaneously

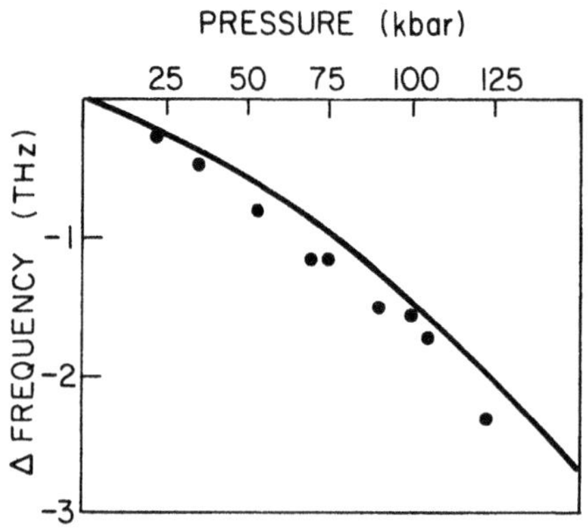

Fig. 8.21. DMP calculation of pressure induced softening of the TA(X) frequency in Si (*solid curve*) compared with experiment (*points* [8.27]). After [8.66]

satisfied charge neutrality by excluding all off-diagonal elements except ε_{OG} and ε_{GO}. These elements were found to be unimportant for TO(Γ) but very important for LO(Γ) and thus e^* (the specific calculation for the pressure dependence of e^* by this method was discussed earlier in Sect. 8.3.1). Reasonable mode γ values were obtained for the tetrahedral semiconductors, e.g., $\gamma_{TO(\Gamma)}=1.1$ for GaAs compared to 1.4 from experiment [8.42].

The second microscopic approach to predicting pressure-dependent phonon properties in tetrahedral semiconductors employs the DFP formalism to calculate the total energy as a sum of kinetic, electron-ion, electron-electron, ion-ion and exchange-correlation terms [8.75]. At each volume the total energy is computed for the equilibrium atomic configuration, and for the configuration distorted by a given phonon eigenvector (frozen phonon). The harmonic expansion of this energy difference gives the phonon frequency at that volume.

Wendel and *Martin* [8.77] calculated TO(Γ), TA(Γ) (corresponding to $c_{11}-c_{12}$) and TA(X) for Si by this method. A local pseudopotential carried to partial self-consistency was used. The γ_i show the correct trends with $\gamma_{TA(X)}<0$; this calculation links the softening of TA(X) to the increasing dominance with pressure of ion-ion forces over the electron-electron contribution. In fact, electron-electron interactions (i.e., forces between bond charges) are found to be essential for stabilizing the lattice against the TA(X) distortion.

Yin and *Cohen* [8.68] employed the DFP formalism to predict several static and dynamic properties of group IV materials. For a given structure the only input to their totally self-consistent nonlocal pseudopotential scheme was the atomic number. They also computed γ_i for Si at the Γ and X points. The predicted and measured γ_i agree *within* the limits of experimental accuracy (Table 8.1). In particular, the success for $\gamma_{TA(X)}$ shows that covalent forces acting via the bond charges and arising from off-diagonal dielectric screening are properly included.

A self-consistent DFP calculation using local pseudopotentials was very recently reported by *Kunc* and *Martin* [8.50] for GaAs. The results also agree well with experiment except for the calculated $\gamma_{TA(X)}=-3.48$, which is negative but too large (Table 8.1). These authors found that anharmonic contributions are very important for TA(X) in GaAs.

8.4 Changes in Phonon Line Shape with Pressure

8.4.1 Phonon-Phonon Interactions

So far our discussion of mode γ's has assumed the quasi-harmonic approximation. This applies when anharmonic terms are small perturbations to the atomic potential. Then $\Gamma_i/2 \ll \omega_i$ in (8.4), and there will be a well-defined Raman peak at the quasi-harmonic frequency ω_i. However, ω_i differs slightly from its harmonic value (appropriate for $T=0$ and $P=0$) by the real part of the phonon self-energy due to anharmonic interactions; it is this term which causes ω_i to shift

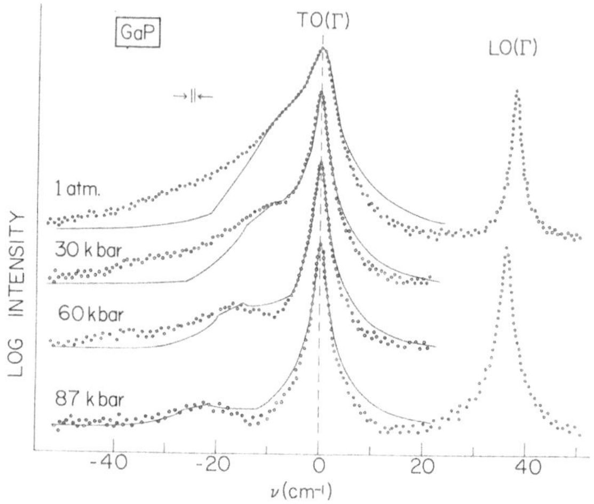

Fig. 8.22. Semilog plot of measured TO(Γ) and LO(Γ) Raman spectra (300 K) of GaP at various pressures (*points*). Calculated TO(Γ) line shape using (8.4, 16) (*solid curves*). Spectra were shifted horizontally to coincide at TO(Γ) and displaced vertically. After [8.78]

with pressure. The imaginary part of the self-energy gives the peak half width $\Gamma_i/2$. We now consider cases where $\Gamma_i/2$ also shows pronounced pressure dependence.

One of the clearest examples is the TO(Γ) Raman peak of GaP, shown as a function of pressure in Fig. 8.22 after [8.78]. At $P=0$ this peak is asymmetrically broadened toward lower energy, whereas LO(Γ) is sharp and symmetric. Under pressure the TO(Γ) peak becomes symmetric, and a weak sideband appears about 23 cm^{-1} below $\omega_{\mathrm{TO}(\Gamma)}$. The width of LO($\Gamma$) is relatively unaffected by pressure. *Barker* [8.79] conjectured that the $P=0$ broadening was due to the anharmonic decay of TO(Γ) into a nearly degenerate continuum of TA+LA combinations near the X-point, TA+LA($\sim X$). He described this by a frequency-dependent Γ_i in (8.4). *Weinstein* [8.78] proposed that the different pressure shifts of TO(Γ) and TA+LA($\sim X$) would decrease this resonant interaction, thereby altering $\Gamma_i/2$ for TO(Γ) sufficiently to explain the induced line shape changes.

The theory of anharmonic phonon decay has been discussed by several authors [8.80, 81]. The contribution of the decay TO(Γ)→TA+LA($\sim X$) to $\Gamma_i/2$, a width parameter that is now expected to be frequency dependent, can be calculated in perturbation theory. The result is [8.78]

$$\frac{\Gamma_i(\omega)}{2}=18\pi(n_{\mathrm{TA}}+n_{\mathrm{LA}}+1)|W_{\mathrm{av}}|^2\varrho_2(\omega). \tag{8.16}$$

Here W_{av} is the effective third-order interaction strength, and the main frequency dependence arises from the two-phonon density of states $\varrho_2(\omega)$ due to

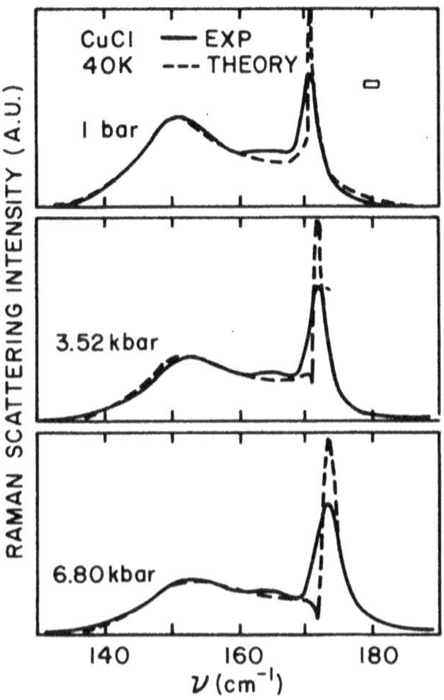

Fig. 8.23. Measured (*solid curve*) and calculated (*dashed curve*) Raman spectra in region of the TO phonon of CuCl as a function of pressure. Box to the right defines uncertainty. After [8.83]

TA+LA($\sim X$) combinations. The thermal factor in parenthesis comes from considering the competition between the decay TO(Γ)→TA+LA($\sim X$) and the reverse process. Using (8.16) and computing $\varrho_2(\omega)$ from the measured dispersion curves at $P=0$ [8.82], the solid line fits shown in Fig. 8.22 were obtained [8.78]. W_{av} was found to be $\sim 2\%$ of $\omega_{TO(\Gamma)}$ at $P=0$, and W_{av} decreased by $\sim 0.4\%$/kbar. A shift of $+0.2\,cm^{-1}$/kbar was deduced for TA+LA($\sim X$), from which $\gamma_{LA(X)} \approx 1.0$ was obtained [using the $\gamma_{TA(X)}$ value from Table 8.1]. Since TO(Γ) shifts by $0.45\,cm^{-1}$/kbar, the near degeneracy responsible for the asymmetric broadening is broken at high pressure as expected. Note also [see (8.2–4, 16)], if $|\omega_{TO(\Gamma)} - \omega| > \Gamma_i/2$, then $d^2\sigma/d\Omega d\omega_s \propto \varrho_2(\omega)$, which explains the density of states-like sideband that persists at high pressure.

Shand et al. [8.83] applied a similar treatment to the pressure dependence of the anomalous TO(Γ) Raman line shape in CuCl. This is shown in Fig. 8.23. The anharmonicity is much larger for CuCl than for GaP, causing a greater distortion of the TO(Γ) peak. These authors calculated both the real and imaginary parts of the self-energy due to the interaction of TO(Γ) with an assumed two-phonon continuum having a M_3-type square root singularity. Their 7 parameter fits (see dashed curve in Fig. 8.23) reproduce the 40K line shape quite well at each pressure, and account for the induced decrease in the relative strength of the broad low-energy band compared to the sharp high-energy peak.

Despite the apparent success of this analysis, *Vardeny* and *Brafman* [8.84a] have maintained that CuCl is so anharmonic that a perturbation treatment is inappropriate. They proposed a disorder model in which the anharmonic potential for the Cu^+ ion has two types of minima. CuCl would then be an "alloy" with Cu^+ ions occupying both sites, giving two distinct systems of optical modes. This would account for the dual-peaked TO(Γ) Raman spectrum at $P=0$. It was argued that under pressure, increased repulsive forces would skew the population of sites against the broad low-energy peak, thereby explaining the observed intensity changes (Fig. 8.23).

In the past, pressure-Raman studies of frequency shifts have often taken precedence over measurements of (usually more subtle) line-shape changes. These examples illustrate that more work in this area could be quite fruitful. (For example, see the very recent work by *Schmeltzer* and *Beserman* [8.84b] invoking fourth order anharmonic coupling between 2TA(X) and TO-TA(X) to explain pressure-induced changes in the two-phonon Raman line shape and intensity of ZnSe).

8.5 Phase Changes

Under compression all solids eventually exhibit one or more phase transformations. These span the range from drastic changes of volume, symmetry and electronic properties to subtle shifts of lattice parameters. Raman scattering can help identify the symmetry of a high-pressure phase through the assignment of new structure. It can also investigate behavior antecedent to a transition by studying modes which soften under pressure. Here we discuss Raman investigations of pressure transformations in the tetrahedral semiconductors. We focus on two illustrative cases: the zincblende→rhombohedral→tetragonal changes of CuI, which offer an excellent example of structural identification, and the metallization transition in groups IV, III–V and II–VI materials, which demonstrates possible antecedent behavior.

Unfortunately, space does not permit us to discuss pressure-Raman studies of transitions in many ionic, ferroelectric and ferroelastic solids; the interested reader is referred to several excellent recent reviews [8.11 – 15].

8.5.1 Transitions in CuI

Often the pressure transitions in tetrahedral materials occur via atomic displacements along one of the initial high-symmetry directions. The resulting reduced-symmetry structures exhibit changes in the Raman selection rules determining the number of active modes. For example, one way (among others) to achieve the zincblende (zb)→NaCl structural change (known for many group II–VI materials [8.85, 86]) is to increase the cation-anion sublattice displacement along ⟨111⟩ to 1/2 of a cubic body diagonal. Whereas the zb phase had two Raman-active lines, the new NaCl phase has none, owing to the presence of inversion symmetry.

The structural sequence zincblende ($P=0$)→rhombohedral (14 kbar)→tetragonal (41 kbar) of CuI illustrates a case for which the Raman spectra change but do not disappear. Figure 8.24 shows the spectra corresponding to these structures and the pressure shifts of the observed peaks after *Brafman* et al.

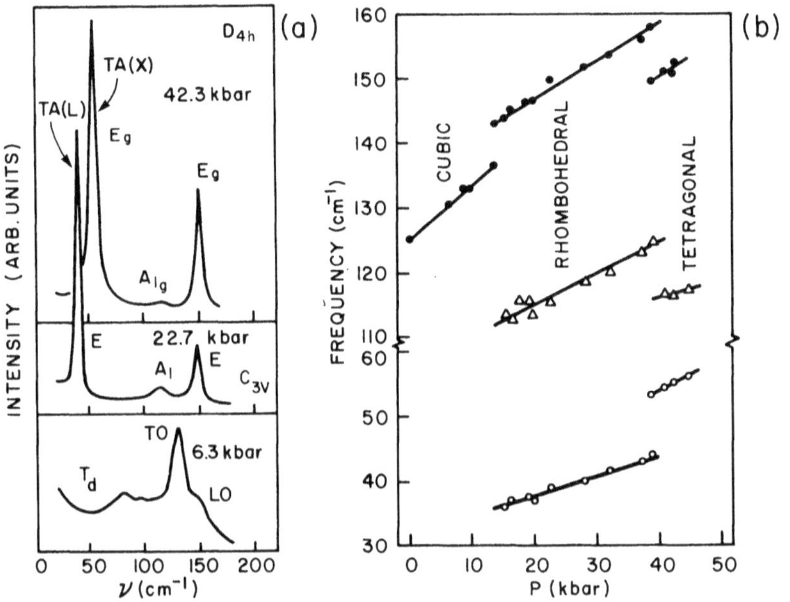

Fig. 8.24. (**a**) Raman spectra (300 K) of CuI in the zincblende (T_d), rhombohedral (C_{3v}) and tetragonal (D_{4h}) phases. TA(L) and TA(X) designate phonon origins in the zincblende BZ. (**b**) Frequencies of observed peaks versus pressure. After [8.58]

[8.58]. In that work, LO phonons could not be followed as a function of pressure because LO scattering was either hidden in the indicated shoulder for zb, or it was simply unobserved for the other phases.

The rhombohedral (trigonal) phase can be achieved by a deformation along the cubic (111) axis in which alternate planes of Cu atoms and alternate planes of I atoms are displaced. The new unit cell is a rhombohedron exactly twice the volume (scaled by the changed lattice parameter) of the original primitive rhombohedron of the fcc lattice. In particular the repeat distance along (111) is doubled. This transformation is depicted in Fig. 8.25a.

The doupling along (111) in real space implies a halving of the BZ along the corresponding (111) direction in reciprocal space such that the cubic L-point is brought to $\boldsymbol{q}=0$. We then expect four new $\boldsymbol{q}=0$ frequencies corresponding to TA(L), LA(L), TO(L) and LO(L) in the zb phase. Since there are now four atoms per unit cell, the trigonal symmetry (C_{3v}) mandates $3E+3A_1$ optic modes that are both Raman and ir active. The original cubic TO(Γ) and LO(Γ) account for $1E$ and $1A_1$, while the folded TA(L), TO(L), and LA(L), LO(L) phonons account for the other $2E$ and $2A_1$ modes, respectively. Given the nonobservation of weak LO scattering and the near degeneracy of TO(Γ) and TO(L) in the zb phase (known from $P=0$ inelastic neutron data [8.87]), three Raman peaks are expected in agreement with experiment (Fig. 8.24a) [8.58].

The relationship between the tetragonal and zb phases of CuI can be seen in Fig. 8.25b. There the primitive tetragonal cell of the high-pressure phase is

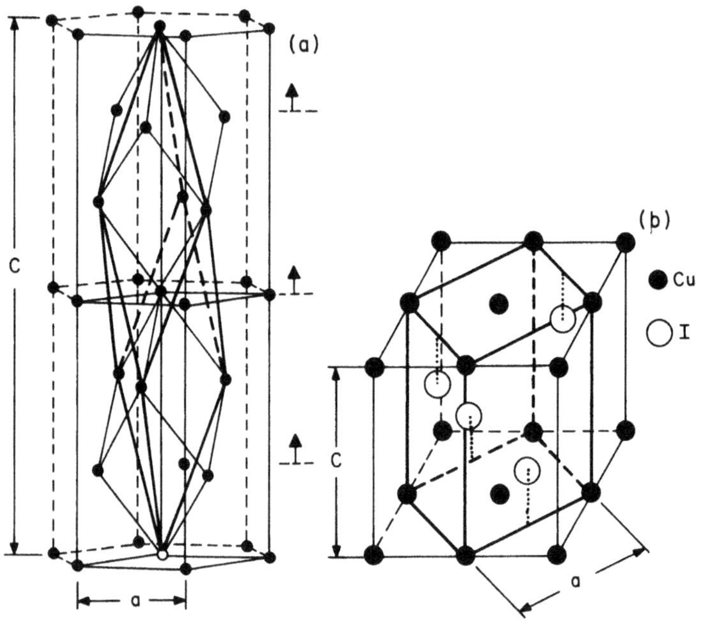

Fig. 8.25. (**a**) Rhombohedral cell for high pressure phase of CuI (*heavy lines*) in a hexagonal supercell (*thin lines*). Original *fcc* rhombohedrons (also *thin lines*) are shown. Arrows indicate displacement of atomic planes during transition. (**b**) Tetragonal cell for CuI (*heavy lines*) inside cubic zincblende cell. After [8.58]

inscribed in the unit cube of zb. The main change is a (001) displacement of the Cu atoms originally at the center of the side faces; in the new structure these atoms lie in the basal plane. The c/a ratio of the tetragonal cell changes little from its cubic value of $\sqrt{2}$. This is quite different from the tetragonal transition in Si and InSb to be discussed below (Sect. 8.5.2). The tetragonal cell also has twice the scaled volume of the primitive fcc rhombohedron. This time the BZ is halved along the (001) c-axis so that the corresponding cubic X-point is folded to $\boldsymbol{q}=0$.

With four atoms per tetragonal cell, the D_{4h} symmetry requires $1A_{1g}$, $1B_{1g}$, $2E_g$ (Raman active) and $1A_{2u}$, $1E_u$ (ir active) optical modes. The cubic LA(X), LO(X), TA(X), and TO(X) phonons transform as A_{1g}, B_{1g}, E_g, and E_g, respectively, while the original TO(Γ) and LO(Γ) modes acquire the ungerade label and become Raman inactive. Given again the nonobservation of LO scattering, three Raman peaks are expected and observed (Fig. 8.24a). They are all derived from phonons at the X-point of the fcc BZ.

In Fig. 8.24b the low-frequency modes corresponding to TA(X) and TA(L) have $\gamma_i>0$ for the rhombohedral and tetragonal phases of CuI, in exception to most other tetrahedral semiconductors (Tables 8.1, 2). *Brafman* et al. [8.58] concluded that $\gamma_{\mathrm{TA}(L)}$ and $\gamma_{\mathrm{TA}(X)}$ are probably positive in zb CuI also because no negative low temperature thermal expansion is observed. A reason for this may be unusually strong anharmonicity in copper halides [8.83, 84a]. In contrast, $\gamma_{\mathrm{TA}(L)}<0$ in the zb phases of CdS, ZnO [8.38, 43], ZnS [8.88] and AgI [8.89] was deduced from the observed negative γ_i of the low frequency E_2 mode in the corresponding wurtzite phases; likewise, $\gamma_{\mathrm{TA}(X)} < 0$ is implied by the negative γ_i measured for low frequency Γ_5 phonons in chalcopyrite structure $CuGaS_2$ and $AgGaS_2$ [8.90]. This raises the question as to what extent the γ_i of Raman forbidden BZ boundary phonons can be surmised from the γ_i of corresponding allowed BZ center modes in a lower symmetry phase of the same material. It is not clear that γ_i (or ω_i) can be extrapolated across a jump discontinuity (Fig. 8.24b) at a phase transition when the atomic positions (and therefore the atomic forces) change substantially, as for the zb→tetragonal transition of CuI.

8.5.2 Pressure-Induced Metallization – Possible Antecedent Behavior

The prototypical transformation of this class is the gray tin [semiconducting, diamond structure (fcc)] to white tin [metallic, body-centered tetragonal (bct)] transition, which actually occurs in tin at $P=0$ and $T=286$ K. In finite band-gap semiconductors, similar phase changes occur only at higher pressure [8.86, 91–93]. The transitions are first order with large volume decreases of 10–20 %. *Klement* and *Jayaraman* [8.85] have reviewed the complicated phase diagrams and structures that apply. For Si, Ge, and Sn (for diamond it is unknown), the final phase has the bct structure; it is metallic, due most likely to overlap of the X-point conduction band minimum (now folded back to Γ) with the valence band [8.94a, b, c]. For those III–V and II–VI materials that have been studied, the bct, NaCl, and several orthorhombic structures have all been reported at high pressure [8.85, 86]. Several of the more ionic compounds – those with Phillips' ionicity >0.3 – remain semiconducting (even if visibly opaque) after the lowest pressure phase change [8.95]; however, the transition still results in a reduced bandgap, signaling a pressure-induced approach toward the metallic state [8.94b, c].

The central question here is to what extent does the softening of BZ boundary TA phonons under pressure constitute antecedent behavior to this metallization transition? The pressure-Raman studies that bear on this issue are those of *Weinstein* and *Piermarini* [8.27] on Si and GaP, *Weinstein* [8.54, 61] on ZnS, ZnSe and ZnTe, *Trommer* et al. [8.42] on InP and GaAs, *Carlone* et al. [8.90] on $CuGaS_2$ and $AgGaS_2$, and *Olego* and *Cardona* [8.96] on Ge.

In [8.27] the phonon dispersion curves of Si and gray-Sn were compared at their respective transition pressures, $P_T \approx 125$ kbar and $P_T=0$. To do this the ω_i were scaled by the ion-plasma frequency appropriate for each material. The scaling insures that any remaining differences are mainly due to the valence electrons. It was found that at P_T the flat TA branch near the BZ boundary was rather similar in both materials, indicating parallel behavior of the valence electrons for these modes. However, a similar comparison of GaP to α-Sn and to InSb showed no analogous result.

An empirical linear relationship between $\gamma_{\mathrm{TA}(X)}$ and P_T that does apply to several compound semiconductors as well as to Si and Ge was first proposed by *Weinstein* [8.54]. Figure 8.26 shows this correlation for Si, GaP, ZnS, ZnSe,

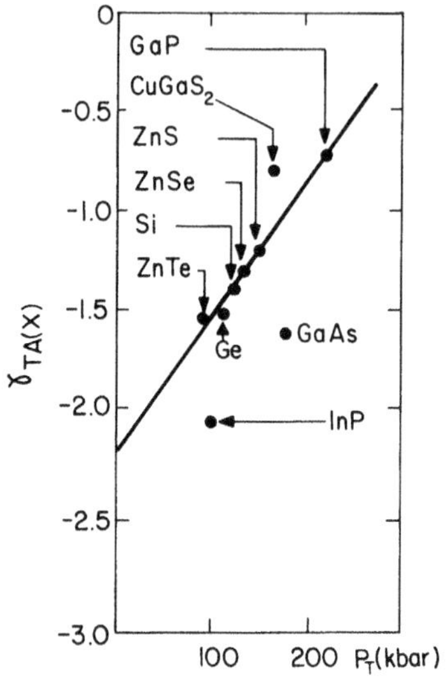

Fig. 8.26. Empirical relationship between the TA(X) mode γ and the metallization transition pressure P_T suggested by *Weinstein* [8.54] (see also [8.42a, 90, 96])

ZnTe [8.54], Ge [8.96], InP, GaAs [8.42, 97] and $CuGaS_2$ [8.90]. (For $CuGaS_2$ the lowest Γ_5-phonon was used; see Sect. 8.5.1). The linearity is remarkable for the first six solids, but an unshown point for $AgGaS_2$ ($\gamma_{\Gamma_5} \approx -4.5$, $P_T = 42$ kbar) falls far afield. Still the trend is clear: the faster the BZ boundary TA frequencies decrease with pressure, the lower is P_T. An alternative empirical correlation for these solids was subsequently suggested by *Carlone* et al. [8.90]. They found that the lowest pressure transition occurred when $\omega_{\mathrm{TA}(X)}$ (or its equivalent) had decreased to 0.7 times its $P=0$ value. Although no ω_i actually vanish prior to this first order transition, these correlations demonstrate some antecedent behavior which we now try to relate to the structural and electronic changes that occur.

The fcc→bct structural transition is depicted in Fig. 8.27 for InSb [8.98]. It corresponds to a contraction along the c-axis (001) and an expansion along the normal a-axes such that the bct cell becomes primitive with $c/a < \sqrt{2}$. The (scaled) first BZ is halved, and the X-point is folded back to Γ as for the tetragonal phase of CuI. *Musgrave* [8.99] pointed out that the $\boldsymbol{q}=0$ shear wave governed by $(c_{11}-c_{12})/2$ could drive this distortion. However, *Demarest* et al. [8.100] discounted a simple strain mechanism because large electronic changes (i.e., metallization) are involved. Although the γ_i corresponding to $(c_{11}-c_{12})/2$, from (8.11), are generally small and sometimes negative [8.60], BZ boundary TA phonons have by far the largest negative γ_i (Table 8.1). This indicates that short wavelength TA modes are the primary source of pressure-induced instability in the diamond and zb lattices as suggested by *Phillips* [8.101]. Because the TA branch is flat near X, *i.e.*, $\nabla_{\boldsymbol{q}}\omega_{\mathrm{TA}(X)} \approx 0$, the TA($X$) mode sets up an approximate standing wave pattern. Since the group of the X-point in diamond and zb (D_{4h} and D_{2d}) exhibits tetragonal symmetry, this pattern is compatible with a $\boldsymbol{q}=0$ optical phonon in a tetragonal lattice. The observation $\omega_{\mathrm{TA}(X)}$ (at P_T) $\approx 0.7\,\omega_{\mathrm{TA}(X)}$ (at $P=0$) shows that the standing wave period undergoes a 30% increase by P_T. Nevertheless, the corresponding distortion

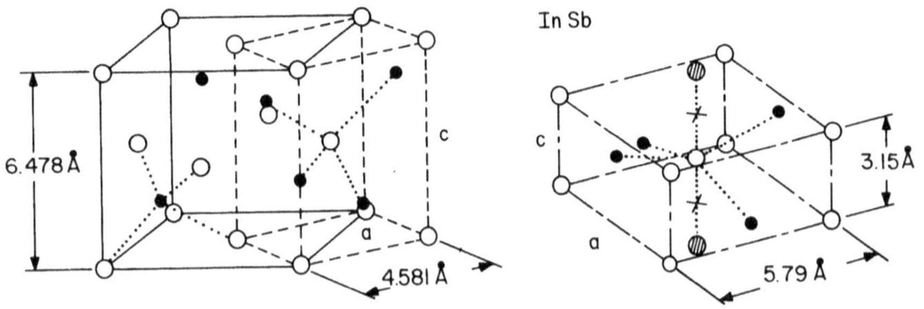

Fig. 8.27. Cubic-to-tetragonal structural change that accompanies the metallization transition in InSb. The nonprimitive bct cell (*dashed lines*) in the fcc lattice becomes primitive after the indicated distortions. After [8.98]

remains dynamic, and the mechanism by which the unstable BZ boundary TA modes are linked to the static structural change is not understood.

The connection between metallization and the softening of BZ boundary TA phonons is likewise unclear. Some insight is provided by the bond-charge model. In *Weber*'s [8.74] treatment, the TA(X) mode corresponds to the ions vibrating within a nearly rigid lattice of *bond charges*. Under pressure, $\omega_{\mathrm{TA}(X)}$ decreases because ion bond-charge forces decrease, so that the valence electrons (within the bond charges) become increasingly *decoupled* from the ions in this mode. Within the dielectric matrix formalism [8.65–67] this is caused by a decrease in the relative magnitude of off-diagonal versus diagonal screening. This corresponds to metallization since diagonal screening is strongly predominant in metals.

In principle, the DFP method [8.50, 68, 69, 77] should be able to elucidate any antecedent role for BZ boundary TA modes in the metallization transition. Nevertheless, a clear picture has not emerged. *Wendel* and *Martin* [8.77] showed for Si that $\gamma_{\mathrm{TA}(X)} < 0$ resulted from the decreasing importance with pressure of electron-ion and electron-electron energies relative to ion-ion terms. *Maschke* and *Andreoni* [8.69] found that the decrease of the electron-ion interaction caused the transition in ZnSe – this time to a semiconducting NaCl structure. *Yin* and *Cohen* [8.68] demonstrated that the high-pressure structures of Si and Ge should be bct and closely predicted P_T; they also calculated reasonable negative values for $\gamma_{\mathrm{TA}(X)}$ (Sect. 8.3.4). However, in each treatment the relationship between the decrease of BZ boundary TA frequencies in the initial phase and the structure, electronic character and onset P_T of the final phase was not explored. Some progress was made in a recent calculation for GaAs by *Kunc* and *Martin* [8.50]. They showed that the total energy as a function of the TA(X) eigenvector amplitude acquired a new minimum at elevated pressure, but they did not examine the uniqueness of this minimum. Also the most recent work of *Cohen* et al. [8.94a, 8.94b] has produced some qualitative insights in terms of pressure-induced changes in the competition between the covalent-bond energy that favors the semiconducting phase, and the Ewald energy (core-core Coulomb interaction) that favors the metallic phase. However, more work along these latest directions is needed to understand the relation to TA(X) mode softening.

8.6 Pressure-Tuned Resonant Raman Scattering

Usually resonant Raman scattering is observed by tuning the incident laser frequency ω_l through various bandgaps which act as poles for the Raman cross section according to (8.2, 5). Alternatively, the bandgap can be tuned through ω_l (or ω_s) by some perturbation. Pressure tuning is an attractive technique for semiconductors in frequency regions not spanned by conventional lasers because the bandgap pressure coefficients are often large – typically ~ 10 meV/kbar. A pressure of 100 kbar can tune a gap through 1.0 eV (~ 2000 Å in the visible). Balancing this advantage is the difficulty of controlling factors that determine the observed intensity during a pressure change (e.g., window and medium depolarization and refractive index, sample orientation, etc.).

There have been very few resonant Raman experiments employing hydrostatic pressure. Our goal here is to illustrate the power of this technique in the hope that it will be more frequently applied. We shall discuss the most studied case, the E_0 gap resonance in GaAs [8.44, 97]. Uniaxial stress has more often been used in resonance measurements, as for the E_1-gaps of Ge, InSb [8.102] and

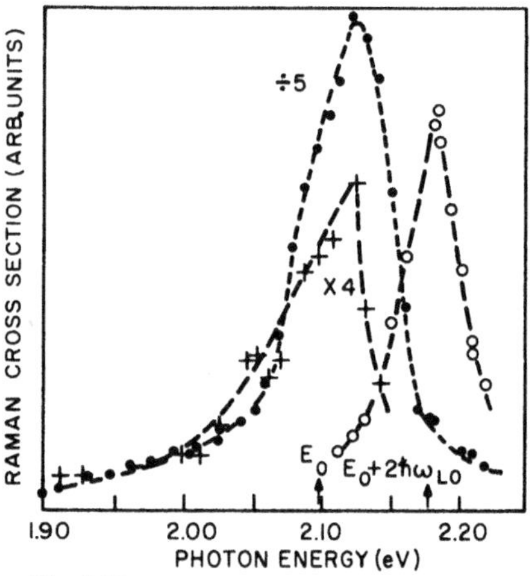

Fig. 8.28

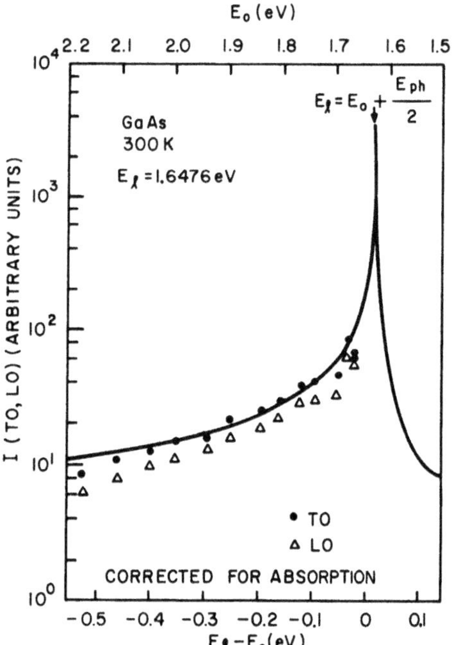

Fig. 8.29

Fig. 8.28. Raman cross section for TO(Γ) (*crosses*), LO(Γ) (*solid circles*) and 2 LO(Γ) (*open circles*) in GaAs at 62 kbar (300 K) versus incident laser energy. After [8.97]

Fig. 8.29. Resonance profiles for TO(Γ) and LO(Γ) near E_0 in GaAs. *Solid curve* was calculated using dielectric response theory. After [8.44]

InAs [8.103]. In keeping with the design of our chapter, we do not discuss these, but refer the reader instead to the recent review by *Anastassakis* [8.104].

First we note that in both [8.44, 97], pressure was exploited to make GaAs transparent to the visible laser frequencies used. This increased the Raman signal (over the diamond-cell background) to acceptable limits by virtue of increased sample penetration depth. (The direct E_0 bandgap of GaAs occurs at 1.43 eV at $P=0$, $T=300$ K, and it shifts by +11.7 meV/kbar [8.105]). This technique could be used to similar advantage for other opaque semiconductors.

Yu and *Welber* [8.97] combined both pressure and frequency (dye laser) tuning to study the GaAs E_0 resonance. Figure 8.28 shows their resonance profiles at 62 kbar for TO(Γ), LO(Γ) and 2LO(Γ). No attempt was made to analyze the profiles in terms of the dielectric response theory, see (8.2, 5), and the known band structure. Combined tuning allowed these authors to measure the pressure shifts of the profile peaks; these shifts were shown to correspond closely to that of the E_0 bandgap. Thus the shift of an *electronic energy* was found from pressure measurements of *phonon* Raman *intensities*.

Trommer et al. [8.44] did study the resonance profile of TO(Γ) and LO(Γ) in the vicinity of E_0 for GaAs. Their data, obtained by pressure tuning with the laser fixed at $E_l=1.648$ eV, are shown in Fig. 8.29. The fitted solid curve was calculated including 2-band (as in Fig. 8.2), 3-band (similar to Fig. 8.2, but the phonon mixes spin-orbit-split valence band states at $\boldsymbol{q}\approx 0$), and nonresonant contributions to the Raman tensor. This is equivalent to summing terms as in (8.5) over the allowed electronic transitions near the E_0 gap [8.2]. The good agreement indicates that pressure did not substantially alter the electron-phonon coupling strength (i.e., the deformation potential) responsible for one-phonon scattering.

8.7 Molecular Solids

8.7.1 Rationale for Pressure-Raman Studies of Molecular Crystals

Because the forces which bind them are so weak, molecular solids are very soft. Because they are so soft, pressure is ideally suited as an investigate probe for these solids. It is practical to attain very substantial (>20%) compressions of the crystal volume and to cause thereby very large (>50%) changes in phonon frequencies. Moreover, because it primarily influences the mushy intermolecular volume, *pressure selectively enhances those effects which are specifically associated with the interactions between molecules.*

The simplest conceivable vibrational model of a molecular solid is illustrated in Fig. 8.30. This elementary figure will serve to introduce a few overall aspects of phonon spectra in molecular solids. Later, with the superposition of a simple ansatz to allow for the anharmonicity of the bonds, this simple model will also provide a framework for understanding the overall effect of pressure on the mode frequencies [8.106].

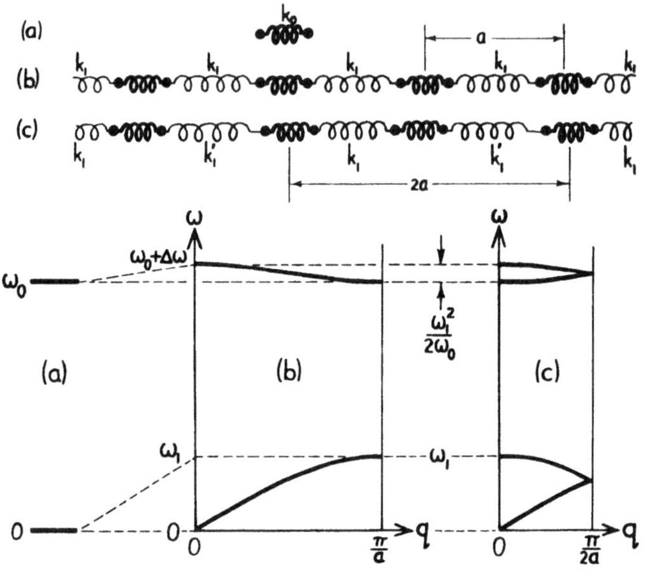

Fig. 8.30a–c. Elementary vibrational models for molecular crystals: (**a**) free molecule, (**b**) linear lattice with one molecule per unit cell, (**c**) dimerized lattice with doubled unit cell. The corresponding phonon dispersion curves are displayed in the lower part of the figure. After [8.106]

Figure 8.30 displays three one-dimensional spring models and their corresponding vibrational spectra (phonon dispersion curves). Figure 8.30a represents a single *isolated* diatomic molecule (in a one-dimensional space) with atoms of mass m held to their equilibrium separation by a spring of force constant k_0. The corresponding vibrational spectrum contains just two frequencies since there are but two degrees of freedom: $\omega=\omega_0$, where $\omega_0=(2k_0/m)^{1/2}$ is the frequency of the *stretching mode*, and $\omega=0$, the vanishing frequency of the rigid-molecule *free translation*. Figure 8.30b depicts a simple linear lattice (one molecule per unit cell) of periodicity a in which many such molecules are *weakly coupled* by a set of soft intermolecular springs of force constant k_1 ($k_1 \ll k_0$). For this condensed phase, the simplest model of a molecular solid, the discrete two-line $(0, \omega_0)$ spectrum of the isolated molecule is replaced by a continuous spectrum composed of two narrow and well-separated bands. These correspond to the two phonon branches indicated in (b) in the lower part of Fig. 8.30; $\omega(q)$ in the right half $(0<q<\pi/a)$ of the one-dimensional Brillouin zone is shown for each branch. The $\omega=0$ free translation of the isolated molecule gives birth to a low-lying acoustic branch of phonon modes in the crystal, rising up to encompass a low-frequency regime from 0 (at $q=0$) to ω_1 (at $q=\pi/a$). The ω_0 stretch vibration of the free molecule gives rise to a flat optical-phonon branch spanning the narrow frequency range from ω_0 (at the zone boundary) to $\omega_0+\Delta\omega$ (at the zone center).

The stretch frequency ω_0 sets the scale for the spectrum of this simple model of a molecular crystal. In the weak-coupling regime $k_1 \ll k_0$, the ratios of the characteristic frequencies of this elementary model are given by

$$\omega_1/\omega_0 = (k_1/k_0)^{1/2}, \tag{8.17}$$

$$\Delta\omega/\omega_0 = (k_1/k_0)/2. \tag{8.18}$$

These relations reveal the way in which both the flatness ($\Delta\omega \ll \omega_0$) of the optical-phonon branch and the lowliness ($\omega_1 \ll \omega_0$) of the acoustic-phonon branch depend upon the weakness ($k_1 \ll k_0$) of the intermolecular bonding. Consider $\varepsilon \equiv (k_1/k_0)^{1/2}$ as a parameter of smallness in the weak-coupling limit. Equations (8.17, 18) show that ω_1 is of first order in ε while $\Delta\omega$ is of second order.

Real molecular crystals are, of course, far more complex than the oversimplified picture of Fig. 8.30b, but many essential aspects are captured by the model. It is useful to introduce the next stage of complication, shown in Fig. 8.30c, in which the unit cell is doubled in size so as to include two molecules. Introduction of this modest amount of complexity is important because it allows us to make contact with spectroscopy, since the unit-cell doubling places the coupling effects (ω_1, $\Delta\omega$) at the zone center (where they must be in order to be accessible to first-order Raman or far-infrared experiments). In the model of Fig. 8.30c, the pairing-off ("dimerization") is accomplished in the gentlest way by allowing alternate intermolecular force constants to adopt a value k_1' slightly different from k_1.

The situation depicted in Fig. 8.30c, in which a wide disparity exists between the upper set of crystalline frequencies near $\omega \approx \omega_0$ and the lower set of frequencies at $\omega \leq \omega_1$, corresponds to the validity of the *separation approximation* [8.107]. In this approximation, which provides the lowest-order view of a molecular crystal, the intermolecular forces are so weak relative to the intramolecular (covalent) forces that there is a very clear distinction between the high-frequency *internal modes* or *intramolecular modes* (or, simply, *molecular modes*) on the one hand and the low-frequency *external modes* or *intermolecular modes* or *lattice modes* on the other. (The internal-mode/external-mode terminology is the most widely-adopted usage for this dichotomy and will usually be employed here. The lattice-mode usage is misleading, and will be avoided). For external modes, the separation approximation is equivalent to the rigid-molecule approximation. The intermolecular forces are regarded as too weak to appreciably deform the internal structure (covalent bond length, bond angles) within the molecule. Within the context of Fig. 8.30c, k_0 is regarded as infinitely stiff and the molecule may be treated as a single particle of mass $2m$ for the external mode motions described by the two lower ($\omega \leq \omega_1$) branches. The maximum external-mode frequency ω_1 occurs at $q=0$ when nearest-neighbor intermolecular (k_1, k_1') springs are stretched/compressed in opposite phase to each other. In our simple model, the two bands below ω_1 correspond, for a real molecular crystal, to acoustic phonons and the lowest-lying optical phonons which are constructed of motions similar to the translations and rotations of the free molecule.

A comparison between the $q=0$ phonon spectrum of Fig. 8.30c and the noninteracting-molecule spectrum of Fig. 8.30a thus reveals the main qualitative spectral features which appear when the intermolecular interaction is turned on. In addition to the very low-frequency external modes (ω_1) discussed above, *Davydov-splitting fine structure* ($\Delta\omega$) is imposed on the intramolecular modes (ω_0). Splittings such as this, in which the degeneracy of corresponding excitations on a set of equivalent molecules (equivalent in the context of the crystal-symmetry factor group) is lifted by their mutual interaction, are characteristic of molecular crystals. Variously referred to as *Davydov* or *factor-group* or *correlation-field* splittings, they are simply (as in Fig. 8.30) the classical frequency differences between normal modes of a set of weakly-coupled oscillators. For some situations, it is possible to make a distinction between Davydov splittings and *crystal-field* splittings in which molecular degeneracies of a given molecule are lifted by the low symmetry of the crystalline environment. Such a distinction is a rather artificial one because the same intermolecular interactions provide the root cause of both effects.

One principal rationale for a program of pressure-Raman studies of molecular crystals is provided by the fact that pressure enhances the above-mentioned spectroscopic consequences of intermolecular interactions in the solid state. This effect of pressure is sketched in Fig. 8.31 in a schematic line-spectrum illustration for Raman-active and infrared-active phonons. These vibrational consequences of the condensation into the solid state include the lifting of the free-molecule translations and rotations from zero frequency into the finite-frequency bands of acoustic phonons and (for crystals with several molecules per unit cell – the usual case) low-lying optical phonons, as well as the subtler shifts and splittings of the molecular modes as they develop into the narrow high-frequency optical-phonon branches. With the application of hydrostatic pressure, as indicated in Fig. 8.31, these intermolecular-induced

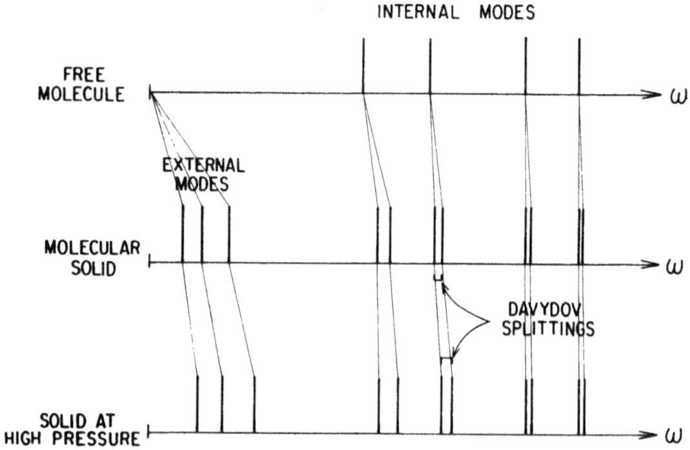

Fig. 8.31. Schematic line-spectrum illustration of the pressure-induced enhancement of intermolecular-interaction effects in a molecular crystal. After [8.108]

effects are enhanced and play a more important role in determining the form of the phonon spectrum.

The value of pressure-Raman experiments for revealing information about intermolecular forces and about the entire hierarchy of interactions which coexist in molecular solids, has been demonstrated in recent studies [8.106, 108–111]. This utility of pressure as a probe of bonding in soft solids provides the main focus of this section of the chapter; it will be discussed in detail in Sect. 8.7.3. An important aspect of this use of pressure-Raman studies is the revelation that the traditional Grüneisen relation which connects phonon frequency with crystal volume breaks down in a dramatic and systematic fashion for molecular solids [8.106, 108] and must be replaced by a more fundamental bond-stiffness/bond-length scaling law.

The search for new solid phases at high pressure is another rationale for pressure-Raman studies on molecular crystals, one which continues the proud tradition pioneered by the vastly extensive volume-versus-pressure measure-

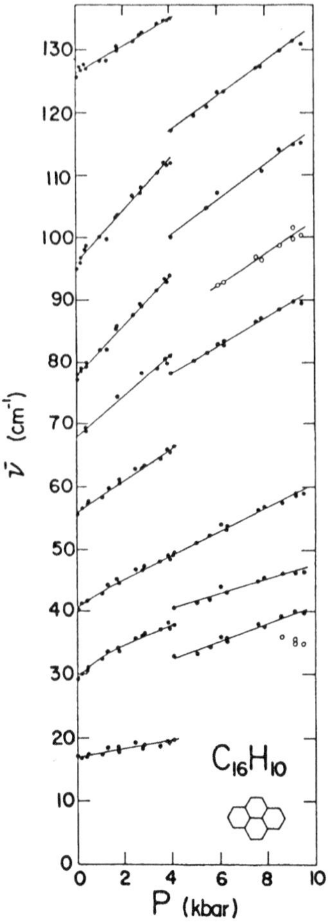

Fig. 8.32. Pressure dependence of Raman-active external-mode frequencies in crystalline pyrene. A first-order transition intervenes at 4.0 kbar. After [8.113]

ments of *Bridgman* [8.112]. Because the forces which are responsible for holding them together are so gentle, molecular solids often undergo solid-solid polymorphic transformations at quite modest pressures. Raman-scattering measurements are very effective in uncovering such phase transitions; an example [8.113] is shown in Fig. 8.32. Pyrene is an organic solid composed of planar ring-system ("aromatic") molecules. The chemical formula and carbon-atom skeleton of the individual molecule is shown at the lower right of the figure; it may be viewed as a graphite fragment bounded by hydrogen on the perimeter. Pyrene is interesting because it is the simplest of a class of organic molecular crystals in which the basic crystallographic building block is a closely-spaced parallel pair (dimer) of planar molecules. The pressure-Raman results shown in Fig. 8.32, along with analogous results obtained with temperature (rather than pressure) as the variable parameter, provided the first direct evidence of a subtle solid-state transition in this material. This figure follows the frequencies, as a function of pressure, of the Raman-active external modes. The discontinuous transformation of the lattice phonon spectrum, observed at 4 kbar, attests to the occurrence of a first-order phase transition. The subtlety of the transition may be appreciated by the fact that dimer pairs survive the phase change, which evidently corresponds to a slight sliding relative displacement of the two molecules within each dimer along with a change in the dimer-dimer conformation. Such a gentle rearrangement of molecules, although nearly indiscernible in the internal-mode high-frequency regime, is easily observed (as shown in Fig. 8.32) when the external-mode frequencies are tracked as a function of pressure.

Pressure-induced molecular-rearrangement solid-solid transitions in molecular crystals, such as the one exhibited by pyrene, are too ubiquitous to cover in this review. (*Bridgman* [8.112] long ago pointed out the high frequency of the occurrence of transformations in this class of solids). Instead, Sect. 8.7.5 will

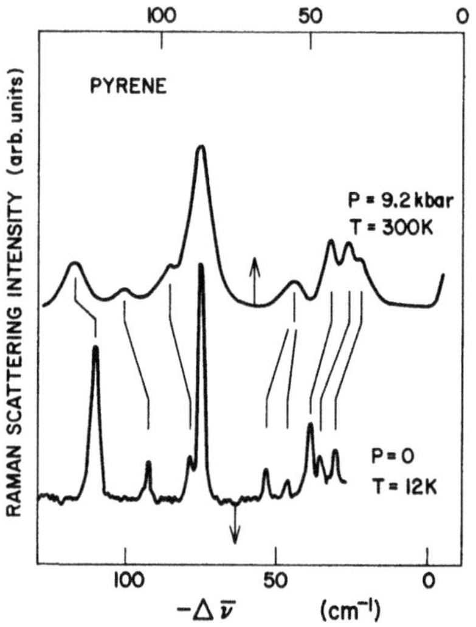

Fig. 8.33. Comparison of the resolution-enhancing aspects of high pressure and low temperature on the external-mode Raman spectrum of pyrene II. After [8.113]

discuss experimental results which bear on the general question of a crossover from molecular to nonmolecular solid-state properties at high pressure.

Before leaving the example provided by pyrene, we show in Fig. 8.33 another illustrative aspect of that study [8.113]. Figure 8.33, which proved the equivalence of the high-pressure form with the form obtained (at $P=0$) at low temperature, displays the correspondence between the high-pressure and low-temperature spectra. At high pressure, the Raman lines are spread over a greater spectral range than at low pressure. Note the low-frequency triplet resolved in both spectra of Fig. 8.33 at around 30 cm^{-1}. At room temperature and low pressure, this triplet is not resolved. High resolution at low temperature is a familiar spectroscopic approach in which the enhanced resolution is achieved primarily via line narrowing at constant splitting. As shown in Fig. 8.33, *high pressure can provide a complementary approach in which enhanced resolution is achieved via increased splitting at constant linewidth.* Other examples of this useful feature of pressure-Raman experiments will appear in the following section. Note that we are dealing throughout this chapter with *isotropic* (hydrostatic) pressure, not with uniaxial stress which is often associated with symmetry-breaking splittings.

A survey of the effect of pressure on the Raman spectra of some simple organic and inorganic molecular solids is presented next in Sect. 8.7.2. The systematics of the response to pressure, and its implications with respect to the extended hierarchy of interaction strengths which coexist in a molecular crystal, are analyzed in Sect. 8.7.3. Here, reference is again made to the simple (but useful) picture of Fig. 8.30 in order to help develop the idea of a bond-stiffness/bond-length scaling law which helps explain the dramatic deviation of molecular solids from the Grüneisen behavior observed for network solids such as the germanium family of semiconductors. In Sect. 8.7.4 we use the pressure-Raman results to dissect the effect of *temperature* into volume-driven ("implicit") and phonon-occupation driven ("explicit") components, and we compare the explicit/implicit mix for modes in molecular crystals with the mix characteristic of covalent and ionic crystals. Section 8.7.5 discusses the search for molecular→nonmolecular transitions at high pressure and touches briefly on a few exotic topics such as the quest for metallic hydrogen and dimensionality (e.g., $2d \rightarrow 3d$) effects.

8.7.2 Pressure-Induced Raman Line Shifts in Simple Organic and Inorganic Molecular Solids

The organic crystal pyrene ($C_{16}H_{10}$), introduced in Figs. 8.32, 33 to illustrate two aspects of pressure effects in molecular solids, is one of a large group of hydrocarbon molecular crystals in which the molecule is built of linked benzene rings. Benzene itself (C_6H_6) is the most basic molecular unit in the aromatic series of organic molecular materials. Figure 8.34 displays the pressure-Raman results of *Ellenson* and *Nicol* [8.114] on crystalline benzene at room temperature.

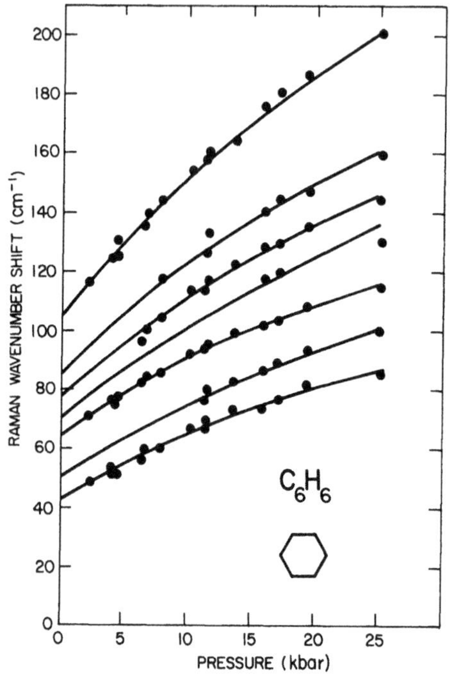

Fig. 8.34. Plot of the wave numbers of the Raman-active lattice bands of benzene I versus pressure. After [8.114]

Shown in Fig. 8.34 are the wave numbers of the observed Raman-active bands of benzene in the external-mode regime, followed up to 25 kbar. Note immediately that this pressure is sufficient to *double* the intermolecular-mode frequencies (i.e., *quadruple* the corresponding force constants). Contrast this enormous effect to that seen in diamond, for which the effect of 25 kbar on the single Raman band is to increase its frequency by a minuscule amount, about one half of one per cent [8.115]. Although this comparison is a bit extreme because, among the hard covalent crystals, diamond is the hardest, it nevertheless conveys a sense of the relatively great sensitivity to pressure of phonons in molecular crystals.

Crystalline benzene has four molecules per unit cell ("unit cell" refers throughout to the primitive unit cell, the smallest translational unit) and each molecule sits on a center of symmetry. The Raman-active modes are therefore the 12 rotational modes (rotational modes are sometimes called librational modes, or *librons*). Note that, in Fig. 8.34, only three lines are resolved at low pressure. As the spectrum spreads out with increasing pressure, more lines are seen via the previously-noted enhanced-resolution aspect of pressure-Raman work. In this study [8.114], a new line was uncovered at high pressure which was never seen earlier (even at low temperature) at $P=0$.

Thus far, organic crystals have been discussed; we now move on to inorganic molecular crystals. A prototypical example is orthorhombic sulfur, an *elemental* molecular crystal composed of eight-atom rings. Unlike the ring molecule in

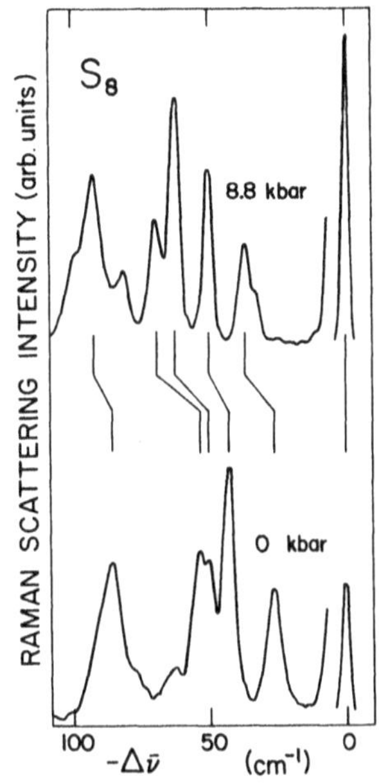

Fig. 8.35. Low-frequency portion of the sulfur Raman spectrum at $P=0$ and at 8.8 kbar. The pressure-induced increases in phonon frequency are indicated for several of the lines. After [8.106]

benzene, the S_8 ring is not planar but is puckered. There are four S_8 rings per unit cell. Figure 8.35 shows the low-frequency region of the Raman spectrum of crystalline sulfur at zero pressure and at a pressure close to 9 kbar [8.106]. Note the two lines near 50 cm^{-1}, barely resolved at $P=0$, are cleanly separated at 9 kbar. Pressure also reveals that the lowest-frequency external-mode band is actually a doublet.

Its extreme chemical simplicity, with only a single type of atom present, has elicited for rhombic sulfur the most thorough *theoretical* treatment of the effect of pressure on the vibrational spectrum of any molecular crystal. *Kurittu* [8.116] has carried out detailed calculations for this crystal, employing a "6–8-exp" atom-atom potential for the van der Waals intermolecular interactions. His analysis included a deformable-molecule model for the S_8 rings. (The latter refinement is necessary for a realistic treatment of sulfur because the rigid-molecule separation approximation is *inadequate* for this crystal: in Fig. 8.35, the modes seen near 85 cm^{-1} have *mixed* internal-mode/external-mode character). The results of these calculations have been compared to the pressure-Raman data of [8.106] and were found to yield satisfactory agreement with experiment.

Lesar et al. [8.117] have succeeded in performing Raman experiments to very high pressure on a very simple molecular crystal, solid nitrogen (N_2). The low-frequency region containing the external modes could not be seen in their experiments, but internal modes (with mode eigenvectors made up of N–N stretching motions of the N_2 molecules) *were* observed and their frequencies are shown in Fig. 8.36. Note the relatively slight influence which pressure has on

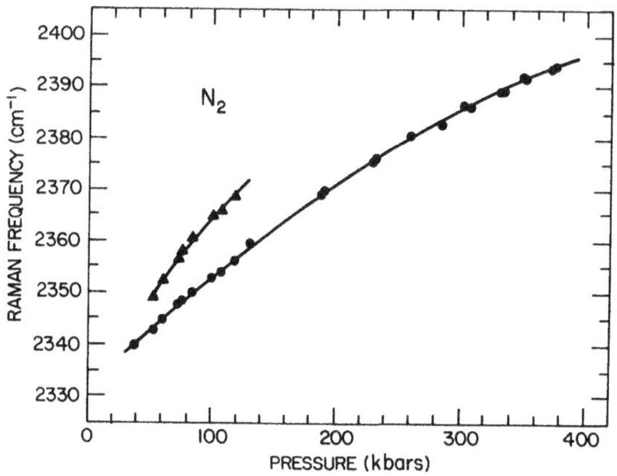

Fig. 8.36. The pressure dependence of internal-mode frequencies in crystalline N_2. After [8.117]

internal-mode frequencies. Over part of the pressure range, a Davydov doublet is observed whose separation (N_2-N_2 intermolecular interaction splitting) increases with pressure.

At this point, a few comments are in order about experimental techniques which have gone unmentioned thus far in our survey of results on molecular crystals. The spectroscopically-accurate but relatively low-pressure data of Figs. 8.32, 33, 35 were obtained with a Bridgman-type optical bomb in conjunction with a Daniels-type single-stage hydraulic press [8.29]. The data of Fig. 8.34 were obtained with a Drickamer-type optical cell [8.118]. The impressive measurements represented in Fig. 8.36 were made possible by recent advances in the use of the gasketed, ruby-calibrated, diamond-anvil cell (DAC) technique. The rest of the experimental results to be shown in subsequent figures have all been obtained by various versions of the DAC technique.

Thus far, we have given a sampling of some of the interesting recent results for this class of solid. In order to provide a degree of insight into how pressure acts upon the vibrational spectrum of a molecular solid and how this action differs significantly from that discussed earlier for covalent solids, it is necessary to pause in our survey and pick a specific material to discuss in some detail. A well-chosen concrete example allows us to display the salient points and also serves to lead into the physical model for these effects which is discussed in the following section.

The crystal chosen to serve this purpose is the molecular chalcogenide As_4S_4; its unit-cell structure is shown in Fig. 8.37. Over 30 phonon frequencies have been tracked under pressure for this inorganic molecular crystal, providing one of the most complete pictures thus far available for the effect of compression on the vibrational spectrum of *any* solid [8.108]. Within each As_4S_4 molecule, the bonding topology is readily understood and is concordant with the 8-*n* rule for covalent coordination. Each sulfur atom is bonded to two nearest-neighbor arsenics while each arsenic is bonded to three nearest neighbors, two of them sulfurs and the third an arsenic. The role of the As–As covalent bonds in the intramolecular structure is to cross-brace the As_4S_4 molecule to form a three-

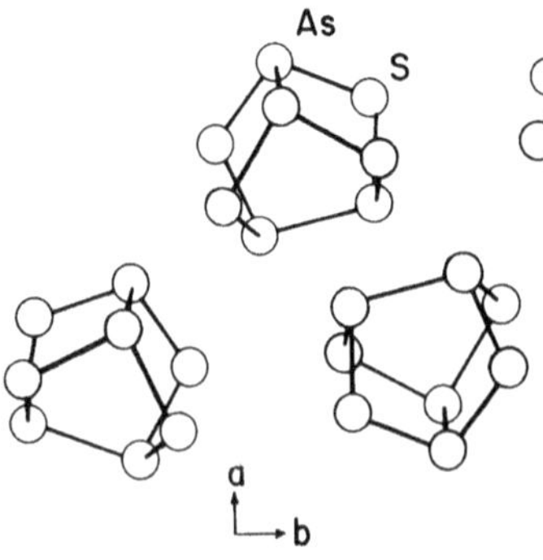

Fig. 8.37. Unit-cell structure of crystalline As_4S_4. Within each tightly-bound cage-like molecule, each As is covalently bonded to three neighbors and each S to two. When the atoms are fleshed out with their van der Waals radii, the As_4S_4 molecules appear nearly globular in shape

dimensional *cage-like* structure. This ensures a high degree of intramolecular rigidity; unlike sulfur with its open ring-like S_8 molecules, the rigid-molecule separation approximation is highly appropriate for crystalline As_4S_4.

As indicated in Fig. 8.37, there are four As_4S_4 molecules per unit cell. While this level of unit-cell complexity (32 atoms/cell) is not untypical for a molecular crystal, it is of course much more complex than the covalent semiconductors of Sect. 8.3. The following point must be strongly emphasized: *in contradiction to the conventional wisdom which holds that physicists must study simple systems to make progress, the unit-cell complexity of crystalline* As_4S_4 *is, in the present context, a blessing and not a curse*. The reason that this complexity is so welcome here is that, in combination with the low crystal symmetry ($P2_1/n$) and consequent generous selection rules for Raman activity, it allows us to monitor many modes in our pressure-Raman experiments. Direct optical access (via first-order Raman scattering) to numerous phonons of diverse frequencies distributed throughout the spectrum enables us to uncover global aspects of the response to pressure, as will be discussed in the following section.

The rich Raman spectrum of crystalline As_4S_4, at $P=0$ and at $P=39$ kbar, is shown in Fig. 8.38 [8.108]. These data, taken with the sample in a Bassett-type DAC [8.119] and using a near-infrared incident beam to which the crystal was transparent (the optical absorption edge of As_4S_4 is about 2.5 eV), show that quite clean spectra can be obtained with the anvil-cell technique even for low-frequency external-mode phonons. Unlike the spectra shown earlier, the spectra of Fig. 8.38 are complete in that they include both the external-mode and internal-mode regimes.

At $P=0$, the external modes in As_4S_4 fall at frequencies below 65 cm^{-1}. The internal modes fall into two clusters near 350 and 200 cm^{-1}, which may be roughly classified as intramolecular (covalent) bond-stretching and bond-bending modes, respectively. Some of the pressure-induced frequency shifts are indicated by horizontal arrows in Fig. 8.38, and the zero-pressure positions of the lines are given in parentheses above the frequency values observed at 39 kbar.

By far the most dramatic changes are those seen in the intermolecular regime. At 39 kbar, nearly all of the external modes have increased in frequency by about

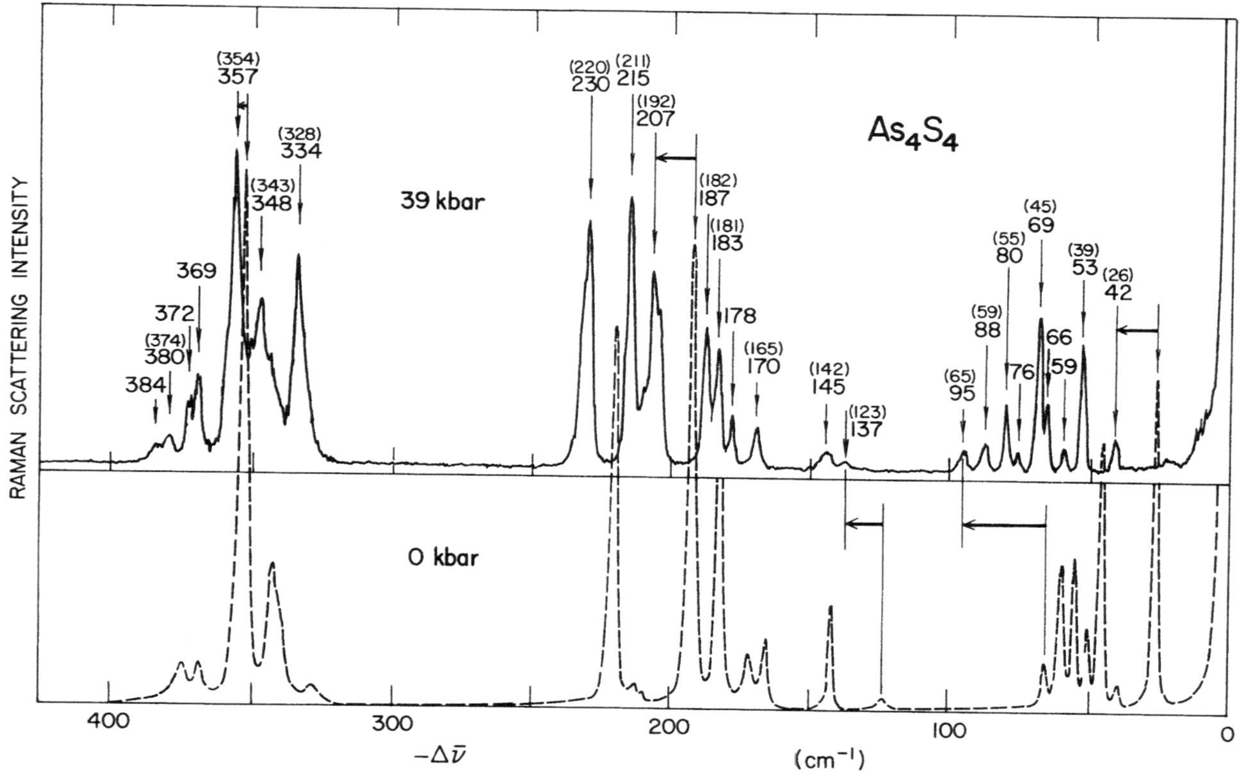

Fig. 8.38. The effect of pressure on the Raman spectrum of crystalline As_4S_4. For a few lines, the pressure-induced frequency shift is indicated by a heavy horizontal arrow. Line positions at 39 kbar are given in cm^{-1}, with the corresponding zero-pressure positions given in parentheses. After [8.108]

50 %. A by-now familiar consequence of this spectral expansion of the lattice-phonon regime is that new lines, hidden at $P=0$ by their proximity to stronger bands, are uncovered at high pressure. The response to pressure of the internal modes is markedly milder. The fractional increase for the bond-bending modes near 200 cm^{-1} is typically 5 %, and for the bond-stretching modes near 350 cm^{-1} the shift is a mere per cent or two.

These changes of frequency with pressure – very swift for the external-mode phonons, sluggish for the lower-frequency group of internal-mode phonons, and very sluggish for the upper group of internal-mode phonons – conspire to close the gaps in the spectrum. In particular, the important gap which separates the intermolecular from the intramolecular vibrations is strongly affected as the highest external mode rapidly gains on the lowest internal mode. As discussed in the following section which continues the analysis of these data, this gap closing under pressure is naturally interpreted in terms of the action of pressure in *selectively* stiffening (via bond shortening) the softest springs in the system. Pressure makes a molecular crystal *less molecular* in character by acting to diminish the intramolecular/intermolecular force-constant disparities.

8.7.3 Vibrational Scaling and the Systematics of the Response to Pressure

The pressure dependence of the phonon spectrum of As_4S_4, as exhibited in Fig. 8.38, is vastly different from the behaviour seen in network solids such as the covalent semiconductors. It is well to step back briefly and mention the usual expectation for the effect of pressure. First recall that, in the harmonic approximation, there is no effect whatsoever; ideal springs shorten, but do not stiffen, under compression. But because of the anharmonicity of the potential acting on atoms in real crystals, phonon frequencies do of course change with pressure. The simplest description for this is in terms of the Grüneisen approximation [8.19, 20] and a simple sketch of this picture is shown in Fig. 8.39.

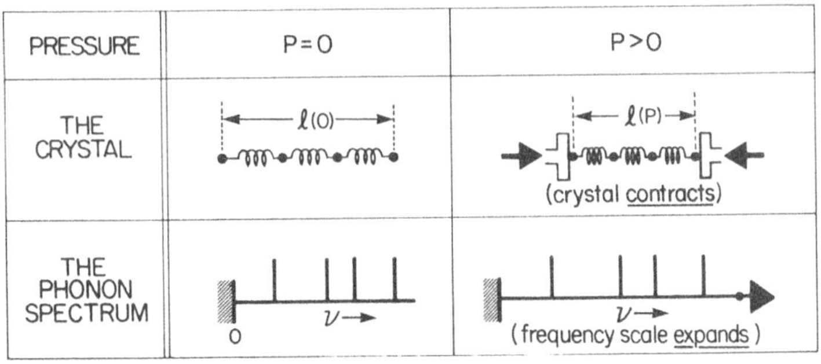

Fig. 8.39. Schematic representation of the scaling relation embodied by the Grüneisen approximation

For mode i in a crystal under compression, the mode-Grüneisen parameter γ_i is the dimensionless quantity connecting the volume dilatation $\Delta V/V$ with the consequent fractional change in phonon frequency, $\Delta\nu_i/\nu_i = \gamma_i(\Delta V/V) = \gamma_i \beta P$. In the Grüneisen approximation, all of the γ_i are assumed to be equal: $\gamma_i = \gamma$, independent of i. This is what is pictorially represented in Fig. 8.39. Pressure produces not only a contraction of the crystal in real space, but also a *uniform expansion of the spectrum in frequency space*. It is as if the spectrum were on a rubber band which experiences a tensile stress when the crystal is subjected to compressive stress. The Grüneisen picture amounts to a phenomenological scaling relationship between frequency and volume: $\nu \sim V^{-\gamma}$.

The Grüneisen model works fairly well as a zeroth-order view of the overall effect of pressure on vibrational frequencies in many materials, especially solids (such as the 3*d*-network covalent semiconductors) in which a single type of bond controls the crystal dynamics. But we shall now show that it does *not* work for molecular solids, for which it fails in very striking fashion.

Figure 8.40 displays a compact representation of the pressure-Raman results of *Zallen* and *Slade* [8.108] on crystalline As_4S_4. *Each individual point* in this plot corresponds to a Raman-active zone-center phonon whose frequency has been followed as a function of pressure. The horizontal coordinate of the point is set by the zero-pressure frequency $\nu(P=0)$. The vertical coordinate is set by the initial $(P=0)$ logarithmic pressure derivative $(1/\nu)(d\nu/dP)$, with the value of $d\nu/dP$ at $P=0$ determined from $\nu(P)$ plots of the type shown in Figs. 8.32, 34. There are 31 points in Fig. 8.40, representing a wealth of information about the

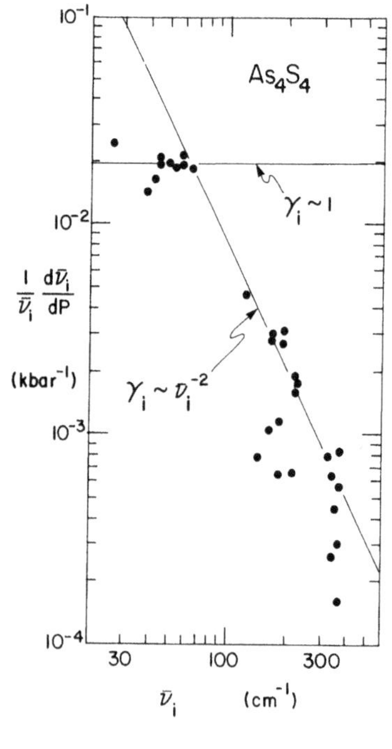

Fig. 8.40. Correlation between pressure sensitivity and phonon frequency for crystalline As_4S_4. Each point represents a phonon whose frequency has been tracked under pressure. After [8.108]

pressure dependence of the phonon spectrum. This wealth is based, as noted earlier, on the structural complexity of crystalline As_4S_4, which generously provides optical accessibility to many phonons. Since a search for systematic trends is the main object of Fig. 8.40, this *benefit derived from complexity* is of appreciable value here. (Contrast Fig. 8.40 with the corresponding plots for the first-order spectrum in diamond and NaCl. For diamond, the plot contains a single point, making it difficult to discern spectral trends. For NaCl, of course, the field is totally blank).

The Grüneisen approximation predicts, for Fig. 8.40, a set of points all lying on or near a horizontal line. Instead, we see that the mode-Grüneisen parameters span an extended range covering two orders of magnitude. Only within the confines of the external-mode manifold at low frequencies do the γ_i values cluster about the same level. As soon as the intramolecular regime is entered at higher frequency, the values observed for $(1/\nu)(d\nu/dP)$ rapidly plummet.

This drastic and systematic departure from Grüneisen scaling is a direct reflection of the *disparity of force constants which coexist in a molecular solid.* When pressure is applied to a solid in which very stiff springs and very soft springs coexist (schematically, as in Fig. 8.30), it is the soft springs which give. Most of the compression which occurs in a molecular solid is borne by the weak intermolecular bonds; the strong bonds within the molecule are relatively unaffected. Both types of bonds stiffen as they shorten (via their intrinsic anharmonicity), but since the strong bonds are shortened only slightly, their stiffening is also relatively slight.

A simple model for all of this, obtained with the aid of the elementary molecular-crystal model of Fig. 8.30, has been described earlier by one of the authors [8.106]. The basic assumption is that, although we must abandon $\nu_i \sim V^{-\gamma}$ (with γ independent of i) for molecular crystals, the idea of a basic vibrational scaling law can be salvaged in the form of the *bond-stiffness/bond-length* microscopic-scale statement $k \sim r^{-6\gamma}$. Here k is the force constant, r is the bond length and γ is a bond-scaling parameter of order unity which is presumed to apply to *both* intermolecular and intramolecular bonds. The last point treats the scaling exponent γ as a "universal" quantity, valid throughout the extended hierachy of bond strengths.

For the elementary model of Fig. 8.30c, $k \sim r^{-6\gamma}$ yields the following mode-Grüneisen parameters for the three zone-center phonons: for ω_1, $\gamma_i \approx 2\gamma$; for ω_0, $\gamma_i \approx (k_1/k_0)2\gamma$; for $\omega_0 + \Delta\omega$, $\gamma_i \approx (k_1/k_0)4\gamma$ [Ref. 8.106, Fig. 7]. The main point to observe is that for the internal modes, γ_i is reduced by a factor of order k_1/k_0 relative to γ_i of the external mode. A subsidiary point is that pressure increases the Davydov splitting of the internal-mode doublet, upshifting the higher line somewhat more than the lower one.

The generalization of the overall behavior displayed by the model is indicated by means of the two lines shown on Fig. 8.40 [8.108]. Within the low-frequency region populated by the external modes, Grüneisen scaling prevails (γ_i constant and of order unity) because the dynamics are dominated by a single bond type, namely, van der Waals. The flagrant fall-off which follows for the

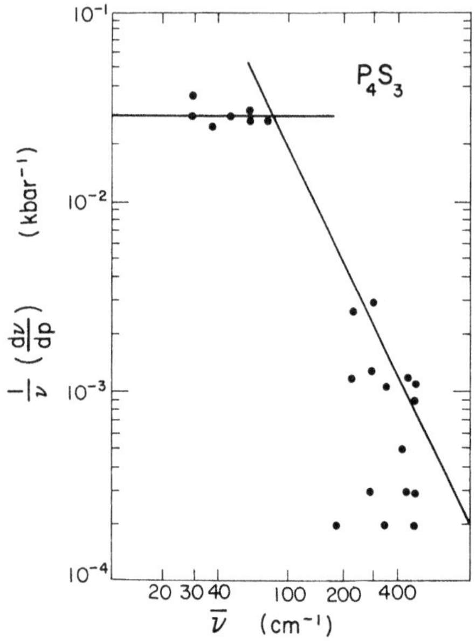

Fig. 8.41. Plot similar to that of Fig. 8.40, observed for crystalline P_4S_3. After [8.110]

internal modes is approximately bounded by an inverse-square correlation between pressure sensitivity and mode frequency: $\gamma_i \sim \nu_i^{-2}$. This follows from $\nu_i \sim k_i^{1/2}$ and $\gamma_i \sim k_i^{-1}$. It also follows that *the range of magnitudes spanned by the observed mode-Grüneisen parameters in a molecular crystal directly reflects the range of force constants which coexist in such a solid.* Thus, pressure-Raman experiments probe the hierarchy of bonding interactions which characterizes a molecular solid.

The broad features of the correlation between γ_i and ν_i, as described above and illustrated by As_4S_4 in Fig. 8.30, has now been seen in a variety of molecular crystals. Figure 8.41 shows recent results obtained by *Chattopadhyay* et al. [8.110] on crystalline P_4S_3. Other crystals for which this overall behavior has been confirmed include S_8 [8.106], S_4N_4 [8.108], $C_{10}F_8$ [8.109] and several organic compounds based on TCNQ [8.111].

8.7.4 The Connection Between the Effects of Pressure and Temperature

A valuable by-product of pressure-Raman experiments is the opportunity to separate the volume-driven ("implicit") and phonon-occupation driven ("explicit") components of the shifts in phonon frequencies which occur when *temperature* changes. Temperature is more subtle in its effect than pressure. A change in pressure alters the equilibrium interatomic spacings. A change in temperature, in addition to its effect on the vibrational excursions of the atoms about their equilibrium positions, *also* alters the interatomic spacings because of thermal expansion. The influence of temperature at constant pressure (normally

$P=0$) consists of two distinct contributions:

$$\left(\frac{\partial v}{\partial T}\right)_P=\left(\frac{\partial v}{\partial T}\right)_V-\left(\frac{\alpha}{\beta}\right)\left(\frac{\partial v}{\partial P}\right)_T \tag{8.19}$$

Here α and β are, respectively, the volume expansivity $(1/V)(\partial V/\partial T)_P$ and the compressibility $-(1/V)(\partial V/\partial P)_T$.

The left side of this equation is the total observed temperature coefficient, usually written simply dv/dT. The first term on the right is the "explicit" contribution at constant volume; it reflects the effect of the change in vibrational amplitudes, i.e., the phonon occupation numbers, at fixed equilibrium positions. The second term $-(\alpha/\beta)(dv/dP)$, known as the "implicit" contribution, reflects the effect of the change in equilibrium interatomic spacings which accompanies, via thermal expansion, a change in temperature. This equation has been discussed in some depth in [Ref. 8.108, Sect. VI.A]. One point worth emphasizing is that the equation evaporates in the absence of anharmonicity. For a perfectly harmonic solid (of which there are none), every quantity vanishes except for β.

The explicit/implicit mix in dv/dT is a matter of intrinsic interest since it is needed for a proper interpretation of the observed influence of temperature on the phonon spectrum. For example, the limiting situation in which the implicit effect is totally dominant corresponds to the validity of the "quasi-harmonic approximation" [8.120] in which V is viewed as the single essential variable controlling v. Exploiting recent pressure-Raman results, an analysis of the explicit/implicit dissection of temperature derivatives has been given by *Zallen* and *Conwell* [8.121]. Their compilation of experimental information on external modes in molecular crystals is presented in Figs. 8.42, 43.

Figures 8.42, 43 display the correlation between the temperature and pressure sensitivities of external-mode frequencies in three inorganic and three organic molecular crystals. Each point represents a Raman-active mode; it is positioned according to the pressure (x-axis) and temperature (y-axis) derivatives of the mode frequency near $P=0$ and $T=300$ K. Experimental uncertainties are typically $\pm 10\%$ for dv/dP, $\pm 20\%$ for dv/dT. The reason that pressure derivatives are usually better determined than temperature derivatives lies in the greater scope available for pressure-induced shifts in v, relative to temperature-induced shifts, making it easier to determine the slope of $v(P)$ than that of $v(T)$. To see this, note that it typically takes a temperature decrease of about 50 K to produce the same upshift in frequency as that produced by a pressure increase of 1 kbar.

The straight lines passing through the origin in these figures are lines of constant explicit/implicit mix. Each of these "isomix" lines is labeled by the value of the dimensionless parameter η, the *implicit fraction* specifying the ratio of the volume-driven term $-(\alpha/\beta)(dv/dP)$ to the total temperature dependence dv/dT. Among the various possibilities for η, two cases are of special interest. Lines corresponding to the volume driven "purely-implicit" $\eta=1$ case have slopes equal to α/β in Figs. 8.42, 43, the expansivity/compressibility ratio for the crystal

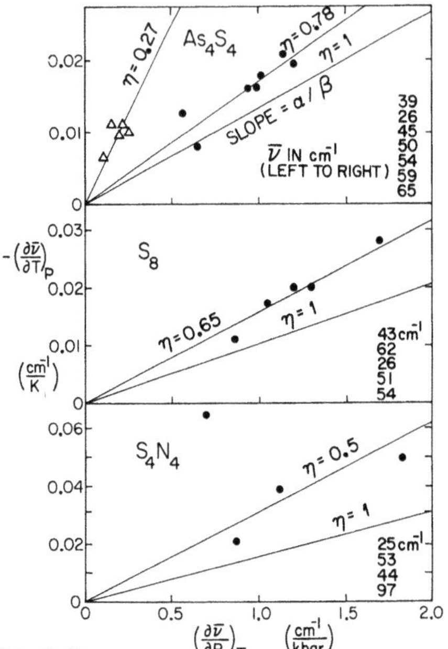

Fig. 8.42.

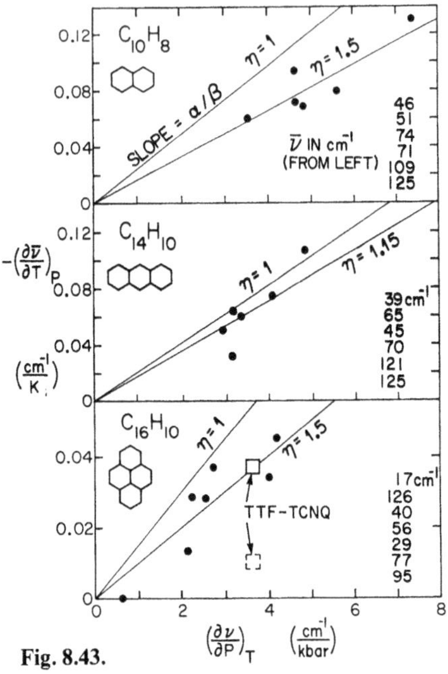

Fig. 8.43.

Fig. 8.42. The correlation between the temperature and pressure derivatives of phonon frequencies in three inorganic molecular crystals. Each point represents an external mode whose frequency has been followed as a function of temperature (at $P \approx 0$) and of pressure (at $T \approx 300$ K). The mode frequencies $\nu(P=0,\ T=300\ \mathrm{K})$, corresponding to the points in sequence from left to right, are listed for each solid. Straight lines define loci of constant explicit/implicit mix in $d\nu/dT$, and are labeled by the implicit fraction η. For As_4S_4, the behavior of several bond-stretching internal modes (shown by the open triangles) is included for comparison. After [8.121]

Fig. 8.43. Temperature and pressure derivatives of intermolecular phonon frequencies in three organic crystals. Format is the same as in Fig. 8.42, but the scales are more extended because these modes are more sensitive to T and P. After [8.121]

in question. The opposite situation, in which the phonon-occupation-driven explicit effect dominates ($\eta = 0$), corresponds to the vertical axes. Other types of behavior, such as exact cancellation between effects of opposite sign ($\eta = \infty$), are indicated in Table 8.4 [8.121].

Figures 8.42, 43 indicate that, for external modes in molecular crystals, a definite correlation exists between the effects of temperature and pressure. The points scatter about lines with η values varying between 0.65 for sulfur to 1.5 for naphthalene and pyrene. The overall tendency certainly corresponds better to $\eta \approx 1$ than to either $\eta \approx 0$ or $\eta \approx \infty$. The data demonstrate that, to a reasonable approximation, *the volume-driven implicit effect dominates the temperature dependence of external-mode frequencies in molecular crystals.*

At the present time, it is not possible to compile a collection of results for internal modes as extensive as that presented for external modes in Figs. 8.42, 43.

Table 8.4. The physical composition of the temperature derivative of phonon frequency, and provisional conclusions for the connection with crystal bonding [8.121]

Physical situation	Composition of $-d\nu/dT$: Total = exp + imp	Implicit fraction η	Crystal-bonding class and mode type
(a) Phonon-occupation driven explicit effect dominates	$1=1+0$	0	Covalent crystals, internal modes in molecular crystals
(b) Comparable contributions of the same sign	$1=0.5+0.5$	0.5	
(c) Thermal-expansion-driven implicit effect dominates, validity of the quasiharmonic approx. ("volume is everything")	$1=0+1$	1	Ionic crystals, external modes in molecular crystals
(d) Implicit effect dominates, small explicit term of the opposite sign	$1=-0.5+1.5$	1.5	
(e) Cancellation, no net effect of temperature	$0=-1+1$	∞	

There is relatively little reliable information about the small temperature coefficients of the high-frequency phonons in molecular crystals. However, we may again make use of the comprehensive study [8.108] available for that paradigmatic molecular solid, crystalline As_4S_4. The relevant data are contained in the top panel of Fig. 8.42. In addition to the results on seven external modes (the solid dots, which lie quite close to the isomix line $\eta=0.8$), results are also shown for five bond-stretching internal modes represented by the open triangles in this figure. These intramolecular modes reveal different behavior than the intermolecular ones; they are characterized by an implicit fraction of about $\eta=0.3$. Another well-documented case is SnI_4 [8.122], which behaves similarly. Thus, unlike the situation for external modes, for internal modes the phonon-occupation driven explicit effect makes a significant contribution to the total temperature effect.

It is of significant interest to compare the conclusions arrived at here for molecular solids, for the explicit/implicit mix in $d\nu/dT$, with the temperature behavior of phonon frequencies in the two other major classes of insulating solids: ionic and covalent. This comparison, for the physical composition of the temperature derivative of phonon frequency, is summarized in Table 8.4 in somewhat oversimplified fashion.

For ionic solids, a substantial body of data [8.123] exists. It demonstrates that the thermal-expansion-dominated "volume-is-everything" limit gives an accurate description of $d\nu/dT$. This is the same conclusion as that reached here for intermolecular modes in molecular solids. There is a key physical similarity between vibrations in ionic crystals and external-mode ($\approx$rigid-molecule) vibrations in molecular crystals; vibrations in both cases involve relative motions of electronically-nonoverlapping closed-shell entities.

Relatively few analyses of the makeup of dv/dT have been carried out for covalent semiconductors [8.124]. The sparse data suggest that the explicit and implicit effects are comparable in size, perhaps with a tendency for the explicit effect to be the larger one. The similarity here is evidently to the behavior of internal modes in molecular crystals, a similarity which seems natural enough because the intramolecular bonding is, after all, covalent. In both cases there is substantial charge overlap between the vibrating units.

8.7.5 Molecular-Nonmolecular Transitions at High Pressure

This chapter has been largely divided along class lines, with two distinct classes of solids (covalent and molecular) being separately treated and discussed. In a very real sense, this closing section bridges that division. Very high pressure severely blurs the distinction between covalent interactions and intermolecular ones, and this section addresses that phenomenon.

Eventually, all materials should go metallic at sufficiently high pressure. In the high density limit, electronic delocalization signals the triumph of kinetic energy over potential energy in a now-familiar story in condensed-matter physics (the Wigner lattice, the Mott transition, Anderson localization, etc.). One well-known scenario for this, although thus-far unrealized (at least on this planet), is the predicted appearance at very high pressure of a metallic solid based on monatomic hydrogen. Metallic hydrogen is the motivation of an active search in several high-pressure laboratories.

Figure 8.44 displays recent pressure-Raman results of *Sharma* et al. [8.28] on solid hydrogen up to 630 kbar. As in the case of the work on solid nitrogen discussed earlier, lattice modes could not be seen and the study focused on a covalent-bond-stretching internal mode. However, there is a remarkable difference between the behavior seen in Fig. 8.44 for solid H_2 and that shown earlier in Fig. 8.36 for solid N_2. The H–H intramolecular-stretch frequency turns over above 300 kbar and begins to *decrease*. This softening of the covalent bond within each H_2 molecule, as the molecules are forced to tightly pack together at high pressure, is suggestive of precursor behavior providing advance notice of the predicted transition to metallic hydrogen (which may take place in the 2–5 Mbar range [8.125]).

Molecular→nonmolecular transition phenomena under pressure are nicely seen in two types of molecular solids which have not yet been mentioned here. In all of the molecular crystals discussed thus far, the covalently-bonded unit (H_2, S_8, As_4S_4, etc.) is microscopic, i.e., finite on an atomic scale. There are many molecular crystals composed of macromolecules, covalently-bonded networks that are *macroscopically* extended in one dimension (*chain-structure* polymeric crystals such as polyethylene or trigonal Se) or in two dimensions (*layer-structure* crystals such as MoS_2 or GaSe). From the viewpoint of the macroscopic *network*

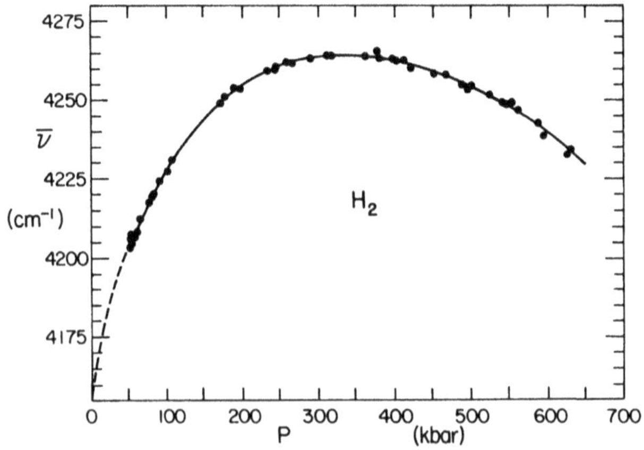

Fig. 8.44. The pressure dependence of an internal-mode frequency in crystalline H_2. Note the softening of the H–H intramolecular bond above 300 kbar. After [8.28]

dimensionality [8.126] of the covalently-bonded molecular unit, chain-structure Se is an example of a 1*d network solid* and layer-structure MoS_2 is an example of a 2*d network solid*. These two types of solids are *intermediate* in bonding topology between the two types of solids discussed earlier: 0*d network* molecular solids (H_2, S_8) and 3*d network* covalent solids (Si, Ge).

Trigonal selenium is an elemental polymer composed of helical chains (three atoms per turn, the simplest helix) packed in hexagonal array. Because of the extreme structural and chemical simplicity of this 1*d* network solid, its vibrational properties have been theoretically investigated in several studies, of which those most relevant to our discussion are the studies of *Martin* and coworkers [8.127]. Already at $P=0$, the separation approximation is invalid and the intermolecular (i.e., *interchain*) coupling contains some appreciable covalent-bonding character. Pressure increases the mixing between the intrachain and interchain bonding, with the result that the chain-chain bonds develop substantial covalent character and at the same time the primary (intrachain) bonding substantially *weakens*. Packing the chains more closely together causes a transfer of bonding charge from the intrachain to the interchain bonding [8.127, 128]. Thus pressure effects a molecular→nonmolecular crossover in this system.

Experimental support for the above picture is seen in the effect of pressure on the three Raman-active phonons in Se. Pressure-Raman experiments were first reported by *Richter* et al. [8.57], and then extended to 140 kbar (where Se undergoes a phase transition to a metallic form) by *Aoki* et al. [8.129] whose data are shown in Fig 8.45. The interesting feature is the pronounced *decrease* with pressure of the frequency of the A_1 internal mode, which is the symmetric-stretch "breathing" mode of the helical chain. The softening of this mode under pressure demonstrates the large pressure-induced decrease of the force constant associated with the intrachain covalent bond.

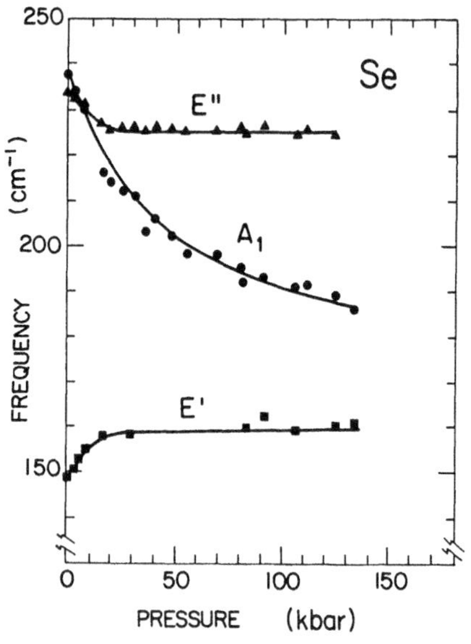

Fig. 8.45. The pressure dependence of the three Raman modes in trigonal Se. Note the pronounced drop in the frequency of the chain-breathing A_1 mode, which announces the pressure-induced weakening of the intrachain covalent bond. After [8.129]

Several two-dimensional network layer-structure crystals have been the subject of pressure-Raman investigations, including As_2S_3 [8.106, 130], GeS [8.131], GeSe [8.131], BN [8.132]. GaS [8.133] and GaSe [8.134]. In layer crystals, the vibrations corresponding to the external modes of ordinary (0*d* network) molecular solids are *rigid-layer* modes [8.135]. *Polian* et al. [8.133] have observed a rigid-layer mode in GaS from zero pressure up to 150 kbar. It experiences an enormous fractional change in frequency (from 23 to 76 cm^{-1}), corresponding to an order-of-magnitude enhancement of the interlayer interaction.

The pressure-induced molecular→nonmolecular crossover described above for chain-structure selenium amounts to a 1*d*→3*d* transition from the viewpoint of the covalent-bonding network dimensionality [8.126]. We close this chapter with a brief description of an analogous phenomenon in a layer-structure crystal, namely, 2*d*→3*d* dimensionality effects observed at high pressure by *Besson* et al. [8.130] for crystalline As_2S_3. This material, whose optical properties at zero pressure have been intensively investigated, is notable as the layer crystal for which the crucial role of the *diperiodic symmetry* (the proper factor-group symmetry of an individual 2*d* network macromolecule) was first appreciated and analyzed [8.136]. The dominance of the diperiodic symmetry at $P=0$ (and, conversely, the minor *subsidiary* role played by the conventional, triperiodic, crystal symmetry) is now well established for both the vibrational and electronic optical properties of this material. In brief, the observation of many degenerate or near-degenerate Raman-infrared line pairs, incomprehensible from the viewpoint of the crystal symmetry, is well explained on the basis of the diperiodic layer symmetry [8.136].

Besson et al. [8.130], in pressure-optical experiments on crystalline As_2S_3 to 100 kbar, have observed an intralayer Raman quadruplet which disperses rapidly as the dominance of the diperiodic symmetry is broken at high pressure and the admixture repulsion of modes of like *crystal* symmetry forces them apart. They also saw a forbidden-crossover repulsion, enforced by crystal symmetry, of another pair of intralayer modes which were well separated in frequency at $P=0$ but attempted to cross at high pressure. These observations, along with the rapid rise in frequency of the rigid-layer modes (which become inextricably admixed with intralayer modes by 100 kbar), are clear evidence of $2d \rightarrow 3d$ molecular→nonmolecular pressure-induced effects.

Acknowledgements. The authors wish to take this opportunity to express their appreciation to Michael L. Slade for his essential contributions to much of the work reviewed here on molecular solids. In addition, we are greatly indebted to Cathyrn A. Horeth for her able and patient help in the preparation of this manuscript.

References

8.1 C.H.Whitfield, E.M.Brody, W.A.Bassett: Rev. Sci. Instrum. **47**, 942 (1976)

8.2a M.Cardona: Surf. Sci. **37**, 100 (1973)
M.Cardona: In *Atomic Structure and Properties of Solids*, ed. by E.Burstein (Academic Press, New York 1972) pp. 514–580

8.2b M.Cardona, G.Güntherodt (eds.): *Light Scattering in Solids II*, Topics Appl. Phys., Vol. 50 (Springer, Berlin, Heidelberg, New York 1982)

8.3 W.Hayes, R.Loudon: *Scattering of Light by Crystals* (Wiley, New York 1978) Chap. 1

8.4 B.A.Weinstein, M.Cardona: Phys. Rev. B**7**, 2545 (1973); Solid State Commun. **10**, 961 (1972)

8.5 P.A.Temple, C.E.Hathaway: Phys. Rev. B**7**, 3685 (1973)

8.6 L.N.Ovander: Opt. Spectrosc. **9**, 302 (1960)

8.7 R.Loudon: Advan. Phys. **13**, 423 (1964)

8.8 R.M.Martin, L.M.Falicov: In *Light Scattering in Solids I*, 2nd. ed., ed. by M.Cardona, Topics Appl. Phys., Vol. 8 (Springer, Berlin, Heidelberg, New York 1983) Chap. 3

8.9 H.Wendel: Solid State Commun. **31**, 423 (1979)

8.10 S.S.Mitra: In *Optical Properties of Solids*, ed. by S.Nudelman, S.S.Mitra (Plenum Press, New York 1969) pp. 378–451

8.11 M.Nicol, J.R.Kessler, Y.Ebisuzaki, W.D.Ellenson, M.Fong, C.S.Gratch: Dev. Appl. Spectrosc. **10**, 79 (1972)

8.12 P.S.Peercy: In *High-Pressure and Low-Temperature Physics*, ed. by C.W.Chu, J.A.Woollam (Plenum Press, New York 1977) p. 279
P.S.Peercy: In *Light Scattering in Solids*, ed. by M.Balkanski, R.C.C.Leite, S.P.S.Porto (Flammarion, Paris 1976) p. 782

8.13 J.R.Ferraro: Coordination Chem. Rev. **29**, 1 (1979)

8.14 W.F.Sherman, G.R.Wilkinson: In *Advances in Infrared and Raman Spectroscopy*, Vol. 6, ed. by R.J.H.Clark, R.E.Hester (Heyden, London 1980) Chap. 4

8.15 G.Martinez: In *Handbook on Semiconductors*, Vol. 2, ed. by T.S.Moss, M.Balkanski (North-Holland, Amsterdam 1980) Chap. 4c

8.16 M.Blackman: Proc. Phys. Soc. Lond. B**70**, 827 (1957)

8.17 W.B.Daniels: In *Lattice Dynamics*, ed. by R.F.Wallis (Pergamon Press, Oxford 1965) p. 273

8.18 P.Debye: Phys. Z. **14**, 259 (1913)

8.19 E.Grüneisen: In *Handbuch der Physik*, Vol. 10, ed. by Geiger and Scheels (Springer, Berlin 1926) Chap. 1

8.20 J.C.Slater: *Introduction to Chemical Physics*, (McGraw-Hill, New York 1939) pp. 217–240

8.21 J.S.Dugdale, D.K.Macdonald: Phys. Rev. **89**, 832 (1953)

8.22 T.Soma: Solid State Commun. **34**, 375 (1980); **34**, 927 (1980)

8.23 O.L.Anderson: Science **213**, 76 (1981)
8.24 G.J.Piermarini, S.Block: Rev. Sci. Instrum. **46**, 973 (1975)
J.D.Barnett, S.Block, G.J.Piermarini: Rev. Sci. Instrum. **44**, 1 (1973)
8.25 C.E.Weir, E.R.Lippincott, A.van Valkenburg, E.N.Bunting: J. Res. Natl. Bur. Stand. A**63**, 55 (1959)
8.26 R.A.Forman, G.J.Piermarini, J.D.Barnett, S.Block: Science **176**, 284 (1972)
8.27 B.A.Weinstein, G.J.Piermarini: Phys. Rev. B**12**, 1172 (1975)
8.28 S.K.Sharma, H.K.Mao, P.M.Bell: Phys. Rev. Lett. **44**, 886 (1980)
8.29 O.Brafman, S.S.Mitra, R.K.Crawford, W.B.Daniels, C.Postmus, J.R.Ferraro: Solid State Commun. **7**, 449 (1969)
8.30 J.M.Besson, J.P.Pinceaux: Rev. Sci. Instrum. **50**, 541 (1979)
8.31 D.M.Adams, S.J.Payne, K.Martin: Appl. Spectrosc. **27**, 377 (1973)
8.32 M.Nicol, Y.Ebisuzaki, W.D.Ellenson, A.Karim: Rev. Sci. Instrum. **43**, 1368 (1972)
8.33a H.K.Mao, P.M.Bell: Science **200**, 1145 (1978)
H.K.Mao, P.M.Bell, K.J.Dunn, R.M.Chrenko, R.C.Devries: Rev. Sci. Instrum. **50**, 1002 (1979)
8.33b A.Jayaraman: Rev. Mod. Phys. **55**, 65 (1983)
8.34 R.P.Lowndes: In *Proc. 4th Intern. Conf. on High Pressure*, Kyoto, 1974, ed. by J.Osugi (The Physico-Chemical Society of Japan, Kyoto 1975) p. 805
8.35 R.A.Noack, W.B.Holzapfel: In *High Pressure Science and Technology*, Vol. 1, ed. by K.D.Timmerhaus, M.S.Barber (Plenum Press, New York 1979) p. 748
8.36 M.Born, K.Huang: *Dynamical Theory of Crystal Lattices* (Oxford University Press, New York 1954) Chap. 2
8.37 C.J.Buchenauer, F.Cerdeira, M.Cardona: In *Light Scattering in Solids*, ed. by M.Balkanski (Flammarion, Paris 1971) p. 280
8.38 S.S.Mitra, O.Brafman, W.B.Daniels, R.K.Crawford: Phys. Rev. **186**, 942 (1969)
8.39 B.J.Parsons, C.D.Clark: In *Light Scattering in Solids*, ed. by M.Balkanski, R.C.C.Leite, S.P.S.Porto (Flammarion, Paris 1976) p. 414
8.40 K.Asaumi, S.Minomura: J. Phys. Soc. Japan-Lett. **45**, 1061 (1978)
8.41 B.A.Weinstein, G.J.Piermarini: Phys. Lett. **48**A, 14 (1974)
8.42a R.Trommer, H.Muller, M.Cardona, P.Vogl: Phys. Rev. B**21**, 4869 (1980)
8.42b D.Olego, M.Cardona, P.Vogl: Phys. Rev. B**25**, 3878 (1982);
J.A.Sanjurjo, E.Lopez-Cruz, P.Vogl, M.Cardona: (in press)
8.43a O.Brafman, S.S.Mitra: In *Light Scattering in Solids*, ed. by M.Balkanski (Flammarion, Paris 1971) p. 284
8.43b S.S.Mitra, K.V.Namjoshi: J.Chem. Phys. **55**, 1817 (1971)
8.44 R.Trommer, E.Anastassakis, M.Cardona: In *Light Scattering in Solids*, ed. by M.Balkanski, R.C.C.Leite, S.P.S.Porto (Flammarion, Paris 1976) p. 396
8.45 B.A.Weinstein, R.Zallen, M.L.Slade, A.deLozanne: Phys. Rev. B**24**, 4652 (1981)
8.46 W.A.Harrison, S.Ciraci: Phys. Rev. B**10**, 1516 (1974)
W.A.Harrison: Phys. Rev. B**8**, 4487 (1973); **14**, 702 (1976)
8.47 P.Vogl: J. Phys. C**11**, 251 (1978)
8.48 J.C.Phillips: Rev. Mod. Phys. **42**, 317 (1970)
J.A.VanVechten: Phys. Rev. **182**, 891 (1969)
8.49 V.Heine, R.O.Jones: J. Phys. C**2**, 719 (1969)
8.50 K.Kunc, R.M.Martin: Phys. Rev. B**24**, 2311 (1981); Phys. Rev. Lett. **48**, 406 (1982)
R.M.Martin, K.Kunc: Phys. Rev. B**24**, 2081 (1981)
8.51 H.Bilz, W.Kress: *Phonon Dispersion Relations in Insulators*, Springer Ser. Solid-State Sci., Vol. 10, (Springer, Berlin, Heidelberg, New York 1979) pp. 95–121
8.52 W.Richter, J.B.Renucci, M.Cardona: Solid State Commun. **16**, 131 (1975)
8.53 B.A.Weinstein, J.B.Renucci, M.Cardona: Solid State Commun. **12**, 473 (1973)
8.54 B.A.Weinstein: Solid State Commun. **24**, 595 (1977)
B.A.Weinstein: In *High Pressure Science and Technology*, Vol. 1, ed. by K.D.Timmerhaus, M.S.Barber (Plenum Press, New York 1979) p. 141
8.55 M.Zigone, R.Beserman, H.D.Fair,Jr.: In *Light Scattering in Solids*, ed. by M.Balkanski, R.C.C.Leite, S.P.S.Porto (Flammarion, Paris 1976) p. 597

8.56 D.N.Talwar, M.Vandevyver, K.Kunc, M.Zigone: Phys. Rev. B**24**, 741 (1981)
8.57 W.Richter, J.B.Renucci, M.Cardona: phys. stat. sol. (b) **56**, 223 (1973)
8.58 O.Brafman, M.Cardona, Z.Vardeny: Phys. Rev. B**15**, 1081 (1977)
8.59 Z.Vardeny, O.Brafman: Phys. Rev. B**19**, 3290 (1979)
8.60 D.S.Rimai, R.J.Sladek: Solid State Commun. **30**, 591 (1979)
8.61 B.A.Weinstein: In *Proc. 13th Intern. Conf. on the Physics of Semiconductors*, Rome, 1976, ed. by F.G.Fumi (North-Holland, Amsterdam 1976) p. 326
8.62 Y.S.Touloukian, R.K.Kirby, R.E.Taylor, T.Y.R.Lee: *Thermal Expansion*, Thermophysical Properties of Matter, Vol. 13, ed. by Y.S.Touloukian, C.Y.Ho (Plenum Press, New York 1977)
8.63 J.F.Vetelino, S.S.Mitra, K.V.Namjoshi: Phys. Rev. B**2**, 967 (1970)
8.64 G.Dolling, R.A.Cowley: Proc. Phys. Soc. Lond. **88**, 463 (1966)
8.65 R.M.Pick, M.H.Cohen, R.M.Martin: Phys. Rev. B**1**, 910 (1970)
R.M.Martin: Phys. Rev. **186**, 871 (1969)
8.66 R.D.Turner, J.C.Inkson: J. Phys. C**11**, 3961 (1978)
8.67 W.Porod, P.Vogl, G.Bauer: J. Phys. Soc. Japan (Suppl. A) **49**, 649 (1980)
8.68 M.T.Yin, M.L.Cohen: Phys. Rev. Lett. **45**, 1004 (1980); J. Phys. Soc. Japan (Suppl. A) **49**, 13 (1980); Solid State Commun. **38**, 625 (1981)
8.69 K.Maschke, W.Andreoni: J. Phys. Soc. Japan (Suppl. A) **49**, 745 (1980)
8.70 H.Jex: phys. stat. sol. (b) **45**, 343 (1971)
8.71 A.Bienenstock: Philos. Mag. **9**, 755 (1964)
8.72 J.C.Phillips: Phys. Rev. **166**, 832 (1968)
8.73 F.Herman: J. Phys. Chem. Sol. **8**, 405 (1959)
8.74 W.Weber: Phys. Rev. Lett. **33**, 371 (1974); Phys. Rev. B**15**, 4789 (1977)
8.75 J.Ihm, A.Zunger, M.L.Cohen: J. Phys. C**12**, 4409 (1979)
8.76 R.W.Shaw: Phys. Rev. **174**, 769 (1968)
8.77 H.Wendel, R.M.Martin: Phys. Rev. B**19**, 5251 (1979)
8.78 B.A.Weinstein: Solid State Commun. **20**, 999 (1976)
8.79 A.S.Barker,Jr.: Phys. Rev. **165**, 917 (1968)
8.80 W.Cochran, R.A.Cowley: In *Handbuch der Physik*, Vol. 25/2a, ed. by S.Flügge, L.Genzel (Springer, Berlin, Heidelberg, New York 1967) p. 118
8.81 J.Ruvalds, A.Zawadowski: Phys. Rev. B**2**, 1172 (1970)
8.82 J.L.Yarnell, J.L.Warren, R.G.Wenzel, P.J.Dean: In 4th *I.A.E.A. Symp. on Neutron Inelastic Scattering*, Vol. 1 (Intern. Atomic Energy Agency, Vienna 1968) p. 301
8.83 M.L.Shand, H.D.Hochheimer, M.Krauzman, J.E.Potts, R.C.Hanson, C.T.Walker: Phys. Rev. B**14**, 4637 (1976)
R.C.Hanson, M.L.Shand: In *High Pressure Science and Technology*, Vol. 1, ed. by K.D.Timmerhaus, M.S.Barber (Plenum Press, New York 1979) p. 453
8.84a Z.Vardeny, O.Brafman: Phys. Rev. B**19**, 3276 (1979)
8.84b D.Schmeltzer, R.Beserman: J. Phys. C**15**, 4173 (1982)
8.85 W.Klement, A.Jayaraman: In *Progress in Solid State Chemistry*, Vol. 3, ed. by H.Reiss (Pergamon Press, London 1966) p. 289
8.86 S.C.Yu, I.L.Spain, E.F.Skelton: Solid State Commun. **25**, 49 (1978)
M.A.Baublitz, Jr., A.L.Ruoff: J. Appl. Phys. **53**, 6179 (1982)
8.87 B.Henion, F.Moussa, B.Prevot, C.Carabatos, C.Schwab: Phys. Rev. Lett. **28**, 964 (1972)
8.88 Y.Ebisuzaki, M.Nicol: J. Phys. Chem. Sol. **33**, 763 (1971)
8.89 R.C.Hanson, T.A.Fjeldly, H.D.Hochheimer: Phys. Stat. Sol. (b) **70**, 567 (1975)
8.90 C.Carlone, D.Olego, A.Jayaraman, M.Cardona: Phys. Rev. B**22**, 3877 (1980)
8.91 H.G.Drickamer: In *Solid State Physics*, **17**, 1 (Academic Press, New York 1965) and references therein
S.Minomura, G.A.Samara, H.G.Drickamer: J. Appl. Phys. **33**, 3196 (1962)
8.92 J.C.Jamieson: Science **139**, 762 (1963); **139**, 845 (1963)
8.93 J.S.Kasper, S.M.Richards: Acta Crystallogr. **17**, 752 (1964)
R.H.Wentorf, Jr., J.S.Kasper: Science **139**, 338 (1963)
8.94a M.T.Yin, M.L.Cohen: Phys. Rev. B**26**, 5668 (1982); Phys. Rev. Lett. **50**, 1172 (1983); J.Ihm, M.L.Cohen: Phys. Rev. B**23**, 1576 (1981)

8.94b S.Froyen, M.L.Cohen: Phys. Rev. B**28**, 3258 (1983); Solid State Commun. **43**, 447 (1982)
8.94c E.J.Mele, J.D.Joannopoulos: Phys. Rev. B**24**, 3145 (1981)
8.95 J.C.Phillips: Phys. Rev. Lett. **27**, 1197 (1971); Phys. Rev. B**25**, 2310 (1982)
8.96 D.Olego, M.Cardona: Phys. Rev. B**25**, 1151 (1982)
8.97 P.Y.Yu, B.Welber: Solid State Commun. **25**, 209 (1978)
8.98 R.E.Hanneman, M.D.Banus, H.C.Gatos: J. Phys. Chem. Sol. **25**, 293 (1964)
8.99 M.J.P.Musgrave: Proc. R. Soc. A**272**, 503 (1963)
8.100 H.H.Demarest,Jr., R.Ota, O.L.Anderson: In *High-Pressure Research*, ed. by M.H.Manghnani, S.Akimoto (Academic Press, New York 1977) p. 281
8.101 J.C.Phillips: *Bonds and Bands in Semiconductors* (Academic Press, New York 1973) p. 93
8.102 W.Richter, R.Zeyher, M.Cardona: Phys. Rev. B**18**, 4312 (1978)
8.103 E.Anastassakis, F.H.Pollak, G.W.Rubloff: Phys. Rev. B**9**, 551 (1974)
8.104 E.M.Anastassakis: In *Dynamical Properties of Solids*, Vol. 4, ed. by G.K.Horton, A.A.Maradudin (North-Holland, New York 1980) pp. 221–227
8.105 B.Welber, M.Cardona, C.K.Kim, S.Rodriguez: Phys. Rev. B**12**, 5729 (1975)
8.106 R.Zallen: Phys. Rev. B**9**, 4485 (1974)
8.107 G.S.Pawley, S.J.Cyvin: J. Chem. Phys. **52**, 4073 (1970)
8.108 R.Zallen, M.L.Slade: Phys. Rev. B**18**, 5775 (1978)
8.109 D.M.Adams, A.C.Shaw, G.A.Mackenzie, G.S.Pawley: J. Phys. Chem. Sol. **41**, 149 (1980)
8.110 T.Chattopadhyay, C.Carlone, A.Jayaraman, H.G. von Schnering: Phys. Rev. B**23**, 2471 (1981)
8.111 C.Carlone, N.K.Hota, H.J.Stolz, M.Elbert, H.D.Hochheimer: J. Chem. Phys. **75**, 3220 (1981)
8.112 P.W.Bridgman: Proc. Am. Acad. Arts Sci. **72**, 227 (1938)
8.113 R.Zallen, C.H.Griffiths, M.L.Slade, M.Hayek, O.Brafman: Chem. Phys. Lett **39**, 85 (1976)
8.114 W.D.Ellenson, M. Nicol: J. Chem. Phys. **61**, 1380 (1974)
8.115 E.Whalley, A.Lavergne, P.T.T.Wong: Rev. Sci. Instrum. **47**, 845 (1976)
8.116 J.V.E.Kurittu: Physica Scripta **21**, 194 (1980); **21**, 200 (1980)
8.117 R.LeSar, S.A.Ekberg, L.H.Jones, R.L.Mills, L.A.Schwalbe, D.H.Schiferl: Solid State Commun. **32**, 131 (1979)
8.118 An excellent review of the various pressure-Raman techniques has been given in [8.14]
8.119 W.A.Bassett, T.Takahashi, P.W.Stook: Rev. Sci. Instrum. **38**, 37 (1967)
8.120 G.Liebfried, W.Ludwig: Solid State Phys. **12**, 275 (1961)
8.121 R.Zallen, E.M.Conwell: Solid State Commun. **31**, 557 (1979)
8.122 P.S.Peercy, G.A.Samara, B.Morosin: J. Phys. Chem. Sol. **36**, 1123 (1975)
8.123 S.S.Mitra, C.Postmus, J.R.Ferraro: Phys. Rev. Lett. **18**, 455 (1967)
C.Postmus, J.R.Ferraro, S.S.Mitra: Phys. Rev. **174**, 983 (1968)
J.F.Asell, M.Nicol: J. Chem. Phys. **49**, 5395 (1968)
R.P.Lowndes: J. Phys. C**4**, 3083 (1971)
P.S.Peercy, B.Morosin: Phys. Rev. B**7**, 2779 (1972)
J.A.Taylor, M.S.Haque, J.B.Page,Jr., C.T.Walker: Phys. Rev. B**12**, 5969 (1975)
8.124 H.D.Hochheimer, M.L.Shand, J.E.Potts, R.C.Hanson, C.T.Walker: Phys. Rev. B**14**, 4630 (1976)
D.G.Mead, G.R.Wilkinson: J. Raman Spectrosc. **6**, 123 (1977)
8.125 J. van Straaten, R.J.Wijngaarden, I.F.Silvera: Phys. Rev. Lett. **48**, 97 (1982)
8.126 R.Zallen: In *Proc. Enrico Fermi Summer School on Lattice Dynamics and Intermolecular Forces*, Varenna, 1972 (Academic Press, New York 1975) p. 159; and in *Proc. 12th Intern. Conf. Physics of Semiconductors*, Stuttgart (Teubner, Stuttgart 1974) p. 621
8.127 R.M.Martin, G.Lucovsky, K.Helliwell: Phys. Rev. B**13**, 1383 (1976)
R.M.Martin, T.A.Fjeldly, W.Richter: Solid State Commun. **18**, 865 (1976)
R.M.Martin, G.Lucovsky: In *Proc. 12th Intern. Conf. Physics of Semiconductors*, Stuttgart (Teubner, Stuttgart 1974) p. 184
8.128 J.D.Joannopoulos, M.Schluter, M.L.Cohen: Phys. Rev. B**11**, 2186 (1975)
8.129 K.Aoki, O.Shimomura, S.Minomura, N.Koshizuka, T.Tsushima: J. Phys. Soc. Japan **48**, 906 (1980)
8.130 J.M.Besson, J.Cernogora, R.Zallen: Phys. Rev. B**22**, 3866 (1980)
8.131 H.R.Chandrasekhar, R.G.Humphreys, M.Cardona: Phys. Rev. B**16**, 2981 (1977)

8.132 T.Kuzuba, Y.Sato, S.Yamaoka, K.Era: Phys. Rev. **B18**, 4440 (1978)
8.133 A.Polian, J.C.Chervin, J.M.Besson: Phys. Rev. **B22**, 3049 (1980)
8.134 E.A.Vinogradov, G.N.Zhizhin, N.N.Melnik, S.I.Subbotin, V.V.Panfilov, K.R.Allakhverdiev, S.S.Babaev, V.F.Zhitar: Phys. Stat. Sol. (b) **99**, 215 (1980)
8.135 R.Zallen, M.L.Slade: Phys. Rev. **B9**, 1627 (1974)
8.136 R.Zallen, M.L.Slade, A.T.Ward: Phys. Rev. **B3**, 4257 (1971)

Bibliography of References Concerning Raman Scattering by Phonons Under Uniaxial Stress

Stress-induced shifts of first-order Raman frequencies of diamond- and zinc-blende-type semiconductors:
F.Cerdeira, C. J.Buchenauer, F. H.Pollak: Phys. Rev. **B5**, 580 (1972)

Piezospectroscopic study of the Raman spectrum of α-quartz:
V. J.Tekippe, A.K.Ramdas, S.Rodriguez: Phys. Rev. **B8**, 706 (1973)

Effect of uniaxial stress on the Raman spectra of cubic Crystals: CaF_2, BaF_2, and $Bi_{12}GeO_{20}$:
S.Venugopalan, A.K.Ramdas: Phys. Rev. **B8**, 717 (1973)

Effects of uniaxial stress on resonance Raman scattering near the E_1-gaps in InSb and InAs:
E.Anastassakis: 11th Intern. Conf. on Physics of Semiconductors (PAN, Warsaw 1972) p. 227

Elastic constants and Raman frequencies of heavily-doped Si under uniaxial stress:
T.A.Fjeldly, F.Cerdeira, M.Cardona: Solid State Commun. **12**, 553 (1973)

Effect of uniaxial stress and doping on the one-phonon Raman spectrum of GaP:
I.Balslev: Phys. Stat. Sol. (b) **61**, 207 (1974)

Effect of free carriers on zone-center vibrational modes in heavily doped p-type Si. II. Optical modes:
F.Cerdeira, T.A.Fjeldly, M.Cardona: Phys. Rev. **B8**, 4734 (1973)

Uniaxial stress dependence of the Raman-active phonons in TiO_2:
P.S.Peercy: Phys. Rev. **B8**, 6018 (1973)

Effects of uniaxial stress on the Raman frequencies of Ti_2O_3 and Al_2O_3:
S.H.Shin, F.Pollak, P.M.Raccah: J. Solid State Chem. **12**, 294 (1975)

Temperature and pressure dependences of the properties and phase transition in paratellurite: ultrasonic, dielectric, and Raman and Brillouin scattering results:
P.S.Peercy, I.J.Fritz, G.A.Samara: J. Phys. Chem. Solids **36**, 1105 (1975)

Effect of uniaxial stress on the unstable phonon in ferroelectric gadolinium molybdate:
B.N.Ganguly, F.G.Ullman, R.D.Kirby, J.R.Hardy: Phys. Rev. **B12**, 3783 (1975)

Stress-induced ferroelectricity and soft phonon modes in $SrTiO_3$:
H. Uwe, T. Sakudo: Phys. Rev. B**13**, 271 (1976)

Resonant Raman scattering under uniaxial stress: $E_1 - E_1 + \Delta_1$ gaps:
W. Richter, R. Zeyher, M. Cardona: In *Light Scattering in Solids*, ed. by M. Balkanski, R.C.C. Leite, S.P.S. Porto (Flammarion, Paris 1976) p. 63

Effects of uniaxial stress on the Raman frequencies of Ti_2O_3 and Al_2O_3:
S.H. Shin, F.H. Pollak, P.M. Raccah: In *Light Scattering in Solids*, ed. by M. Balkanski, R.C.C. Leite, S.P.S. Porto (Flammarion, Paris 1976) p. 401

Piezospectroscopic study of the Raman spectrum of cadmium sulfide:
R.J. Briggs, A.K. Ramdas: Phys. Rev. B**13**, 5518 (1976)

Effects of stress on the Raman spectra of Mg_2Si and Mg_2Sn:
S. Onari, M. Cardona, E. Schönherr, W. Stetter: Phys. Stat. Sol. (b) **79**, 269 (1977)

Raman-scattering study of stress-induced ferroelectricity in $KTaO_3$:
H. Uwe, T. Sakudo: Phys. Rev. B**15**, 337 (1977)

Piezospectroscopy of Raman lines exhibiting linear wave-vector dependence. Quartz:
M.H. Grimsditch, A.K. Ramdas, S. Rodriguez, V.J. Tekippe: Phys. Rev. B**15**, 5869 (1977)

Piezospectroscopy of the Raman spectrum of α-quartz:
R.J. Briggs, A.K. Ramdas: Phys. Rev. B**16**, 3815 (1977)

Effects off interband excitations on Raman phonons in heavily-doped n-Si:
M. Chandrasekhar, J.B. Renucci, M. Cardona: Phys. Rev. B**17**, 1623 (1978)

Self-energy of phonons interacting with free electrons in silicon:
M. Chandrasekhar, M. Cardona: In *Lattice Dynamics*, ed. by M. Balkanski (Flammarion, Paris 1978) p. 186

Lattice dynamics of paratellurite under uniaxial stress:
M.A.F. Scarparo, V. Lemos, R.S. Katiyar, F. Cerdeira: In *Lattice Dynamics*, ed. by M. Balkanski (Flammarion, Paris 1978) p. 707

Raman-scattering measurements and the effect of uniaxial stress on the ferroelectric transition in $Gd_2\,(MoO_4)_3$:
Q. Kim, F.G. Ullman, R.D. Kirby, and J.R. Hardy: In *Lattice Dynamics*, ed. by M. Balkanski (Flammarion, Paris 1978) p. 664

Linear wave-vector dependence of an optical phonon frequency in $Bi_{12}GeO_{20}$ in the vicinity of the Brillouin zone center:
W. Imaino, A.K. Ramdas, S. Rodriguez: Solid State Commun. **28**, 211 (1978)

Anomalous damping of phonons in ferroelectric $Gd_2\,(MoO_4)_3$:
Q. Kim, and F.G. Ullman: Phys. Rev. B**18**, 3579 (1978)

Resonance Raman scattering in semiconductors under uniaxial stress. E_1 (E_1 $+\Delta_1$) Gaps:
W. Richter, R. Zeyher, M. Cardona: Phys. Rev. B**18**, 4312 (1978)

Stress dependence of the zone-center optical phonons of LaF_3:
F. Cerdeira, V. Lemos, R.S. Katiyar: Phys. Rev. B**19**, 5413 (1979)

Intra- and interband Raman scattering in heavily-doped p-Si:
M. Chandrasekhar, U. Rössler, M. Cardona: In *The Physics of Semiconductors*, ed. by B.L.H. Wilson (Institute of Physics, London 1979) p. 961

Uniaxial-stress dependence of the first-order Raman spectrum of rutile. I. Experiments:
P. Merle, J. Pascual, J. Camassel, H. Mathieu: Phys. Rev. B**21**, 1617 (1980)

Uniaxial-stress dependence of the first-order Raman spectrum of rutile. II. Model calculation:
J. Pascual, J. Camassel, P. Merle, H. Mathieu: Phys. Rev. B**21**, 2439 (1980)

Raman scattering under uniaxial and hydrostatic stresses in $Ba_2NaNb_5O_{15}$ crystals:
J. Sapriel, A. Boudou, G. Martinez: Ferroelectrics **29**, 15 (1980)

Study of the localized vibrations of boron in heavily-doped Si:
M. Chandrasekhar, H.R. Chandrasekhar, H. Grimsditch, M. Cardona: Phys. Rev. B**22**, 4825 (1980)

Linear wave-vector dependence of optical phonon frequencies in bismuth germanium oxide in the vicinity of the Brillouin zone center:
W. Imaino, A.K. Ramdas, S. Rodriguez: Phys. Rev. B**22**, 5679 (1980)

Inelastic light scattering in the presence of uniaxial stresses:
E. Anastassakis: J. Raman Spectroscopy **10**, 64 (1981)

Uniaxial-stress dependence of the first-order Raman spectrum of rutile-type crystals. III. MgF_2:
J. Pascual, J. Camassel, P. Merle, B. Gil, H. Mathieu: Phys. Rev. B**24**, 2101 (1981)

Uniaxial stress dependence of the amplitude mode of K_2SeO_4 at 82 K:
N.E. Massa, F.G. Ullman, J.R. Hardy: In *Proc. 7th Intern. Conf. on Raman Spectroscopy* (North-Holland, Amsterdam 1980) p. 64

Effect of uniaxial stress on the zone-center phonons of diamond:
M.H. Grimsditch, E. Anastassakis, M. Cardona: Phys. Rev. B**18**, 901 (1978)

Errata for Light Scattering in Solids II (TAP 50)

Page	Line	Is	Should be
39	4	Sect. 2.1.2	Sect. 2.1.3
47		Trigonal	move down 7 lines
49	Table 2.2	$a^2+\frac{1}{3}d^2(\mathrm{LO})+\frac{2}{3}d^2(\mathrm{TO})$	$a^2+\frac{1}{3}d^2(\mathrm{LO})+\frac{2}{3}d^2(\mathrm{TO})+b^2$
49	Table 2.2	$a^2+\frac{2}{3}d^2(\mathrm{TO})$	$\frac{2}{3}d^2(\mathrm{TO})+b^2$
63	Eq. (2.97)	$\lvert\ldots\rvert$	$\lvert\ldots\rvert^2$
93	Table 2.8	Silicon 1.68	Silicon 168
		GaAs 13 ± 3	GaAs 13 ± 5
		GaP 30 ± 5	GaP 30 ± 10
		GaP 39 ± 4	GaP 39 ± 7
		ZnSe 2.2 ± 0.2	ZnS 2.2 ± 0.4
		ZnTe 22 ± 4	ZnTe 22 ± 10
		2.2.18.1, 2.2.18.2, …	2.1.18a, 2.1.18b, …
110	4	Laguerre	Hermite
113	Eq. (2.170)	$(\omega-\omega_0)^{-1/2}$	$(\omega-\omega_0)^{1/2}$
	Eq. (2.173)	$x^{1/2}$	2
119	Eq. (2.185)	$-\frac{4\sqrt{3}(\omega_1+\Delta/3)}{9\pi a_0\omega_0^2}$	$+\frac{4\sqrt{3}(\omega_1+\Delta/3)}{9\pi a_0\omega_1^2}$
		$\cdot F^{(2)}(1-x^2)$	$\cdot\ln(1-x^2)$
121	Eq. (2.189)	$\frac{2P^2}{\omega_0}$	$\frac{\omega_0}{2P^2}$
		$\frac{m_e^*}{2}\frac{P^2}{\omega_0}$	$\frac{m_e^*}{2}\simeq\frac{\omega_0}{4P^2}$
125	Eq. (2.201)	$=-$	$=+$
127	8	Sect. 2.2.10	Sect. 2.1.10
129	Eq. (2.211, 213)	V_c	V
129	Eq. (2.212)	$+$	$-$
129	Eq. (2.212)	$V^{-1/2}$	delete
129	13	(2.111)	(2.211)

Errata for Light Scattering in Solids II (TAP 50) (continued)

Page	Line	Is	Should be
130	Eq. (2.215a)		add equation $\frac{d\chi'}{d\xi}=\sqrt{\frac{2\omega_v V}{\hbar}}\,\underset{\sim}{\chi}'_s$
131	Eq. (2.217)	$\simeq -$	$\simeq +$
132	Eq. (2.221a)	$= \dots$	$= i \dots$
133	Eq. (2.222)	$= \dots$	$= i \dots$
133	Eq.		
133	Eq. (2.224)	$= \frac{C_F}{24} \dots$	$= -\frac{C_F}{24} \dots$
139	Eq. (2.235)	$\simeq \frac{1}{3} \dots$	$\simeq -\frac{1}{3} \dots$
139	Eq. (2.236)	$\simeq \frac{4\pi}{3} \dots$	$\simeq -\frac{4\pi}{3} \dots$
140	Eq. (2.237)	$= 2\pi^2 \dots$	$= -2\pi^2 \dots$
140	14	-12	$+12$,
140	15	-468	$+468$
140	15	$+456$	-456
142	Eq. (2.243)	$\simeq 5$	$\simeq -5$
157	Table 2.10	add to (^{f})	reference
	E. Anastassakis, F.H. Pollak, G. Rubloff: Phys. Rev. B**9**, 551 (1974)		
177	Ref. 2.195	6284	5284

Subject Index

GPSR Compliance

The European Union's (EU) General Product Safety Regulation (GPSR) is a set of rules that requires consumer products to be safe and our obligations to ensure this.

If you have any concerns about our products, you can contact us on ProductSafety@springernature.com

In case Publisher is established outside the EU, the EU authorized representative is:

Springer Nature Customer Service Center GmbH
Europaplatz 3
69115 Heidelberg, Germany

Batch number: 10844669

Printed by Printforce, the Netherlands